Communications
in Computer and Information Science 81

Eyke Hüllermeier Rudolf Kruse
Frank Hoffmann (Eds.)

Information Processing and Management of Uncertainty in Knowledge-Based Systems

Applications

13th International Conference, IPMU 2010
Dortmund, Germany, June 28 – July 2, 2010
Proceedings, Part II

Springer

Volume Editors

Eyke Hüllermeier
Philipps-Universität Marburg
Marburg, Germany
E-mail: eyke@mathematik.uni-marburg.de

Rudolf Kruse
Otto-von-Guericke-Universität Magdeburg
Magdeburg, Germany
E-mail: kruse@iws.cs.uni-magdeburg.de

Frank Hoffmann
Technische Universität Dortmund
Dortmund, Germany
E-mail: frank.hoffmann@tu-dortmund.de

Library of Congress Control Number: 2010929196

CR Subject Classification (1998): I.2, H.3, F.1, H.4, I.5, I.4

ISSN 1865-0929
ISBN-10 3-642-14057-2 Springer Berlin Heidelberg New York
ISBN-13 978-3-642-14057-0 Springer Berlin Heidelberg New York

springer.com

Printed in Germany

Typesetting: Camera-ready by author, data conversion by Scientific Publishing Services, Chennai, India
Printed on acid-free paper 06/3180

Preface

The International Conference on Information Processing and Management of Uncertainty in Knowledge-Based Systems, IPMU, is organized every two years with the aim of bringing together scientists working on methods for the management of uncertainty and aggregation of information in intelligent systems. Since 1986, this conference has been providing a forum for the exchange of ideas between theoreticians and practitioners working in these areas and related fields. The 13th IPMU conference took place in Dortmund, Germany, June 28–July 2, 2010.

This volume contains 77 papers selected through a rigorous reviewing process. The contributions reflect the richness of research on topics within the scope of the conference and represent several important developments, specifically focused on applications of methods for information processing and management of uncertainty in knowledge-based systems.

We were delighted that Melanie Mitchell (Portland State University, USA), Nihkil R. Pal (Indian Statistical Institute), Bernhard Schölkopf (Max Planck Institute for Biological Cybernetics, Tübingen, Germany) and Wolfgang Wahlster (German Research Center for Artificial Intelligence, Saarbrücken) accepted our invitations to present keynote lectures. Jim Bezdek received the Kampé de Fériet Award, granted every two years on the occasion of the IPMU conference, in view of his eminent research contributions to the handling of uncertainty in clustering, data analysis and pattern recognition.

Organizing a conference like this one is not possible without the assistance and continuous support of many people and institutions. We are particularly grateful to the organizers of sessions on dedicated topics that took place during the conference—these 'special sessions' have always been a characteristic element of the IPMU conference. Frank Klawonn and Thomas Runkler helped a lot to evaluate and select special session proposals. The special session organizers themselves rendered important assistance in the reviewing process, that was furthermore supported by the Area Chairs and regular members of the Programme Committee. Thomas Fober was the backbone on several organizational and electronic issues, and also helped with the preparation of the proceedings. In this regard, we would also like to thank Alfred Hofmann and Springer for providing continuous assistance and ready advice whenever needed.

Finally, we gratefully acknowledge the support of several organizations and institutions, notably the German Informatics Society (Gesellschaft für Informatik, GI), the German Research Foundation (DFG), the European Society for Fuzzy Logic and Technology (EUSFLAT), the International Fuzzy Systems Association (IFSA), the North American Fuzzy Information Processing Society (NAFIPS) and the IEEE Computational Intelligence Society.

April 2010

Eyke Hüllermeier
Rudolf Kruse
Frank Hoffmann

Organization

Conference Committee

General Chair	Eyke Hüllermeier (Philipps-Universität Marburg)
Co-chairs	Frank Hoffmann (Technische Universität Dortmund)
	Rudolf Kruse (Otto-von-Guericke Universität Magdeburg)
	Frank Klawonn (Hochschule Braunschweig-Wolfenbüttel)
	Thomas Runkler (Siemens AG, Munich)
Web Chair	Thomas Fober (Philipps-Universität Marburg)
Executive Directors	Bernadette Bouchon-Meunier (LIP6, Paris, France)
	Ronald R. Yager (Iona College, USA)

International Advisory Board

G. Coletti, Italy
M. Delgado, Spain
L. Foulloy, France
J. Gutierrez-Rios, Spain
L. Magdalena, Spain
C. Marsala, France
M. Ojeda-Aciego, Spain
M. Rifqi, France
L. Saitta, Italy
E. Trillas, Spain
L. Valverde, Spain
J.L. Verdegay, Spain
M.A. Vila, Spain
L.A. Zadeh, USA

Special Session Organizers

P. Angelov
A. Antonucci
C. Beierle
G. Beliakov
G. Bordogna
A. Bouchachia
H. Bustince
T. Calvo
P. Carrara
J. Chamorro Martínez
D. Coquin
T. Denoeux
P. Eklund
Z. Elouedi
M. Fedrizzi
J. Fernandez
T. Flaminio
L. Godo
M. Grabisch
A.J. Grichnik
F. Hoffmann
S. Kaci
J. Kacprzyk
G. Kern-Isberner
C. Labreuche
H. Legind Larsen
E. William De Luca
E. Lughofer
E. Marchioni
N. Marin
M. Minoh
G. Navarro-Arribas
H. Son Nguyen
V. Novak
P. Melo Pinto
E. Miranda
V.A. Niskanen
D. Ortiz-Arroyo
I. Perfilieva
O. Pons
B. Prados Suárez
M. Preuß
A. Ralescu
D. Ralescu
E. Reucher
W. Rödder
S. Romaní
G. Rudolph
G. Ruß
D. Sanchez
R. Seising
A. Skowron
D. Slezak
O. Strauss
E. Szmidt
S. Termini
V. Torra
L. Valet
A. Valls
R.R. Yager

International Programme Committee

Area Chairs

P. Bosc, France
O. Cordon, Spain
G. De Cooman, Belgium
T. Denoeux, France
R. Felix, Germany
L. Godo, Spain
F. Gomide, Spain
M. Grabisch, France
F. Herrera, Spain
L. Magdalena, Spain
R. Mesiar, Slovenia
D. Sanchez, Spain
R. Seising, Spain
R. Slowinski, Poland

Regular Members

P. Angelov, UK
J.A. Appriou, France
M. Baczynski, Poland
G. Beliakov, Australia
S. Ben Yahia, Tunisia
S. Benferat, France
H. Berenji, USA
J. Bezdek, USA
I. Bloch, France
U. Bodenhofer, Austria
P.P. Bonissone, USA
C. Borgelt, Spain
H. Bustince, Spain
R. Casadio, Italy
Y. Chalco-Cano, Chile
C.A. Coello Coello, Mexico
I. Couso, Spain
B. De Baets, Belgium
G. De Tré, Belgium
M. Detyniecki, France
D. Dubois, France
F. Esteva, Spain
M. Fedrizzi, Italy
J. Fodor, Hungary
D. Fogel, USA
K. Fujimoto, Japan
P. Gallinari, France
B. Gerla, Italy
M.A. Gil, Spain
S. Gottwald, Germany
S. Grossberg, USA
P. Hajek, Czech Republic
L. Hall, USA
E. Herrera-Viedma, Spain
C. Noguera, Spain
K. Hirota, Japan
A. Hunter, UK
H. Ishibuchi, Japan
Y. Jin, Germany
J. Kacprzyk, Poland
A. Kandel, USA
G. Kern-Isberner, Germany
E.P. Klement, Austria
L. Koczy, Hungary
V. Kreinovich, USA
T. Kroupa, Czech Republic
C. Labreuche, France
J. Lang, France
P. Larranaga, Spain
H. Larsen, Denmark
A. Laurent, France
M.J. Lesot, France
C.J. Liau, Taiwan
W. Lodwick, USA
J.A. Lozano, Spain
T. Lukasiewicz, UK
F. Marcelloni, Italy
J.L. Marichal, Luxembourg
N. Marin, Spain
T. Martin, UK
L. Martinez, Spain
J. Medina, Spain
J. Mendel, USA
E. Miranda, Spain
P. Miranda, Spain
J. Montero, Spain
S. Moral, Spain
M. Nachtegael, Belgium
Y. Nojima, Japan
V. Novak, Czech Republic
H. Nurmi, Finland
E. Pap, Serbia
W. Pedrycz, Canada
F. Petry, USA
V. Piuri, Italy
O. Pivert, France
P. Poncelet, France
H. Prade, France
A. Ralescu, USA
D. Ralescu, USA
M. Ramdani, Morocco
M. Reformat, Canada
D. Ruan, Belgium
E. Ruspini, USA
R. Scozzafava, Italy
P. Shenoy, USA
G. Simari, Argentina
P. Sobrevilla, Spain
U. Straccia, Italy

Table of Contents – Part II

Data Analysis Applications

Intelligent Databases

Information Fusion

Decision Support

Modelling Collective Choices

Fuzzy Decision Theory

Applications in Finance

Fuzzy Systems

Philosophical Aspects

Fuzzy Numbers

Fuzzy Arithmetic

Fuzzy Equations

Soft Computing Applications

Image Processing

Privacy and Security

The Sensing Web

Manufacturing and Scheduling

Table of Contents – Part I

Reasoning with Uncertainty

Decomposable Models

Imprecise Probabilities

Logics for Reasoning

Preference Modeling

Data Analysis and Knowledge Processing

Belief Functions

Rough Sets

Machine Learning

Probabilistic Methods

Fuzzy Methods

Fuzzy Sets and Fuzzy Logic

Fuzzy Measures and Integrals

Fuzzy Inference

Intuitionistic Fuzzy Sets

Aggregation Functions

Evolutionary Agorithms

Data-Driven Design of Takagi-Sugeno Fuzzy Systems for Predicting NOx Emissions*

Edwin Lughofer[1], Vicente Macián[2],
Carlos Guardiola[2], and Erich Peter Klement[1]

[1] Department of Knowledge-based Mathematical Systems/Fuzzy Logic Laboratorium
Linz-Hagenberg, Johannes Kepler University of Linz, Austria
[2] CMT-Motores Térmicos/Universidad Politécnica de Valencia, Spain

Abstract. New emission abatement technologies for the internal combustion engine, like selective catalyst systems or diesel particulate filters, need of accurate, predictive emission models. These models are not only used in the system calibration phase, but can be integrated for the engine control and on-board diagnosis tasks. In this paper, we are investigating a data-driven design of prediction models for NOx emissions with the help of (regression-based) Takagi-Sugeno fuzzy systems, which are compared with analytical physical-oriented models in terms of practicability and predictive accuracy based on high-dimensional engine data recorded during steady-state and dynamic engine states. For training the fuzzy systems from data, the *FLEXFIS* approach (short for *FLEXible Fuzzy Inference Systems*) is applied, which automatically finds an appropriate number of rules by an incremental and evolving clustering approach and estimates the consequent parameters with the local learning approach in order to optimize the weighted least squares functional.

Keywords: Combustion engines, NOx emissions, physical models, data-driven design of fuzzy systems, steady-state and dynamic engine data.

1 Introduction and Motivation

Automotive antipollution legislation are increasingly stringent, which boost technology innovations for the control of engine emissions. A combination of active methods (which directly address the pollutant formation mechanism) and passive methods (which avoid the pollutant emission) is needed. Between the first, innovations in fuel injection and combustion systems, and also exhaust gas recirculation [13], have been successfully applied to spark ignited and compressed ignited engines. In this frame, pollutant emission models (in particular NOx models) are currently under development to be included in the engine control system and the on-board diagnostic system for optimizing the control of NOx after-treatment devices as NOx traps and selective reduction catalyst.

* This work was supported by the Upper Austrian Technology and Research Promotion. Furthermore, we acknowledge PSA for providing the engine and partially supporting our investigation. Special thanks are given to PO Calendini, P Gaillard and C. Bares at the Diesel Engine Control Department.

E. Hüllermeier, R. Kruse, and F. Hoffmann (Eds.): IPMU 2010, Part II, CCIS 81, pp. 1–10, 2010.

There are several ways for estimating the amount of a given pollutant that reaches the after-treatment device [1]: 1.) a direct mapping of the pollutant emitted by a reference engine as a function of rotation speed and torque implemented as a series of look-up tables; 2.) a physical-based model developed by engine experts, based on some engine operating parameters continuously registered by the engine control unit (ECU) can be used; 3.) a direct measurement of the pollutant emission in the exhaust gases. Although the latter option is ideal because is the only that fully addresses the diagnosis function, the technology in order to be able to produce low cost, precise and drift-free sensors, however, is still under development depending on the considered pollutant [11]. Hence, emission models are of great interest, leaving the first two options.

Direct engine maps are usually unable to compensate for production variations and variations in the operating conditions (e.g., warming-up of the engine, altitude, external temperature, etc.) of the engine along the vehicle lifetime. Hence, they are usually not flexible enough to predict the NOx content with sufficient accuracy. Physical-based models compensate this weakness of direct engine maps by including a deeper knowledge of experts about the emission behavior of an engine. However, the deduction of physical-based models often require significant development time and is usually very specific. Reviews and different model implementations can be found in the literature [16] [6] [18].

1.1 Our Fuzzy Modelling Approach

Our modelling approach tries to find a compromise between a physical-oriented and a pure mapping approach by extracting automatically high-dimensional nonlinear fuzzy models from static as well as dynamic measurements recorded during the test phases of an engine. These measurements reflect the emission behavior of the corresponding engine and hence provide a representation of the intrinsic relations between some physical measurement channels (such as temperatures, pressures, engine speed, torque etc.) and the NOx concentration in the exhaust gases in the emission. Our methodology of a machine-learning driven building up of fuzzy models is able to recognize this relation and hence to map input values (from a subset of measurement channels) onto the NOx concentration (used as target) appropriately, fully automatically and with high precision. The learning algorithm consists of two phases, the first phase estimates the clusters = rules in the product space with an iterative batch learning variant of the evolving vector quantization method *eVQ* [8]; the second phase completes the learning by estimating linear weights in the consequent hyper-planes of the models with a weighted least squares approach.

The fuzzy model generation process proposed in this paper benefits of automated model generation with very low human intervention. On the other hand, physical-based models usually need a long setting-up phase where physical relations and boundary conditions are specified. In the case of higher order CFD models, this includes laborious tasks as geometry definition, grid generation, etc. while in simpler look-up table mapping alternatives the definition of the number of tables, their size and input signals, the general model structure and how the different tables outputs are combined, sums up a considerable development time. Presented automated model generation can shorten this process, and also the data fitting process. Another advantage of the presented methodology is that the model structure and the automated model training can simultaneously deal with both steady and dynamical data, thus shortcoming the existence of two different engine states. Main drawback of pure data-driven approach is that resulting models are of very low physical interpretability; this issue also affects fine tuning capabilities and manual fine-tuning and model adaptation, which is usually appreciated by the designers for correcting problems during the engine development process. However, this deficiency is weakened when using fuzzy systems in the data-driven modelling process as these are providing some insight into and understanding of the model components in form of linguistic rules (if-then causualities).

The paper is organized in the following way: Section 2 provides an insight into the experimental setup we used at an engine test bench for performing steady and transient tests; Section 3 describes the fuzzy modelling component, Section 4 provides an extensive evaluation of the fuzzy models trained from steady state and transient measurements and a mixture of these; Section 5 concludes the paper with a summary of achieved and open issues.

2 Experimental Setup and DoE

2.1 Experimental Setup

The engine was a common-rail diesel engine, equipped with a variable geometry turbine (VGT) and an exhaust recirculation (EGR) system. The engine control was performed by means of an externally calibratable engine control unit (ECU) in a way that boost pressure, exhaust gas recirculation rate and injection characteristics could be modified during the tests. Temperature conditioning systems were used for the control of the engine coolant temperature and of the intake air mass flow. An eddy current dynamometer was used for loading the engine, which was able to perform transient tests, and thus to replicate the driving tests measured in real-life driving conditions.

Different acquisition frequencies ranging from 10 Hz to 100 Hz were used depending on the signal characteristic. Since most engine signals fluctuate with the engine firing frequency, antialiasing filters were used when needed for mitigating this effect. Two different test campaigns, covering steady and transient operation, were performed. A test design comprising 363 steady operation tests was done. Tests ranged from full load to idle operation, and different repetitions varying EGR rate (i.e. oxygen concentration in the intake gas), boost pressure, air charge temperature and coolant temperature were done. Test procedure for each one of the steady tests was as follows: 1.) Operation point is fixed, and stability of the signals is checked; 2.) data is acquired during 30 s; 3.) data is averaged for the full test; 4.) data is checked for detecting errors, which are corrected when possible. The last two steps are usually done offline. As a result of this procedure, steady test campaign produced a data matrix were each row

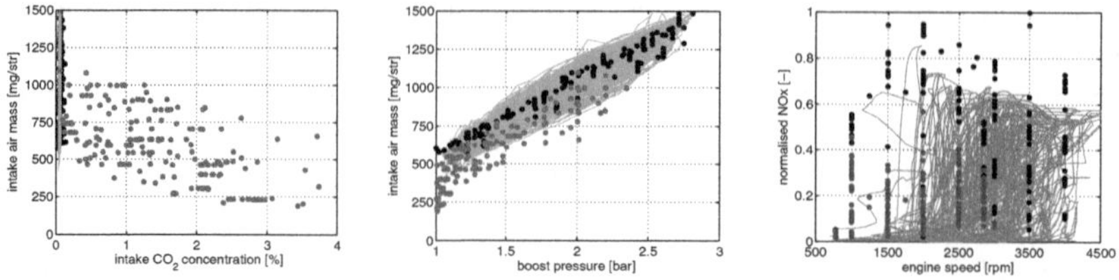

Fig. 1. Comparison of the range of several operating variables during the steady tests without EGR (black points), those with EGR (grey points) and during the transient test (light grey line)

corresponds to a specific test, while each column contained the value from a measured or calculated variable (such as engine speed, intake air mass, boost pressure etc.). A second test campaign covering several engine transient tests was performed. Tested transient covered European MVEG homologation cycle and several driving conditions, including dan MVEG cycle, a sportive driving profile in a mountain road and two different synthetic profiles. Several repetitions of these tests were done varying EGR and VGT control references, in a way that EGR rate and boost pressure are varied from one test to another. In opposition to steady state tests, where each test provides an independent row of averaged values, here a matrix of dynamically dependent measurements is provided. In addition, during dynamical operation the engine reaches states that are not reachable in steady operation. In Figure 1 boost and exhaust pressures are represented for the steady tests and for a dynamical driving cycle, note that the range of the variation during the transient operation clearly exceeds that of the steady operation. Furthermore, steady tests do not show the dynamical (i.e. temporal) effects.

3 Fuzzy Model Identification

3.1 Pre-processing the Data

Our fuzzy modelling component is applicable to any type of data, no matter whether they were collected from steady-state or from dynamic processes. The only assumption is that the data is available in form of a data matrix, where the rows represent the single measurements and the columns represent the measured variables. This is guaranteed by the data recording and pre-processing phase as described in the previous section. In case of dynamic data, the matrix (ev. after some down-sampling procedure) has to be shifted in order to include time delays of the measurement variables and hence to be able to identify dynamic relationships in form of k-step ahead prediction models. In case of a mixed data set (steady-state and dynamic data) for achieving a single model, in order to prevent time-intensive on-line checks and switches between two different models, the static data is appended at the end of the dynamic data matrix, by copying the same (static) value of the variables to all of their time delays applied in the dynamic data matrix.

3.2 Model Architecture

For the fuzzy modelling component (based on the pre-pared data sets), we exploit the Takagi-Sugeno fuzzy model architecture [14] with Gaussian membership functions and product operator, also known as fuzzy basis function networks [17] and defined by:

$$\hat{f}(\boldsymbol{x}) = \hat{y} = \sum_{i=1}^{C} l_i \Psi_i(\boldsymbol{x}) \qquad \Psi_i(\boldsymbol{x}) = \frac{e^{-\frac{1}{2}\sum_{j=1}^{p}\frac{(x_j-c_{ij})^2}{\sigma_{ij}^2}}}{\sum_{k=1}^{C} e^{-\frac{1}{2}\sum_{j=1}^{p}\frac{(x_j-c_{kj})^2}{\sigma_{kj}^2}}} \tag{1}$$

with consequent functions

$$l_i = w_{i0} + w_{i1}x_1 + w_{i2}x_2 + \dots + w_{ip}x_p \tag{2}$$

The symbol x_j denotes the j-th input variable (static or dynamically time-delayed), c_{ij} the center and σ_{ij} the width of the Gaussian fuzzy set in the j-th premise part of the i-th rule. Often, it is criticized that the consequents have a poor interpretable power [4] as represented by hyper-planes instead of fuzzy partitions for which linguistic labels can be applied. However, it depends on the application which variant of consequents is preferred. For instance, in control or identification problems it is often interesting to know in which parts the model behaves almost constant or which influence the different variables have in different regions [12].

3.3 Model Training Procedure

Our model training procedure consists of two main phases: the first phase estimates the number, position and range of influence of the fuzzy rules and the fuzzy sets in their antecedent parts; the second phase estimates the linear consequent parameters by applying a local learning approach [19] with the help of a weighted least squares optimization function. The first phase is achieved by finding an appropriate cluster partition in the product space with the help of evolving vector quantization (*eVQ*) [8], which is able to extract the required number of rules automatically by evolving new clusters on demand. The basic steps of this algorithm are:

- Checking whether a newly loaded sample (from the off-line data matrix) fits into the current cluster partition; this is achieved by checking whether an already existing cluster is close enough to the current data sample.
- If yes, update the nearest cluster center $\boldsymbol{c}_{win}$ by shifting it towards the current data sample:

$$\boldsymbol{c}_{win}^{(new)} = \boldsymbol{c}_{win}^{(old)} + \eta(\boldsymbol{x} - \boldsymbol{c}_{win}^{(old)}) \tag{3}$$

  — by using a decreasing learning gain η over the number of samples forming this cluster.

- If no, a new cluster is born in order to cover the input/output space sufficiently well; its center is set to the current data sample and the algorithm continues with the next sample.
- Estimating the range of influence of all clusters by calculating the variance in each dimension based on those data samples responsible for forming the single clusters.

Once the local regions (clusters) are elicited, they are projected from the high-dimensional space to the one-dimensional axes to form the fuzzy sets as antecedent parts of the rules. Hereby, one cluster is associated with one rule.

The (linear) consequent parameters are estimated by local learning approach, that is for each rule separately. This is also because in [7] it is reported that local learning has some favorable advantages over global learning (estimating the parameters from all rules in one sweep) such as smaller matrices to be inverted (hence more stable and faster), or providing a better interpretation of the consequent functions. With the weighting matrix

$$Q_i = \begin{bmatrix} \Psi_i(\boldsymbol{x}(1)) & 0 & \dots & 0 \\ 0 & \Psi_i(\boldsymbol{x}(2)) & \dots & 0 \\ \vdots & \vdots & \vdots & \vdots \\ 0 & 0 & \dots & \Psi_i(\boldsymbol{x}(N)) \end{bmatrix}$$

a weighted least squares method is achieved in order to estimate the linear consequent parameters $\hat{w}_i$ for the ith rule:

$$\hat{\boldsymbol{w}}_{\boldsymbol{i}} = (R_i^T Q_i R_i)^{-1} R_i^T Q_i \boldsymbol{y} \tag{4}$$

with R_i the regression matrix containing the original variables (+ some time delays in case of dynamic data).

In case of an ill-posed problem, i.e. the matrix $R_i^T Q_i R_i$ singular or nearly singular, we apply the estimation of consequents by including a Tichonov regularization [15] step, that is we add αI to $R_i^T Q_i R_i$ with α a regularization parameter. In literature there exists a huge number of regularization parameter choice methods, a comprehensive survey can be found in [3]. Here, we use an own developed heuristic method, proven to be efficient in both, computational performance and accuracy of the final obtained fuzzy models [10]. For further details on our fuzzy modelling approach, called *FLEXFIS* which is short for *FLEXible Fuzzy Inference Systems*, see also [9].

4 Evaluation

4.1 Setup

Measurements. Experimental tests presented in Section 2 were used for evaluating the performance of our fuzzy modelling technique presented. For that, data was rearranged in different data sets, namely:

- A steady-state data set including all 363 measurements.
- A dynamic data including the 42 independent tests delivering 217550 measurements in total, down-sampled to 21755 measurements, approx. three quarters taking as training data, the remaining quarter as test data.
- Mixed data which appends the steady-state data to the dynamic measurements to form one data set where the fuzzy models are trained from.

Physical Model. In order to establish a baseline, fuzzy model results were compared with those obtained with a simple physical-oriented model. This physical-based model is composed by a mean value engine model (MVEM) similar to the one presented in [5] and a NOx emission model which correlates the NOx emissions with several operating variables, mainly engine speed, load and oxygen concentration at the intake manifold. The applied NOx model used several look-up tables which provided NOx nominal production and corrective parameters which depended on the operative conditions. For identifying the relevant operative condition and fixing the NOx model structure, several thousands of simulation of a Zeldovich mechanism-based code [2] were used.

Fuzzy Model. For the fuzzy model testing, we used the nine essential input channels used in the physical-oriented model which were extended by a set of additional 30 measurement channels, used as intermediate variables in the physical model (such as EGR rate, intake manifold oxygen concentration, etc.). For the dynamic and mixed data set all of these were delayed up to 10 samples. We split the measurements into two data sets, one for model evaluation and final training and one for model testing (final validation). Model evaluation is performed within a 10-fold cross-validation procedure coupled with a best parameter grid search scenario. In order to choose appropriate inputs, i.e. the inputs which are most important for achieving a model with high accuracy, we apply a modified version of forward selection as filter approach before the actual training process. Once the 10-fold cross-validation is finished for all the defined grid points, we take those parameter setting for which the CV error, measured in terms of the mean absolute error between predicted and real measured target (NOx), is minimal; and train a final model using all training data.

4.2 Some Results

Figure 2 shows the results of this model when tested on the steady data set, including the correlation between predicted and measured values (left plot), the absolute error over all samples (middle plot) and the histogram of the errors normalized to the unit interval (right plot). A major portion of the errors lie in the 10% range, see the histogram plot at the right side in Figure 2. The same is the case when applying the data-driven fuzzy component, compare Figure 3 with Figure 2. However, the error performance is slightly worse in case of physical modeling: the normalized mean absolute error (MAE) is about 20% above the normalized mean absolute error (MAE) in case of the Takagi-Sugeno fuzzy models. A more clear improvement of the fuzzy modeling approach over the

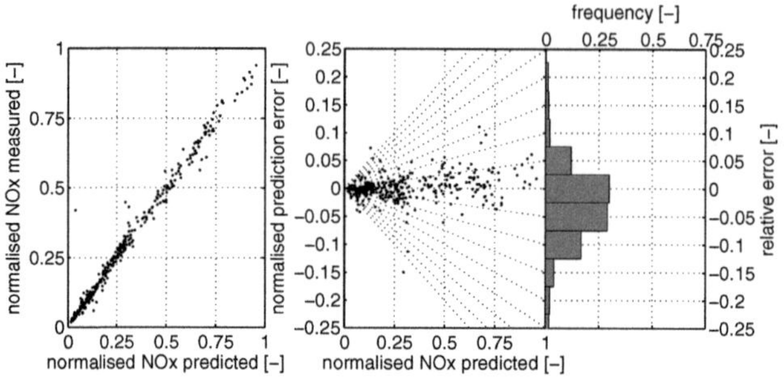

Fig. 2. Physical-based model results when applied to steady tests

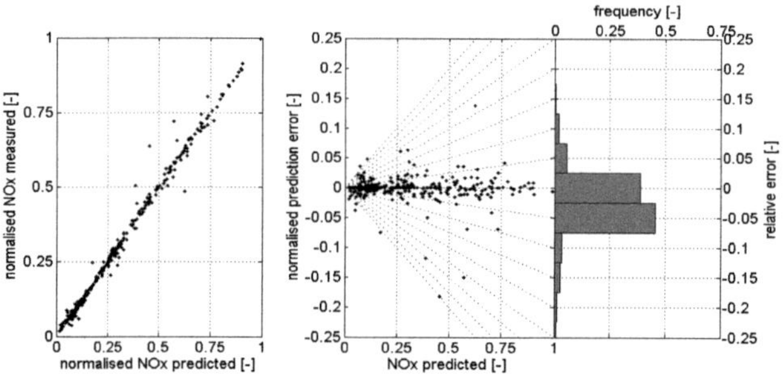

Fig. 3. Fuzzy model results when applied to steady tests

physical-based model can be realized when comparing the two right most plots in Figures 3 and 2: significantly more samples are distributed around 0 error. Figure 4 illustrates the results of the physical-based model applied to two fragments of the transient tests. Figure 5 below shows the results obtained from the fuzzy model (trained based on the optimal parameter setting) on the same two fragments. Obviously, our model is able to follow the highly fluctuating trend of the measured NOx content during dynamic system states quite well and similarly as the physical-based model (compare lines in dark predicted values with light measured values). In total 8 such fragments were available as independent test set, the normalized MAE of the physical model over all these fragments was 2.23%, of the fuzzy model slightly better: 2.04%.

One major problem in the physical modelling approach is that the static model must be modified for being applied to dynamic data. In fact, someone may simply use the dynamic model for static measurements or vice versa. However, this is somewhat risky, as significant extrapolation situation may arise (compare Figure 1). Hence, it is a big challenge to have one single model for transient and steady states available. This can be accomplished with our fuzzy modelling approach by using the data set extension as demonstrated in Section 3.1 and apply the *FLEXFIS* (batch) modelling procedure. Similar error plots as in Figures 3 and 5 could be obtained, the resulting normalized MAE worsened only slightly: from

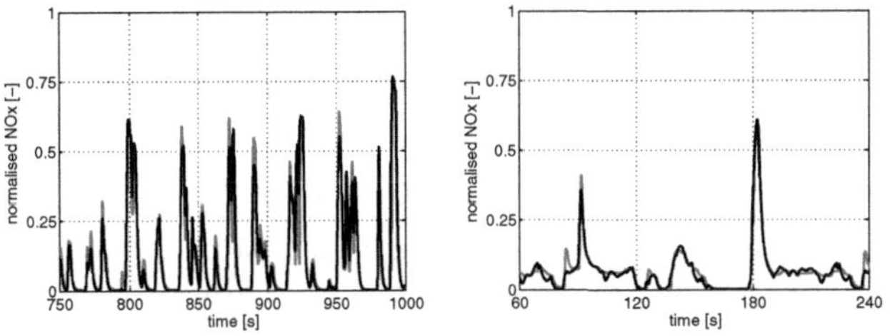

Fig. 4. Physical-based model results (black) when applied to two fragments of the transient tests and experimental measurement (grey)

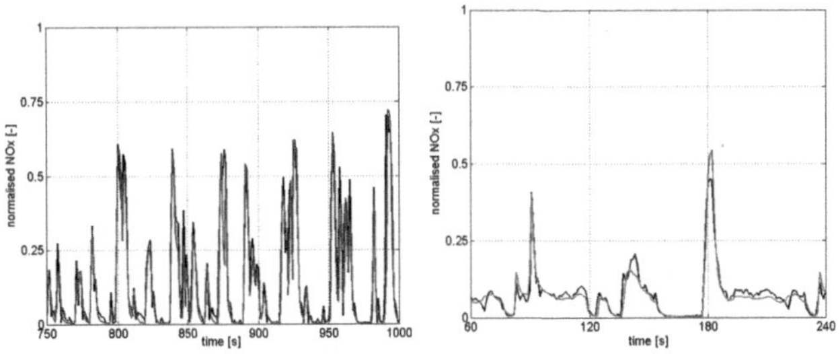

Fig. 5. Fuzzy model results (black) when applied to two fragments of the transient tests and experimental measurement (grey), compare with Figure 4

1.32% to 1.61% accuracy for the static and from 2.04% to 2.46% accuracy for the dynamic data sets. The complexities of our models measured in terms of the number of rules stayed in a reasonable range: 14 rules in case of static data, 11 in case of dynamic data and 22 for the mixed data set; the number of finally used inputs was between 5 and 9.

5 Conclusions

In this paper, we presented an alternative to conventional NOx prediction models by training fuzzy models directly from measurement data, representing static and dynamic operation modes. The used fuzzy systems modelling method was the *FLEXFIS* approach. The fuzzy models could slightly outperform physical-based models, no matter whether using static and dynamic data sets. Together with the aspects that 1.) it was also possible to set up a mixed model with high accuracy, which is able to predict new samples either from static or dynamic operation modes, and 2.) to have a kind of plug-and-play method available for setting up new models, we can conclude that our fuzzy modelling component is in fact a reliable and good alternative to physical-based models.

References

1. Arrègle, J., López, J., Guardiola, C., Monin, C.: Sensitivity study of a nox estimation model for on-board applications. SAE paper 2008-01-0640 (2008)
2. Arrègle, J., López, J., Guardiola, C., Monin, C.: On board NOx prediction in diesel engines. A physical approach. In: del Re, L., Allgöwer, F., Glielmo, L., Guardiola, C., Kolmanovsky, I. (eds.) Automotive Model Predictive Control: Models, Methods and Applications, pp. 27–39. Springer, Heidelberg (2010)
3. Bauer, F.: Some considerations concerning regularization and parameter choice algorithms. Inverse Problems 23, 837–858 (2007)
4. Casillas, J., Cordon, O., Herrera, F., Magdalena, L.: Interpretability Issues in Fuzzy Modeling. Springer, Heidelberg (2003)
5. Eriksson, L., Wahlström, J., Klein, M.: Physical modeling of turbocharged engines and parameter identification. In: del Re, L., Allgöwer, F., Glielmo, L., Guardiola, C., Kolmanovsky, I. (eds.) Automotive Model Predictive Control: Models, Methods and Applications, pp. 59–79. Springer, Heidelberg (2010)
6. Kamimoto, T., Kobayashi, H.: Combustion processes in diesel engines. Progress in Energy and Combustion Science 17, 163–189 (1991)
7. Lughofer, E.: Evolving Fuzzy Models — Incremental Learning, Interpretability and Stability Issues, Applications. VDM Verlag Dr. Müller, Saarbrücken (2008)
8. Lughofer, E.: Extensions of vector quantization for incremental clustering. Pattern Recognition 41(3), 995–1011 (2008)
9. Lughofer, E.: FLEXFIS: A robust incremental learning approach for evolving TS fuzzy models. IEEE Trans. on Fuzzy Systems 16(6), 1393–1410 (2008)
10. Lughofer, E., Kindermann, S.: Improving the robustness of data-driven fuzzy systems with regularization. In: Proc. of the IEEE World Congress on Computational Intelligence, WCCI 2008, Hongkong, pp. 703–709 (2008)
11. Moos, R.: A brief overview on automotive exhaust gas sensors based on electroceramics. International Journal of Applied Ceramic Technology 2(5), 401–413 (2005)
12. Piegat, A.: Fuzzy Modeling and Control. Physica Verlag. Physica Verlag, Springer Verlag Company, Heidelberg (2001)
13. Riesco, J., Payri, F., Molina, J.B.S.: Reduction of pollutant emissions in a HD diesel engine by adjustment of injection parameters, boost pressure and EGR. SAE paper 2003-01-0343 (2003)
14. Takagi, T., Sugeno, M.: Fuzzy identification of systems and its applications to modeling and control. IEEE Trans. on Systems, Man and Cybernetics 15(1), 116–132 (1985)
15. Tikhonov, A., Arsenin, V.: Solutions of ill-posed problems. Winston & Sonst, Washington D.C. (1977)
16. Gartner, U., Hohenberg, G., Daudel, H., Oelschlegel, H.: Development and application of a semi-empirical nox model to various hd diesel engines. In: Proc. of THIESEL, Valencia, Spain, pp. 487–506 (2002)
17. Wang, L., Mendel, J.: Fuzzy basis functions, universal approximation and orthogonal least-squares learning. IEEE Trans. Neural Networks 3(5), 807–814 (1992)
18. Weisser, G.: Modelling of combustion and nitric-oxide formation for medium-speed DI Diesel engines: a comparative evaluation of zero- and three-dimensional approaches. Ph.D. thesis, Swiss Federal Institute of Technology (Zuerich 2001)
19. Yen, J., Wang, L., Gillespie, C.: Improving the interpretability of TSK fuzzy models by combining global learning and local learning. IEEE Trans. on Fuzzy Systems 6(4), 530–537 (1998)

Coping with Uncertainty in Temporal Gene Expressions Using Symbolic Representations

Silvana Badaloni and Marco Falda*

Dept. of Information Engineering
University of Padova
Via Gradenigo 6/A - 35131 Padova, Italy
{silvana.badaloni,marco.falda}@unipd.it

Abstract. DNA microarrays can provide information about the expression levels of thousands of genes, however these measurements are affected by errors and noise; moreover biological processes develop in very different time scales. A way to cope with these uncertain data is to represent expression level signals in a symbolic way and to adapt sub-string matching algorithms (such as the Longest Common Subsequence) for reconstructing the underlying regulatory network. In this work a first simple task of deciding the regulation direction given a set of correlated genes is studied. As a validation test, the approach is applied to four biological datasets composed of Yeast cell-cycle regulated genes under different synchronization methods.

1 Introduction

A significant challenge in dealing with genomic data comes from the enormous number of genes involved in biological systems (for example the human Genome has 30.000 genes). Furthermore, uncertainty in the data, represented by the presence of noise, enhances the difficulty in distinguishing real from random patterns and increases the potential of misleading analyses. To overcome these problems, some studies proposed to identify symbolic features of the series; examples include temporal abstraction-based methods that define trends (i.e., increasing, decreasing and steady) over subintervals [1], or a difference-based method that uses the first and second order differences in expression values to detect the direction and rate of change of the temporal expressions for clustering [2].

In this paper a recently started study about symbolic representations for gene temporal profiles affected by uncertainty will be presented. Tests performed on a simple fragment of a real biological regulatory network seem to show that such qualitative representations could be useful for finding the correct regulation directions, since they have the further advantage to be able to abstract delays among genes due to biological reactions, and therefore be less penalized by the diverse temporal scales typical of biological systems.

* Corresponding author.

E. Hüllermeier, R. Kruse, and F. Hoffmann (Eds.): IPMU 2010, Part II, CCIS 81, pp. 11–20, 2010.

2 Symbolic Representations

Interactions among genes can be formalized as a directed graph $\langle G, A\rangle$ where G represents the set of genes and A the set of relations between genes; the graph can be weighted by associating a number to each arc $a_{ij} \in A$, but in a simpler scenario each arc a_{ij} will assume the value 1 or 0 depending on the fact that gene i influences gene j or not. The temporal evolution of a single gene in a regulatory network, that is its *time series*, is usually represented as a sequence of K samples $\mathcal{V} = \{v_k, k \in \{1, \ldots, K\}\}$, where $k \in \mathbb{N}^+$ is the index of the discrete sampling time and $v_k \in \mathbb{R}$ its value at index k.

2.1 Preprocessing of Data

When one is measuring a variable that is both slowly varying and also corrupted by random noise, as in the case of gene temporal profiles, it can sometimes be useful to replace each data point by some kind of local average of surrounding data points. Since nearby points measure very nearly the same underlying value, averaging can reduce the level of noise without (much) biasing the value obtained.

A particular type of low-pass filter, well-adapted for data smoothing is Savitzky-Golay filters family, initially developed to render visible the relative widths and heights of spectral lines in noisy spectrometric data [3]. The simplest type of digital filter replaces each data value $v_k \in \mathcal{V}$ by a linear combination of itself and some number of nearby neighbors:

$$\widetilde{v}_k = \sum_{n=-n_L}^{n_R} c_n \cdot v_{k+n}$$

Here n_L is the number of points used "to the left" of a data point k, i.e., earlier than it, while n_R is the number used to the right, i.e., later.

The algorithm of Savitzky-Golay applies the least-squares principle to determine an improved set of kernel coefficients c_n for use in a digital convolution; these improved coefficients are determined using polynomials rather than, as for the case of simple averaging, a constant value determined from a sub-range of data. Indeed, the Savitzky-Golay method could be seen as a generalization of averaging data, since averaging a sub-range of data corresponds to a Savitzky-Golay polynomial of degree zero. The idea of this kind of filtering is to find coefficients that preserve higher moments.

2.2 Features

To reason about the temporal evolution of each gene, a symbolic representation can be developed starting from quantitative data and applying simple Discrete Calculus definitions; in this way, it is possible to describe a time series $\mathcal{V}$ as sequence of symbols $\mathcal{S}_{\mathcal{V}}$ representing significant features. The features that have been considered are: maxima, minima, inflection points and points where the series becomes stationary, zero or saturates:

Definition 1 (Symbolic features). *The significant features of a time series are defined over the set* $\mathcal{F} = \{M, m, f, s, z, S\}$.

Definition 2 (Symbolic representation). *A time series* $\mathcal{V}$ *can be represented as a sequence of symbols* $\mathcal{S}_{\mathcal{V}} = \{\sigma_j, j \in \{1, \ldots, J\}\}$ *where each symbol* σ_j *belongs to the set of features* $\mathcal{F}$.

To maintain a link with the original series a mapping function $m_{\mathcal{S}}$ between $\mathcal{S}_{\mathcal{V}}$ and $\mathcal{V}$ is defined:

Definition 3 (Mapping function). *Given a symbolic representation* $\mathcal{S}_{\mathcal{V}}$ *and its original time series* $\mathcal{V}$, $m_{\mathcal{S}} : \mathbb{N}^+ \rightarrow \mathbb{N}^+$ *is a function that maps the index* j *of a symbol* $\sigma_j \in \mathcal{S}_{\mathcal{V}}$ *in the index* k *of the corresponding time series element* $v_k \in \mathcal{V}$.

2.3 Enriching the Symbolic Representation

In the symbolic sequence it is possible to add further information, namely the intensity, both relative and absolute, of the time series at a given point, and the width of the feature itself. To do this, it is necessary to define how this kind of information will be represented, and a natural way is to express it in terms of time series parameters.

Definition 4 (Range of a time series). *The range of a time series* $\mathcal{V} = \{v_k, k \in \{1, \ldots, K\}\}$ *is provided by the function* $ext : \mathbb{R}^K \longrightarrow \mathbb{R}^+$ *defined as* $ext(\mathcal{V}) = |\max_k(v_k) - \min_k(v_k)|$.

Definition 5 (Range of a set of time series). *The range of a set of time series* $\mathcal{W} = \{\mathcal{V}_h, h \in \{1, \ldots, H\}\}$ *is defined as* $set\_ext : (\mathbb{R}^K)^H \longrightarrow \mathbb{R}^+$, $set\_ext(\mathcal{W}) = |\max(v_k) - \min(v_k)|, v_k \in \mathcal{W}$.

Definition 6 (Length of a time series). *The length of a time series is the cardinality of the set* $\mathcal{V}$ *and it will be written as* $|\mathcal{V}|$.

Given these basic parameters which allow to have a reference w.r.t. a specific time series and w.r.t. the whole set of time series, it is possible to describe more intuitively the properties of the features identified.

Definition 7 (Absolute height of a feature). *Given a set of time series* $\mathcal{W} = \{\mathcal{V}_h, h \in \{1, \ldots, H\}\}$ *and a symbolic sequence* $\mathcal{S}_{\mathcal{V}h}$, *the absolute height of the feature represented by the symbol* $\sigma_j \in \mathcal{S}_{\mathcal{V}h}$ *is defined by the function* $ha_{\mathcal{S}} : \mathbb{N}^+ \longrightarrow \mathbb{R}^+$

$$ha_{\mathcal{S}}(j) = \left| \frac{v_{m_{\mathcal{S}}(j)}}{set\_ext(\mathcal{W})} \right|$$

Definition 8 (Relative height of a feature). *Given time series* $\mathcal{V} = \{v_k\}$ *and its symbolic sequence* $\mathcal{S}_{\mathcal{V}}$, *the relative height of the feature represented by the symbol* $\sigma_j \in \mathcal{S}_{\mathcal{V}}$ *is defined by the function* $hr_{\mathcal{S}} : \mathbb{N}^+ \longrightarrow \mathbb{R}^+$

$$hr_{\mathcal{S}}(j) = \left| \frac{v_{m_{\mathcal{S}}(j)} - v_{m_{\mathcal{S}}(j-1)}}{ext(\mathcal{V})} \right|$$

Definition 9 (Width of a feature). *Given time series $\mathcal{V}$ and its symbolic sequence $\mathcal{S}_\mathcal{V}$, the width of the feature represented by the symbol $\sigma_j \in \mathcal{S}_\mathcal{V}$ is defined by the function $wr_\mathcal{S} : \mathbb{N}^+ \longrightarrow \mathbb{R}^+$*

$$wr_\mathcal{S}(j) = \left| \frac{m_\mathcal{S}(j) - m_\mathcal{S}(j-1)}{|V|} \right|$$

These functions can be associated to the symbols of a sequence $\mathcal{S}$ by means of a function $q_\mathcal{S}$ that describes the properties of a feature.

Definition 10 (Properties of a symbol). *Given a symbolic sequence $\mathcal{S}_\mathcal{V}$, the properties of a symbol $\sigma_j \in \mathcal{S}_\mathcal{V}$ are defined by the function $q_\mathcal{S} : \mathbb{N}^+ \longrightarrow \langle \mathbb{R}^+, \mathbb{R}^+, \mathbb{R}^+ \rangle$*

$$q_\mathcal{S}(j) = \langle ha_\mathcal{S}(j), hr_\mathcal{S}(j), wr_\mathcal{S}(j) \rangle$$

Example 1. The series $\mathcal{V}$ in Figure 1 can be represented by $\mathcal{S}_\mathcal{V} = \{m, f, M, f, m, \ldots\}$, and the properties of its symbols are $q_\mathcal{S}(1) = \langle 0.63, 0, 0 \rangle$, $q_\mathcal{S}(2) = \langle 0.12, 0.51, 0.06 \rangle$, $q_\mathcal{S}(3) = \langle 0.33, 0.45, 0.09 \rangle$, $q_\mathcal{S}(4) = \langle 0.08, 0.25, 0.07 \rangle$, $q_\mathcal{S}(5) = \langle 0.17, 0.25, 0.09 \rangle$ *et c.* .

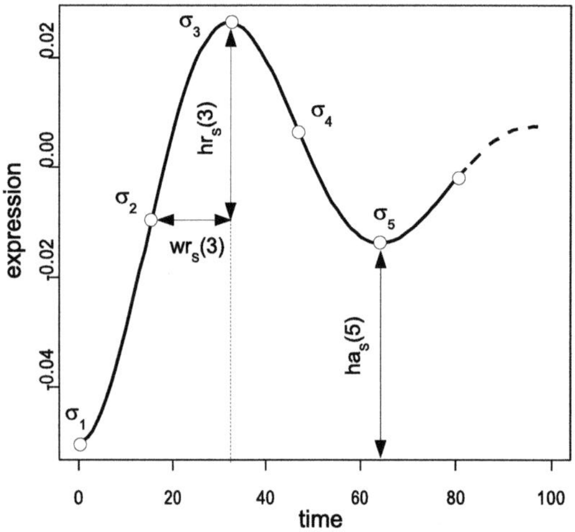

Fig. 1. Example of a numerical time series and its symbolic representation

3 Reasoning about Regulation Directions

The symbolic representation of time series allows reasoning about strings in which each symbol representing a feature is linked to a point of the real series (through an index given by the function $m_\mathcal{S}$). In the following five methods proposed by [4] have been considered. In all cases the hypothesis that in a causal process the cause always precedes its consequence is assumed and exploited. By now just the basic symbolic representations have been used.

Reverse Engineering of a gene regulatory network means inferring relations among genes starting from experimental data, in this specific case from time series data. It can be solved by providing a "similarity measure" function $f : \mathbb{N}^{|G|} \longrightarrow \mathbb{R}$ from a set of indices, which identify the genes, to a real number; $|G|$ represents the cardinality of the set G. Since the focus of this work is the symbolic processing of time series, the domain of the measures will be $\mathcal{F}^J \times \mathcal{F}^J$, that is pairs of symbolic sequences whose length is J. In this work these measures will be used just to establish whether two genes are correlated or not, so the resulting real number will be eventually compared with a given threshold to obtain a Boolean value.

3.1 Shifted Correlation (sC)

The simplest metric that can be applied on two symbol sequences x and y is a Pearson correlation:

$$r = \frac{\sum_i (x_i - \bar{x}) \cdot (y_i - \bar{y})}{\sqrt{\sum_i (x_i - \bar{x})^2} \cdot \sqrt{\sum_i (y_i - \bar{y})^2}}$$

where $\bar{x}$ and $\bar{y}$ are the means of the sequences.

The aim is to identify directions, so this measure has been made asymmetric by shifting the series by one temporal sample (the cause precedes the effect); it will be called "Shifted Correlation" (sC). The correlation is applied to the original time series points identified by the mapping function $m_{\mathcal{S}}$. In the case the series have different lengths the shorter length is considered.

3.2 Matching between Maxima and Minima with Temporal Precedences (tMM)

A second easy idea is to find a one-to-one correspondence between maxima and minima, direct or inverse, in the symbolic representation taking into account the fact that the regulator gene should always precede the regulated one, and to evaluate the relative length of the matching features with respect to the shorter sequence:

$$tMM(\mathcal{S}_1, \mathcal{S}_2) = \frac{\max(|\mathcal{M}_{1,2}|, |\mathcal{M}_{\overline{1},2}^{-}|)}{\min(|\mathcal{M}_1|, |\mathcal{M}_2|)}$$

$\mathcal{M}_1$ and $\mathcal{M}_2$ are sub-sequences containing only maxima and minima of the sequences $\mathcal{S}_1$ and $\mathcal{S}_2$, $\mathcal{M}_{1,2} = \{\sigma_{j1} \in \mathcal{S}_1 : \exists \sigma_{j2} \in \mathcal{S}_2 \wedge \sigma_{j1} = \sigma_{j2} \wedge m_{\mathcal{S}_1}(j_1) < m_{\mathcal{S}_1}(j_2)\}$ and $\mathcal{M}_{\overline{1},2}^{-}$ is defined as $\mathcal{M}_{1,2}$ but matching in an inverse fashion (e.g.: maxima with minima and vice versa).

3.3 Temporal Longest Common Substring ($tLCStr$)

A further step is to notice that noise could alter the series, therefore it could be the case that just some segments of the temporal expressions match, therefore looking for the longest segment should help. The longest segment shared between two symbolic sequences can be found using the Longest Common Substring algorithm, which exploits Dynamic Programming techniques and has a $O(J^2)$ asymptotic complexity in the worst case [5]. As for the precedence criterion, the algorithm matches only the features of the regulator gene which precede the corresponding features of the regulated one (the "t" in the name *tLCStr*). The formula is:

$$tLCStr(\mathcal{S}_1, \mathcal{S}_2) = \frac{\max(|tLCStr_{1,2}|, |tLCStr^-_{1,2}|)}{\min(|\mathcal{S}_1|, |\mathcal{S}_2|)}$$

where $tLCStr^-_{1,2}$ is the Longest Common Substring matching in an inverse fashion.

3.4 Temporal Longest Common Subsequence (*tLCS*)

It is possible to hypothesize that the effects of a gene could be hidden by saturation effects, and therefore trying to identify the longest non-contiguous subsequence shared between two symbolic sequences could be useful. Also in this case there exists a $O(J^2)$ algorithm based on Dynamic Programming techniques [5]; the formula is analogous to the previous one and so it has not been reported here.

The precedence criterion has been added as in the previous case:

$$tLCS(\mathcal{S}_1, \mathcal{S}_2) = \frac{\max(|tLCS_{1,2}|, |tLCS^-_{1,2}|)}{\min(|\mathcal{S}_1|, |\mathcal{S}_2|)}$$

where $tLCS^-_{1,2}$ is the Longest Common Subsequence matching in an inverse fashion.

3.5 Directional Dynamic Time Warping (*dDTW*)

The last algorithm, adapted to take into account the asymmetry of the time arrow, is the Dynamic Time Warping, a procedure coming from the Speech Recognition field [6]; it is a "elastic" alignment that allows similar shapes to match even if they are out of phase in the time axis; the algorithm complexity is again $O(J^2)$. The precedence criterion has been added by matching features of regulated genes with preceding features of the regulator ones. The computations are performed on the original time series points identified by the mapping function $m_{\mathcal{S}}$.

3.6 Adding Qualitative Properties

In the symbolic sequence it is possible to add further information, namely the intensity, both relative and absolute, of the time series at a given point, and the relative width of the feature itself; the definitions are not reported in this paper, we will simply postulate the existence of the functions $hr_{\mathcal{S}}(j)$, $ha_{\mathcal{S}}(j)$ and $wr_{\mathcal{S}}(j)$ respectively.

These functions can be associated to the symbols of a sequence $\mathcal{S}$ by means of a function $q_{\mathcal{S}}$ that describes the properties of a feature.

Definition 11 (Qualitative properties). *Given a symbolic sequence $\mathcal{S}_\mathcal{V}$, the properties of a symbol $s_j \in \mathcal{S}_\mathcal{V}$ are given by the function $q_\mathcal{S} : \mathbb{N}^+ \longrightarrow \langle \mathbb{R}^+, \mathbb{R}^+, \mathbb{R}^+ \rangle$*

$$q_\mathcal{S}(j) = \langle ha_\mathcal{S}(j), hr_\mathcal{S}(j), wr_\mathcal{S}(j) \rangle$$

Functions $ha_\mathcal{S}(j)$, $hr_\mathcal{S}(j)$ and $wr_\mathcal{S}(j)$ has been quantized in a fixed number n of levels:

Definition 12 (Quantized functions). *Given a number $n \in \mathbb{N}$, the quantized version of a function $f : \mathbb{N}^+ \longrightarrow \mathbb{R}^+$ is a function $\varphi_n : (\lambda : \mathbb{N}^+ \longrightarrow \mathbb{R}^+) \longrightarrow \mathbb{N}^+$*

$$\varphi_n[f] = \left\lceil \frac{f}{|\max[f]|} \cdot n \right\rceil$$

In this way the properties can be fuzzified in n levels and compared in an approximated way using a function $\widetilde{eq}_{\mathcal{S}_1,\mathcal{S}_2}$ defined as follows.

Definition 13 (Approximately equal). *Given two symbol sequences $\mathcal{S}_1$ and $\mathcal{S}_2$ the function $\widetilde{eq}_{\mathcal{S}_1,\mathcal{S}_2} : \mathbb{N}^+ \times \mathbb{N}^+ \rightarrow \{0,1\}$ is defined as*

$$\begin{aligned}\widetilde{eq}_{\mathcal{S}_1,\mathcal{S}_2}(j_1, j_2) = g(&(\varphi_n[ha_{\mathcal{S}1}](j_1)) = \varphi_n[ha_{\mathcal{S}2}](j_2)),\\ &(\varphi_n[hr_{\mathcal{S}1}](j_1)) = \varphi_n[hr_{\mathcal{S}2}](j_2)),\\ &(\varphi_n[wr_{\mathcal{S}1}](j_1)) = \varphi_n[wr_{\mathcal{S}2}](j_2)))\end{aligned}$$

where $g : \{0,1\}^3 \rightarrow \{0,1\}$ is a function that weights the relevance of each qualitative property and can be defined using heuristics.

4 Results

To test the five measures discussed above, time series coming from the Yeast cell cycle under four different synchronization conditions [7] have been considered; each series has 26 time samples. To validate the results the simplified Yeast network topology from [8], which represents interactions among 29 genes, has been chosen (Figure 2).

Also the algorithms which exploit the qualitative properties of the features have been implemented but, by now, no extensive tests have yet been done.

As a performance criterion, the precision of the above algorithms in recognizing the regulation directions has been taken into account in the hypothesis that another algorithm gave the correct undirected pairs (for example the state-of-the-art ARACNe algorithm [9] has good performances but it does not compute directions). Let a_{ij} be the arc between two genes i and j in the graph G and $f(i,j)$ be a function that estimates how much they are correlated, then the definitions for *true positives* (TP), *false positives* (FP) and *false negatives* (FN) become:

$$TP \Leftarrow (a_{ij} = 1) \wedge f(i,j) > \vartheta$$

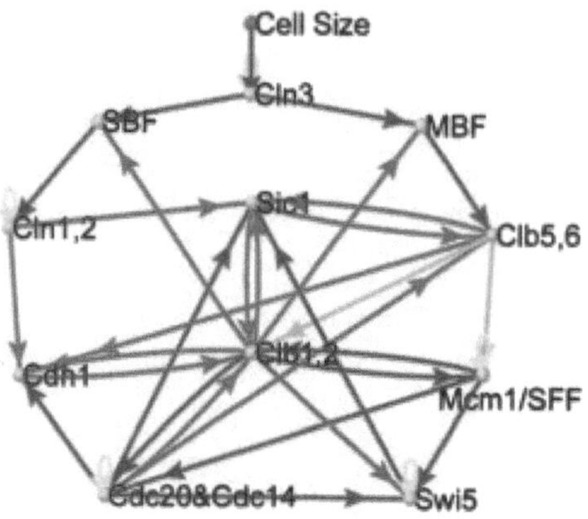

Fig. 2. Simplified cell-cycle network with only one checkpoint [Li *et al.*, 2004]

$$FP \Leftarrow (a_{ij} = 0 \wedge a_{ji} = 1) \wedge f(i,j) > \vartheta$$

$$FN \Leftarrow (a_{ij} = 1) \wedge f(i,j) \leq \vartheta$$

where ϑ is a threshold, in this work set to zero; this value has been chosen because it gave good results, but an in-depth parametric analysis has not yet been performed.

In particular, two common indices have been calculated[1]: the positive predictive value (PPV), called also precision, which refers to the fraction of returned true positives that are really positives:

$$PPV = \frac{TP}{TP + FP}$$

and the sensitivity (also known as recall), which gives the proportion of real positives which are correctly identified:

$$Sensitivity = \frac{TP}{TP + FN}$$

In order to have an idea of the performances obtained, a software based on Dynamic Bayesian Networks (Banjo [10]) has been applied with default parameters and 7 quantization levels to the same datasets: it seems to be precise but not very sensitive; the upper bound time for computation has been set to 10 minutes, a reasonable time, considering that the other four algorithms take seconds.

The mean performances of the five measures over the four different synchronization experiments have been reported in Table 1; for sC and $dDTW$ there are also numerical versions, computed on the original time series (their scores

[1] The performances of ARACNe algorithm on the considered dataset for the problem of identifying *undirected* pairs are: PPV = 65.2 % and Sensitivity = 13.9 %.

Table 1. Mean PPV and sensitivity values for the measures discussed in the paper over different synchronization experiments (in parentheses the performances on the numerical series)

	PPV	**Sensitivity**
sC	67.5 % (64.6 %)	40.0 % (38.5 %)
tMM	61.1 %	32.8 %
lcstr	63.4 %	42.0 %
lcs	64.0 %	32.1 %
dDTW	52.2 % (51.3 %)	2.6 % (2.2 %)
Banjo (DBN)	59.3 %	3.6 %

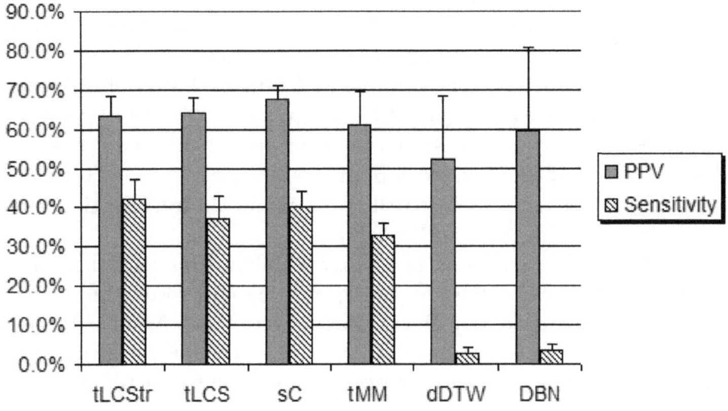

Fig. 3. Positive Predictive Value and Sensitivity of the four algorithms proposed in the paper compared with those of Dynamic Bayesian Networks

have been reported in parentheses). In Figure 3 the results of the algorithms operating on symbolic data have been plotted with their standard deviation as error bar.

It is possible to notice that the symbolic versions of the Shifted Correlation *sC* and the Directional Dynamic Time Warping *dDTW* both improve with respect to their numerical counterparts. Besides, all the measures provide a PPV above the 50% threshold; this means that they could be useful for deciding regulation directions.

5 Conclusions

In this work a first simple task of deciding the regulation direction of correlated genes has been studied by representing expression profiles in a symbolic way and by designing 5 new sub-string matching algorithms. In this way it is possible to reason more flexibly about temporal profiles affected by uncertainty due to noise or variable delays typical of biological systems. The next step will be to perform extended tests, possibly on larger datasets, with series enriched by qualitative properties of the features estimated using fuzzy quantization levels; hopefully, this should enhance the recall index, that are still under the threshold of a random choice, in particular the recall of $dDTW$, the most recently studied among the five measures proposed.

References

1. Sacchi, L., Bellazzi, R., Larizza, C., Magni, P., Curk, T., Petrovic, U., Zupan, B.: TA-clustering: Cluster analysis of gene expression profiles through temporal abstractions. Int. J. Med. Inform. 74, 505–517 (2005)
2. Kim, J., Kim, J.H.: Difference-based clustering of short time-course microarray data with replicates. BMC Bioinformatics 8, 253 (2007)
3. Savitzky, A., Golay, M.J.E.: Smoothing and differentiation of data by simplified least squares procedures. Analytical Chemistry 36, 1627–1639 (1964)
4. Falda, M.: Symbolic representations for reasoning about temporal gene profiles. In: Proc. of IDAMAP 2009 workshop, pp. 9–14 (2009)
5. Cormen, T.H., Leiserson, C.E., Rivest, R.L., Stein, C.: Introduction to algorithms, 2nd edn. McGraw-Hill, New York (2005)
6. Sakoe, H., Chiba, S.: Dynamic programming algorithm optimization for spoken word recognition. IEEE Trans. Acous. Speech Signal Process. 26, 43–49 (1978)
7. Spellman, P.T., Sherlock, G., Zhang, M.Q., Iyer, V.R., Anders, K., Eisen, M.B., Brown, P.O., Botstein, D., Futcher, B.: Comprehensive identification of cell cycle-regulated genes of the yeast saccharomyces cerevisiae by microarray hybridization. Mol. Biol. of the Cell 9, 3273–3297 (1998)
8. Li, F., Long, T., Lu, Y., Ouyang, Q., Tang, C.: The yeast cell-cycle network is robustly designed. PNAS 101, 4781–4786 (2004)
9. Margolin, A.A., Nemenman, I., Basso, K., Wiggins, C., Stolovitzky, G., Dalla Favera, R., Califano, A.: Aracne: an algorithm for the reconstruction of gene regulatory networks in a mammalian cellular context. BMC Bioinformatics 7 (2006)
10. Yu, J., Smith, V., Wang, P., Hartemink, A., Jarvis, E.: Advances to bayesian network inference for generating causal networks from observational biological data. Bioinformatics 20, 3594–3603 (2004)

Olive Trees Detection in Very High Resolution Images

Juan Moreno-Garcia[1], Luis Jimenez Linares[2], Luis Rodriguez-Benitez[2], and Cayetano Solana-Cipres[2]

[1] Escuela Universitaria de Ingenieria Tecnica Industrial, Universidad de Castilla-La Mancha, Toledo, Spain
juan.moreno@uclm.es
http://oreto.esi.uclm.es/

[2] Escuela Superior de Informatica, Universidad de Castilla-La Mancha, Ciudad Real, Spain
{luis.jimenez,luis.rodriguez,cayetanoj.solana}@uclm.es
http://oreto.esi.uclm.es/

Abstract. This paper focuses on the detection of olive trees in Very High Resolution images. The presented methodology makes use of machine learning to solve the problem. More concretely, we use the K-Means clustering algorithm to detect the olive trees. K-Means is frequently used in image segmentation obtaining good results. It is an automatic algorithm that obtains the different clusters in a quick way. In this first approach the tests done show encouraging results detecting all trees in the example images.

1 Introduction

The remote sensing is a very important factor inside the management and control of the Common Agricultural Policy [10]. The European Commission (EC) has been very interested in the use of remote sensing with Very High Resolution (VHR) images (airborne and satellite) to identify orchards and the position of fruit trees, as well as the measurement of the area of the orchards from GIS and remote sensing. The subsidies are totally or partially based in the orchards area and there are three types of permanent crops getting subsidies: olive, vineyards and more recently nuts. For this reason, remote sensing by using VHR images takes relevance within this scope. Remote sensing and GIS techniques help measuring the parcel area based on orthoimages, counting the trees by using automatic or semi-automatic methods, calculating the position of the trees, selecting the tree species and, detecting changes in olive parcels.

Olive production is very important for the economy of the European Union (EU) since EU is the main olive producer in the world. In 1997 the EC proposes a reform of the olive scheme oil based in the number of the olive trees or in the area of the orchards [10]. There is not reliable information about the olive trees number and the olive growing areas during years, and the new schema provided a boost for research in these areas. In the EC the research responsibility

E. Hüllermeier, R. Kruse, and F. Hoffmann (Eds.): IPMU 2010, Part II, CCIS 81, pp. 21–29, 2010.

was transferred to the Joint Research Centre (JRC) of EC under the research projects OLISTAT and OLIAREA [10]. OLISTAT stands for the estimate of the number of olive trees in the EU and OLIAREA makes an estimation of the olive area and the number of maintained trees in the EU. OLISTAT is based on aerial image acquisition; it computes the number of olive trees in a selected sample and makes an extrapolation to national levels using statistical estimators. A semi-automatic counting tool called OLICOUNT was created to count the olive trees by using aerial orthophotos [7]. OLIAREA tool uses the position of the olive trees to calculate the olive area.

In Spain, the research motivation in this area is very important since our country is a big producer of olive oil and wine, and our region, Castilla-La Mancha, is the second olive producer and the first wine producer of Spain. Due to it our research group participates in a project entitled "Automatic Vectorization of Cartographic Entities from Satellite Images" with the aim of to design and to implement an automatic method to cartographic entities vectorization in a GIS by means of Data Mining techniques. In this work, we present a first approach to detect olive trees in VHR images using a K-Means clustering algorithm.

The paper is organized as follows. Section 2 briefly reviews some related works. Section 3 describes the proposed approach that is based on the use of the K-Means algorithm. Later, in Section 4 experimental results are shown. Finally, conclusions and future works are described in Section 5.

2 Previous Works

The first approach to the automatic identification of individual trees by using remote sensing was developed to forestry applications [12,3,2]. These approaches show a good behavior detecting fruit trees, but they do not have the expected behavior detecting olive trees. That is because this kind of tree is usually less dense than forestry and besides is characterized by the fact that the fruit tree crown (plus shadow) is locally darker than its surrounding background [10].

The most popular method to count olive trees is the OLICOUNT tool by the Joint Research Centre [4]. OLICOUNT [7] is based on a combination of image threshold (i.e. using the spectral characteristics of trees), region growing and tests based on tree morphological parameters (i.e. using the morphology of individual trees). It operates with four parameters: (1) Grey value threshold, (2) Tree diameter, (3) Crown shape and (4) Crown compactness. OLICOUNT is a semi-automatic approach; an operator is required for tuning the parameters per parcel during the training step and for manually checking the results (trees can be manually added or deleted). The problem is that these manual tasks are time-consuming. OLICOUNT supports VHR images, and works with a single band of 8 bits. For this reason, the OlicountV2 tool [11] improved OLICOUNT tool to be able to handle various types of image file formats, with various pixel size (8 bits or 16 bits) and resolutions. The following upgrades were implemented in OlicountV2:

- TIFF file format support for 8/16 bits using LibTiff library.
- GEOTIFF file format support for 8/16 bits using LibGeoTiff library.
- 16 bits support for Olive Tree detection in OTCOUNT and OTVALUES libraries.
- Upgrade ArcView OLICOUNT project in order to be able to display the 16 bits images.
- Bugs revision and correction in ArcView OLICOUNT project.

The tests confirm that the results obtained by OlicountV2 do not improve the result of OLICOUNT [11]. OLICOUNT and OlicountV2 do not work with multispectral images that require another type of approach.

The method of regional minima was tested with the intent of reducing the manual work required by OLICOUNT. It is based on the principle that since crowns are dark objects and usually contain a regional minimum. A regional minimum is defined as a connected component of pixels whose neighbors all have a strictly higher intensity value [13]. The whole image is processed and a mask with the regional minima is built, and then it is clipped based on the parcel boundary layer in order to keep only the minima within the test parcels.

Karantzalos and Argialas in [6] developed and applied an image processing scheme towards automatic olive tree extraction using satellite imagery. The processing scheme is composed of two steps. In the first step, enhancement and smoothing was performed using nonlinear diffusion. In the second step, olive tree extraction was accomplished after extracting the local spatial maxima of the Laplacian. The algorithm has fixed parameters and is fast and efficient. The algorithm pays attention to the pre-processing step meanwhile image selective smoothing is accomplished. The algorithm output is a list of identified and labeled olive trees and it can be extended to include geographic coordinates.

3 Proposed Method

We propose the use of a clustering algorithm to solve this problem. More concretely, we will apply the K-Means algorithm to VHR images of olive fields. A preprocessing phase will be done before running the clustering algorithm. In this phase, we perform smoothing on the input image. We consider that the information contained in a pixel is not only represented by its value. Besides, it must be taken into account *the information of its neighbors*. Because of this, before to apply the K-Means algorithm, each pixel component value is recalculated by means of Equation 1, this process is done separately for each component. This equation calculates the mean between the pixel value $p(i,j)_c$ and its neighbors of the component c (R, G or B) with a distance d. Figure 1 shows the position of the pixels of the neighbors with distance $d = 1$, and the neighbors with distance $d = 2$, the central pixel is the pixel $p(i,j)_c$.

$$p(i,j)_c = \frac{\sum_{a=i-d}^{i+d} \sum_{b=j-d}^{j+d} p(a,b)_c}{(d+1)^2} \tag{1}$$

where $p(i,j)_c$ represents the pixel value of the component c in the position (i,j), and d is the distance between the central pixel and the neighbors.

The K-Means clustering algorithm is commonly used in image segmentation obtaining successful results. The results of the segmentation are used later to border detection and object recognition. The term *K-Means* was first used by James MacQueen in 1967 [9]. The standard algorithm was first proposed by Stuart Lloyd in 1957 as a technique for pulse-code modulation, though it was not published until 1982 [5]. K-Means clustering is a method of cluster analysis which aims to partition n observations into k clusters in which each observation belongs to the cluster with the nearest mean. It attempts to find the centers of natural clusters in the data.

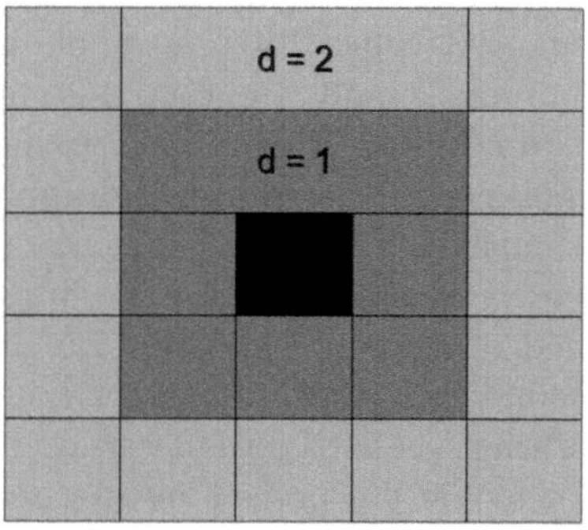

Fig. 1. Neighbors with distance $d = 1$ and $d = 2$

Let be $(x_1, x_2, \ldots, x_n)$ a set of observations where each observation is a d-dimensional real vector, then K-Means clustering aims to partition the n observations into k sets $(k < n)$ $S = S_1, S_2, \ldots, S_k$ so as to minimize the within-cluster sum of squares 2 (Equation 2).

$$argmin_S \sum_{i=0}^{k} \sum_{x_j \in S_i} |x_j - \mu_i|^2 \quad (2)$$

where μ_i is the mean of S_i.

The most common algorithm uses an iterative refinement technique. The behavior of the algorithm is shown in Algorithm 1. The result depends on the initial clusters and there is no guarantee that it will converge to the global optimum. The algorithm is usually very fast, although there are certain point sets on which the algorithm takes superpolynomial time [1].

Algorithm 1. K-means Algorithm

{Let be $m_1^{(1)},\dots,m_k^{(1)}$ an initial set of k means, which may be specified randomly or by some heuristic; the algorithm proceeds by alternating between two steps: [8]}

1. Assignment step: Assign each observation to the cluster with the closest mean (i.e. partition the observations according to the Voronoi diagram generated by the means). $S_i^{(t)} = \left\{x_j : \left|x_j - m_i^{(t)}\right| \leq \left|x_j - m_{i*}^{(t)}\right| \textit{ for all } i^* = 1,\dots,k\right\}$

2. Update step: Calculate the new means to be the centroid of the observations in the cluster. $m_i^{(t+1)} = \frac{1}{\left|S_i^{(t)}\right|} \sum_{x_j \in S_i^{(t)}} x_j$

{The algorithm is deemed to have converged when the assignments no longer change.}

4 Results and Discussion

Aerial images are used to do the test. These images are obtained by using the SIGPAC viewer of the Ministry of the Environment and Rural and Marine Affairs (*http://sigpac.mapa.es/fega/visor/*). This viewer offers VHR aerial images of the Spanish territory with a spatial resolution of 1 meter. In this first approach we obtain images in JPEG format with three bands. The difference with other methods is that they do not allow the use of three bands, i.e., they work with a single band. The images used belong to the municipality of Cobisa in the province of Toledo, Spain. The parameters that define the center of the image by using the Universal Transverse Mercator (UTM) coordinate system correspond to the *Huso* 30 with $X = 411943.23$ and $Y = 4406332.6$.

The open source software issued under the GNU General Public License *Weka* [14] has been used for the accomplishment of the tests. Weka is a collection of machine learning algorithms for data mining tasks, data pre-processing, classification, regression, clustering, association rules, and visualization [15]. The algorithms can either be applied directly to a dataset or called from Java code. This software contains an implementation with different options to carry out the tests, for example see Figure 2.

These tests are our first approach to this subject. Two tests are done by using little parts of an aerial image. We use the distance d assigned to 1 for all tests. Three tests are done where the number of clusters k is assigned to 2, 3 and 4; only one cluster represents the olive tree and the rest of the clusters are the field. The cluster that represents to the olive class has been manually selected when $k > 2$ in the test done in this paper. In order to automate this process, we are currently working in a similarity function between the *typical RGB values* that represent the olive trees and the RGB values that represent each obtained cluster. The metrics used to analize the results are *the omission rate* and *the commission rate* (also used in other works [10]). The omission rate represents the trees present on the field which were not identified by the method, and the commission rate is the number of trees which were identified by the method but which are not present on the field.

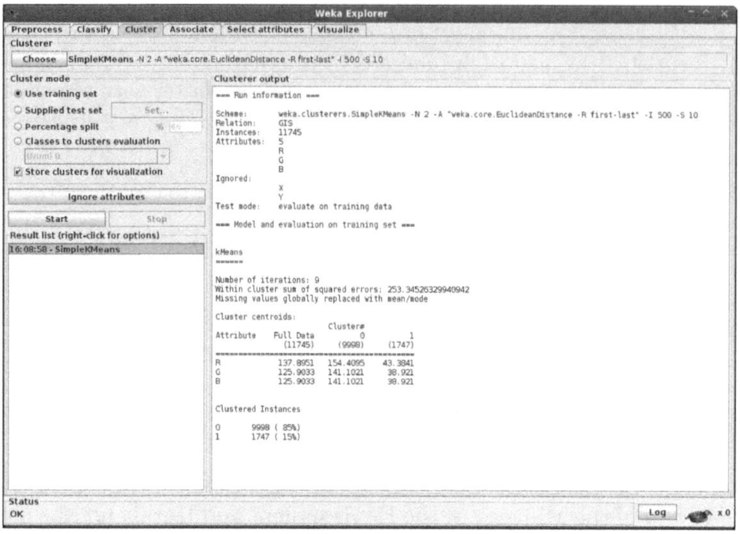

Fig. 2. Weka's window used for the test

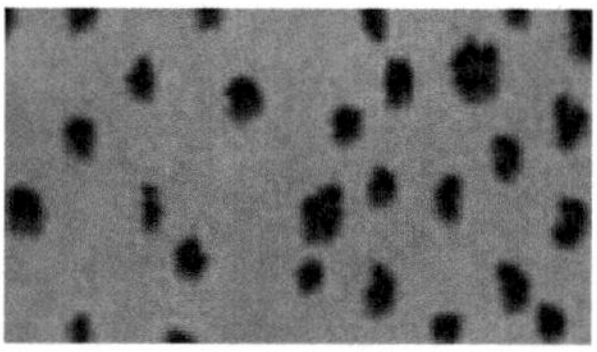

Fig. 3. Used image for the first test

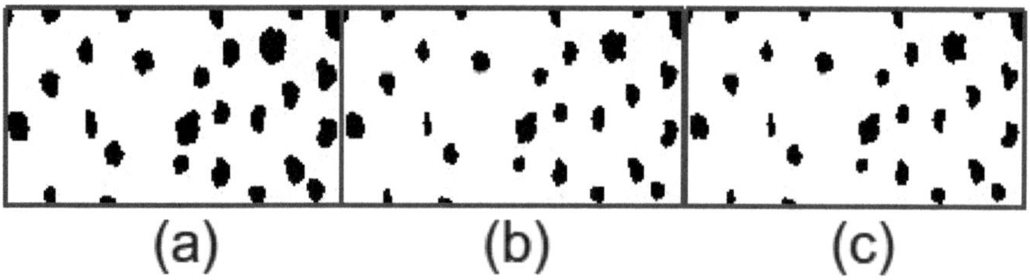

Fig. 4. Images of the results for the first test. (a) $k = 2$, (b) $k = 3$, (c) $k = 4$.

For the first test, a little part of an aerial image with 28 olive trees is selected (Figure 3). This test shows the possibilities of the K-Means algorithm to detect the olive trees. Neighbors with a distance 1 ($d = 1$) are used for this test (Equation 1). The number of clusters k are assigned to 2, 3 and 4, that is, three proofs are done. Figure 4 shows the obtained results. The omission rate is 0 in the three tests, that is, all the trees are detected whatever k used. Respect to the comission rate, it happens just like with the omission rate, it is 0 for the three tests, nothing is detected that is not tree. The size of the detected trees is greater for the smaller values of the variable k. The execution time is smaller when smaller is the value of k since the less classes, the less iterations. As a conclusion of this test we can say that the K-Means algorithm obtains good results, the ideal number of clusters are 2 since it obtains the trees with a good size, all the detected items are trees, and finally it has the smaller run time.

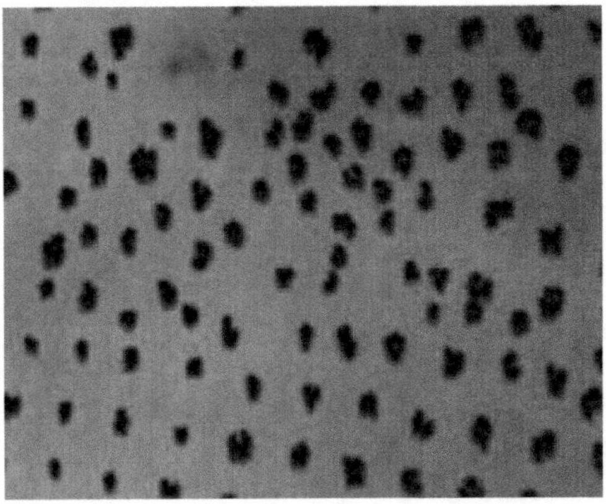

Fig. 5. Used image for the second test

Figure 5 is the input image of the second test. This part is greater than the image used in Test 1. The used image contains a part of a field with 100 olive trees with the next features:

- There are olive trees well-aligned, and there are olive trees without alignment.
- Ground of different tones.
- Different sizes of the tree crown.

Figure 6 shows the obtained results. The omission rate is 0 in the three test, so all the trees are detected for the three values of k. The comission rate are 1 for $k = 2$ (mark with a red circle in Figure 6), 0 for $k = 3$ and 0 for $k = 4$. The size of the detected trees is greater for the smaller values of the variable k what causes that the crowns of two trees are being touched (indicated with a blue rectangle).

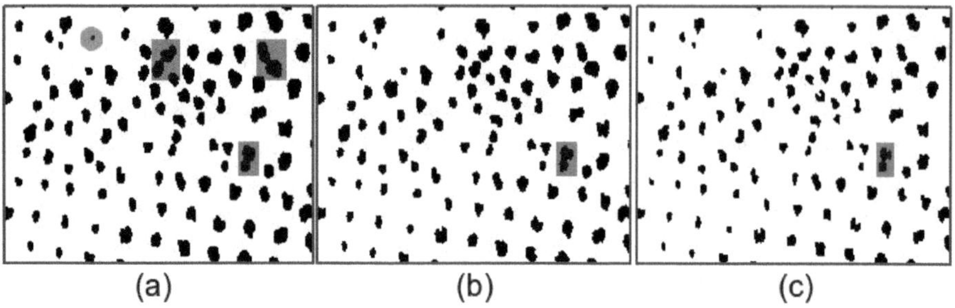

Fig. 6. Segmented images for the second test. (a) $k = 2$, (b) $k = 3$, (c) $k = 4$.

For $k = 2$ there are three cases of "joint crowns", one case is motivated by the input image, see Figure 5, and the rest of cases are a consequence of the detected trees size (this situation only occurs for $k = 2$ since the detected tree size is greater than the cases $k = 3$ and $k = 4$). Respect to the execution time occurs the same situation. As a conclusion of this test we can say that the K-Means algorithm obtains good results in this case, the ideal number of clusters are 3 since it obtains the trees with a good size but avoiding that two trees crowns appear like one crown.

5 Conclusions and Future Work

In this work, the K-Means clustering algorithm has been used to detect olive trees in Very High Resolution images. This is our first approach, but the obtained results allow to infer that it is a valuable method to detect olive trees in VHR images. The omission rate is 0 in 5 of the 6 proofs, and only 1 in the other case. The commission rate is 0 in all tests. The K-Means method is a fast method to detect olive trees since the number of clusters is small (2 or 3) and because it can avoid the different ground tones through the number of clusters.

The K-Means algorithm is an automatic method, meanwhile the reference work in this subject (OLICOUNT) needs an operator for tunning its four parameters [10]. In addition, the presence of parameters in image segmentation has a negative impact in the behavior of the method.

As future work, we must do more tests with images with joint crowns, very irregular parcels, parcels not well maintained (presence of shrubs or weeds), young trees, and so on. Also, we would like to probe the algorithm to other type of trees (nuts, citrus and associated fruit trees); the results could be good. Another line of work consists of to improve the smoothing phase by means of sub-pixel accuracy. Finally, we must to test other methodologies to know the real improvement when our approach is using.

Acknowledgments

This work was supported by the Council of Science and Technology of Castilla-La Mancha under FEDER Projects PIA12009-40, PII2I09-0052-3440 and PII1C09-0137-6488.

References

1. Arthur, D., Vassilvitskii, S.: How Slow is the K-Means Method? In: Proceedings of the 2006 Symposium on Computational Geometry (SoCG), pp. 144–153 (2006)
2. Brandberg, T., Walter, F.: Automated Delineation of Individual Tree Crowns in High Spatial Resolution Aerial Images by Multi-Scale Analysis. Machine Vision and Applications 11, 64–73 (1998)
3. Gougeon, F.: A crown following approach to the automatic delineation of individual tree crowns in high spatial resolution aerial images. Canadian Journal of Remote Sensing 3(21), 274–284 (1995)
4. European Comission, Joint Research Center, `http://ec.europa.eu/dgs/jrc/index.cfm` (last visit January 25, 2010)
5. Lloyd, S.P.: Least square quantization in PCM. Bell Telephone Laboratories Paper (1982); Least squares quantization in PCM. IEEE Transactions on Information Theory 28(2), 129–137 (1957)
6. Karantzalos, K., Argialas, D.: Towards the automatic olive trees extraction from aerial and satellite imagery. International Archives of the Photogrammetry, Remote Sensing and Spatial Information Sciences 35(5), 1173–1177 (2004)
7. Kay, S., Leo, P., Peedel, S., Giordino, G.: Computer-assisted recognition of olive trees in digital imagery. In: Proceedings of International Society for Photogrammetry and Remote Sensing Conference, pp. 6–16 (1998)
8. MacKay, D.: An Example Inference Task: Clustering. In: Information Theory, Inference and Learning Algorithms, ch. 20, pp. 284–292. Cambridge University Press, Cambridge (2003)
9. MacQueen, J.B.: Some Methods for classification and Analysis of Multivariate Observations. In: Proceedings of 5th Berkeley Symposium on Mathematical Statistics and Probability, pp. 281–297. University of California Press (1967)
10. Masson, J.: Use of Very High Resolution Airborne and Spaceborne Imagery: a Key Role in the Management of Olive, Nuts and Vineyard Schemes in the Frame of the Common Agricultural Policy of the European Union. In: Proceedings of the Information and Technology for Sustainable Fruit and Vegetable Production (FRUTIC 2005), pp. 709–718 (2005)
11. Bagli, S.: Olicount v2, Technical documentation, Joint Research Centre IPSC/G03/P/SKA/ska D (5217) (2005)
12. Pollock, R.J.: A model-based approach to automatically locating tree crowns in high spatial resolution images. In: Desachy (ed.) Image and Signal Processing for Remote Sensing. SPIE, vol. 2315, 526–537 (1994)
13. Soille, P.: Morphological Image Analysis: Principles and Applications, 2nd edn. Springer, Heidelberg (2004)
14. Weka Software, `http://www.cs.waikato.ac.nz/~ml/weka/` (last visit January 25, 2010)
15. Witten, I.H., Frank, E.: Data Mining: Practical Machine Learning Tools and Techniques, 2nd edn. Morgan Kaufmann, San Francisco (2005)

A Fast Recursive Approach to Autonomous Detection, Identification and Tracking of Multiple Objects in Video Streams under Uncertainties

Pouria Sadeghi-Tehran[1], Plamen Angelov[1], and Ramin Ramezani[2]

[1] Department of Communication Systems, Infolab21, Lancaster University Lancaster, LA1 4WA, United Kingdom
p.sadeghi-tehran@lancaster.ac.uk, p.angelov@lancs.ac.uk
[2] Department of Computing, Imperial College London, United Kingdom
raminr@doc.ic.ac.uk

Abstract. Real-time processing the information coming form video, infra-red or electro-optical sources is a challenging task due the uncertainties such as noise and clutter, but also due to the large dimensionalities of the problem and the demand for fast and efficient algorithms. This paper details an approach for automatic detection, single and multiple objects identification and tracking in video streams with applications to surveillance, security and autonomous systems. It is based on a method that provides recursive density estimation *(RDE)* using a Cauchy type of kernel. The main advantage of the RDE approach as compared to other traditional methods (e.g. KDE) is the low computational and memory storage cost since it works on a frame-by-frame basis; the lack of thresholds, and applicability to multiple objects identification and tracking. A robust to noise and clutter technique based on spatial density is also proposed to autonomously identify the targets location in the frame.

1 Introduction

Uncertainties are inherently related to video streams and can broadly be categorised as;

i) noise (rather probabilistic disturbances and errors);
ii) clutter (correctly identifying objects that are however of no interest to the observer – e.g. not a target that we want to track etc.).

Processing in real-time information that is coming form image, infra-red (IR) or electro-optical (EO) sources is a challenging task due to these uncertainties, but also due to the large dimensionalities of the problem (the resolution nowadays allow having millions of pixels and the rates of collecting information in order of dozens or more frames per second). At the same time the demand from applications that are related to surveillance, security and autonomous systems require fast and efficient algorithms. Recently, the use of security and surveillance systems is the centre of attention due to growing insecurity and terrorism activities

E. Hüllermeier, R. Kruse, and F. Hoffmann (Eds.): IPMU 2010, Part II, CCIS 81, pp. 30–43, 2010.

around the world. A pressing demand is the problem of automating the video analytical processes which require short processing time and low memory and storage requirement to enable real-time autonomous applications.

Traditional visual surveillance systems are not very efficient since they require a large amount of computer storage to archive video streams for further batch mode processing [1-3, 16]. They also often rely on manual (as opposed to automatic) and off-line target/object identification. One of the most widely used approaches for novelty detection is based on so called background subtraction [4-7]. This approach is based on building a representation of the scene background and compares new frames with this representation to detect unusual motions [4] . Instead of using window of consecutive frames to build background and keep them in the memory for off-line processing [4, 5], we propose a fully autonomous analysis on a per frame basis which is using recursive calculations and removes the need of computer storage to archive video frames. Additionally, the introduced approach is threshold-independent and minimises the processing time by discarding the unnecessary data. The main idea of the proposed approach is to approximate the probability density function (pdf) using a Cauchy type of kernel (as opposed to Gaussian one used in KDE technique), and then in order to update this estimation we apply a recursive expression using the colour intensity of each pixel. In this manner, only the accumulated information which represents the colour intensity of each pixel is stored in the memory and there is no need to keep huge volumes of data in the memory. As a result, the proposed technique is considerably (in an order of magnitude) faster and more computationally efficient.

The second innovation that is introduced in this paper is the automatic single and multiple object(s) identification in the frame. For the newly proposed multi-object detection we use a novel clustering technique to group the foreground pixels which represents objects/targets and distinguish them from the noise (due to luminance variation) and clutter. The proposed approach can be extended for tracking objects using Kalman Filter (KF) or evolving Takagi-Sugeno fuzzy model [8] , and landmark detection used in robotics [9, 10].

The remainder of the paper is organised as follows. In section two, the RDE novelty detection in video streams method is introduced. First, the widely used method KDE is explained and then its recursive version, RDE is introduced. The problem of single and multi-objects tracking and the mechanism for approaching this problem is explained in section 3. Section 4 represents the tracking technique based on eTS fuzzy system. Section 5 displays the experimental results. At the end, section 6 provides conclusion and discussion.

2 Novelty Detection in Video Streams through Recursive Densirty Estimation

2.1 Background Subtraction

One of the most popular and widely used methods for visual novelty detection in video stream is background subtraction method (BS) [4, 7]. Background subtraction is a method used to detect unusual motion in the scene by comparing each new frame to a model of the scene background. It is based on statistical modelling the background of the scene to achieve a high sensitivity to detect a moving object and robust to the noise. Robustness is required to distinguish fluctuations in the statistical characteristic due to non-rigid objects and noise, such as tree branches and bushes movements, luminance change, etc.

In [6] the absolute difference between every two frames is calculated and a threshold is used for decision making and model a foreground. As result, this method has low robustness to noise (e.g. luminance, variations, movement of tree branches, etc.) and clutter.

In order to cope with this problem a window of frames with length N (usually $N > 10$ is defined and analyzed in an off-line mode. Each pixel in the video frame is modelled separately as a random variable in a particular feature space and estimates its probability density function (pdf) across the window of Nframes [4, 5] (Fig. 1). The pdf is usually modelled as Gaussian. A more advanced approach is based on mixture of Gaussian (rather than a simple Gaussian) which is more realistic [11]. A drawback of this method is also using a threshold to selecting the proper distribution as a background model.

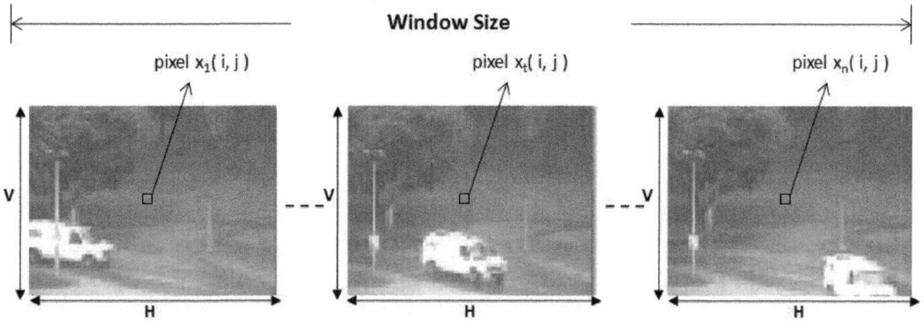

Fig. 1. Window of N frames used in KDE approach, H denotes the number of pixel in the horizontal and V – the number of pixels in the vertical

2.2 Kernel Density Estimation

Some of the most common techniques for modelling the background in video stream processing are non-parametric techniques such as the well known Kernel Density Estimation (KDE) [4]. Different types of kernels can be used to represent the pdf of each pixel of the frame each one having different properties. Typically, the Gaussian kernel is used for its continuity, differentiability, and locality properties [4].

$$p(z_t^{ij}) = \frac{1}{N}\sum_{r=1}^{N}\prod_{l=1}^{n} k_\sigma(z_{tl}^{ij} - z_{rl}^{ij}) \tag{1}$$

where k_σ denotes the kernel function (sometimes called a "window" function) with bandwidth (scale) σ; n denotes the colour channel (R,G,B or H,S,V) or, more generally, the number of input features; $z^{ij} = \left[z_1^{ij}, z_2^{ij}, \ldots, z_t^{ij}, \ldots, z_N^{ij}\right]^T; z \in R^n$ denotes the colour intensity values of N consecutive frames of a video stream that have a specific $(i,j)^{th}$ position in each frame (Fig. 1); $i=[1,H]$; $j=[1,V]$. If Gaussian is chosen to be a kernel function k_σ, then the colour intensity can be estimated as:

$$p(z_t^{ij}) = \frac{1}{N}\sum_{r=1}^{N}\prod_{l=1}^{n}\frac{1}{\sqrt{2\pi\sigma_l^2}}e^{-\frac{1}{2}\frac{(z_{tl}^{ij}-z_{rl}^{ij})^2}{\sigma_l^2}} \tag{2}$$

This can be simplified as:

$$p(z_t^{ij}) = \frac{1}{N\sqrt{2\pi\sigma_l^2}}\sum_{r=1}^{N}e^{-\sum_{l=1}^{n}\frac{(z_{tl}^{ij}-z_{rl}^{ij})^2}{2\sigma_l^2}} \tag{3}$$

Once the pdf is estimated by calculating the kernel function, it should be classified as a background (BG) or foreground (FG) by comparing to the pre-defined threshold [4].

$$IF\ (\ p(z_t^{ij}) < threshold)\ THEN\ (\ z_t^{ij}\ is\ foreground)\ ELSE\ (\ z_t^{ij}\ is\ background) \tag{4}$$

Although non-parametric kernel density estimation is very accurate, it is computationally expensive and the significant disadvantage of this method is the need to use a threshold. A wrong choice of the value of the threshold may cause a low performance of the whole system in difference outdoor environment. Another major problem/difficulty is to define a proper bandwidth for the kernel function. Practically, since only a finite number of samples are used and the computation must be performed in real time, the choice of suitable bandwidth is essential. Too small value of the bandwidth may lead the density estimation to be over-sensitive, while a wide bandwidth may cause the density estimation to be over-smoothed.

2.3 The Concept of the Proposed RDE Approach

The main idea of the proposed RDE approach is to estimate the pdf of the colour intensity given by equation (1)-(3) using a Cauchy type kernel (instead of Gaussian kernel) and calculate it recursively [12]. Such a recursive technique removes the dependence of a threshold and parameters (such as bandwidth) and allows the image frame to be discarded once they have been processed and not to be kept in the memory. Instead, information concerning the colour intensity per pixel is accumulated and is being kept in the memory. In this way, the amount of information kept in the memory is significantly smaller than original KDE approach, namely *(n+1)*H*V* or *(n+1)* per pixel compare to KDE which needs *(n*N*H*V)* data stored in the memory.

The Gaussian kernel can be approximated by a Cauchy function since the Cauchy function has the same basic properties as the Gaussian [13]. *a)* It is monotonic; *b)* its maximum is unique and of value 1; *c)* it asymptotically tends to zero when the argument tends to plus or minus infinity.

In RDE approach with using Cauchy type function the density of a certain (ij^{th}) pixel is estimated based on the similarity to **all** previously image frames (unless some requirements impose this to be limited to a potentially large window, N) at the same ij^{th} position.

$$D(z_t^{ij}) = \frac{1}{1 + \sum_{l=1}^{N}\sum_{r=1}^{n} \frac{(z_{tl}^{ij} - z_{rl}^{ij})^2}{2\sigma_r^2}} \tag{5}$$

It is very important that the density, D can be calculated recursively as demonstrated for other types of problems in [12, 13]. In a vector form (for all pixels in the frame) we have:

$$D(z_{tl}) = \frac{t-1}{(t-1)(z_t^T z_t + 1) - 2c_t + b_t} \tag{6}$$

Value c_t can be calculated from the current image frame only:

$$c_t = z_t^T d_t \tag{7}$$

where d_t is calculated recursively:

$$d_t = d_{(t-1)} + z_{(t-1)};\ d_1 = 0 \tag{8}$$

The value b_t is also accumulated during the processing of the frames one by one as given by the following recursive expression:

$$b_t = b_{t-1} + \|z_{t-1}\|^2;\ b_1 = 0 \tag{9}$$

As mentioned earlier, to identify a foreground (novelty) the density of each ij^{th} pixel of the image frame is compared to pixels at the same ij^{th} position in **all** previous frames. In this way, the expression (10) should be applied for each pixel, (Fig. 2). It should be highlighted that in RDE approach there is no need to pre-define any threshold since we estimate the statistical properties of the density:

$$IF\left(D(z_t^{ij}) < \min_{l=1}^{t} D(z_l^{ij} - std(D(z_l^{ij}))\right) THEN\ (\ z_l^{ij}\ is\ FG)\ ELSE\ (\ z_l^{ij}\ is\ BG)\ \ i=[1,H];j=[1,V] \tag{10}$$

where $std\left(D\left(z_l^{ij}\right)\right)$ is the standard deviation of the densities of image frames seen so far.

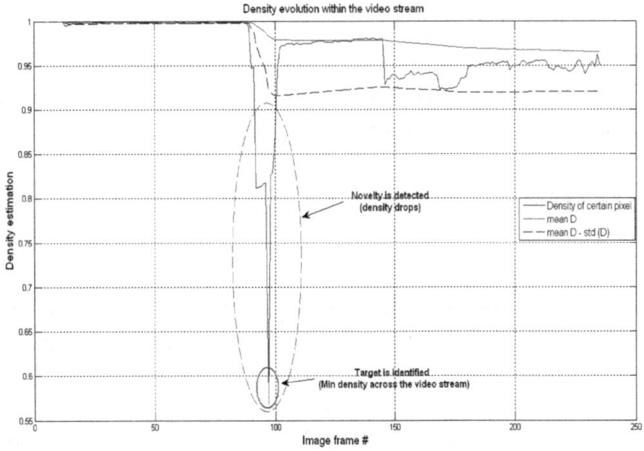

Fig. 2. The frames for which the value of the density drops below the value of meanD-std(D) are denoted by red circle and a novelty is detected there

3 Single/Multi Object(S) Identification Using RDE

3.1 Single Object Identification

After applying condition (11) to each pixel and detecting a novelty at a pixel level, the standard way to identify the object for tracking purposes is to find the spatial **mean value** of all pixels that have been identified to be background [5]. The drawback of this technique is the influence of the noise caused by change of illumination, move of tree branches and bushes, clutter, etc. This may lead to locating the object in a wrong position which might be misleading for the tracking. An alternative that is also often used for target tracking in video streams is the manual identification of the target/object which is obviously an off-line process [5]. In this paper we propose two alternative techniques to cope with this problem:

a) Based on the minimum density in the feature (colour) space.
b) Based on maximum value of the density inside the current frame.

In the first proposed technique, the same colour density which is calculated recursively by equation (6) is used to identify a novel object. In this technique, out of the F_t pixels identified as a foreground in the current frame, t the one, $O_t^* = [h_t^*, v_t^*]$ which has minimum density (D), will be the most different from the background and most likely to represent the novel object/target on the image frame:

$$O_t^* = \underset{t=1,i=1,j=1}{\overset{N;H;V}{\arg\min}} D(z_t^{ij}) \tag{11}$$

It is a very fast technique and free of computational complexity. It is also guarantees a better lock on the object for tracking purposes (Fig. 3).

In the second alternative technique, we use again the density, but this time in terms of the spatial position of the pixels inside the current frame (for $i=1,2,\ldots,H$; $j=1,2,\ldots,V$) which were identified already to be susceptive foreground, F_t. The pixel with maximum value of the density inside the current frame can be chosen to represent the novel object/target on the scene.

$$O_t^* = \arg\max_{i,j=1}^{F} \{D_t^{ij}\}, \quad O_t^* = [h_t^*, v_t^*] \tag{12}$$

where O_t^* denotes the vector of the object position in the current frame with its horizontal and vertical components. F denoted the number of pixels in a frame classified as foreground ($F<<H*V$).

The rationale of this technique is that this point represents the pixel that has higher spatial density if only foreground (F_t) pixels are taken into account in the frame (Fig. 3).

The spatial density can be calculated recursively in a vector form similarly to (6)-(9):

$$D(O_t^*) = \frac{l-1}{(l-1)(f^T f+1)-2\gamma+\beta}; \; l=[1,F] \tag{13}$$

$$\gamma = f^T \delta \tag{14}$$

$$\beta(l) = \beta(l-1) + \|f(l-1)\|^2; \quad \beta(1) = 0 \tag{15}$$

$$\delta(l) = \delta(l-1) + f(l-1); \quad \delta(1) = 0 \tag{16}$$

where $f\in R^F$ denotes the vector of the foreground pixels in a frame

This method can be extended for image segmentation [14] , and landmark detection [9] used in self-localisation in robotics [10] . As result it is more robust to locate the position of the object in the current image frame compare to the standard mean value technique (Fig. 3).

3.2 A New Method for Multiple Objects Identification in Video Frames by Real-Time Clustering

Multiple objects tracking always has been a challenging part in computer vision. Several methods are used to identify and track the fixed number of objects [17]. Many of them are only applicable to tracking humans or vehicles [18-20]. The method that is proposed in this paper can be applied to tracking multiple objects whose number is unknown and varies during tracking. The proposal is for real-time on-line fast non-iterative clustering that does not require the number of clusters to be specified beforehand. This clustering approach is applied only to those pixels (F_t) in a frame that were identified as a novelty/foreground. In this approach the number of the clusters is not pre-specified and generated based on the position of the novelties in each frame. Each novelty/foreground is assigned

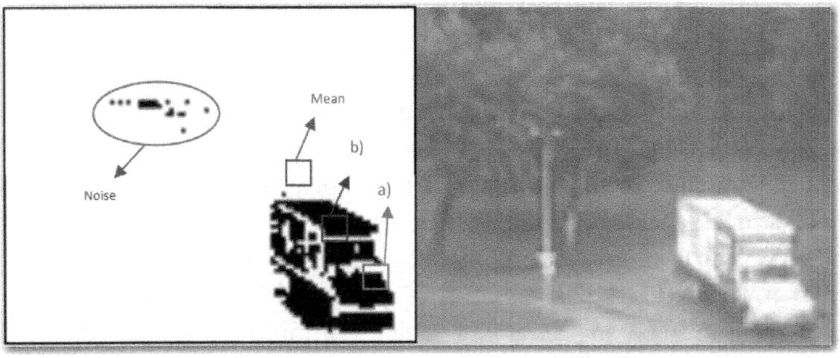

Fig. 3. Pixels detected as novelty; left hand side of the scene are due to noise and clutter; right hand side is modelled one. Note the pixels on the left hand side of the modelled scene are due to noise and clutter. The green and red square denotes the centre of the target as identified by the proposed techniques a) and b). Brown square denotes the centre as identified by the mean.

Fig. 4. Multi objects identification. Right hand side scenes are original frames; left hand side scenes are modelled ones. The red square denotes the focal point of the foreground.

to the cluster with the nearest mean. Initially a single cluster (object) is formed around the pixel identified as described in the previous section. Its radius, $r_1 = \sigma_1$ is determined based on the spatial variance of the positions of the pixels that are associated to it. After that, if the distance between the pixels that is a susceptive novelty/foreground and the centre of the cluster that is already formed is less than r_1, then a new cluster/target is created. At the same time in a pursuit to avoid noise and clutter we ignore the new clusters that are formed around a small number of pixels, $\underline{S}$. This number is determined in such a way that the size of an object/target that is expected to be detected to be comparable with the size of a regular (square) blob formed by $\underline{S}$ pixels. If any of the newly formed clusters has less than $\underline{S}$ pixels as members, it will be not specified as an object/target and will be ignored, see Fig. 4.

4 Real-Time Tracking Using Evolving Takagi-Sugeno eTS Fuzzy Systems

After detecting all the foreground pixels in an image frame and identifying the object/target often the problem is to track it. Therefore, the efficient tracking algorithm can be vital. In this paper we propose to use evolving Takagi-Sugeno (eTS) fuzzy model [12, 22] which represents a fuzzy mixture of locally active Kalman Filter (KF) [21] where the number of the local regions is not pre-specified and fixed. eTS is an on-line self-developing version of the widely used Takagi-Sugeno fuzzy systems [23] which combine a linguistic fuzzy *IF* part and a functional/linear *THEN* consequents part.

The proposed algorithm is non-linear with an evolving structure. This means that the number of local regions can grow or be reduced, the eTS structure – fuzzy rules, input variables/features, fuzzy sets – can expand or shrink according to the data pattern in the joint input-output (current – next frame) data space.

In a nutshell, learning eTS consists of two stages [24]:

1) Decomposing the data (pixel locations in the current and next/predicted images) space into local sub-areas.
2) Adapting the parameter of the consequent parts of the fuzzy rules.

Both stages are performed in real-time for an interval of time shorter than the time of arrival of the next image frame (less than *40ms* if assume *25fps* rate of the video).

In the tracking problem the aim is to predict the position of the object/target in the next, $(t+1)^{th}$ frame:

$$\hat{O}^*_{t+1} = eTS(O^*_t) \tag{17}$$

where $\hat{O}^*_{t+1} = eTS(O^*_t)$ is the predicted position of the target in the $(t+1)^{th}$ frame.

Another advantage of eTS is that it can be represented by linguistically tractable fuzzy rules of the following type:

$$Rule: IF(h^*_t \text{ is about } \chi^*) \text{ AND } (v^*_t \text{ is about } w^*) \text{ THEN } \begin{cases} \hat{h}_{t+1} = a_0 + a_1 h_t + a_2 v_t \\ \hat{v}_{t+1} = b_0 + b_1 h_t + b_2 v_t \end{cases} \tag{18}$$

where χ^*, w^* are the prototypes (centres of the membership functions); a and b are the parameters of the (linear) consequents.

The fuzzy sets can defined by their membership functions, e.g. of a Gaussian type:

$$\mu^l_\chi(h) = e^{\frac{\|h^l_t - \chi^*\|}{2\sigma^l_t}} \tag{19}$$

where $\mu_\chi^l(h)$ denotes the membership to the fuzzy set (h_t^* *is about* χ^*) form the lth fuzzy rule.

Similar membership functions can be defined for the vertical component, v for each fuzzy rule, $l=[1,R]$. The overall prediction of the position of the target in the next frame is produced using centre of gravity type defuzzification:

$$\hat{O}^*_{t+1} = \sum_{l=1}^{R} \lambda^l O_t^{l*} \tag{20}$$

where $\lambda^l = \frac{\mu_\chi^l(h)\mu_\omega^l(v)}{\sum_{r=1}^{R}\mu_\chi^r(h)\mu_\omega^r(v)}$ is the normalized firing level of the lth rule; T_t^{l*} is the prediction by the lth rule.

Since eTS (same as TS) is linear in the consequents part it renders the use of well established learning approaches such as RLS. In fact, this is partially correct, because the linearity is correct only locally and eTS (same as TS) is non-linear as a whole. Therefore a fuzzily weighted version of RLS (wRLS) [24] is necessary to be applied, not the standard RLS. wRLS can be applied locally (per fuzzy rule and per cluster) or globally [13] . The local implementation has a number of advantages, including better convergence properties, better interpretability, smaller computational demands etc. [13] and is described as:

$$\hat{a}_t^l = \hat{a}_{t-1}^l + C_{t-1}^l \tau_{t-1}^* \lambda^l \left(\tau_t^* - \tau_{t-1}^{*T} \hat{a}_{t-1}^l \right); \; a_1^l = 0 \tag{21}$$

$$C_t^l = C_{t-1}^l - \frac{\lambda_i(\tau_{t-1})\; C_{t-1}^l\; \tau_{t-1}^*\; \tau_{t-1}^{*T}\; C_{t-1}^l}{1+\lambda_i(\tau_{t-1}^*)\; \lambda_{t-1}^{*T}\; C_{t-1}^l\; \tau_{t-1}^*}; \; C_1^l = \Omega I \tag{22}$$

where $\tau = [1;T^*]^T$ denotes the extended position vector; C is the co-variance matrix, λ is the normalized firing strength of the lth fuzzy rule; Ω is a large value and $I$is an identity matrix.

5 Experimental Results

In this section, we present some real-time novelty detection results using the proposed RDE method. The RDE approach is applied to two video streams in different environment and illumination conditions (Fig. 5). The results are compared with the well known KDE approach in Table 1. Note that, the proposed algorithm is implemented in MATLAB, however the time can be significantly reduced using C language.

The first video sequence has 237 frames of size 176×144 while the second one is 320×240 with 195 frames. In both clips, the RDE method is applied for

Fig. 5. Background Subtraction using the proposed RDE method, Left hand side scenes are original frames; right hand side scenes are modelled ones. The red square denotes the focal point of the foreground.

autonomous novelty detection. After novelty/foreground of the image frame is determined, two alternative novelty techniques known as minimum density in colour space and maximum spatial density (explained in section 3) are used to identify the single target, while the real-time clustering is implemented for automatic identification of multiple targets.

Despite the noise caused by illumination and camera oscillation in both video streams, the proposed approach (RDE) has a superior performance as compared to the original KDE approach. As opposed to KDE, the proposed algorithm is significantly faster and requires significantly less memory storage (Table 1). It should be stressed that RDE approach can be applied in real-time; while KDE approach is limited by the size of the window. If the size of the window is too large the sensitivity of the approach diminishes, on the contrary if the window size is too short it may lead to an oversensitive realization. Also, it is important to note that RDE can be realised on hardware (an n-line clustering approach using recursive Cauchy formula was implemented on FPGA [15] and proved to work extremely fast) which paves the way to various practical real-time implementations.

For tracking part, eTS was applied to predict the position of the target in the next t^{th} frame. At the end, the performance of the eTS was compared with KF in terms of two dimensions, h and v. Table 2 displays that eTS provides the smaller root mean error (RMSE) and non-dimensional index (NDEI) in estimating the true location of the target and overall has a better performance than KF.

In addition eTS provides the closer predicted values to the actual values comparing to KF (Fig. 6)

Table 1. Comparison of the performance of RDE and KDE in two different video streams

	Method	window size	Frame size (pixels)	Time per frame (sec)	Memory used Per pixel (bites)	Memory calculation
Video Clip (1)	KDE	20	176×144	15.57	60	O(60)
	RDE	1	176×144	0.43	4	O(4)
Video Clip (2)	KDE	20	320×240	44.56	60	O(60)
	RDE	1	320×240	1.17	4	O(4)

Table 2. Tracking precision using eTS vs. KF

	Method	RMSE	NDEI	VAF (%)
v, pixel	KF	15.58	0.7	50.57
	eTS	13.2	0.56	69.14
h, pixel	KF	23.78	0.66	55.58
	eTS	21.31	0.56	68.56

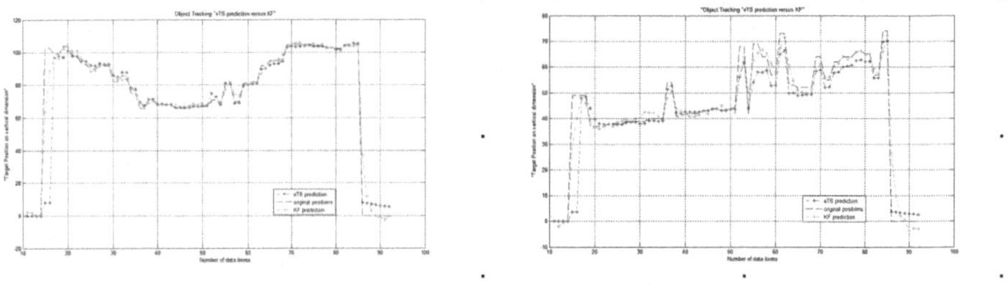

Fig. 6. Tracking performance of eTS vs. KF (Left plot – vertical component, Right plot – horizontal component)

6 Conclusion and Discussion

In this paper, we introduced novel techniques for detection and automatic object/target identification and tracking in video streams under uncertainties. We compared the results with the best known used methods such as kernel density estimation (KDE) for novelty detection and Kalman filter for tracking. The key innovation of the proposed approach is the use of a recursively calculated Cauchy type of kernel for the density estimation (as opposed to widely used Gaussian one) and in the tracking part of the problem – the use of evolving fuzzy Takagi-Sugeno model. The proposed approach is particularly suitable for real-time autonomous applications and is very fast and robust to uncertainties in the video stream (it does not need to be tuned to different environments and has built-in robustness to noise and clutter.

For single object/target identification, two alternative techniques based on minimum density in the colour across the video stream and the maximum spatial density inside the current frame was introduced and compared with are used and compared with spatial mean method. The results show a better lock on the target and more robust recognition comparing to the standard spatial mean technique. To autonomously identify multiple objects in a frame we proposed a real-time on-line clustering method which is computationally fast and the number of the clusters does not need to be pre-defined. For tracking autonomously identified objects/targets we proposed to use evolving Takagi-Sugeno fuzzy model (eTS) which provides real-time high prediction, is fast and human-interpretable. The overall proposed approach is fully autonomous and suitable for video-analytical tasks in surveillance and autonomous systems design.

References

1. Hampapur, A.: Smart Video Surveillance, Exploring the concept of multi-scale spatiotemporal tracking. IEEE Signal Processing Magazine, 38–51 (2005)
2. Han, B., Comaniciu, D., Davis, L.: Sequential kernel density approximation through mode propagation: applications to background modeling. In: Proc. ACCV - Asian Conf. on Computer Vision (2004)
3. Gonzalez, R.C., Woods, R.E.: Digital Image Processing, 2nd edn. Prentice Hall, Englewood Cliffs (2002) ISBN: 0201180758
4. Elgammal, A., Duraiswami, R., Harwood, D., Davis, L.: Background and Foreground modeling using nonparametric Kernel Density Estimation for visual surveillance KDE. Proc. 2002 IEEE 90(7), 1151–1163 (2002)
5. Cheung, S.-C., Kamath, C.: Robust techniques for background subtraction in urban traffic video. In: Proc. SPIE, Electronic Imaging Video Comm. and Image Proc., San Jose, pp. 881–892 (2004)
6. Cucchiara, R., Grana, C., Piccardi, M., Prati, A.: Detecting moving objects, ghosts and shadows in video streams. IEEE Trans. on Pattern Analysis and Machine Intelligence, 1337–1342 (2003)
7. Han, B., Comaniciu, D., Davis, L.: Sequential kernel density approximation through mode propagation: applications to background modeling. In: Proc. ACCV - Asian Conf. on Computer Vision (2004)
8. Memon, A., Angelov, P., Ahmed, H.: An Approach to Real-Time Color-based Object Tracking. In: Proc. 2006 Intern. Symp. on Evolving Fuzzy Systems, Ambleside, Lake District, UK, pp. 81–87. IEEE Press, Los Alamitos (2006)
9. Zhou, X., Angelov, P.: Real-Time joint Landmark Recognition and Classifier Generation by an Evolving Fuzzy System. In: Proc. 2006 World Congress on Computational Intelligence, Vancouver, Canada, pp. 6314–6321 (2006)
10. Zhou, X., Angelov, P.: Autonomous Visual Self-localization in Completely Unknown Environment using Evolving Fuzzy Rule-based Classifier. In: Proc. IEEE Intern. Conf. on Comp. Intel. Applic. for Defense and Security, Honolulu, USA, pp. 131–138 (2007)
11. Zhivkovic, Z., Van der Heijden, F.: Efficient adaptive density estimation per image pixel for the task of background subtraction. Pattern Recognition Letters, 773–780 (2006)
12. Angelov, P.: An Approach for Fuzzy Rule-base Adaptation using On-line Clustering. Intern. Journal of Approximate Reasoning, 275–289 (2004)

13. Angelov, P., Filev, D.: An Approach to On-line Identification of Takagi-Sugeno Fuzzy Models. IEEE Trans. on System, Man, and Cybernetics, part B - Cybernetics, 484–498 (2004)
14. Alzate, C., Suykens, J.: Image segmentation using Weighted Kernel PCA Approach to Spectral Clustering. In: Proc. IEEE Intern. Conf. on Comp. Intell. Applicat. for Image and Signal Processing, CIISP 2007, Honolulu, HI, USA, pp. 220–225 (2007)
15. Everett, M., Angelov, P.: EvoMap: On-Chip Implementation of Intelligent Information Modelling using EVOlving MAPping. Technical Report, Lancaster University, UK, pp. 1–15 (2005)
16. Cucchiara, R., Grana, C., Piccardi, M., Prati, A.: Detecting moving objects, ghosts and shadows in video streams. IEEE Trans. on Patt. Anal. and Machine Intell. 25, 1337–1342 (2003)
17. MacCormick, P.J., Blake, A.: A probabilistic exclusion principle for tracking multiple objects. In: ICCV 1999, pp. 572–578 (1999)
18. Mao, X., Qi, F., Zhu, W.: Multiple-part based Pedestrian Detection using Interfering Object Detection. In: Third International Conference on Natural Computation, pp. 165–169 (2007)
19. Cai, Q., Aggarwal, J.K.: Tracking human motion using multiple cameras. In: 13th International Conference on Pattern Recognition, pp. 68–72 (1996)
20. Yang, L., Johnstone, J., Zhang, C.: A Multi-camera Approach to Vehicle Tracking Based on Features. In: IEEE International Symposium on Multimedia, pp. 79–80 (2007)
21. Kalman, R.E.: A New Approach to linear filtering and prediction problem. Trans. of the ASME, Ser. D, Journal of Basic Engineering, 34–45 (1960)
22. Angelov, P., Zhou, X.: Evolving fuzzy systems from data streams in Real time. In: Proc. 2006 Intern. Symp. on Evolving Fuzzy Systems, Ambleside, Lake District, UK, pp. 29–35. IEEE Press, Los Alamitos (2006)
23. Takagi, T., Sugeno, M.: Fuzzy identification of systems and its application to modelling and control. IEEE Trans. on Systems, Man and Cybernetics, 116–132 (1985)
24. Angelov, P.: Evolving Rule-based Models: A Tool for Design of Flexible Adaptive Systems. Springer, Heidelberg (2002)

Soft Concept Hierarchies to Summarise Data Streams and Highlight Anomalous Changes

Trevor Martin[1,2], Yun Shen[3], and Andrei Majidian[1,2]

[1] Artificial Intelligence Group, University of Bristol BS8 1TR UK
firstname.lastname@bristol.ac.uk
[2] BT Innovate, Adastral Park, Ipswich IP5 3RE, UK
[3] Hewlett-Packard Laboratories, Stoke Gifford, Bristol, BS34 8QZ, UK
yun.shen@hp.com

Abstract. A hierarchical approach is natural when managing large volumes of information, from both static (database) and dynamic (datastream) sources. Hierarchies allow progressively finer division into more specific categories, but frequently the categories are fuzzy rather than crisp. In this paper, we use fuzzy formal concept analysis to extract soft hierarchies from data. The hierarchies are used to classify data and to monitor changes over time by means of a fuzzy confidence measure for association analysis. A (simulated) stream of terrorism incident data is used as proof of concept.

Keywords: fuzzy formal concept hierarchies, fuzzy association rules, fuzzy confidence, dynamic data streams.

1 Introduction

Information overload or "digital obesity" [1] is widely recognised as a cause of major difficulties in the knowledge economy. It arises from too much stored data combined with inadequate methods to access the right data, and profoundly affects the productivity of knowledge workers. Many decisions are made on the basis of incomplete knowledge simply because the relevant information is scattered over different repositories (personal filestores, web, networked databases, etc). These problems are exacerbated by changes in data, which make static analysis difficult or impossible. (We argue that we can treat a database as effectively static if it is not updated frequently and most of the data are unchanged after update. The terms "most" and "frequent" are deliberately left open to interpretation.) The extreme case of non-static data is a data stream which may contain numerical data (such as network traffic flow, stock prices, output from weather sensors, etc.), text and symbolic data (e.g. short text reports, emails, newsfeeds, etc.), or a mixture of both. Values may be updated on an almost continuous basis, or at discrete intervals, typically short. The dynamic nature of data streams and limits on storage and processing may mean that only summary measures can be examined when analysing the data or using it to support a decision - it is often not possible to consider the entire historic dataset. In this paper, we describe a new approach to summarising symbolic data (static and non-static) using fuzzy formal

E. Hüllermeier, R. Kruse, and F. Hoffmann (Eds.): IPMU 2010, Part II, CCIS 81, pp. 44–54, 2010.

concept hierarchies, and outline a method of tracking changes in the summary structures, suitable for application to data streams.

2 Fuzzy Formal Concept Analysis

When faced with a large volume of information, a natural human approach is to "divide and conquer" - we look for similarities and group together roughly equivalent items, often in a taxonomic or hierarchical fashion. Frequently, this analysis can lead to formal or informal hierarchical classification schemes - for example, items in a supermarket or web-based retailer, product catalogues, topics in an online discussion forum, software component libraries, scientific literature, etc, all follow "divide and conquer" style taxonomies, gradually becoming more specific as one ventures deeper.

Classification schemes are typically static structures. To illustrate, consider a consumer electronics retailer which devised a scheme some years ago using separate categories such as camera, personal organiser, mobile phone, portable music player, etc. Although there are now devices that fit into all these categories, the effort involved in conceptual re-design can mean it is easier to force the products into the closest category (rather than the most accurate category). We accept this limitation but propose fuzzy membership and the ability of an object to belong to multiple categories as one way to address the problem.

Fuzzy set theory is also applicable in modelling the approximations made in classification - particularly in deciding whether an item belongs or doesn't belong to a category. A fundamental tenet underlying fuzzy set theory [2] is the idea that humans work with groups of entities (or conceptual categories) that are loosely defined. Music and film genres are clear examples- there is a shared understanding but no precise definition of (say) comedy or blues. Many categorisations are actually based on subjective criteria and judgments, rather than objective definitions.

Finally we note that data often contains implicit taxonomies - a relational table may flatten hierarchical data into one or more attributes, and XML tags may hide

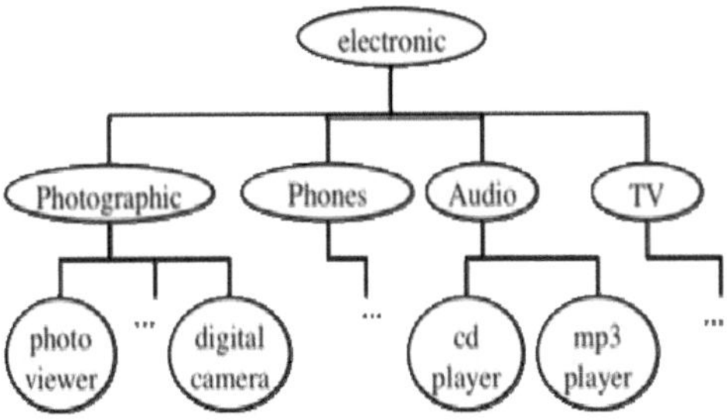

Fig. 1. A possible hierarchical organisation for electronic goods

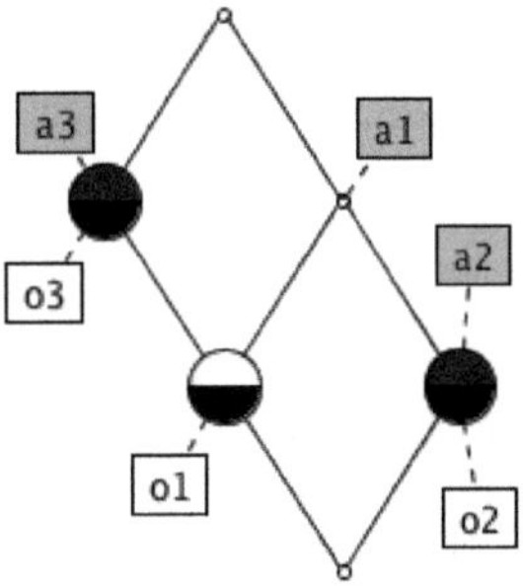

Fig. 2. concept lattice corresponding to the formal context in Table 1. The lattice is drawn using the *conexp* tool (conexp.sourceforge.net).

structure. Data often relies on human interpretation for its "semantics" - a programmer can take advantage of the fact that <*iPod*> and <*walkman*> are subtypes of <*music Player*>, but a program has no way of knowing this unless it is made explicit by means of a taxonomy.

These arguments hold for both static and dynamic data sources.

2.1 Formal Concept Analysis

Formal concept analysis (FCA) [3, 4] is a way of extracting hidden structure (specifically, lattice-based structure) from a dataset that is represented in object-attribute-value form. In its simplest form, FCA considers a binary-valued table, where each row corresponds to an object and each column to an attribute (property). The table contains 1 (true) in cell *i, j* if object *i* has attribute *j* and 0 (false) otherwise. Formally, we consider a set of objects *O* and attributes *A*, together with a relation

$$R \subseteq O \times A$$

The structure *(O, A, R)* is called a formal context. Given *X*, a subset of objects and *Y*, a subset of attributes, i.e. $X \subseteq O, \quad Y \subseteq A$ we define operators ↑ and ↓ as follows:

$$X^{\uparrow} = \{y \in Y \mid \forall x \in X : (x, y) \in R\} \quad (1)$$

$$Y^{\downarrow} = \{x \in X \mid \forall y \in Y : (x, y) \in R\} \quad (2)$$

Table 1. A simple formal context ; the concepts are shown on the right

	a1	a2	a3
o1	1	0	1
o2	1	1	0
o3	0	0	1

({o2}, {a1, a2})
({o1}, {a1, a3})
({o1, o2}, {a1})
({o1, o3}, {a3})

Then any pair (*X, Y*) such that $X^{\uparrow}=Y$ and $Y^{\downarrow}=X$ is a formal concept.

Table 1 and Fig. 2 show the relation between three objects *o1, o2, o3* and attributes *a1, a2* and *a3*. The resulting concepts are shown on the right of Table 1. In larger tables, this is less obvious to inspection.

A partial order, $\leq$, is defined on concepts such that

$$(X1, Y1) \leq (X2, Y2) \text{ means } X1 \subseteq X2 \text{ and } Y2 \subseteq Y1$$

The higher concept contains more objects / fewer conditions than the lower concept. This leads to a lattice (Fig 2). Attributes *a2* and *a3* are mutually exclusive, since no object has both attributes - the least upper bound is the top element and the greatest lower bound is the bottom. A node drawn as a large circle represents an object (or objects); each object has all the attributes attached to its node and all higher nodes linked directly or indirectly to it. A node with a black lower half represents at least one object; a blue upper half shows the highest node corresponding to an attribute. This convention allows the diagram to be simplified by omitting labels.

2.2 Conceptual Scaling

For attributes which take a range of values (rather than true / false as above), the idea of "conceptual scaling" is introduced [5]. This transforms a many-valued attribute (e.g. a number) into a symbolic attribute - for example, an attribute such as "*height in centimetres*", given an integer or real value between 0 and 200 could be transformed to attributes "*heightlessthan50*", *heightfrom50to99*, etc. These derived attributes have true/false values and can thus be treated within the framework described above.

2.3 Fuzzy FCA

Clearly the idea of conceptual scaling is ideally suited to a fuzzy treatment, which reduces artefacts introduced by drawing crisp divisions between the categories. The notion of a binary-valued relation is easily generalised to a fuzzy relation, in which an object can have an attribute to a degree. In the context *(O, A, R)* each tuple of R, $(o,a)\in R$ has a membership value in [0, 1]. Several approaches have been proposed in the literature, starting with [6]; the reader is referred to [7] for discussion of fuzzy FCA. Fuzzy FCA starts from a fuzzy relation and (broadly speaking) we can distiguish approaches that convert to crisp FCA by means of alpha cuts or similar mechanisms [8-10], and approaches that use various fuzzy implications to generalise crisp FCA to the fuzzy case [7, 11].

In this work, we take a different approach and define a fuzzy formal concept as a pair X, Y where X is a fuzzy set of objects and Y is a crisp set of properties such that $X^{\uparrow}$=Y and $Y^{\downarrow}$=X where

$$X^{\uparrow} = \{y \in Y \mid \forall x \in X : \mu_R(x,y) \geq \mu_X(x)\} \tag{3}$$

$$Y^{\downarrow} = \left\{ x / \mu_X(x) \middle| \mu_X(x) = \min_{y \in Y}(\mu_R(x,y)) \right\} \tag{4}$$

This is linked to approaches based on fuzzy implication; the precise relationship will be discussed in future work. Table 2 shows a fuzzy context, where the table cells represent the degree to which an attribute (column) is relevant to each object (listed in first column). The objects are messages posted on a forum, and the fuzzy memberships are assumed to be created by automated analysis of the message content and expert judgment. Pairs such as

({a/1, e/0.8 }, {message-content=answer, author-expertise=expert})

and *({a/0.3, b/1, f/0.2, h/0.1, i/0.9 }, {device-type=music player, message-content=complaint})*

are examples of fuzzy concepts derived from this table.

The aim of formal concept analysis is to expose dependencies between attributes in a dataset. In the crisp case these dependencies are generally implications of the form $A1 \rightarrow A2$. An implication holds when *A1* and *A2* are sets of attributes such that every object having all attributes from *A1* also has all attributes from *A2*.

$$A1 \rightarrow A2 \textit{ iff } A2 \subseteq \left(A1^{\downarrow}\right)^{\uparrow} \qquad \textit{where } A1, A2 \subseteq A$$

Table 2. A fuzzy context describing messages posted on a forum

		device type				message content					author expertise		
message-id	subject of message	camera	phone	music player	personal organiser	comment	complaint	question	answer	other content	expert	intermediate	novice
a	iphone	0.6	1	0.8	0.7	1	0.3	0	1	0.6	1	0	0
b	ipod touch	0	0	1	0.7	0.7	1	0	0	0	0.8	0.6	0
c	blackberry	0	0.8	0	1	0	0.4	1	0	0	1	0.6	0
d	blackberry	0	0.8	0	1	0.6	0.1	1	0	0.6	0.4	1	0
e	blackberry	0	0.8	0	1	0.8	0	0	1	0.8	0.8	0.4	0
f	zune	0	0	1	0.2	0	0.2	0.6	0.1	1	0	0.4	0.9
g	camera phone	1	1	0	0.6	1	0	0.8	0	0	0	0.6	0.9
h	mp3 phone	0.3	1	1	0	0	0.1	1	0	0.3	0	0.3	1
i	mp3 phone	0.3	1	1	0	0.3	0.9	1	0	1	1	0	0

Within a crisp concept lattice, any upward link from a concept to its parent (trivially) represents an implication relation. It is also possible to derive association rules with high confidence - typically this is restricted to rules of the form *parent* → *child* (termed the Luxenberger basis, see [12] for discussion; also [13]).

Within a fuzzy concept lattice, the definitions of Eqs. (3) and (4) mean that the set of elements contained within a node is a fuzzy subset of the elements contained in its parent node(s), that is, all properties true of elements in the parent node are also true of elements in the child. In some cases there can also be a strong association in the opposite direction, i.e. most elements in the parent node are also in the child node. We discuss the extension of association rule analysis to fuzzy concepts in the next section.

The fuzzy context in table 2 is derived from three separate tables, as indicated by the headings at the top (*device type, message content, author expertise*). Rather than deriving a large fuzzy formal concept lattice from the entire context, it is possible to derive smaller lattices from the individual tables and look for associations between concepts in different lattices.

3 Association Rules in Fuzzy Categorised Data

Taxonomies and, more generally, ontologies, are essential tools in the knowledge discovery process. Classification of data in taxonomic form is useful to enable subsequent searching but is not just an end in itself. The ability to group multiple entities together into an (approximately) uniform whole allows us to efficiently represent an entire group as a single concept, enabling us to reason, and to derive knowledge, about groups of entities. A simple form of derived knowledge is association - essentially, that the extensions of two concepts overlap significantly.

Association rules (in their crisp form) are a well-established technique for knowledge discovery in databases, enabling "interesting" relations to be discovered. For two concepts *C1* and *C2*,

$$Support(C1 \rightarrow C2) = |C1 \cap C2|$$

$$Confidence(C1 \rightarrow C2) = \frac{|C1 \cap C2|}{|C1|}$$

where the cardinality of a concept is the number of objects in its extension.

Consider the database of sales employees, salaries and sales figures in Fig 3. A mining task might be to find out whether the good sales figures are achieved by the highly paid employees. We can obtain rule confidences ranging from 1/3 up to 1 by different crisp definitions of *"good sales"* and *"high salary"*, as shown on the right of Fig. 3. Although this is a contrived example, such sensitivity to the cut-off points adopted for crisp definitions is a good indication that a fuzzy approach is more in line with human understanding of the categories.

name	sales	salary
a	100	1000
b	80	400
c	50	800
d	20	700

definition of *good sales*	definition of *high salary*	*rule confidence*
sales≥80	*high≥400*	1
sales≥50	*high≥500*	0.667
sales>50	*high>500*	0.5
sales≥50	*high>800*	0.333

Fig. 3. A simple database of names (a, b, c, d), sales and salary figures (left) and (right) the confidences for an association rule *good sales => high salary* arising from different crisp definitions of the terms *good sales* and *high salary*

The associations corresponding to these crisp definitions of *good sales* and h*igh salary* can also be found from the formal concept lattices. Fig. 4 shows the lattices corresponding to the crisp definitions listed in Fig 3. The cardinality of a concept is easily found by counting the elements at the concept node plus the elements at its descendant nodes; the cardinality of the intersection of two concepts is the cardinality of their greatest lower bound. So for the third set of definitions above (*sales>50, high>500*), the cardinality of the *good sales* concept is 2 (objects *a* and *b*) and the cardinality of the intersection of *good sales* and *high salary* is 1 (object *a*), giving the rule confidence ½.

An alternative to this approach is to build two smaller lattices corresponding to the individual attribute sets - in this case, one corresponding to *good sales* and one to *high salary* - and then look for associations between pairs of concepts drawn respectively from the two lattices. For example, if the lattice *C1* contains sets *C1i* and the lattice *C2* contains sets *C2j*, we would look for associations *C1i* → *C2j*. In the work described below, we have adopted the second method but are actively examining the first approach in other research.

There have been a number of proposals to extend association rules to the fuzzy case, that is, to discover the degree of association between fuzzy concepts. A good overview is given in [14]. Our recent work takes a different approach, using mass assignment theory [15-17] to find a point valued association confidence between fuzzy categories [18] or a fuzzy-valued version [19, 20]. We argue that, in looking for association strengths between fuzzy categories, it is better to propagate the fuzziness through the calculation and produce a fuzzy value rather than a single value. Other association rule measures (e.g. lift) can be treated in the same framework.

In streaming applications, it is not sufficient to take a dataset at a specific time and extract the interesting relations. Because data streams are continually updated, the strength of relations (i.e. association confidence) is continually changing. We address this issue by considering the change in fuzzy association confidence over a specified time window as more data are added. The window can be crisp or fuzzy.

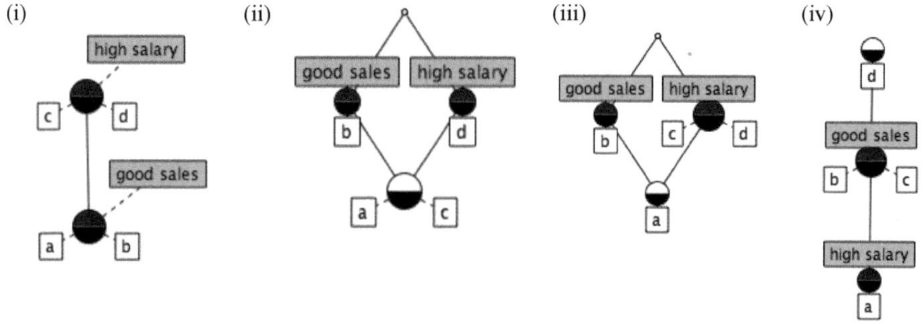

Fig. 4. Concept lattices derived from the four definitions in Fig 3

Various associations can be extracted by consideration of fuzzy categories in different taxonomies. Since the dataset is not static, we cannot assume that significant associations will remain significant – indeed, valuable insight arises from detecting changes in association levels relative to other associations, and trends in the strength of an association. Our approach has been to average association strengths over a user-specifiable time window, and highlight anomalous values.

In this work, we focus on two classes of anomalous values, illustrated in the next section. Assume *A*, *B* are general fuzzy concepts in different taxonomies deemed to be of interest. The first class monitors the evolution of fuzzy association levels for rules $A \rightarrow B$. If the rule confidence at two user-specified times t_1 and t_2 does not conform to

$$Conf(A \rightarrow B, t_1) \approx Conf(A \rightarrow B, t_2)$$

then the association is flagged as potentially anomalous. The second class of interesting associations is not time-dependent but involves parent/child discrepancies. Let A_P, A_{C1} ... A_{Cn} (resp B_P, B_{Ci}) be parent and child concepts in two different taxonomies. In the absence of further information, we would expect similar confidence for $A_P \rightarrow B$ and $A_{Ci} \rightarrow B$ since the confidences are given by

$$Confidence(A_P \rightarrow B) = \frac{|A_P \cap B|}{|A_P|} \qquad Confidence(A_{Ci} \rightarrow B) = \frac{|A_{Ci} \cap B|}{|A_{Ci}|}$$

and there is no reason to expect attribute set B to be dependent on Ci. Similarly, assuming more than one child , we would expect $A \longrightarrow B_P$ to be larger than $A \longrightarrow B_{Ci}$ Any deviation from these expectations can be flagged as potentially interesting.

4 Case Study

We have applied these methods to the calculation of associations in a database of terrorist incidents (Worldwide Incidents Tracking System (WITS) [21] augmented by information from the MIPT Terrorism Knowledge Base[1]. As described in [19] we integrate data from these sources and recategorise it according to various taxonomic views, e.g. fuzzy regions ("*Middle East*" or "*in/near Iraq*") or fuzzy categories based on the casualty levels, *low*, *medium*, *high*, *very-high*, or perpetrator, weapon-type, etc. Each incident represents one object; the attributes of interest include location details, (city, country, region), and weapon type. The concept hierarchies are generally quite simple, and are a mixture of automatically extracted taxonomies and manually refined taxonomies based on fuzzy FCA. Although the entire dataset is available, we have simulated a data stream by updating the known data on an incident-by-incident basis. The examples are presented as a proof of concept in a reasonably large dataset (tens of thousands of incidents). To date, they have not been compared to other analyses.

To illustrate the first class of anomalous values, Fig. 5 shows changes in fuzzy association confidence over time; these would be drawn to the attention of human experts for interpretation. The associations in Fig. 5 relate incidents in the (fuzzy) geographic region *Israel / near Israel* to *weapon type*, and show a recent sharp increase in *weapon type= missile / rocket*. The plot shows point-valued confidences for the association, averaged over 2 months, with the bars showing the minimum / maximum of the fuzzy set associated with each point value..

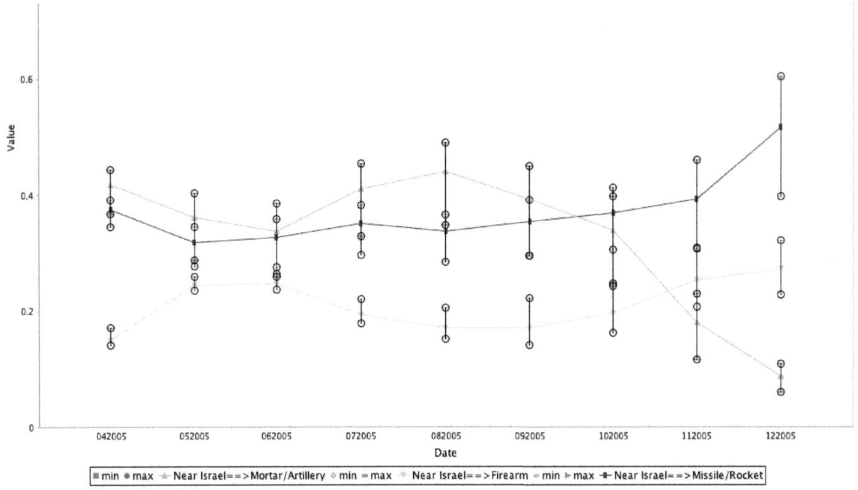

Fig. 5. Change in fuzzy association strength between incidents occurring in / near Israel and incidents involving specified weapons

[1] http://www.start.umd.edu/data/gtd/

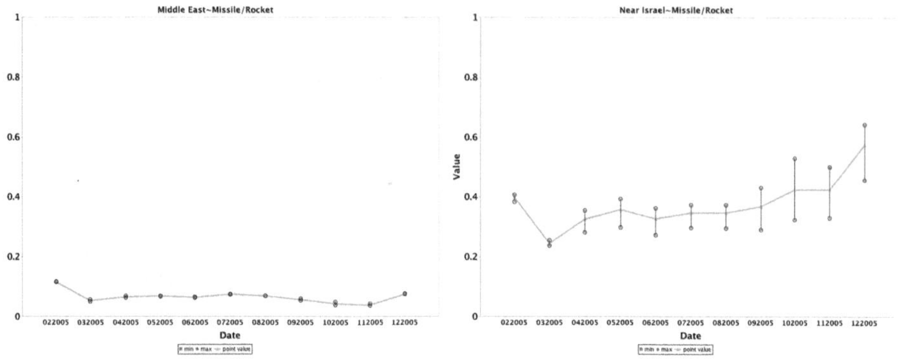

Fig. 6. Strength of association for *geographical region*:X => *weapon type*:missile / rocket. The left figure uses X=Middle East as the (fuzzy) region, whereas the right figure shows the child category *Israel/near Israel* which has a much higher association strength than its parent category (or any of its siblings).

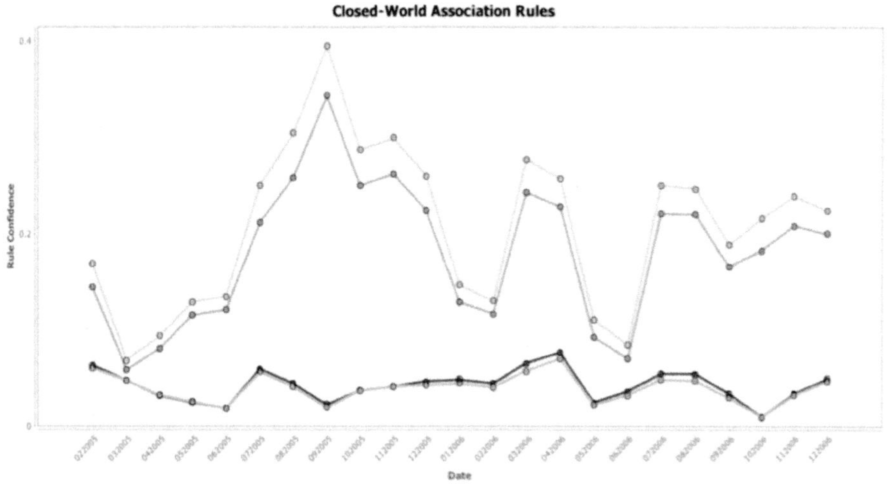

Fig. 7. Change in association confidence between incidents classified as kidnap and incidents occurring in South Asia (top line), Nepal (second line), Middle East (third line) and in/near Iraq (fourth line, mostly obscured by third line). Uncertainty is negligible in these cases.

The second class of interesting associations involves parent/child discrepancies. We compare the association between incidents in the fuzzy geographical regions

(i) *Middle East* (parent category) (ii) *in/near Israel* (child category)

and the set of incidents involving *missile/rocket* attack. Fig. 6 shows a clear difference between the two cases. Although all incidents in the child category are included in the parent category, the association with this particular child category is much stronger. Associations with other child categories (*near Iraq*, *near Iran*, etc) are similar to the association with the parent category.

Fig. 7 considers *incident-type* → *country* (where *incident type* includes *kidnap, assault, hijack, explosion* etc). The anomalous values show that almost all recorded incidents of kidnap in the considered time periods occur in Nepal (for South Asia mainland region) and in Iraq (for Middle East region).

5 Summary

The contribution of this paper is two-fold. We use fuzzy FCA to extract taxonomies, either to be used unchanged or to form a starting point for further refinement. We use a novel form of fuzzy association analysis to detect potentially interesting static and dynamic relations between concepts in different taxonomies. The feasibility of these methods has been illustrated by application to a (simulated) stream of reports concerning incidents of terrorism.

References

[1] Martin, T.P.: Fuzzy sets in the fight against digital obesity. Fuzzy Sets and Systems 156, 411–417 (2005)

[2] Zadeh, L.A.: Fuzzy Sets. Information and Control 8, 338–353 (1965)

[3] Ganter, B., Wille, R.: Formal Concept Analysis: Mathematical Foundations. Springer, Heidelberg (1998)

[4] Priss, U.: Formal Concept Analysis in Information Science. Annual Review of Information Science and Technology 40, 521–543 (2006)

[5] Prediger, S.: Logical Scaling in Formal Concept Analysis. In: Delugach, H.S., Keeler, M.A., Searle, L., Lukose, D., Sowa, J.F. (eds.) ICCS 1997. LNCS, vol. 1257, pp. 332–341. Springer, Heidelberg (1997)

[6] Burusco, A., Fuentes-Gonzalez, R.: The study of the L-fuzzy concept lattice. Mathware and Soft Computing 1, 209–218 (1994)

[7] Belohlavek, R., Vychodil, V.: What is a fuzzy concept lattice? In: Proceedings of International Conference on Concept Lattices and their Applications, CLA 2005, Aalborg, Denmark, July 16-21, pp. 34–45. Czech Republic (2005)

[8] Quan, T.T., Hui, S.C., Cao, T.H.: Ontology-Based Fuzzy Retrieval for Digital Library. In: Goh, D.H.-L., Cao, T.H., Sølvberg, I.T., Rasmussen, E. (eds.) ICADL 2007. LNCS, vol. 4822, pp. 95–98. Springer, Heidelberg (2007)

[9] Zhou, W., Liu, Z., Zhao, Y.: Ontology Learning by Clustering Based on Fuzzy Formal Concept Analysis. In: 31st Computer Software and Applications Conference (COMPSAC 2007), pp. 204–210 (2007)

[10] Ceravolo, P., Damiani, E., Viviani, M.: Extending Formal Concept Analysis by Fuzzy Bags. In: IPMU 2006, Paris, France (2006)

[11] Djouadi, Y., Prade, H.: Interval-Valued Fuzzy Formal Concept Analysis. In: Rauch, J., Raś, Z.W., Berka, P., Elomaa, T. (eds.) Foundations of Intelligent Systems. LNCS, vol. 5722, pp. 592–601. Springer, Heidelberg (2009)

[12] Valtchev, P., Missaoui, R., Godin, R.: Formal Concept Analysis for Knowledge Discovery and Data Mining: The New Challenges. In: Formal concept analysis; Concept lattices, Singapore, pp. 352–371 (2004)

[13] Lakhal, L., Stumme, G.: Efficient Mining of Association Rules Based on Formal Concept Analysis. In: Ganter, B., Stumme, G., Wille, R. (eds.) Formal Concept Analysis. LNCS (LNAI), vol. 3626, pp. 180–195. Springer, Heidelberg (2005)
[14] Dubois, D., Hullermeier, E., Prade, H.: A systematic approach to the assessment of fuzzy association rules. Data Mining and Knowledge Discovery 13, 167–192 (2006)
[15] Baldwin, J.F.: The Management of Fuzzy and Probabilistic Uncertainties for Knowledge Based Systems. In: Shapiro, S.A. (ed.) Encyclopedia of AI, 2nd edn., pp. 528–537. John Wiley, Chichester (1992)
[16] Baldwin, J.F.: Mass Assignments and Fuzzy Sets for Fuzzy Databases. In: Fedrizzi, M., Kacprzyk, J., Yager, R.R. (eds.) Advances in the Shafer Dempster Theory of Evidence. John Wiley, Chichester (1994)
[17] Baldwin, J.F., Martin, T.P., Pilsworth, B.W.: FRIL - Fuzzy and Evidential Reasoning in AI. Research Studies Press (John Wiley), U.K. (1995)
[18] Martin, T.P., Azvine, B., Shen, Y.: Finding Soft Relations in Granular Information Hierarchies. In: 2007 IEEE International Conference on Granular Computing, Fremont, CA, USA (2007)
[19] Martin, T.P., Shen, Y.: TRACK - Time-varying Relations in Approximately Categorised Knowledge. International Journal of Computational Intelligence Research 4, 300–313 (2008)
[20] Martin, T.P., Shen, Y.: Fuzzy Association Rules to Summarise Multiple Taxonomies in Large Databases. In: Laurent, A., Lesot, M.-J. (eds.) Scalable Fuzzy Algorithms for Data Management and Analysis: Methods and Design, pp. 273–301. IGI-Global (2009)
[21] WITS, WITS - Worldwide Incidents Tracking System, National Counterterrorism Center, Office of the Director of National Intelligence 2007 (2007)

Using Enriched Ontology Structure for Improving Statistical Models of Gene Annotation Sets

Frank Rügheimer

Institut Pasteur, Laboratoire Biologie Systémique
Département Génomes et Génétique
F-75015 Paris, France
CNRS, URA2171, F-75015 Paris, France
frueghei@pasteur.fr

Abstract. The statistical analysis of annotations provided for genes and gene products supports biologists in their interpretation of data from large-scale experiments. Comparing, for instance, distributions of annotations associated with differentially expressed genes to a reference, highlights interesting observations and permits to formulate hypotheses about changes to the activity pathways and their interaction under the chosen experimental conditions. The ability to reliably and efficiently detect relevant changes depends on properties of the chosen distribution models. This paper compares four methods to represent statistical information about gene annotations and compares their performance on a public dataset with respect to a number of evaluation measures. The evaluation results demonstrate that the inclusion of structure information from the Gene Ontology enhances overall approximation quality by providing suitable decompositions of probability distributions.

1 Introduction

The Gene Ontology (GO) [1] establishes standardized sets of annotation terms for genes and gene products. Terms are grouped in three separate sub-ontologies that are concerned with intracellular location, molecular functions and associated biological processes of gene products respectively. In addition, the ontology provides a network of relations that formalize the relationships between annotation terms. This is used, for example, to associate a general category with more specific terms subsumed under that category, so annotations on different levels of detail may be applied concurrently. The resulting formal description of domain knowledge has been a key contribution to expanding the use of computational methods in the analysis and interpretation of large scale biological data. For instance, the GO enables the definition of semantic similarity measures, which in turn can be used to compare or cluster gene products and groups thereof [7] or to implement semantic search strategies for retrieval of information from domain-specific text collections [5].

E. Hüllermeier, R. Kruse, and F. Hoffmann (Eds.): IPMU 2010, Part II, CCIS 81, pp. 55–64, 2010.

The utility of the term relations is further increased when the ontology is combined with an annotation database. In the case of the GO term relations have been combined with databases of annotations for gene products to identify statistically significant term enrichment [2] within subset of genes undergoing expression changes in large scale experiments (microarrays, ChIP-chip etc.). For this reason plug-ins for GO annotations have been integrated into standard data visualization and analysis software for systems biology [3]. In a similar way annotation sources and term relations can be combined with data from experiments investigating effects of interventions, e.g. from knockout or RNAi studies. The relational information provided by GO contributes to the integration of observations on different levels of detail so a subsequent statistical analysis of the results becomes possible. Beyond this role in data fusion, the ontology structure can guide the construction of statistical models by providing decompositions of probability distributions over annotations. As an additional benefit this approach establishes consistency between distributions regardless of the level of detail under which data is viewed for the purpose of the analysis.

In annotation databases following the GO standard each entry assigns an annotation term to a particular gene. Both genes and annotation terms may occur in several entries. Therefore several terms may be annotated to the same gene, and such combinations of annotations terms are often used to indicate roles in the interaction of pathways associated with different biological functions. While an analysis of the annotation sets as a whole is desirable, a direct representation via empirical distributions is usually impractical as the theoretical size of the sample space for annotations with n possible terms is on the order of 2^n. To construct probabilistic models of annotation frequencies a number of representations are employed, which differ in their inherent modeling assumptions and the simplifications applied. In this paper I compare such strategies using publicly available data sets and discuss their differences in the light of the resulting properties.

Section 2 provides a brief exposition of the data set used in the comparison, its connection to the Gene Ontology and the preprocessing applied to it. In section 3 the three different types of distribution models are presented and pointers to the relevant literature given. This is followed by details of the evaluation method and the evaluation measures employed (section 4). Results are summarized and discussed in section 5.

2 Data Sets and Preprocessing

The data set used throughout the experiment was constructed from a collection of annotations on the function of the genes and gene products of the baker's and brewer's yeast *Saccharomyces cerevisiae* – one of the most well-studied eukariotic model organisms. The collection is maintained by the Saccharomyces Genome Database project [11] and will be referred to as the SGD. The annotations provided by that source are compliant with the GO annotation standard.

Within the GO, terms are organized into three non-overlapping term sets. Each of the three sets covers one annotation aspect, namely biological processes,

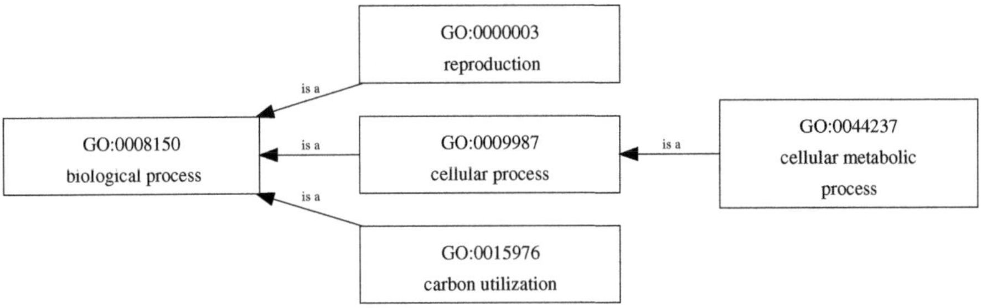

Fig. 1. Extract from the quasi-hierarchical term structure as specified by the Gene Ontology relations

molecular functions or cellular components. For the "cellular components" the term set is structured by a quasi-hierarchical partial ordering defined via the "`part_of`" relation whereas the "`is_a`" relation fulfills the same role for the two other aspects (Figure 1). The "molecular function" annotation refers to biochemical, e.g. enzymatic, activity and is closely linked to the protein (and therefore gene) sequence. However, many proteins that posses very similar functions or even a shared evolutionary history are found in largely unrelated pathways. The "cellular component" annotation provides information on cell compartments in which the gene products are known to occur. On the biological side this information allows to restrict a search for potential additional players in a partially known mechanism. This enables experimenters to look specifically at, for instance, candidates for a postulated membrane bound receptor. Finally, the "biological process" annotations provide an idea of the overall functionality to which a gene product contributes. The terms occurring in the ontology fragment of Figure 1 are examples of this annotation type. Because the targeted interactions in large scale expression studies are focused on overall biological processes only annotations for this aspect were considered in the experiments.

It should also be noted that the full term set of the GO (at the time of this writing >30,000 terms) provides a very high level of detail. Only a small subset of the available annotation terms are actually used in the SGD. Even among the terms that are used, many have a low coverage in the database. To obtain a standardized, broader perspective on the data that lends itself to a statistical analysis, less specific versions of the ontology can be employed. These so-called "slim ontologies" define subsets of comparatively general Gene Ontology terms. In the case of *S. cerevisiae* a species-specific slim version of the ontology has been released together with the full annotation data [12]. For the study described in this paper any annotation terms from the SGD that were not already included in the slim version of the ontology were mapped to their most specific ancestors in the reduced term set. Note that for the selected sub-ontology the corresponding term subset of the Yeast Slim GO has tree structure.

The resulting term sets consists mainly of leaf nodes of the slim ontology, but still contains elements representing coarser descriptions. For the evaluation these

```
tY(GUA)M1 {GO_0006412}
YAL004W   {GO_0008150}
YAL005C   {GO_0006412 GO_0006457 GO_0006810 GO_0006950 GO_0006996}
YAL010C   {GO_0006810 GO_0006996 GO_0016044}
YAL014C   {GO_0006810}
YAL017W   {GO_0005975 GO_0006464 GO_0007047}
YAL018C   {GO_0008150}
YAL019W   {GO_0006996}
YAL026C   {GO_0006810 GO_0006996 GO_0016044 GO_0016192 GO_0042254}
```

Fig. 2. Fragment of constructed gene list with associated GO term identifiers

terms were considered as competing with their more specific hierarchical children, reflecting the GO annotation policy of assigning the most specific suitable term supported by the observations. For the analysis the example database was constructed by aggregating the mapped annotations for each of the known genes into gene-specific annotation sets. The resulting file summarizes the known biological processes for each of 6849 genes using a total of 909 distinct annotation sets (Figure 2).

In parallel, the preprocessing assembled information about the annotation scheme employed. To that end the term hierarchy was extracted from the ontology and converted into a domain specification. This specification serves to describe how the annotation on different levels of details relate to each other and was later used to by one of the models to integrate the ontology information during the learning phase.

3 Distribution Models

In order to cover a broad spectrum of different strategies four representations for distributions on annotation sets were implemented:

a) A model using binary random variables to encode presence or absence of elements in a set. The variables are treated as independent, so the distribution of set-instantiations is obtained as a product of the proabilities for the presence or absence of the individual elements of the underlying carrier set.
b) A condensed distribution [8] model using an unstructured attribute domain
c) An enriched term hierarchy using condensed random sets for the representation of branch distribution [10,9]
d) A random set representation [6]

The representation task is formalized as follows: Let Ω denote the set of available annotation terms. The preprocessed annotation database is rendered as a list $D = \{S_1, \dots, S_m\}$ $m \in \mathbb{N}$, $S_i \in 2^\Omega$, where the S_i represent annotation term sets associated with individual genes. The representation task is to model relevant properties of the generating probability distribution characterized via its probability mass function $p_{\text{Annot}} : 2^\Omega \to [0, 1]$. To this end distribution models are trained from a non-empty training set $D_{\text{trn}} \subset D$ and subsequently tested using the corresponding test set $D_{\text{tst}} = D \setminus D_{\text{trn}}$. To increase the robustness of evaluation results several training and test runs are embedded into a cross-validation framework (cf. section 4).

The independence assumptions in (a) allow a compact representation of probability distributions by decomposing them into a small set of binary factor distributions $\hat{p}_{\omega_i} : \{+,-\} \to [0,1]$, where the outcomes + and − denote presence or absence of the term ω_i in the annotation set. This results in the decomposition

$$\forall S \subseteq \Omega : \quad \hat{p}_{\text{Annot}}(S) = \left(\prod_{\omega \in S} \hat{p}_\omega(+)\right)\left(\prod_{\omega \in \Omega \setminus S} \hat{p}_\omega(-)\right) \tag{1}$$

$$= \left(\prod_{\omega \in S} \hat{p}_\omega(+)\right)\left(\prod_{\omega \in \Omega \setminus S} 1 - \hat{p}_\omega(+)\right) . \tag{2}$$

Because only one value per term needs to be stored this results in a very compact model. Moreover, the approach allows to rapidly compute probability estimates and is thus popular in text mining and other tasks involving large term sets. The strong independence assumptions, however, are also a potential source of errors in of the representation of probability distribution over the set domain 2^Ω.

Approach (d) represents the opposite extreme: Each possible combination of terms is represented in the sample space, which for this model is the power set 2^Ω of the term set. Therefore the the target distribution is estimated directly from observations of the samples. Due to the size of the distribution model, and the sparse coverage of the sample space no explicit representation of the model was provided. Instead all computations were conducted at evaluation time based on counts of annotation term sets shared by the training and test database applying a subsequent modification for the Laplace correction (see page 60). Nevertheless, computation time for evaluating the random set model exceeded that of the other models by several orders of magnitude and, giving its scaling behavior, is not considered an option for application in practice.

Finally approach (b) and (c) represent two variants of the condensed random set model introduced in [8] and [10] respectively. The central idea of these approaches is to use a simplified sample space that represents annotations consisting of single terms separately, but groups those for non-singleton instantiations. The probability mass assigned to the non-singleton instantiations is then further distributed according to a re-normalized conditional probability distribution, which is encoded using the method proposed in (a). This two-step approach allows to better reflect the singletons (which are overrepresented in GO-Annotations), while retaining the performance advantages of (a). Approach (c) additionally integrates structure information from the ontology by associating a condensed random set with each branch of the ontology structure. Because condensed random sets use the probabilistic aggregation operation observations on coarsened versions of the enriched ontology remain consistent with aggregated

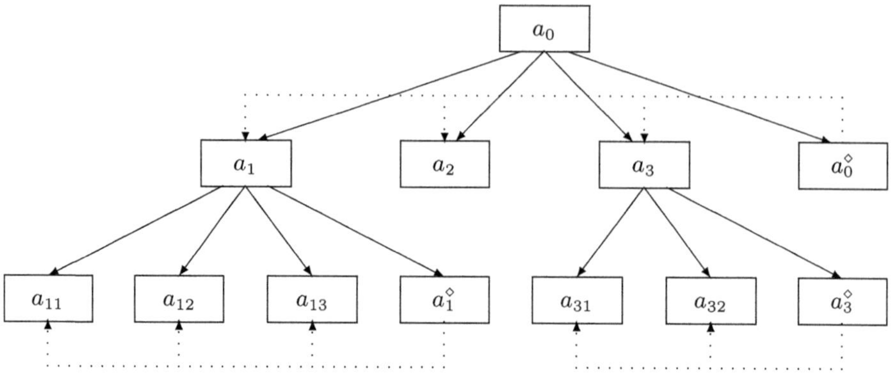

Fig. 3. Decomposition principle for the hierarchical version of the Condensed Random Set model. Conditional probabilities and coverage factors are indicated by solid and dotted arrows respectively (image from [10]).

results from observations on higher levels of detail. For an in-depth discussion of parameters and the model induction algorithm see [9].

In all cases, the parameters were estimated from the observed frequencies in the training data with a Laplace correction applied. The value of the Laplace correction was set to of 0.5 for models (a), (b), (c) and to $2.5 \cdot 10^{-9}$ for model (d), contributing similar portions of the total probability mass for all models.

4 Evaluation

Preprocessing resulted in a database of annotation sets for 6849 genes. To limit sampling effects, the evaluation measures were computed in a 5-fold cross-validation process [4]. To this end the data set was split into five partitions with genes randomly assigned (4 partition with 1370 genes each and one partition with 1369 genes). In each of the five runs a different partition was designated as a test data set whereas the remaining partitions used in the role of training data. Evaluation measures were chosen to provide complementary information on how well different aspects of the set-distribution are captured by each model type. All measures are described with respect to and evaluation against a test data set $D_{\text{tst}} \subset D$.

Log-Likelihood. A common way to evaluate the fit of a probability-based model M is to consider the likelihood of the observed test data D_{tst} under the model, that is, the conditional probability estimate $\hat{p}(D_{\text{tst}} \mid M)$. The closer the agreement between test data and model, the higher that likelihood will be. The likelihood is also useful to test model generalization, as models that overfit the training data tend to predict low likelihoods for test datasets drawn from the same background distribution as the training data. To circumvent technical limitations concerning the representation of and operations with small numbers in the computer, the actual measure used in practice is based on the logarithm of the likelihood:

$$\log L(D_{\text{tst}}) = \log \prod_{S \in D_{\text{tst}}} \hat{p}(S \mid M) \tag{3}$$

$$= \sum_{S \in D_{\text{tst}}} \log \hat{p}(S \mid M). \tag{4}$$

In that formula the particular term used to estimate the probabilities $\hat{p}(S \mid M)$ of the records in D are model-dependent. Since the likelihood takes values from $[0, 1]$ the values for the log-transformed measure are from $(-\infty, 0]$ with larger values (closer to 0) indicating better fit. The idea of the measure is that the individual cases (genes) in both the training and the test sets are considered as independently sampled instantiations of a multi-valued random variable drawn from the same distribution. The likelihood of a particular test database D_{tst} is computed as the product of the likelihoods of its $|D_{\text{tst}}|$ elements. Due to the low likelihood of individual sample realizations even for good model approximation, the Log-Likelihood is almost always implemented using the formula given in Equation 4, which yields intermediate results within the bounds of standard floating point format number representations.

One particular difficulty connected with the Log-Likelihood, resides in the treatment of previously unobserved cases in the test data set. If such values are assigned a likelihood of zero by the model then this assignment entails that the whole database is considered as impossible and the Log-Likelihood becomes undefined. In the experiment this undesired behavior was countered by applying a Laplace correction during the training phase. The Laplace correction ensures that all conceivable events that have not been covered in the training data are modeled with a small non-zero probability estimate and allow the resulting measures to discriminate between databases containing such records.

Average Record Log-Likelihood. The main idea of the log-likelihood measure is to separately evaluate the likelihood of each record in the test database with respect to the model and consider the database construction process a sequence of a finite number of independent trials. As a result log-likelihoods obtained on test databases of different sizes are difficult to compare. By correcting for the size of the test database one obtains an average record log-likelihood measure that is better suited to a comparative study:

$$\text{arLL}(D_{\text{tst}}) = \frac{\log L(D_{\text{tst}})}{|D_{\text{tst}}|}. \tag{5}$$

Note that in the untransformed domain the mean of the log-likelihoods corresponds to the geometric mean of the likelihoods, and is thus consistent with the construction of the measure from a product of evaluations of independently generated instantiations.

Singleton and Coverage Rate Errors. In addition to the overall fit between model and data, it is desirable to characterize how well other properties of a set-distribution are represented. In particular it has been pointed out that the condensed distribution emphasizes the approximation of both singleton probabilities and the values of the element coverage. To assess how well these properties are preserved by the investigated representation methods, two additional measures – $\mathrm{d_{sglt}}$ and $\mathrm{d_{cov}}$ – have been employed. These measures are based on the sum of squared errors for the respective values over all elements of the base domain:

$$\mathrm{d_{sglt}} = \sum_{\omega \in \Omega} \left(p'(\omega) - p(\omega)\right)^2 , \tag{6}$$

$$\mathrm{d_{cov}} = \sum_{\omega \in \Omega} \left(\mathrm{opc}'(\omega) - \mathrm{opc}(\omega)\right)^2 . \tag{7}$$

In this equation the function opc computes the one-point coverage of an element by a random set, defined as the cumulative probability of every instantiation containing the argument.

5 Results

For the assessment and comparison of the different methods, a 5-fold cross-validation was conducted. All approaches were applied with the same partitioning of the data. Evaluation results of the individual cross-validation runs were collected and – with the exception of the logL measure – averaged. These results are summarized in Table 1.

The two condensed random set-based models (b) and (c) achieve a better overall fit to the test data (higher value of arLL-measure) than the model assuming independence of term coverages (a) indicating that those assumptions are not well suited for annotation data. The highest accuracy for all models and variants is achieved using the hierarchical version of the CRS model. This is interpreted as a clear indication of the benefits provided due to additional structure information. Despite its large number of parameters the full random set representation (d) does not achieve acceptable approximation results. Due to the large sample space that model is prone to overfitting.

For the prediction of singleton annotations model (a) exhibits large prediction errors. This again is explained by the independence assumption in that representation being too strong. In contrast, with their separate representation of singleton annotation sets, the CRS-based models show only small prediction errors for the singleton frequencies, though the incomplete separation between real singletons and single elements in local branch distributions appear to leads to a slightly increased error for the hierarchical version.

Table 1. Evaluation results for individual runs and result summaries; best and second best results highlighted in dark and light gray respectively (from top left to bottom right: Model using independent binary variables (a) with Laplace correction of 0.5, Condensed Random Sets on unstructured domain (b) with Laplace correction of 0.5, Condensed Random Sets on hierarchically structured domain (c) with Laplace correction of 0.5, Random Set representation (d) with Laplace correction of $2.5 \cdot 10^{-9}$)

a)

$\log L$	arLL	d_{sglt}	d_{cov}
-9039.60	-6.60	0.067856	0.001324
-8957.19	-6.54	0.064273	0.001524
-9132.09	-6.67	0.060619	0.001851
-8935.82	-6.52	0.074337	0.001906
-9193.44	-6.72	0.059949	0.001321
	-6.61	0.065406	0.001585

b)

$\log L$	arLL	d_{sglt}	d_{cov}
-7992.76	-5.83	0.000241	0.001342
-7885.19	-5.76	0.000222	0.001531
-8045.31	-5.87	0.000411	0.001838
-7839.16	-5.72	0.000612	0.001895
-8195.49	-5.99	0.000268	0.001316
	-5.83	0.00035	0.001584

c)

$\log L$	arLL	d_{sglt}	d_{cov}
-7629.66	-5.57	0.000539	0.008293
-7559.38	-5.52	0.000457	0.011652
-7752.21	-5.66	0.000857	0.006998
-7529.83	-5.50	0.001014	0.004767
-7828.44	-5.72	0.000567	0.009961
	-5.59	0.000686	0.008334

d)

$\log L$	arLL	d_{sglt}	d_{cov}
-8259.66	-6.03	0.001823	0.098462
-8346.13	-6.09	0.001311	0.100860
-8651.12	-6.31	0.000964	0.095850
-8288.68	-6.05	0.003105	0.103200
-8534.30	-6.23	0.000671	0.094305
	-6.14	0.001574	0.098536

This is consistent with the higher error d_{cov} of that model in the prediction of coverage factors. The non-hierarchical models (a) and (b) represent one-point coverages directly and therefore achieve identical prediction error[1]. Large deviations for coverage rate predicted by the Random Set model (d) are explained by the cumulative effect of the Laplace correction after aggregating over the a large number of combinations.

6 Conclusions

The presented contribution analyzed the effect of different modeling assumptions for representing distributions over annotation sets. Although parsimonious models should be preferred whenever justified from the data, the often applied independence assumption for term occurrence do not seem to hold for annotation data in biology. It could be shown that the inclusion of background information on relations between annotation terms contributes to improving the overall accuracy of the representation at some cost for the accuracy of coverage rates and singleton frequencies. In combination with the additional benefit of consistent aggregation operations the results indicate that the probabilistic

[1] The minor differences between the tables are merely artifacts of the two-factor decomposition of coverage factors in the condensed distribution.

enrichment of ontologies provides an both effective approach to the statistical modeling of distributions over annotation sets and integrates well with already available resources for data analysis in biology.

References

1. Ashburner, M., Ball, C.A., Blake, J.A., Botstein, D., Butler, H., Cherry, J.M., Davis, A.P., Dolinski, K., Dwight, S.S., Eppig, J.T., Harris, M.A., Hill, D.P., Issel-Tarver, L., Kasarskis, A., Lewis, S., Matese, J.C., Richardson, J.E., Ringwald, M., Rubin, G.M., Sherlock, G.: Gene Ontology: tool for the unification of biology. Nature Genetics 25, 25–29 (2000)
2. Boyle, E.I., Weng, S., Gollub, J., Jin, H., Botstein, D., Cherry, J.M., Sherlock, G.: GO::TermFinder—open source software for accessing Gene Ontology information and finding significantly enriched Gene Ontology terms associated with lists of genes. Bioinformatics 20(18), 3710–3715 (2004)
3. Garcia, O., Saveanu, C., Cline, M., Fromont-Racine, M., Jacquier, A., Schwikowski, B., Aittokallio, T.: GOlorize: a Cytoscape plug-in for network visualization with Gene Ontology-based layout and coloring. Bioinformatics 23(3), 394–396 (2006)
4. Kohavi, R.: A study of cross-validation and bootstrap for accuracy estimation and model selection. In: Proc. of the 14th Int. Joint Conference on Artificial Intellligence (IJCAI 1995), pp. 1137–1145 (1995)
5. Müller, H.M., Kenny, E.E., Sternberg, P.W.: Textpresso: An ontology-based information retrieval and extraction system for biological literature. PLoS Biology 2(11) (2004)
6. Nguyen, H.T.: On random sets and belief functions. Journal Math. Anal. Appl. 65, 531–542 (1978)
7. Ovaska, K., Laakso, M., Hautaniemi, S.: Fast Gene Ontology based clustering for microarray experiments. BioData Mining 1(11) (2008)
8. Rügheimer, F.: A condensed representation for distributions over set-valued attributes. In: Proc. 17th Workshop on Computational Intelligence. Universitätsverlag Karlsruhe, Karlsruhe (2007)
9. Rügheimer, F., De Luca, E.W.: Condensed random sets for efficient quantitative modelling of gene annotation data. In: Proc. of the Workshop "Knowledge Discovery, Data Mining and Machine Learning 2009" at the LWA 2009, pp. 92–99. Gesellschaft für Informatik (2009) (published online)
10. Rügheimer, F., Kruse, R.: An uncertainty representation for set-valued attributes with hierarchical domains. In: Proceedings of the 12th International Conference on Information Processing and Management of Uncertainty in Knowledge-Based Systems (IPMU 2008), Málaga, Spain (2008)
11. SGD Curators: Saccharomyces genome database, `http://www.yeastgenome.org`, (accessed 2008/11/16)
12. SGD Curators: SGD yeast gene annotation dataset (slim ontology version). via Saccharomyces Genome Database Project [11], `ftp://genome-ftp.stanford.edu/pub/yeast/data_download/literature_curation/go_slim_mapping.tab` (accessed November 16, 2008)

Predicting Outcomes of Septic Shock Patients Using Feature Selection Based on Soft Computing Techniques*

André S. Fialho[1,2,3], Federico Cismondi[1,2,3], Susana M. Vieira[1,3], João M.C. Sousa[1,3], Shane R. Reti[4], Michael D. Howell[5], and Stan N. Finkelstein[1,2]

[1] MIT–Portugal Program, 77 Massachusetts Avenue, E40-221, 02139 Cambridge, MA, USA
[2] Massachusetts Institute of Technology, Engineering Systems Division, 77 Massachusetts Avenue, 02139 Cambridge, MA, USA
[3] Technical University of Lisbon, Instituto Superior Técnico, Dept. of Mechanical Engineering, CIS/IDMEC – LAETA, Av. Rovisco Pais, 1049-001 Lisbon, Portugal
[4] Division of Clinical Informatics, Department of Medicine, Beth Israel Deaconess Medical Centre, Harvard Medical School, Boston, MA, USA
[5] Silverman Institute for Healthcare Quality and Safety, Beth Israel Deaconess Medical Center, Harvard Medical School, Boston, MA, USA

Abstract. This paper proposes the application of new knowledge based methods to a septic shock patient database. It uses wrapper methods (bottom-up tree search or ant feature selection) to reduce the number of features. Fuzzy and neural modeling are used for classification. The goal is to estimate, as accurately as possible, the outcome (survived or deceased) of these septic shock patients. Results show that the approaches presented outperform any previous solutions, specifically in terms of sensitivity.

1 Introduction

A patient is considered to be in septic shock when the hypotensive state related to a sepsis condition persists, despite adequate fluid resuscitation [1]. This advanced stage of sepsis carries a high burden, which translates into a high mortality rate (about 50%) and high costs of treatments compared with other intensive care unit (ICU) patients [2].

With regard to clinical predictors based on knowledge discovery techniques, previous works have applied knowledge-based neural networks and neuro-fuzzy techniques in the domain of outcome prediction for septic shock patients [3,4]. This paper uses these same clinical predictors to determine the outcome (deceased or survived) of septic shock patients.

Our main goal is the application of soft computing techniques to a publicly available septic shock patient database, and compare our results with the ones obtained in [4]. As with other real-world databases, the septic shock patient dataset dealt with here, involves a relatively large number of non-linear features. Thus, input selection is a crucial step, in order to reduce model's complexity and remove inputs which do not improve

* This work is supported by the Portuguese Government under the programs: project PTDC/SEM-ENR/100063/2008, Fundação para a Ciência e Tecnologia (FCT), and by the MIT-Portugal Program and FCT grants SFRH/43043/2008 and SFRH/43081/2008.

E. Hüllermeier, R. Kruse, and F. Hoffmann (Eds.): IPMU 2010, Part II, CCIS 81, pp. 65–74, 2010.

the prediction performance of the model. In this paper, four different combinations of modeling and feature selection approaches are proposed and compared: artificial neural networks and fuzzy modeling with ant colonies and bottom-up tree search feature selection.

The paper is organized as follows. Section 2 briefly describes modeling techniques. Proposed feature selection algorithms are presented in Section 3. Section 4 describes the database used and presents the obtained results after use of the described techniques. Conclusions are drawn in Section 5.

2 Modeling

A large number of systems are complex and only partially understood, making simple rules difficult to obtain. For these complex systems, nonlinear models based on artificial intelligence techniques can be used. This paper uses fuzzy modeling and neural modeling, as they represent highly nonlinear problems effectively due to their universal function approximation properties. These modeling techniques are described in more detail below.

2.1 Fuzzy Modeling

Fuzzy modeling is a tool that allows an approximation of nonlinear systems when there is few or no previous knowledge of the system to be modeled [5]. The fuzzy modeling approach has several advantages compared to other nonlinear modeling techniques. In general, fuzzy models provide not only a more transparent model, but also a linguistic interpretation in the form of rules. This is appealing when dealing with clinical related classification systems. Fuzzy models use rules and logical connectives to establish relations between the features defined to derive the model. A fuzzy classifier contains a rule base consisting of a set of fuzzy if–then rules together with a fuzzy inference mechanism. There are three general methods for fuzzy classifier design that can be distinguished [6]: the regression method, the discriminant method and the maximum compatibility method. In the discriminant method the classification is based on the largest discriminant function, which is associated with a certain class, regardless of the values or definitions of other discriminant functions. Hence, the classification decision does not change by taking a monotonic transformation of the discriminant function. The utility of this property is the reason we focus on this method here [7].

In the discriminant method, a separate discriminant function $d_c(\mathbf{x})$ is associated with each class ω_c, $c = 1, \ldots, C$. The discriminant functions can be implemented as fuzzy inference systems. In this work, we use Takagi-Sugeno (TS) fuzzy models [8], which consist of fuzzy rules where each rule describes a local input-output relation. When TS fuzzy systems are used, each discriminant function consists of rules of the type

$$\begin{aligned}\text{Rule } R_i^c : &\text{ If } x_1 \text{ is } A_{i1}^c \text{ and } \ldots \text{ and } x_M \text{ is } A_{iM}^c \\ &\text{ then } d_i^c(\mathbf{x}) = f_i^c(\mathbf{x}),\ i = 1, 2, \ldots, K,\end{aligned}$$

where f_i^c is the consequent function for rule R_i^c. In these rules, the index c indicates that the rule is associated with the output class c. Note that the antecedent parts of the rules can be for different discriminants, as well as the consequents. Therefore, the output of each discriminant function $d_c(\mathbf{x})$ can be interpreted as a score (or evidence) for the associated class c given the input feature vector $\mathbf{x}$. The degree of activation of the ith rule for class c is given by: $\beta_i = \prod_{j=1}^{M} \mu_{A_{ij}^c}(\mathbf{x})$, where $\mu_{A_{ij}^c}(\mathbf{x}) : \mathbb{R} \rightarrow [0,1]$. The discriminant output for each class c, with $c = 1, \ldots, C$, is computed by aggregating the individual rules contribution: $d_c(\mathbf{x}) = \frac{\sum_{i=1}^{K} \beta_i f_i^c(\mathbf{x})}{\sum_{i=1}^{K} \beta_i}$. The classifier assigns the class label corresponding to the maximum value of the discriminant functions, i.e

$$\max_{c} d_c(\mathbf{x}). \tag{1}$$

The number of rules K, the antecedent fuzzy sets A_{ij}, and the consequent parameters $f_i^c(\mathbf{x})$ are determined in this step, using fuzzy clustering in the product space of the input and output variables. The number of fuzzy rules (or clusters) that best suits the data must be determined for classification. The following criterion, as proposed in [9], is used in this paper to determine the optimum number of clusters:

$$S(c) = \sum_{k=1}^{N} \sum_{i=1}^{c} (\mu_{ik})^m (\| \mathbf{x}_k - v_i \|^2 - \| v_i - \bar{\mathbf{x}} \|^2), \tag{2}$$

2.2 Neural Networks

Artificial Neural Networks (ANN) have been largely used as input-output mapping for different applications including modelling and classification [10]. The main characteristics of a neural network are parallel distributed structure and ability to learn, which produce excellent outputs for inputs not encountered during training. Moreover, the structure can be set to be simple enough to compute the output(s) from the given input(s) in very low computational time.

The basic processing elements of neural networks are called artificial neurons, or simply neurons or nodes. Each processing unit is characterized by an activity level (representing the state of polarization of a neuron), an output value (representing the firing rate of the neuron), a set of input connections (representing synapses on the cell and its dendrites), a bias value (representing an internal resting level of the neuron), and a set of output connections (representing a neuron's axonal projections).

The processing units are arranged in layers. There are typically three parts in a neural network: an input layer with units representing the input variables, one or more hidden layers, and an output layer with one or more units representing the output variables(s). The units are joined with varying connection strengths or weights. Each connection has an associated weight (synaptic strength) which determines the effect of the incoming input on the activation level of the unit. The weights may be positive (excitatory) or negative (inhibitory). The neuron output signal is given by the following relationship:

$$\sigma = f\left(\mathbf{w}^T x\right) = f\left(\sum_{j=1}^{n} (w_j x_j)\right) \tag{3}$$

where $\mathbf{w} = (w_1, \ldots, w_n)^T \in \mathbb{R}^n$ is the weight vector, and $x = (x_1, \ldots, x_n)^T \in \mathbb{R}^n$ is the vector of neuron inputs. The function $f\left(\mathbf{w}^T x\right)$ is often referred to as the activation (or transfer) function. Its domain is the set of activation values, net, of the neuron model, and is often represented by $f(net)$. The variable net is defined as a scalar product of the weight and input vectors: $net = \sum_{j=1}^{n}(w_j x_j) = w_1 x_1 + \ldots + w_n x_n$.

Training a neural network can be defined as the process of setting the weights of each connection between units in such a way that the network best approximates the underlying function, thus turning it into an optimization problem. In this work the Levenberg-Marquardt optimization method is used.

3 Feature Selection

Feature selection is generally used for a set of reasons: to reduce computational complexity, to eliminate features with high mutual correlation [11] as well as due to required generalization properties of the model [12].

In this paper, two methods are used for feature selection: a wrapper method – bottom–up; and a hybrid method – ant colony metaheuristic combined with Fisher's rank as the guiding heuristic.

3.1 Bottom–Up

A detailed description of the bottom-up approach used here may be encountered in [9]. However, it is importnat to note that a more recent algorithm that minimizes the computational time with similar performance was already developed and proposed in [13] and further detailed in [5].

According to [14], a Receiver–Operator Characteristics (ROC) curve can be used to study the behavior of two–class classifiers. This is a function of the true positive ratio versus the false positive ratio. Consequently, in order to compare the performance of various classifiers, the Area Under the ROC Curve (AUC) can be used [14]. It corresponds to the total performance of a two-class classifier integrated over all thresholds:

$$AUC = 1 - \int_0^1 FP(FN)dFN \tag{4}$$

where FP and FN represent, respectively, the false positive rate the false negative rate. This AUC measure was used to evaluate the fuzzy and neural models produced by our algorithm. The bottom-up approach looks for single inputs that may influence the output, and combines them in order to achieve the model with the best performance. Two subsets of data are used in this stage, T (train) and V (validation). Using the train data set, a model is built for each of the n features in consideration, and evaluated using the described performance criterion - AUC (4) - upon the validation data set. The feature that returns the best value of AUC is the one selected. Next, other feature candidates are added to the previous best model, one at a time, and evaluated. Again, the combination of features that maximizes the AUC value is selected. When this second stage finishes, the model has two features. This procedure is repeated until the value of the performance criterion stops increasing. In the end, one should have all the relevant features for the considered process.

3.2 Ant Feature Selection

Ant Colony Optimization (ACO) methodology is an optimization method suited to find minimum cost paths in optimization problems described by graphs [15]. This paper uses the ant feature selection algorithm (AFS) proposed in [16], where the best number of features is determined automatically. In this approach, two objectives are considered: minimizing the number of features (*features cardinality* - N_f) and maximizing the prediction performance. Two cooperative ant colonies optimize each objective. The first colony determines the number (cardinality) of features and the second selects the features based on the cardinality given by the first colony. Thus, two pheromone matrices and two different heuristics are used.

The objective function of this optimization algorithm aggregates both criteria, the maximization of the model performance and the minimization of the features cardinality:

$$J^k = w_1(1 - PC^k) + w_2\frac{N_f^k}{n} \tag{5}$$

where $k = 1, \ldots, g$, PC is the performance criterion defined in 4 , N_n is the number of used data samples and n is the total number of features. The weights w_1 and w_2 are selected based on experiments.

To evaluate performance, both fuzzy and neural classifier are built for each solution following the procedure described in Section 2.

Heuristics. The heuristic value used for each feature (ants visibility) for the second colony, is computed as $\eta_{f_j} = PC_j$ for $j = 1, \ldots, n$. PC_j quantifies the relevance of each feature in the prediction model, which is measured using the AUC criteria defined in (4).

For the features cardinality (first colony), the heuristic value is computed using the Fisher discriminant criterion for feature selection [17]. The Fisher discriminant criterion is described as

$$F(i) = \frac{|\mu_1(i) - \mu_2(i)|^2}{\sigma_1^2 + \sigma_2^2} \tag{6}$$

where $\mu_1(i)$ and $\mu_2(i)$ are the mean values of feature i for the samples in class 1 and class 2, and σ_1^2 and σ_2^2 are the variances of feature i for the samples in class 1 and 2. The score aims to maximize the between-class difference and minimize the within a given class spread. Other currently proposed rank-based criteria generally come from similar considerations and show similar performance [17].The score is used to limit the number of features chosen by the AFS algorithm, particularly when the number of available features is large.

4 Results

This paper uses the public available MEDAN database [18], containing data of 382 patients with abdominal septic shock, recorded from 71 German intensive care units (ICUs), from 1998 to 2002. This database holds personal records, physiological parameters, procedures, diagnosis/therapies, medications, and respective outcomes (survived or deceased). For the purpose of the present work, we will focus exclusively in physiological parameters, which include a total of 103 different variables/features.

Several drawbacks were found within this database. First, from the 382 abdominal septic shock patients, only 139 were found to effectively have data. Second, not all 139 patients have entries for the 103 features. Third, short breaks (missing data) exist in these records together with outliers, similarly to other data acquisition systems.

4.1 Data Preprocessing

The first step in preprocessing the data, consisted of replacing outliers by missing values. An entry was considered an outlier whenever the difference between its value and the mean value for the entries in that feature was larger than three times the standard deviation. The second step dealt with missing values. From the several available methods, we chose linear regression. Despite its drawbacks, namely the fact that it increases the sample size and reduces the standard error (underestimates the error) by not adding more information, it has the advantage of being easily implemented and of imputing values that are in some way conditional to existent entries.

The original database has inhomogeneous sampling times between different variables and for each individual variable. For modeling purposes, all variables are required to have the same sampling time and to be uniformly sampled during the whole recording period. In order to overcome this incongruity, an important medical-logical criterion was used:

- The value of the variable is zero-order held until a new value appears. If the variable was originally sampled once per hour, each value would be hold for the other 59 minutes until the new value appears.
- At the starting time, if there are no values for a specific variable, the algorithm looks for the closest value in time and hold it back again at zero order.

After this preprocess, the whole dataset is normalized in time span and sampling frequency, with no missing data or outliers present. The sampling rate was set to 24 hours, so that it would be consistent with the sampling used in [4]. In this way, no major differences would exist between our preprocessed data and the the preprocessed data used in [4].

4.2 Simulations and Results

From the initial set containing a total of 103 features, two different subsets of features were chosen as inputs for our models. One was defined as in [4] for purposes of comparison and contains the 12 most frequently measured variables. A total of 121 patients were found to have data including these features. However, as our group considered the previous subset of features too narrow, a second data subset was defined including a total of 28 variables. These were found to be present in a total of 89 patients. Bottom-up and ant feature selection algorithms were then applied to each of these subsets, using fuzzy and neural models. The reasoning behind the choice to apply feature selection techniques to these smaller subsets associates to the clinical relevance of finding the specific variables that relate the most with the prediction of a patient's outcome. In order to evaluate the performance of the developed prediction models, upon the described

Table 1. % of correct classifications for fuzzy and neural models

	12 features						28 features					
	Fuzzy models			Neural models			Fuzzy models			Neural models		
	Mean	Std	NF	Mean	Std	NF	Mean	Std	NF	Mean	Std	NF
[4]	–	–	–	69.00	4.37	–	–	–	–	–	–	–
Bottom–up	74.10	1.31	2–6	73.24	2.03	2–8	82.27	1.56	2–7	81.23	1.97	4–8
AFS	72.77	1.44	2–3	75.67	1.37	2–7	78.58	1.44	3–9	81.90	2.15	5–12

Table 2. AUC values for fuzzy and neural models using feature selection with 12 and 28 features

	12 features				28 features			
	Fuzzy models		Neural models		Fuzzy models		Neural models	
	Mean	Std	Mean	Std	Mean	Std	Mean	Std
Bottom–up	75.01	1.06	71.94	1.17	81.79	1.97	80.78	1.28
AFS	73.48	0.01	72.61	0.01	78.74	0.02	78.07	0.03

subsets of data, four different criteria were used: correct classification rate, AUC, sensitivity and specificity.

To reiterate, the goal of this paper is to apply new knowledge based methods to a known septic shock patient database and compare the outcome prediction capabilities of these methods with the ones developed in [4].

Table 1 presents classification rates both for the developed methods and for [4]. From the analysis of this table, it is apparent not only that all developed models perform better than the ones used in [4] but also that use of the subset with 28 features leads to the best results. However, the correct classification rate is not always the best way to evaluate the performance of the classifier. For this particular application, the goal is to correctly classify which patients are more likely to decease, in order to rapidly act in their best interest. Bearing this in mind the classifier should classify as accurate as possible the cases that result in death, or true positives, and aim to have a low number of false negatives. In other words, the classifier should maximize sensitivity. To perform this analysis, Table 2 with the obtained results for AUC, as well as the correspondent values of specificity and sensitivity are presented (Table 3 and 4, respectively)

Looking at Table 2, one can see that again, the use of the subset with 28 features leads to better results when considering AUC values. Fuzzy models perform slightly better than neural networks. Namely, the model with highest value of AUC arises from the combination of fuzzy models with bottom–up feature selection. It is also possible to observe that the standard deviation of AUC values obtained using bottom-up are higher than the ones obtained with ant feature selection, which might suggest that the variability on the number of features selected by the bottom-up algorithm is also higher. Table 3 shows results obtained for sensitivity and specificity using the 12 feature subset, while Table 4 shows same results using the 28 features subset. A few observations can be made from these tables. When comparing the obtained values of sensitivity and specificity with [4], it is clear that our models have much better sensitivity, but slightly worse specificity (Table 3). This means that the models developed in [4], accurately

Table 3. Mean sensitivity and specificity using feature selection with 12 features

	12 features							
	Sensitivity				Specificity			
	Fuzzy models		Neural models		Fuzzy models		Neural models	
	Mean	Std	Mean	Std	Mean	Std	Mean	Std
[4]	–	–	15.01	–	–	–	92.26	–
Bottom–up	79.89	2.60	54.53	5.42	71.16	2.86	81.65	3.61
AFS	76.49	0.03	59.64	0.02	70.46	0.02	85.59	0.02

Table 4. Mean sensitivity and specificity sing feature selection with 28 features

	28 features							
	Sensitivity				Specificity			
	Fuzzy models		Neural models		Fuzzy models		Neural models	
	Mean	Std	Mean	Std	Mean	Std	Mean	Std
Bottom–up	82.26	1.56	64.16	3.92	83.30	2.62	90.33	2.05
AFS	79.23	0.04	66.98	0.05	78.24	0.03	90.16	0.02

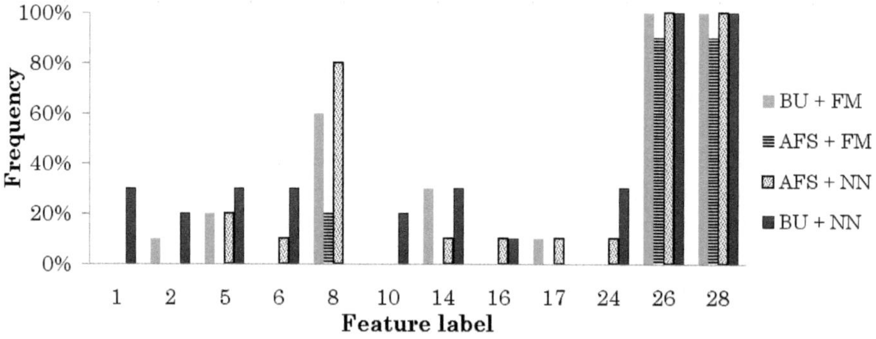

Fig. 1. Histogram with the selection rate of each feature for the 12 features subset. BU - bottom–up; AFS - ant feature selection; FM - fuzzy modelling; NN - neural networks.

predict which patients survive, but have poor confidence when predicting which of them will decease. Conversely, our models are very accurate in predicting which patients are in risk of death, which is the goal of the paper. Additionally, fuzzy models present higher values for sensitivity, while neural networks present higher values for specificity, and bottom-up feature selection leads to better results than ant feature selection. Lastly, these tables also point that by using 28 features, results are substantially improved, suggesting that important features exist within the 28 feature subset that were not included in the 12 feature one. This can be confirmed comparing Figure 1 and Figure 2.

Figure 1 presents an histogram of the selected features, for all the tested approaches, when using the 12 features subset. From this figure, it is apparent that three features are more commonly selected: 8, 26 and 28, corresponding, respectively, to pH, Calcium

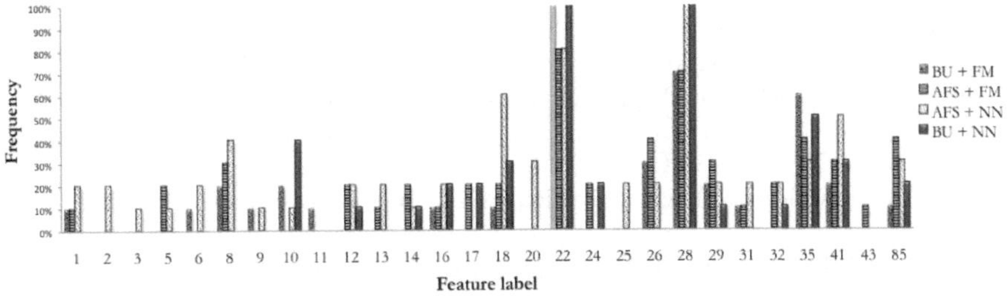

Fig. 2. Histogram with the selection rate of each feature for the 28 features subset. BU - bottom–up; AFS - ant feature selection; FM - fuzzy modelling; NN - neural networks.

and Creatinine. Moreover, it is possible to confirm what was previously mentioned: ant feature selection with fuzzy modelling selects the smallest number of variables, i.e. the variability of features selected is minimum.

Figure 2 shows the histogram for the 28 features subset. The same three features mentioned above are selected. Additionally, four more features are commonly selected, which may be responsible and explain the substantial improvements in the obtained results for correct classification rate, AUC, sensitivity and specificity. These features are number 18, 35, 41 and 85, which correspond respectively to thrombocytes, total bilirubin, CRP (C-reactive protein) and FiO2.

5 Conclusions

This paper applied wrapper feature selection based on soft computing methods to a publicly available ICU database. Fuzzy and neural models were derived and features were selected using a tree search method and ant feature selection. The proposed approaches clearly outperformed previous approaches in terms of sensitivity, which is the most important measure for the application in hands.

In the future, these techniques will be applied to larger health care databases which have more available features. To initially reduce the number of features, a filter method will be applied in order to alleviate the computational burden in the wrapper methods. Once, fuzzy models perform better in terms of sensitivity and neural networks perform better in terms of specificity, an hybrid approach will be considered in future work combining both advantages.

References

1. American College of Chest Physicians/Society of Critical Care Medicine Consensus Conference: Definitions for sepsis and organ failure and guidelines for the use of innovative therapies in sepsis. Crit. Care Med. (20), 864–874 (1992)
2. Burchardi, H., Schneider, H.: Economic aspects of severe sepsis: a review of intensive care unit costs, cost of illness and cost effectiveness of therapy. Pharmacoeconomics 22(12), 793–813 (2004)

3. Paetza, J., Arlt, B., Erz, K., Holzer, K., Brause, R., Hanisch, E.: Data quality aspects of a database for abdominal septic shock patients. Computer Methods and Programs in Biomedicine 75, 23–30 (2004)
4. Paetza, J.: Knowledge-based approach to septic shock patient data using a neural network with trapezoidal activation functions. Artificial Intelligence in Medicine 28, 207–230 (2003)
5. Mendonça, L.F., Vieira, S.M., Sousa, J.M.C.: Decision tree search methods in fuzzy modeling and classification. International Journal of Approximate Reasoning 44(2), 106–123 (2007)
6. Kuncheva, L.I.: Fuzzy Classifier Design. Springer, Heidelberg (2000)
7. van den Berg, J., Kaymak, U., van den Bergh, W.M.: Fuzzy classification using probability-based rule weighting. In: Proceedings of the 2002 IEEE International Conference on Fuzzy Systems, FUZZ-IEEE 2002, vol. 2, pp. 991–996 (2002)
8. Takagi, T., Sugeno, M.: Fuzzy identification of systems and its applications to modelling and control. IEEE Transactions on Systems, Man and Cybernetics 15(1), 116–132 (1985)
9. Sugeno, M., Yasukawa, T.: A fuzzy-logic-based approach to qualitative modeling. IEEE Transactions on Fuzzy Systems 1(1), 7–31 (1993)
10. Haykin, S.: Neural Networks and Learning Machines, 3rd edn. Prentice-Hall, Upper Saddle River (2008)
11. Jensen, R., Shen, Q.: Are more features better? a response to attributes reduction using fuzzy rough sets. IEEE Transactions on Fuzzy Systems 17(6), 1456–1458 (2009)
12. Guyon, I., Elisseeff, A.: An introduction to variable and feature selection. Journal of Machine Learning Research 3, 1157–1182 (2003)
13. Vieira, S.M., Mendonça, L., Sousa, J.M.C.: Modified regularity criterion in dynamic fuzzy modeling applied to industrial processes. In: Proc. of 2005 IEEE International Conference on Fuzzy Systems, FUZZ-IEEE 2005, Reno, Nevada, May 2005, pp. 483–488 (2005)
14. Pekalska, E., Duin, R.P.W.: The Dissimilarity Representation for Pattern Recognition: Foundations And Applications (Machine Perception and Artificial Intelligence). World Scientific Publishing Co., Inc., River Edge (2005)
15. Dorigo, M., Birattari, M., Stützle, T.: Ant colony optimization. IEEE Computational Intelligence Magazine 1(4), 28–39 (2006)
16. Vieira, S.M., Sousa, J.M.C., Runkler, T.A.: Two cooperative ant colonies for feature selection using fuzzy models. Expert Systems with Applications 37(4), 2714–2723 (2010)
17. Duda, R.O., Hart, P.E., Stork, D.G.: Pattern Classification, 2nd edn. Wiley–Interscience Publication, Chichester (2001)
18. Hanisch, E., Brause, R., Arlt, B., Paetz, J., Holzer, K.: The MEDAN Database (2003), `http://www.medan.de` (accessed October 20, 2009)

Obtaining the Compatibility between Musicians Using Soft Computing

Teresa Leon* and Vicente Liern

University of Valencia, Spain
{teresa.leon,vicente.liern}@uv.es

Abstract. Modeling the musical notes as fuzzy sets provides a flexible framework which better explains musicians' daily practices. Taking into account one of the characteristics of the sound: the pitch (the frequency of a sound as perceived by human ear), a similarity relation between two notes can be defined. We call this relation compatibility. In the present work, we propose a method to asses the compatibility between musicians based on the compatibility of their interpretations of a given composition. In order to aggregate the compatibilities between the notes offered and then obtain the compatibility between musicians, we make use of an OWA operator. We illustrate our approach with a numerical experiment.

Keywords: Musical note, Trapezoidal Fuzzy Number, Similarity relation, OWA operators.

1 Introduction

In this work, we are concerned with the compatibility between musicians. As an example we can imagine a situation in which the staff of an orchestra needs to be augmented. Some new instrumentalists have to be hired and they should tune well but also be compatible with the members of the orchestra. Certainly, a decision based on experts' subjective judgement can be made by listening them playing together. However if a list of candidates for the position need to be casted it can be useful to quantify their compatibilities.

Firstly, it is important to remark that we are only taking into account one of the characteristics of the sound: the pitch. Pitch is the frequency of a sound as perceived by human ear, more precisely: the pitch of a sound is a psychological counterpart of the physical phenomenon called "frequency" of that sound. In the middle zone of the audible field the sensation of pitch changes approximately according to the logarithm of the frequency i.e. follows the Weber-Fechtner's law: "As a stimulus is increased multiplicatively, sensation is increased additively."

The word tone is used with different meanings in music. In [1] we can read that a tone is "a sound of definite pitch and duration, as distinct from noise and from less definite phenomena, such as the violin portamento. " In this dictionary we find that the notes are "the signs with which music is written on a staff.

* Corresponding author.

E. Hüllermeier, R. Kruse, and F. Hoffmann (Eds.): IPMU 2010, Part II, CCIS 81, pp. 75–84, 2010.

In British usage the term also means the sound indicated by a note". A pure tone can be defined as the sound of only one frequency, such as that given by an electronic signal generator. The fundamental frequency of a tone has the greatest amplitude. The other frequencies are called overtones or harmonics and they determine the quality of the sound. Loudness is a physiological sensation. It depends mainly on sound pressure but also on the spectrum of the harmonics and the physical duration. Although timbre and loudness are very important, we are focusing on pitch.

A tuning system is the system used to define which tones to use when playing music, these tones are the tuned notes. Some examples of tuning systems are: the Pythagorean intonation, the Tempered, the Just Intonation or the 12 Tone Equal Temperament. Most tuning systems have been obtained through mathematical arguments, this facilitates their transmission and the manufacture of instruments. However, many musicians feel that these mathematical arguments are impractical. Different musicians play together in a classical orchestra and they must adjust their instruments to tune well, in particular, continuous pitch instruments such as the violin, do not limit them to particular pitches, allowing to choose the tuning system "on the fly". It is fair to say that some music traditions around the world do not use our type of precision tuning because of an aesthetic preference for wide tuning. In these traditions, the sound of many people playing precisely the same pitch is considered a thin, uninteresting sound; the sound of many people playing near the same pitch is heard as full, lively, and more interesting.

As musicians need flexibility in their reasoning, the use of fuzzy logic to connect music and uncertainty is appropriate. In fact musical scores can be also seen as fuzzy systems, for instance the tempo of some compositions by J.S. Bach was not prescribed on the corresponding scores.

A musical note should be understood as a fuzzy number. This idea is fundamental in [6] where the following paragraph can be read:

> *"It seems that N.A. Garbuzov was the first who applied the fuzzy theory (in a naive form) when considering interval zones with regard to tuning in music".*

In [10], a note is considered as a triangular fuzzy number which reflects the sensation that a frequency f produces; the definition of compatibility between two notes and a formula to compute it are also given. Notes are modeled as trapezoidal fuzzy numbers in [3] and [8], because this approach better reflects the fact that the human ear perceives notes with very close frequencies as if they were the same note.

In this work we model the pitches of the tones as fuzzy numbers and subsequently analyze the compatibility between them and between different music players. Our method to asses the compatibility between musicians is based on the compatibility of their interpretations of the notes written on a staff. We assume that the notes are not distinguishable in the traditional music notation. In order to aggregate the compatibilities between the notes offered and obtain the compatibility between musicians we make use of an OWA operator [13]. We illustrate our approach with a numerical experiment.

2 Preliminaries

2.1 Some Basic Music Theory Concepts

Each note is usually identified with the frequency of its fundamental harmonic, i.e. the frequency that tuners measure. The usual way to relate two frequencies is through their ratio and this number is called the interval. Given two sounds with frequencies f_1 and f_2, we say that f_2 is one octave higher than f_1 if f_2 is twice f_1. Two notes one octave apart from each other have the same letter-names. This naming corresponds to the fact that notes an octave apart sound like the same note produced at different pitches and not like entirely different notes. As our interest is in the pitch sensation, we should work with the exponents of the frequencies. To be precise, when the diapason is fixed at 440 Hz, the note C_4 is identified with $2^{-\frac{3}{4}}.440$ Hz and the reference interval is

$$[f^0, 2f^0[:= \left[2^{-\frac{3}{4}}.440, 2^{-\frac{1}{4}}.440\right[.$$

Once we have established f^0, with the aim of translating each frequency f to the interval $[1, 2[$, and subsequently take the exponent corresponding to 2, we make use of the following expression:

$$t = \log_2(\frac{f}{f^0}) - \left\lfloor \frac{f}{f^0} \right\rfloor . \tag{1}$$

Therefore, a natural way to measure the distance between two notes f_1 and f_2 is the following:

$$d(f_1, f_2) = 1200. \left| \log_2 \frac{f_1}{f_2} \right| .$$

An equal temperament is a tuning system in which every pair of adjacent notes has an identical frequency ratio. In these tunings, an interval (usually the octave) is divided into a series of equal frequency ratios. For modern Western music, the most common tuning system is twelve-tone equal temperament which divides the octave into 12 (logarithmically) equal parts. An electronic tuner is a device used by musicians to detect and display the pitch of notes played on musical instruments. Chromatic tuners allow tuning to an equal temperament scale.

2.2 Recalling Some Definitions

Let us include the well-known definition of a trapezoidal fuzzy number for notational purposes.

Definition 1. *$\tilde{p} = (m, M, \alpha, \beta)$ is a trapezoidal fuzzy number if*

$$\mu_{\tilde{p}}(x) = \begin{cases} 1 - \frac{m-x}{\alpha} & x < m \\ 1 & m \le x \le M \\ 1 - \frac{x-M}{\beta} & x > M \end{cases}$$

where $[m, M]$ is the modal interval and α (resp. β) is the left (resp. right) spread.

In Section 4 we study the compatibility between musicians and we make use of an OWA operator [13].

Definition 2. *An OWA operator of dimension n is a mapping $f : \mathbb{R}^n \to \mathbb{R}$ with an associated weighting vector $W = (w_1, \ldots, w_n)$ such that $\sum_{j=1}^{n} w_j = 1$ and where $f(a_1, \ldots, a_n) = \sum_{j=1}^{n} w_j b_j$ being b_j the jth largest element of the collection of aggregated objects $a_1, \ldots, a_n$.*

As OWA operators are bounded by the Max and Min operators, Yager introduced a measure called *orness* to characterize the degree to which the aggregation is like an *or* (Max) operation:

$$\text{orness}(W) = \frac{1}{n-1} \sum_{i=1}^{n} (n-i) w_i \tag{2}$$

A key issue in defining an OWA operator is in the choice of the vector of weights. Several approaches can be found in the literature including learning from data, exponential smoothing or aggregating by quantifiers. Our proposal (already detailed in [9]) is to construct a parametric family of weights as a mixture of a binomial and a discrete uniform distribution. Let us recall the most important points of this approach.

From the definition of orness is direct to prove that:

1. If $W = (w_1, \ldots, w_k)$ and $W' = (w'_1, \ldots, w'_k)$ are two vectors of weights such that orness $(W) = \alpha$ and orness $(W') = \beta$ with $\alpha, \beta \in [0, 1]$, then for all $\lambda \in [0, 1]$ we have that
$$\text{orness}(\lambda W + (1-\lambda)W') = \lambda\alpha + (1-\lambda)\beta,$$
where $\lambda W + (1-\lambda)W' = (\lambda w_1 + (1-\lambda)w'_1, \ldots, \lambda w_k + (1-\lambda)w'_k)$.
2. If $W = (w_1, \ldots, w_k) = (\frac{1}{k}, \ldots, \frac{1}{k})$, then orness $(W) = 0.5$.

Let X a random variable which follows the binomial distribution with parameters $k-1$ (number of trials) and $1-\alpha$ (probability of success), i.e. $X \sim B(k-1, 1-\alpha)$, and let us put $\phi_j = P(X = j)$ for $j \in \{0, 1, \ldots, k-1\}$, then from the properties of the binomial probability distribution it is easy to check that:

1. Let $W = (w_1, \ldots, w_k)$ be a vector of weights such that $w_i = \phi_{i-1}$, then orness$(W) = \alpha$.

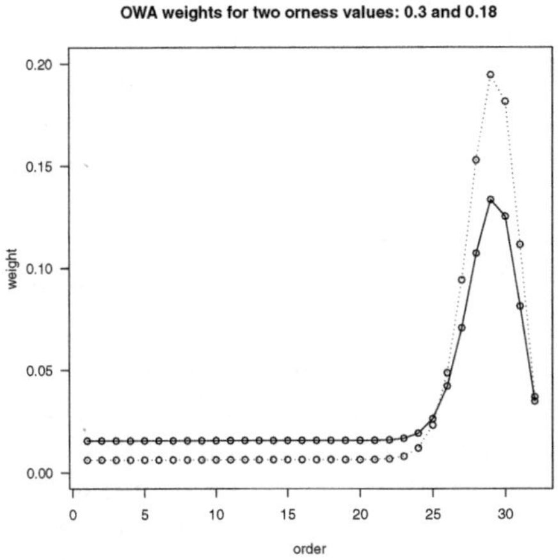

Fig. 1. Aggregation weights with n=32 obtained using a mixture of probabilities of a Bi(31, 0.95) and a discrete uniform distribution. Two different values for the mixture parameter λ have been used. For $\lambda = 0.5$, the orness value equals 0.3 (weights connected with a continuous line) and for $\lambda = 0.8$ we have that $\alpha = 0.18$ (weights connected with a dotted line).

2. Let $W = (w_1, \ldots, w_k) = (\lambda\phi_0 + (1-\lambda)\frac{1}{k}, \ldots, \lambda\phi_{k-1} + (1-\lambda)\frac{1}{k})$, then $\text{orness}(W) = \lambda(1-p) + (1-\lambda)0.5$.

For a given orness value, α, we can obtain a vector of weights as a mixture of a binomial $B(k-1,p)$ and a discrete uniform distribution with support $\{1,\ldots,k\}$. The relationship between the parameters α, λ, and p is $2\alpha - 1 = \lambda(1-2p)$.

We have chosen this way of defining the weights for aggregation because it is simple and intuitive. Figure 1 may help us to justify this. It shows two vectors of weights for two different orness values. The y-coordinates of the points correspond to the weights and the x-coordinates to the order. The points have been connected for a better visualization. The binomial distribution concentrates the higher values of the weights around $\mu = (k-1)\alpha$, while the discrete uniform component of the mixture keeps all the weights different to zero and then the information from all the objects is taken into account in the aggregation process. The parameter value selection allows us to control the distribution of the weights.

3 Notes and Compatibility

Firstly, let us justify that it is appropriate to model a note as a trapezoidal fuzzy number. It is well known that the human ear perceives notes with very close frequencies as if they were the same note (see [2]).

In 1948, N. A. Garbuzov made thousands of experiments and used them to assign band frequencies to every musical interval, called the Garbuzov zones (see

[6] and [7]). According to his studies, we perceive as the same note, unison, two frequencies that are 12 cents apart[1]. It deserves to be mentioned that the Garbuzov zones were obtained as arithmetical means from hundreds of measurements to an accuracy of 5 cents due to the imprecise measuring equipments of that time. Other authors reduce this interval to 5 or 6 cents [11]. It seems that the accuracy of an instrumentalist is not better than 5 cents and that this accuracy is between 10 and 20 cents for non-trained listeners.

In any case, the modal interval corresponding to the pitch sensation should be expressed as $[t-\varepsilon, t+\varepsilon]$.

Next, let us focus on its support. If the amount of notes per octave is q, the octave can be divided into q intervals of widths $1200/q$ cents. So, if we represent it as a segment, the (crisp) central pitch would be in the middle, and the extremes would be obtained by adding and subtracting $1200/(2\times$ q) cents. In fact, chromatic tuners assign $q = 12$ divisions per octave suggesting that a tolerance of $\delta = 50/1200 = 1/24$ is appropriate. Therefore, the support of the pitch sensation should be expressed as $[t-\delta, t+\delta]$, where $\delta = 1/2q$. Certainly, a semi-tone is a large interval for the human ear and other choices for the width of the support can be made; however we prefer to take $\delta = \frac{1}{2q}$ because it is consistent with the traditional practice of using a tuner.

In [10], the compatibility between two fuzzy notes is defined as the Zadeh consistency index between their pitch sensations. The pitch sensations were modeled as triangular fuzzy numbers in the original definition therefore let us adapt it to consider the case in which pitch sensations are trapezoidal fuzzy numbers.

Definition 3. *Let $2^{\tilde{t}}$ and $2^{\tilde{s}}$ be two notes where $\tilde{t} = (t-\varepsilon, t+\varepsilon, \delta-\varepsilon, \delta-\varepsilon)$ and $\tilde{s} = (s-\varepsilon, s+\varepsilon, \delta-\varepsilon, \delta-\varepsilon)$. The compatibility between them is defined as*

$$\mathrm{comp}(2^{\tilde{\mathrm{t}}}, 2^{\tilde{\mathrm{s}}}) := \mathrm{consistency}(\tilde{\mathrm{t}}, \tilde{\mathrm{s}}) = \max_{\mathrm{x}} \mu_{\tilde{\mathrm{s}} \cap \tilde{\mathrm{t}}}(\mathrm{x}).$$

And we say that they are p-compatible for $p \in [0,1]$, if $\mathrm{comp}(2^{\tilde{t}}, 2^{\tilde{s}}) \geq p$.

Remark 1. The Zadeh consistency index is a similarity measure and therefore the p-compatibility is an equality at level p, see [4]. Although similarity measures are very numerous in the literature and other definitions could have been considered, we find that Definition 3 better reflects our approach.

Remark 2. Notice that, if notes are modeled as triangular fuzzy numbers with $\Delta = 50$ cents, then the compatibility between two notes in the Garbuzov zone (i.e. 12 cents apart) would be equal to 0.88 and the compatibility between two notes which are 5 cents apart is 0.95. However it makes more sense to have a definition in which notes that are indistinguishable have the maximum compatibility.

It is easy to check that following expression provides the compatibility between two notes.

[1] The Garbuzov zones are much larger for other musical intervals than in the case of the unison, and depend on the modality, the key, the instrumentalist and the composition (tempo, style).

$$\text{comp}\left(2^{\tilde{t}}, 2^{\tilde{s}}\right) = \begin{cases} 1 & \text{if} \quad |t-s| < 2\varepsilon \\ 1 - \frac{|t-s|-2\varepsilon}{2(\delta-\varepsilon)} & \text{if} \quad 2\varepsilon \le |t-s| \le 2\delta \\ 0 & \text{if} \quad |t-s| > 2\delta \end{cases} \tag{3}$$

4 Compatibility between Musicians

In the previous section we have defined the compatibility between two notes, we are now concerned with the compatibility between instrumentalists.

1. A "difficult" composition which allows the decision maker to evaluate the expertness of the musicians $P_1, P_2, \ldots P_k$ should be selected.
   Let us assume that it is represented by
   $$j_1, ..., j_n \in \{C, C\sharp, D, D\sharp, E, F, F\sharp, G, G\sharp, A, A\sharp, B\}.$$
2. Each musician performs the composition a reasonable number of times, m (which depends on the composition length and its difficulty). Notice that too many times can be counterproductive. Each performance is recorded and the frequencies corresponding to each note are obtained.
3. Denote by f_{il} (resp.F_{il}) the lowest (resp. highest) interpretation of j_l performed by P_i , for $l \in \{1, \ldots, n\}$ and $i \in \{1, \ldots, k\}$.
4. As our interest is in the pitch sensation, we work with "their exponents" computed according to Equation 1 and denoted by s_{il}, and S_{il} respectively, where $l \in \{1, \ldots, n\}$ and $i \in \{1, \ldots, k\}$.
5. For simplicity, let us suppose that we want to compare P_1 with P_2. Compute $a_{12}^l = \text{consistency}(\tilde{s_{1l}}, \tilde{s_{2l}})$ and $A_{12}^l = \text{consistency}\left(\tilde{S_{1l}}, \tilde{S_{2l}}\right)$, for l taking values in $\{1, \ldots, n\}$.
6. Making use of an OWA operator, aggregate the quantities $\{a_{12}^l, A_{12}^l\}_{l=1}^n$ into a single value C_{12}. We suggest using a small value for the orness when professional musicians are being compared because they probably will tune quite well most of the notes.

5 Numerical Results

Three saxophonists helped us to perform our experiment. One of the instrumentalists is a jazz musician who plays the tenor saxophone, the other is a musician and teacher of the alto saxophone and the third one is a student of his. Each musician interpreted the fragment represented in Figure 2 five times using four different saxophones. The saxophone brand names are the following: Selmer Superaction II (alto and tenor), Amati (soprano) and Studio (baritone).

Therefore our data set comprises 60 ($3 \times 5 \times 4$) recordings made using the software *Amadeus II*©. It is a sound editor for Macintosh which allows to make

Fig. 2. Score of the excerpt interpreted by the musicians

Table 1. Intervals containing the exact frequencies performed by the players with the alto and the soprano saxophones

	SOPRANO				ALTO		
Note	P1	P2	P3	Note	P1	P2	P3
$G_4^\sharp$	[414.1, 417.81]	[414.33, 416.68]	[418.66, 423.46]	$C_4^\sharp$	[276.8, 277.22]	[276.45, 277.83]	[275.73, 276.97]
$C_5^\sharp$	[552.32, 555.56]	[555.08, 555.97]	[556.77, 560.5]	$F_4^\sharp$	[371.84, 373.41]	[371.57, 373.42]	[369.86, 372]
$F_5^\sharp$	[730.93, 732.31]	[726.56, 731.41]	[731.33, 736.52]	B_4	[498.09, 499.01]	[495.75, 498.23]	[492.23, 493.36]
C_5	[526.44, 529.36]	[527.6, 530.72]	[530.03, 532.41]	F_4	[353.56, 354.2]	[353.38, 354.96]	[350.37, 352.37]
E_5	[646.3, 651.01]	[649.02, 650.36]	[651.29, 653.38]	A_4	[441.45, 443.24]	[441.07, 443.53]	[437.85, 440.04]
$A_4^\sharp$	[464.52, 465.68]	[466.18, 469.98]	[468.18, 470.32]	$D_4^\sharp$	[310.55, 313.15]	[308.81, 311.45]	[308.23, 309.33]
$F_5^\sharp$	[727.57, 732.01]	[729.38, 733.61]	[726.63, 734.31]	B_4	[497.82, 498.7]	[496.36, 497.47]	[491.27, 494.73]
$A_5^\sharp$	[923.18, 932.55]	[922.07, 928.17]	[922, 933.51]	$D_5^\sharp$	[628.7, 630.76]	[625.77, 630.24]	[619.57, 622.79]
$G_5^\sharp$	[822.45, 827.03]	[818.28, 828.43]	[811.19, 823.72]	$C_5^\sharp$	[555.55, 557.11]	[553.82, 557.03]	[551.1, 553]
D_6	[1153.2, 1169.2]	[1154.5, 1167.9]	[1152.2, 1159]	G_5	[794.02, 796.13]	[787.71, 793.74]	[787.15, 786.92]
$A_5^\sharp$	[926.21, 929.73]	[926.49, 929.31]	[925.66, 930.1]	$D_5^\sharp$	[628.41, 629.95]	[626.86, 628.22]	[616.4, 622.8]
F_5	[683.8, 690.01]	[687.8, 691]	[687.96, 689.48]	$A_4^\sharp$	[468.27, 471.97]	[466.96, 468.16]	[462.9, 464.16]
E_5	[649.67, 661.08]	[646.63, 649.01]	[649.86, 652.51]	A_4	[441.07, 445.01]	[441.19, 442.49]	[439.09, 440.45]
C_5	[524.98, 526.59]	[528.97, 529.78]	[529.52, 531.37]	F_4	[352.93, 354.42]	[352.98, 353.8]	[350.38, 351.23]
B_4	[489.22, 486.54]	[489.95, 490.92]	[487.97, 488.84]	E_4	[323.71, 324.91]	[323.96, 324.91]	[322.87, 323.15]
$A_4^\sharp$	[463.54, 469.99]	[467.62, 469.79]	[468.02, 470.63]	$D^\sharp$	[310.8, 311.65]	[309.23, 311.89]	[308.78, 309.19]

a detailed spectral analysis of signals. All subsequent data manipulations were performed using the spreadsheet Microsoft Office Excel© and R© software ([12]).

Table 1 displays the exact frequencies offered by the players for the alto and soprano saxophones. The lower limit of the intervals correspond to the lowest interpretations of the musical notes while the upper limits reflect their highest interpretations.

We have considered different orness values: $0.3, 0.18, 0.095$ and 0.075. The four vectors of weights have been constructed as mixtures of binomial and discrete uniform distributions.

The support of the discrete uniform distribution is $\{1, \ldots, 32\}$ because the excerpt interpreted by the musicians contains 16 notes. The values of the parameter p of the binomial distribution and the values of the mixture parameter λ can be found in Table 2 ($n = 31$ in all cases).

Table 3 contains the compatibility between the three musicians for different orness values. We have set $\varepsilon = \frac{6}{1200}$ and $\delta = \frac{50}{1200}$ for the calculations.

Let us comment the results for the alto saxophone. As expected, the most compatible are $P1$ and $P2$ (they are teacher and student). The same results are attained for the tenor saxophone which is the main instrument of $P3$. For the

Table 2. Orness, parameter values of the binomial distributions and the mixture parameter value

orness	p	λ
0.3	0.9	0.5
0.18	0.9	0.8
0.095	0.95	0.9
0.0725	0.95	0.95

Table 3. Compatibility between musicians with different orness

	orness 0.0725			orness 0.095		
	P1-P2	P2-P3	P2-P3	P1-P2	P2-P3	P1-P3
alto	0.98895	0.82896	0.90291	0.98946	0.83518	0.90680
baritone	0.73674	0.84487	0.73968	0.74812	0.85101	0.75079
soprano	0.92645	0.88858	0.93048	0.92973	0.89317	0.93360
tenor	0.86634	0.77636	0.75271	0.87236	0.78456	0.76143
	orness 0.018			orness 0.3		
	P1-P2	P2-P3	P1-P3	P1-P2	P2-P3	P1-P3
alto	0.99686	0.87973	0.94262	0.99752	0.90503	0.95549
baritone	0.86598	0.87769	0.84845	0.89861	0.90919	0.88684
soprano	0.97055	0.92573	0.97248	0.97739	0.94447	0.97901
tenor	0.96660	0.83379	0.82685	0.97187	0.87063	0.86117

baritone and the soprano saxophones we can see that the most compatible are $P2$ with $P3$ and $P1$ with $P3$ respectively.

We can also observe that the orness values are not influential in the sense that the relative order of the compatibilities between the musicians is the same.

6 Conclusions

In the present work we have given a method to numerically asses the compatibility between musicians. We have only taken into account the pitch of the notes. Other characteristics of the sound, as its quality, are very important and should be considered. Some of them can be subjectively evaluated during the auditions: an instrumentalist which is not good in terms of sound quality and richness should not pass the selection process. In any case, as a future work some other aspects of the sound could be incorporated to our methodology.

We have presented a small size example in which we were comparing only three musicians; however a table displaying the pairwise compatibility measures $\{C_{ij}\}_{1\leq i<j\leq k}$ between the k different musicians can be useful for $k > 3$ and also a set of selection criteria for potential decision-makers.

Acknowledgments. The authors acknowledge the kind collaboration of José Martínez-Delicado, Jorge Sanz-Liern, and Julio Rus-Monge in making the recordings used in computational tests and would also like to thank the financial support of research projects TIN2008-06872-C04-02 and TIN2009-14392-C02-01 from the Science and Innovation Department of the Spanish government.

References

1. Apel, W.: Harvard Dictionary of Music, 2nd edn., Revised and Enlarged. The Belknap Press of Harvard University Press, Cambridge (1994)
2. Borup, H.: A History of String Intonation, `http://www.hasseborup.com/ahistoryofintonationfinal1.pdf`
3. Del Corral, A., León, T., Liern, V.: Compatibility of the Different Tuning Systems in an Orchestra. In: Chew, E., Childs, A., Chuan, C.-H. (eds.) Communications in Computer and Information Science, vol. 38, pp. 93–103. Springer, Heidelberg (2009)
4. Dubois, D., Prade, H.: Fuzzy Sets and Systems: Theory and Applications. Academic Press, New York (1980)
5. Goldáraz Gaínza, J.J.: Afinación y temperamento en la música occidental. Alianza Editorial, Madrid (1992)
6. Halušca, J.: Equal Temperament and Pythagorean Tuning: a geometrical interpretation in the plane. Fuzzy Sets and Systems 114, 261–269 (2000)
7. Halušca, J.: The Mathematical Theory of Tone Systems. Marcel Dekker, Inc., Bratislava (2005)
8. Leon, T., Liern, V.: Mathematics and Soft Computing in Music (2009), `http://www.softcomputing.es/upload/web/parrafos/00694/docs/lastWebProgram.pdf`
9. Leon, T., Zuccarello, P., Ayala, G., de Ves, E., Domingo, J.: Applying logistic regression to relevance feedback in image retrieval systems. Pattern Recognition 40, 2621–2632 (2007)
10. Liern, V.: Fuzzy tuning systems: the mathematics of the musicians. Fuzzy Sets and Systems 150, 35–52 (2005)
11. Piles Estellés, J.: Intervalos y gamas. Ediciones Piles, Valencia (1982)
12. R Development Core Team. R: A language and environment for statistical computing. In: R Foundation for Statistical Computing, Vienna, Austria (2009), `http://www.R-project.org`, ISBN 3-900051-07-0
13. Yager, R.R.: On ordered weighted averaging aggregation operators in multi-criteria decision making. IEEE Trans. Systems Man Cybernet. 18, 183–190 (1988)

Consistently Handling Geographical User Data
Context-Dependent Detection of Co-located POIs

Guy De Tré[1], Antoon Bronselaer[1], Tom Matthé[1],
Nico Van de Weghe[2], and Philippe De Maeyer[2]

[1] Department of Telecommunications and Information Processing, Ghent University, Sint-Pietersnieuwstraat 41, B-9000 Ghent, Belgium
{Guy.DeTre,Antoon.Bronselaer,Tom.Matthe}@UGent.be

[2] Department of Geography, Ghent University, Krijgslaan 281 (S8), B-9000 Ghent, Belgium
{Nico.VandeWeghe,Philippe.DeMaeyer}@UGent.be

Abstract. In the context of digital earth applications, points of interest (POIs) denote geographical locations which might be of interest for some user purposes. Examples are nice views, historical buildings, good restaurants, recreation areas, etc. In some applications, POIs are provided and inserted by the user community. A problem hereby is that users can make mistakes due to which the same POI is, e.g., entered multiple times with a different location and/or description. Such POIs are coreferent as they refer to the same geographical object and must be avoided because they can introduce uncertainty in the map. In this paper, a novel soft computing technique for the automatic detection of coreferent locations of POIs is presented. Co-location is determined by explicitly considering the scale at which the POI is entered by the user. Fuzzy set and possibility theory are used to cope with the uncertainties in the data. An illustrative example is provided.

Keywords: GIS, POIs, duplication detection, soft computing.

1 Introduction

Digital earth applications are characterized by a tremendous amount of data, which must be collected, processed and represented by a geographical information system. Moreover, some of these data must regularly be actualised as geographic objects like roads, buildings or borderlines often change. A commonly used approach is to allow users to add, update and delete their own data. This approach is especially useful in cases where detailed, not commonly known data must be maintained.

A specific kind of information are descriptions of geographic locations or entities at geographic locations. In general, such information is modelled by objects which are called *points of interest* (POIs). Examples of POIs are objects that describe historical buildings, public services, hotels, restaurants and bars, panoramic views, interesting points to visit, etc. Usually, POIs contain information about location (coordinates) and a short textual description, but also other

E. Hüllermeier, R. Kruse, and F. Hoffmann (Eds.): IPMU 2010, Part II, CCIS 81, pp. 85–94, 2010.

information such as the category the POI belongs to, multimedia like pictures and video and meta-data like the creator's name, the timestamp of creation, the file size, etc. can be provided.

If POIs are created and added by users, one should especially take care about the consistency of the data. Indeed, user data is extremely vulnerable to errors, which might among others be due to uncertainty, imprecision, vagueness or missing information. A problem that seriously decreases the data quality occurs when multiple descriptions, which all refer to the same POI are entered in a geographical information system (GIS) as different POIs. Such POIs are called *coreferent POIs*: they all refer to the same geographic location or object at a geographic location. Coreferent POIs introduce uncertainty and storage overhead in a GIS and hence must be avoided [7].

Two basic strategies for avoiding coreferent POIs are possible. In the first strategy, the existence of coreferent POIs is *prevented* with techniques that, e.g., inform users about POIs that are in the neighbourhood of a new POI or by assigning different levels of trustworthiness to different users. In the second strategy, it is assumed that coreferent POIs can exist and must be *detected*. After detection, the problem has to be *solved* by removing the detected duplicates or by merging them into one single, consistent POI. The research in this paper contributes to the automatic detection of coreferent POIs. More specifically the sub problem of determining the (un)certainty about the co-location of two POIs is studied. This study is motivated by the observation that the detection of duplicated POIs is important in every application where a user community is assumed to participate actively in the deliverance of data.

In the remainder of the paper a novel soft computing approach for the detection of co-location of POIs is presented. In Section 2, a brief overview of related work and some preliminary definitions and notations with respect to POIs are given. Next, in Section 3, the problem of determining the uncertainty about the co-location of two POIs in a two-dimensional space is dealt with. Firstly, a basic technique is proposed. Secondly, this technique is enhanced in order to explicitly cope with the scale at which the POI is entered by the user. The presented technique is illustrated in Section 4. Furthermore, in Section 5 it is briefly described how the presented technique can be used to determine coreference of POIs. Hereby, also linguistic and semantic characteristics of POIs are considered in order to estimate the (un)certainty that two POIs are indeed coreferent. Finally, some conclusions and indications for further work are given in Section 6.

2 Preliminaries

2.1 Related Work

The topic of coreferent POI detection has already been studied from different perspectives. A basic work is [6]. Both traditional and fuzzy approaches exist.

In traditional approaches, a clustering method is typically used. An example is the DBSCAN algorithm [5], where clusters of coreferent POIs are expanded by adding similar POIs. Similarity between POIs is usually determined by means of a multidimensional similarity measure, which can be a weighted linear combination of spatial, linguistic and semantic measures. Spatial similarity is usually measured by calculating the distance between two POIs [9] and map this to inverse values in the interval $[0, 1]$ where 1 denotes an identical location and 0 represents the maximal distance. Linguistic similarity is usually measured by applying the Jaro-Winkler string comparison metric [8,14] and semantic similarity can be computed by comparing the relative positions of the concepts under consideration in a taxonomic ontology structure [11].

In fuzzy approaches, the problem of detecting coreferent POIs is usually addressed by considering that duplicates are due to uncertainty and by explicitly handling this uncertainty by means of fuzzy set theory [15] and its related possibility theory [16,4] (see e.g., [13]). Fuzzy ranges are then used to model spatial uncertainty about the co-location of two POIs. In [13], rectangular ranges are used, but other approaches are possible. Fuzzy rectangular ranges are interesting from a practical point of view because their α-cuts can be efficiently computed by using indexes.

In the presented approach, fuzzy set theory [15] is used to further enhance spatial similarity measures so that these better cope with imperfections in the descriptions (of the locations) of the POIs.

2.2 Basic Definitions and Notations

POIs are assumed to be described in a structured way. In the remainder of this paper, it is assumed that the structure t of a POI is defined by

$$t(c_1 : t_1, c_2 : t_2, \ldots, c_n : t_n), \text{ with } n \in \mathbb{N}. \tag{1}$$

As such, the structure is composed by a finite number of characteristics $c_i : t_i$, $i = 1, 2, \ldots, n$ that are all characterized by a name c_i and a data type t_i. The data types are either atomic or complex. An atomic data type is characterized by atomic domain values that are further processed as a whole. Examples are data types for the modelling of numbers, text, character strings, truth values, etc. Complex data types are themselves considered to be structured and hence have a structure like in Eq. (1).

Each POI with structure t is then characterized by a unique identifier id and a structured value that is composed of the values of each of its characteristics $c_i : t_i$, $i = 1, 2, \ldots, n$. It is denoted by

$$id(c_1 : v_1, c_2 : v_2, \ldots, c_n : v_n), \text{ with } v_i \in dom_{t_i}, i = 1, 2, \ldots, n. \tag{2}$$

Hence, the value v_i of characteristic c_i has to be an element of the domain dom_{t_i} of the data type t_i that is associated with c_i.

Example 1. An example of a POI structure with three characteristics is:

$$POI(loc : pos(lat : real,\ \ lon : real),\ \ descr : text,\ \ cat : text)$$

The first characteristic denotes the location of the POI, which is modelled by two real numbers that respectively express the latitude and longitude of the POI in decimal degrees (where 0.000001 degrees corresponds to 0.111 metres). The second characteristic denotes a free description, provided by the user and modelled by full text, whereas the third characteristic denotes the category with which the POI is labeled. It is assumed that this label is chosen from a given list. Examples of POIs with this structure are:

$$POI_1(loc : loc_{POI_1}(lat : 51.056934,\ \ lon : 3.727112),$$
$$descr : \text{“Friday Market, Ghent”},\ \ cat : \text{“market place”})$$

$$POI_2(loc : loc_{POI_2}(lat : 51.053036,\ \ lon : 3.727015),$$
$$descr : \text{“St-Baafs Kathedraal, Ghent”},\ \ cat : \text{“church”})$$

$$POI_3(loc : loc_{POI_3}(lat : 51.053177,\ \ lon : 3.726382),$$
$$descr : \text{“Saint-Bavo's Cathedral - Ghent”},\ \ cat : \text{“cathedral”})$$

$$POI_4(loc : loc_{POI_4}(lat : 51.033333,\ \ lon : 3.700000),$$
$$descr : \text{“St-Bavo - Ghent”},\ \ cat : \text{“cathedral”})$$

POI_2, POI_3 and POI_4 are examples of coreferent POIs. All four POIs have a different location.

3 Co-location of POIs

In the context of GIS, coreferent POIs are points that refer to the same geographic location or geographic entity. Geographic entities in general are located at a given geographic area which might consist of several locations. Consider for example all locations of the surface of a building, bridge, or recreation area. Thus, even in a perfect world, coreferent POIs can be denoted by different locations. In the case of imperfect data, coreferent POIs can also be assigned to different locations due to imprecision or uncertainty.

In the remainder of this section a novel technique for estimating the uncertainty about the co-location of two POIs is presented. Firstly, a basic technique commonly used in fuzzy geographic applications is presented. Secondly, this basic technique is further enhanced.

3.1 Basic Technique

As illustrated in Example 1, the location of a POI in a two-dimensional space is usually modelled by means of a latitude *lat* and longitude *lon*. Consider two POIs POI_1 and POI_2 with locations (lat_1, lon_1) and (lat_2, lon_2). In geographic applications, the distance (in metres) between the two locations is usually approximately computed by

$$d(POI_1, POI_2) = 2R \arcsin(h) \tag{3}$$

where $R = 6367000$ is the radius of the earth in metres and

$$h = \min\left(1, \sqrt{\sin^2\left(\frac{lat_2^r - lat_1^r}{2}\right) + \cos(lat_1^r)\cos(lat_2^r)\sin^2\left(\frac{lon_2^r - lon_1^r}{2}\right)}\right)$$

with $lat_j^r = \frac{\pi}{180} lat_j$ and $lon_j^r = \frac{\pi}{180} lon_j$, for $j = 1, 2$, being the conversions in radians of lat_j and lon_j [12]. The higher the precision of the measurement of the latitude and longitude, the higher the precision of this distance.

From a theoretical point of view, POIs are considered to be locations. Hence, two POIs are considered to be co-located if their distance equals zero. In practice however, POIs can refer to geographic areas (or entities located at geographic areas). Therefore, it is more realistic to consider two POIs as being co-located if they refer to the same area and are thus *close enough*. In traditional approaches *'close enough'* is usually modelled by a threshold $\epsilon > 0$, such that two POIs POI_1 and POI_2 are ϵ-close if and only if

$$d(POI_1, POI_2) \leq \epsilon. \tag{4}$$

The problem with such a single threshold is that it puts a hard constraint on the distance, which implies an 'all or nothing' approach: depending on the choice for ϵ, two POIs will be considered as being co-located or not. If an inadequate threshold value is chosen, this will yield in a bad decision.

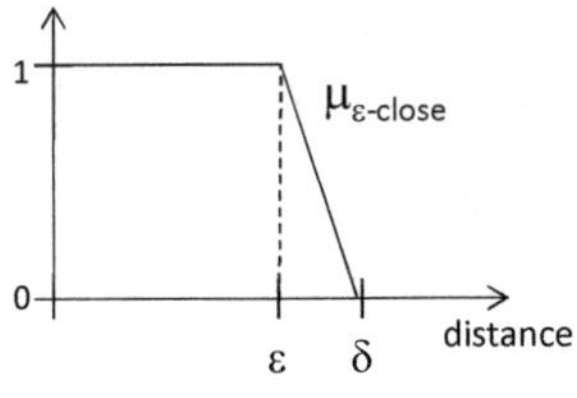

Fig. 1. Fuzzy set for representing 'close enough'

Fuzzy sets [15] have been used to soften the aforementioned constraint. In general, a fuzzy set with a membership function $\mu_{\epsilon-close}$, as presented in Figure 1, is used to model *'close enough'*. This membership function is defined by

$$\mu_{\epsilon-close} : [0, +\infty] \rightarrow [0, 1]$$

$$d \mapsto \begin{cases} 1 & \text{if } d \leq \epsilon \\ \dfrac{\delta - d}{\delta - \epsilon} & \text{if } \epsilon < d \leq \delta \\ o & \text{if } d > \delta. \end{cases} \tag{5}$$

The extent to which two POIs POI_1 and POI_2 are considered to be co-located is then given by $\mu_{\epsilon-close}(d(POI_1, POI_2))$. Hence, for distances below ϵ, $\mu_{\epsilon-close}$ denotes co-location, for distances larger than δ no co-location is assumed, whereas for distances between ϵ and δ, there is a gradual transition from co-location to no co-location. Other membership functions can be used.

3.2 Enhanced Technique

A practical problem with fuzzy approaches as described in the previous subsection, is that the membership function has to reflect reality as adequate as possible. This implies that adequate values for ϵ and δ must be chosen. Values that are too stringent (too small) will result in false negatives, i.e., some POIs will falsely be identified as not being co-located, whereas values that are too soft (too large) will result in false positives, i.e., some POIs will falsely be identified as being co-located. In this subsection, it is considered that POIs are created and added by users. This situation often occurs in applications where the user community helps with maintaining the data sets. In such a case, it makes sense to study how the parameters ϵ and δ are influenced by the *context* in which the user inserted the POI. Eq. (5) can then be further enhanced in order to better reflect the imprecision in the placement of the POI.

In practice, users work with maps on computer screens or screens of mobile devices when entering, searching, or maintaining POIs. Each map is a representation of the real world and is drawn on a given *scale* $1 : s$, which means, e.g., that 1 cm on the scale corresponds to s cm in reality. For example, a map of Europe on a computer screen can be drawn at scale $1 : 15000000$, a map of Belgium at scale $1 : 1000000$ and a map of Ghent at scale $1 : 125000$. Is is clear that the precision with which a user can place a POI on a map depends on the scale of the map. Denoting a POI that represents the Eiffel tower on a map of Europe will be less precise than on a map of France, which on its turn will be less precise than on a map of Paris. On the other hand, depending on his or her knowledge about the location of the new POI the user can zoom-in or zoom-out on the map to enter the POI at the map with the most appropriate detail for the user. In practice the scales supported by a given GIS will be within the range $1 : s_{min}$ (corresponding to the most detailed level) and $1 : s_{max}$ (corresponding to the least detailed level). Hence, $s_{min} \leq s \leq s_{max}$.

Another aspect to take into account is the *precision* with which the user can denote the location of a POI on the screen. Usually, when working at an appropriate scale $1 : s$, the user will be able to place a point on the screen with a precision of a couple of centimetres, i.e., the exact location of the point will be within a circle with the denoted point as centre and radius d_s. This radius can be considered to be a parameter that depends on the scale $1 : s$ and the user's abilities for accurately denoting the POI on the screen. Therefore, in practical applications, d_s could be adjustable by the user or by a user feedback mechanism.

The scales $1 : s$, $s_{min} \leq s \leq s_{max}$, and corresponding radiuses d_s can now be used to further enhance the definition of the membership function $\mu_{\epsilon-close}$ that is used in the basic technique presented in the previous subsection.

Estimating the Value of ϵ. In order to approach reality, ϵ should reflect the maximum distance for which two POIs are indistinguishable and hence must be considered as being co-located. If no further information about the geographical area of the POI is available, then the POI is positioned at the location that is entered by the user and modelled by its latitude and longitude. Two POIs are then indistinguishable if they are mapped by the GIS to the same latitude and longitude. The maximum precision of the GIS, which can be approximated by the dot pitch of the screen can then be used to estimate the value of ϵ. The dot pitch dp of a screen is defined as the diagonal distance between two pixels on the screen and usually has a standard value of 0.28mm. Considering the minimum scale $1 : s_{min}$, the value of ϵ can then be approximated by

$$\epsilon = d \cdot s_{min} \tag{6}$$

If information about the geographical area of the POI is given, then the length l of the diagonal of the minimum bounding rectangle that surrounds this area can be used to approximate ϵ. Indeed, all POIs that are placed in the rectangle can reasonably be considered as being co-located. If the POI information for POI_1 and POI_2 is respectively entered at a scale $1 : s_1$ and $1 : s_2$, the value of ϵ can be approximated by

$$\epsilon = \max(\frac{l}{2} \cdot s_1, \frac{l}{2} \cdot s_2) \tag{7}$$

where the maximum operator is used to take the roughest, largest approximation (which is due to the least precise scale) in cases where both POIs were entered at a different scale.

Estimating the Value of δ. Taking into account the scale $1 : s_1$ and precision d_{s_1} with which a user entered POI_1 and the scale $1 : s_2$ and precision d_{s_2} with which POI_2 was entered, the value of δ can be defined by

$$\delta = \epsilon + \max(s_1 \cdot d_{s_1}, s_2 \cdot d_{s_2}) \tag{8}$$

where the maximum operator is again used to take the roughest approximation in cases where both POIs were entered at a different scale. With this definition the precisions d_{s_1} and d_{s_2} are handled in a pessimistic way. Alternative definitions for δ are possible.

Possibilistic Interpretation. The problem of determining whether two POIs are co-located or not can be approached as a problem of estimating the uncertainty of the Boolean proposition

$$p = \text{“}POI_1 \textit{ is co-located with } POI_2\text{”}.$$

Possibility theory [16,4] can be used to express this uncertainty. More specifically, in our approach, a possibilistic truth value (PTV) is used for this purpose. A PTV is a normalized [10] possibility distribution

$$\tilde{p} = \{(T, \mu_{\tilde{p}}(T)), (F, \mu_{\tilde{p}}(F))\} \tag{9}$$

over the set of Boolean values true (T) and false (F), representing the possibility that $p = T$ and the possibility that $p = F$.

The membership function $\mu_{\epsilon-close}$ can then be used to estimate the PTV $\tilde{p}$ of p. A simple approach to do this is given by

$$\mu_{\tilde{p}}(T) = \frac{\mu_{\epsilon-close}(d(POI_1, POI_2))}{\max(\mu_{\epsilon-close}(d(POI_1, POI_2)), 1 - \mu_{\epsilon-close}(d(POI_1, POI_2)))} \quad (10)$$

$$\mu_{\tilde{p}}(F) = \frac{1 - \mu_{\epsilon-close}(d(POI_1, POI_2))}{\max(\mu_{\epsilon-close}(d(POI_1, POI_2)), 1 - \mu_{\epsilon-close}(d(POI_1, POI_2)))}. \quad (11)$$

4 An Illustrative Example

The enhanced technique, presented in Section 3 is illustrated in Example 2.

Example 2. Consider the four POIs of Example 1. POI_1, POI_2 and POI_3 are entered at scale $1 : 10000$ which corresponds to a street map of Ghent, whereas POI_4 is entered at scale $1 : 1000000$ which corresponds to a map of Belgium. The latitude, longitude, scale, radius of screen precision, and parameter value for ϵ of these POIs are summarized in Table 1. The minimum scale supported by the GIS is assumed to be $1 : 10000$. For all POIs, the same precision $d_s = 0.01m$ is used. This precision is assumed to be provided by the user (or could alternatively be set by default in the system). The PTVs representing the uncertainty about the co-location of these POIs are given in Table 2. These results reflect that POI_1 is not co-located with POI_2 and POI_3 which is denoted by the PTV $\{(F, 1)\}$. Due to the fact that POI_4 is entered at scale $1 : 1000000$, which is less precise than scale $1 : 10000$ it is either possible with possibility 1 that POI_4 is co-located with POI_1, POI_2 and POI_3, or either to a lesser extent (resp. 0.48, 0.41 and 0.40) possible that it is not co-located with these POIs. Likewise, it is either

Table 1. Information about POIs

POI	**lat**	**lon**	$1 : s$	d_s	$\epsilon = dp.s_{min}$
POI_1	51.056934	3.727112	1:10000	0.01m	2.8m
POI_2	51.053036	3.727015	1:10000	0.01m	2.8m
POI_3	51.053177	3.726382	1:10000	0.01m	2.8m
POI_4	51.033333	3.700000	1:1000000	0.01m	2.8m

Table 2. Information about co-location of POIs

x	**y**	$d(x, y)$	$\delta = \epsilon + \max(s_1.d_{s_1}, s_2.d_{s_2})$	$\mu_{\tilde{p}}(T)$	$\mu_{\tilde{p}}(F)$
POI_1	POI_2	433.2m	102.8m	0	1
POI_1	POI_3	420.6m	102.8m	0	1
POI_1	POI_4	3235.2m	10002.8m	1	0.48
POI_2	POI_3	46.9m	102.8m	1	0.79
POI_2	POI_4	2890.8m	10002.8m	1	0.41
POI_3	POI_4	2874.1m	10002.8m	1	0.40

possible with possibility 1 that POI_2 and POI_3 are co-located, or to an extent 0.79 possible that these are not co-located. This rather high value of 0.79 is due to the pessimistic estimation of ϵ being only 2.8m, where Saint-Bavo cathedral has a diagonal of about 110m. Using Eq. (7), $\epsilon = 55m$ and $\delta = 155m$, which yields the PTV $\{(T, 1)\}$ that corresponds to true.

5 Coreference of POIs

The presented technique can be used as a component of a technique to determine whether two POIs are coreferent or not. The resulting PTVs as obtained by Eq. (10) and (11), then denote a measure for the uncertainty about the co-location or spatial similarity of the POIs.

Considering the other relevant characteristics in the structure of the POIs (Eq. (1)), other techniques can be constructed to estimate the uncertainty about the linguistic and semantic similarity of two POIs [1,2]. Applying such a technique for each characteristic, will then yield in a PTV that reflects the uncertainty that the values of this characteristic (in both POIs) are coreferent or not. All resulting PTVs can then be aggregated, using a technique as, e.g., described in [3]. The resulting PTV then represents the overall possibility that the POIs are coreferent or not.

6 Conclusions and Further Work

In this paper, a novel soft computing technique to estimate the uncertainty about the potential co-location of two POIs is described. The technique is a further enhancement of a traditional fuzzy technique where fuzzy ranges are used to determine in a flexible way whether two POI locations can be considered to be close enough to conclude that they are coreferent, i.e., they refer to the same geographic entity or area. Typical for the technique is that it is context-dependent as it explicitly copes with the precision and scale at which a given POI is entered in a GIS. Furthermore, the estimated uncertainty is modelled by a possibilistic truth value (PTV). This makes the technique especially suited for the detection of coreferent POIs in applications where POIs are provided and inserted by a user community.

The technique allows for a human consistent estimation and representation of the uncertainty about the co-location of POIs, which is induced by the imprecision in the POI placement and is due to the physical limitations of computer screens and handheld devices. If used to detect coreferent POIs, the technique allows for a semantic justifiable, direct comparison of POIs. This opens new perspectives to enhance existing clustering algorithms for the detection of coreferent POIs, but also offers opportunities for new detection techniques which are based on direct comparisons of POIs. Integration of the technique in a real GIS application is planned.

Further research is required and planned. An important aspect that will be further investigated is the further development and optimization of the POI comparison technique. Optimization is possible as not all pairs in a set of POIs must be checked to detect all coreferent POIs. Moreover, not all characteristics must necessarily in all cases be evaluated to come to a conclusion regarding coreference. Another aspect concerns the use of advanced indexing techniques, which might speed up the comparison process. Finally, a last mentioned research topic is related to the further processing of coreferent POIs. More specifically, in view of the deduplication of coreferent POIs, it is worth to study how information of two coreferent POIs could be merged and further processed.

References

1. Bronselaer, A., De Tré, G.: A Possibilistic Approach to String Comparison. IEEE Trans. on Fuzzy Systems 17, 208–223 (2009)
2. Bronselaer, A., De Tré, G.: Semantical evaluators. In: Proc. of the 2009 IFSA/EUSFLAT Conference, pp. 663–668 (2009)
3. Bronselaer, A., Hallez, A., De Tré, G.: Extensions of Fuzzy Measures and Sugeno Integral for Possibilistic Truth Values. Int. Journal of Intelligent Systems 24, 97–117 (2009)
4. Dubois, D., Prade, H.: Possibility Theory. Plenum Press, New York (1988)
5. Ester, M., Kriegel, H.P., Sander, J., Xu, X.: A Density-Based Algorithm for Discovering Clusters in Large Spatial Databases with Noise. In: Proc. of the 2nd Int. Conf. on Knowledge Discovery and Data Mining (1996)
6. Fellegi, I., Sunter, A.: A Theory for Record Linkage. American Statistical Association Journal 64(328), 1183–1210 (1969)
7. Federal Geographic Data Committee: Content standard for digital geospatial metadata. FGDC-STD-001-1998, Washington D.C., USA (1998)
8. Jaro, M.: Unimatch: A record linkage system: User's manual. US Bureau of the Census, Tech. Rep. (1976)
9. National Imagery and Mapping Agency (NIMA): Department of Defence World Geodetic System 1984: Its Definitions and Relationships with Local Geodetic Systems. NIMA Technical Report 8350.2 (2004)
10. Prade, H.: Possibility sets, fuzzy sets and their relation to Lukasiewics logic. In: Proc. of the Int. Symposium on Multiple-Valued Logic, pp. 223–227 (1982)
11. Rodríguez, M.A., Egenhofer, M.J.: Comparing Geospatial Entity Classes: An Asymmetric and Context-Dependent Similarity Measure. Int. Journal of Geographical Information Science 18 (2004)
12. Sinnott, R.W.: Virtues of the Haversine. Sky and Telescope 68(2), 159 (1984)
13. Torres, R., Keller, G.R., Kreinovich, V., Longpré, L., Starks, S.A.: Eliminating Duplicates under Interval and Fuzzy Uncertainty: An Asymptotically Optimal Algorithm and Its Geospatial Applications. Reliable Computing 10(5), 401–422 (2004)
14. Winkler, W.E.: The State of Record Linkage and Current Research Problems. R99/04, Statistics of Income Division, U.S. Census Bureau 1999 (1999)
15. Zadeh, L.A.: Fuzzy Sets. Information and Control 8(3), 338–353 (1965)
16. Zadeh, L.A.: Fuzzy sets as a basis for a theory of possibility. Fuzzy Sets and Systems 1, 3–28 (1978)

A Model Based on Outranking for Database Preference Queries

Patrick Bosc, Olivier Pivert, and Grégory Smits

Irisa – Enssat, University of Rennes 1
Technopole Anticipa 22305 Lannion Cedex France
{bosc,pivert,smits}@enssat.fr

Abstract. In this paper, we describe an approach to database preference queries based on the notion of outranking, suited to the case where partial preferences are incommensurable. This model constitutes an alternative to the use of Pareto order. Even though outranking does not define an order in the strict sense of the term, we describe a technique which yields a complete pre-order, based on a global aggregation of the outranking degrees computed for each pair of tuples.

keywords: Preference queries, outranking, incommensurability.

1 Introduction

The last decade has witnessed an increasing interest in expressing preferences inside database queries. This trend has motivated several distinct lines of research, in particular fuzzy-set-based approaches and Pareto-order-based ones.

Fuzzy-set-based approaches [1,2] use fuzzy set membership functions that describe the preference profiles of the user on each attribute domain involved in the query. Then, individual satisfaction degrees associated with elementary conditions are combined using a panoply of fuzzy set connectives, which may go beyond conjunction and disjunction. Let us recall that fuzzy-set-based approaches rely on a *commensurability* hypothesis between the degrees pertaining to the different attributes involved in a query.

Approaches based on Pareto order aim at computing non Pareto-dominated answers (viewed as points in a multidimensional space, their set constitutes a so-called skyline), starting with the pioneering works of Bőrzsőnyi *et al.* [3]. Clearly, the skyline computation approach does not require any commensurability hypothesis between satisfaction degrees pertaining to elementary requirements that refer to different attribute domains. Thus, some skyline points may represent very poor answers with respect to some elementary requirements while they are excellent w.r.t. others. Let us emphasize that Pareto-based approaches yield a strict partial order only, while fuzzy set-based approaches yield a complete pre-order. Kießling [4,5] has provided foundations for a Pareto-based preference model for database systems. A preference algebra including an operator called *winnow* has also been proposed by Chomicki [6].

The present paper proposes an alternative to the use of Pareto order in the case where preferences are incommensurable. Our goal is not to show that this approach is "better" than those based on Pareto order, but that it constitutes a different way to deal with preferences inside database queries, that some users may find more suitable and intuitive (at

E. Hüllermeier, R. Kruse, and F. Hoffmann (Eds.): IPMU 2010, Part II, CCIS 81, pp. 95–104, 2010.

least in some given contexts). The situation considered is that of queries involving preferences on several attributes, which use different ordinal scales and/or different scoring measures (expressed either by fuzzy set membership functions or by *ad hoc* functions as in [7]). Then, for a given tuple, each atomic preference leads to the computation of a score. It is assumed, however, that the user does not authorize any trade-off between the atomic preference criteria. In other terms, contrary to the assumption underlying fuzzy-set-based approaches, the scores associated with the different partial preferences *cannot be aggregated.*

The approach we advocate rests on the concept of outranking, which was introduced in the context of decision-making [8] but has never been used so far in a database context, to the best of our knowledge. However, the way we define outranking differs on some aspects from the definition given in [8]. In particular, we choose to have a strict symmetry between the concepts of concordance and discordance, and we introduce the additional notion of an indifferent criterion (which is absent from [8] where indifference is only considered a binary relation between values). Besides, the mechanism we propose to order the tuples on the basis of their global "quality" yields a *total order* whereas no such mechanism exists in [8] where only a partial order is obtained.

In this paper, we deal mainly with the *query model.* As to the processing aspect, we just outline a simple evaluation method but it is clear that query optimization should be tackled in future work, as it has been done for skyline queries (see, e.g., [9]).

The remainder of the paper is organized as follows. Section 2 presents some related work, in particular the Pareto-based approach to preference handling in databases. In Section 3, we present a preference model based on the notion of outranking. In Section 4, we describe the way such preference queries can be expressed by means of an SQL-like language and we briefly deal with query evaluation. Finally, Section 5 concludes the paper and outlines some perspectives for future work.

2 Related Work

Let us first recall the general principle of the approaches based on Pareto order. Let $\{G_1, G_2, ..., G_n\}$ be a set of the atomic preferences. We denote by $t \succ_{G_i} t'$ (resp. $t \succeq_{G_i} t'$) the statement "tuple t satisfies preference G_i better than (resp. as least as good as) tuple t'". Using Pareto order, a tuple t dominates another tuple t' iff

$$\forall i \in \{1, \ldots, n\},\ t \succeq_{G_i} t' \text{ and } \exists k \in \{1, \ldots, n\},\ t \succ_{G_k} t'.$$

i.e., if t is at least as good as t' regarding every preference, and is strictly better than t' regarding at least one preference. The following example uses the syntax of the language *Preference SQL* [5], which is a typical representative of a Pareto-based approach.

Example 1. Let us consider a relation *car* of schema (make, category, price, color, mileage) whose extension is given in Table 1, and the query:

select * **from** car **where** mileage ≤ 20,000
preferring
  (category = 'SUV' **else** category = 'roadster') **and**
  (make = 'VW' **else** make = 'Ford' **else** make = 'Opel');

The idea is to retain the tuples which are not dominated in the sense of the "preferring" clause. Here, t_1, t_4, t_5, t_6 and t_7 are discarded since they are Pareto-dominated by t_2 and t_3. On the other hand, t_2 and t_3 are incomparable and the final answer is $\{t_2, t_3\}$.⋄

When the number of dimensions on which preferences are expressed gets high, many tuples may become incomparable. Several approaches have been proposed to define an order for two incomparable tuples in the context of skylines, based on:

- the number of other tuples that each of the two tuples dominates (notion of k-representative dominance proposed by Lin *et al.* [10]) or
- some preference order of the attributes; see for instance the notions of k-dominance and k-frequency introduced by Chan *et al.* [11,12].

Even if these approaches make it possible to some extent to avoid incomparable elements, they are all based on a *Boolean* notion, namely that of dominance. What we propose is an alternative semantics to the modeling of preference queries involving incommensurable criteria, which takes into account the *extent* to which an element is better than another for a given atomic preference. In other words, the approach we propose is fundamentally gradual, unlike those based on Pareto order such as Skyline and its different variants.

Unlike the family of approaches based on Pareto order, score-based approaches (including those based on fuzzy set theory as well as the quantitative approach proposed by Agrawal and Wimmers [7] and top-k queries [13]) do not deal with incommensurable preferences. Besides Pareto-order-based approaches, CP-nets [14,15] also handle incommensurable preferences, but they do so only within a restrictive interpretation setting. Indeed, CP-nets deal with *conditional* preference statements and use the *ceteris paribus* semantics, whereas we deal with *non-conditional* preference statements and consider the *totalitarian* semantics (i.e., when evaluating the preference clause of a query, one *ignores* the values of the attributes which are not involved in the preference statement). This latter semantics is implicitly favored by most of the authors in the database community, including those who advocate a Pareto-based type of approach.

3 Principle of the Approach

3.1 Basic Notions

Atomic preference modeling. An ordinal scale is specified as:

$$S_1 > S_2 > ... > S_m$$

such that the elements from S_1 get score m while those from S_m get score 1. In other words, an ordinal scale involving m levels is associated with a mapping: $level \rightarrow \{1, , \ldots, m\}$ such that the preferred level corresponds to score m and the less preferred one to score 1. A value absent from the scale gets score zero. The scale may include the special element *other* as a bottom value so as to express that any value non explicitly specified in the list is an acceptable choice but is worse than the explicitly specified ones: it then corresponds to score 1. Notice that this way of doing "freezes" the distance between the elements of the list. For instance, with the ordered list {VW, Audi}

$\succ$ {BMW} $\succ$ {Seat, Opel} $\succ$ {Ford}, the distance between, e.g., VW and BMW is assumed to be the same as, e.g., that between Opel and Ford. If the user wants to avoid this phenomenon, he/she can elicitate the scores in an explicit manner, specifying for instance: {1/{VW, Audi}, 0.8/{BMW}, 0.5/{Seat, Opel}, 0.3/{Ford}}. This has no impact on the interpretation model described further.

As to scoring functions concerning numerical attributes, they model flexible conditions of the form $attribute \leq \alpha$, $attribute \approx \alpha$ and $attribute \geq \alpha$ where α is a constant. In the following examples, we assume that they take their values in the unit interval $[0, 1]$ but this is not mandatory.

Concordance, indifference, discordance. The outranking relation relies on two basic notions, concordance and discordance. Concordance represents the proportion of preferences which validate the assertion "t is preferred to t'", denoted by $t \succ t'$, whereas discordance represents the proportion of preferences which contradict this assertion.

Let $A_1, A_2, ..., A_n$ be the attributes concerned respectively by the set of preferences $G = \{G_1, G_2, ..., G_n\}$. Let $g_1, g_2, ..., g_n$ be the scoring functions associated to preferences $G_1, G_2, ..., G_n$ respectively.

Indifferent preferences: Each preference G_j may be associated with a threshold q_j. Preference G_j is indifferent with the statement "t is preferred to t'" iff $|g_j(t.A_j) - g_j(t'.A_j)| \leq q_j$. This notion makes it possible to take into account some uncertainty or some tolerance on the definition of the elementary preferences.

Concordant preferences: G_j is concordant with the statement "t is preferred to t'" iff $g_j(t.A_j) > g_j(t'.A_j) + q_j$.

Discordant preferences: Preference G_j is discordant with the statement "t is preferred to t'" iff $g_j(t'.A_j) > g_j(t.A_j) + q_j$.

In the following, we denote by $C(t, t')$ (resp. $I(t, t')$, resp. $D(t, t')$) the set of concordant (resp. indifferent, discordant) preferences from G w.r.t. $t \succ t'$. One may also attach a weight w_j to each preference G_j expressing its importance. It is assumed that the sum of the weights equals 1.

3.2 The Preference Model

First, let us define:

$$conc(t, t') = \sum_{G_j \in C(t, t')} w_j,$$

$$disc(t, t') = \sum_{G_j \in D(t, t')} w_j$$

$$ind(t, t') = \sum_{G_j \in I(t, t')} w_j$$

where w_j denotes the importance attached to preference G_j (recall that $\sum_{j=1}^{n} w_j = 1$ is assumed).

Theorem 1. One has: $\forall(t, t')$, $conc(t, t') + ind(t, t') + disc(t, t') = 1$. □

Lemma 1. One has:

$conc(t, t') = 1 \Rightarrow disc(t, t') = 0$ and $disc(t, t') = 1 \Rightarrow conc(t, t') = 0$. □

The outranking degree attached to the statement $t \succeq t'$ (meaning "t is at least as good as t'"), denoted by $out(t, t')$, reflects the truth of the statement: most of the important criteria are concordant *or indifferent* with $t \succeq t'$ and few of the important criteria are discordant with $t \succeq t'$. It is evaluated by the following formula:

$$out(t, t') = conc(t, t') + ind(t, t') = 1 - disc(t, t'). \quad (1)$$

Theorem 2. $\forall (t, t'), conc(t, t') = disc(t', t)$. □

Theorem 3. $\forall (t, t'), out(t, t') \geq 1 - out(t', t)$. □

Theorem 4. $\forall t, out(t, t) = 1$. □

Theorems 1 to 4 are straightforward and their proofs are omitted. From Equation 1 and Theorem 2, one gets:

$$out(t, t') = 1 - conc(t', t). \quad (2)$$

Example 2. Let us consider the extension of the relation *car* from Table 1 and the preferences:
for make:
$\{\text{VW}\} \succ \{\text{Audi}\} \succ \{\text{BMW}\} \succ \{\text{Seat}\} \succ \{\text{Opel}\} \succ \{\text{Ford}\} \succ$ *other*; $q_{make} = 1$; $w_{make} = 0.2$.
for category:
$\{\text{sedan}\} \succ \{\text{roadster}\} \succ \{\text{coupe}\} \succ \{\text{SUV}\} \succ$ *other*; $q_{category} = 1$; $w_{category} = 0.3$.
for price:
score(price) = 1 if price $\leq$ 4000, 0 if price $\geq$ 6000, linear in-between; $q_{price} = 0.2$; $w_{price} = 0.2$.
for color:
$\{\text{blue}\} \succ \{\text{black}\} \succ \{\text{red}\} \succ \{\text{yellow}\} \succ \{\text{green}\} \succ \{\text{white}\} \succ$ *other*; $q_{color} = 1$; $w_{color} = 0.1$.
for mileage:
score(mileage) = 1 if mileage $\leq$ 15,000, 0 if mileage $\geq$ 20,000, linear in-between; $q_{mileage} = 0.2$; $w_{mileage} = 0.2$.

Table 1. An extension of relation *car*

	make	category	price	color	mileage
t_1	Opel	roadster	4500	blue	20,000
t_2	Ford	SUV	4000	red	20,000
t_3	VW	roadster	5000	red	10,000
t_4	Opel	roadster	5000	red	8,000
t_5	Fiat	roadster	4500	red	16,000
t_6	Renault	sedan	5500	blue	24,000
t_7	Seat	sedan	4000	green	12,000

Table 2. Scores obtained by the values from *car*

	make	category	price	color	mileage
t_1	3	4	0.75	7	0
t_1	2	2	1	5	0
t_1	7	4	0.5	5	1
t_1	3	4	0.5	5	1
t_1	1	4	0.75	5	0.8
t_1	1	5	0.25	7	0
t_1	4	5	1	3	1

Table 2 gives the scores obtained by each tuple for every preference. Notice that in the sense of Pareto order, t_3 dominates t_4 and tuples t_1, t_2, t_3, t_5, t_6, t_7 are incomparable. Thus the result of a Pareto-based system such as *Preference SQL* [5] would be the "flat" set $\{t_1, t_2, t_3, t_5, t_6, t_7\}$, whereas the approach we propose yields a much more discriminated result, as we will see below.

Let us compute the degree $out(t_1, t_2)$. The concordant criteria are *category* and *color*; the indifferent ones *make* and *mileage*; the only discordant one is *price*. We get: $conc(t_1, t_2) = w_{category} + w_{color} = 0.4$, $ind(t_1, t_2) = w_{make} + w_{mileage} = 0.4$, $disc(t_1, t_2) = w_{price} = 0.2$, hence: $out(t_1, t_2) = 0.4 + 0.4 = 0.8$.

Table 3. Concordance degrees

	t_1	t_2	t_3	t_4	t_5	t_6	t_7
t_1	0	0.4	0.3	0.3	0.3	0.4	0.1
t_2	0.2	0	0.2	0.2	0.2	0.2	0.1
t_3	0.4	0.7	0	0.2	0.2	0.6	0.3
t_4	0.2	0.5	0	0	0.2	0.6	0.1
t_5	0.2	0.5	0.2	0.2	0	0.4	0.1
t_6	0	0.4	0.1	0.1	0.1	0	0.1
t_7	0.4	0.7	0.2	0.2	0.4	0.6	0

Table 3 gives the concordance degree obtained for every pair of tuples (t, t') from relation *car* (a row corresponds to a t and a column to a t'). Table 4 — which can be straightforwardly computed from Table 3 thanks to Equation 2 — gives the outranking degree of $t \succeq t'$ for every pair of tuples (t, t') from relation *car*. Table 4 includes an extra column μ_1 whose meaning is given hereafter.◇

Notice that the degree of outranking does not define an order since the notion of outranking is not transitive (there may exist cycles in the outranking graph). However, several ways can be envisaged to rank the tuples, based on different aggregations of the outranking degrees, thus on a *global* evaluation of each tuple. We suggest the following:

1. for every tuple t, one computes the degree:

$$\mu_1(t) = \frac{\Sigma_{t' \in r \setminus \{t\}} out(t, t')}{|r| - 1}$$

Table 4. Outranking degrees

	t_1	t_2	t_3	t_4	t_5	t_6	t_7	μ_1
t_1	1	0.8	0.6	0.8	0.8	1	0.6	0.77
t_2	0.6	1	0.3	0.5	0.5	0.6	0.3	0.47
t_3	0.7	0.8	1	1	0.8	0.9	0.8	0.83
t_4	0.7	0.8	0.8	1	0.8	0.9	0.8	0.8
t_5	0.7	0.8	0.8	0.8	1	0.9	0.6	0.77
t_6	0.6	0.8	0.4	0.4	0.6	1	0.4	0.53
t_7	0.9	0.9	0.7	0.9	0.9	0.9	1	0.87

where $|r|$ denotes the cardinality of r. Degree $\mu_1(t)$ expresses the extent to which t is better to (or as good as) most of the other tuples from r (where the fuzzy quantifier *most* is assumed to be defined as $\mu_{most}(x) = x, \forall x \in [0, 1]$). These degrees appear in the last column of Table 4.

2. one ranks the tuples in increasing order of $\mu_1(t)$.

The data from Table 1 leads to:

$$0.87/t_7 > 0.83/t_3 > 0.8/t_4 > 0.77/\{t_1, t_5\} > 0.53/t_6 > 0.47/t_2.$$

It is interesting to notice that $\mu_1(t)$ also captures the extent to which t is not worse than most of the other tuples. Indeed, let us consider

$$\mu_2(t) = \frac{\Sigma_{t' \in r \setminus \{t\}} conc(t', t)}{|r| - 1}$$

Degree $\mu_2(t)$ expresses the extent to which t is worse than most of the other tuples from r. Due to Equation 2, one has: $\forall t, \mu_1(t) = 1 - \mu_2(t)$. Thus, ranking the tuples according to μ_1 or to $1 - \mu_2$ leads to the same ordering.

3.3 Relation with Pareto Order

Let t and t' be two tuples such that t is better than t' in the sense of Pareto order (denoted by $t >_P t'$). It can be proven that in the case where $\forall j, q_j = 0$ (case of usual Skyline or *Preference SQL* queries), one has:

$$t >_P t' \Rightarrow$$
$$\forall t'', out(t, t'') \geq out(t', t'') \text{ and } out(t'', t) \leq out(t'', t').$$

The proof is straightforward and is omitted here due to space limit. This result guarantees that t will be ranked before t' in the final result. In other terms, the outranking-based model applied to classical Skyline queries refines the order produced by the Pareto-based model.

4 About Query Expression and Processing

4.1 Syntactical Aspects

Let us consider the SQL language as a framework. We introduce a new clause aimed at expressing preferences, which will be identified by the keyword *preferring* as in the

Preference SQL approach. This clause can come as a complement to a *where* clause, and then only the tuples which satisfy the condition from the *where* clause are concerned by the preference clause.

The preference clause specifies a list of preferences, and each element of the list includes:

- the name of the attribute concerned,
- an ordered scale or the definition of a scoring function,
- the optional weight associated with the preference,
- the optional threshold q.

We assume that scoring functions take their values in [0, 1]. A simple way to define them, when they concern numerical attributes, is to specify their core (ideal values) and support (acceptable values) and to use trapezoidal functions:

- $attribute \leq \alpha : \langle ideal : \leq \alpha,\ acceptable : < \alpha + \beta \rangle$,
- $attribute \approx \alpha : \langle ideal : \in [\alpha - \beta,\ \alpha + \beta],$
  $acceptable : \in]\alpha - \beta - \lambda,\ \alpha + \beta + \lambda[\rangle$
- $attribute \geq \alpha : \langle ideal : \geq \alpha,\ acceptable : > \alpha - \beta \rangle$.

When scoring functions concern categorical attributes (case where the user wants to avoid the "distance freezing" phenomenon induced by an ordinal scale, cf. Subsection 3.1), they have to be given in extension, as in: {1/{VW, Audi}, 0.8/{BMW}, 0.5/{Seat, Opel}, 0.3/{Ford}}.

As to the weights, their sum must be equal to 1, and if none is given by the user, each weight is automatically set to $1/m$ where m is the number of preferences (sets) in the list. In order to make the system more user-friendly, one can also think of letting the user specify the weights by means of a linguistic scale such as {very important, rather important, medium, not very important, rather unimportant}, assuming that the system automatically translates these linguistic terms into numerical weights and normalizes the set of weights obtained in such a way that their sum equals 1.

The optional threshold q must be consistent with the ordinal scale used or with the unit interval in the case of a scoring function. If q is not specified, its default value is zero, which means that indifference corresponds to equality.

The preference concerning an attribute can be either *strict* (then one uses the keywords *strict*) or *tolerant*. If it is strict, it means that a tuple which gets the score zero for the preference concerned is discarded. If it is tolerant (as in the previous examples), even the tuples which get a zero degree on that preference are ranked. The notion of a strict preference frees the user from the tedious task of specifying an additional condition in the *where* clause.

Example 3. An example of such a query is:

select * **from** car **preferring**
color: ⟨(blue) > (black) > (red, orange) > (green) > (black) > other | w = 0.1 | q = 1⟩
make: ⟨(VW, Audi) > (BMW) > (Seat, Opel) > (Ford) > other | w = 0.2 | q = 1⟩
category **strict**: ⟨(sedan) > (roadster) > (coupe) > (SUV) > other | w = 0.3 | q = 1⟩
price **strict**: ⟨ideal: ≤ 4000 | acceptable: ≤ 6000 | w = 0.2 | q = 0.2⟩
mileage: ⟨ideal: ≤ 15,000 | acceptable: ≤ 20,000 | w = 0.2 | q = 0.2⟩.◇

4.2 On a "Naive" Query Evaluation Technique

Let us denote by n the cardinality of the relation concerned. The data complexity of a preference query based on outranking, if a straightforward evaluation technique is used, is in $\theta(n^2)$ since all the tuples have to be compared pairwise. But it is important to emphasize that this is also the case of "naive" evaluation methods for Pareto-order-based preference queries (as in the approaches *Skyline*, *Preference SQL*, etc), even though some recent works have proposed more efficient processing techniques (see, e.g., [9]). On the other hand, fuzzy queries have a linear data complexity, but let us recall that they can be used only when the preferences are commensurable. Even though outranking-based preference queries are significantly more expensive than regular selection queries (n^2 instead of n), they remain tractable (they belong to the same complexity class as self-join queries in the absence of any index).

Notice that when the result of the SQL query on which the preferences apply is small enough to fit in main memory, the extra cost is small (data complexity is then linear).

5 Conclusion

In this paper, we have proposed an alternative to the use of Pareto order for the modeling of preference queries in the case where preferences on different attributes are not commensurable. The approach we defined is based on the concept of outranking, which was initially introduced in a decision-making context (but its definition was revisited here so as to fit our purpose). Outranking makes it possible to compare tuples pairwise, and even though it does not define an order (it is not transitive), we showed how a complete preorder could be obtained by aggregating the outranking degrees in such a way that the aggregate characterizes the global "quality" of a tuple (regarding a given set of preferences) w.r.t. the others.

As perspectives for future research, we notably intend to deal with query optimization, in order to see whether some suitable techniques could reduce the data complexity associated with a "naive" evaluation of such queries (in the spirit of what has been done for skyline queries). Furthermore, it is desirable to perform a user evaluation of the approach. Still another perspective concerns the extension of the model in such a way that smooth transitions between the concepts of concordance, indifference and discordance are taken into account (as suggested in [16] in a decision-making context).

References

1. Bosc, P., Pivert, O.: SQLf: A relational database language for fuzzy querying. IEEE Trans. Fuzzy Syst. 3(1), 1–17 (1995)
2. Dubois, D., Prade, H.: Using fuzzy sets in flexible querying: Why and how? In: Proc. of FQAS 1996, pp. 89–103 (1996)
3. Bőrzsőnyi, S., Kossmann, D., Stocker, K.: The skyline operator. In: Proc. of ICDE 2001, pp. 421–430 (2001)
4. Kießling, W.: Foundations of preferences in database systems. In: Proc. of VLDB 2002, pp. 311–322 (2002)

5. Kießling, W., Köstler, G.: Preference SQL — Design, implementation, experiences. In: Proc. of VLDB 2002, pp. 990–1001 (2002)
6. Chomicki, J.: Preference formulas in relational queries. ACM Trans. Database Syst. 28(4), 427–466 (2003)
7. Agrawal, R., Wimmers, E.: A framework for expressing and combining preferences. In: SIGMOD 2000, pp. 297–306 (2000)
8. Roy, B.: The outranking approach and the foundations of ELECTRE methods. Theory and Decision 31, 49–73 (1991)
9. Bartolini, I., Ciaccia, P., Patella, M.: Efficient sort-based skyline evaluation. ACM Trans. Database Syst. 33(4), 1–49 (2008)
10. Lin, X., Yuan, Y., Zhang, Q., Zhang, Y.: Selecting stars: the k most representative skyline operator. In: ICDE 2007, pp. 86–95 (2007)
11. Chan, C., Jagadish, H., Tan, K., Tung, A., Zhang, Z.: Finding k-dominant skylines in high dimensional space. In: Proc. of SIGMOD 2006, pp. 503–514 (2006)
12. Chan, C., Jagadish, H., Tan, K., Tung, A., Zhang, Z.: On high dimensional skylines. In: Ioannidis, Y., Scholl, M.H., Schmidt, J.W., Matthes, F., Hatzopoulos, M., Böhm, K., Kemper, A., Grust, T., Böhm, C. (eds.) EDBT 2006. LNCS, vol. 3896, pp. 478–495. Springer, Heidelberg (2006)
13. Bruno, N., Chaudhuri, S., Gravano, L.: Top-k selection queries over relational databases: mapping strategies and performance evaluation. ACM Trans. Database Syst. 27(2), 153–187 (2002)
14. Boutilier, C., Brafman, R., Domshlak, C., Hoos, H., Poole, D.: CP-nets: A tool for representing and reasoning with conditional ceteris paribus preference statements. J. Artif. Intell. Res. (JAIR) 21, 135–191 (2004)
15. Brafman, R., Domshlak, C.: Database preference queries revisited. Technical report, TR2004-1934, Cornell University (2004)
16. Perny, P., Roy, B.: The use of fuzzy outranking relations in preference modelling. Fuzzy Sets and Systems 49(1), 33–53 (1992)

Incremental Membership Function Updates

Narjes Hachani, Imen Derbel, and Habib Ounelli

Faculty of Science of Tunis

Abstract. Many fuzzy applications today are based on large databases that change dynamically. Particularly, in many flexible querying systems this represents a huge problem, since changing data may lead to poor results in the absence of proper retraining. In this paper we propose a novel incremental approach to represent the membership functions describing the linguistic terms for a dynamically changing database. It exploits fuzzy knowledge models previously determined in order to simplify the modelling process. Experiments testing the method's efficiency are also reported.

1 Introduction

A fuzzy database is a database which is able to deal with uncertain or incomplete information using fuzzy logic [1997, 2006]. Basically, imprecise information in a fuzzy database is stored and/or retrieved using linguistic terms. For applications applying fuzzy database (Fuzzy querying, Fuzzy data mining) one crucial part is the proper design of the proper membership function of each linguistic term. Hence, the problem of fuzzy membership function generation is of fundamental importance [1998].The problem of modelling fuzzy knowledge from data has been widely investigated in the last decade. Earlier works focused mostly on the determination of membership functions that respect subjective perceptions about vague or imprecise concepts [1978, 1984] and summarized by Turksen [1991] under the framework of measurement theory.

Medasani [1998] provided a general overview of several methods for generating membership functions from domain data for fuzzy pattern recognition applications. However, addressing the incremental updating of data sets hasn't been reported by any work. In these approaches each update of the database, insertion or deletion of a new numeric value requires the modelling of fuzzy knowledge of the new data from scratch. This might be a huge computational workload. This problem becomes even more severe in very large databases characterized by frequent updates. In this application context, in order to speed up the query execution time, it makes sense to exploit the membership functions already generated.

In this paper, we propose a new approach for incremental and automatic generation of trapezoidal membership functions. In other words, we generate the new membership function of the new data set, starting from the fuzzy knowledge model of the previous data. The remainder of this paper is organized as follows. In section 2, we discuss the basic idea of our approach. In section 3,

E. Hüllermeier, R. Kruse, and F. Hoffmann (Eds.): IPMU 2010, Part II, CCIS 81, pp. 105–114, 2010.

we present the incremental algorithm. In section 4, we report the experimental results conducting to evaluate our algorithm. In section 5, we recall the main points of this paper.

2 Basic Idea of the Approach

2.1 Approach Requirements

The fuzzy set theory provides the most appropriate framework to model vague linguistic terms and to evaluate vague queries. Therefore, building the proper membership functions, of each linguistic term, is a key in almost all systems based on fuzzy set theory, including fuzzy databases systems. We submit the following requirements as long term goals for our model:

- *Automatic generation of membership function.* Given the quantitative data describing a linguistic term, we want to automatically identify the optimal number of clusters and the corresponding membership functions.
- *Incremental updates of membership functions after insertions.* Given membership functions that model a specified linguistic term , we want to exploit them in order to deal with the insertion of new values in the data relative to the linguistic term.
- *Incremental updates of membership functions after deletions.* Given membership functions that model a specified linguistic term , we want to exploit them in order to deal with the deletion of some values in the data relative to the linguistic term.

The First goal was the purpose of a previous paper [2008]. For the purposes of this paper, we content ourselves with the second goal. The third one will be the purpose of a future paper.

2.2 Basic Concepts

In this section, we define the basic concepts used in our approach, for detailed definitions, the reader can see papers [2007, 2008].

- The validity index DB* [2001] is a criterion measuring the quality of a partition. This index decides about the number of clusters of the fuzzy partition. The index DB* is defined as follows:

$$DB^*(nc) = \frac{1}{nc}\sum_{i=1}^{nc}\left(\frac{max_{k=1,\ldots,nc,k\neq i}\{S_i + S_k\}}{min_{l=1,\ldots,nc,l\neq i}\{d_{il}\}}\right) \quad (1)$$

  Where nc is the number of clusters, d_{il} is calculated as the distance between the centers of two clusters. Given a centroid c_i of a cluster C_i, S_i is a scatter distance. It is defined by:

$$S_i = \frac{1}{|C_i|}\sum_{x\in C_i} \parallel x - c_i \parallel \quad (2)$$

- Let x_i and x_j be two vertices of a cluster C. x_j is a direct neighbor of x_i if it exists an edge connecting x_i and x_j.
- A density function of a vertex $x_i \in C$ is defined by the following expression [2004]:

$$De(x_i) = \frac{Diam_C - \frac{1}{Dg(x_i)} \sum_{x_j \in V(x_i)} d(x_i, x_j)}{Diam_C} \tag{3}$$

  Where $V(x_i)$ is the set of direct neighbors of a vertex x_i, $Dg(x_i)$ is the cardinality of $V(x_i)$, $d(x_i, x_j)$ is the distance between x_i and x_j and $Diam_C$ is the diameter of C and it is defined as the maximum distance between two cluster objects.
  $De(x_i)$ has a high value when the elements of $V(x_i)$ are close to x_i.
- The density threshold is defined as follows:

$$thresh = \frac{(Dmin_C + Dmax_C)}{2}$$

  Where $Dmin_C$ represents the minimal density in C and $Dmax_C$ is the maximal density in C.
- A dense vertex of a cluster C is an object having a density value greater than the density's threshold of C.

3 Handling with Incremental Insertion

In this section, we start by introducing the characteristics of a consistent partition. Then, we present the algorithm for incremental updates of membership functions parameters after inserting data in the context of dynamic databases. We recall first that according to our clustering method [2007], a cluster C can be defined as a subgraph (X, E) where X is a set of vertices and E is a set of edges.

3.1 A Consistent Partition

Let CK be a partition resulted from the clustering algorithm $CLUSTERDB^*$. We suppose that CK is composed of k clusters $CK = \{C_1, C_2, ..., C_k\}$ and suppose that each cluster C_i contains the objects x_i^j, such that $j = 1..|C_i|$. We define a consistent partition as a partition that satisfies the following properties:

1. **Property 1 (P1):** The objects of two neighboring clusters C_i and C_{i+1}, $\forall i \in [1, k-1]$ are contiguous. In other words $\forall i \in [1, k-1]$ $x_i^{|Ci|} < x_{i+1}^1$
2. **Property 2 (P2):** if two objects x_i^j and x_{i+1}^k belong to two distinct clusters respectively C_i and C_{i+1}, then $d(x_i^j, x_{i+1}^k) > d(x_i^j, x_i^l)$ $\forall l \in [1, |C_i|], l \neq i$. Where $d(x, y)$ is the distance between two objects x and y. This property can also be expressed by the two equations:

$$\forall i \in [1, k-1] D_{C_iC_{i+1}} > max\{d(x_i^j, x_i^{j+1}) \forall j \in [1, (|C_i| - 1)]\} \tag{4}$$

$$D_{C_iC_{i+1}} > max\{d(x_{i+1}^j, x_{i+1}^{j+1}) \forall j \in [1, (|C_{i+1}| - 1)]\} \tag{5}$$

Where $D_{C_iC_{i+1}}$ is the distance between two clusters.

We consider that we want to insert an object p into a database D. Starting from a consistent initial partition CK and the corresponding trapezoidal membership functions (TMF), we propose to determine the necessary changes on this partition as well as on TMF. To this end, the problem is summarized in two stages:

1. determine the appropriate cluster for p
2. determine the new membership functions parameters.

To meet such needs, we proceed incrementally. The incremental aspect of the proposed approach lies in the fact that following the insertion of an object, initially generated clusters and their membership functions will evolve according to the new element.

3.2 Identification of the Appropriate Cluster

The main objective of this step is to determine the cluster associated to the new inserted object p while maintaining the consistency of the partition. Indeed, the resulting partition must be consistent so that it verifies properties P1 and P2. We note that we insert p so that P1 is satisfied. According to the value of p, we distinguish three cases:

First case: The value of p is between the lower and the upper bound of a cluster $C_i, i \in [1, k]$.

Second case: The value of p is equidistant from the upper bound of a cluster C_i and the lower bound of a cluster $C_{i+1}, i \in [1, k-1]$.

Third case: The value of p is between the upper bound of the cluster C_i and the lower bound of a cluster C_{i+1} and the distance between p and C_i is different from that between p and C_{i+1}.

The problem amounts to identify the cluster which will contain p so that we obtain a partition verifying the property P2. We consider the following rules associated to the cases already enumerated:

Rule 1: The value of p is between lower and upper bound of a cluster C_i. In this case, if we suppose that p belongs to the cluster $C_i, i \in [1, k]$, then p will be attributed to the cluster C_i such that $x_i^j < p < x_i^{j+1}$ with $j \in [1, (|C_i| - 1)]$.

Justification: The partition obtained after the insertion of p in $C_i, i \in [1, k]$ verify the property P2 for two raisons. First, the insertion of p does not affect the inter-cluster distances between C_i and the direct neighboring clusters C_{i-1} and C_{i+1}, if they exist. Furthermore, if the maximum distance in the cluster C_i changes, then it will decrease. Consequently, the equations 4 and 5of the subsection 3.1 remain available.

Rule 2: Let suppose that the value of p is equidistant of the upper bound of a cluster C_i and the lower bound of a cluster $C_{i+1}, i \in [1, k-1]$. In such case, we should re-apply the clustering to the whole data set using the algorithm CLUSTERDB*. Figure 1 illustrates this rule.

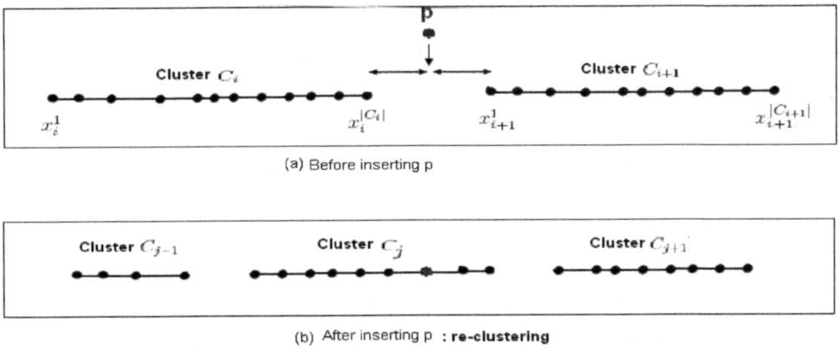

Fig. 1. Illustration of the second case of insertion

Justification: The partition resultant of the insertion of p in the cluster C_i or in the cluster C_{i+1} does not confirm the property P2. If we suppose that p belong to C_i, then p is equidistant from his left neighbor in the cluster C_i and the first object of the cluster C_{i+1}. Thus, the property $P2$ is not confirmed.

Rule 3: Let suppose that the value of p is between the upper bound of a cluster cluster C_i and the lower bound of a cluster C_{i+1} and the distance separating p from C_i is different from that separating it from C_{i+1}, then we check if the partition, resultant of the insertion of p in the nearest cluster (C_i or C_{i+1}), will confirm the property P2. If P2 is satisfied, then the partition is coherent (figure 2-b1), else the resultant partition is not consistent and we propose to re-apply the clustering (figure 2-b2).

Justification: Let suppose that p is assigned to the most distant cluster, let C_{i+1}. The distance between p and its right neighbor in C_{i+1} is then

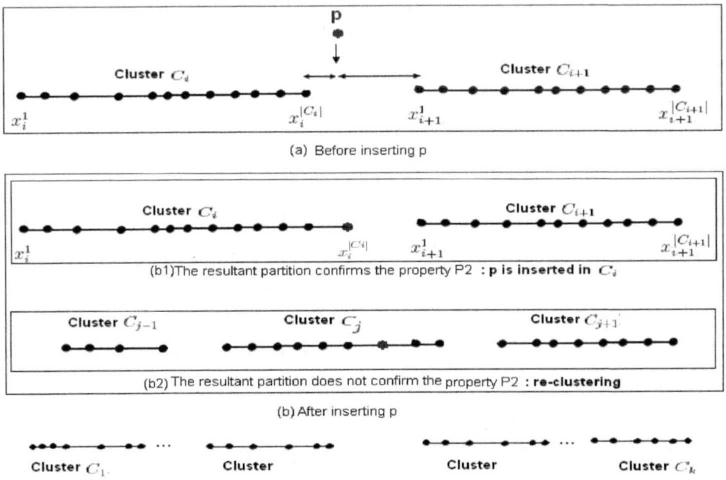

Fig. 2. Illustration of the third case of insertion

greater than the distance separating the two clusters C_i et C_{i+1}. This contradicts the property P2. Therefore the inclusion of p at the most distant cluster led always to an inconsistent partition.

3.3 Incremental Generation of Membership Functions

Inserting a new object p in the appropriate cluster C_i, will disturb the membership function parameters of C_i and those of clusters C_{i-1} and C_{i+1}. Indeed, some elements in the neighborhood of p will have their density changed. This change of density value may induce in turn not only an adjust in the core of the membership function related to C_i but also a change in the supports of clusters C_{i-1} and C_{i+1}. The algorithm 1, named *Coreupdating*, describes the changes that may occur on the core of the cluster C_i after the insertion of p. We submit the following notations:

- C: the cluster including the inserted object p.
- Nc: the centroid of C after insertion of p.
- $Oinf$, $Osup$: the lower bound and the upper bound of the core before inserting P.
- $Ninf$, $Nsup$: the lower bound and the upper bound of the core after inserting p.
- *Othresh*, *Nthresh*: the density threshold before and after the insertion of p.
- $LOinf$: the left direct neighbor of $Oinf$.
- *ROsup*: the right direct neighbor of $Osup$.

Coreupdating returns the new bounds of the core associated to a cluster C. Indeed, when inserting a new object p, updates depends on threshold's density value, centroid's position and the position of the inserted object p. In case the cluster's centroid remains in the old core of the cluster, then the new core will either be extended or reduced depending on threshold's density. Hence, one of the three cases may occur:

1. The threshold is constant. In such case, the extension of the core is performed only in two cases:
   - p is the direct left neighbor of $Oinf$. In this case, the function Newlbound(C, p) extends the core with p and $LOinf$ if they are dense.
   - p is the direct right neighbor of $Osup$. In this case, the function NewRbound(C, p) extends the core with p and $ROsup$ if they are dense.
2. The threshold decreases. Our algorithm is based on the functions *Lneighbors* and *Rneighbors* to update the core. The function *Lneighbors* allows to identify the dense objects at the left neighborhood of $Oinf$. Thus, it determines the new lower bound of the core. As the same, *Rneighbors* searches the dense objects at the right neighborhood of $Osup$.
3. The threshold increases. In such case, we can not compute the new core based on the old core. Thus, the core generation algorithm is re-applied.

Algorithm 1. Coreupdating
Input: the cluster C, Nc, p, $Oinf$, $Osup$
Output: $Ninf$,$Nsup$

```
begin
    Ninf ← Oinf
    Nsup ← Osup
    if Nc ∈ [Oinf, Osup] then
        if Nthresh = Othresh then
            if p not ∈ [Oinf, Osup]  then
                if p = LOinf then
                    Ninf ← Newlbound(C, p)
                else
                    if p = ROsup then
                        Nsup ← Newupbound(C,p)
        else
            if Nthresh < Othresh then
                Ninf ← Lneighbors(C, Oinf, Nthresh)
                Nsup ← Rneighbors(C, Osup, Nthresh)
            else
                Coregeneration(C)
    else
        Coregeneration(C)
end
```

4 Experiments

4.1 Experimental Setup

As part of our experiments, we conducted tests on the data set *Books* taken from the web site "www.amazon.com" and on data sets *Census Income* and *Hypothyroid* issued from the UCI Machine Learning Repository [2009]. All these data sets are labelled and contain only numeric data. Below we present a description of each of these data sets in the number of objects and clusters.

1. *Books* includes over 400 prices of books. It contains two clusters.
2. *Census Income* DB [2009] includes 763 objects. We are interested in the value of age attribute which allows to identify three clusters.
3. *Hypthyroid* DB [2009] includes 1000 objects. We are interested in the values of the TSH which allows to identify two clusters.

4.2 Experiments on Incremental Generation of MF

The following experiments were conducted to validate our approach of the incremental integration of elements in data sets. Indeed, these tests involve inserting successively the elements listed in the tables below in the order of their

Table 1. Updates after insertion in *Books*

p	Cluster	InitialCore	FinalCore	MF.Parameters
87	$C2$	134.5..135	115..165	$C1 : 15.61, 36, 115$ $C2 : 36, 115, 165$
70	$C1$	15.61..36	15..55	$C1 : 15, 55, 155$ $C2 : 55, 115, 165$

Table 2. Re-clustering after insertion in *Books*

p	Partition 1	DB^*1	Partition 2	DB^*2
80	$C1 : [15, 70]$ $C2 : [80, 165]$	0.253	$C1 : [15, 87]$ $C2 : [110, 165]$	0.232

Table 3. Updates after insertion in *CensusIncome*

p	Cluster	InitialCore	FinalCore	MF.Parameters
77	$C1$	1..76	1..78	$C1 : 1, 78, 83$ $C2 : 78, 83, 87, 89$ $C3 : 87, 89, 90$
87	$C2$	83..86	83..87	$C1 : 1, 76, 83$ $C2 : 76, 83, 87, 89$ $C3 : 87, 89, 90$

presentation. While the results in tables 1, 3 and 5 are relative to the case of inserting elements in the initial partition, tables 2, 4 and 6 illustrate the case of re-clustering. Tables 1, 3 and 5 present the parameters of Membership Functions (MF.Parameters) after the insertion of a new element. Tables 2, 4 and 6 present the results of inserting the value (p) indicating the resultant partition, its quality evaluated using the validity index DB^* before the re-clustering (Partition 1, DB^*1) and after the re-clustering (Partition 2, DB^*2).

To judge the usefulness of the decision re-clustering, we propose to compare the qualities of two partitions. One obtained after the insertion of p in the current partition and the other is the result of the re-application of clustering on the new data set. In case the second partition is better, we can say that re-clustering is the right choice. To make this comparison, we proceed as follows:

1. Inserting p in the current partition in order to obtain a partition P1. Then, we ought to determine the cluster that will receive p. Since the case of re-clustering can arise when the position is p between the upper end of a cluster C_i and the lower end of C_{i+1}, $i \in [1, k-1]$, we distinguish the following cases:
   - p is equidistant of C_i and C_{i+1} ,$i \in [1, k-1]$: we use the silhouette index defined in [2001]in order to choose the cluster that will contain p. Let's recall that a positive value of the index silhouette indicates that p is well placed while a negative value of this index shows that p is misplaced

Table 4. Reclustering after insertion in *CensusIncome*

p	Partition 1	DB^*1	Partition 2	DB^*2
88	$C1 : [1, 78]$ $C2 : [81, 87]$ $C3 : [88, 90]$	3.145	$C1 : [1, 78]$ $C2 : [81, 90]$	0.470

Table 5. Updates after insertion in *Hypothyroid*

p	Cluster	InitialCore	FinalCore	MF.Parameters
50	$C1$	0.005..18	0.005..27	$C1 : 0.005, 27, 143$ $C2 : 27, 143, 199$
100	$C2$	143..160	143..199	$C1 : 0.005, 27, 143$ $C2 : 27, 143, 199$

Table 6. Reclustering after insertion in *Hypothyroid*

p	Partition 1	DB^*1	Partition 2	DB^*2
85	$[0.005, 50]$ $[85, 199]$	0.267	$[0.005, 100]$ $[143, 199]$	0.157

and should be assigned to the nearest cluster. The appropriate cluster allowing to have a positive value of the index silhouette is the one in which the average distance between p and all its elements is minimal.
   - p is closer to one of the clusters: assign p to the nearest cluster.
2. Compute the index of validity DB^* (DB^*1) of the new partition $P1$.
3. Repeat the clustering on the new data set and we compute the index DB* (DB^*2) of the obtained partition $P2$.
4. Compare the two values DB^*1 and DB^*2. The lower index of validity is related to the best partition. In the case P2 is the best partition, we affirm that the decision of re-clustering is adequate.

Results shown in tables 2, 4 and 6 are interesting on two levels. We note first that the re-clustering has generated a new partition with new clusters. Then, we notice that in all cases presented, the value of validity index DB^*2 is lower than DB^*1. Hence, the quality of the partition $P2$, resulting from re-clustering, is much better than the one of the partition $P1$ obtained after inserting p in the initial partition. Accordingly, we conclude the usefulness of re-clustering on the quality of the partition. Tables 1, 3 and 5 were interested in the case of inserting the object in the partition requires a re-adjustment of the core of membership functions. The new cores are generated incrementally. They are defined in most cases by extending the initial core.

5 Conclusion

Methods aiming to generate automatically membership functions are attractive. However, in these approaches each insertion of a new numeric value in the database requires the modelling of fuzzy knowledge of the new data from scratch. This might be a huge computational workload. In this paper we proposed a novel incremental approach to updating membership functions after insertion of new values. The application of this approach with very large databases remains a main point for future work.

References

[1978] MacVicar-Whelan, P.J.: Fuzzy sets, the concept of height, and the hedge VERY. IEEE Trans. Syst. Man Cybern. 8, 507–511 (1978)

[1984] Norwich, A.M., Turksen, I.B.: Model for the measurement of membership and the consequences of its empirical implementation. Int. J. Fuzzy Sets and Syst. 12, 1–25 (1984)

[1991] Turksen, I.B.: Measurement of membership functions and their acquisition. Int. J. Fuzzy Sets and Syst. 40, 5–38 (1991)

[1997] Dubois, D., Prade, H.: Using fuzzy sets in flexible querying: why and how? Flexible Query Answering Systems, 45–60 (1997)

[1998] Medasani, S., Kim, J., Krishnapuram, R.: An overview of Membership Function Generation Techniques for Pattern Recognition. Int. J. Approx. Reason. 19, 391–417 (1998)

[2001] Halkidi, M., Batistakis, Y., Vazirgiannis, M.: On clustering validation techniques. J. Int. Inf. Syst., 107–145 (2001)

[2004] Guénoche, A.: Clustering by vertex density in the graph. In: Proceeding of IFCS congress classification, pp. 15–24 (2004)

[2006] Galindo, J., Urrutia, A., Piattini, M.: Fuzzy Databases: Modeling, Design, and Implementation, vol. 12, pp. 1–25. IGI Publishing, Hershey (2006)

[2007] Hachani, N., Ounelli, H.: Improving Cluster Method Quality by Validity Indices. In: Flairs Conference, pp. 479–483 (2007)

[2008] Derbel, I., Hachani, N., Ounelli, H.: Membership Functions Generation Based on Density Function. In: International Conference on Computational Intelligence and Security, pp. 96–101 (2008)

[2009] Blake, C., Merz, C.: UCI repository of machine learning databases, http://www.ics.uci.edu/~mlearn/MLRepository.html

A New Approach for Comparing Fuzzy Objects

Yasmina Bashon, Daniel Neagu, and Mick J. Ridley

Department of Computing, University of Bradford,
Bradford, BD7 1DP, UK
{Y.Bashon,D.Neagu,M.J.Ridley}@bradford.ac.uk

Abstract. One of the most important issues in fuzzy databases is how to manage the occurrence of vagueness, imprecision and uncertainty. Appropriate similarity measures are necessary to find objects which are close to other given fuzzy objects or used in a user vague query. Such similarity measures could be also utilized in fuzzy database or even classical relational database modeling. In this paper we propose a new family of similarity measures in order to compare two fuzzy objects described with fuzzy attributes. This is done by comparing the corresponding attributes of the objects using a generalization of Euclidean distance measure into fuzzy sets. The comparison is achieved for two cases: fuzzy attribute/fuzzy attribute comparison and crisp attribute/fuzzy attribute comparison. Each case is examined with experimental examples.

Keywords: Similarity measures, fuzzy objects, fuzzy attributes, crisp attributes, fuzzy database.

1 Introduction

Similarity is a comparison measure that is widely known and commonly used, but it is still increasingly difficult to assess and quantify. Assessing the degree to which two objects in a query statement are similar or compatible is an essential factor of human reasoning. Consequently, the assessment of similarity has become important in the development of classification, information retrieval and decision systems [1].

For object comparison, three different types of data can be considered to describe an object: numerical (e.g. *175 cm tall*), categorical (e.g. colour: *white, black*) and attributes with fuzzy values (e.g. age: *young, middle-aged, old*). We can call the latter examples fuzzy attributes. In this paper the third attribute type is addressed in depth. Accordingly, the attributes of any object can be classified as follows:

1. Attributes with crisp (or numerical) values. Each attribute has a well-defined basic domain (called the universe of discourse) $U \subseteq \mathbb{R}$ which is usually a bounded set.
2. Categorical attributes: take values in a discrete and fixed set of linguistic terms.
3. Fuzzy attributes (or attributes with fuzzy values). Each attribute value is a set of linguistic labels, which can be easily described by a set $F(U) = \{A|A \subseteq U\}$ of fuzzy subsets defined over the basic domain U. Each fuzzy subset $A \in F(U)$ is characterized by a membership function $\mu_A: U \to [0,1]$. There are many different ways to represent the semantic of the fuzzy labels.

E. Hüllermeier, R. Kruse, and F. Hoffmann (Eds.): IPMU 2010, Part I, CCIS 81, pp. 115–125, 2010.

Distance and similarity measures are well defined for numerical data, and also, there exist some extensions to categorical data [2], [3]. However, it is possible sometimes to have additional semantic information about the domain. Computing with words (e.g. dealing with both numerical and symbolic attributes) especially in complex domains adds a more natural facet to modern data processing [4].

In this paper, we propose a family of similarity measures using the geometric distance metric and their application for fuzzy objects comparison. We focus on the study of the object comparison problem by offering both an abstract analysis and a simple and clear method to use our theoretical results in practice. We consider hereby the objects that are described with both crisp and fuzzy attributes. We define the semantics of attribute values by means of fuzzy sets, and propose a similarity measure of the corresponding attributes of the fuzzy objects. Then the overall similarity between two fuzzy objects is calculated in two different ways (weighted average and minimum) to finally decide on how similar two fuzzy objects are.

The organization of the paper is as follows. In section 2 we discuss related work, and introduce some other similarity measures proposed by various authors. In section 3 the motivation of our research is discussed. In section 4 we define our proposed similarity measures and give some experimental examples for the similarity between fuzzy objects with fuzzy attributes. Section 5 explains the case of crisp attribute/fuzzy attribute comparison. Finally, conclusions and further work are provided in section 6.

2 Related Work

There are several ways in which fuzzy set theory can be applied to deal with imprecision and uncertainty in databases [5]. Using similarity and dissimilarity measurement for comparing fuzzy objects is the approach we present in this paper. This approach and many different approaches are based on comparing fuzzy sets that are defined using membership functions. In this section, we will review some of them.

George et al. [6] proposed in early 1990s a generalization of the equality relationship with a similarity relationship to model attribute value imperfections. This work was continued by Koyuncu and Yazici in IFOOD, an intelligent fuzzy object-oriented data model, developed using a similarity-based object-oriented approach [7].

Possibility and necessity measures are two fuzzy measures proposed by Prade and Testemale [8]. This is one of the most popular approaches, and is based on possibility theory. Each possibility measure associated with a necessity measure is used to express the degree of uncertainty to whether a data item satisfies a query condition.

Semantic measures of fuzzy data in extended possibility-based fuzzy relational databases have been presented by Zongmin Ma et al. in [9]. They introduced the notions of semantic space and semantic inclusion degree of fuzzy data, where fuzzy data is represented by possibility distributions. Marin et al. [10] have proposed a resemblance generalized relation to compare two sets of fuzzy objects. This relation that recursively compares the elements in the fuzzy sets has been used by Berzal et al. [4], [11] with a framework proposed to build fuzzy object-oriented capabilities over an existing database system. Hallez et al. have presented in [12] a theoretical framework for constructing an objects comparison schema. Their hierarchical approach aimed to be able to choose appropriate operators and an evaluation domain in which the comparison results are expressed. Rami Zwick et al. have briefly reviewed some of these measures and compared their performances in [13]. In [14] and [15] the authors continue these studies, focused on fuzzy database models.

Geometric models dominate the theoretical analysis of similarity relations [16]. Objects in these models are represented as points in a coordinate space, and the metric distance between the respective points is considered to be the dissimilarity among objects. The Euclidean distance is used to define the dissimilarity between two concepts or objects. The same distance approach has been adopted in this paper. However, in our proposal we consider the problem of fuzzy object comparison where the Euclidean distance is applied to fuzzy sets, rather than just to points in a space.

3 Motivation

To introduce the motivation behind our research, we consider the following example: a student is looking to book a room and wants to compare the existing rooms in order to choose the most suitable one. Every room is described by its quality, price, as well as the distance to the University (DTU). He found two rooms as described in Fig. 1. He asks himself "*How can one compare these two rooms?*"

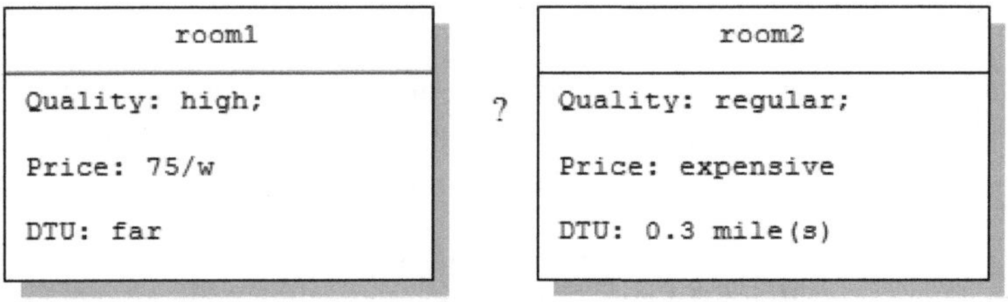

Fig. 1. A case study of fuzzy objects comparison

As shown in Fig. 1, the description of the two rooms is imprecise and mixed, since their features or attributes are expressed using linguistic labels as well as numerical values. In other words, `room1` and `room2` are fuzzy objects of the class `Room` (that is; at least one of their attributes has a fuzzy value). We represent firstly any value in a fuzzy format to be consistent and also because, when fuzzy sets describe the linguistic labels of a particular domain and their membership functions are defined, the terms can be compared in a more accurate way.

Before introducing the similarity measure to compare two fuzzy objects (e.g. two rooms as described in Fig. 1), we should be able to:

1. define the basic domain for each fuzzy attribute.
2. define the semantics of the linguistic labels by using fuzzy sets (or fuzzy terms which are characterized by membership functions) built over the basic domains.
3. calculate the similarity among the corresponding attributes; then
4. aggregate or calculate the average over all similarities in order to give the final judgement on how similar the two objects (rooms) are.

When comparing two fuzzy objects, we consider the following cases:

Case I: the comparison of two fuzzy attributes; and
Case II: the comparison of a crisp attribute with a fuzzy one and vice versa.

4 Fuzzy Similarity Measures

In this section we address *Case I*, and also explain in detail our methodology for comparing objects described with fuzzy attributes. Initially we define the similarity between two corresponding fuzzy attributes of the fuzzy objects and then we calculate the overall similarity between two fuzzy objects using two different definitions (7) and (8). Fig. 2 illustrates the way of calculating similarity between two fuzzy objects.

Let o_1 and o_2 be any two objects with sets of attributes $at_{o_1} = \{a_1, a_2, \dots, a_n\}$ and $at_{o_2} = \{b_1, b_2, \dots, b_n\}$, respectively. The similarity $S\colon at_{o_1} \times at_{o_2} \to [0,1]$ between any two corresponding attributes a_j , b_j is defined as:

$$S(a_j , b_j) = \frac{1 - d(a_j , b_j)}{1 + k_j d(a_j , b_j)}; \qquad \text{for some } k_j \geq 0 \tag{1}$$

where $j = 1,2, \dots, n$, n stands for the number of attributes, and the distance metric d is represented by a mapping $\oplus_j\colon [0,1]^{m_j} \to [0,1]$ as follows:

$$d(a_j , b_j) = \oplus_j \left(dis(A_{1j}, B_{1j}), dis(A_{2j}, B_{2j}), \dots, dis\left(A_{m_j j}, B_{m_j j}\right) \right) \tag{2}$$

where m_j stands for the number of fuzzy sets that represent the value of jth attribute over basic domain U_j. $\oplus_j$ can be defined by the generalization of the Euclidean distance metric into fuzzy subsets divided by the number of fuzzy sets m_j:

$$d(a_j , b_j) = \left[\frac{\sum_{i=1}^{m_j} dis(A_{ij}, B_{ij})^2}{m_j} \right]^{1/2} \tag{3}$$

The distance metric $dis\colon F(U_j) \times F(U_j) \to [0,1]$ describes the dissimilarity among fuzzy sets and it can be defined in the following two cases:

a) if attributes a_j and b_j are characterized by linguistic labels having their semantics defined using fuzzy sets represented by the same membership function (i.e. $\mu_{A_{ij}}(x) = \mu_{B_{ij}}(x)$ for all $x \epsilon U_j$), for example comparing two student rooms in the same country say UK (see Fig. 4 in example 1 below for more clarity), then:

$$dis(A_{ij}, B_{ij}) = \left| \mu_{A_{ij}}(x) - \mu_{A_{ij}}(y) \right|; \text{for any } x, y \epsilon U_j \tag{4}$$

b) if the attributes a_j , and b_j are characterized by linguistic labels represented by different membership functions $\mu_{A_{ij}}(x)$, $\mu_{B_{ij}}(x)$ respectively, e.g. comparing a student room in UK with a student room in Italy (see Fig. 6 in example 2 below) then:

$$dis(A_{ij}, B_{ij}) = \left| \mu_{A_{ij}}(x) - \mu_{B_{ij}}(y) \right|; \text{ for any } x, y \epsilon U_j \tag{5}$$

The proposed similarity definition in equation (1) guarantees normalization and allows us to determine to which extent the attributes of two objects are similar.

Parameter $k_j$in eq (1) is used for tuning the similarity by adjusting the contribution of distance d in the similarity measure. As a consequence, k_j can be calculated in terms of the distance d according to the user application or can be estimated.

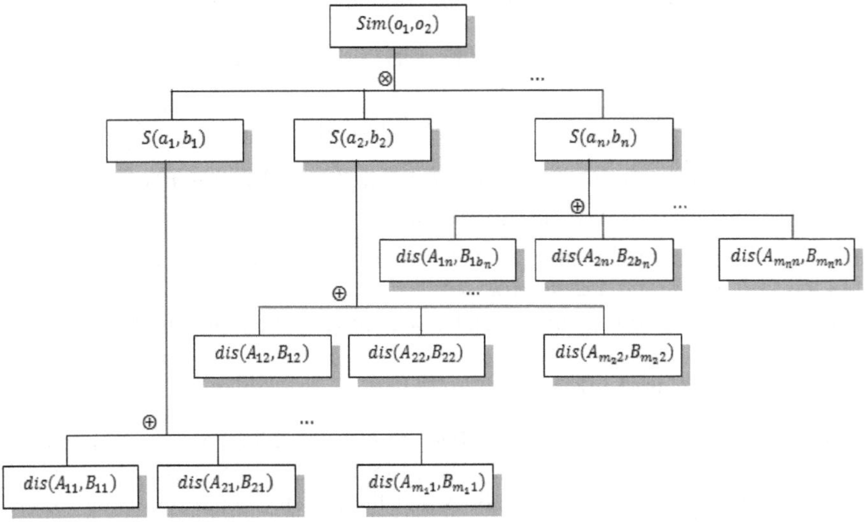

Fig. 2. Calculating similarity between two fuzzy objects o_1 and o_2

We define similarity $Sim(o_1, o_2)$ between the two fuzzy objects o_1 and o_2 as:

$$Sim(o_1, o_2) = \otimes\big(S(a_1, b_1), S(a_2, b_2), \dots, S(a_n, b_n)\big) \tag{6}$$

where the mapping $\otimes: [0,1]^n \to [0,1]$ is defined as an aggregation operator such as the weighted average or the minimum function:

1) the weighted average of the similarities of attributes:

$$\otimes\big(S(a_1, b_1), S(a_2, b_2), \dots, S(a_n, b_n)\big) = \frac{\sum_{j=1}^{n} \alpha_j S(a_j, b_j)}{\sum_{j=1}^{n} \alpha_j}; \quad \alpha_j \epsilon [0,1] \tag{7}$$

2) the minimum of the similarities of attributes:

$$\otimes\big(S(a_1, b_1), S(a_2, b_2), \dots, S(a_n, b_n)\big) = \min[S(a_1, b_1), S(a_2, b_2), \dots, S(a_n, b_n)] \tag{8}$$

An assessment for our similarity approach is provided below. The justification of similarity measures will help us to guarantee that our model respects the main properties of similarity measures between two fuzzy objects. In the following paragraphs we examine some metric properties satisfied by our similarity measure:

Proposition: The definition of similarity $Sim(o_1, o_2)$ between the two fuzzy jects o_1 and o_2 (as in eqs 7, 8) satisfies the following properties:

a) Reflexivity: the similarity between an object o_1 to itself is equal to one: $Sim(o_1, o_1) = 1$, for any object o_1.
b) The similarity between two different objects o_1 *and* o_2 must be less than the similarity between the object o_1 and itself: $Sim(o_1, o_2) \leq Sim(o_1, o_1)$.
c) Symmetry: $Sim(o_1, o_2) = Sim(o_2, o_1)$, for any two fuzzy objects o_1 and o_2.

Proof: Since $dis(A_{ij}, A_{ij}) = \left|\mu_{A_{ij}}(x) - \mu_{A_{ij}}(x)\right|$ for all $x \epsilon U_j$, $i = 1,2, \ldots, m_j$, then: $d(a_j, a_j) = \left[\frac{\sum_{i=1}^{m_j} dis(A_{ij}, B_{ij})^2}{m_j}\right]^{1/2} = 0$. Thus, $S(a_j, a_j) = \frac{1-d(a_j,a_j)}{1+k_j d(a_j,a_j)} = \frac{1-0}{1+k_j(0)} = 1$. Therefore we get a). From a) since for $a_j \neq b_j, d(a_j, b_j) > 0$ implies $S(a_j, b_j) < 1$. Hence b) is true. Since $dis(A_{ij}, B_{ij}) = \left|\mu_{A_{ij}}(x) - \mu_{B_{ij}}(y)\right| = \left|\mu_{B_{ij}}(y) - \mu_{A_{ij}}(x)\right| = dis(B_{ij}, A_{ij})$ for $x, y \epsilon U_j$, then $d(a_j, b_j) = d(b_j, a_j)$ and thus $S(a_j, b_j) = S(b_j, a_j)$. Consequently $Sim(o_1, o_2) = Sim(o_2, o_1)$.

The two cases mentioned above can be illustrated by the following examples:

4.1 Example 1: Case I (a)

Let us consider two rooms in the UK. Each room is described by its quality and price as shown in Fig. 3. In order to know how similar the two rooms are, we will first measure similarity between the qualities and the prices of both rooms by following the previous procedure. Let us define basic quality domain $D_Q = [0,1]$ of each room as the interval of degrees between 0 and 1. We can determine a fuzzy domain of room quality by defining fuzzy subsets (linguistic labels) $FD_Q = \{\text{Low}, \text{Regular}, \text{High}\}$. over the basic domain$D_Q$. Here we assumed only three fuzzy subsets ($m_j = 3$).

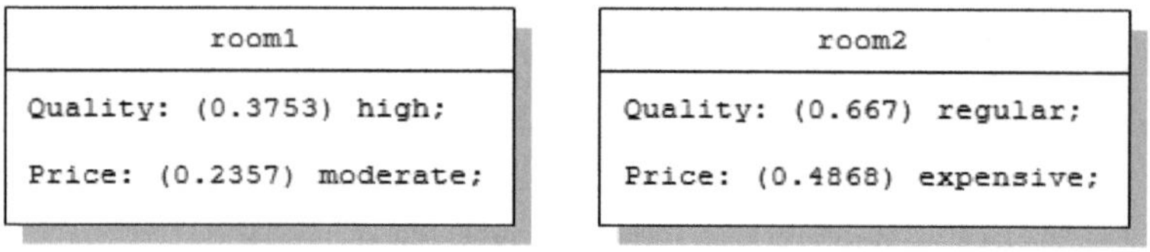

Fig. 3. Case I (a) comparing two rooms in UK

Accordingly, quality of `room1` and quality of `room2` are respectively defined as:

$$Q(room1) = \{0.0/\text{Low}, 0.198/\text{Regular}, 0.375/\text{High}\}$$

$$Q(room2) = \{0.0497/\text{Low}, 0.667/\text{Regular}, 0.0/\text{High}\}$$

using the membership functions shown in Fig. 4. Now similarity between these attributes can be measured by:

$$d(a_j, b_j) = \left[\frac{\sum_{i=1}^{m_j}\left|\mu_{A_{ij}}(x) - \mu_{A_{ij}}(y)\right|^2}{m_j}\right]^{1/2} \quad ; x, y \in D_Q \qquad (9)$$

Let attributes $Q(room1)$ and $Q(room2)$ be denoted by a_1 and b_1, respectively, and let A_{11}, A_{21}, and A_{31} stand for Low, Regular, and High, respectively. Thus we have:

$$d(a_1, b_1) = \left[\frac{\left|\mu_{A_{11}}(x) - \mu_{A_{11}}(y)\right|^2 + \left|\mu_{A_{21}}(x) - \mu_{A_{21}}(y)\right|^2 + \left|\mu_{A_{31}}(x) - \mu_{A_{31}}(y)\right|^2}{3}\right]^{1/2}$$

$$= \left[\frac{|0.0 - 0.0497|^2 + |0.1979 - 0.667|^2 + |0.3753 - 0.0000|^2}{3}\right]^{1/2} \cong 0.35$$

Hence, the similarity between a_1 and b_1: $S(a_1, b_1) = \frac{1-0.3480}{1+k_1(0.3480)}$; for some $k_1 \geq 0$.

We can get different similarity measures between the attributes, by assuming different values for k_1, for example, when $k_1 = 1$, we get: $S(a_1, b_1) \cong 0.4836$ and when $k_1 = 2$, we get: $S(a_1, b_1) \cong 0.3844$. Similarly, we can measure similarity between the prices of the two rooms. Let $D_P = [0,600]$. The fuzzy domain $FD_P = \{\text{Cheap}, \text{Moderate}, \text{Expensive}\}$. The prices for *room1* and *room2* are respectively:

$$P(room1) = \{0.2353/\text{Cheap}, 0.726/\text{Moderate}, 0.0169/\text{Expensive}\}$$

$$P(room2) = \{0.0/\text{Cheap}, 0.2353/\text{Moderate}, 0.4868/\text{Expensive}\}$$

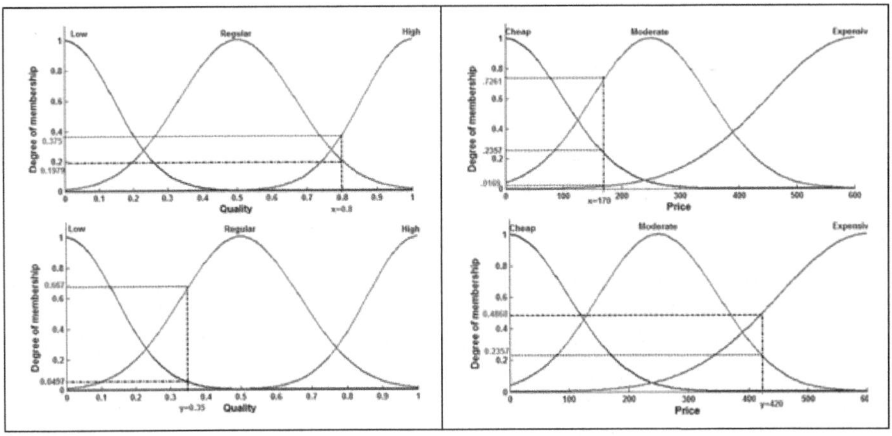

Fig. 4. Case I (a) Fuzzy representation of quality and price of two rooms in UK (using the same membership function)

Let $P(room1)$ and $P(room2)$ be denoted by the attributes a_2 and b_2 respectively. Let also A_{12}, A_{22}, and A_{32} stand for `Cheap`, `Moderate`, and `Expensive`, respectively (Fig. 4 shows a fuzzy representation of qualities and prices for the two rooms). Distance $d(a_2, b_2) \cong 0.4151$. For $k_2 = 1$, we get: $S(a_2, b_2) \cong 0.4133$, and when $k_2 = 2$, we get: $S(a_2, b_2) \cong 0.3196$. Thus, the overall similarity $Sim(o_1, o_2) = Sim(room1, room2) = \otimes\big(S(a_1, b_1), S(a_2, b_2)\big)$ is calculated as:

1) the weighted average of the similarities of attributes:
Let us assume that $\alpha_1 = 0.5$ and $\alpha_2 = 0.8$. When $k_1 = k_2 = 1$ we get:

$$Sim(o_1, o_2) = \frac{\sum_{j=1}^{2} \alpha_j S(a_j, b_j)}{\sum_{j=1}^{2} \alpha_j} = \frac{0.5 * S(a_1, b_1) + 0.8 * S(a_2, b_2)}{0.5 + 0.8} \cong 0.4403$$

and when $k_1 = k_2 = 2$: $Sim(o_1, o_2) \cong 0.3445$; or

2) the minimum of the similarities of attributes:
When $k_1 = k_2 = 1$ we get: $Sim(o_1, o_2) = \min_{n=2}[0.4836, 0.4133] = 0.4133$, and when $k_1 = k_2 = 2$ we get: $Sim(o_1, o_2) = 0.3196$.

4.2 Example 2: Case I (b)

In the case of comparing a room in UK with a room in Italy as described in Fig. 5, e.g. when the membership functions of fuzzy sets are different, we can use:

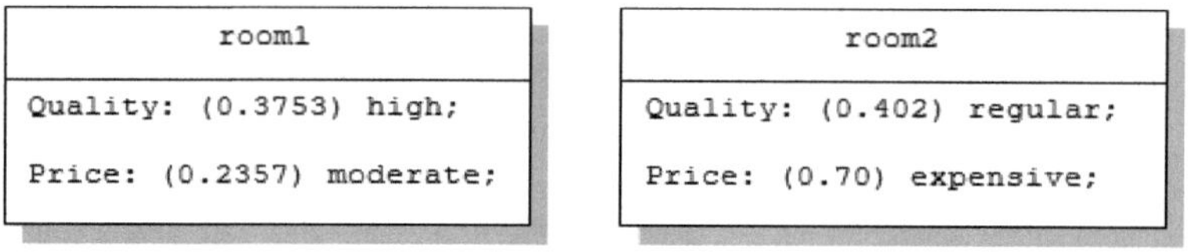

Fig. 5. Case I (b) Comparing a room in UK with a room in Italy

$$d(a_j, b_j) = \left[\frac{\sum_{i=1}^{m_j} \left| \mu_{A_{ij}}(x) - \mu_{B_{ij}}(y) \right|^2}{m_j} \right]^{1/2} \quad ; x, y \in D \tag{10}$$

Let A_{11}, B_{11} stand for *Low*, A_{21}, B_{21} stand for *Regular*, and A_{31}, B_{31} stand for *High*, respectively. Thus $d(a_1, b_1) \cong 0.2469$. Hence, the similarity between a_1 and b_1 when $k_1 = 1$, is: $S(a_1, b_1) \cong 0.6039$, and we get: $S(a_1, b_1) \cong 0.5041$ when $k_1 = 2$. We can also compare the prices a_2 and b_2 by the same way, where A_{12}, B_{12} stand for *Cheap*, A_{22}, B_{22} stand for *Moderate* and A_{32}, B_{32} stand for *Expensive*, respectively (fuzzy representations of quality and price for both rooms are shown in Fig. 6). Thus we have: $d(a_2, b_2) \cong 0.5979$ and when $k_2 = 1$, we get: $S(a_2, b_2) \cong 0.2566$ and when $k_2 = 2$, we get: $S(a_2, b_2) \cong 0.1871$.

The similarity $Sim(o_1, o_2)$ between the two rooms is calculated as follows:

1) the weighted average of attributes' similarities: let $\alpha_1 = 0.5$ and $\alpha_2 = 0.8$. Then, when $k_1 = k_2 = 1$ we get: $Sim(o_1, o_2) \cong 0.3902$, and when $k_1 = k_2 = 2$ we get: $Sim(o_1, o_2) \cong 0.3090$.
2) the minimum of the similarities of attributes: when $k_1 = k_2 = 1$ we get: $Sim(o_1, o_2) \cong 0.2566$, and when $k_1 = k_2 = 2$ we get: $Sim(o_1, o_2) = 0.1871$.

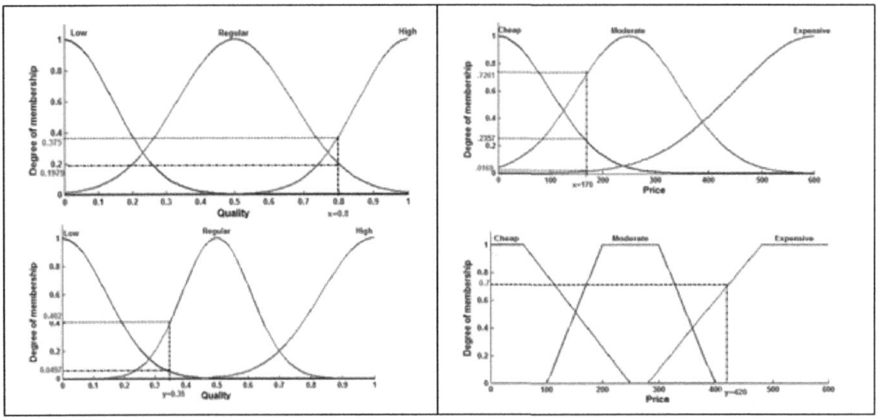

Fig. 6. Case I (b) Fuzzy representation of quality and price of a room in UK and quality and price of a room in Italy (using different membership functions)

Consequently similarity among fuzzy sets defined by using same membership function is greater than similarity among the same fuzzy sets defined by using different membership functions. This means that assessment of similarity is relative to definition of membership functions and interpretation of linguistic values.

5 Comparing a Crisp Attribute with a Fuzzy Attribute

In this section we address the second case: comparing a crisp attribute value (numerical) of a fuzzy object (that is, an object that has one or more fuzzy attribute(s)) with a corresponding fuzzy attribute of another fuzzy object. First, we have fuzzified the crisp value into fuzzy or linguistic label [17], [18], then the comparison has been made following the same procedure as used in *Case I.* For the sake of consistency, we have used (as shown in Fig. 5 above) the Gaussian membership function without restricting the generality of our proposal. This is illustrated by the following example.

5.1 Example 3: Case II

Let us consider the same two rooms in Example 1, but now the value of attribute quality of `room1` and the value of attribute price of `room2` are crisp (see Fig.7)

room1	room2
Quality: 0.8;	Quality: (0.667) regular;
Price: (0.2357) moderate;	Price: £420;

Fig. 7. Case II Comparing rooms described by both crisp and fuzzy attributes

After the fuzzification for both crisp values assuming the same membership functions as in Example1, we get the following:

$$Q(room1) = 0.8 \equiv \{0.0/\text{Low}, 0.1979/\text{Regular}, 0.3753/\text{High}\}$$

$$P(room2) = 420 \equiv \{0.0/\text{cheap}, 0.2353/\text{Moderate}, 0.4868/\text{Expensive}\}$$

Using the procedure above, we will get the same results as in Example 1 above.

6 Conclusions and Further Work

In this paper we propose a new approach to compare two fuzzy objects by introducing a family of similarity measures. This approach employs fuzzy set theory and fuzzy logic coupled with the use of Euclidean distance measure. Since some objects of the same class may have fuzzy values and some may have crisp values in the same set for the same attribute, our approach is suitable to represent and process attribute values of such objects, as has been discussed in the two cases mentioned above.

When we define the domain for both compared attributes, we should use the same unit, even if they are in different contexts. The parameter k_j allows us to balance the impact of fuzzification in equation (1) and can be obtained by user estimation or can be inferred because of the distance *d*. Also similarity between any two objects in the same context should be greater than similarity between objects in different contexts, as noted in the examples given above. Our similarity measures are applied when fuzzy values describing some object's attributes are supported by a degree of confidence.

The two previous cases do not constitute a complete assessment of our approach. Further work on other cases is required, such as comparing objects described with non supported fuzzy attributes. For further research we also intend to introduce a general framework that allows programmers as well as database designers to deal with fuzzy information when developing some applications without need for any more treatment. Our work is motivated by the need for an easy-to-use mechanism to develop applications that deal with vague, uncertain, and fuzzy information. We consider our similarity measure easy to be implemented and thus a basis for further data mining applications to process vague information.

References

1. Cross, V.V., Sudkamp, A.T.: Similarity and Compatibility in Fuzzy Set Theory: Assessment and Applications. Studies in Fuzziness and Soft Comp. Physica-Verlag, Heidelberg (2002)
2. Lourenco, F., Lobo, V., Bacao, F.: Binary-based similarity measures for categorical data and their application in Self-Organizing Maps. In: Procs. JOCLAD 2004, Lisbon (2004)
3. Boriah, S., Chandola, V., Kumar, V.: Similarity measures for categorical data: A comparative evaluation. In: Procs. 8th SIAM Int'l. Conf. on Data Mining, pp. 243–254 (2008)
4. Berzal, F., Cubero, J.C., Marin, N., Vila, M.A., Kacprzyk, J., Zadrozny, S.: A General Framework for Computing with Words in Object-Oriented Programming. Int'l. Journal of Uncertainty, Fuzziness and Knowledge-Based Systems 15(suppl.1), 111–131 (2007)
5. De Caluwe, R. (ed.): Fuzzy and Uncertain Object-Oriented Databases: Concepts and Models. Advances in Fuzzy Systems, Applications and Theory, 13 (1998)

6. George, R., Buckles, B.P., Petry, F.E.: Modelling Class Hierarchies in the Fuzzy Object-Oriented Data Model. Fuzzy Sets and Systems 60, 259–272 (1993)
7. Koyuncu, M., Yazici, A.: IFOOD: an intelligent fuzzy object-oriented database architecture, Knowledge and Data Engineering. IEEE Trans. KDE 15/5, 1137–1154 (2003)
8. Prade, H., Testemale, C.: Generalizing database relational algebra for the treatment of incomplete/uncertain information and vague queries. Inf. Sciences 34, 115–143 (1984)
9. Ma, Z.M., Zhang, W.J., Ma, W.Y.: Extending object-oriented databases for fuzzy information modeling. Information Systems 29/5, 421–435 (2004)
10. Marin, N., Medina, J.M., Pons, O., Sánchez, D., Vila, M.: Complex object comparison in a fuzzy context. In: Information and Software Technology. Elsevier, Amsterdam (2003)
11. Berzal, F., Pons, O.: A Framework to Build Fuzzy Object-Oriented Capabilities over an Existing Database System. In: Ma, Z. (ed.) Advances in fuzzy OODB, pp. 177–205. IGI (2005)
12. Hallez, A., De Tre, G.: A Hierarchical Approach to Object Comparison. In: Melin, P., Castillo, O., Aguilar, L.T., Kacprzyk, J., Pedrycz, W. (eds.) IFSA 2007. LNCS (LNAI), vol. 4529, pp. 191–198. Springer, Heidelberg (2007)
13. Zwick, R., Carlstein, E., Budescu, D.V.: Measures of similarity among fuzzy concepts: A comparative analysis. Int'l. Journal of Approx. Reasoning 1, 221–242 (1987)
14. Lee, J., Kuo, J.-Y., Xue, N.-L.: A note on current approaches to extending fuzzy logic to object-oriented modeling. Int. Journal Intelligent Systems 16(7), 807–820 (2001)
15. Ma, Z.M., Li, Y.: A Literature Overview of Fuzzy Database Models. Journal of Information Science and Engineering 24, 189–202 (2008)
16. Kaufmann, A.: Introduction to the theory of fuzzy subsets. Academic Press, New York (1975)
17. Fuzzification Techniques, `http://enpub.fulton.asu.edu/powerzone/fuzzylogic/`
18. Zadeh, L.: Fuzzy Sets. Information and Control 8, 335–353 (1965)

Generalized Fuzzy Comparators for Complex Data in a Fuzzy Object-Relational Database Management System

Juan Miguel Medina[1], Carlos D. Barranco[2],
Jesús R. Campaña[1], and Sergio Jaime-Castillo[1]

[1] Department of Computer Science and Artificial Intelligence, University of Granada
C/ Periodista Daniel Saucedo Aranda s/n, 18071 Granada, Spain
{medina,jesuscg,sjaime}@decsai.ugr.es
[2] Division of Computer Science, School of Engineering, Pablo de Olavide University
Utrera Rd. Km. 1, 41013 Seville, Spain
cbarranco@upo.es

Abstract. This paper proposes a generalized definition for fuzzy comparators on complex fuzzy datatypes like fuzzy collections with conjunctive semantics and fuzzy objects. This definition and its implementation on a Fuzzy Object-Relational Database Management System (FORDBMS) provides the designer with a powerful tool to adapt the behavior of these operators to the semantics of the application considered.

Keywords: Fuzzy Databases, Fuzzy Object Oriented Databases, Complex Fuzzy Objects Comparison.

1 Introduction

Fuzzy database models and systems have evolved from being extensions of the relational model to be extensions of the object-oriented and object-relational database models. These two last approaches deal with complex fuzzy datatypes and the semantics of the fuzzy operators involved in complex object retrieval is dependent on the application considered. The work [1] uses a FORDBMS to represent and store dominant color descriptions extracted from images stored in the database. To perform flexible retrieval of the images based on their dominant colors, it is necessary to use implementations of fuzzy operators that compute the inclusion degree of a set of dominant colors into another set. Also, if we are interested in the retrieval of images with a similar dominant color description, the system must provide an implementation for the resemblance operator for conjunctive fuzzy collections and for fuzzy objects. In [2] a FORDBMS is used to represent description of curves in spines suffering a deformation called scoliosis. To obtain appropriate results in the fuzzy search, it is necessary to use an implementation of the operators involved in complex data retrieval that is different from the one used in the previously mentioned application. These facts prove that, for complex objects, it is necessary to provide a parameterized approach to adapt the behavior of the comparison operations on them to the semantics of the considered application. This paper proposes a general definition for the operators involved in comparison operations on fuzzy collections of elements with conjunctive semantics, and on complex fuzzy objects. Also,

E. Hüllermeier, R. Kruse, and F. Hoffmann (Eds.): IPMU 2010, Part II, CCIS 81, pp. 126–136, 2010.

changes on FORDBMS structures and methods to provide the designer with the necessary mechanisms to set the desired behavior of these operators in accordance to the semantics of his/her application are proposed.

The paper is organized as follows. Section 2 describes the general structure of our FORDBMS. Section 3 introduces the new definition proposed for the comparators on complex fuzzy datatypes. The extension of the catalog to parameterize these comparators is described in Section 4. An example illustrating a real world application of the proposal is shown in Section 5. Finally, main conclusions and future work are summarized in Section 6.

2 The Fuzzy Object-Relational Database Management System

In [3,4] we introduced the strategy of implementation of our FORDMS model, that is based on the extension of a market leader RDBMS (Oracle®) by using its advanced object-relational features. This strategy let us take full advantage of the host RDBMS features (high performance, scalability, etc.) adding the capability of representing and handling fuzzy data provided by our extension.

2.1 Fuzzy Datatypes Support

Our FORDBMS is able to handle and represent a wide variety of fuzzy datatypes, which allow to easily model any sort of fuzzy data. These types of fuzzy data, that are shown in Fig. 1, are the following:

- Atomic fuzzy types (AFT), represented as possibility distributions over ordered (OAFT) or non ordered (NOAFT) domains.
- Fuzzy collections (FC), represented as fuzzy sets of objects with conjunctive (CFC) or disjunctive (DFC) semantics.
- Fuzzy objects (FO), whose attribute types could be crisp or fuzzy, and where each attribute is associated with a degree to weigh its importance in object comparison.

All fuzzy types define a Fuzzy Equal operator (FEQ) that computes the degree of fuzzy equality for each pair of instances. Each fuzzy datatype has its own implementation

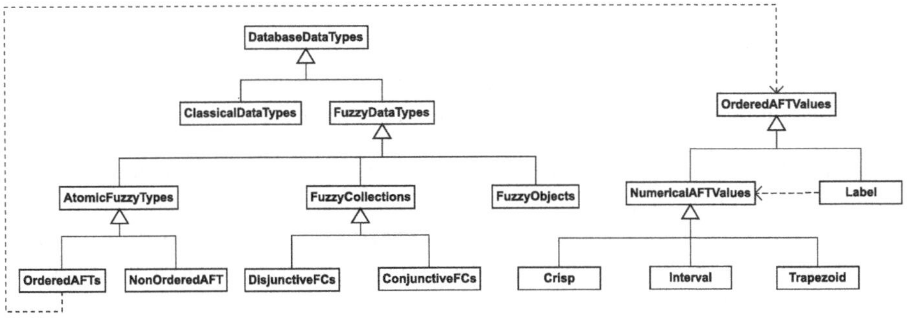

Fig. 1. Datatype hierarchy for the FORDBMS

of this operator in accordance with its nature. Moreover, the FORDBMS provides parameters to adjust the fuzzy equality computation to the semantics of the data handled. For OAFTs the system uses the possibility measure to implement FEQ and implements other fuzzy comparators such FGT (Fuzzy Greater Than), FGEQ (Fuzzy Greater or Equal), FLT (Fuzzy Less Than) and FLEQ (Fuzzy Less or Equal), using this measure. Also, OAFTs implement the necessity based measure for these operators (NFEQ, NFGT, NFEGQ, NFLT and NFLEQ).

3 Comparators for Complex Fuzzy Datatypes

Our FORDBMS provides complex fuzzy datatype structures to model complex problems from the real world. In order to properly capture the rich semantics present in these real objects, it is necessary to provide a flexible mechanism to model the way the system retrieves instances of the datatypes. In other words, the FORDBMS must provide a parameterized way to adapt the behavior of the flexible comparators on complex fuzzy datatypes instances to the semantics of the real object modeled.

The complex fuzzy datatypes that the FORDBMS provides are fuzzy collections (FC) and fuzzy objects (FO). The implementation of flexible comparators on these datatypes is not straight-forward as they must return a degree that represents the whole resemblance for each pair of instances of the datatype considered. The problem is that each instance has a complex structure and a fuzzy equality degree must be computed for each of their component values, and then, perform an aggregation of all these degrees. There are several options available for each step in the computation of the resemblance degree between a pair of complex fuzzy datatype instances. Depending on the alternative used, the semantics of the comparison may vary substantially. In this section we will provide a general definition for the flexible comparators for these datatypes.

3.1 Comparators for Conjunctive Fuzzy Collections

This fuzzy datatype models collections of elements with the same type, where each element can belong to the collection with a degree between 0 and 1. The semantics of the collection is conjunctive. The FORDBMS must provide an operator that computes to which degree an instance of a CFC data type is included into another.

Fuzzy Inclusion Operator. The operator `FInclusion(A,B)` calculates the inclusion degree of $A \subseteq B$, where A and B are instances of CFC. There are some proposals for this operator like the *Resemblance Driven Inclusion Degree* introduced in [5]:

Definition 1. (Resemblance Driven Inclusion Degree). *Let A and B be two fuzzy sets defined over a finite reference universe $\mathcal{U}$, μ_A and μ_B the membership functions of these fuzzy sets, S the resemblance relation defined over the elements of $\mathcal{U}$, $\otimes$ be a t-norm, and I an implication operator. The inclusion degree of A in B driven by the resemblance relation S is calculated as follows:*

$$\Theta_S(B|A) = \min_{x \in \mathcal{U}} \max_{y \in \mathcal{U}} \theta_{A,B,S}(x, y) \tag{1}$$

where

$$\theta_{A,B,S}(x,y) = \otimes(I(\mu_A(x), \mu_B(y)), \mu_S(x,y)) \quad (2)$$

For some applications, like [2], this definition using the min as t-norm and the Gödel implication works fine. However, in others applications like [1], we obtain a better result if we substitute in equation 1 the minimum aggregation operator by a weighted mean aggregation operator, whose weight values are the membership degrees in A of the elements of $\mathscr{U}$, divided by the number of elements of A.

As this kind of operations require the use of an implication operator, a t-norm and a aggregation function, we will propose a definition that provides freedom in the choice of the two first elements, and that includes the use of the well known OWA operators [6] to model the aggregation task. This way, the particular semantics of each application can be taken into account choosing the right operators (i.e. implication, t-norm and aggregation) to compute the resemblance. This is the proposed definition:

Definition 2. (Generalized Resemblance Driven Inclusion Degree). *Let A and B be two fuzzy sets defined over a finite reference universe $\mathscr{U}$, μ_A and μ_B the membership functions of these fuzzy sets, S the resemblance relation defined over the elements of $\mathscr{U}$, $\otimes$ be a t-norm, I an implication operator, F an OWA operator, and $K(A)$ an aggregation correction factor depending on A. The generalized inclusion degree of A in B driven by the resemblance relation S is calculated as follows:*

$$\Theta_S(B|A) = K(A) \cdot F_{x\in\mathscr{U}}(\mu_A(x) \cdot \max_{y\in\mathscr{U}} \theta_{A,B,S}(x,y)) \quad (3)$$

where

$$K(A) : \mathscr{P}(\mathscr{U}) \to \mathbb{R}^+, K(A) > 0, A \in \mathscr{P}(\mathscr{U}) \quad (4)$$

and,

$$\theta_{A,B,S}(x,y) = \otimes(I(\mu_A(x), \mu_B(y)), \mu_S(x,y)) \quad (5)$$

Note that if in equation 3, the min OWA operator F_* is chosen and $K(A) = 1$, the result is the following:

$$\Theta_S(B|A) = F_{*x\in\mathscr{U}}(\mu_A(x) \cdot \max_{y\in\mathscr{U}} \theta_{A,B,S}(x,y)) \quad (6)$$

which has a similar behavior to equation 1 when $\mu_A(x) = 1, \forall x \in \mathscr{U}$.

With the use of OWA operators we can model "orness" (F^*), "andness" (F_*), average (F_{Ave}), and other semantics for the aggregation.

Fuzzy Equality Operator. When A and B are two instances of CFC, this resemblance degree is calculated using the concept of double inclusion.

Definition 3. *(Generalized resemblance between fuzzy sets). Let A and B be two fuzzy sets defined over a finite reference universe $\mathscr{U}$, over which a resemblance relation S is defined, and $\otimes$ be a t-norm. The generalized resemblance degree between A and B restricted by $\otimes$ is calculated by means of the following formula:*

$$\beta_{S,\otimes}(A,B) = \otimes(\Theta_S(B|A), \Theta_S(A|B)) \quad (7)$$

Therefore, the implementation of the operator `FEQ(A,B)`, when A and B are instances of CFC, aggregates the results of `FInclusion(A,B)` and `FInclusion(B, A)` using a t-norm.

3.2 Comparators for Fuzzy Objects

For this kind of fuzzy datatypes, our FORDBMS provides the operator `FEQ(A,B)`, that computes the resemblance of two instances of the same subclass of *FO*. The definition of the operator proposed in this section aims to provide the designer with a flexible framework to express the specific semantics of the considered problem. First, we will introduce a parameterized version of the FEQ operator for OAFT datatypes, to allow to fuzzify crisp values and to relax fuzzy values in flexible comparisons.

Definition 4. *(Relaxation Function) Let A be a trapezoidal possibility distribution defined on a numerical domain $\mathscr{U}$ whose characteristic function is given by $[\alpha, \beta, \gamma, \delta]$, let $s, k \geq 0$ be two real numbers that represent support and kernel increments, respectively. Then, we define the relaxation function, $r_{k,s}(A)$, as follows:*

$$r_{k,s}(A) = \mu_{(k,s)}A(x) \tag{8}$$

where $\mu_{(k,s)}A(x)$ is a trapezoidal possibility distribution described by the values:
$[\min(\alpha \cdot (1-s), \beta \cdot (1-k)), \beta \cdot (1-k), \gamma \cdot (1+k), \max(\gamma \cdot (1+k), \delta \cdot (1+s))]$

Note that $r_{0,0}(A) = A$.

Definition 5. *(Relaxed numerical resemblance) Let A and B be two trapezoidal possibility distributions defined on a numerical domain $\mathscr{U}$, with membership functions $\mu_A(x)$ and $\mu_B(x)$, respectively, and $k, s \geq 0$ two real numbers that represent the kernel and support increments, respectively, then we define the relaxed numerical resemblance, $FEQ_{k,s}(A,B)$, as follows:*

$$FEQ_{k,s}(A,B) = FEQ(r_{k,s}(A), r_{k,s}(B)) = \sup_{x \in \mathscr{U}} \min(\mu_{(k,s)}A(x), \mu_{(k,s)}B(x)) \tag{9}$$

Note that $FEQ_{0,0}(A,B) = FEQ(A,B)$ and, because of the use of possibility measure, $FEQ_{k,s}(A,B) = FEQ_{k,s}(B,A)$.

Definition 6. *(Parameterized Object Resemblance Degree). Let C be a Class, n the number of attributes defined in the class C, $\{a_i : i \in 1,2,...,n\}$ the set of attributes of C that does not generate cycles in subclass definition, o_1 and o_2 be two objects of the class C, $o_j.a_i$ the value of the i-th attribute of the object o_j, $rel(a_i) \in [-1,1]$, a value whose absolute value means the relevance degree of the i-th attribute of the class C, and that indicates that the attribute is discriminant in the resemblance comparison if it has a negative sign, $n_{>0}$ a parameter that establishes the minimum of attributes whose comparison degree must be greater than 0. The Parameterized Object Resemblance*

Degree is recursively defined as follows:

$$OR(o_1, o_2) = \begin{cases} 1 & \textit{if} \quad o_1 = o_2 \\ FEQ_{k,s}(o_1, o_2) & \textit{if} \quad o_1, o_2 \in \textit{subClassOf(OAFT)} \\ FEQ(o_1, o_2) & \textit{if} \quad o_1, o_2 \in \textit{subClassOf(NOAFT)} \vee \\ & \quad\;\; o_1, o_2 \in \textit{subClassOf(CFC)} \\ 0 & \textit{if} \quad (\exists i \in [1, n] : OR(o1.a_i, o2.a_i) = 0 \\ & \quad\;\; \wedge rel(a_i) < 0) \\ & \quad\;\; \vee(|\{OR(o1.a_i, o2.a_i) > 0 : \\ & \quad\;\; i \in [1, n]\}| < n_{>0}) \\ K(C) \cdot F(OR(o1.a_i, o2.a_i).|rel(a_i)|) & \textit{otherwise} \end{cases} \tag{10}$$

where, $k, s \geq 0$ *are two real numbers that represent the kernel and support increments for the considered class (if not defined, both take 0 as value),* F *is an OWA operator, that aggregates the comparisons of attributes and has an associated vector* $W = [w_1, \cdot\cdot, w_n]^T$*, and* $K(C) > 0$ *an aggregation correction factor depending on* C*.*

The previous definition provides the designer with a parameterized rich framework to model the semantics of complex object comparisons. The definition offers the following design alternatives:

- To set a relaxation percentage for elements of a given subclass of OAFT, using the k and s parameters. This allows to perform flexible comparisons on crisp values that are not exactly equal, as well as relax fuzzy data in these kind of comparisons.
- The possibility to determine that a given object attribute is discriminant in the whole comparison of two objects, in the sense that if the comparison of two objects return a 0 value for this attribute, the whole object comparison must also return 0.
- To set the number of comparisons of attributes for an object that needs to be distinct from 0 to return a whole object comparison distinct from 0. For some kinds of problems it is better to return a 0 value for the whole object comparison if there are a certain number of comparisons of attributes that return 0.
- To set the relevance of each object attribute, $rel(a_i)$, in the whole object comparison.
- To choose the OWA operator F and the aggregation correction factor $K(C)$ that best matches the semantics of the problem modeled.
- To set the FEQ parameters and behavior for each subclass involved in a complex object comparison.

4 FORDBMS Elements to Control the Comparison Behavior

Our FORDBMS implements the complex object comparison introduced in the previous section by means of the datatype structure shown in Fig. 1, where the definition and implementation of methods, constructors and operators take into account a set of parameters, stored in a specific database catalog, to determine their behavior. This section will briefly describe the database elements involved in the comparisons of complex object.

4.1 Catalog for Parameterized Comparison

The FORDBMS has a database catalog extension which allows to set parameters to control the behavior of comparisons on complex fuzzy datatypes. We will describe this structure in relation with each kind of datatype considered.

Conjunctive Fuzzy Collections. The following are the tables with the parameters that provide the behavior of the FInclusion operator:

`CFC_FInclusion(type_name,oper_name)`, where the first attribute identifies the subtype of CFC and the second stores the associated implementation. There is a predefined implementation labeled as: "min" (the default value) that implements the definition 1, using the Gödel implication. The designer can provide other implementations whose definitions are parameterized in the following tables.
`CFC_FInclusion_def(type_name,oper_name,tnorm,implication,owa,ka)`, where the first two attributes identify the FInclusion operator whose parameters must be set, `tnorm` identifies the t-norm used (the min t-norm is used by default), the user can also select the product t-norm; the attribute `implication` sets the implication operator used, by default Gödel implication is used, others operators are available in the implementation; the last attributes are usually related and define the OWA operator and the aggregation correction factor used, by default the FORDBMS provides an implementation with the values used in the equation 5. If the designer wants to provide his own OWA operator and aggregation correction factor, the adequate values must be set in the following tables:

`OWA_def(owa_name,weight)`, this table stores the weight for each OWA operator defined.

`Ka_def(type_name,ka_name,user_function)`, the user must define and implement a user function in the FORDBMS that computes the value of the aggregation correction factor, and sets the identifier of this function in the column `user_function`.

To parameterize the FEQ operator on CFC, FORDBMS provides the following table: `CFC_FEQ(type_name,tnorm,same_size)`, where the second attribute sets the t-norm used for the double fuzzy inclusion (by default the min), the third is boolean and, if it is set to 'true', then the FEQ operator returns 0 if the compared CFC instances have different number of elements.

Fuzzy Objects. The following catalog tables are created to store the parameters that establish the behavior of the FEQ operator on instances of FO:

`FTYPE_Relax(type_name,k,s,active)`, by means of this table the designer sets the parameters `k` and `s` that relax the instances of OAFTs subtypes in FEQ comparisons; if the attribute active is set to 'true' then this relaxation is applied in further FEQ comparisons, if it is set to 'false', this relaxation is not considered.

`FO_FEQ_Aggr(type_name,owa,k_a,min_gt_0)`, this table stores the identificator of the OWA operator and the aggregation correction factor used for the subclass identified in the column `type_name`. By default, the FORDBMS implements and uses the F_{Ave} OWA operator and the aggregation correction factor: $K(A) = n/\sum_{i=1}^{n} |rel(a_i)|$. The description and implementation for other operators must be set in the tables `OWA_def` and `Ka_def` described above. Also this table allows to establish the minimum number of attributes whose comparison degree must be greater than 0 so that the whole object comparison does not return 0.

`FO_FEQ_Attrib(type_name,name,relevance)`, by means of this table, the designer can set the relevance values for each attribute of the subclass of FO considered, if the relevance value for a given attribute is < 0, this means that this attribute is discriminant.

5 An Example of Modeling Complex Fuzzy Datatypes and Flexible Comparators

To illustrate the use of the complex fuzzy datatypes handled by the proposed FORDBMS and the way the designer can adjust the behavior of comparators for them, we will use an example based on a flexible representation of the structure of a spine with scoliosis (see more details in [2]). This pathology consists of a three-dimensional deformation of the spine. An anteroposterior(AP) X-ray of a spine with scoliosis projects an image that shows several curves on the spine. To measure the severity of this disease, physicians measure, on the AP X-ray, the Cobb angle [7] for each curve in the spine. Each Cobb angle measurement is characterized by means of four values: angle value, superior vertebra, inferior vertebra and direction of the curve (left or right). Another parameter that characterizes a spine curve is the apical vertebra. The whole spine measurement comprises a set of curves, each one represented by the previously mentioned parameters. Figure 2 shows an example of Cobb angle measurement performed on an AP X-ray and the values obtained. For the diagnosis and treatment of scoliosis it is useful to have medical files of other similar cases. In order to gather this data, it is interesting the possibility to retrieve X-rays of patients with similar parameters for the deformation of the spine. Therefore, a physician should be able to formulate queries to the FORDBMS looking for images of spines that include a given curve (or a set of them). Note that the queries must be flexible in the sense of retrieving images with similar values for the parameters, but not exactly the same values.

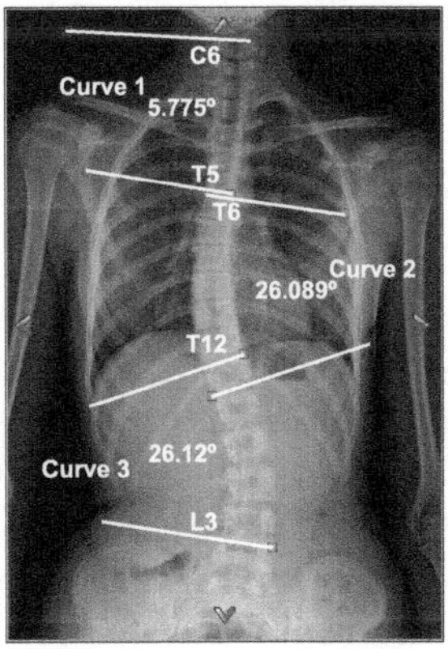

Parameters of Curve 1:
- Direction of the curve: "Right",
- Cobb angle: 5.775,
- Superior vertebra: "C6",
- Apex vertebra: "T2",
- Inferior vertebra: ""T5"

Parameters of Curve 2:
- Direction of the curve: "Left",
- Cobb angle: 26.089,
- Superior vertebra: "T6",
- Apex vertebra: "T9",
- Inferior vertebra: ""T12"

Parameters of Curve 3:
- Direction of the curve: "Right",
- Cobb angle: 26.12,
- Superior vertebra: "T12",
- Apex vertebra: "L3",
- Inferior vertebra: ""L4"

Fig. 2. Example of Cobb angle measurement for a spine with three curves. On the right, the parameter values for each curve.

5.1 Data Definition and Setting the Behavior of the Comparators

First we need to create all subtypes needed to represent the structure of data. According to Fig. 2, we use a subtype of OAFT to model Cobb angle, subtypes of NOAFT to model the boundary vertebrae of the Cobb angle and its direction. To model a whole curve measurement we select a FO subtype, and to model the whole spine measurement we use a CFC subtype. Below, these definitions, using the DDL provided by the FORDBMS, are shown:

```
EXECUTE OrderedAFT.extends('CobbAngleT');
EXECUTE NonOrderedAFT.extends('CurvDirectionT');
EXECUTE NonOrderedAFT.extends('VertbSetT');
-- This type has two values: LEFT and RIGHT
EXECUTE NonOrderedAFT.defineLabel('CurvDirectionT','LEFT');
EXECUTE NonOrderedAFT.defineLabel('CurvDirectionT','RIGHT');
-- This set of sentences defines labels for the 24 vertebrae
EXECUTE NonOrderedAFT.defineLabel(user,'VertbSetT','L5');
..
EXECUTE NonOrderedAFT.defineLabel(user,'VertbSetT','C1');
-- Creates the subtype for a whole Cobb angle measurement
CREATE OR REPLACE TYPE CobbCurvT UNDER FO (Direction CurvDirectionT,
Angle CobbAngleT,SupVertb VertbSetT,ApexVertb VertbSetT,InfVertb VertbSetT););
-- Creates the subtype for the whole spine measurement
EXECUTE ConjunctiveFCs.extends('SpineCurves','CobbCurvT',4);
```

The type `VertbSetT` represents the 24 vertebrae of the spine considered, to perform comparisons it is necessary to provide an order relation for this set, to do this we define the following mapping: 'L5' $\rightarrow$ 1, 'L4' $\rightarrow$ 2 $\cdot\cdot$ 'C1' $\rightarrow$ 24. With this order relation we define a static function on the type `VertbSetT` to relax the proximity value of two vertebrae in FEQ comparisons. This function has the form `create_nearness_vert(k,s)` and generates and stores a nearness relation based on the parameters k and s. The former extends *k vertebrae* the kernel for a `VertbSetT` value, and the second extends *s vertebrae* the support. In this example, we will model that a given vertebra when compared with the same and adjacent ones return 1, and decreases comparison degree to 0 when there are four vertebrae between the compared vertebrae. To get this behavior we need to execute: `VertbSetT.create_nearness_vert(1,3)`.The following statements set the behavior for the comparison of instances of the data structure defined:

```
-- Extends the kernel and support in FEQ comparisons
execute orderedAFT.setRelax('CobbAngleT', 0.4, 0.7, 'Y');
-- Set the relevance for CobbCurvT attributes, the first two are discriminant
execute fo.setAttributeRelevance('CobbCurvT','Direction',-1);
execute fo.setAttributeRelevance('CobbCurvT','Angle',-1);
-- All the vertebrae attributes have relevance value equal to 1
execute fo.setMin_gt_0('CobbCurvT',3); -- Set the min attributes nonzero
```

For the aggregation on the attributes of `CobbCurvT` we select the default implementation, then we do not need to set any values. The same is valid for the implementation used for the FInclusion operator. For FEQ comparisons on whole spine instances we want that spines with different number of curves return a 0 value. To do this, we execute the sentence: `EXECUTE ConjunctiveFCs.set('SpineCurves',null,'true')`.Now, we can create a table that stores instances of `SpineCurves` with the X-rays images:

```
create table APXRay( image# number, xray bfile, SpineDescription SpineCurves);
```

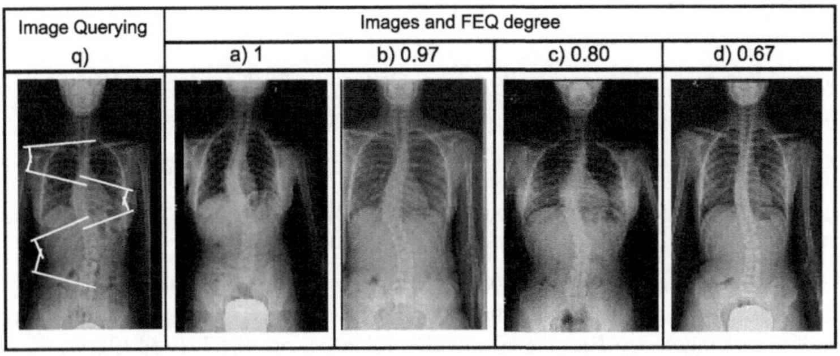

Fig. 3. Searching images that present similar spine curvature to image q)

5.2 Querying

Using the behavior configured for FEQ on `CobbCurvT` and `SpineCurves` subtypes, we can retrieve images that present a similar spine curve pattern to a given one. To do this we execute the following sentence:

```
SELECT ap1.image#,ap1.xray,ap1.cdeg(1) FROM apxray ap1, apxray ap2
WHERE ap1.image='q' AND
FCOND(FEQ(ap1.spinedescription, ap2.spinedescription),1)>0 order by cdeg(1) desc;
```

As can be seen in Fig. 3, the query evaluated holds that, the higher is the curve pattern matching, the higher the computed compliance degree is.

6 Concluding Remarks and Future Work

This paper proposes a parameterized behavior for the operators FInclusion on CFC and FEQ on CFC and FO. This idea is motivated by the need to adapt the behavior of these comparators to the specific semantics of the modeling applications. The FORDBMS is designed to support these changes through the implementation of a catalog that stores the parameters that define the behavior of these comparators, and defining and implementing the types, methods and operators that provides that functionality. Some examples of applications prove the usefulness of this proposal. Although some alternatives for the operators are implemented by default into the FORDBMS, future work will be oriented to extend the number of variants of operators supported and to implement techniques to improve the performance of retrieval operations based on these operators.

Acknowledgment

This work has been supported by the "Consejería de Innovación Ciencia y Empresa de Andalucía" (Spain) under research projects P06-TIC-01570 and P07-TIC-02611, and the Spanish (MICINN) under grants TIN2009-08296 and TIN2007-68084-C02-01.

References

1. Chamorro-Martínez, J., Medina, J., Barranco, C., Galán-Perales, E., Soto-Hidalgo, J.: Retrieving images in fuzzy object-relational databases using dominant color descriptors. Fuzzy Sets and Systems 158, 312–324 (2007)
2. Medina, J.M., Jaime-Castillo, S., Barranco, C.D., Campaña, J.R.: Flexible retrieval of x-ray images based on shape descriptors using a fuzzy object-relational database. In: Proc. IFSA-EUSFLAT 2009, Lisbon (Portugal), July 20-24, pp. 903–908 (2009)
3. Cubero, J., Marín, N., Medina, J., Pons, O., Vila, M.: Fuzzy object management in an object-relational framework. In: Proc. 10th Int. Conf. on Information Processing and Management of Uncertainty in Knowledge-Based Systems, IPMU 2004, pp. 1767–1774 (2004)
4. Barranco, C.D., Campaña, J.R., Medina, J.M.: Towards a fuzzy object-relational database model. In: Galindo, J. (ed.) Handbook of Research on Fuzzy Information Processing in Databases, pp. 435–461. IGI Global (2008)
5. Marín, N., Medina, J., Pons, O., Sánchez, D., Vila, M.: Complex object comparison in a fuzzy context. Information and Software Technology 45(7), 431–444 (2003)
6. Yager, R.: Families of owa operators. Fuzzy Sets and Systems 59, 125–148 (1993)
7. Cobb, J.: Outline for the study of scoliosis. Am. Acad. Orthop. Surg. Inst. Course Lect. 5, 261–275 (1948)

The Bipolar Semantics of Querying Null Values in Regular and Fuzzy Databases
Dealing with Inapplicability

Tom Matthé and Guy De Tré

Ghent University, Dept. of Telecommunications and Information Processing,
St.-Pietersnieuwstraat 41, B-9000 Ghent, Belgium
{TOM.MATTHE,GUY.DETRE}@UGent.be

Abstract. Dealing with missing information in databases, either because the information is unknown or inapplicable, is a widely addressed topic in research. Most of the time, null values are used to model this missing information. This paper deals with querying such null values, and more specifically null values representing inapplicable information, and tries to come up with semantically richer, but still correct, query answers in the presence of null values. The presented approach is based on the observation that, when used in the context of a query, an inapplicable value can be treated semantically equivalent to some other regular domain value, resulting in a query criteria satisfaction being either '*true*', '*false*' or '*unknown*'. So, the data itself can be inapplicable, but the criteria (and query) satisfaction is not inapplicable.

Keywords: fuzzy database querying, null values, inapplicability.

1 Introduction

Since many years, an emphasis has been put on research that aims to make database systems more flexible and better accessible. An important aspect of flexibility is the ability to deal with imperfections of information, like imprecision, vagueness, uncertainty or missing information. Imperfection of information can be dealt with at the level of the data modeling, the level of database querying or both.

In literature, fuzzy set theory [28] and its related possibility theory [19,29] have been used as underlying mathematical frameworks for those enhancement approaches that are called 'fuzzy': integrating imperfection at the level of the data modeling leads to what is usually called 'fuzzy' databases, whereas dealing with imperfections at the level of database querying leads to what is usually called fuzzy querying [3,4,5,11,21,27], which is a special kind of flexible querying.

This paper deals with the handling of missing information, and more specifically the querying of such missing information. The treatment of missing information in traditional database models has been widely addressed in research and continues to be explored. The most commonly adopted technique is to model

E. Hüllermeier, R. Kruse, and F. Hoffmann (Eds.): IPMU 2010, Part II, CCIS 81, pp. 137–146, 2010.

missing data with a pseudo-description, called *null*, that denotes 'missing' [6]. Once null values are admitted into a database, it is necessary to define their impact on database querying and manipulation.

In his approach, Codd extends the relational calculus based on an underlying three-valued logic [6,7] in order to formalize the semantics of null values in traditional databases. Even further extensions of this approach have been made, among them a four-valued logic (4VL) model containing two types of 'null' values, respectively representing 'missing but applicable' and 'missing and inapplicable' [8], and an approach based on 'marked' null values, which can be interpreted as named variables [22].

In this paper, the querying of null values, and more specifically inapplicable values, will be dealt with. The remainder of the paper is structured as follows: The following section gives an overview of how null values are dealt with in databases, in general. Next, Section 3 describes how this paper tries to deal with inapplicable information which is being queried, both in regular and in fuzzy databases. The last section states the conclusions of the paper.

2 Null Values in Databases

Missing information in databases could be indicated and handled by using a 'null' value, which can be seen as a special mark that denotes the fact that the actual database value is missing. In order to assign correct semantics to such a mark, it is important to distinguish between two main reasons for information being missing. As originally stated by Codd [7], information is either missing because:

- data is unknown to the users, but the data is applicable and can be entered whenever it happens to be forthcoming;
- data is missing because it represents a property that is inapplicable to the particular object represented by the database record involved.

As an illustration of these two cases, consider a database with information about birds. For each bird, among other things, the flying speed is registered in the database. The first case, unknown information, occurs for example in a situation where it is known that a bird can fly, but the bird still has to be better observed in order to obtain its speed. The second case, inapplicable information, occurs in a situation where it is known that the bird cannot fly, as for example is the case with a penguin (flying speed information is inapplicable for penguins).

2.1 Traditional Approaches

In many traditional database approaches a 'null' value mark is used to handle both mentioned cases of missing data. In most approaches no explicit distinction is made and a single kind of null values is used to handle both cases. In [8], Codd introduces the idea of making an explicit distinction in the modeling of both cases by using two different kinds of null values, one meaning "value unknown" and the other "value not defined" (to be interpreted as "value which cannot exist").

In formal definitions of database models null values are represented by some special symbol, e.g., by the bottom symbol '$\bot$' [1,26]. In some formal approaches, null values are considered to be domain dependent [26]: the domain dom_t of each data type t supported by the database model contains a domain specific null value $\bot_t$, which implies that an explicit distinction is made between for example a missing integer value, a missing string value, etc. The idea behind this, is that the situation of a missing integer value differs from the situation of a missing string value. In both cases information is missing, but the known information about the data type of the expected values —if these exist— should not be neglected (at least from a theoretical point of view).

In order to define the impact of null values on database querying and manipulation, a many-valued logic [25] has been used. This logic is three-valued if only a single kind of null values is used [2,7] and four-valued if two distinct kinds of null values are considered [8]. The truth values of Codd's four-valued logic are respectively true (T), false (F), unknown ($\bot_U$) and inapplicable ($\bot_I$). In Codd's three-valued logic, the latter two values have been combined into one truth value $\bot_{U/I}$, which stands for 'either unknown or inapplicable'.

An important drawback of using these approaches based on many-valued logics is that the law of excluded middle and the law of non-contradiction may not hold in all cases. For example, by using a three-valued Kleene logic [25] with truth values T, F and $\bot_{U/I}$, $\bot_{U/I} \wedge \neg(\bot_{U/I}) = \bot_{U/I} \neq F$ and $\bot_{U/I} \vee \neg(\bot_{U/I}) = \bot_{U/I} \neq T$. Moreover, a truth value $\bot_U$ or $\bot_{U/I}$ induces a problem of truth functionality (the degree of truth of a proposition can no longer be calculated from the degrees of truth of its constituents). 'Unknown' stands for the *uncertainty* of whether a proposition is true or false, which differs from the idea of 'many-valuedness' in a logical format what many-valued logics are intended for: degrees of uncertainty and degrees of truth are different concepts [20]. These observations explain some of the rationales behind Date's criticism on the use of null values [9,10]. In this paper, although the criticism is understood, it is assumed that 'null' values can never be ruled out, certainly not when working with existing databases, and therefore they should be dealt with in an adequate way.

2.2 'Fuzzy' Approaches

In fuzzy database approaches the situation with respect to null values is different, because almost all fuzzy database models, at least in a theoretical view, allow to model unknown information without a need for an extra specific null value. For example, in the possibilistic database approach, as originally presented in [24], unknown information with respect to an attribute of a given database record is represented by means of a possibility distribution [19,29] that has been defined over the domain of the attribute. The modeling of inapplicability however, still requires a specific element, $\bot_I$ in the domain of the attribute. With respect to null values this implies that, in fuzzy databases, it is sufficient to have only one kind of null values in order to be able to represent 'inapplicability', because 'unknown' can be represented by a uniform (normalized) possibility distribution.

An adequate logic can be obtained by imposing possibilistic uncertainty on a three-valued logic with truth values 'True' (T), 'False' (F) and 'Inapplicable' ($\perp_I$, or $\perp$ for short). Such a logic, based on a three-valued Kleene logic [25], has been developed in [14]. The resulting truth values have been called 'extended possibilistic truth values' (EPTV's), and are defined as an extension of the concept 'possibilistic truth value' (PTV) which was originally introduced in [23] and further developed in [12,13]. Each EPTV $\tilde{t}^*(p)$ of a proposition p is defined as a normalized possibility distribution over the set of truth values 'True', 'False' and 'Inapplicable', i.e.

$$\tilde{t}^*(p) = \{(T, \mu_T), (F, \mu_F), (\perp, \mu_\perp)\}$$

with μ_T, μ_F and $\mu_\perp$ the respective membership degrees in the possibility distribution. As illustrated in [16,18], EPTV's can be used to express query satisfaction in flexible database querying: the EPTV representing the extent to which it is (un)certain that a given database record belongs to the result of a flexible query can be obtained by aggregating the calculated EPTV's denoting the degrees to which it is (un)certain that the record satisfies the different criteria imposed by the query. The arithmetic rules for the aggregation operations can be found in [15,17]. This final, aggregated, EPTV expresses the extent to which it is possible that the record satisfies the query, the extent to which it is possible that the record does not satisfy the query and the extent to which it is possible that the record is inapplicable for the query. The logical framework based on these EPTV's extends the approach presented in [24] and allows to explicitly deal with the inapplicability of information during the evaluation of the query conditions: if some part of the query conditions are inapplicable, this will be reflected in the resulting EPTV.

Using a special truth value to handle inapplicable information brings along with it the same problem as mentioned above. The law of non-contradiction is not always valid, e.g. if you look at "'Inapplicable' AND NOT('Inapplicable')":

$$\begin{aligned}&\{(T,0),(F,0),(\perp,1)\} \wedge \neg\{(T,0),(F,0),(\perp,1)\}\\&= \{(T,0),(F,0),(\perp,1)\} \wedge \{(T,0),(F,0),(\perp,1)\} = \{(T,0),(F,0),(\perp,1)\}\\&\neq \{(T,0),(F,1),(\perp,0)\}\end{aligned}$$

And analogous for the disjunction.

Falling back to PTV's ($\tilde{t}(p) = \{(T, \mu_T), (F, \mu_F)\}$, with μ_T or μ_F, or both, equal to 1) does not entirely solve this problem (e.g. if $\tilde{t}(p) = \{(T,1),(F,1)\} = UNK$, i.e. 'Unknown', $\tilde{t}(p \wedge \neg p) = UNK \neq F$, and $\tilde{t}(p \vee \neg p) = UNK \neq T$), but at least the possibility will always be 1 to come up with a correct result (i.e. $\mu_F = 1$ for "$\tilde{t}(p \wedge \neg p)$", and $\mu_T = 1$ for "$\tilde{t}(p \vee \neg p)$"), which is not the case in the above example. Therefore, in this paper, PTV's will be used to express query satisfaction when querying fuzzy databases. It is understood that this has the drawback of losing some information, namely that the result might be based on the inapplicability of some of the criteria. Therefore, it will be tried to deal with inapplicable information in a way that results are still semantically correct.

3 Dealing with Inapplicability in Querying

Certain attributes in a database can be inapplicable for certain records. This is the case when the value of the attribute cannot be determined for the concerning record, simply because it does not exist. An example that was already given above, is the flying speed of birds, which is inapplicable for a penguin since a penguin cannot fly. Other examples are the results of tests in a recruitment database (if it is known that a person did not take a test, the test score is inapplicable for the concerning test), or the pregnancy of people in an employee database (the attribute pregnant –yes or no– is inapplicable for all male employees).

Remark that inapplicable information is not the same as unknown information, e.g., a test result is inapplicable in case the test was not taken, so there is no score for it, while a test result is unknown in case the test was taken but the score itself is unknown. So, in the data representation itself there should be a major difference between unknown information and inapplicable information. In the approach taken in this paper however, when querying such (imperfect) data, this differences should be handled and reflected transparently in the query result. It is assumed that, when a user poses a query to a database system, the possible answers to his/her question ("Do the records satisfy my query?") are: '*yes*', '*no*' or '*maybe*' (to a certain degree, in case of flexible querying), but not '*inapplicable*'. A record is either (partly) satisfactory with respect to the query or not. So the satisfaction degree itself cannot be inapplicable.

So, since the satisfaction of a query result should not be inapplicable, should it be forbidden to use inapplicable information in databases? No, it should not! It is still better to know that information is inapplicable than to know nothing at all. However, the database design should be optimized, to avoid the need for inapplicability as much as possible. This is not always feasible though. So, we need ways to deal with this inapplicable information in querying. In traditional systems, only records for which all query criteria evaluate to 'True' are considered in the query result. Because 'inapplicable' is modeled by 'NULL' and the evaluation of a criterion on 'NULL' always evaluates to 'False' (unless the criterion specifically searches for inapplicable data, e.g., with the "IS NULL" operator in SQL), the records with inapplicable data simply do not show up in the result if there is a query criterion on the concerning attribute. This is not always the best option though. Sometimes, depending on the context, it is better to let an inapplicable value satisfy the imposed criterion. The reason for this is that, in querying, an inapplicable value can be viewed as being semantically equivalent to another value in the domain (or an unknown value in case of fuzzy databases). Indeed, e.g., in the examples above, considering query results:

- an inapplicable flying speed of a bird can be treated exactly the same as a flying speed of 0
- if a test was not taken, and hence the score is inapplicable, this can be treated in the same way as an unknown test score
- in the employee database, the attribute 'pregnant' is inapplicable for all male employees, but, for querying, can be treated semantically equivalent to all females not being pregnant

Remark that it is not because the satisfaction degree cannot be inapplicable, that the query result itself (i.e. the data in the records of the query result) may not contain inapplicable information. E.g., when querying an employee database for all non pregnant employees, the result should return all male employees as well, with full satisfaction, but with an inapplicable value for the 'pregnant' attribute.

3.1 Traditional Querying of Regular Databases

Traditional querying of regular databases results in boolean truth values, 'True' or 'False'. A record is either part of the result or not. As mentioned above, most traditional database systems use a single 'null' value for representing both unknown and inapplicable information. So, there is no real distinction between them. When evaluating queries, a criterion on an attribute containing a 'null' value will always result in the truth value 'False' (i.e. the record will not appear in the result), unless, as also mentioned above, the "IS NULL" operator is used. Even if one would use distinct 'null' values for 'unknown' and 'inapplicable', or even marked 'null' values, the result would be the same. A criterion will always evaluate to 'False' if either of those 'null' values was encountered, again with the exception when using the "IS NULL" operator.

For an 'unknown' value, this is a good approach, because by lack of more or better knowledge, you have no choice but to discard such values from the result, because the only other option, accepting it, would mean that the value is totally satisfactory for the query, which it clearly is not since the value is not even known.

For an 'inapplicable' value on the other hand, the situation is totally different. It is also possible that an 'inapplicable' value is totally unacceptable, but it might as well be the case that it should be accepted in the result. Which one of the two it should be, acceptable or not, depends on the context and the query. E.g. when querying the birds database for birds with flying speed below 20km/h, the penguin should be in the result, while when querying the same database for birds with maximum flying speed above 20km/h, the penguin clearly should not be in the result. Hence, when used in the context of a query, 'inapplicable' has bipolar semantics: in some situations 'inapplicable' should contribute in a positive way, in others in a negative way.

Of course, even in traditional querying systems it is possible to let inapplicable values (or 'null' values in general) appear in the result. E.g., the query for birds with flying speed below 20km/h, could be formulated as "speed < 20 OR speed IS NULL". However, this solution treats unknown and inapplicable values in the same way, which, as explained above, is not the best approach. Trying to solve this by making an explicit distinction between 'unknown' and 'inapplicable' (and hence also between two operators "IS UNKNOWN" and "IS INAPPLICABLE") is only a partial solution. Although this would make a distinction between 'unknown' and 'inapplicable', it still treats all inapplicable values in the same way. However, as seen above, it is necessary to distinguish between the semantics of different inapplicable values (depending on the context). Furthermore, in this way, the decision of whether or not to search for 'null' values is the responsibility of the user who is formulating the query. Since a user might not even be aware of the presence of 'null' values in the database, it would be better if this could be handled transparently to the user as much as possible. Therefore, we should try to come with (semantically correct) query answers, without the need for users to explicitly take into account the possible presence of 'null' values in the database.

When we want to deal with 'null' values as described in the semantics above, it is clear that using one single 'null' value will not be enough. In this paper, we propose to use a form of 'marked' null values [22], to make the distinction between 'unknown' ($\perp_U$) and 'inapplicable' ($\perp_I$) on the one hand, but on the other hand also between different interpretations of 'inapplicable'. We propose to add an extra "mark", stating to which domain value (including 'unknown') 'inapplicable' should be treated semantically equivalent in case the value is queried. E.g., if we look back at the examples above:

- $\perp_{I,0}$ for the 'inapplicable' value which, for querying, can be treated semantically equivalent to 0, like in the case of the flying speed
- $\perp_{I,\perp_U}$ for the 'inapplicable' value which, for querying, can be treated semantically equivalent to 'unknown', like in the case of the test score
- $\perp_{I,False}$ for the 'inapplicable' value which, for querying, can be treated semantically equivalent to 'False', like in the case of the pregnancy of men

When evaluating query criteria on these kind of 'null' values, the semantics of 'unknown' ($\perp_U$) is equal to the semantics of a regular 'null' value as explained above, i.e. regular criteria will always return 'False' and the "IS NULL" or "IS UNKNOWN" operator will return 'True'. On the other hand, when evaluating a criterion on a marked 'inapplicable' value, the evaluation can be done using the value stated in the additional "mark" of the marked value.

3.2 Flexible Querying of Regular Databases

Flexible querying of regular databases results in a satisfaction degree $s \in [0, 1]$. This makes it possible that a record *partly* satisfies some query criteria (and as a consequence the entire query). $s = 1$ means total satisfaction, while $s = 0$ means total dissatisfaction. Although the concept of *partial satisfaction* can be modeled by this, there is still no way to adequately deal with 'unknown' or 'inapplicable' information, or 'null' values in general. In fact, the situation is quite similar to that of traditional querying of regular databases, described above. Although the satisfaction degree s can take values in $[0, 1]$, the evaluation of a query criterion on a 'null' value will still result in $s = 0$ (unless the "IS NULL" operator is used), because one cannot speak of partial satisfaction for a 'null' value: in that case, the satisfaction is 'unknown', not 'partial'.

Because the handling of 'null' values in flexible querying of regular databases is analogous to traditional querying of regular databases, the solution proposed in this paper for achieving answers that are semantically more correct, is the same as in the previous case of traditional querying of regular databases. So, here also we propose to introduce marked 'null' values to make the distinction between 'unknown' and 'inapplicable' information, with, in case of 'inapplicable' information, an additional "mark", stating to which domain value (including 'unknown') the value 'inapplicable' should be treated semantically equivalent in case this value is queried. The semantics when evaluating query criteria on these kind of 'null' values is the same as described above (where a truth value 'False' corresponds to $s = 0$).

3.3 Fuzzy Querying of Fuzzy Databases

Different frameworks for dealing with fuzzy querying of fuzzy databases exist. As mentioned above, Possibillistic Truth Values (PTV's) will be used in this paper. In that case, the evaluation of a criterion will lead to a PTV $\tilde{t}(p)$, which has the advantage over regular satisfaction degrees $s \in [0,1]$ of also being able to model an unknown satisfaction degree ($\tilde{t}(p) = \{(T,1),(F,1)\}$), or even a partially unknown satisfaction degree (e.g. $\tilde{t}(p) = \{(T,1),(F,0.5)\}$. So, in case of unknown information in the database (which in a fuzzy databases is a possibility distribution over the domain of the attribute), this can be handled very easily using PTV's. On the other hand, inapplicable information, which in fuzzy databases still requires a special 'null' value, still can't be handled in a natural way, and if nothing else is done, will lead to a satisfaction degree expressed by the PTV $\{(T,0),(F,1)\}$ (i.e. 'False'). Again, this is similar to the two situations above, and is not what is really desired when we want to deliver semantically correct answers to the user.

So, even when using PTV's, inapplicable information still requires a special approach. Again, it is proposed to use marked 'null' values, but this time only for the handling of inapplicable information because in a fuzzy database there is no need for a 'null' value to express unknown information. As in the previous cases, an additional "mark" will be used to indicate to which domain value (including 'unknown') the value 'inapplicable' should be treated semantically equivalent in case this value is queried. As examples of such marked 'inapplicable' values, and the evaluation of them when processed in queries, consider the following:

- $\perp_{I,0}$ stands for an 'inapplicable' value which, for querying, can be treated semantically equivalent to 0. This can, e.g., be used for the flying speed of a penguin. If a fuzzy criterion on the attribute 'flying speed' would be "speed < *moderate*" (with *moderate* a linguistic label for the possibility distribution representing, e.g., '*around 20*'), the evaluation would lead to the PTV $\{(T,1),(F,0)\}$, i.e. 'True'. If the fuzzy criterion would be "speed > *moderate*", the evaluation would lead to the PTV $\{(T,0),(F,1)\}$, i.e. 'False'.
- $\perp_{I,UNKNWON}$ stands for an 'inapplicable' value which, for querying, can be treated semantically equivalent to 'unknown', where 'unknown' represents a uniform possibility distribution over the domain. This can, e.g., be used for test scores in case it is known the test was not taken. For any criterion on the attribute 'test score' (e.g. "score = *very high*", with *very high* a linguistic label for the possibility distribution representing very high test scores), not taking into account an "IS NULL" operator, the evaluation for such value will lead to the PTV $\{(T,1),(F,1)\}$, i.e. 'unknown', which indicates that it is unknown whether the concerning person is satisfactory for the query.

- $\perp_{I,False}$ stands for an 'inapplicable' value which, for querying, can be treated semantically equivalent to domain value 'False'. This can, e.g., be used in the case of the pregnancy attribute for men. If the criterion on the attribute 'pregnancy' would be "pregnancy = 'False' ", the evaluation would lead to the PTV $\{(T, 1), (F, 0)\}$, i.e. 'True'. If the criterion would be "pregnancy = 'True' ", the evaluation would lead to the PTV $\{(T, 0), (F, 1)\}$, i.e. 'False'.

Remark that the evaluation of these marked 'inapplicable' values does not always lead to the same result, but could differ depending on the context and the query. E.g. the first and last example above can evaluate to either 'True' or 'False'.

4 Conclusion

In this paper, a new approach for querying 'null' values in databases, and more specifically 'inapplicable' values, has been presented. It is shown that *marked 'null' values* can be used to come up with (semantically correct) answers, even if the user is not aware of the presence of any 'null' or 'inapplicable' values. The evaluation of such *marked 'null' values* in case they are being queried, is different than in the traditional approaches. The "mark" in a *marked 'null' value* is used to denote a domain value to which the inapplicable value should be treated semantically equivalent in case the value is queried. As a result of this, the evaluation, of such 'null' values will not always lead to the same result, but can differ depending on the context and the query.

Acknowledgements

The authors would like to thank Prof. Dr. Patrick Bosc for the fruitful discussions which, amongst others, led to the realization of this paper.

References

1. Abiteboul, S., Hull, R., Vianu, V.: Foundations of databases. Addison-Wesley Publishing Company, Reading (1995)
2. Biskup, J.: A formal approach to null values in database relations. In: Gallaire, H., Minker, J., Nicolas, J. (eds.) Advances in Data Base Theory, pp. 299–341. Plenum, New York (1981)
3. Bordogna, G., Pasi, G. (eds.): Recent Issues on Fuzzy Databases. Physica-Verlag, Heidelberg (2000)
4. Bosc, P., Pivert, O.: SQLf: A Relational Database Language for Fuzzy Querying. IEEE Transactions on Fuzzy Systems 3, 1–17 (1995)
5. Bosc, P., Kacprzyk, J. (eds.): Fuzziness in Database Management Systems. Physica-Verlag, Heidelberg (1995)
6. Codd, E.F.: RM/T: Extending the Relational Model to capture more meaning. ACM Transactions on Database Systems 4(4) (1979)
7. Codd, E.F.: Missing information (applicable and inapplicable) in relational databases. ACM SIGMOD Record 15(4), 53–78 (1986)

8. Codd, E.F.: More commentary on missing information in relational databases (applicable and inapplicable information). ACM SIGMOD Record 16(1), 42–50 (1987)
9. Date, C.J.: Null Values in Database Management. In: Relational Database: Selected Writings, pp. 313–334. Addisson-Wesley Publishing Company, Reading (1986)
10. Date, C.J.: NOT is Not 'Not'! (notes on three-valued logic and related matters). In: Relational Database Writings 1985–1989, pp. 217–248. Addison-Wesley Publishing Company, Reading (1990)
11. De Caluwe, R. (ed.): Fuzzy and Uncertain Object-oriented Databases: Concepts and Models. World Scientific, Singapore (1997)
12. De Cooman, G.: Towards a possibilistic logic. In: Ruan, D. (ed.) Fuzzy Set Theory and Advanced Mathematical Applications, pp. 89–133. Kluwer Academic Publishers, Boston (1995)
13. De Cooman, G.: From possibilistic information to Kleene's strong multi-valued logics. In: Dubois, D., et al. (eds.) Fuzzy Sets, Logics and Reasoning about Knowledge, pp. 315–323. Kluwer Academic Publishers, Boston (1999)
14. De Tré, G.: Extended Possibilistic Truth Values. International Journal of Intelligent Systems 17, 427–446 (2002)
15. De Tré, G., De Caluwe, R., Verstraete, J., Hallez, A.: Conjunctive Aggregation of Extended Possibilistic Truth Values and Flexible Database Querying. In: Andreasen, T., Motro, A., Christiansen, H., Larsen, H.L. (eds.) FQAS 2002. LNCS (LNAI), vol. 2522, pp. 344–355. Springer, Heidelberg (2002)
16. De Tré, G., De Caluwe, R.: Modelling Uncertainty in Multimedia Database Systems: An Extended Possibilistic Approach. International Journal of Uncertainty, Fuzziness and Knowledge-Based Systems 11(1), 5–22 (2003)
17. De Tré, G., De Baets, B.: Aggregating Constraint Satisfaction Degrees Expressed by Possibilistic Truth Values. IEEE Transactions on Fuzzy Systems 11(3), 361–368 (2003)
18. De Tré, G., De Caluwe, R., Prade, H.: Null Values in Fuzzy Databases. Journal of Intelligent Information Systems 30(2), 93–114 (2008)
19. Dubois, D., Prade, H.: Possibility Theory. Plenum, New York (1988)
20. Dubois, D., Prade, H.: Possibility Theory, Probability Theory and Multiple-Valued Logics: A Clarification. Annals of Mathematics and Artificial Intelligence 32(1-4), 35–66 (2001)
21. Galindo, J., Medina, J.M., Pons, O., Cubero, J.C.: A Server for Fuzzy SQL Queries. In: Andreasen, T., Christiansen, H., Larsen, H.L. (eds.) FQAS 1998. LNCS (LNAI), vol. 1495, pp. 164–174. Springer, Heidelberg (1998)
22. Imieliǹski, T., Lipski, W.: Incomplete Information in Relational Databases. Journal of the ACM 31(4), 761–791 (1984)
23. Prade, H.: Possibility sets, fuzzy sets and their relation to Lukasiewicz logic. In: 12th International Symposium on Multiple-Valued Logic, pp. 223–227 (1982)
24. Prade, H., Testemale, C.: Generalizing Database Relational Algebra for the Treatment of Incomplete or Uncertain Information and Vague Queries. Information Sciences 34, 115–143 (1984)
25. Rescher, N.: Many-Valued Logic. Mc Graw-Hill, New York (1969)
26. Riedel, H., Scholl, M.H.: A Formalization of ODMG Queries. In: 7th Working Conf. on Database Semantics, Leysin, Switzerland, pp. 90–96 (1997)
27. Yazici, A., George, R.: Fuzzy Database Modeling. Physica-Verlag, Heidelberg (1999)
28. Zadeh, L.A.: Fuzzy Sets. Information and Control 8(3), 338–353 (1965)
29. Zadeh, L.A.: Fuzzy sets as a basis for a theory of possibility. Fuzzy Sets and Systems 1, 3–28 (1978)

Describing Fuzzy DB Schemas as Ontologies: A System Architecture View

Carmen Martínez-Cruz[1], Ignacio J. Blanco[2], and M. Amparo Vila[2]

[1] Dept. Computers, 035-A3, University of Jaen
Las Lagunillas Campus, 23071, Jaen, Spain
cmcruz@ujaen.es

[2] Dept. Computer Science and Artificial Intelligence, University of Granada,
C/ Periodista Daniel Saucedo Aranda S/N, 18071, Granada, Spain
{iblanco,vila}@decsai.ugr.es

Abstract. Different communication mechanisms between ontologies and database (DB) systems have appeared in the last few years. However, several problems can arise during this communication, depending on the nature of the data represented and their representation structure, and these problems are often enhanced when a Fuzzy Database (FDB) is involved. An architecture that describes how such communication is established and which attends to all the particularities presented by both technologies, namely ontologies and FDB, is defined in this paper. Specifically, this proposal tries to solve the problems that emerge as a result of the use of heterogeneous platforms and the complexity of representing fuzzy data.

1 Introduction

Fuzzy Relational Databases (FRDBs) can be used to represent fuzzy data and perform flexible queries [10,20,18,17,7,15]. Indeed, once a fuzzy data management system has been implemented, any other application can use this flexible information to perform logical deductions [3], data mining procedures [9], data warehouse representations, etc. [4]. However, a FRDB requires complex data structures that, in most cases, are dependent on the platform in which they are implemented. Such drawbacks mean that these FRDB systems are poorly portable and scalable, even when implemented in standard Relational Databases (RDBs).

A proposal that involves representing an FRDB using an ontology as an interface has been defined to overcome these problems [6]. This ontology, whose definition is extended in this paper, provides a frame where fuzzy data are defined in a platform-independent manner and which is Web-understandable because it is represented in OWL. An implementation layer, which is responsible for parsing and translating user requests into the corresponding DB implementations in a transparent manner, is required to establish communication between the ontology and the relational database management system (RDBMS). This paper presents an architecture that shows the flow of information from the data/schema

E. Hüllermeier, R. Kruse, and F. Hoffmann (Eds.): IPMU 2010, Part II, CCIS 81, pp. 147–157, 2010.

definition, the ontology definition, to its representation in any DBMS platform. This architecture describes both the elements involved in the ontology definition process and the communication mechanisms with three different RDMBS implementations. The main differences between these implementations include their ability to manage information using internal procedures and their ability to execute developed Fuzzy SQL (FSQL) procedures which allow fuzzy data to be managed efficiently.

A brief review of databases and ontologies is presented in the 2 section, which is followed by a description of the proposed ontology in the 3 section. The system architecture that describes how communication is established between the ontology and the heterogeneous DB technologies is presented in the 4 section. Finally, an example is shown in the 5 section and conclusions are discussed in the 6 section.

2 Antecedents

Since the concept of RDBMSs first appeared in the 1970s [10,20,17], these systems have been extended in order to increase their functionality in different ways, for example, to manage time, spatial data, multimedia elements, objects, logic information, etc. One of these extensions consists of including fuzzy data management capabilities [17,18,7,15] to store and query classical or fuzzy information. A specific representation of a Fuzzy RDMBS, known as GEFRED [18], describes a complete system where fuzzy data can be represented and managed efficiently [6]. Consequently, this representation has been chosen to develop this proposal.

Ontologies allow the knowledge of any domain to be represented in a formal and standardised way [14,21]. Ontologies can represent the knowledge of a domain in a similar manner to a DB, although there are also several differences. For example, the main purpose of an ontology is to describe the reality semantically and make this semantics machine-readable, whereas the majority of databases have a commercial purpose and try to represent a data system as efficiently as possible in order to manage an organization's information. Thus, both technologies can be used to describe the same reality in different ways. Relational Databases are the subject of this paper as, despite the fact that Object-Oriented and Object-Relational Databases are more similar to ontologies interms of their representation structures, they are currently the most widely used DB mode. Both technologies currently exist alongside each other and can exchange information in order to take advantage of the information that they represent. Several proposals have been developed to establish a communication between ontologies and databases [22]. Some of these involve creating ontologies from database structures [1], whereas others populate ontologies using database data [2] and others represent databases as ontologies [8,19]. The latter are used herein to solve the problems presented.

3 Ontology Description

The ontology that describes a Fuzzy Database Schema, as defined previously [6,5], consists of a fuzzy DB schema and the fuzzy data stored in the DB (the tuples). This ontology cannot, however, represent schemas and data simultaneously as ontology classes cannot be instantiable twice, therefore two ontologies are defined, one of which describes fuzzy schema as instances of a DB catalog ontology and the other of which describes the same schema as a domain ontology which will allow the data instatiating it to be defined.

3.1 Fuzzy Catalog Ontology

An ontology, hereinafter referred to as the *Fuzzy Catalog Ontology*, has been defined to represent the Fuzzy RDBMS[18,4] catalog. This ontology contains all the relational concepts defined in the ANSI Standard SQL [12,8], for example the concept of Schema, Table, Column, Constraints, Data Types, etc., and the fuzzy structures extension to manage flexible information: *Fuzzy Columns, fuzzy labels, fuzzy discrete values* and *fuzzy data types* [6,5]. Moreover, the following fuzzy structures have been added to this ontology to complete the fuzzy RDBMS description:

- *Fuzzy Constraints.* These restrictions, which are described in table 1, can only be applied to fuzzy domains and are used either alone or in combination to generate domains such as, for example, *not unknown, undefined or null values are allowed*, or *only labels are allowed.*

Table 1. Fuzzy Constraints

Constraint	*Description*
Label_Const	The use of labels for this attribute is not allowed.
Crisp_Const	Crisp values are not allowed for this attribute.
Interval_Const	Interval values are not allowed for this attribute.
Trapezoid_Const	Non-trapezoidal values are allowed for this attribute.
Approx_Const	Non-approximated values are allowed for this attribute.
Nullability_Const	Null values cannot be defined for this attribute.
Unknown_Const	Undefined values cannot be defined for this attribute.
Undefined_Const	Unknown values cannot be defined for this attribute

- *Fuzzy Domains.* These represent a set of values that can be used by one or more attributes. They are defined by a fuzzy datatype, one or more fuzzy constraints, and those labels or discrete values that describe this domain. For example, *Temperature of the land is a Fuzzy Type 2 attribute (because it is defined over a numerical domain) which has the Not Null and Only Label constraints and the only labels it can use are High, Medium, Low.*

3.2 Fuzzy Schema Ontology

A *Fuzzy Schema Ontology* is a domain Ontology [11] that represents a specific FDB schema. This ontology is generated using the previously defined fuzzy schema as instances of the *Fuzzy Catalog Ontology*. The generation process for this ontology has been described previously [6] and consists of translating the *Table* instances into classes, and the *Attributes* instances into properties. The *constraints* restrictions establish the conections between the *Attribute* properties and the Fuzzy Data Structures. For example, the object property of the *Tall* attribute allows *Trapezoidal, Approximate, Crispt, Interval, Label values* to be defined but not *Null, Unknown or Undefined* ones. Moreover, the previously defined structures, namely *Fuzzy Labels* and *Discrete Tags*, are included in the *Schema Ontology*. The result is a mixed ontology containing the *Fuzzy Schema Ontology* and the *Fuzzy Catalog Ontology* instances or a new ontology where the fuzzy values are imported as a separate ontology.

4 Architecture

As mentioned above, the representation of fuzzy data in an RDBMS using the described ontologies is not trivial due, first of all, to the complexity of the data structure. Secondly, the schema and data representation are dependent on the characteristics and functionality added to the RDBMS implementation. A fuzzy database representation ontology simplifies the process of representing this information because the data structure is defined independently of any RDBMS implementation. However, when represented as ontologies, the data are defined without any need for constraints due to the flexibility of the data model, which means that a data-definition process where the sequence of steps and the elements involved in the ontology definition process are explicitly declared is necessary. Moreover, despite their noticeable differences, most RDBMS platforms are catalogued and predefined to make the data representation particularities invisible to the user. These systems have been categorised accordingly to their fuzzy management capabilities in the system architecture shown in figure 1. This architecture specifies the flow of information from the user to a DBMS and the different elements involved when defining a classic or fuzzy DB. This architecture is divided into two subarchitectures, the first of which manages the representation of the information into ontologies and the second one of which manages the communication between a schema ontology and any DBMS implementation.

4.1 Ontology Architecture

This architecture guides the definition of a database schema as an ontology by the user. The elements involved in this process are:

- *User Interface.* This interface allows the user to represent all the elements that constitute a fuzzy database model as an ontology independently of the particularities of any RDBMS system.

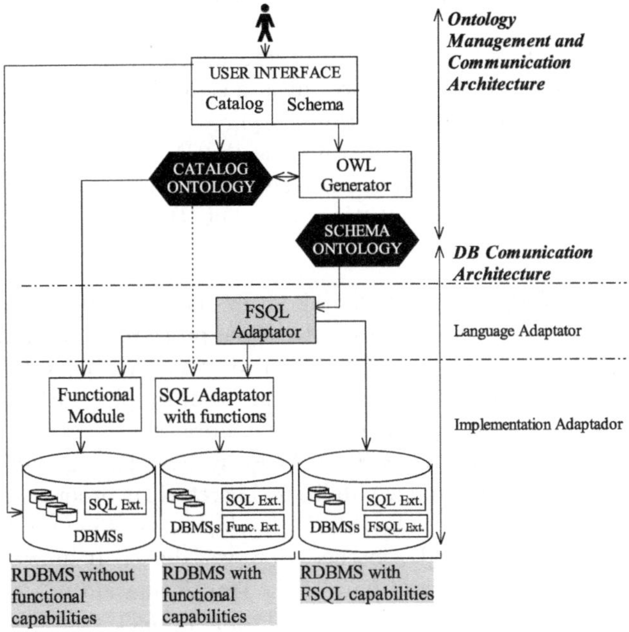

Fig. 1. System Architecture

- *Catalog Ontology.* This is the ontology that represents a Fuzzy RDBMS. A schema definition is performed to instantiate this ontology.
- *OWL Generator.* This module allows the *Schema Ontology* to be generated automatically from the previously defined instances of the *Catalog Ontology.*
- *Schema Ontology.* This is the equivalent ontology to the *Catalog Ontology* instances that describes a specific fuzzy schema. Consequently, this ontology can be instantiated to store fuzzy data.

The following applications has been developed to accomplish this functionality:

- A Protégé [16] plug-in which allows any fuzzy database schema to be defined in an intuitive and platform-independent manner. The domain ontology can be generated automatically in this plug-in using the definition of the fuzzy database schema. A screen shot of this application is shown in Figure 3.
- A Protégé [16] plug-in that allows the user to define fuzzy data in a domain ontology. These data can be loaded from the ontology, stored in it, loaded from a database or stored in one or more database systems. Moreover, this plug-in provides the user with the fuzzy domain information previously defined in the schema, thus making the definition process as easy as possible. A screen shot of this application is shown in Figure 4.

All these applications provide a DB connection as described below.

4.2 Database Communication Architecture

This architecture guides the process of generating fuzzy database schemas and data, previously defined as ontologies, in a specific RDBMS. First of all, the DBMS must have the catalog structures required to define the fuzzy information installed in it. The connection with the RDBMS can then be established in accordance with the fuzzy data management implementation that the DB system had installed. At this point, three different RDBMS implementations are identified:

- *RDBMS with FSQL capabilities.* These systems are able to execute any FSQL[13] sentence as they can execute stored procedures and have the developed FSQL management libraries installed. This capability makes interaction with the DB faster and the communication simpler, although currently only *Oracle* © systems can manage these libraries.
- *RDBMS with functional capabilities.* These systems have no access to FSQL libraries but have functional capabilities. Therefore, in order to manage fuzzy data, an external module, called the SQL adapter, is defined to translate any FSQL sentence into an SQL sentence. However, due to its functional capabilities, this system can execute part of the transformation process internally, thus making the system more efficient. *PostgreSQL*© systems are included in this modality.
- *RDBMS without functional capabilities.* This system only implements basic SQL functions, which means that any procedure to manage fuzzy data must be done outside of this system, thus delaying the process. These external functions are defined using Java procedures. *MySQL* © is an example of this kind of system.

The procedures to manage communication with the different implementations include the development of a parser that considers the particularities of each RDBMS. Finally, a translator converts a fuzzy domain ontology into the appropriate database query depending on the capabilities of the RDBMS implementation.

As a result, this architecture allows several connections to different database implementations to be established simultaneously. Indeed, the same database schema can be replicated in heterogeneous RDBMSs. This functionality has been implemented in the plug-ins developed and shown in figures 3, 4, where a simultaneous connexion to Oracle © and MySql © systems has been established.

5 Example

One part of a fuzzy database schema concerning *"Land Characteristics"* is shown in figure 2 item A) as an example. The definition process of this schema involves instantiating the *Ontology Catalog* described in the 3.1 section. For example, the relation *Location* has the attribute *Tavg*, which represents the average temperature. *Tavg* values can be fuzzy or classical numbers or one of the labels *low, high*

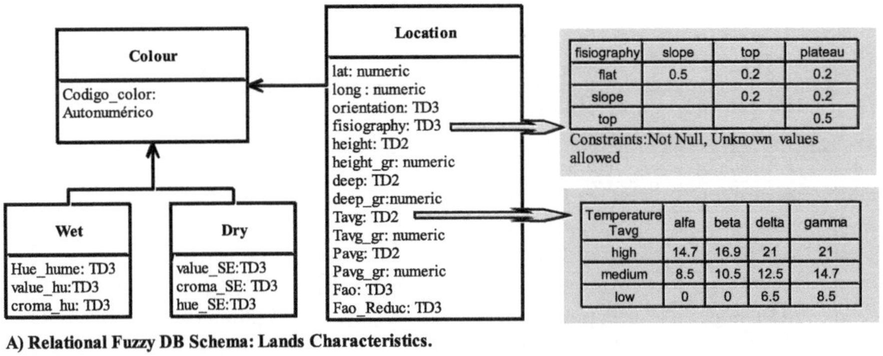

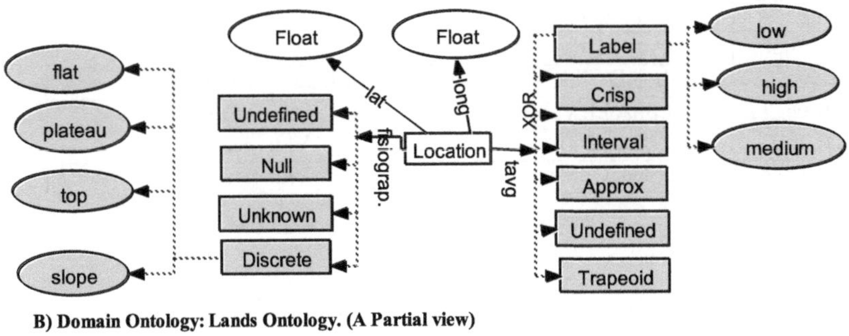

Fig. 2. Example of a land analysis

or medium, but never *Null* or *Unknown*. A subset of the resulting instances is shown in table 2.

Once the instances have been completely defined, the corresponding domain ontology can be generated. One part of the domain ontology generated is shown in figure 2 item B). The set of classes of this ontology, which are represented as white rectangles, can be instantiated to define fuzzy data (tuples). All the fuzzy attributes are properties, represented as arrows, that point towards their allowed representation structures. For example, the *Tavg* property points towards *Label, Crisp, Interval Approx, Undefined or Trapezoid* data structures. These fuzzy structures are classes imported from the *Catalog Ontology* and are marked as grey ovals in the figure 2. *Label* values, namely *low, high, medium*, have previously been defined and imported for the *Tavg* domain.

A screen shot of the application used to define fuzzy schemas is shown in figure 3. The definition of the entire *Land Characteristics* schema is shown as an ontology in the application. Moreover, all the attributes and constraints defined for the class *Location*, along with the labels and constraints defined for the attribute *Tavg*.

The plug-in that represents the domain ontology is shown in the screen shot in figure 4. This application shows the structure of a concrete relation, namely

Table 2. *Land* Ontology Schema Instances

ID	instance of	values
Localization	Table	Ref: Lat, fisiography, Tav, ...
Lat	Base_Column	-
PK_Loc	Primary_Key	Ref: Lat, Long
fisiography	Fuzzy_Column	Ref: Dom_fisiog
Tavg	Fuzzy_Column	-
Dom_fisiog	Fuzzy_Domain	Ref: TD_fisiog, XX
Dom_Tavg	Fuzzy_Domain	Ref: TD_Tavg, FC1_Tavg, FC2_Tavg,..
TD_fisiog	FType2_Struct	Value: 1
TD_Tavg	FType3_Struct	Value: 3,4 Ref: Float
Flat	Discrete_Definition	-
Slope	Discrete_Definition	-
Flat-Slope	Discrete_Relation	Val: 0.5
Low	Label_Definition	Ref: Low_Tavg_TD
Low_Tavg_TD	Trapezoid_Value	[0,0,6.5,8.5]
FC1_Tavg	Nullability_Constraint	Val: true
FC2_Tavg	Unknown_Constraint	Val: true
...	...	...

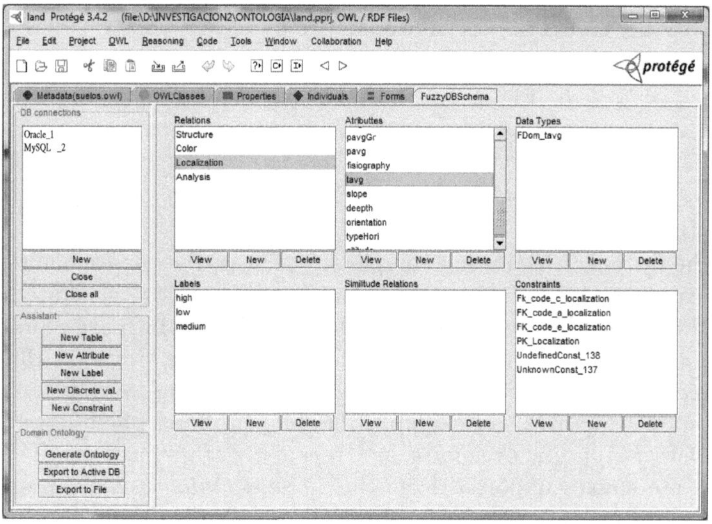

Fig. 3. Protégé plug-ins for defining fuzzy tuples in a Fuzzy DB

Location, for defining or showing the data associated with it. Furthermore, the interface restricts the values that can be inserted into the environment on the basis of the schema data constraints and provides the imported labels for the appropriate attributes to make the data-definition process easier.

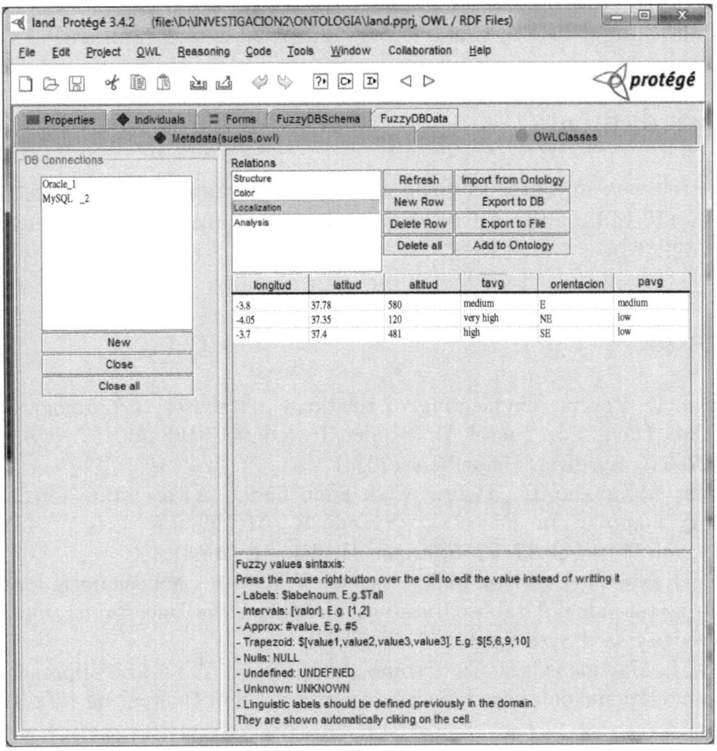

Fig. 4. Protégé plug-ins for defining fuzzy tuples in a Fuzzy DB

6 Conclusions

A fuzzy relational database has been isolated from any specific database implementation and fuzzy database schemas have been presented as OWL ontologies, thereby making this knowledge platform-independent and accessible through the Semantic Web. Consequently, DB information can be shared among different database systems and heterogeneous data representations. The architecture defined in this paper defines the flow of information within the system and helps to identify the elements involved in the process of communication between the user and the DB systems. In this process, the fuzzy capabilities that a RDBMS can execute are detected to choose the most suitable platform to serve any request. Moreover, the implementation of this proposal allows any data or schema to be defined in several and heterogeneous RDBMS implementations simultaneously. Finally, this architecture presents a highly scalable system where the FRDBMS can be extended easily with other functionalities already implemented in an RDBMS [3,9], such as logical databases and data mining operations using fuzzy data. Both these extensions will be implemented in the near future. Moreover, an extension to represent object functionalities required to complete the ANSI 2003 standard described in this ontology proposal is also planned.

Acknowledgments

This research was supported by the Andalusian Regional Government (projects TIC1570, TIC1433 and TIC03175) and the Spanish Government (project TIN2008-02066).

References

1. Astrova, I.: Reverse engineering of relational databases to ontologies. In: Bussler, C.J., Davies, J., Fensel, D., Studer, R. (eds.) ESWS 2004. LNCS, vol. 3053, pp. 327–341. Springer, Heidelberg (2004)
2. Barrasa, J., Corcho, O., Perez, A.G.: Fund finder: A case study of database to ontology mapping. In: Fensel, D., Sycara, K., Mylopoulos, J. (eds.) ISWC 2003. LNCS, vol. 2870, pp. 17–22. Springer, Heidelberg (2003)
3. Blanco, I., Martin-Bautista, M.J., Pons, O., Vila, M.A.: A mechanism for deduction in a fuzzy relational database. International Journal of Uncertainty, Fuzziness and Knowledge-Based Systems 11, 47–66 (2003)
4. Blanco, I., Martinez-Cruz, C., Serrano, J.M., Vila, M.A.: A first approach to multipurpose relational database server. Mathware and Soft Computing 12(2-3), 129–153 (2005)
5. Blanco, I., Martínez-Cruz, C., Vila, M.A.: Looking for Information in Fuzzy Relational Databases accessible via the Web. In: Handbook of Research on Web Information Systems Quality, pp. 300–324. Idea Group Ref. (2007)
6. Blanco, I.J., Vila, M.A., Martinez-Cruz, C.: The use of ontologies for representing database schemas of fuzzy information. International Journal of Intelligent Systems 23(4), 419–445 (2008)
7. Bosc, P., Galibourg, M., Hamon, G.: Fuzzy querying with sql: Extensions and implementation aspects. Fuzzy Sets and Systems 28, 333–349 (1988)
8. Calero, C., Piattini, M.: Ontological Engineering: Principles, Methods, Tools and Languages. In: An Ontological Approach to SQL 2003, pp. 49–102. Springer, Heidelberg (2006)
9. Carrasco, R.A., Vila, M.A., Galindo, J.: Fsql: a flexible query language for data mining. In: Enterprise information systems IV, pp. 68–74 (2003)
10. Codd, E.F.: Extending the database relational model to capture more meaning. ACM Transactions on Database Systems 4, 262–296 (1979)
11. Corcho, O., FernándezLópez, M., GómezPérez, A.: Ontological Engineering: Principles, Methods, Tools and Languages. In: Ontologies for Software Engineering and Software Technology, pp. 49–102. Springer, Heidelberg (2006)
12. International Organization for Standardization (ISO). Information Technology. Database language sql. parts 1 to 4 and 9 to 14. 9075-1:2003 to 9075-14:2003 International Standards Standard, No. ISO/IEC 9075: 2003 (September 2003)
13. Galindo, J., Medina, J.M., Pons, O., Cubero, J.C.: A server for fuzzy sql queries. In: Proceedings of the Third International Conference on Flexible Query Answering Systems, pp. 164–174 (1998)

14. Gómez-Pérez, A., Férnandez-López, M., Corcho-García, O.: Ontological Engineering. Springer, New york(2003)
15. Kacprzyk, J., Zadrozny, S.: Sqlf and fquery for access. In: IFSA World Congress and 20th NAFIPS International Conference. Joint 9th, vol. 4, pp. 2464–2469 (2001)
16. H. Knublauch. An ai tool for the real world. Knowledge modeling with protègè. Technical report, http://www.javaworld.com/javaworld/jw-06-2003/jw-0620-protege.html.
17. Ma, Z.: Fuzzy Database Modeling of Imprecise and Uncertain Engineering Information. Springer, Heidelberg (2006)
18. Medina, J.M., Pons, O., Vila, M.A.: Gefred. a generalized model of fuzzy relational databases. Information Sciences 76(1-2), 87–109 (1994)
19. de Laborda Perez, C., Conrad, S.: Relational.owl: a data and schema representation format based on owl. In: CRPIT '43: Proceedings of the 2nd Asia-Pacific conference on Conceptual modelling, pp. 89–96 (2005)
20. Raju, K.V.S.V.N., Majumdar, A.K.: Fuzzy functional dependencies and lossless join decomposition of fuzzy relational database systems. ACM Transactions on Database Systems 13(2), 129–166 (1988)
21. Studer, R., Benjamins, V.R., Fensel, D.: Knowledge engineering: Principles and methods. IEEE Transactions on Data and Knowledge Eng. 25(1-2), 161–197 (1998)
22. Vysniauskas, E., Nemuraite, L.: Transforming ontology representation from owl to relational database. Information Technology and Control 35(3A), 333–343 (2006)

Using Textual Dimensions in Data Warehousing Processes

M.J. Martín-Bautista[1], C. Molina[2], E. Tejeda[3], and M. Amparo Vila[1]

[1] University of Granada, Spain
[2] University of Jaen, Spain
[3] University of Camagüey, Cuba

Abstract. In this work, we present a proposal of a new multidimensional model handling semantical information coming from textual data. Based of a semantical structures called AP-structures, we add new textual dimensions to our model. This new dimension aloow the user to enrich the data analisis not only using lexical information (a set or terms) but the meaning behind the textual data.

1 Introduction

The information and knowledge management is a strategic activity for the success of the companies. Textual information takes part of this information, specially from the coming of the Internet. However, it is complex to process this kind of data due to the lack of structure and its heterogeneity. For this reason, there exist not many integrated tools processing this textual information together with other processes such as Data Mining, Data Warehouse, OLAP, etc.

In particular, and as far as we know, there exists no implementations of Data Warehousing and OLAP able to analyze textual attributes in databases from a semantical point of view. The proposal in this work try to solve this problem.

This work shows a multidimensional model with semantical treatment of texts to build the data cubes. In this way, we can implement a Data Warehousing with OLAP processing using this model. That is, a process that be able to get useful information from textual data coming from external files or from textual attributes in a database.

For this purpose, this paper is organized as follows: in the next section, we review the literature related to our proposal, specially those works about Data Warehousing with texts. In section 3, we presents a classical multidimensional model as a base for the extension without textual dimensions. Section 4 collects the formal model proposed and in 5 an example is shown. The papers finishes with the main conclusions.

2 Related Work

In this section, we include some of the most relevant works about Data Warehousing related to processing of textual data. In most of them, different techniques

E. Hüllermeier, R. Kruse, and F. Hoffmann (Eds.): IPMU 2010, Part II, CCIS 81, pp. 158–167, 2010.

are used to manage textual data and to incorporate them in a multidimensional model, but the source of texts are usually XML documents or texts with some internal structure.

The creation of a Data Warehouse of Words (WoW) is proposed in [3]. This proposal extracts a list of words from plain text and XML documents and stores the result in DataCubes. The proposal in [9] is based on XML too, and propose a distribute system to build the datacubes in XML. In [10], short texts such as emails and publications are transformed into a multidimensional models and can be queried.

Obviously, the restriction of using XML or structured texts implies generally the intervention of the user to generate them and structure them. In our proposal, the entry can be either a set of external files in plain text, XML, or any other format. However, our approach also considers the textual attributes in a database; in fact, whatever the entry data, they are transformed into an attribute in a database which will be a textual dimension in the future. This transformed textual attribute has two main advantages with respect to other textual representations. First, it takes the semantic of the text. In this process althought statistic a method is applied, the resulting structure is directly understandable by the user. Second, it can be obtained automatically and without the user's intervention. The process to perform this transformation is shown in section 4.4. Due to the semantical treatment of the textual data, a semantical dimension in the data cube is generated. Data Warehousing and OLAP processes are then performed.

3 Background

3.1 The *Classical* Multidimensional Model

The model presented here is a resume of the characteristics of the first models proposed in the literature of Data Warehousing and OLAP [1], [4], since we do not consider that there be a standard one [8]. This model is the base of most of the proposals reviewed in Section 2, and also the starting point to achieve our goal: a new multidimensional model with a more powerful textual processing.

In a classical multidimensional model we can consider the following elements:

- A set of *dimensions* $d_1, ..d_n$ defined in a database. That is, attributes with a discrete domain belonging to the database scheme. The data are grouped attending these attributes. Each dimension d_i has associated:
  - A basic domain $D_i = \{x_1....x_{m_i}\}$ of discrete values so, each tuple t of the database takes an unique and well determined value x_i in the attribute d_i. Let us note $d_i[t] = x_i$.
  - A grouping hierarchy that allow us to consider different values for the analysis. Such a hierarchy $\mathcal{H}_i = \{\mathcal{C}_{i1}...\mathcal{C}_{il}\}$ is formed by partitions D_i in a way that:

$$\forall k \in \{1, 2...l\}\ \mathcal{C}_{ik} \subseteq \mathcal{P}(D_i)\ \mathcal{C}_{ik} = \{X^1_{ik}, .., X^h_{ik}\}$$

being $\forall j, r\ X_{ik}^{j} \bigcap X_{ik}^{r} = \emptyset$ and $\bigcup_{j=1}^{h} X_{ik}^{j} = D_i$.
The hierarchy $\mathcal{H}_i$ is an inclusion reticulum which minimal element is D_i, considering element by element, and the maximal is D_i considering a partition of just one element.

- A numeric measure V associated to these dimensions, so we can always obtain $V = f(Y_1, Y_2..Y_n)$ where $Y_1..Y_n$ are values of the dimensions considered above. We must point out that these values may not be exactly the same as the ones in the domain, but the ones in some partition of the hierarchy. That is, if we consider the level $\mathcal{C}_{ik}$ in the dimension d_i, then $Y_i \in \mathcal{C}_{ik}$. This measure V can be:
  - A count measure which gives us the number of tuples in the database that verify $\forall i \in \{1,..n\}\ d_i[t] \in Y_i\}$
  - Any other numerical attribute that is semantically associated to the considered dimensions.
- There exits also an aggregation criterion of V, AGG, which is applied when 'set' values are considered in any of the dimensions. That is,

$$V = f(x_1, ..Y_k, ..x_n) = AGG_{x_k \in Y_k} f(x_1, ..x_k, ..x_n)$$

  AGG can be a sum, SUM, or any other statistical function like the average, AVG, the standard deviation, STD, etc. Obviously, is the measure is the count one, the aggregation function is SUM.

From the concept of data cube, the normal operations are defined. They correspond to the different possibilities of analysis on the dimensions (roll-up, drill-down, slice and dice).

We must also remark that there are other approaches in the literature where there are no explicit hierarchies defined on the dimensions, like the one in [2].

4 Formal Model

Due to space limitation, in this paper we only present the main aspect needed to understand de proposal. The complete model can be found in [7, 6, 5].

4.1 AP-Set Definition and Properties

Definition 1. AP-Set
Let be $X = \{x_1...x_n\}$ any referential and $\mathcal{R} \subseteq \mathcal{P}(X)$ we will say $\mathcal{R}$ is and AP-Set if and only if:

1. $\forall Z \in \mathcal{R} \Rightarrow \mathcal{P}(Z) \subseteq \mathcal{R}$
2. $\exists Y \in \mathcal{R}$ *such that :*
   (a) $card(Y) = max_{Z \in \mathcal{R}}(card(Z))$ *and not exists* $Y' \in \mathcal{R}$ *such that* $card(Y') = card(Y)$
   (b) $\forall Z \in \mathcal{R}; Z \subseteq Y$

The set Y of maximal cardinal characterizes the AP-Set and it will be called *spanning set of* $\mathcal{R}$. We will denote $\mathcal{R} = g(Y)$, that is $g(Y)$ will be the AP-Set with spanning set Y.

We will call *Level of* $g(Y)$ to the cardinal of Y. Obviously, AP-Set of level equal to 1 are the elements of X, we will consider the empty set $\emptyset$ as the AP-Set of zero level.

It should be remarked that the definition 8 implies that any AP-Set $g(Y)$ is in fact the reticulum of $\mathcal{P}(Y)$

Definition 2. AP-Set Inclusion
Let be $\mathcal{R} = g(R)$ *and* $\mathcal{S} = g(S)$ *two AP-Sets with the same referential:*

$$\mathcal{R} \subseteq \mathcal{S} \Leftrightarrow R \subseteq S$$

Definition 3. Induced sub-AP-Set: *Let be* $\mathcal{R} = g(R)$ *and* $Y \subseteq X$ *we will say* $\mathcal{S}$ *is the* sub-AP-Set induced by Y *iff:*

$$\mathcal{S} = g(R \bigcap Y)$$

Definition 4. Induced super-AP-Set: *Let be* $\mathcal{R} = g(R)$ *and* $Y \subseteq X$ *we will say* $\mathcal{V}$ *is the* super-AP-Set induced by Y *iff:*

$$\mathcal{V} = g(R \bigcup Y)$$

4.2 AP-Structure Definition and Properties

Once we have established the AP-Set concept we will use it to define the information structures which appear when frequent itemsets are computed. It should be considered that such structures are obtained in a constructive way, by initially generating itemsets with cardinal equal to 1, next these ones are combined to obtain those of cardinal equal 2, and by continuing until getting itemsets of maximal cardinal, with a fixed minimal support. Therefore the final structure is that of a set of AP-Sets, which formally is defined as follows.

Definition 5. AP-Structure
Let be $X = \{x_1 ... x_n\}$ *any referential and*
$S = \{A, B, ...\} \subseteq \mathcal{P}(X)$ *such that:*

$$\forall A, B \in S\,;\; A \nsubseteq B\,,\; B \nsubseteq A$$

We will call AP-Structure of spanning S,
$\mathcal{T} = g(A, B, ...)$, *to the set of AP-Set whose spanning sets are* $A, B, ...$

Now we will give some definition and properties of these new structures.

Definition 6. *Let be* $\mathcal{T}_1$, $\mathcal{T}_2$, *two AP-Structures with the same referential:*

$$\mathcal{T}_1 \subseteq \mathcal{T}_2 \;\Leftrightarrow\; \forall \mathcal{R} \textit{ AP-Set of } \mathcal{T}_1\,,$$

$$\exists \mathcal{S} \textit{ AP-Set of } \mathcal{T}_2 \textit{ such that } \mathcal{R} \subseteq \mathcal{S}$$

It should be remarked that the inclusion of AP-Set is that which is given in the definition 2.

Extending the definitions 3 and 4 we can defined the *Induced AP-Substructure* and *Induce AP-Superstructure* (see [6] for details).

4.3 Matching Sets with AP-Structures

Now we will establish the basis for querying in a database where the AP-structure appears as data type. The idea is that the users will express their requirements as sets of terms and in the database will be AP-structures as attribute values, therefore some kind of matching has to be given.

Two approaches are proppopsed: *weak* and *strong matching*. A detail definition can be found in [5, 6]. The idea behind the matching is compare the spanning sets for the AP-struture and the set of terms given by the user. The *strong matching* consider that the set of terms by tue user and the AP-structure match if all the terms are include in a spanning set. The *weak matching* relaxes the condition and return *true* if at least on of the term is included ina spanning set.

These matching criterias can be complemented by giving some measures or indexes which quantify these matchings. The idea is to consider that the matching of a long set of terms will have an index greater than other with less terms, additionally if some term set match with more than one spanning set will have an index greater than that of the other one which only match with one set. Obviously two matching indexes can be established, but both two have similar definitions.

Definition 7. strong(weak) matching index

Let be an AP-structure $\mathcal{T} = g(A_1, A_2, ..., A_n)$ *with referential* X *and* $Y \subseteq X$, *we define the* strong(weak) matching index between Y and $\mathcal{T}$ *as follows:*

$\forall A_i \in \{A_1, A_2, ..., A_n\}$ *we denote* $m_i(Y) = card(Y \bigcap A_i)/card(A_i)$, $S = \{i \in \{1, ..., n\} | Y \subseteq A_i\}$, $W = \{i \in \{1, ..., n\} | Y \bigcap A_i \neq \emptyset\}$.

Then we define the strong and weak matching indexes between Y and $\mathcal{T}$ *as follows:*

$$\textbf{Strong index} = S(Y|\mathcal{T}) = \sum_{i \in S} m_i(Y)/n$$

$$\textbf{Weak index} = W(Y|\mathcal{T}) = \sum_{i \in W} m_i(Y)/n$$

Obviously:

$$\forall Y \text{ and } \mathcal{T}\ ,\ S(Y|\mathcal{T}) \in [0,1]\ ,\ W(Y|\mathcal{T}) \in [0,1] \text{ and } W(Y|\mathcal{T}) \geq S(Y|\mathcal{T})$$

4.4 Transformation into an AP-Attribute

In this section we briefly describe the process to transform a textual attribute in an AP-structure valuated attribute, what we call an *AP-attribute.*

1. The frequent terms associated to the textual attribute are obtained. This process includes cleaning process, empty words deleting process, synonymous management process using dictinaries, etc. Then we get a set of basic terms T to work with. In this point the value of textual attribute on each tuple t is subset of basic terms T_t. This consideration allow us to work with the tuples as in a transactional database regarding the textual attribute.

2. Maximal frequent itemsets are calculated. Been $\{A_1, .., A_n\}$ the itemsets, the AP-structure $S = g(A_1, .., A_n)$ includes all the frequent itemsets, so we can consider the AP-structure to cover the semantic of the textual attribute.
3. Once we have the global AP-structure, we obtain the AP-structure associated to tuple t: if T_t is the set of terms associated to t, the value of AP-attribute for the tuple is:

$$S_t = g(A_1, .., A_n) \bigwedge T_t$$

This process obtains the domain for any AP-attribute.

Definition 8. Domain of an AP-attribute: *Considering a database to build the AP-attribute A with global structure $(A_1, ..., A_n)$, the domain of attribute A is*

$$D_A = \{R = g(B_1, ..B_m), /, \forall i \in \{1, .., m\}, \exists, j \in \{1, .., n\} such\ that B_i \subseteq A_j\}$$

So D_A is the set of all sub-AP-structures of the global AP-structure associated to the attribute, because these are all the possible values for attribute A according to previous constraint.

As an example let consider a simplification of data of patient in an emergencies service at an hospital. Table 1 shows some records stored in the database. Attributes *Patient number (no)*, *Waiting time*, *Town* are classical attributes. *Diagnosis* is textual attribute that stores the information given by the medical doctor for the patient.

After applying the propposed process, we tranform the textual attribute into an AP-attribute. Figure 1 shows the AP-structure obtained for the diagnosis attribute. The sets at the top of the structure are the spanning set of the attribute. The other are all the possible subsets with the elements in the spanning sets.

Then the database is transformed to stores the spanning sets associated to each records as shown in Table 2.

4.5 Dimension Associated to an AP-Attribute

To use the AP-attribute on a multidimensional model we need to define a concept hierarchy and the operations over it. We need first some considerations.

Table 1. Example of database with a textual attribute

No.	Waiting time	Town	Diagnosis
1	10	Granada	pain in left leg
2	5	Gojar	headache and vomit
3	10	Motril	voimit and headache
4	15	Granada	rigth arm fractured and vomit
5	15	Armilla	intense headache
...	...	...	...

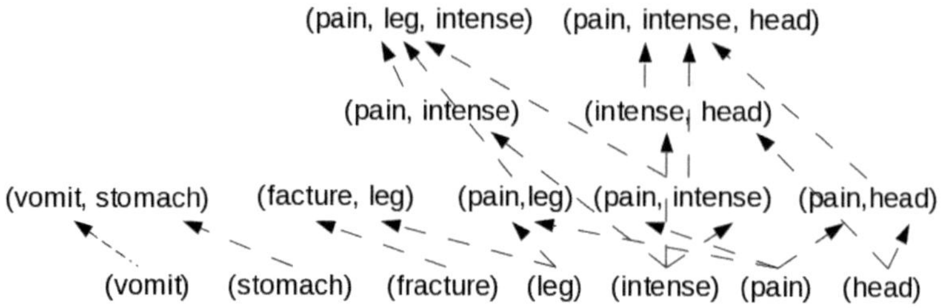

Fig. 1. Global AP-structure

Table 2. Database after the process

No.	Waiting time	Town	spanning set of AP-attribute
1	10	Granada	(pain,leg)
2	5	Gojar	(pain head), (vomit)
3	10	Motril	(pain,head), (vomit)
4	15	Granada	(fracture) (vomit)
5	15	Armilla	(pain,intense, head)
6	5	Camaguey	(pain, intense, leg)
7	5	Málaga	(pain, leg)
8	5	Sevilla	(pain,head)
9	10	Sevilla	(pain), (stomach)
10	5	Gojar	(fracture)
11	10	Granada	(fracture leg)
12	5	Santafé	(fracture) (head)
13	5	Madrid	(vomit, stomach)
14	5	Madrid	(vomit, stomach)
15	12	Jaen	(pain, intense, leg)
16	15	Granada	(pain, intense, leg)
17	5	Motril	(pain, intense, head)
18	10	Motril	(pain, intense)
19	5	London	(fracture, leg)
20	15	Madrid	(pain, intense), (vomit, stomach)

- Although the internal representation of a AP-attribute are structures, the input and output for the user is carry out by means of terms sets ("sentences"), which are spanning for the AP-structures.
- This will be the same case for OLAP. The user will give as input a set of sentences, as values of the dimension, although these sentences are values of the AP-attribute domain.
- According to definition 8 we are working with a structure domain and closed when we consider the union. So, a set of elements of the domain is include in the domain. Then, the basic domain for a dimension associated to an AP-structure and the domain of the hierarchies is the same.

According to these considerations we have the following definition.

Definition 9. AP-structure partition associated to a query
Let $C = \{T_1, .., T_q\}$ where $T_i \subseteq X$ is subset of "sentences" given by an user for a dimension of a AP-attribute. Been S the global AP-structure associated to that attribute. We define the **AP-structure partition associated to** C *as:*

$$\mathcal{P} = \{S_1, .., S_q, S_{q+1}\}$$

where

$$S_i = \begin{cases} S \bigwedge T_i & \textit{if } i \in \{1, .., q\} \\ S \bigwedge (X - \bigcup_{i=1}^{q} T_i) & \textit{otherwise} \end{cases}$$

Now we can introduce a multidimensional model as define in section 3.1 that use an AP-dimension:

- $\forall i \in \{1, .., q\}$ $f(.., S_i, ..)$ is an aggregation (count, or other numeric aggregation) associated with the tuples that satisfy T_i in any way.
- $f(..., S_q, ...)$ is an aggregation associated to the tuples not matching any sentences in T_i, or part of them. That means, the sentences that are not related with the sentences given by the user.

Obviously, the matching concept and the considerate aggregations have to be adapted to the characteristics of an AP-dimension.

5 Example

Let consider the example introduced in Section 4.4 about an emergencies service at an hospital to show how queries are answered in a datacube with the AP-attribute.

Let suppose the partition for the following query:

$$C = \{(pain, intense), (vomit)\}$$

If we choose the count aggregation and the *weak matching* (definition 7) the results are shown in Table 3. On the other hand, if we use the strong matching (definition 7) the results are the one collected in Table 4. As it was expected, when considering the *weak matching* more records satisfy the constraint than for the very strict *strong matching*

We can use classical dimensions for the query and the AP-attribute at the same time. Let suppose we have an hierarchy over *home town* attribute and we grouped the values as follows:

{Granada county, Malaga county, Jaen county, Rest of Spain, Abroad}

If we choose again the *count* aggregation the result for weak matching and strong matching are shown in Tables 5 and 6 respectively. A example using a different aggregation function is shown in Table 7, using the *average* to aggregate the *waiting time*.

Table 3. One dimension datacube using weak matching

(pain intense)	(vomit)	Other	Total
13	6	4	23

Table 4. One dimension datacube using strong matching

(pain intense)	(vomit)	Other	Total
7	6	8	21

Table 5. Two dimensions datacube using weak matching

	(pain intense)	(vomit)	Other	Total
Granada c.	7	3	3	13
Malaga c.			1	1
Jaen c.	1			1
Rest of Spain	3	3	0	6
Abroad	1		1	2
Total	13	6	4	23

Table 6. Two dimensions datacube using strong matching

	(pain intense)	(vomit)	Other	Total
Granada c.	4	3	4	11
Malaga c.			1	1
Jaen c.	1			1
Rest of Spain	1	3	2	6
Abroad	1		1	2
Total	7	6	8	21

Table 7. Two dimensional datacube using strong matching and average time aggregation

	(pain intense)	(vomit)	Other	Total
Granada c.	11.5	10	7.5	9.3
Malaga c.			5	5
Jaen c.	12			12
Rest of Spain	15	6.5	7.5	9.6
Abroad	5		5	5
Total	10.8	8.3	6.6	

6 Conclusions

In this paper we have presented a multidimensional model that supports the use of textual information in the dimensions by means of a semantical structures called AP-structures. To build these structure, a process is carried out so these AP-structure represent the meaning behind the text instead of a simple set of terms. The using of the AP-structure inside the multidimensional model enrich the OLAP analisys so the user may introduce the sematic of textual attribute in the queries over the datacube.

To complete the model we need to provide the dimension associated to the AP-attribute with the normal operation over a hierarchy allow the user to choose different granularities in the detail levels. All these extension to the multidimensional will be integrated inside an OLAP system to build a prototype a test the behaviour of the proposal with real databases.

References

[1] Agrawal, R., Gupta, A., Sarawagi, S.: Modeling multidimensional databases (1995)
[2] Datta, A., Thomas, H.: The cube data model: A conceptual model and algebra for on-line analytical processing in data warehouses. Decision Support Systems 27, 289–301 (1999)
[3] Keith, S., Kaser, O., Lemire, D.: Analyzing large collections of electronic text using olap. In: APICS 2005 (2005); Technical report
[4] Kimball, R.: The Data Warehouse Toolkit. Wiley, Chichester (1996)

[5] Marín, N., Martín-Bautista, M.J., Prados, M., Vila, M.A.: Enhancing short text retrieval in databases. In: Larsen, H.L., Pasi, G., Ortiz-Arroyo, D., Andreasen, T., Christiansen, H. (eds.) FQAS 2006. LNCS (LNAI), vol. 4027, pp. 613–624. Springer, Heidelberg (2006)
[6] Martín-Bautista, M.J., Martínez-Folgoso, S., Vila, M.A.: A new semantic representation for short texts. In: Song, I.-Y., Eder, J., Nguyen, T.M. (eds.) DaWaK 2008. LNCS, vol. 5182, pp. 347–356. Springer, Heidelberg (2008)
[7] Martín-Bautista, M.J., Prados, M., Vila, M.A., Martínez-Folgoso, S.: A knowledge representation for short texts based on frequent itemsets. In: Proceedings of IPMU, Paris, France (2006)
[8] Molina, C., Rodríguez-Ariza, L., Sánchez, D., Vila, M.A.: A new fuzzy multidimensional model. IEEE T. Fuzzy Systems 14(6), 897–912 (2006)
[9] Niemi, T., Niinimki, M., Nummenmaa, J., Thanisch, P.: Applying grid technologies to xml based olap cube construction. In: Proc. DMDW 2003, pp. 2003–2004 (2003)
[10] Tseng, F.S.C., Chou, A.Y.H.: The concept of document warehousing for multidimensional modeling of textual-based business intelligence. Decision Support Systems 42(2), 727–744 (2006)

Uncertainty Estimation in the Fusion of Text-Based Information for Situation Awareness

Kellyn Rein, Ulrich Schade, and Silverius Kawaletz

Fraunhofer FKIE, Neuenahrer Straße 20, 53343 Wachtberg-Werthhoven, Germany
{rein,schade,kawaletz}@fkie.fraunhofer.de

Abstract. Rapid correlation and analysis of new and existing information is essential to recognizing developing threats and can easily be considered one of the most important challenges for military and security organizations. Automatic preprocessing (fusion) of information can help to "connect the dots" but arguably as important to the intelligence process is the assessment of the combined reliability of the information derived from diverse sources: how can one estimate the overall reliability of a collection of information which may be only partially correct, and is incomplete, vague or imprecise, or even intentionally misleading? In this paper we present a simple heuristics-based model for fusing text-based information from multiple diverse sources which exploits standardized representation of underlying source information. The strength of the model lies in the analysis and evaluation of the uncertainty at three levels: data (source and content), fusion (correlation and evidence), and model (accuracy of representation).

Keywords : Evaluation of uncertainty, diverse sources, information fusion, situation awareness.

1 Introduction

One of the great challenges facing military and security analysts is finding a way to deal with the overwhelming volume of incoming data. This is particularly so in today's world dominated by asymmetric terrorist operations, where the battlefields are not clearly defined, the enemy is difficult to identify, and where subterfuge and long timelines make it difficult to recognize potentially threatening activities. Similarly, non-war activities such as peacekeeping or disaster relief require the coordination of information gathered and provided by multiple (often multinational) agencies both military and civilian to deal effectively with the crisis, avoid duplication of effort or to eliminate the chance of oversight in an area requiring attention. Time and again the inability to "connect the dots" has resulted in missed opportunities or embarrassing failures of the intelligence, security and support services. Even in the best of situations, the sheer enormity of the information available makes correlation and complex analysis of this information nearly intractable. Clearly the most promising solution to this problem is to automate the pre-processing of inflowing information – automating the process of sifting, sorting and collating incoming messages with existing information – would be

E. Hüllermeier, R. Kruse, and F. Hoffmann (Eds.): IPMU 2010, Part II, CCIS 81, pp. 168–177, 2010.

of great benefit to alleviate the load on analysts, reduce timelines and allow decision-makers to react more quickly to potential situations.

Finding a useful way to identify and connect various pieces of information with one another to make sense of them is not the only challenge for analysts. A second, and at least as important, challenge is how to assess the overall reliability of a "picture" which is composed of "dots" of varying reliability. The underlying information is imperfect. Some of the individual pieces of information forming the "dots" may be vague or imprecise, others are from sources which are less than completely reliable. Some of the correlations which connect the individual pieces to each other may be strong, other correspondences more uncertain and indirect. Still other "dots" may be missing. And yet the analyst must attempt to evaluate the overall credibility of the collection at hand so that decisions are made with an understanding of the validity of the underlying information.

However, it turns out that uncertainty arises in many ways and on a variety of levels. One must first understand the various manifestations of uncertainty in order to come up with an appropriate method for analysis.

In this paper we will discuss the types of uncertainty which arise in the information fusion process. We then briefly discuss Battle Management Language (BML) as a language to represent information. After that we outline a simple heuristics-based system for threat-modeling which uses a simplified computer linguistic algorithm exploiting the BML representations and which incorporates the elements of uncertainty analysis to provide decision-makers with a "rating" of the uncertainty inherent in the threat analysis.

2 Information Fusion and Uncertainty

Situation awareness, according to Dominguez et al., is "the continuous extraction of environmental information, the integration of this information with previous knowledge to form a coherent mental picture, and the use of that picture in directing further perception and anticipating future events."[1] One cannot simply rely on single items of information in isolation, but rather must attempt to identify the larger picture which individual elements may form when combined. This means, an important part of situation awareness is the fusing of individual pieces of information into a cohesive whole [2].

Among the numerous definitions for information fusion, the one which best encapsulates the focus of our work comes from the University of Skövde's Information Fusion Web site [3]: "[the] combining, or fusing, of information from different sources in order to facilitate understanding or provide knowledge that is not evident from individual sources." There are several steps in this fusion process:

- collection of data from various diverse sources,
- where necessary, the conversion of data from its original form to a standardized format in preparation for pre-processing,
- selection and correlation of potentially related individual pieces of information,
- mapping of individual pieces of data to existing threat models,
- evaluation of the credibility of the results of the correlation and mapping process, and
- assessment of the accuracy of models used as the basis for fusion. [4]

However, simply combining individual pieces of information is not in and of itself sufficient. In an ideal world, the information which we have available through various sources would be reliable, precise and unambiguous in its content and would come from unimpeachable sources. Further it would be unambiguously clear which information can be clustered together, and which pieces corroborate or contradict each other. And in an ideal world, the underlying threat models to which this information is mapped are accurate mirrors of reality.

Unfortunately, the world in which we live is imperfect. Our sources provide information which may be vague, ambiguous, misleading or even contradictory. Reports coming in from the field may contain speculation which is phrased used modal expressions ("possibly", "probably") which convey uncertainty. The sources themselves vary in reliability: information available through open sources such as internet sources and the press may vary widely in credibility, HUMINT sources such as refugees may be well-intentioned but less well-informed, or prisoners of war may deliver disinformation in the form of half-truths. Tips may be accurate or the result of gossip. Sensors provide readings which may vary with environmental conditions and thus are accurate only within a given range. Information may be missing.

Even if all underlying information were truthful and from completely reliable sources, sorting and correlating various pieces within the flood of available information into useful patterns poses problems. Connections between individual bits of information are not always direct. They are indeed more often achieved through a chain of mappings, each link of which introduces more uncertainty. And finally, despite our best attempts to model actions and situations, such models are seldom completely accurate. The enemy tries to hide his activities from us. We attempt to guess what he is doing by tracking observable events. We are thus attempting to discern the shape and size of the iceberg through the peaks which we can see above the surface.

In other words, at all stages of the information gathering, analysis and fusion process uncertainty creeps in. In [5] we identified three levels of uncertainty in the information fusion process: *data*, *fusion* and *model*:

- *Data level* consists of uncertainties involving the source and content of information;
- *Fusion level* is concerned with the correlation of individual bits of information and the strength of evidence for a given threat;
- *Model level* concerns the reliability of the model itself.

In the following sections we discuss these levels in more depth. Following this discussion we will introduce briefly BML and our fusion model. At least, we demonstrate how we implement a methodology for calculation of uncertainty into this model.

2.1 Data Level Uncertainty

At the data level we attempt to evaluate the quality of the information available to us, that is, the perceived competence of the source of the information as well as the perceived truth of the information delivered by this source. For device-based information, we may have known statistics available concerning the reliability of the device which we can use as a basis. Other sources are less easy to quantify. Open source information such as web-based, media such as television, newspaper, etc., may vary widely from instance to instance.

HUMINT sources and reports are routinely assigned rankings based upon the reporter's (or analyst's) belief in the credibility of the source of the information as well as the validity of the content. In general, it is next to impossible for humans to evaluate independently the perceived reliability of source and content. As former CIA analyst Richards J. Heuer, Jr. [6] notes, "Sources are more likely to be considered reliable when they provide information that fits what we already think we know." Further, we humans tend to put greater trust in the information that is delivered by a source that we trust and, of course, the converse: we mistrust the information from a source we are suspicious of. Other factors play a role in the perception of truth. Nisbett and Ross offer a thorough discussion of the fallibility of human perception in [7].

2.2 Fusion Level Uncertainty

At the fusion level uncertainty arises as to how individual pieces of information are identified as belonging together (correlation-based uncertainty) and as to how clearly a given piece of information indicates the presence of (i.e., provides evidence for) a potential developing threat (evidential uncertainty).

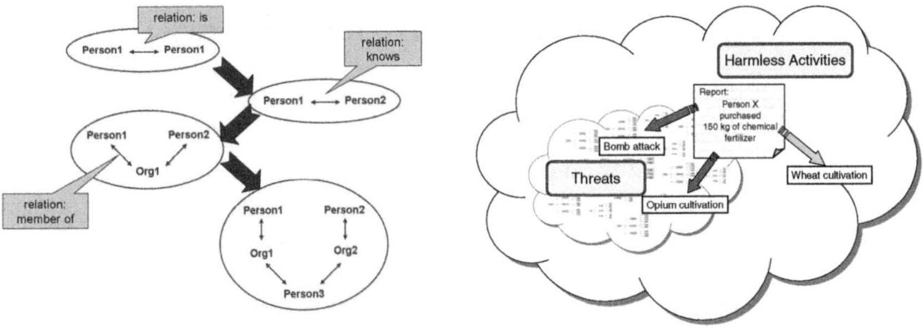

Fig. 1. Left: Correlation uncertainty reflects the mapping of connections between pieces of information: each interim step in the connection introduces more uncertainty. Right: Evidential uncertainty arises because a single action may provide evidence for more than one potential threat, or even be completely innocent.

Correlation uncertainty arises when the mapping of connections between two pieces of information is not direct. For example, when the same individual is referenced in two pieces of information, the correlation of the two reports is obvious. When two reports may be linked because our ontology indicates that the person in the first message knows the person in the second, the connection between the two is weaker. Every additional link in the chain of connections creates more uncertainty in the correlation.

Evidential uncertainty quantifies the likelihood that a given piece of information signals a particular threat. Observed actions may be indicative of more than one single threat or even be perfectly innocent; for example, the purchase of 20 kg of chemical fertilizer may (in Afghanistan) indicate a direct threat (intention to build an explosive device), an indirect threat (opium cultivation for funding terrorist activity) or simply indicate that the purchaser will be assisting his aged parents in planting wheat on the family farm.

2.3 Model Level Uncertainty

In threat recognition we are attempting to create a recognizable picture based upon incomplete and fragmentary information. A threat model is a formalized description of a conjecture, which is designed to assist us to predict potential threats by organizing information such that a recognizable pattern arises. The designers of the model have selected, organized and weighted the various components contained in the model. Nisbett and Ross[7] make a powerful argument against expert neutrality (or, more precisely, the lack thereof) in selecting information and assigning weights to data.

How certain can we be, even when all of our identified indicators for a specific threat are present, that this prediction is accurate? Sometimes what we think we see is in fact not there; in fact, it may well be that experience tells us that only in one instance in ten does this constellation of factors truly identify an impending threat.

Thus a final assessment must be made: the uncertainty associated at the model level. This can be seen as a sort of "reality check" and may be based upon heuristics derived from observation over time in the field.

3 Battle Management Language (BML) for Consistency

Information needed for accurate situation analysis is generated by or obtained from a variety of sources. Each source generally has its own format, which can pose a significant hurdle for automatic fusion of the different pieces of information. In multinational endeavors there are often even multiple languages involved. Preprocessing available information by converting it into a standardized format would greatly support fusion.

Originally designed for commanding simulated units, BML is supposed to be a standardized language for military communication (orders, requests and reports) [8]. Under the aegis of the NATO MSG-048 "Coalition BML", a BML version based on the formal grammar C2LG [9][10] has been developed, implemented and tested in multinational experiment [11][12][13]. There are also expansions to BML being developed such as AVCL (Autonomous Vehicle Command Language) [15] which will facilitate communications (tasking and reporting) with autonomous devices. Currently, we are expanding BML to task robots.

BML is lexically based upon the Joint Consultation, Command and Control Information Exchange Data Model (JC3IEDM) which is used by all participating NATO partners. As NATO standard, JC3IEDM defines terms for elements necessary for military operations, whether wartime or non-war. To take these terms as BML lexicon means that BML expressions consists of words which meanings are defined in the standard.

Another particular interesting feature of MSG-048's BML version is that its statements can be unambiguously transformed into feature-value matrices that represent the meaning of the expression. These matrices can be fused through unification, a standard algorithm in computational linguistics [16]. Since data retrieved from databases and ontologies may also be easily represented as feature-value pairs, BML structure facilitates the fusion of not only field reports from deployed soldiers and intelligence sources, it also supports fusion of these reports with previous information stored as background information.

BML reports may be input via a BML-GUI [12].) or converted from free text using natural language processing techniques [16]. In either case, the information contained in the reports is ultimately converted to feature-value matrices. Additionally, operational and background information from the deployment area which is stored in databases or ontologies is essentially also represented as feature-value matrices, thus providing the common format necessary for fusion [17].

As described in [10], a basic report in BML delivers a "statement" about an individual task, event or status. A task report is about a military action either observed or undertaken. An event report contains information on non-military, "non-perpetrator" occurrences such as flooding, earthquake, political demonstrations or traffic accidents. Event reports may be important background information for a particular threat: for example, a traffic accident may be the precursor of an IED detonation. Status reports provide information on personnel, materiel, facilities, etc., whether own, enemy or civilian, such as number of injured, amount of ammunition available, condition of an airfield or bridge.

4 BML and Uncertainty

There are several important elements to BML basic reports for the fusion process. First is the fact that each BML "report" is a statement representing a single (atomic) statement. Second is that each basic report has its own values representing source and content reliability (cf. figure 2). Third is that each report also has a reference label to its origination so that the context is maintained for later use by an analyst.

The first point (atomicity) is essential for the fusion process: each statement of a more complex report may be processed individually.

However, this atomicity is additionally significant for the second point (uncertainty evaluation). Natural language text sources such as HUMINT reports usually contain multiple statements. Some of these statements may be declarative ("three men on foot heading toward the village"), other statements may be speculative ("possibly armed"). While an analyst may assign a complex HUMINT communication an overall rating (e.g., using the familiar "A1"-"F6" system), individual statements contained therein have greater or lesser credibility. Therefore the conversion to BML assigns first the global rating, but adjusts each individual statement according to the uncertainty in its formulation, e.g., on the basis of modality term analysis.

Finally, the label of the BML statement referencing the origin of the statement allows the analyst or decision-maker to easily re-locate the statement in its original form and in its original context, which may be necessary in critical situations.

5 The Threat Model

In practice soldiers and intelligence analysts have mental checklists ("rules of thumb") of events or states that they watch out for as harbingers of potential developing threats or situations. For example, the checklist of the factors which might constitute forewarning of a potential bomb attack on a camp would include such things as the camp appearing to be under surveillance, reports that a local militant group may have acquired bomb materiel, and a direct tip from an informant concerning an attack. Many of these factors may be further broken down into more detail, the matching of which "triggers" the activation of the factor. For example, the acquisition of blasting caps would activate the factor "bomb materiel". The result is a simple tree-like structure as shown in Figure 2.

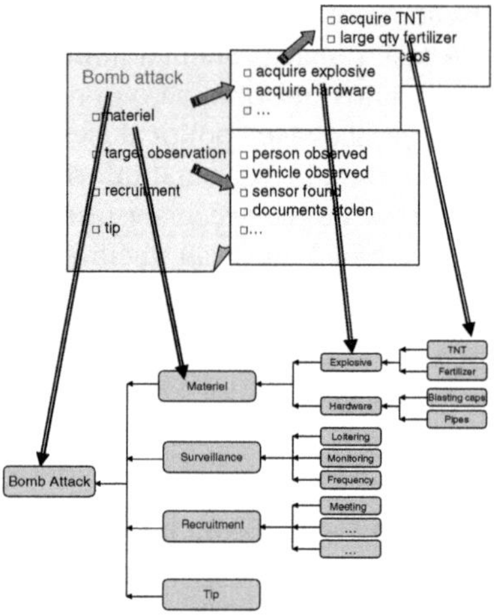

Fig. 2. Converting heuristic "checklist" to tree structure for threat "bomb attack"

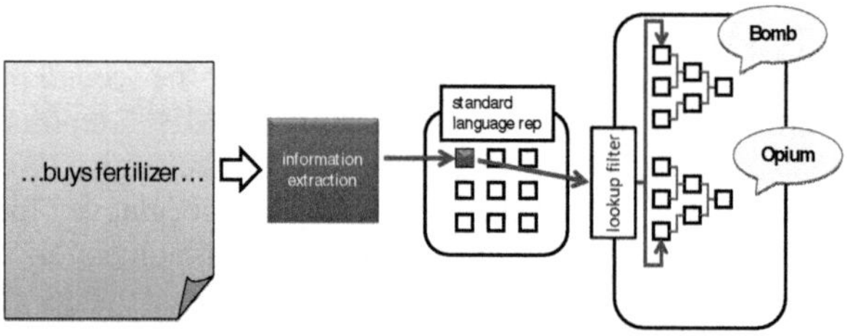

Fig. 3. Mapping new information into instances of threat structures. In this example, we have a report of a large amount of fertilizer having been acquired by suspicious actors. Within the area of interest, this may be indicative of two threats (bomb or opium cultivation).

Within a given structure different elements are weighted as to how significant an indicator of the threat they are (local evidence, i.e., significance within a given threat structure). For example, while "fertilizer" may be a trigger for bomb materiel, it may not be as strong an indicator as, say, blasting caps, would therefore have a relatively low weighting. A direct tip from a reliable local source may be a better indicator of an attack than the fact that there may be indications that the camp is being watched and the two factors therefore likewise weighted accordingly.

Global evidential weighting, which describes the likelihood solely within the set of described threat situations, is assigned to each trigger and factor. For example, within our area of interest, the acquisition of chemical fertilizer is indicative of two threats (Figure 3). However, our experience is that this only weakly indicates bomb construction, but strongly indicates opium cultivation. The analyst creating the model would assign weights which reflect the relative likelihood of the observed activity predicting each type of threat..

Within the model we also define which elements need to be correlated and which can or should be ignored. There may clustering based upon a common set of features for the triggers, but a different set of connections between factors. Within the model the correlating attributes are identified at different levels. As previously discussed in this paper, the uncertainty will be calculated based upon whether the correlations between the two branches are strong or weak.

Finally, the last assigned uncertainty weight is at the model level and is the assessment that, even when all apparent indications are there, how likely is it that this threat will actually materialize.

The threat model as well as its weights for analysis of the various uncertainties is designed and populated by an analyst based upon heuristics. These values stay static during running. The only "dynamic" element in the evaluation of uncertainty is at the data level: incoming information which activates the given structure arrives with a (numerical) weighting reflecting our assessment of the reliability of the information.

This reliability, in essence a basically trivial accumulation of weights, is then propagated through the model to produce a value which reflects cumulative result. As more supporting information arrives, the greater the likelihood that the threat is real. The various weights -- the credibility (source, content) of the initial information, the evidential weighting between and within models – interact to assure that there is a certain amount of checks and balances: unreliable information (assigned a low credibility) may trigger a strong indicator for a threat, but doubt is covered through the balance of the weights.

As more information flows into a model instance, the cumulative result increases and eventually reaches a predefined threshold, at which point the information contained in the checklist is passed on to analysts and decision-makers for final determination.

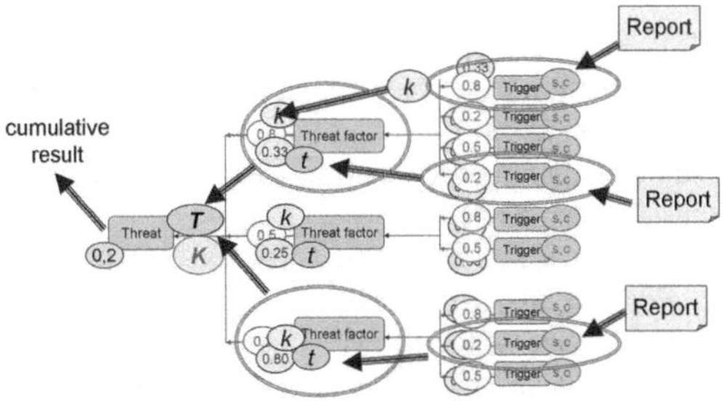

Fig. 4. Flow of calculation through a threat tree

6 Summary and Future Work

In this paper we have discussed the sources of uncertainty in information fusion for situation awareness and presented a simple heuristics-based model for fusing text-based information from multiple diverse sources which exploits standardized representation of underlying source information using BML. The model is designed around the analysis and evaluation of uncertainty of at three levels, in a manner which is open to scrutiny (no black box), while at the same time providing a mechanism through referencing to reconfirm the context of the original information.

This model been conceived to provide rapid (linear, near real-time), first pass warning of potential developing threats, based upon heuristic knowledge of the area of operations and upon observed behavior of the enemy. It has no mechanism at this time for analysis of constraints; for example, since it only accumulates information, it is not able to recognize or take into account contradictory information. It is not intended to replace more sophisticated deeper processing, but rather signal potential developing threats to allow for quick reaction from decision-makers.

Many of the parts of the system as described above have been or are currently being implemented (e.g., natural language parsing and conversion module), others are still in the design phase.

For the future, an extension of the model which would allow for the processing of more complex constraints is being investigated.

References

1. Dominguez, C., et al.: Situation awareness: Papers and annotated bibliography. Armstrong Laboratory, Human System Center, ref. AL/CF-TR-1994-0085 (1994)
2. Biermann, J.: Understanding Military Information Processing – An Approach To Supporting the Production of Intelligence in Defence and Security. In: Shahbazian, E., Rogova, G. (eds.) NATO Science Series: Computer & Systems Sciences: Data Fusion Technologies on Harbour Protection. IOS Press, Amsterdam (2006)
3. University of Skövde Information Fusion, `http://www.his.se/templates/vanligwebbsida1.aspx?id=16057`
4. Kruger, K., Schade, U., Ziegler, J.: Uncertainty in the fusion of information from multiple diverse sources for situation awareness. In: Proceedings Fusion 2008, Cologne, Germany (July 2008)
5. Kruger, K.: Two 'Maybes', One 'Probably' and One 'Confirmed' Equals What? Evaluating Uncertainty in Information Fusion for Threat Recognition. In: Proceedings MCC 2008, Cracow, Poland (September 2008)
6. Heuer Jr., R.J.: Limits of Intelligence Analysis. Seton Hall Journal of Diplomacy and International Relations (Winter 2005)
7. Nisbett, R., Ross, L.: Human Inference: Strategies and Shortcomings of Social Judgment. Prentice-Hall, Inc., Englewood Cliffs (1980)
8. Carey, S., Kleiner, M., Hieb, M.R., Brown, R.: Standardizing Battle Management Language – A Vital Move Towards the Army Transformation. In: Paper 01F-SIW-067, Fall Simulation Interoperability Workshop (2001)

9. Schade, U., Hieb, M.: Formalizing Battle Management Language: A Grammar for Specifying Orders. In: Paper 06S-SIW-068, Spring Simulation Interoperability Workshop, Hunts-ville, AL (2006)
10. Schade, U., Hieb, M.R.: Battle Management Language: A Grammar for Specifying Reports. In: Spring Simulation Interoperability Workshop (= Paper 07S-SIW-036), Norfolk, VA (2007)
11. De Reus, N., de Krom, P., Pullen, M., Schade, U.: BML – Proof of Principle and Future Development. In: I/ITSEC, Orlando, FL (December 2008)
12. Pullen, M., Carey, S., Cordonnier, N., Khimeche, L., Schade, U., de Reus, N., LeGrand, N., Mevassvik, O.M., Cubero, S.G., Gonzales Godoy, S., Powers, M., Galvin, K.: NATO MSG-048 Coalition Battle Management Initial Demonstration Lessons Learned and Follow-on Plans. In: 2008 Euro Simulation Interoperability Workshop (= Paper 08E-SIW-064), Edinburgh, UK (June 2008)
13. Pullen, M., Corner, D., Singapogo, S.S., Clark, N., Cordonnier, N., Menane, M., Khimeche, L., Mevassvik, O.M., Alstad, A., Schade, U., Frey, M., de Reus, N., de Krom, P., LeGrand, N., Brook, A.: Adding Reports to Coalition Battle Management Language for NATO MSG-048. In: 2009 Euro Simulation Interoperability Workshop (= Paper 09E-SIW-003), Istanbul, Turkey (July 2009)
14. Huijsen, W.-O.: Controlled Language – An Introduction. In: Proc. of the Second International Workshop on Controlled Language Applications (CLAW 1998), May 1998, pp. 1–15. Language Technologies Institute, Carnegie Mellon University, Pittsburgh (1998)
15. Shieber, S.M.: An Introduction to Unification-Based Approaches to Grammar. = CSLI Lecture Notes 4. CSLI, Stanford (1986)
16. Jenge, C., Kawaletz, S., Schade, U.: Combining Different NLP Methods for HUMINT Report Analysis. In: NATO RTO IST Panel Symposium. Stockholm, Sweden (October 2009)
17. Jenge, C., Frey, M.: Ontologies in Automated Threat Recognition. In: MCC 2008, Krakau, Polen (2008)

Aggregation of Partly Inconsistent Preference Information

Rudolf Felix

F/L/S Fuzzy Logik Systeme GmbH,
Joseph-von-Fraunhofer Straße 20, 44227 Dortmund, Germany
Tel.:+49 231 9700 921; Fax: +49 231 9700 929
felix@fuzzy.de

Abstract. Traditionally the preference information is expressed as a preference relation defined upon the power set of the set of the decision alternatives. The preference information required for the method described in this paper is significantly less complex and is simply defined on the decision set. For every decision goal the preference of the decision alternatives for this goal is defined upon the set of decision alternatives as a linear ranking of the decision alternatives. It is discussed in which way a decision making approach based on interactions between goals is applied using this kind of preference information even if it is partly inconsistent. In a recent work for the case of consistent preference information a link to the theory of matroids was found. In this paper the case of partly inconsistent preference information is considered and a new link to the theory of matroids is given.

Keywords: Aggregation complexity, decision making, interactions between goals, reduced preference relation, inconsistent preferences, weighted sum.

1 Introduction

Many decision making approaches are limited with respect to the management of inconsistency [4], [6], [9], [10], [11]. The consequence of this limitation is that these models are rather not applicable for real world problems. Many aggregation approaches require a consistent preference relation as input. For more complex decision problems such preference relations are difficult to obtain because a preference relation is defined upon the power set of the set of decision alternatives. If the number of decision alternatives increases it becomes almost impossible to provide the exponentially increasing preference statements and to supervise their consistency. In a former work [5] it is discussed in which way a decision making approach based on interactions between goals is adapted to the aggregation of inconsistent preference information in complex decision situations. However, the required input preference information is not defined upon the power set of the decision alternatives. Instead of this, for every single decision goal only a linear preference ranking of the decision alternatives with respect to that goal is required. In [5] it has already been shown that in the case of consistent preference information there is a link to the theory of matroids and that in such a case the aggregation based upon the preference information may be done by weighted sums. Of course, single goal preference rankings may rank the decision alternatives in

E. Hüllermeier, R. Kruse, and F. Hoffmann (Eds.): IPMU 2010, Part II, CCIS 81, pp. 178–187, 2010.

different ways for different goals as the goals usually are partly conflicting. Therefore, the preference information may be partly inconsistent. The aggregation approach based on relationships between decision goals presented in former papers [1],[2],[3],[4] ascertains for each pair of goals their conflicts (and correlations) by computing the so called interactions between the goals from the initial input single goal rankings.

The only additional information needed for calculating the interactions between goals is a linear importance information of each goal, which is expressed in terms of so called goal priorities. The goal priorities are numbers of [0,1] and are comparable with a fuzzy measure that ranks not the decision alternatives but the decision goals themselves with respect to their importance (priority).

It turned out that decision making based on interactions between goals is less limited because the complexity of both the input information required and the aggregation process is not higher than polynomial. Since the model has successfully been applied to many real world problems [3] it clearly helped to manage many complex aggregation processes. It also turned out that in the case of consistent preference information, the model may be linked to the theory of matroids [5] and therefore could be related to decision approaches based on weighted sums. In this paper we show a new result that describes the link to the theory of matroids in the case that the preference information is partly inconsistent. We show that explicit reasoning upon relationships between decision goals helps to see that aggregation based on weighted sums may only be applied to these parts of the decision set which are consistent in terms of the preference information and not to the decision set as a whole.

For better readability of the paper, in the subsequent sections first we repeat the description of the decision making approach based on interactions between decision goals. Then we repeat how the approach is applied to single goal preference rankings and we also repeat under which conditions decision models based on weighted sums help and how they are related to decision situations with interacting decision goals. After this we show how the approach behaves in case of inconsistent preference information. Finally, the consequences of the new result are discussed.

2 Decision Making Based on Interactions between Goals

In the following it is shown how an explicit modeling of interaction between decision goals that are defined as fuzzy sets of decision alternatives helps to manage complexity of the decision making and aggregation. This modeling of the decision making and aggregation process significantly differs from the related approaches and the way they manage complex decision situations. First the notion of positive and negative impact sets is introduced. Then different types of interaction between goals are defined. After this it is shown how interactions between goals are used in order to aggregate pairs of goals to the so called local decision sets. Then it is described how the local decision sets are used for the aggregation of a final decision set. The complexity of the different steps is discussed.

2.1 Positive and Negative Impact Sets

Before we define interactions between goals as fuzzy relations, we introduce the notion of the positive impact set and the negative impact set of a goal. A more detailed discussion can be found in [1],[2] and [3].

Def. 1a). Let A be a non-empty and finite set of decision alternatives, G a non-empty and finite set of goals, $A \cap G = \emptyset, a \in A, g \in G, \delta \in (0,1]$. For each goal g we define the two fuzzy sets S_g and D_g each from A into [0, 1] by:1. Positive impact function of the goal g: $S_g(a) := \delta$, if a affects g positively with degree δ, $S_g(a):=0$ else. 2. Negative impact function of the goal g: $D_g(a) := \delta$, if a affects g negatively with degree δ, $D_g(a):=0$ else.

Def. 1b). Let S_g and D_g be defined as in Def. 1a). S_g is called the positive impact set of g and D_g the negative impact set of g.

The set S_g contains alternatives with a positive impact on the goal g and δ is the degree of the positive impact. The set D_g contains alternatives with a negative impact on the goal g and δ is the degree of the negative impact.

2.2 Interactions between Goals

Let $P(A)$ be the set of all fuzzy subsets of A. Let X, Y ∈ $P(A)$, x and y the membership functions of X and Y respectively. Assume now that we have a binary fuzzy inclusion ∈ and a fuzzy non-inclusion ∈, such that ∈. In such a case the degree of inclusions and non-inclusions between the impact sets of two goals indicate the degree of the existence of interaction between these two goals. The higher the degree of inclusion between the positive impact sets of two goals, the more cooperative the interaction between them. The higher the degree of inclusion between the positive impact set of one goal and the negative impact set of the second, the more competitive the interaction. The non-inclusions are evaluated in a similar way. The higher the degree of non-inclusion between the positive impact sets of two goals, the less cooperative the interaction between them. The higher the degree of non-inclusion between the positive impact set of one goal and the negative impact set of the second, the less competitive the relationship. The pair (S_g, D_g) represents the whole known impact of alternatives on the goal g. Then S_g is the fuzzy set of alternatives which satisfy the goal g. D_g is the fuzzy set of alternatives, which are rather not recommendable from the point of view of satisfying the goal g.

Based on the inclusion and non-inclusion between the impact sets of the goals as described above, 8 basic fuzzy types of interaction between goals are defined. The different types of interaction describe the spectrum from a high confluence between goals (analogy) to a strict competition (trade-off) [1].

Def. 2). Let S_{g_1}, D_{g_1}, S_{g_2} and D_{g_2} be fuzzy sets given by the corresponding membership functions as defined in Def. 1). For simplicity we write S_1 instead of S_{g_1} etc.. Let $g_1, g_2 \in G$ where G is a set of goals. *T* is a t-norm.

The fuzzy types of interaction between two goals are defined as relations which are fuzzy subsets of as follows:

1. g_1 *is independent of* g_2: <=> $T(N(S_1, S_2), N(S_1, D_2), N(S_2, D_1), N(D_1, D_2))$
2. g_1 *assists* g_2: <=> $T(I(S_1 S_2), N(S_1, D_2))$

3. g_1 *cooperates with* g_2*:* <=> $T(\mathrm{I}(S_1, S_2), \mathrm{N}(S_1, D_2), \mathrm{N}(S_2, D_1))$
4. g_1 *is analogous to* g_2*:* <=> $T(\mathrm{I}(S_1, S_2), \mathrm{N}(S_1, D_2), \mathrm{N}(S_2, D_1), \mathrm{I}(D_1, D_2))$
5. g_1 *hinders* g_2*:* <=> $T(\mathrm{N}(S_1, S_2), \mathrm{I}(S_1, D_2))$
6. g_1 *competes with* g_2*:* <=> $T(\mathrm{N}(S_1, S_2), \mathrm{I}(S_1, D_2), \mathrm{I}(S_2, D_1))$
7. g_1 *is in trade-off to* g_2*:* <=> $T(\mathrm{N}(S_1, S_2), \mathrm{I}(S_1, D_2), \mathrm{I}(S_2, D_1), \mathrm{N}(D_1, D_2))$
8. g_1 *is unspecified dependent from* g_2*:* <=>
   $T(\mathrm{I}(S_1, S_2), \mathrm{I}(S_1, D_2), \mathrm{I}(S_2, D_1), \mathrm{I}(D_1, D_2))$

The interactions between goals are crucial for an adequate orientation during the decision making process because they reflect the way the goals depend on each other and describe the pros and cons of the decision alternatives with respect to the goals. For example, for cooperative goals a conjunctive aggregation is appropriate. If the goals are rather competitive, then an aggregation based on an exclusive disjunction is appropriate.

Note that the complexity of the calculation of every type of interaction between two goals is $O(\mathrm{card}(A) * \mathrm{card}(A)) = O((\mathrm{card}(A))^2)$ [4].

2.3 Two Goals Aggregation Based on the Type of Their Interaction

The assumption, that cooperative types of interaction between goals imply conjunctive aggregation and conflicting types of interaction between goals rather lead to exclusive disjunctive aggregation, is easy to accept from the intuitive point of view. It is also easy to accept that in case of independent or unspecified dependent goals a disjunctive aggregation is appropriate. For a more detailed formal discussion see for instance [1],[2]. Knowing the type of interaction between two goals means to recognize for which goals rather a conjunctive aggregation is appropriate and for which goals rather a disjunctive or even exclusively disjunctive aggregation is appropriate. This knowledge then in connection with information about goal priorities is used in order to apply interaction dependent aggregation policies which describe the way of aggregation for each type of interaction. The aggregation policies define which kind of aggregation operation is the appropriate one for each pair of goals. The aggregation of two goals gi and gj leads to the so called local decision set $L_{i,j}$. For each pair of goals there is a local decision set $L_{i,j} \in P(A)$, where A is the set of decision alternatives (see Def 1 a)) and P(A) the power set upon A. For conflicting goals, for instance, the following aggregation policy which deduces the appropriate decision set is given: if (g_1 is in trade-off to g_2) and (g_1 is slightly more important than g_2) then $L_{1,2} := S_1 / D_2$. In case of very similar goals (analogous or cooperative goals) the priority information even is not necessary:
if (g_1 cooperates with g_2) then $L_{1,2} := S_1 \cap S_2$ because $S_1 \cap S_2$ surely satisfies both goals. if (g_1 is independent of g_2) then $L_{1,2} := S_1 \cup S_2$ because $S_1 \cup S_2$ surely do not interact neither positively nor negatively and we may and want to pursue both goals.

In this way for every pair of goals g_i and g_j, i,j ∈ {1,...,n} decision sets are aggregated. The importance of goals is expressed by the so called priorities. A priority of a goal g_i is a real number $P_i \in [0,1]$. The comparison of the priorities is modeled based on the linear ordering of the real interval [0,1]. The statements like g_i *slightly more important than* g_j are defined as linguistic labels that simply express the extend of the difference between P_i and P_j.

2.4 Multiple Goal Aggregation as Final Aggregation Based on the Local Decision Sets

The next step of the aggregation process is the final aggregation. The final aggregation is performed based on a sorting procedure of all local decision sets $L_{i,j}$. Again the priority information is used to build a semi-linear hierarchy of the local decision sets by sorting them. The sorting process sorts the local decision sets with respect to the priorities of the goals. Subsequently an intersection set of all local decision sets is built. If this intersection set is empty then the intersection of all local decision sets except the last one in the hierarchy is built. If the resulting intersection set again is empty then the second last local decision set is excluded from the intersection process. The process iterates until the intersection is not empty (or more generally speaking until its fuzzy cardinality is big enough with respect to a given threshold). The first nonempty intersec-tion in the iteration process is the final decision set and the membership values of this set give a ranking of the decision alternatives that is the result of the aggregation process (for more details see [2]).

2.5 Complexity Analysis of the Aggregation Process

As already discussed for instance in [4] the complexity of the aggregation process is $O((card(A))^2 * (card(G))^2)$ and the complexity of the information required for the description of both the positive and the negative impact functions is $O(card(A) * card(G))$.

3 Application of the Aggregation in Case of Reduced Preference Relations

In preference based decision making the input preference information has to be defined upon the power set of the decision alternatives [7]. This means that the complexity of the input information required is exponential with respect to the cardinality of the set of decision alternatives. This also means that the input information required is very difficult to obtain. Especially if the number of decision goals increases and the goals are partly conflicting the required preference information has to be multidimensional and the provider of the preference information has to express all the multidimensional interactions between the goals through the preference relation. With increasing number of goals and interactions between them the complexity of the required input preference relation possesses the same complexity as the decision problem itself. But, if the complexity of the required input is the same as the solution of the underlying decision problem itself then the subsequent aggregation of the input does not really help to solve the problem and is rather obsolete.

Therefore we propose to reduce the complexity of the input preference relation. Instead of requiring a preference relation defined upon the power set of the set of the decision alternatives which expresses the multidimensionality of the impacts of the decision alternatives on the goals for every single decision goal a linear preference ranking of the decision alternatives with respect to that goal is required. This means that for every goal a preference ranking defined on the set of decision alternatives is required instead of a ranking defined on the power set of the alternatives. The multidimensionality of the goals is then computed from all the single goal preference rankings using the concept of interactions between the goals as defined in section X.2.

In the sequent we extend the definition Def. 1. The extension defines how a single goal preference ranking defined on the set of the decision alternatives is transformed into positive and negative impact sets:

Def. 1c). Let A be a non-empty and finite set of decision alternatives, G a non-empty, finite set of goals as defined in Def. 1a), $A \cap G = \emptyset, a \in A, g \in G, \delta \in (0,1]$.

Let $>p_g$ be a preference ranking defined upon A with respect to g defining a total order upon A with respect to g, such that $a_{i1} >p_g a_{i2} >p_g a_{i3} >p_g \dots >p_g a_{im}$, where m=card(A) and $\forall a_{ij}, a_{ik} \in A, a_{ij} >p_g a_{ik} :\Leftrightarrow a_{ij}$ is preferred to a_{ik} with respect to the goal g. The preference relation $>p_g$ is called the *reduced single goal preference relation* of the goal g.

For simplicity, instead of $>p_g$ we also equivalently write RSPR of the goal g. All the RSPRs for all $g \in G$ are called the *reduced preference relation* RPR for the whole set of goals *G*. In order to avoid complete redundancy within the RPR we additionally define that the RSPRs of all the goals are different.

Let us assume that there is a decision situation with *n* decision goals where *n=card(G)* and *m* decision alternatives where *m=card(A)*. In the subsequent we propose an additional extension of the original definition Def. 1b with the aim to transform the single goal preference relations RSPR of every goals $g \in G$ into the positive and negative impact sets *Sg* and *Dg*:

Def. 1d). Let again A be a non-empty and finite set of decision alternatives, G a non-empty, finite set of goals as defined in Def. 1) a), $A \cap B \neq \emptyset$, $a_i \in A$, *m=card(*A*)*, $g \in G$, i, $c \in \{1, \dots, m\}$. For any goal g we obtain both the positive and the negative impact sets Sg and Dg by defining the values of δ according to Def. 1a) and 1b) as follows:

Def. 1d1). For the positive impact set:

$Sg(a_i)=\delta:=1/i$ iff $i \in [1,c-1]$, $Sg(a_i)=\delta:=0$ iff $i \in [c,m]$.

Def. 1d2). For the negative impact set:

$Dg(a_i)=\delta:=0$ iff $i \in [1,c-1]$, $Sg(a_i) = \delta:= 1/(m-i+1)$ iff $i \in [c, m]$.

Using the definition Def. 1d1 and Def. 1d2 for any goal $g \in G$ we obtain a transformation of the RPR into positive and negative impact sets of all the goals and can evaluate the interactions of the goals using Def. 2 that are implied by the RPR.

Compared to classical preference based decision models this transformation helps to reduce the complexity of the input preference information required without losing modeling power for complex real world problems. The advantage is that using Def. 2.

the interactions between goals that are implied by the RPR expose the incompatibilities and compatibilities that may be hidden in the RPR. The exposed incompatibilities and compatibilities are used adequately during the further calculation of the decision sets. Note that the exposition is calculated with a polynomial number of calculation steps with degree 2. The only additional information required from the decision maker is the priority information for each goal which has to be expressed as a weight with a value between 0 and 1. Many real world applications show that despite the reduced input complexity there is no substantial loss of decision quality [3].

Statement 1: *In particular this means that it is not necessary to have classical preference relations defined upon the power set of the decision alternatives in order to handle complex decision problems with both positively and negatively interacting decision goals.*

Another interesting question is how the decision making based on interactions between decision goals is formally related to aggregation methods based on weighted sums. In order to investigate this we introduce the notion of r-consistency of RPRs and will consider the question under which conditions the weighted sum aggregation may be appropriate from the point of view of the application of the decision making based on interactions between goals if we have an RPR as input. For this we define the following:

Def.3). Given a discrete and finite set A of decision alternatives. Given a discrete and finite set G of goals. Let $r \in (0,1]$. The reduced preference relation RPR is called r-consistent $:\Leftrightarrow \exists\, c_l \in \{1, \dots ,m\}$, $m=card(A)$ such that $\forall\, (g_i, g_j) \in G \times G$, $i,j \in \{1, \dots ,n\}$, $n=card(G)$, $(g_i\, cooperates\ with\ g_j) \geq r$.

Let us now consider an important consequence of the interaction between goals using the notion of r-consistency. This notion will imply a condition under which an aggregation based on a weighted sum may lead to an appropriate final decision.

For this let us assume that the quite sophisticated final aggregation process of the iterative intersections as described in section 2.4 is replaced by the following rather intuitive straight forward consideration of how to obtain an optimal final decision.

Again let us identify the local decision sets $L_{i,j}$ obtained after the application of the local decision policies for each pair of goals (g_i, g_j) as the first type of decision subsets of the set of all decision alternatives A which, according to the decision model are expected to contain an optimal decision alternative a_k. Thus we define the set of sets $T1 := \{ L_{i,j} \mid i, j \in \{1, .. n\}, n=card(G)\}$ and expect that the optimal decision alternative has a positive membership in at least one of the sets of T1. Since we want to consider multiple goals we may also expect that an optimal decision alternative may have a positive membership in at least one of the intersections of pairs of the local $L_{i,j}$. Thus we define $T2 := \{ L_{k,l} \cap L_{p,q} \mid k,l,p,q \in \{1, .. n\}, n=card(G)\}$. In order to simplify the subsequent explanation we concentrate on the crisp case and replace all membership values > 0 in all $L_{i,j}$, and $L_{k,l} \cap L_{p,q}$ by the membership value 1. Now we define the system of these crisp sets GDS as follows:

Def.4). GDS:=$\{\emptyset$,T1,T2$\}$, T1,T2 are the sets of crisp sets that we construct as described above by replacing all membership values > 0 in all $L_{i,j}$, and $L_{k,l} \cap L_{p,q}$ by the membership value 1.

With this definition we are able to formulate the following theorem that describes a property of the crisp GDS in the case that the underlying decision situation stems from a reduced preference relation RPR to which the decision model based on interactions between decision goals is applied. This property will enable us to relate this decision making to the calculation of optimal decisions by concepts based on weighted sums that are strongly connected with the notion of a matroid [8] and optimal decisions obtained by Greedy algorithms.

Theorem 1: If the reduced preference relation RPR is r-consistent then the system ($\mathcal{P}(A)$,GDS) is a matroid.

Sketch of the Proof: The proof will show that the following matroid conditions [8] hold: 1. $\emptyset \in$ GDS, 2. $X \subseteq Y$, $Y \in$ GDS $\Rightarrow X \in$ GDS and 3. X,Y$\in$GDS card(X)<card(Y) $\Rightarrow \exists x \in Y \setminus X$ and $X \cup \{x\} \in$ GDS.

Ad1.: Holds by definition of GDS. Ad2.: If X is one of the $L_{i,j}$s then Y must be one of the another $L_{k,l}$ with $i \neq k$, $j \neq l$. If X is one of the $L_{k,l} \cap L_{p,q}$ then Y must be $L_{k,l}$ or $L_{k,l}$. Ad3.: This condition holds because the RPR is r-consistent and therefore we have $\forall$ k,l,p,q $\in \{1, .. n\}$, $n=card(G)$, card($L_{i,j}$)=card($L_{k,l} \cap L_{p,q}$)= c_1.

With this result we obtain the following corollary that relates decision making based on interactions between decision goals to calculating optimal decisions by concepts based on weighted sums.

Corollary 1: If the reduced preference relation RPR is r-consistent then there exists a greedy algorithm based on the maximization of the weighted sum $\Sigma_{i\in\{1,\ldots,n\},k\in\{1,\ldots,m\}}$ $P_i * S_i(a_k)$, where S_i is defined as in Def. 1d1 and $a_k \in$ A being a decision alternative, n=card(G), m=card(A). P_i is the priority of the goal g_i as used for the calculation of both the local *and the global decision sets (see sections 4.3 and 4.4).*

Sketch of proof: The proof is an immediate result of Theorem 1 and the theory of matroids used in the field of combinatorial optimization [8].

Let us now consider the case of inconsistency of the RPR as follows:

Def.5. Given a discrete and finite set *A* of decision alternatives. Given a discrete and finite set *G* of goals. Let r $\in$ (0,1]. The reduced preference relation RPR is called r-inconsistent :$\Leftrightarrow \exists\, c_2 \in \{1, \ldots, m\}$, $m=card(A)$ such that $\forall\, (g_i, g_j) \in G \times G$, $i,j \in \{1, \ldots, n\}$, $n=card(G)$, (g_i *competes with* g_j) $\geq r$.

Let us now consider that a given RPR is inconsistent in the following way:

RPR is *r-consistent* for a c_1 and RPR is *r-inconsistent* for a c_2 and c_2=m and $c_2 > c_1$. In such a case we can conclude that the RPR is r-consistent only for a real subset A_1 of *A*.

Corollary 2: If the reduced preference relation RPR is r-inconsistent for c_2 *and* c_2*=m and RPR is r-consistent for* c_1 and $c_2 > c_1$ *then there exists a set* A_1, $A_2 \subset A$ *such that the system* ($\mathcal{P}(A_1)$,GDS) *is a matroid.*

Scetch of proof: Since $c_2 > c_1$, the condition that RPR is *r-consistent* can only hold for a real subset A_1 of *A*. Therefore *Theorem1* holds only for real subset A_1 of *A*.

As a consequence we see that *Corollary 1* and the fact that we can calculate the decisions using a greedy algorithm based on weighted sums cannot be proofed for the whole set A but only for a part of it. According to Corollary 2 we only know that the weighted sums are appropriate to A_1 but not to *A* as a whole.

4 Discussion of the Consequences

As already mentioned in [5], the *Corollary 1* relates the decision making based on interactions between decision goals to the calculation of optimal decisions by concepts based on weighted sums. It shows that weighted sums both as aggregation and optimization concept are rather appropriate if the goals or criteria are cooperative e.g. if they interact positively (or at least do not interact at all being independent). In contrast to this, the decision making based on interactions between decision goals is more general and it reflects both positive and negative interactions between the goals. *Corollary 2* shows that even better because based on *Corollary 2* we can see that the decision making based on relationships between goals helps to understand that in case of partly inconsistent preferences weighted sums as aggregation method are only adequately applicable to the consistent parts of the decision set but not to the decision set as a whole. This is important in the context of real world decision making and optimization problems, which usually possess partly conflicting goal structures and partly inconsistent preferences. If the aggregation is performed by weighted sums, like for instance in case of Choquet integrals, it becomes evident that the aggregation will only work for these parts of the decision set which are free of conflicts with respect to the preference information given. The last-mentioned statement has already been formulated in [5] based on *Corollary 1*. The new result expressed by the *Corollary 2* supports the statement even better.

5 Conclusions

Both the *Corollary 1* and the new *Corollary 2* relate the decision making based on interactions between decision goals to the calculation of optimal decisions by concepts based on weighted sums. Both corollaries show that weighted sums both as aggregation and optimization concept are rather appropriate if the decision goals are not partly inconsistent, that is if they interact positively or are at least independent. In contrast to this the decision making based on interactions between decision goals is more general. It is able to handle partly inconsistent preference information. *Corollary 2* shows that in case of partly inconsistent preferences weighted sums as aggregation method are only adequately applicable to the consistent parts of the decision set but not to the decision set as a whole. The last-mentioned statement has already been formulated formerly based on *Corollary 1*. The new result expressed by *Corollary 2* supports the statement better. The statement is important in the context of real world decision making and optimization problems, which usually possess partly conflicting goal structures and where partly inconsistent preferences are rather the normal case than the exception.

References

1. Felix, R.: Relationships between goals in multiple attribute decision making. Fuzzy Sets and Systems 67, 47–52 (1994)
2. Felix, R.: Decision-making with interacting goals. In: Ruspini, E., Bonissone, P.P., Pedrycz, W. (eds.) Handbook of Fuzzy Computation. IOP Publishing Ltd. (1998)

3. Felix, R.: Real World Applications of a Fuzzy Decision Model Based on Relationships between Goals (DMRG). In: Forging the New Frontiers, Fuzzy Pioneers I (1965-2005), October 2007. Studies in Fuzziness and Soft Computing. Springer, Heidelberg (2007)
4. Felix, R.: Multicriterial Decision Making (MCDM): Management of Aggregation Complexity Through Fuzzy Interactions Between Goals or Criteria. In: Proceedings of the 12th International IPMU Conference, Málaga, Spain (2008)
5. Felix, R.: Multi-Goal Aggregation of Reduced Preference Relations Based on Fuzzy Interactions between Decision Goals. In: Proceedings of the IFSA World Congress, Lisbon, Portugal (2009)
6. Modave, F., Dubois, D., Grabisch, M., Prade, H.: A Choquet integral representation in multicriteria decision making. In: AAAI Fall Symposium, Boston, MA (November 1997)
7. Modave, F., Grabisch, M.: Preference representation by a Choquet integral: Commensurability hypothesis. In: Proceedings of the 7th International IPMU Conference, Paris, France, pp. 164–171 (1998)
8. Oxley, J.: Matroid Theory. Oxford University Press, Oxford (1992)
9. Saaty, T.L.: The Analytic Hierarchy Process. McGraw-Hill, New York (1980)
10. Torra, V.: Weighted OWA operators for synthesis of information. In: Proceedings of the fifth IEEE International Conference on Fuzzy Systems, New Orleans, USA, pp. 966–971 (1996)
11. Yager, R.R.: Families of OWA operators. Fuzzy Sets and Systems 59, 125–148 (1993)

Risk Neutral Valuations Based on Partial Probabilistic Information

Andrea Capotorti, Giuliana Regoli, and Francesca Vattari

Dip. Matematica e Informatica, Università degli Studi di Perugia - Italy

Abstract. In a viable single-period model with one stock and $k \geq 2$ scenarios the completeness of the market is equivalent to the uniqueness of the risk neutral probability; this equivalence allows to price every derivative security with a unique fair price. When the market is incomplete, the set of all possible risk neutral probabilities is not a singleton and for every non attainable derivative security we have a *bid-ask* interval of possible prices. In literature, different methods have been proposed in order to select a unique risk neutral probability starting with the *real world probability* $\mathbf{p}$. Contrary to the complete case, in all these models $\mathbf{p}$ is really used for the option pricing and its elicitation is a crucial point for every criterion used to select a risk neutral probability. We propose a method for the valuation problem in incomplete markets which can be used when $\mathbf{p}$ is a partial conditional probability assessment as well as when we have different expert opinions expressed through conditional probability assessments. In fact, it is not always possible to elicit a probability distribution $\mathbf{p}$ over all the possible states of the world: the information that we have could be partial, conditional or even not coherent. Therefore we will select a risk neutral probability by minimizing a discrepancy measure introduced in [2] and analized in [3] between $\mathbf{p}$ and the set of all possible risk neutral probability, where $\mathbf{p}$ can be a partial conditional probability assessments or it can be given by the fusion of different expert opinions.

1 Introduction

In literature, different methods have been proposed in order to select a risk neutral probability starting with the *real world probability* $\mathbf{p}$ (see for example [8], [9] and [10]); contrary to the complete case where this assessment is not used at all for the valuation problem, in all these methods $\mathbf{p}$ is really used and its elicitation is a crucial point for every criterion used to select a risk neutral probability. Usually the information that we have about the possible states of the world can be partial, it can be given on conditional events or it can even be incoherent. Therefore we propose a method for the valuation in incomplete markets which can be used when $\mathbf{p}$ is a partial conditional probability assessments as well as when there are more partial conditional probability assessments given by different expert opinions.

E. Hüllermeier, R. Kruse, and F. Hoffmann (Eds.): IPMU 2010, Part II, CCIS 81, pp. 188–197, 2010.

The paper is organised as in the following: in the next subsections 1.1 and 1.2 we give a brief overview of the technical results on the risk neutral valuation and we will analyse a discrepancy measure. In section 2 the selection procedure of a risk neutral probability is described, with the aid of some examples. Section 3 generalizes the procedure when disparate probabilistic opinions are given. Finally, section 4 closes the contribution with a short conclusion.

1.1 Basic Notion on Risk Neutral Valuation

Let us recall some basic notions to deal with incomplete markets in line with [1], [7] and [12].

A *risk neutral probability* $\boldsymbol{\alpha}$ in a single-period single-stock model with $k \geq 2$ scenarios is a probability $(\alpha_1, \ldots, \alpha_k)$ such that the price S_0 of the stock can be computed as the expected value of the stock price S_1 at time 1 discounted with the risk free interest rate r. Denoting by $\overline{S}_t := S_t/(1+r)^t$, $t = 0, 1$, we have

$$S_0 = \frac{1}{1+r}\mathbb{E}_{\boldsymbol{\alpha}}(S_1) = \mathbb{E}_{\boldsymbol{\alpha}}(\overline{S}_1),$$

that means $\boldsymbol{\alpha}$ is a martingale measure for the discounted stock price process $\overline{S}_t$.

A market model is said to be *viable* if there are no arbitrage opportunities and it is said to be *complete* if every derivative security admits a replicating portfolio; a $fair$ price for a derivative is any price for it that prevents arbitrages. We know that a single-period model with one stock and $k \geq 2$ scenarios is viable if and only if there is a risk neutral probability and the model is viable and complete if and only if this probability is unique. When the market is viable and complete, the fair price π of any derivative security D is given by

$$\pi = \mathbb{E}_{\boldsymbol{\alpha}}(\overline{D})$$

where $\overline{D}$ is the discounted price of D and $\boldsymbol{\alpha}$ is the risk neutral probability.
When the market is incomplete the set F of all possible fair prices for a derivative D is

$$F = [l, u]$$

where

$$l := \inf\{\mathbb{E}_{\boldsymbol{\alpha}}(\overline{D}) \mid \boldsymbol{\alpha} \text{ is a risk neutral probability}\},$$
$$u := \sup\{\mathbb{E}_{\boldsymbol{\alpha}}(\overline{D}) \mid \boldsymbol{\alpha} \text{ is a risk neutral probability}\}.$$

A derivative security is said to be *attainable* if it admits a replicating portfolio; obviously a derivative security is attainable if and only if $l = u$; otherwise $l < u$ and we have to consider the interval $[l, u]$ of all possible fair prices.
Finally we know that there is a one-to-one correspondence between the set $[l, u]$ of possible prices for a derivative and the convex set of possible risk neutral probabilities, that we will denote by $\mathcal{Q}$ (see [12]). Hence taken $\tilde{\boldsymbol{\alpha}} \in \mathcal{Q}$, for every derivative security there will be a corresponding fair price π given by

$$\pi = \mathbb{E}_{\tilde{\boldsymbol{\alpha}}}(\overline{D}).$$

1.2 Discrepancy Measure

Let $\mathbf{p} = (p_1, \ldots, p_n) \in (0,1)^n$ be a conditional probability assessment given by an expert over the set of conditional events $\mathcal{E} = [E_1|H_1, \ldots, E_n|H_n]$ and let $\Omega = \{\omega_1, \ldots, \omega_k\}$ be the set of all possible states of the world. In the following E_iH_i will denote the logical connection "$E_i \wedge H_i$", E_i^c will be "$\neg E_i$".

We need to define the following hierarchy of probability distributions over Ω:

let $\mathcal{A} := \left\{\boldsymbol{\alpha} = [\alpha_1, \ldots, \alpha_k], \sum_1^k \alpha_i = 1, \alpha_j \geq 0, j = 1, \ldots, k\right\}$;

let $\mathcal{A}_0 := \{\boldsymbol{\alpha} \in \mathcal{A} | \alpha(\bigcup_{i=1}^n H_i) = 1\}$;

let $\mathcal{A}_1 := \{\boldsymbol{\alpha} \in \mathcal{A}_0 | \alpha(H_i) > 0, i = 1, \ldots, n\}$;

let $\mathcal{A}_2 := \{\boldsymbol{\alpha} \in \mathcal{A}_1 | 0 < \alpha(E_iH_i) < \alpha(H_i), i = 1, \ldots, n\}$.

Any $\boldsymbol{\alpha} \in \mathcal{A}_1$ induces a coherent conditional assessment on $\mathcal{E}$ given by

$$\mathbf{q}_{\boldsymbol{\alpha}} := [q_i = \frac{\sum\limits_{j:\,\omega_j \subset E_iH_i} \alpha_j}{\sum\limits_{j:\,\omega_j \subset H_i} \alpha_j}, i = 1, \ldots, n]. \tag{1}$$

Associated to any assessment $\mathbf{p} \in (0,1)$ over $\mathcal{E}$ we can define a scoring rule

$$S(\mathbf{p}) := \sum_{i=1}^n |E_iH_i| \ln p_i + \sum_{i=1}^n |E_i^cH_i| \ln(1 - p_i) \tag{2}$$

with $|\cdot|$ indicator function of unconditional events. This score $S(\mathbf{p})$ is an "adaptation" of the "proper scoring rule" for probability distributions proposed by Lad in [11]. We have extended this scoring rule to partial and conditional probability assessments defining the "discrepancy" between a partial conditional assessment $\mathbf{p}$ over $\mathcal{E}$ and a distribution $\boldsymbol{\alpha} \in \mathcal{A}_2$ through the expression

$$\Delta(\mathbf{p}, \boldsymbol{\alpha}) := E_{\boldsymbol{\alpha}}(S(\mathbf{q}_{\boldsymbol{\alpha}}) - S(\mathbf{p})) = \sum_{j=1}^k \alpha_j [S_j(\mathbf{q}_{\boldsymbol{\alpha}}) - S_j(\mathbf{p})]. \tag{3}$$

It is possible to extend by continuity the definition of $\Delta(\mathbf{p}, \boldsymbol{\alpha})$ in $\mathcal{A}_0$ as

$$\Delta(\mathbf{p}, \boldsymbol{\alpha}) = \sum_{i=1}^n \ln \frac{q_i}{p_i} \alpha(E_iH_i) + \ln \frac{(1-q_i)}{(1-p_i)} \alpha(E_i^cH_i) =$$

$$= \sum_{i=1}^n \alpha(H_i) \left(q_i \ln \frac{q_i}{p_i} + (1-q_i) \ln \frac{(1-q_i)}{(1-p_i)} \right).$$

In [3] is formally proved that $\Delta(\mathbf{p}, \boldsymbol{\alpha})$ is a non negative function on $\mathcal{A}_0$ and that $\Delta(\mathbf{p}, \boldsymbol{\alpha}) = 0$ if and only if $\mathbf{p} = \mathbf{q}_{\boldsymbol{\alpha}}$; moreover $\Delta(\mathbf{p}, \cdot)$ is a convex function on $\mathcal{A}_2$ and it admits a minimum on $\mathcal{A}_0$. Finally if $\boldsymbol{\alpha}, \boldsymbol{\alpha}^0 \in \mathcal{A}_0$ are distributions that minimize $\Delta(\mathbf{p}, \cdot)$, then for all $i \in \{1, \ldots, n\}$ such that $\alpha(H_i) > 0$ and $\alpha^0(H_i) > 0$ we have $(\mathbf{q}_\alpha)_i = (\mathbf{q}_{\alpha^0})_i$; in particular if $\Delta(\mathbf{p}, \cdot)$ attains its minimum value on $\mathcal{A}_1$ then there is a unique coherent assessment $\mathbf{q}_{\underline{\alpha}}$ such that $\Delta(\mathbf{p}, \underline{\boldsymbol{\alpha}})$ is minimum. The discrepancy measure $\Delta(\mathbf{p}, \boldsymbol{\alpha})$ can be used to correct incoherent[1] assessments [2], to aggregate expert opinions [5] and it can be even applied with imprecise probabilities [4]. Here we consider a particular optimization problem involving $\Delta(\mathbf{p}, \boldsymbol{\alpha})$ which will be used to select a risk neutral probability in the set of all possible martingale measures.

2 Selection of a Risk-Neutral Probability

In order to keep the market tractable, we start with a single period model without transaction costs and with the following assumptions:

A1. Trading takes place at time 0 and time 1;

A2. The risk-free interest rate r is given and the value at time 1 of the risk-free asset (*bond*) with initial value B_0 is $B_1 = B_0(1 + r)$;

A3. The set of possible states of the world is $\Omega = \{\omega_1, \ldots, \omega_k\}$;

A4. There is a risky asset (*stock*) with price S_0 and its value at time 1 is

$$S_1 = \begin{cases} S_1(\omega_1) = a_1 S_0 \\ S_1(\omega_2) = a_2 S_0 \\ \ldots \\ S_1(\omega_k) = a_k S_0 \end{cases}, \quad a_1 > a_2 > \ldots > a_k > 0;$$

A5. The model is viable.

In a single-period single-stock model with $k \geq 2$ states of the world, the viability is equivalent to the following condition

$$min\{S_1(\omega), \omega \in \Omega\} < S_0(1 + r) < max\{S_1(\omega), \omega \in \Omega\};$$

therefore with the assumptions $A1 - A4$ the viability is equivalent to

$$a_k < 1 + r < a_1.$$

Moreover, in this model a probability distribution $\boldsymbol{\alpha}$ over Ω is a risk neutral probability if and only if

$$S_0 = \frac{1}{1 + r} [\alpha_1 a_1 S_0 + \ldots + \alpha_k a_k S_0] \Leftrightarrow \boldsymbol{\alpha} \cdot \mathbf{a} = 1 + r$$

[1] For coherence notions we refer to [6].

and we can define the set of all possible martingale measures as:

$$\mathcal{Q} := \left\{ \boldsymbol{\alpha} \in \mathbb{R}^k : \boldsymbol{\alpha} \cdot \mathbf{1} = 1,\ \boldsymbol{\alpha} \cdot \mathbf{a} = 1 + r,\ \boldsymbol{\alpha} \geq 0 \right\}.$$

Notice that $\mathcal{Q}$ is a singleton if and only if $k = 2$; otherwise $\mathcal{Q}$ is a convex set with infinitely many elements.

Finally we assume that $\mathbf{p} = (p_1, \ldots, p_n) \in (0,1)^n$ is a partial conditional probability assessment given by an expert over the set of conditional events $\mathcal{E} = [E_1|H_1, \ldots, E_n|H_n]$.

Let $\mathcal{Q}_0$ be the convex set $\mathcal{Q}_0 := \mathcal{Q} \cap \mathcal{A}_0$; we propose to select a martingale measure in $\mathcal{Q}_0$ starting from the assessment $\mathbf{p}$. In fact, we suggest a selection procedure which uses the discepancy measure $\Delta(\mathbf{p}, \boldsymbol{\alpha})$ and which is based on the following result:

Theorem 1. *Let* $\mathcal{M} := \arg\min\{\Delta(\mathbf{p}, \boldsymbol{\alpha}), \boldsymbol{\alpha} \in \mathcal{Q}_0\}$ *be the set of all martingale measures minimizing* $\Delta(\mathbf{p}, \boldsymbol{\alpha})$*; then* $\mathcal{M}$ *is a non-empty convex set.*

Proof. $\Delta(\mathbf{p}, \boldsymbol{\alpha})$ is a convex function on $\mathcal{Q}_0$ and then there is at least one $\underline{\boldsymbol{\alpha}}$ in $\mathcal{Q}_0$ such that

$$\Delta(\mathbf{p}, \underline{\boldsymbol{\alpha}}) = \min_{\boldsymbol{\alpha} \in \mathcal{Q}_0} \Delta(\mathbf{p}, \boldsymbol{\alpha}) \tag{4}$$

and then $\mathcal{M}$ is non empty. Notice that the convexity of $\Delta(\mathbf{p}, \boldsymbol{\alpha})$ guarantees the existence of this minimum but it is possible that more than one distribution minimize $\Delta(\mathbf{p}, \boldsymbol{\alpha})$ in $\mathcal{Q}_0$ and in this case $\mathcal{M}$ is not a singleton. However, since $\Delta(\mathbf{p}, \boldsymbol{\alpha})$ is a convex function and $\mathcal{M}$ is the set of minimal points of $\Delta(\mathbf{p}, \boldsymbol{\alpha})$ in $\mathcal{Q}_0$, $\mathcal{M}$ is a convex set. □

The previous theorem guarantees the existence of a solution $\underline{\boldsymbol{\alpha}}$ for the optimization problem (4) but it does not assure its uniqueness. Therefore we need another criterion to choose, between the martingale measure minimizing $\Delta(\mathbf{p}, \boldsymbol{\alpha})$, a unique $\boldsymbol{\alpha}^*$ as risk-neutral probability. The idea is to select one distribution in $\mathcal{M}$ which in some sense minimizes the exogenous information. In fact, we will define $\boldsymbol{\alpha}^*$ as

$$\boldsymbol{\alpha}^* := arg \min_{\boldsymbol{\alpha} \in \mathcal{M}} \sum_{j-1}^{k} \alpha_j \ln \alpha_j \tag{5}$$

that is the distribution which minimize the relative entropy with respect to the uniform distribution (i.e. the distribution with maximum entropy).

In the following applicative examples, since extremely simplified, we will encounter single optimal solutions for (4). Anyhow, in more complex situations (e.g. *Example 3* in [2]) multiple optimal solutions appear so that the further selection step (5) gets real significance.

Example 1 (Partial information). A european call option is a contract that gives its owner the right to buy one unit of underlying stock S at time 1 at strike price K. At time 1 the decision of whether to exercise the option or not will depend on the stock price S_1 at time 1: the investor will exercise the call if and only if the option is in the money (i.e. $S_1 > K$) and the payoff at time one can be written as

$$C_1 = [S_1 - K]^+ = \max\{S_1 - K, 0\}.$$

Let us consider the simplest example of incomplete model with one stock and one step, the *trinomial model*:

$$S_1 = \begin{cases} S_0(1+u) \\ S_0 \\ S_0(1+d) \end{cases}$$

where $u, d \in \mathbb{R}$ and $u > r > d > -1$. To price this call option by risk neutral valuation we have to select a risk neutral probability in the convex set

$$\mathcal{Q}_0 = \{\boldsymbol{\alpha} \in \mathbb{R}^3 \mid \alpha_1(1+u) + \alpha_2 + \alpha_3(1+d) = 1 + r,\ \boldsymbol{\alpha} \geq 0\}.$$

Given a risk neutral probability $\boldsymbol{\alpha} \in \mathcal{Q}$ the corresponding fair price is $C_0 = \mathbb{E}_{\boldsymbol{\alpha}}(\overline{C_1})$. Let us take $r = 0$, $u = 0.1$, $d = 0.1$, $S_0 = 100$ and $K = 95$; then

$$S_1 = \begin{cases} 110 \\ 100 \\ 90 \end{cases}, \qquad C_1 = \begin{cases} 15 \\ 5 \\ 0 \end{cases}$$

and the convex set of possible martingale measure is

$$\mathcal{Q}_\lambda = (\lambda, 1 - 2\lambda, \lambda), \quad \lambda \in [0, 1/2].$$

Suppose that we have the probability that the market goes up $p_1 = \frac{2}{3}$. Then

$$\Delta(\mathbf{p}, \boldsymbol{\alpha}) = q_1 \ln \frac{q_1}{p_1} + (1 - q_1) \ln \frac{1 - q_1}{1 - p_1} = q_1 \ln \frac{3}{2} q_1 + (1 - q_1) \ln 3(1 - q_1)$$

is a strictly convex function and the minimum for $\boldsymbol{\alpha} \in \mathcal{Q}_\lambda$ is attained at $\alpha_1 = 1/2$. Therefore the distribution that minimize $\Delta(\mathbf{p}, \boldsymbol{\alpha})$ is $(1/2, 0, 1/2)$ and the corresponding fair price for the call option is $C_0 = 7.5$.

It is important to remark that the initial probability assessment $\mathbf{p}$ can also be incoherent: we can start with an evaluation which is not consistent with the set of all distributions $\mathcal{A}$ or, as described in the next section, we can have several expert opinions and in this case, since it is easy that the experts disagree, the merging of different evaluations can easily give rise to incoherence.

Example 2 (Incoherent assessment). Let us consider the call option of previous example and let us suppose that we have the probabilities

$$p_1 = P(\text{“the market goes up”}) = 1/3,$$

$$p_2 = P(\text{“the market goes down given that the market change”}) = 4/5.$$

This assessment is incoherent; in fact the system

$$\begin{cases} \alpha_1 = 1/3 \\ \alpha_3/(\alpha_1+\alpha_3) = 4/5 \\ \alpha_1+\alpha_2+\alpha_3 = 1 \\ \alpha_j \geq 0, j = 1,2,3 \end{cases}$$

has no solution. In this case

$$\varDelta(\mathbf{p},\boldsymbol{\alpha}) = \alpha_1 \ln 3\alpha_1 + (1-\alpha_1)\ln\frac{3}{2}(1-\alpha_1) + \alpha_3 \ln\frac{5\alpha_3}{4(\alpha_1+\alpha_3)} + \alpha_1 \ln\frac{5\alpha_1}{\alpha_1+\alpha_3}$$

and we will select the distribution $\boldsymbol{\alpha}^*$ which minimize the discrepancy measure in $\mathcal{Q}_\lambda = (\lambda, 1-2\lambda, \lambda)$ with $\lambda \in \left[0, \frac{1}{2}\right]$. So we can write $\varDelta(\mathbf{p},\boldsymbol{\alpha})$ as function of λ and we have

$$\varDelta(\mathbf{p},\lambda) = \lambda \ln 3\lambda + (1-\lambda)\ln\frac{3}{2}(1-\lambda) + \lambda \ln\frac{25}{16}.$$

Since

$$\varDelta'(\mathbf{p},\lambda) = \ln\lambda - ln(1-\lambda) + \ln\frac{25}{8} = 0 \Leftrightarrow \frac{\lambda}{1-\lambda} = \frac{8}{25} \Leftrightarrow \lambda = \frac{8}{33}$$

we get

$$\boldsymbol{\alpha}^* = \left(\frac{8}{33}, \frac{17}{33}, \frac{8}{33}\right), \quad C_0 = \frac{8}{33}\cdot 15 + \frac{17}{33}\cdot 5 = 6.\overline{21}.$$

Obviously the coherence of the initial assessment $\mathbf{p}$ is not sufficient for the compatibility of $\mathbf{p}$ with $\mathcal{Q}_0$. In fact, as shown in the next example, even if the initial assessment is coherent, it is possible that the intersection between the set of probability distributions over Ω which are consistent with $\mathbf{p}$ and the set $\mathcal{Q}_0$ of risk neutral probabilities is empty; therefore also in this case we will select the martingale measure in $\mathcal{Q}_0$ which minimize the discrepancy with respect to $\mathbf{p}$.

Example 3 (Coherent assessment incompatible with $\mathcal{Q}_0$). In the same framework of *Example 2*, let us suppose that we have $p_1 = 1/2$ and $p_2 = 1/3$. In this case $\mathbf{p}$ is a coherent assessment but it is incompatible with $\mathcal{Q}_\lambda$; in fact the system

$$\begin{cases} \alpha_1 = 1/2 \\ \alpha_3/(\alpha_1+\alpha_3) = 1/3 \\ \alpha_1+\alpha_2+\alpha_3 = 1 \\ \alpha_j \geq 0, j = 1,2,3 \end{cases}$$

admits the solution $\left(\frac{1}{2}, \frac{1}{4}, \frac{1}{4}\right) \notin \mathcal{Q}_\lambda$. Then we have to minimize

$$\varDelta(\mathbf{p},\lambda) = \lambda\ln\lambda + (1-\lambda)\ln(1-\lambda) + \lambda\ln\frac{9}{8} + \ln 2, \quad \lambda \in [0, 1/2].$$

Since

$$\varDelta'(\mathbf{p},\lambda) = \ln\lambda - ln(1-\lambda) + \ln\frac{9}{8} = 0 \Leftrightarrow \frac{\lambda}{1-\lambda} = \frac{8}{9} \Leftrightarrow \lambda = \frac{8}{17}$$

we get

$$\boldsymbol{\alpha}^* = \left(\frac{8}{17}, \frac{1}{17}, \frac{8}{17}\right), \quad C_0 = 125/17 \cong 7.3529.$$

In this case we can compare our result with other methods used to select a martingale measure starting with a probability given over all the possible states of the world. For example, by minimizing the *Euclidean Distance* between $\left(\frac{1}{2}, \frac{1}{4}, \frac{1}{4}\right)$ and $\mathcal{Q}_\lambda$ we get

$$\boldsymbol{\alpha}^* = \left(\frac{3}{8}, \frac{1}{4}, \frac{3}{8}\right), \quad C_0 = 6.875;$$

using the *Minimal Entropy Criterion* proposed by Frittelli in [10], we get

$$\boldsymbol{\alpha}^* = \left(\frac{\sqrt{2}}{2\sqrt{2}+1}, \frac{1}{2\sqrt{2}+1}, \frac{\sqrt{2}}{2\sqrt{2}+1}\right), \quad C_0 \cong 6.847.$$

Notice that minimize the relative entropy of $\mathbf{q}$ with respect to $\mathbf{p}$ is equivalent to minimize $E_{\mathbf{q}}(S(\mathbf{q}) - S(\mathbf{p}))$ using the logarithmic scoring rule $S(\mathbf{p}) = \sum_{i=1}^{n} \ln p_i$; we prefer to use the proper scoring rule proposed by Lad because the assessor of $\mathbf{p}$ "loses less" the higher are the probabilities assessed for events that are verified and, at the same time, the lower are the probabilities assessed for those that are not verified. But it is important to remark that the main difference between our method and the previous methods is that the discrepancy measure $\Delta(\mathbf{p}, \boldsymbol{\alpha})$ can be used to select a risk neutral probability when the information that we have is partial or conditional or if we have incoherent assessments even given by different expert opinions, as we will see in the next section.

Finally observe that, as described in [3], when $p_i = 0$ we just take $q_i = 0$ that is $\mathbf{q} \prec\prec \mathbf{p}$ as is usually done in the literature (see for example [10]).

3 Aggregation of Expert Opinions

In this section we generalize the previous procedure to the case of several assessments given by different experts. Suppose that we have S experts which give S partial conditional probability assessments, depending on their partial knowledge and on their informations about the future states of the world. The different sources of information will be indexed by a subscript index s varying on the finite set $\mathcal{S}$.

We formalize the domain of the evaluations through finite families of conditional events of the type $\mathcal{E}_s = \{E_{s,i}|H_{s,i}\,,\, i = 1, \ldots, n_s\}$, $s \in \mathcal{S}$; the numerical part of the different assessments can be elicited through $\mathbf{p}_s = (p_{s,1}, \ldots, p_{s,n_s})$ as evaluation of the probabilities $P(E_{s,i}|H_{s,i})$, $i = 1, \ldots, n_s$. When the different evaluations are merged, we get a unique assessment with repetitions, i.e. conditional events with different absolute frequencies. To distinguish the whole merged assessments by its components we simply ignore the indexes $s \in \mathcal{S}$, so that we deal with the domain $\mathcal{E} = [E_1|H_1, \ldots, E_n|H_n] = \bigcup_{s\in\mathcal{S}} \mathcal{E}_s$ with associated assessment $\mathbf{p} = (p_1, \ldots, p_n) = \bigcup_{s\in\mathcal{S}} \mathbf{p}_s$. The possible multiplicity of some conditional event $E_i|H_i$ in $\mathcal{E}$ can be simply treated as peculiar logical relations. This means that actually we have a unique assessment $\mathbf{p} = (p_1, \ldots, p_n)$ over $\mathcal{E} = [E_1|H_1, \ldots, E_n|H_n]$ and to select a risk neutral probability in $\mathcal{Q}$ we can solve the optimization problem (4). Therefore, the existence and uniqueness of $\boldsymbol{\alpha}^*$ can be proved as in the previous section and the procedure gives one and only one martingale measure to price uniquely all the derivative securities.

Example 4 (Two distributions). In a trinomial one-step model, suppose that we have the opinions of two experts over all the possible states of the world; if we denote with $\boldsymbol{\alpha}' = (\alpha'_1, \alpha'_2, \alpha'_3)$ the distribution of the first expert and with $\boldsymbol{\alpha}'' = (\alpha''_1, \alpha''_2, \alpha''_3)$ the distribution of the second expert we have

$$\Delta(\mathbf{p}, \boldsymbol{\alpha}) = \sum_{j=1}^{3} \alpha_j \ln \frac{\alpha_j^2}{\alpha'_j \alpha''_j} + (1-\alpha_j) \ln \frac{(1-\alpha_j)^2}{(1-\alpha'_j)(1-\alpha''_j)}$$

and from the properties of $\Delta(\mathbf{p}, \boldsymbol{\alpha})$ it follows that there is a unique $\boldsymbol{\alpha}^* \in \mathcal{Q}_0$ such that $\Delta(\mathbf{p}, \boldsymbol{\alpha})$ is minimum.

It is also possible to associate different weights to the elements of the joined assessment $(\mathcal{E}, \mathbf{p})$: we can denote by $\mathbf{w} = [w_1, \ldots, w_n]$ such weights and adjust the expression of $\Delta(\mathbf{p}, \boldsymbol{\alpha})$ as

$$\Delta^{\mathbf{w}}(\mathbf{p}, \boldsymbol{\alpha}) = \sum_{i=1}^{n} \alpha(Hi) \left(q_i \ln \frac{q_i^{w_i}}{p_i^{w_i}} + (1-q_i) \ln \frac{(1-q_i)^{w_i}}{(1-p_i)^{w_i}} \right). \qquad (6)$$

Example 5 (Agriculture Derivative and Different Sources of Information). Let us suppose that the underlying asset is an agricultural product and that its value depends on the weather conditions and on the presence of insect populations which can damage the production, that is it will depend on on the events E_1="favourable weather" and E_2="absence of dangerous insect populations". Let $\omega_1 := E_1^c E_2^c$, $\omega_2 := E_1 E_2^c$, $\omega_3 := E_1^c E_2$ and $\omega_4 := E_1 E_2$ and let $S_0 = 100$ be the value at time 0 of the the underlying asset with payoff

$$S_1 := \begin{cases} S_1(\omega_1) = 1.1 S_0 \\ S_1(\omega_2) = S_1(\omega_3) = S_0 \\ S_1(\omega_4) = 0.9 S_0 \end{cases}$$

The set of all martingale measures is $\mathcal{Q}_{\lambda,\mu} = (\lambda, \mu, 1-2\lambda-\mu, \lambda)$ with $\lambda, \mu \in [0,1]$ and $2\lambda + \mu < 1$.

To price a call option with underlying S and strike price $K = 90$, we ask two different expert: an entomologist give us $p_1 := P(E_2^c|E_1) = 1/3$, $p_2 := P(E_2) = 2/3$ and a meteorologist give us $p_3 := P(E_1) = 2/3$. So we have

$$\Delta(\mathbf{p}, \lambda, \mu) = \mu \ln \frac{3\mu}{\lambda+\mu} + \lambda \ln \frac{3\lambda}{2(\lambda+\mu)} + (1-\lambda-\mu) \ln \frac{9}{2}(1-\lambda-\mu)^2 + (\lambda+\mu) \ln \frac{9}{2}(\lambda+\mu)^2$$

and the optimal distribution $\boldsymbol{\alpha}^* = \left(\frac{1}{3}, \frac{1}{6}, \frac{1}{6}, \frac{1}{3}\right)$ gives $C_0 = 10$.

4 Conclusions

With this paper we want to present a method for the valuation problem in incomplete markets which can be used to select a unique risk neutral probability in a one-step model with a finite number of scenarios starting with partial information given as conditional probability assessment or given by different expert opinions. Further investigations are required to analyse the multi-period model where conditional probabilities play a fundamental role.

References

1. Bingham, N.H., Kiesel, R.: Risk-Neutral Valuation: pricing and hedging of financial derivatives. Springer, London (2004)
2. Capotorti, A., Regoli, G.: Coherent correction of inconsistent conditional probability assessments. In: Proc. of IPMU 2008, Malaga (Es) (2008)
3. Capotorti, A., Regoli, G., Vattari, F.: Theoretical properties of a discrepancy measure among partial conditional probability assessments. Submitted to International Journal of Approximate Reasoning (to appear)
4. Capotorti, A., Regoli, G., Vattari, F.: On the use of a new discrepancy measure to correct incoherent assessments and to aggregate conflicting opinions based on imprecise conditional probabilities. In: Proc. of ISIPTA 2009, Durham (UK) (2009)
5. Capotorti, A., Regoli, G., Vattari, F.: Merging different probabilistic information sources through a new discrepancy measure. In: Proc. of WUPES 2009, Liblice (CR) (2009)
6. Coletti, G., Scozzafava, R.: Probabilistic Logic in a Coherent Setting. Trends in Logic. Kluwer, Dordrecht (2002)
7. Elliot, R.J., Kopp, P.E.: Mathematics of Financial Markets. Springer Finance, New York (2005)
8. Follmer, H., Schied, A.: Stochastic Finance: an introduction in discrete time. Walter de Gruyter, Berlin (2004)
9. Follmer, H., Sondermann, D.: Hedging of Non-Redundant Contingent Claim. In: Contributions to Mathematical Economics. Elsevier Science Publishers, Amsterdam (1986)
10. Frittelli, M.: The minimal entropy martingale measure and the valuation problem in incomplete markets. Mathematical Finance 10, 39–52 (2000)
11. Lad, F.: Operational Subjective Statistical Methods: a mathematical, philosophical, and historical introduction. John Wiley, New York (1996)
12. Musiela, M., Rutkowski, M.: Martingale Methods in Financial Modelling. Springer, New York (2005)

A New Contextual Discounting Rule for Lower Probabilities

Sebastien Destercke

INRA/CIRAD, UMR1208, 2 place P. Viala, F-34060 Montpellier cedex 1, France
sdestercke@gmail.com

Abstract. Sources providing information about the value of a variable may not be totally reliable. In such a case, it is common in uncertainty theories to take account of this unreliability by a so-called discounting rule. A few discounting rules have been proposed in the framework of imprecise probability theory, but one of the drawback of those rules is that they do not preserve interesting properties (i.e. n-monotonicity) of lower probabilities. Another aspect that only a few of them consider is that source reliability is often dependent of the context, i.e. a source may be more reliable to identify some values than others. In such cases, it is useful to consider contextual discounting, where reliability information is dependent of the variable values. In this paper, we propose such a contextual discounting rule that also preserves some of the interesting mathematical properties a lower probability can have.

Keywords: information fusion, reliability, discounting, probability sets.

1 Introduction

When sources providing uncertain information about the value assumed by a variable X on the (finite) domain $\mathscr{X}$ are not fully reliable, it is necessary to integrate information about this reliability in uncertainty representations. In imprecise probability theories (i.e. possibility theory, evidence theory, transferable belief model, lower previsions), where imprecision in beliefs or information is explicitly modelled in uncertainty representations, it is usual to take account of this reliability through the operation commonly called discounting. Roughly speaking, the discounting operation consists in making the information all the more imprecise (i.e. less relevant) as it is unreliable.

Many authors have discussed discounting operations in uncertainty theories [1,2,3]. In most cases, authors consider that reliability is modelled by a single weight (possibly imprecise) λ whose value is in the unit interval, i.e. $\lambda \in [0,1]$. In a few other cases, they consider that different weights can be given to different elements of a partition of the referential $\mathscr{X}$, and in this case reliability information is given by a vector of weights $\lambda = (\lambda_1, \ldots, \lambda_L)$, with L the cardinality of the partition and $\lambda_i \in [0,1]$. The reason for considering such weights is that, in some cases, the ability of the source to recognise the true value of X may depend on this value. For example, a specialised physician will be very reliable when it comes to recognise diseases corresponding to its speciality, but less reliable when the patient has other diseases. A sensor may be very discriminative for some kinds of objects, while often confusing other objects between them.

E. Hüllermeier, R. Kruse, and F. Hoffmann (Eds.): IPMU 2010, Part II, CCIS 81, pp. 198–207, 2010.

Many rules handling more than precise single reliability weight have been proposed in the framework of imprecise probability theory [2,4,5], in which uncertain information is represented by bounds over expectation values or by associated convex probability sets, the two representations being formally equivalent. Both Karlsson *et al.* [4] and Benavoli and Antonucci [5] consider the case where a unique but possibly imprecise reliability weight is given for the whole referential $\mathscr{X}$, but start from different requirements, hence proposing different discounting rules. Karlsson *et al.* [4] require a discounted probability set to be insensitive to Bayesian combination (i.e. using the product) when the source is completely unreliable. It brings them to the requirement that the information provided by a completely unreliable source should be transformed into the precise uniform probability distribution. Benavoli and Antonucci [5] model reliability by the means of coherent conditional lower previsions [6] and directly integrates it to an aggregation process, assuming that the information provided by a completely unreliable source should be transformed into a so-called vacuous probability set (i.e. the probability set corresponding to all probabilities having $\mathscr{X}$ for support). Moral and Sagrado [2] start from constraints given on expectations value and assume that reliability weights are precise but can be contextual (i.e., one weight per element of $\mathscr{X}$) or can translate some (fuzzy) indistinguishability relations.

Each of these rules is justified in its own setting. However, a common defect of all these rules is that when reliability weights are not reduced to a single precise number, the discounted probability set is usually more complex and difficult to handle than the initial one. This is a major inconvenient to their practical use, since using generic probability sets often implies an heavy computational burden. In this paper, we propose a new discounting rule for lower and upper probabilities, inspired from the discounting rule proposed by Mercier et al. [3] in the framework of the transferable belief model [7]. We show that this rule preserves both the initial probability set complexity, as well as some of its interesting mathematical properties, provided the initial lower probability satisfies them.

Section 2 recalls the basics of lower/upper probabilities needed here, as well as some considerations about the properties discounting rules can satisfy. Section 3 then presents our rule, discusses its properties and possible interpretation, and compares its properties with those of other discounting rules.

2 Preliminary Notions

This section recalls both the notion of lower probabilities and of associated sets of probabilities. It then details some properties that may or may not have a given discounting rule.

2.1 Probability Sets and Lower Probabilities

In this paper, we consider that our uncertainty about the value assumed by a variable X on a finite space $\mathscr{X} = \{x_1, \ldots, x_N\}$ is modelled by a lower probability $\underline{P} : \wp(\mathscr{X}) \to [0,1]$, i.e. a mapping from the power set of $\mathscr{X}$ to the unit interval, satisfying the boundary constraints $\underline{P}(\emptyset) = 0$, $\underline{P}(\mathscr{X}) = 1$ and monotonic with respect to inclusion, i.e. for any $A, B \subseteq \mathscr{X}$ such that $A \subseteq B$, $\underline{P}(A) \leq \underline{P}(B)$. To a lower probability can be associated an upper probability $\overline{P}$ such that, for any $A \subseteq \mathscr{X}, \underline{P}(A) = 1 - \overline{P}(A^c)$, with A^c the complement of A. A lower probability induce a probability set $\mathscr{P}_{\underline{P}}$ such that

$$\mathscr{P}_{\underline{P}} := \{p \in \Sigma_{\mathscr{X}} | (\forall A \subseteq \mathscr{X})(P(A) \geq \underline{P}(A)\},$$

with p a probability mass, P the induced probability measure and $\Sigma_{\mathscr{X}}$ the set (simplex) of all probability mass functions on $\mathscr{X}$. A lower probability is said to be *coherent* if and only if $\mathscr{P}_{\underline{P}} \neq \emptyset$ and $\underline{P}(A) = \min\{P(A)|p \in \mathscr{P}_{\underline{P}}\}$ for all $A \subseteq \mathscr{X}$, i.e., if $\underline{P}$ is the *lower envelope* of $\mathscr{P}_{\underline{P}}$ on events.

Inversely, from any probability set $\mathscr{P}$, one can extract a lower probability measure defined, for any $A \subseteq \mathscr{X}$, as $\underline{P}(A) = \min\{P(A)|p \in \mathscr{P}\}$. Note that lower probabilities alone are not sufficient to describe any probability set. Let $\mathscr{P}$ be a probability set and $\underline{P}$ its lower probability, then the probability set $\mathscr{P}_{\underline{P}}$ induced by this lower probability is such that $\mathscr{P} \subseteq \mathscr{P}_{\underline{P}}$ with the inclusion being usually strict. In general, one needs the richer language of expectation bounds to describe any probability set [8].

In this paper, we will restrict ourselves to credal sets induced by lower probabilities alone. Note that such lower probabilities already encompass an important number of practical uncertainty representations, such as necessity measures [9], belief functions [10] or so-called p-boxes [11]. An important classes of probability sets induced by lower probabilities alone and encompassing these representations are the one for which lower probabilities satisfy the property of n-monotonicity for $n \geq 2$. n-monotonicity is defined as follows:

Definition 1. *A lower probability $\underline{P}$ is n-monotone, where $n > 0$ and $n \in \mathbb{N}$, if and only if for any set $\mathscr{A} = \{A_i | i \in \mathbb{N}, 0 < i \leq n\}$ of events $A_i \subseteq \mathscr{X}$, it holds that*

$$\underline{P}(\bigcup_{A_i \in \mathscr{A}} A_i) \geq \sum_{I \subseteq \mathscr{A}} (-1)^{|I|+1} \underline{P}(\bigcap_{A_i \in I} A_i).$$

An ∞-montone lower probability (i.e., a belief function) is a lower probability n-monotone for every n. Both 2-monotonicity and ∞-monotonicity have been studied with particular attention in the literature [12,10,13,14], for they have interesting mathematical properties that facilitate their practical handling. When processing lower probabilities, it is therefore desirable to preserve such properties, if possible.

2.2 Discounting Operation: Definition and Properties

The discounting operation consists in using the reliability information λ to transform an initial lower probability $\underline{P}$ into another lower probability $\underline{P}^{\lambda}$. λ can take different forms, ranging from a single precise number to a vector of imprecise numbers. In order to discriminate between different discounting rules, we think it is useful to list some of the properties that they can satisfy.

*Property 1 (coherence preservation, **CP**).* A discounting rule satisfies coherence preservation **CP** when $\underline{P}^{\lambda}$ is coherent whenever $\underline{P}$ is coherent.

This property ensures some consistency to the discounting rule.

Property 2 (Imprecision monotony, ***IM****).* A discounting rule satisfies Imprecision monotony **IM** if and only if $\underline{P}^{\lambda} \leq \underline{P}$, that is if the discounted information is less precise than the original one.[1]

This property simply means that imprecision should increase when a source is partially unreliable. This may seem a reasonable request, however for some particular cases [4], there may exist arguments against such a property.

Property 3 (n-monotonicity preservation, ***MP****).* A discounting rule satisfies n-monotonicity preservation **MP** when $\underline{P}^{\lambda}$ is n-monotone whenever $\underline{P}$ is n-monotone.

Such a property ensures that interesting mathematical properties of a lower probabilities will be preserved by the discounting operation.

Property 4 (lower probability preservation, ***LP****).* A discounting rule satisfies lower probability preservation, **LP** when the discounted probability set $\mathscr{P}^{\lambda}$ resulting from discounting is such that $\mathscr{P}^{\lambda} = \mathscr{P}_{\underline{P}^{\lambda}}$, provided initial information was given as a lower probability.

This property ensures that if the initial information is entirely captured by a lower probability, so will be the discounted information. It ensures to some extent that the uncertainty representation structure will keep a bounded complexity.

Property 5 (Reversibility, ***R****).* A discounting rule satisfies reversibility **R** if the initial information $\underline{P}$ can be recovered from the knowledge of the discounted information $\underline{P}^{\lambda}$ and λ alone, when $\lambda > 0$.

This property, similar to the de-discounting discussed by Denoeux and Smets [15], ensures that, if one receives as information the discounted information together with the source reliability information, he can still come back to the original information provided by the source. This can be useful if reliability information is revised. This requires the discounting operation to be an injection.

3 The Discounting Rule

We now propose our contextual discounting rule, inspired from the contextual discounting rule proposed by Mercier at al. [3] in the context of the transferable belief model. We show that, from a practical viewpoint, this discounting rule has interesting properties, and briefly discuss its interpretation.

3.1 Definition

We consider that source reliability comes into the form of a vector of weights $\lambda = (\lambda_1, \ldots, \lambda_L)$ associated to elements of a partition $\Theta = \{\theta_1, \ldots, \theta_L\}$ of $\mathscr{X}$ (i.e. $\theta_i \subseteq \mathscr{X}$, $\cup_{i=1}^{L} \theta_i = \mathscr{X}$ and $\theta_i \cap \theta_j = \emptyset$ if $i \neq j$). We denote by $\mathscr{H}$ the field induced by Θ. Value one is given to λ_i when the source is judged completely reliable for detecting element

[1] This is equivalent to ask for $\mathscr{P}_{\underline{P}} \subseteq \mathscr{P}_{\underline{P}^{\lambda}}$.

x_i, and zero if it is judged completely unreliable. We do not consider imprecise weights, simply because in such a case one can still consider the pessimistic case where the lowest weights are retained.

Given a set $A \subseteq \mathscr{X}$, its inner and outer approximations in $\mathscr{H}$, respectively denoted A_* and A^*, are:

$$A_* = \bigcup_{\substack{\theta \in \Theta \\ \theta \subseteq A}} \theta \quad \text{and} \quad A^* = \bigcup_{\substack{\theta \in \Theta \\ \theta \cap A \neq \emptyset}} \theta .$$

We then propose the following discounting rule that transforms an initial information $\underline{P}$ into $\underline{P}^{\lambda}$ such that, for every event $A \subseteq \mathscr{X}$, we have

$$\underline{P}^{\lambda}(A) = \underline{P}(A) \prod_{\theta_i \subseteq (A^c)^*} \lambda_i, \tag{1}$$

with the convention $\prod_{\theta_i \subseteq \emptyset} \lambda_i = 1$, ensuring that $\underline{P}(\mathscr{X}) = \underline{P}^{\lambda}(\mathscr{X}) = 1$ and $\overline{P}(\emptyset) = \overline{P}^{\lambda}(\emptyset) = 0$.

Example 1. Let us illustrate our proposition on a 3-dimensional space $\mathscr{X} = \{x_1, x_2, x_3\}$. Assume the lower probability is given by the following constraints:

$$0.1 \leq p(x_1) \leq 0.3; \quad 0.4 \leq p(x_2) \leq 0.5; \quad 0.3 \leq p(x_3) \leq 0.5.$$

Lower probabilities induced by these constraints (through natural extension [8]) can be easily computed, as they are probability intervals [16]. They are summarised in the next table:

	x_1	x_2	x_3	$\{x_1,x_2\}$	$\{x_1,x_3\}$	$\{x_2,x_3\}$
$\underline{P}$	0.1	0.4	0.3	0.5	0.5	0.7

Let us now assume that $\Theta = \{\{x_1,x_2\} = \theta_1, \{x_3\} = \theta_2\}$ and that $\lambda_1 = 0.5$, $\lambda_2 = 1$. The discounted lower probability $\underline{P}^{\lambda}$ is given in the following table

	x_1	x_2	x_3	$\{x_1,x_2\}$	$\{x_1,x_3\}$	$\{x_2,x_3\}$
$\underline{P}^{\lambda}$	0.05	0.2	0.15	0.5	0.25	0.35

Figure 1 pictures, in barycentric coordinates (i.e. each point in the triangle is a probability mass function over $\mathscr{X}$, with the probability of x_i equals to the distance of the point to the side opposed to vertex x_i), both the initial probability set and the discounted probability set resulting from the application or the proposed rule. As we can see, only the upper probability of $\{x_3\}$ (the element we are certain the source can recognise with full reliability) is kept at its initial value.

3.2 Properties of the Discounting Rule

Let us now discuss the properties of this discounting rule. First, by Equation (1), we have that the results of the discounting rule is still a lower probability, and since $\lambda \in [0,1]$, $\underline{P}^{\lambda} \leq \underline{P}$, hence the property of imprecision monotony is satisfied. We can also show the following proposition:

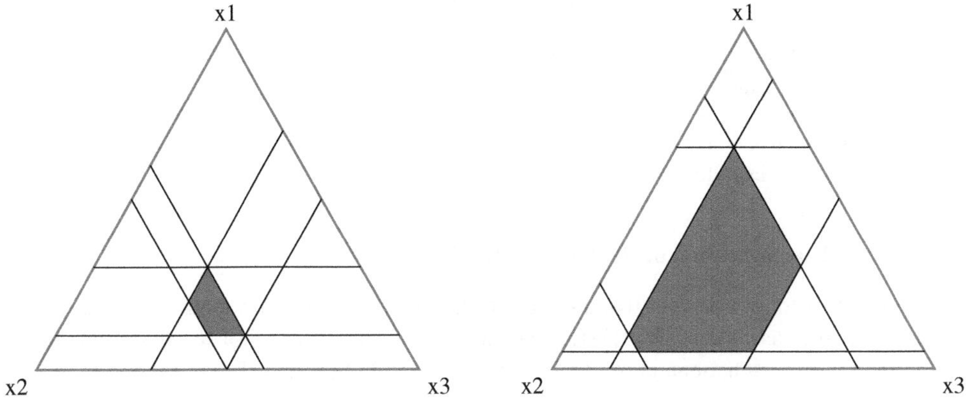

Fig. 1. Initial (right) and discounted (left) probability sets of Example 1

Proposition 1. *Let $\underline{P}$ be a lower probability and λ a strictly positive weight vector. The contextual discounting rule preserves the following properties:*

1. *Coherence*
2. *2-monotonicity*
3. *∞-monotonicity*

See Appendix A for the proof. These properties ensure us that the discounting rule preserves the desirable properties of lower probabilities that are coherence, as well as other more "practical" properties that keep computational complexity low, such as 2-monotonicity. The discounting operator is also reversible.

Property 6 (Reversibility). Let $\underline{P}^{\lambda}$ and λ be the provided information. Then, $\underline{P}$ can be retrieved by computing, for any $A \subseteq \mathscr{X}$,

$$\underline{P}(A) = \frac{\underline{P}^{\lambda}(A)}{\prod_{\theta_i \subseteq (A^c)^*} \lambda_i}.$$

Table 1 summarises the properties of the discounting rule proposed here, together with the properties of other discounting rules proposed in the literature. It considers the following properties and features: whether a discounting can cope with generic probability sets, with imprecise weights and with contextual weights, and if it satisfies or not the properties proposed in Section 2.2. This table displays some of the motivations that have led to the rule proposed in this paper. Indeed, while most rules presented in the literature have been justified and have the advantages that they can be applied to any probability set (not just the ones induced by lower probabilities), applying them also implies losing properties that have a practical interest and importance, especially the properties of 2$-$ and $\infty-$monotonicity. When dealing with lower probabilities, our rule offers a convenient alternative, as it preserves important properties.

Table 1. Discounting rules properties

Paper	Any $\mathscr{P}$	Imp. weights	contextual	CP	IM	MP	LP	R
This paper	×	×	✓	✓	✓	✓	✓	✓
Moral et al. [2]	✓	×	✓	✓	✓	×	×	×
Karlsson et al. [4]	✓	✓	×	✓	×	×	×	✓
Benavoli et al. [5]	✓	✓	×	✓	✓	×	×	×

3.3 Interpretation of the Discounting Rule

In order to give an intuitive interpretation of the proposed discounting rule, let us consider the case where $\Theta = \{x_1,\dots,x_N\}$ and $\mathscr{H}$ is the power set of $\mathscr{X}$, that is one weight is given to each element of $\mathscr{X}$. In this case, Eq. (1) becomes, for an event $A \subseteq \mathscr{X}$,

$$\underline{P}^{\lambda}(A) = \underline{P}(A) \prod_{x_i \in A^c} \lambda_i,$$

and the upper discounted probability $\overline{P}^{\lambda}$ of an event A becomes

$$\overline{P}^{\lambda}(A) = 1 - \underline{P}^{\lambda}(A^c) = 1 - (\underline{P}(A^c) \prod_{x_i \in A} \lambda_i) = 1 - \prod_{x_i \in A} \lambda_i + \overline{P}^{\lambda}(A) \prod_{x_i \in A} \lambda_i.$$

Hence, in this particular case, we have the following lemma:

Lemma 1. *For any event $A \subseteq \mathscr{X}$, we have*

- $\underline{P}^{\lambda}(A) = \underline{P}(A)$ *iff* $\lambda_i = 1$ *for any* $x_i \in A^c$,
- $\overline{P}^{\lambda}(A) = \overline{P}(A)$ *iff* $\lambda_i = 1$ *for any* $x_i \in A$.

This means that our certainty in the fact that the true answer lies in A (modeled by $\underline{P}(A)$) does not change, provided that we are certain that the source is able to eliminate all possible values outside of A. Consider for instance the case $\underline{P}(A) = 1$, meaning that we are sure that the true answer is in A. It seems rational to require, in order to fully trust this judgement, that the source can eliminate with certainty all possibilities outside A. Conversely, the plausibility that the true value lies in A ($\overline{P}(A)$) does not change when the source is totally able to recognise elements of A. Consider again the extreme case $\overline{P}(A) = 0$, then it is again rational to ask for $\overline{P}(A)$ to increase if the source is not fully able to recognise elements of A, and for it to remain the same otherwise, as in this case the source would have recognised an element of A for sure.

Now, consider the case where $\Theta = \{\mathscr{X}\}$ and $\mathscr{H} = \{\emptyset, \mathscr{X}\}$, with λ the associated unique weight. We retrieve the classical discounting rule consisting in mixing the initial probability set with the vacuous one, that is $\underline{P}^{\lambda}(A) = \lambda \underline{P}(A)$ for any $A \subseteq \mathscr{X}$ and we have $\mathscr{P}_{\underline{P}^{\lambda}} = \{\lambda \cdot p + (1-\lambda) \cdot q | p \in \mathscr{P}_{\underline{P}}, q \in \Sigma_{\mathscr{X}}\}$. Note that when $\Theta = \{\theta_1,\dots,\theta_L\}$ with $L > 1$ and $\lambda := \lambda_1 = \dots = \lambda_L$, the lower probability $\underline{P}^{\lambda}$ obtained from $\underline{P}$ is not equivalent to the one obtained by considering $\Theta = \{\mathscr{X}\}$ with λ, contrary to the rule of Moral and Sagrado [2]. However, if one thinks that reliability scores have to be distinguished for some different parts of the domain $\mathscr{X}$, there is no reason that the rule should act like if there was only one weight when the different weights are equal.

4 Conclusion

In this paper, we have proposed a contextual discounting rule for lower probabilities that can be defined on general partitions of the domain $\mathscr{X}$ on which a variable X assumes its values. Compared to previously defined rules for lower probabilities, the present rule have the advantage that its result is still a lower probability (one does not need to use general lower expectation bounds). It also preserves interesting mathematical properties, such as 2– and ∞-monotonicity, which are useful to compute the so-called natural extension.

Next moves include the use of this discounting rule and of others in practical applications (e.g. merging of classifier results, of expert opinions, ...), in order to empirically compare their practical results. From a theoretical point of view, the rule presented here should be extended to the more general case of lower previsions, so as to ensure that extensions of n-monotonicity [17] are preserved. Although preserving n-monotonicity for values others than $n = 2$ and $n = \infty$ has less practical interest, it would also be interesting to check whether it is preserved by the proposed rule (we can expect that it is, given results for 2-monotonicity and ∞-monotonicity). Another important issue is to provide a stronger and proper interpretation (e.g. in terms of betting behaviour) to this rule, as the interpretation given in the framework of the TBM [3] cannot be applied to generic lower probabilities.

A Proof of Proposition 1

Proof. Let $\underline{P}$ be the lower probability given by the source

- Let us start with property 3, as we will use it to prove the other properties. This property has been proved by Mercier et al. [3] in the case of the transferable belief models, in which are included normalized belief functions (i.e. ∞-monotone lower probabilities).
- Let us now show that property 1 of coherence is preserved. First, note that if $\underline{P}$ is coherent, it means that $\mathscr{P}_{\underline{P}} \neq \emptyset$, and since $\underline{P}^{\lambda} \leq \underline{P}$, $\mathscr{P}_{\underline{P}^{\lambda}} \neq \emptyset$ too. Now, consider a particular event A. If $\underline{P}$ is coherent, it means that there exists a probability measure $P \in \mathscr{P}_{\underline{P}}$ such that it dominates $\underline{P}$ (i.e., $\underline{P} \leq P$) and moreover $\underline{P}(A) = P(A)$. P being a special kind of ∞-monotone lower probability, we can also apply the discounting rule to P and obtain a lower probability P^{λ} which remains ∞-monotone (property (3)) and is such that $P^{\lambda}(A) = \underline{P}^{\lambda}(A)$. The fact that $\prod_{\theta_i \subseteq \emptyset} \lambda_i = 1$ ensures us that $P^{\lambda}(\emptyset) = 0$ and $P^{\lambda}(\mathscr{X}) = 1$, hence P^{λ} is coherent. Also note that P^{λ} still dominates $\underline{P}^{\lambda}$, since both P and $\underline{P}$ are multiplied by the same numbers on every event to obtain P^{λ} and $\underline{P}^{\lambda}$. Therefore, $\exists P'$ such that $\underline{P}^{\lambda} \leq P^{\lambda} \leq P'$ and $P'(A) = P^{\lambda}(A) = \underline{P}^{\lambda}(A)$. As this is true for every event A, this means that $\underline{P}^{\lambda}$ is coherent.
- We can now show property 2. If $\underline{P}$ is 2-monotone, it means that $\forall A, B \subseteq \mathscr{X}$, the inequality

$$\underline{P}(A \cup B) \geq \underline{P}(A) + \underline{P}(B) - \underline{P}(A \cap B)$$

holds. Now, considering $\underline{P}^{\lambda}$, we have to show that $\forall A, B \subseteq \mathscr{X}$, the following inequality holds

$$\underline{P}(A \cup B) \prod_{\theta_i \subseteq ((A \cup B)^c)^*} \lambda_i \geq \underline{P}(A) \prod_{\theta_i \subseteq (A^c)^*} \lambda_i + \underline{P}(B) \prod_{\theta_i \subseteq (B^c)^*} \lambda_i - \underline{P}(A \cap B) \prod_{\theta_i \subseteq ((A \cap B)^c)^*} \lambda_i. \quad (2)$$

Let us consider the three following partitions:

$$\begin{aligned}
(A^c)^* &= ((A^c)^* \setminus ((A \cup B)^c)^*) \cup ((A \cup B)^c)^*, \\
(B^c)^* &= ((B^c)^* \setminus ((A \cup B)^c)^*) \cup ((A \cup B)^c)^*, \\
((A \cap B)^c)^* &= ((A^c)^* \setminus ((A \cup B)^c)^*) \cup ((B^c)^* \setminus ((A \cup B)^c)^*) \cup ((A \cup B)^c)^*.
\end{aligned}$$

To simplify notation, we denote by $\mathscr{S} = (A \cup B)^c$. We can reformulate Eq (2) as

$$\begin{aligned}
\underline{P}(A \cup B) \prod_{\theta_i \subseteq \mathscr{S}} \lambda_i \geq \; & \underline{P}(A) \prod_{\theta_i \subseteq ((A^c)^* \setminus \mathscr{S})} \lambda_i \prod_{\theta_i \subseteq \mathscr{S}} \lambda_i + \underline{P}(B) \prod_{\theta_i \subseteq ((B^c)^* \setminus \mathscr{S})} \lambda_i \prod_{\theta_i \subseteq \mathscr{S}} \lambda_i \\
& - \underline{P}(A \cap B) \prod_{\theta_i \subseteq ((A^c)^* \setminus \mathscr{S})} \lambda_i \prod_{\theta_i \subseteq ((B^c)^* \setminus \mathscr{S})} \lambda_i \prod_{\theta_i \subseteq \mathscr{S}} \lambda_i.
\end{aligned}$$

Dividing by $\prod_{\theta_i \subseteq \mathscr{S}}$, we obtain

$$\begin{aligned}
\underline{P}(A \cup B) \geq \; & \underline{P}(A) \prod_{\theta_i \subseteq ((A^c)^* \setminus \mathscr{S})} \lambda_i + \underline{P}(B) \prod_{\theta_i \subseteq ((B^c)^* \setminus \mathscr{S})} \lambda_i \\
& - \underline{P}(A \cap B) \prod_{\theta_i \subseteq ((A^c)^* \setminus \mathscr{S})} \lambda_i \prod_{\theta_i \subseteq ((B^c)^* \setminus \mathscr{S})} \lambda_i.
\end{aligned}$$

Now, using the fact that $\underline{P}$ is 2-monotone and replacing $\underline{P}(A \cup B)$ by the lower bound $\underline{P}(A) + \underline{P}(B) - \underline{P}(A \cap B)$ in the above equation, we must show

$$\begin{aligned}
& \underline{P}(A)(1 - \prod_{\theta_i \subseteq ((A^c)^* \setminus \mathscr{S})} \lambda_i) + \underline{P}(B)(1 - \prod_{\theta_i \subseteq ((B^c)^* \setminus \mathscr{S})} \lambda_i) \\
& - \underline{P}(A \cap B)(1 - \prod_{\theta_i \subseteq ((A^c)^* \setminus \mathscr{S})} \lambda_i \prod_{\theta_i \subseteq ((B^c)^* \setminus \mathscr{S})} \lambda_i) \geq 0.
\end{aligned}$$

Now, we can replace $\underline{P}(A \cap B)$ by $\min(\underline{P}(A), \underline{P}(B))$, considering that $\min(\underline{P}(A), \underline{P}(B)) \geq \underline{P}(A \cap B)$. Without loss of generality, assume that $\underline{P}(A) \leq \underline{P}(B)$, then we have

$$\begin{aligned}
& \underline{P}(A)(\prod_{\theta_i \subseteq ((A^c)^* \setminus \mathscr{S})} \lambda_i \prod_{\theta_i \subseteq ((B^c)^* \setminus \mathscr{S})} \lambda_i - \prod_{\theta_i \subseteq ((A^c)^* \setminus \mathscr{S})} \lambda_i) + \underline{P}(B)(1 - \prod_{\theta_i \subseteq ((B^c)^* \setminus \mathscr{S})} \lambda_i) \geq 0 \\
& - \underline{P}(A)(1 - \prod_{\theta_i \subseteq ((B^c)^* \setminus \mathscr{S})} \lambda_i)(\prod_{\theta_i \subseteq ((A^c)^* \setminus \mathscr{S})} \lambda_i) + \underline{P}(B)(1 - \prod_{\theta_i \subseteq ((B^c)^* \setminus \mathscr{S})} \lambda_i) \geq 0 \\
& - \underline{P}(A)(\prod_{\theta_i \subseteq ((A^c)^* \setminus \mathscr{S})} \lambda_i) + \underline{P}(B) \geq 0
\end{aligned}$$

and, since $\underline{P}(A)(\prod_{\theta_i \subseteq ((A^c)^* \setminus \mathscr{S})} \lambda_i) \leq \underline{P}(A) \leq \underline{P}(B)$, this finishes the proof.

References

1. Dubois, D., Prade, H.: Possibility theory and data fusion in poorly informed environments. Control Engineering Practice 2, 811–823 (1994)
2. Moral, S., Sagrado, J.: Aggregation of imprecise probabilities. In: BouchonMeunier, B. (ed.) Aggregation and Fusion of Imperfect Information, pp. 162–188. Physica-Verlag, Heidelberg (1997)
3. Mercier, D., Quost, B., Denoeux, T.: Refined modeling of sensor reliability in the bellief function framework using contextual discounting. Information Fusion 9, 246–258 (2008)
4. Karlsson, A., Johansson, R., Andler, S.F.: On the behavior of the robust bayesian combination operator and the significance of discounting. In: ISIPTA 2009: Proc. of the Sixth Int. Symp. on Imprecise Probability: Theories and Applications, pp. 259–268 (2009)
5. Benavoli, A., Antonucci, A.: Aggregating imprecise probabilistic knowledge. In: ISIPTA 2009: Proc. of the Sixth Int. Symp. on Imprecise Probability: Theories and Applications, pp. 31–40 (2009)
6. Miranda, E.: A survey of the theory of coherent lower previsions. Int. J. of Approximate Reasoning 48, 628–658 (2008)
7. Smets, P., Kennes, R.: The transferable belief model. Artificial Intelligence 66, 191–234 (1994)
8. Walley, P.: Statistical reasoning with imprecise Probabilities. Chapman & Hall, New York (1991)
9. Dubois, D., Prade, H.: Possibility Theory: An Approach to Computerized Processing of Uncertainty. Plenum Press, New York (1988)
10. Shafer, G.: A mathematical Theory of Evidence. Princeton University Press, New Jersey (1976)
11. Ferson, S., Ginzburg, L., Kreinovich, V., Myers, D., Sentz, K.: Constructing probability boxes and dempster-shafer structures. Technical report, Sandia National Laboratories (2003)
12. Chateauneuf, A., Jaffray, J.Y.: Some characterizations of lower probabilities and other monotone capacities through the use of Mobius inversion. Mathematical Social Sciences 17(3), 263–283 (1989)
13. Miranda, E., Couso, I., Gil, P.: Extreme points of credal sets generated by 2-alternating capacities. I. J. of Approximate Reasoning 33, 95–115 (2003)
14. Bronevich, A., Augustin, T.: Approximation of coherent lower probabilities by 2-monotone measures. In: ISIPTA 2009: Proc. of the Sixth Int. Symp. on Imprecise Probability: Theories and Applications, SIPTA , pp. 61–70 (2009)
15. Denoeux, T., Smets, P.: Classification using belief functions: the relationship between the case-based and model-based approaches. IEEE Trans. on Syst., Man and Cybern. B 36(6), 1395–1406 (2006)
16. de Campos, L., Huete, J., Moral, S.: Probability intervals: a tool for uncertain reasoning. I. J. of Uncertainty, Fuzziness and Knowledge-Based Systems 2, 167–196 (1994)
17. de Cooman, G., Troffaes, M., Miranda, E.: n-monotone lower previsions and lower integrals. In: Cozman, F., Nau, R., Seidenfeld, T. (eds.) Proc. 4th International Symposium on Imprecise Probabilities and Their Applications (2005)

The Power Average Operator for Information Fusion

Ronald R. Yager

Machine Intelligence Institute
Iona College
New Rochelle, NY 10801

Abstract. The power average provides an aggregation operator that allows similar argument values to support each other in the aggregation process. The properties of this operator are described. We see this mixes some of the properties of the mode with mean. Some formulations for the support function used in the power average are described. We extend this facility of empowerment to a wider class of mean operators such as the OWA and generalized mean.

Keywords: information fusion, aggregation operator, averaging, data mining.

1 Introduction

Aggregating information using techniques such as the average is a task common in many information fusion processes. Here we provide a tool to aid and provide more versatility in this process. In this work we introduce the concept of the power average [1]. With the aid of the power average we are able to allow values being aggregated to support each other. The power average provides a kind of empowerment as it allows groups of values close to each other to reinforce each other. This operator is particularly useful in group decision-making [2]. It also helps to moderate the effects of outlier values a problem that can arise with the simple average.

2 Power Average

In the following we describe an aggregation type operator called the **P**ower **A**verage (P–A), this operator takes a collection of values and provides a single value [1]. We define this operator as follows:

$$\text{P-A}(a_1, \ldots, a_n) = \frac{\sum_{i=1}^{n} (1 + T(a_i))a_i}{\sum_{i=1}^{n} (1 + T(a_i))}$$

where $T(a_i) = \sum_{\substack{j=1 \\ j \neq i}}^{n} \text{Sup}(a_i, a_j)$ and is denoted the support for a_i.

E. Hüllermeier, R. Kruse, and F. Hoffmann (Eds.): IPMU 2010, Part II, CCIS 81, pp. 208–220, 2010.

Typically we assume that Sup(a, b) satisfies the following three properties:

1. Sup(a, b) $\in$ [0, 1], 2. Sup(a, b) = Sup(b, a), 3. Sup(a, b) $\geq$ Sup(x, y) if $|a - b| \leq |x-y|$

We see the more similar, closer, two values the more they support each other.

We shall find it convenient to denote $V_i = 1 + T(a_i)$ and $w_i = \frac{V_i}{\sum_{j=1}^{n} V_j}$. Here the w_i are a proper set of weights, $w_i \geq 0$ and $\Sigma_i w_i = 1$. Using this notation we have

$$\text{P-A}(a_1, ..., a_n) = \Sigma_i w_i a_i,$$

it is a weighted average of the a_i. However, this is a non-linear weighted average as the w_i depends upon the arguments.

Let us look at some properties of the power average aggregation operator. First we see that this operator provides a generalization of the simple average, if $Sup(a_i, a_j) = k$ for all a_i and a_j then $T(a_i) = k (n - 1)$ for all i and hence $\text{P-A}(a_1, ..., a_n) = \frac{1}{n} \Sigma_i a_i$. Thus when all the supports are the same the power average reduces to the simple average.

We see that the power average is commutative, it doesn't depend on the indexing of the arguments. Any permutation of the arguments has the same power average.

The fact that $\text{P-A}(a_1, ..., a_n) = \Sigma_i w_i a_i$ where $w_i \geq 0$ and $\Sigma_i w_i = 1$ implies that the operator is bounded, $Min[a_i] \leq \text{P-A}(a_1, a_2, ..., a_n) \leq Max_i[a_i]$. This in turn implies that it is idempotent, if $a_i = a$ for all i then $\text{P-A}(a_1, ..., a_n) = a$

As a result of the fact that the w_i depends upon the arguments, one property that is not generally satisfied by the power average is monotonicity. We recall that monotonicity requires that if $a_i \geq b_i$ for all i then $\text{P-A}(a_1, ..., a_n) \geq \text{P–A}(b_1, ..., b_n)$. As the following example illustrates, the increase in one of the arguments can result in a decrease in the power average.

Example: Assume the support function Sup is such thatSup(2, 4) = 0.5, Sup(2, 10) = 0.3, Sup(2, 11) = 0, Sup(4, 10) = 0.4Sup(4, 11) = 0 the required symmetry means S(a, b) = S(b, a) for these values. Consider first P–A(2 4, 10), T(2) = Sup(2, 4) + Sup(2, 10) = 0.8, T(4) = Sup(4, 2) + Sup(4, 10) = 0.9, T(10) = Sup(10, 2) + Sup(10, 4) = 0.7and therefore P-A(2, 4, 10) = 5.22.

Consider now P-A(2, 4, 11): T(2) = Sup(2, 4) + Sup(2, 11) = 0.5, T(4) = Sup(4, 2) + Sup(4, 11) = 0.5, T(11) = Sup(11, 2) + Sup(11, 2) = 0, and hence-A(2, 4, 11) = 5

Thus we see that P-A(2, 4, 10) > P(2, 4, 11).

As we shall subsequently see, this ability to display non-monotonic behavior provides one of the useful features of this operator that distinguishes it from the usual average. For example the behavior displayed in the example is a manifestation of the ability of this operator to discount outliers. For as we shall see in the subsequent discussion, as an argument moves away from the main body of arguments it will be accommodated, by having the average move in its direction, this will happen up to point then when it gets too far away it is discounted by having its effective weighting factor diminished.

To some degree this power average can be seen to have some of the characteristics of the mode operator. We recall that the mode of a collection of arguments is equal to the value that appears most in the argument. We note that the mode is bounded by the arguments and commutative, however as the following example illustrates it is not monotonic.

Example: Mode(1, 1, 3, 3, 3) = 3. Consider now Mode(1, 1, 4, 7, 8) = 1, here we increased all the threes and obtain a value less than the original.

As we shall subsequently see, while both the power average and mode in some sense are trying to find the most supported value, a fundamental difference exists between these operators. We note that in the case of the mode we are not aggregating, blending, the values we are counting how many of each, the mode must be one of the arguments. In the case of power average we are allowing blending of values.

It is interesting, however, to note a formal relationship between the mode and the power average. To understand this we introduce an operator we call a **P**ower **M**ode. In the case of the power mode we define a support function $Sup_m(a, b)$, indicating the support for a from b, such that: 1) $Sup_m(a, b) \in [0, 1]$, 2) $Sup_m(a, b) = Sup_m(b, a)$, 3) $Sup_m(a, b) \geq Sup_m(x, y)$ if $|a - b| \leq |x - y|$, 4). $Sup_m(a, a) = 1$.

We then calculate $\text{Vote}(i) = \sum_{j=1}^{n} Sup_m(a_i, a_j)$ and define the Power Mode(a_1, ..., a_n) = a_{i^*} where i* is such that Vote(i*) = Max_i[Vote(i)], it is the argument with the largest vote.

If $Sup_m(a, b) = 0$ for $b \neq a$ then we get the usual mode. Here we are allowing some support for a value by neighboring values. It is also interesting to note the close relationship to the mountain clustering method introduced by Yager and Filev [3-4]. and the idea of fuzzy typical value introduced in [5].

3 Power Average with Binary Support Functions

In order to obtain some intuition for the power average aggregation operator we shall consider first a binary support function. Here we assume Sup(a, b) = K if $|a - b| \leq d$ and Sup(a, b) = 0 if $|a - b| > d$. Here two values support each other if they are less than or equal d away, otherwise they supply no support. Here K is the value of support. In the following discussion we say a and b are neighbors if $|a - b| \leq d$. The set of points that are neighbors of x will be denoted N_x. We shall call a set of points such that all points are neighbors and no other points are neighbors to those points a cluster. We note if x and y are in the same cluster then the subset $\{x\} \cup N_x = \{y\} \cup N_y$ defines the cluster.

Let us first assume that we have two disjointed clusters of values A = $\{a_1, \ldots, a_{n_1}\}$ and B = $\{b_1, \ldots, b_{n_2}\}$. Here all points in A support each other but support none in B while the opposite holds for B. In this case for all i and j, $|a_i - a_j| \le d$, $|b_i - b_j| \le d$ and $|a_i - b_j| > d$. Here for each a_i in A, $T(a_i) = K(n_1 - 1)$ and for each b_j in B, $T(b_j) = K(n_2 - 1)$. From this we get $1 + T(a_i) = (1 - K) + n_1 K$ and $1 + T(b_j) = (1 - K) + n_2K$. Using this we have

$$\text{P-A}(a_1, \ldots, a_{n_1}, b_1, \ldots, b_{n_2}) = \frac{\sum_{i=1}^{n_1}((1-K)+n_1K)a_i + \sum_{j=1}^{n2}((1-K)+n_2K)b_j}{n_1(1-K+n_1K)+n_2(1-K+n_2K)}$$

Letting $\bar{a} = \frac{1}{n_1}\sum_{i=1}^{n_1} a_i$ and $\bar{b} = \frac{1}{n_2}\sum_{j=1}^{n_2} b_j$ we have

$$\text{PA}(a_1, \ldots, a_{n_1}, b_1, \ldots, b_{n_2}) = \frac{((1-K)+n_1K)n_1\bar{a} + ((1-K)+n_2K)n_2\bar{b}}{n_1(1-K+n_1K)+n_2(1-K+n_2K)}$$

We get a weighted average of the cluster averages. If we let

$$w_a = \frac{((1-K)+n_1K)n_1}{n_1(1-K+n_1K)+n_2(1-K+n_2K)}$$

$$w_b = \frac{((1-K)+n_2K)n_2}{n_1(1-K+n_1K)+n_2(1-K+n_2K)}$$

then $\text{PA}(a_1, \ldots, a_{n_1}, b_1, \ldots, b_{n_2}) = w_a\,\bar{a} + w_b\,\bar{b}$. We note $w_a + w_b = 1$ and $\frac{w_a}{w_b} = \frac{(1-K+n_1K)n_1}{(1-K+n_2K)n_2}$, We see that if k = 1, then $\frac{w_a}{w_b} = (\frac{n_1}{n_2})^2$, the weights proportional to the square of the number of elements in the clusters. Thus in this case $w_a = \frac{n_1^2}{n_1^2+n_2^2}$ and $w_b = \frac{n_2^2}{n_1^2+n_2^2}$. On the other hand if we allow no support, K = 0, then $\frac{w_a}{w_b} = \frac{n_1}{n_2}$, the weights are just proportional to the number of elements in each cluster. In this case $w_a = \frac{n_1}{n_1+n_2}$ and $w_b = \frac{n_2}{n_1+n_2}$. Thus we see as we move from K = 0 to K = 1 we move from being proportional to number of elements in each cluster to being proportional to the square of the number of elements in each cluster. We now begin to see the effect of this power average. If we allow support then elements that are close gain power. This becomes a reflection of the adage that there is power in sticking together. We also observe that if n_1K and $n_2K >> (1 - K)$, there are a large number of arguments, then again $\frac{w_a}{w_b} = (\frac{n_1}{n_2})^2$. Furthermore we note if $n_1 = n_2$ then we always have $\frac{w_a}{w_b} = 1$, here we take the simple average.

Consider now the case when we have q disjoint clusters, each only supporting elements in its neighborhood. Let a_{ji} for i = 1 to n_j be the elements in the jth cluster.

In this case

$$\text{P-A} = \frac{\sum_{j=1}^{q}(\sum_{i=1}^{n_j}(1-K+n_jK)a_{ji})}{\sum_{j=1}^{q} n_j(1-K+n_jK)}$$

Letting $\frac{1}{n_j}\sum_{i=1}^{n_j} a_{ji} = \bar{a}_j$, the individual cluster averages, we can express this power average as $\text{P-A} = = \frac{\sum_{j=1}^{q}((1-K+n_jK)n_j\bar{a}_j}{\sum_{j=1}^{q} n_j(1-K+n_jK)}$ Again we get a weighted average of the individual cluster averages, $\text{P-A} = \sum_{j=1}^{q} w_j\bar{a}_j$. In this case $\frac{w_i}{w_j} = \frac{(1-K+n_iK)n_i}{(1-K+n_jK)n_j}$.

Again we see if K= 1, then $\frac{w_i}{w_j} = (\frac{n_i}{n_j})^2$, the proportionality factor is the square of the number of elements. Here then $w_i = \frac{n_i^2}{\sum_{j=1}^{q} n_j^2}$. If we allow no support, K = 0, then $\frac{w_i}{w_j} = \frac{n_i}{n_j}$, here we get the usual average. We note that K is the value of support.

Consider a case with small value of support, 1 - K ≈1. Furthermore assume n_i is a considerable number of elements while n_j is a very small number. Here (1 - K) + n_j K ≈ 1 while (1 - K) + n_i K ≈ n_1K then $\frac{w_i}{w_j} = \frac{n_i^2 K}{n_j}$

On the other hand if n_i and n_j are large, n_iK and $n_jK >>> 1$ then $\frac{w_i}{w_j} = (\frac{n_i}{n_j})^2$. We that if $(1 - K) << n_jK$ for all j then $\text{P-A} = \frac{\sum_{j=1}^{q} n_j^2 \bar{a}_j}{\sum_{j=1}^{q} n_j^2}$, the weights in proportion to the square of the number of elements.

Let us observe another interesting property of this P-A. To most clearly illustrate the property we shall assign K = 1. Assume we have two clusters then with K = 1 we have $\text{P-A} = \frac{n_1^2 \bar{a}_1 + n_2^2 \bar{a}_2}{n_1^2 + n_2^2}$. If $n_1 \approx n_2 = \frac{1}{2}$ n, they have the same number of elements then $\text{P-A} = \frac{1}{2}\bar{a}_1 + \frac{1}{2}\bar{a}_2$. Assume now that the second cluster is broken into two equal disjoint clusters. Then $P(A) = \frac{n_1^2 \bar{a}_1 + n_2^2 \bar{a}_2 + n_3^2 \bar{a}_3}{n_1^2 + n_2^2 + n_3^2}$ with $n_1 = \frac{1}{2}$ n, $n_2 = \frac{1}{4}$ n and $n_3 = \frac{1}{4}$ n. From this we see that $P(A) = \frac{4\bar{a}_1 + \bar{a}_2 + \bar{a}_3}{6}$. We see cluster one's influence (power) has greatly increased because of the fragmentation of cluster two..

We now consider a situation in which we have three sets of elements, $A = \{a_1, ..., a_{n_1}\}$, $B = \{b_1, ..., b_{n_2}\}$ and $C = \{c_1, ..., c_{n_3}\}$. We assume all the elements in A are a neighbors with each other as well as with those in B. Those in B are neighbors with each other and also with those in both A and C. The elements in C are neighbors with themselves and B. Thus B is seen to be between A and C. Here we see that for all a_i we have $T(a_i) = K(n_1 + n_2 - 1)$, for all b_i $T(b_i) = K(n_1 + n_2 + n_3 - 1)$ and for all $c_i T(c_i) = K(n_2 + n_3 - 1)$. Let $\bar{a} = \frac{1}{n_1}\sum a_j$, $\bar{b} = \frac{1}{n_2}\sum b_j$ and $\bar{c} = \frac{1}{n_3}\sum b_j$. Using this we have P-A equals

$$\frac{(1 - K + K(n_1 + n_2))n_1\bar{a} + (1 - K + K(n_1 + n_2 + n_3))n_2\bar{b} + (1 - K + K(n_2 + n_3))n_3\bar{c}}{(1 - K + K(n_1 + n_2))n_1 + (1 - K + K(n_1 + n_2 + n_3))n_2 + (1 - K + K(n_2 + n_3))n_3}$$

Again for illustrative purposes we assume K =1 hence

$$\text{P-A} = \frac{(n_1 + n_2)n_1\bar{a} + nn_2\bar{b} + (n - n_1)n_3\bar{c}}{n^2 - 2n_1n_3}$$

We see that relationship between the weights associated A and C is

$$\frac{w_a}{w_c} = \frac{(n_2 + n_1)n_1}{(n_2 + n_3)n_3}$$

If n_2 is large compared with both n_1 and n_3 then $\frac{w_a}{w_c} = \frac{n_1}{n_3}$, their relationship is proportion to the number of elements in A and C. If n_2 is small compared with both n_1 and n_3 then $\frac{w_a}{w_c} = (\frac{n_1}{n_3})^2$. Consider the relationship between A and B, which is analogous to B and C, $\frac{w_a}{w_b} = \frac{n_1(n_1 + n_2)}{nn_2}$. If n_2 is large compared with n_1 and n_3 then $\frac{w_a}{w_b} = \frac{n_1}{n}$

We consider now another situation that exemplifies the possibility for non-monotonicity. Let $\{a_1, ..., a_n, a_{n+1}\}$ be a collection of points in the same cluster, for all a_i and a_j, $|a_i - a_j| \leq d$. In this case P-A$\{a, ..., a_{n+1}\} = \frac{1}{n+1}\sum_{j=1}^{n+1} a_j = \bar{a}$. Assume now that we replace a_{n+1} by $\hat{a}_{n+1}$ where $\hat{a}_{n+1} \geq a_{n+1}$ and $|a_{n+1} - a_j| > d$ for all other a_j. That is we have moved the n+1th observation all the way to the right. In this case we can view the situation having two disjoint clusters one being $\{a_1, ..., a_n\}$ and the other $\{\hat{a}_{n+1}\}$. As we already established the power average of this situation is

$$\text{P-A}(a_1, a_2, ..., a_n, \hat{a}_{n+1}) = w_1 \tilde{a} + w_2 \hat{a}_{n+1}$$

here $\tilde{a} = \frac{1}{n}\sum_{i=1}^{n} a_i$ and $\hat{a}_{n+1} = a_{n+1} + \Delta$. In the situation where K = 1 we have $\frac{w_1}{w_2} = \frac{n^2}{1}$. This gives us $w_1 = \frac{n^2}{n^2+1}$ and $w_2 = \frac{1}{n^2+1}$ .and hence

$$\text{P-A}(a_1, ..., \hat{a}_{n+1}) = a + \frac{\Delta - (n-1)(a_{n+1} - \bar{a})}{n^2+1}$$

Thus we see that if a_{n+1} was the right most element then we get a non-monotonicity as long as Δ is not too big.

4 Forms for the Support Function

The support function is a crucial part of the power average method. The form of the support function is context dependent. Here we describe some useful parameterized formulations for expressing the Sup function. The determination of the values of the parameters may require the use of some learning techniques. We recall if **R** is the range of the values to be aggregated then Sup:$\mathbf{R} \times \mathbf{R} \to [0, 1]$ such that Sup(a, b) = Sup(b, a), and Sup(a, b) $\geq$ Sup(x, c) if $|a - b| \leq |x - c|$

In the preceding we assumed a binary Sup function, Sup(a, b) = K if $|a - b| \leq d$ and Sup(a, b) = 0 if $|a - b| > d$. A natural extension of this is to consider a partitioned type support function. Let K_i for i = 1 to p be a collection of values such that $K_i \in [0, 1]$ and where $K_i > K_j$ if i < j. Let d_i be a collection of values such that $d_i \geq 0$ and where $d_i < d_j$ if i < j. We now can define a support function as

If $|a - b| \leq d_1$ then $Sup(a, b) = K_1$
If $d_{j-1} < |a - b| \leq d_j$ then $Sup(a, b) = K_j$ for j = 2 to p - 1
If $d_{p-1} < |a - b|$ then $Sup(a, b) = K_p$

Inherent in the above type of support function is a discontinuity as we move between the different ranges.

One form of the Sup function with a continuous transition is $Sup(a, b) = K\, e^{-\alpha(a - b)^2}$ where $K \in [0, 1]$ and $\alpha \geq 0$. We easily see that this function is symmetric and lies in the unit interval. We see K is the maximal allowable support and α is acting as a attenuator of the distance. The larger the α the more meaningful differences in distance. We note here that a = b gives us Sup(a, b) = K and as the distance between a and b gets larger, $Sup(a, b) \to 0$.

Using this form for support function we have $P\text{-}A(a_1, ..., a_n) = \dfrac{\sum_{i=1}^{n}(1+T(a_i))a_i}{\sum_{i=1}^{n}(1+T(a_i))}$ where $T(a_i) = \sum_{\substack{j=1\\ j\neq i}}^{n} K e^{-\alpha(a_i - a_j)^2}$. Denoting $V_i = 1 + T(a_i)$ we express $P\text{-}A(a_1, ..., a_n) = \Sigma_i\, w_i\, a_i$ where $w_i = \dfrac{V_i}{\sum_{j=1}^{n} V_j}$. Since $e^{-\alpha(a_i - a_i)^2} = 1$ we can express $V_i = 1 - K + K\, M_i$ where $M_i = \sum_{j=1}^{n} e^{-\alpha(a_i - a_j)^2}$. Noting the similarly of M_i to the mountain function used in mountain clustering [3] we call M_i the support mountain at i. It's clear that if $a_p = a_q$ then $M_q = M_p$ and hence $V_q = V_p$. It is also noted that $M_i \geq 1$ for all i.

We see here that

$$P\text{-}A(a, ..., a_n) = \frac{\sum_{i=1}^{n} n(1-K)a_i + K\sum_{i=1}^{n} M_i a_i}{n(1-K) + K\sum_{i=1}^{n} M_i}$$

In the special case where K = 1 then $V_i = M_i$ and hence

$$\text{P-A}(a_1, \ldots, a_n) = \frac{\sum_{i=1}^{n} M_i a_i}{\sum_{i=1}^{n} M_i}$$

A simple algorithm approach somewhat is in spirit of the mountain method is as follows:

1. For each argument value a_i, i = 1 to n, initialize $M_i = 0$

2. For each data point a_j j = 1 to n augment M_i, $M_i = M_i + e^{-\alpha(a_i - a_j)^2}$. This builds the support mountain.

3. Calculate $V_i = (1 - K) + K M_i$ - linear transformation of mountain values

4. Calculate $w_i = \frac{V_i}{\sum_{j=1}^{n} V_j}$

5. $\text{P-A} = \Sigma_i w_i a_i$

As we have noted an important characteristic of this power average is its possibility for displaying non-monotonicity, a feature that can provide one of the benefits of this method. The following example illustrates the occurrence of non-monotonicity.

Example: Consider the Power average of twenty elements, 10 of which are ten's and 10 of which are five's. In this case the ordinary average evaluates to 7.5 and for any choice of K and α the power average also evaluates to 7.5. The following table shows what happens as we change one of the values originally equal to 10. For illustrative purposes we used K = 1 and $\alpha = 0.3$

Value	AVE	P-A
10	7.5	7.5
9	7.45	7.398
8	7.4	7.278
7	7.35	7.193
6	7.3	7.083
5	7.25	6.982
11	7.55	7.4727
12	7.6	7.346
13	7.65	7.259
14	7.7	7.232
15	7.75	7.22856
16	7.8	7.22828
17	7.85	7.22829
18	7.9	7.22831
19	7.9	7.22832
20	8	7.22834

We see that as we decrease the value and move it towards the cluster of fives our P-A decrease, although more dramatically than the average. Essentially the variable value is beginning to join the cluster of fives and increase its power. In the case of increasing the value, initially the power average instead of increasing as does the average begins to decrease, exhibiting non–monotonicity. This decrease is a reflection of the fragmentation of the cluster at 10, it is losing its power because it lost a member and the cluster at five has gained in power more than compensating for the increase in value. This decreasing in the P-A continues as we increase the element until it reaches eighteen at which time we see a reversal and now the P-A starts increasing At this point the increase in value begins overcoming the loss of power. But still we are favoring the cluster of fives.

We describe another approach to obtaining the support function that combines the partitioning of the first method with the continuity displayed by the exponential function. This approach motivated by Zadeh's idea of computing with words [6] makes use of fuzzy systems modeling technology [7]. We shall briefly describe the possibilities for this approach. Using this approach we can express our support function by a description of its performance in terms of a set of rules using linguistic values. For example.

If difference is *very small* then support is K_1
If difference is *small* then support is K_2
If difference is *moderate* the support is K_3
If difference is *large* the support is K_4
If difference is *very large* the support is K_5

Representing the italic terms as fuzzy sets, VS, S, M, L, and VL respectively and denoting the difference between a and b as Δ than we have a collection of fuzzy if-then rules, a fuzzy systems model:

If Δ is VS then $S(a, b) = K_1$
If Δ is S then $S(a, b) = K_2$
If Δ is M then $S(a, b) = K_3$
If Δ is L then $S(a, b) = K_4$
If Δ is VL then $S(a, b) = K_4$

here $K_i < K_j$ if $i > j$.

To obtain the Sup(a, b) we use the inference mechanism of fuzzy systems modeling . Letting $\Delta = |a - b|$ then the analytic formulation of our support function is

$$\text{Sup}(a, b) = \frac{K_1 VS(\Delta) + K_2 S(\Delta) + K_3 M(\Delta) + K_4 L(\Delta) + VL(\Delta)}{VS(\Delta) + S(\Delta) + M(\Delta) + L(\Delta) + VL(\Delta)}$$

here $VS(\Delta)$ indicates the membership of Δ in the fuzzy subset VS.

We now look at the power average in the special situation in which the arguments that are being aggregated, the a_i, always be in the unit interval [0, 1]. This is a situation that occurs in many environments when the arguments are degrees of belief. We note a particular important situation is in the aggregation of fuzzy subsets.

In the case when the arguments lie in the unit interval a very natural definition for the Sup function is $\text{Sup}(a, b) = K(1 - |a - b|^{\alpha})$ for $\alpha \geq 0$. Here we see that the term $|a - b|$ is a measure of distance between the arguments. We note since a and b are assumed to lie in the unit interval then $|a - b|$ must also lie in the unit interval as well as $|a - b|^{\alpha}$. We see $|a - b| \to 0$ indicates the elements are close and $|a - b| \to 1$ indicates the elements are far. We see that is Sup is related to the negation of the distance.

We notice that because a and b always lie in the unit interval, $|a - b| = 1$ if and only if one of the arguments equal zero and the other equals one. Furthermore we note that α modifies the effects of distance. Since $(a - b) < 1$ then $\alpha > 1$ reduces the effect of distance while $\alpha < 1$ increase the effects of distance. We note $\text{Sup}(a, b) = K$ when $a = b$.

As in the preceding $\text{P-A}(a_1, ..., a_n) = \dfrac{\sum_{i=1}^{n} V_i a_i}{\sum_{i=1}^{n} V_i}$. Let us consider the case when $\alpha =$ 2, $\text{Sup}(a, b) = K(1 - (a - b)^2)$. Here $V_i = 1 + T(a_i)$ with $T(a_i) = K\sum_{\substack{j=1\\ j\neq i}}^{n}(1-(a_i - a_j)^2)$.

Realizing $1 - (a_i - a_i)^2 = 1$ then $V_i = (1 - K) + K\sum_{\substack{j=1\\ j\neq i}}^{n}(1-(a_i - a_j)^2)$. Letting $Q_i =$ $\sum_{j=1}^{n}(a_i - a_j)^2$ we have $V_i = 1 - K + Kn - KQ_i$.

Let us carefully look at the term Q_i. We shall denote $\bar{a} = \sum_{j=1}^{n} a_j$, it is the average, and denote $\text{Var}(a) = \frac{1}{n}\sum_{j=1}^{n}(a_j - \bar{a})^2$. Using these notations we can express

$$Q_i = \sum_{j=1}^{n}(a_i - a_j)^2 = \sum_{i=1}^{n} (a_i - \bar{a})^2 + \sum_{j=1}^{n} (a_j - \bar{a})^2$$

Letting $\Delta_i = |a_i - \bar{a}|$. we have $Q_i = n\ \Delta_i^2 + n\ \text{Var}(a)$.

From this we have $V_i = (1 - K) + Kn - nK(\Delta_i^2 + \text{Var}(a))$. Using this we get that

$$\sum_{i=1}^{n} V_i = n(1 - K) + Kn^2 - n^2 K\ \text{Var}(a) - n\ K\sum_{j=1}^{n}\Delta_i^2$$

Since $\frac{1}{n}\sum_{i=1}^{n}\Delta_i^2 = \text{Var}(a)$ then $\sum_{i=1}^{n} V_i = n(1 - K) + Kn^2 - 2n^2 K \text{Var}(a)$

Let us consider the special case where K = 1, here $V_i = n\,(1 - Var(a) - \Delta_i^2)$ and

$\sum_{i=1}^{n} V_i = n^2\,(1 - 2\,Var(a))$. Using this $P\text{-}A(a_1, ..., a_n) = \bar{a} + \frac{\bar{a}\sum_{i=1}^{n}\Delta_i^2 - \sum_{i=1}^{n}\Delta_i^2 a_i}{n^2 + (1 - 2Var(a))}$

We see that if the arguments are such that there are a few large values far away from the rest of the values mean then the power average tends to pull $\bar{a}$ downwards.

Another interesting case of $Sup(a, b) = K(1 - |a - b|^{\alpha})$ occurs when $\alpha = 1$, here $Sup(a\ b) = K(1 - |a - b|)$. We note that $|a - b| = Max(a, b) - Min(a, b) = (a \vee b) - (a \wedge b)$. Here again $P\text{-}A(a_1, ..., a_n) = \frac{\sum_{i=1}^{n} V_i a_i}{\sum_{i=1}^{n} V_i}$. In this case $V_i = 1 + (\nu - 1)\,K - K\sum_{j=1}^{n} [(a_i \vee a_j) - (a_j \wedge a_i)]$

Without loss of generality let us assume that the a_i have been indexed in descending order, thus a_i is the ith largest of the arguments. In this case

$$a_i = Min[a_i, a_j] \text{ and } a_j = Max[a_i, a_j] \qquad \text{for } j = 1 \text{ to } i - 1$$
$$a_j = Min[a_i, a_j] \text{ and } a_i = Max[a_i, a_j] \qquad \text{for } j = i + 1 \text{ to } n$$
$$a_i = Min[a_i, a_j] = Max[a_i, a_j] \qquad \text{for } j = 1$$

If we denote $Q_i = \sum_{j=1}^{n} |a_i - a_j|$ then after some arithmetic we get

$$Q_i = \sum_{j=1}^{i-1} a_j - \sum_{j=i+1}^{n} a_j + (n - 2i)a_i$$

Denoting $SL(i) = \sum_{j=1}^{i} a_j$ and $SU(i) = \sum_{j=i+1}^{n} a_j$ then $Q_i = SL(i) - SU(i) + (n - 2i)\,a_i$ and

$$V_i = 1 + (n - 1)\,K - K\,(SL(i) - SU(i) + (n - 2i)\,a_i)$$

and $\sum_{i=1}^{n} V_i = n + n\,(n - 1)\,K - K\sum_{i=1}^{n} (SL(i) - SU(i) + (n - 2i)\,a_i)$.

Let us consider the special case where K = 1, we get

$$\sum_{i=1}^{n} V_i = n^2 - \sum_{i=1}^{n} (2n - 4i + 1)\,a_i = n^2 - n(2n - 1)\,\bar{a} + 4\sum_{i=1}^{n} i\,a_i$$

References

[1] Yager, R.R.: The power average operator. IEEE Transaction on Systems, Man and Cybernetics Part A 31, 724–730 (2001)
[2] Xu, Z., Yager, R.R.: Power geometric operators and their use in group decision making. IEEE Transactions on Fuzzy Sets and Systems (to appear)
[3] Yager, R.R., Filev, D.P.: Approximate clustering via the mountain method. IEEE Transactions on Systems, Man and Cybernetics 24, 1279–1284 (1994)
[4] Chiu, S.L.: Fuzzy model identification based on cluster estimation. Journal of Fuzzy and Intelligent Systems 2, 267–278 (1994)
[5] Yager, R.R.: A note on a fuzzy measure of typicality. International Journal of Intelligent Systems 12, 233–249 (1997)
[6] Zadeh, L.A.: Generalized theory of uncertainty (GTU)-principal concepts and ideas. Computational Statistics and Data Analysis 51, 15–46 (2006)
[7] Yager, R.R., Filev, D.P.: Essentials of Fuzzy Modeling and Control. John Wiley, New York (1994)
[8] Beliakov, G., Pradera, A., Calvo, T.: Aggregation Functions: A Guide for Practitioners. Springer, Heidelberg (2007)
[9] Yager, R.R.: On ordered weighted averaging aggregation operators in multi-criteria decision making. IEEE Transactions on Systems, Man and Cybernetics 18, 183–190 (1988)

Performance Comparison of Fusion Operators in Bimodal Remote Sensing Snow Detection

Aureli Soria-Frisch, Antonio Repucci, Laura Moreno, and Marco Caparrini*

Starlab Barcelona S.L.
aureli.soria-frisch@starlab.es
http://www.starlab.es

Abstract. This contribution describes the system developed and implemented for the detection of snow based on the fusion of optical and Synthetic Aperture Radar (SAR) remote sensing modalities. The work is focused on the performance comparison of different fusion operators for the implementation of the fusion stage. In case of the optical signal the so-called Normalized Difference Snow Index (NDSI) is used, whereas in SAR, the binary presence of wet and dry snow are used. We take into account soft data fusion, a framework where several operators are included. The comparison is undertaken on a set of satellite images by computing the standard Receiver Operating Curves (ROC) and the corresponding Area Under the Curves (AUC).

1 Introduction and Background

The detection of snow is a very important intermediate step in a number of earth observation services. One of the most important is the estimation of snow volume in order to feed hydrological models used in several monitoring services. The monitoring of snow melt, snow water equivalent, and soil water content values are key for the management of water resources and for snow melt runoff modeling and forecasting. In the work presented herein we attain the detection of snow by making use of two different remote sensing modalities. The general goal is to improve the robustness in the snow detection by using two complementary information sources.

The system described herein attains the detection of snow based on two different remote sensing modalities, namely SAR and optical. The general problem of the optical modality, which is based on the NDSI [6], is the presence of clouds, which can not be avoided in an easy form. Moreover NDSI delivers a high value in water areas. This is avoidable through the utilization of geographical masks. However this is not easily implementable in an operational service. On the other hand, the detection of snow based on the SAR modality [10] presents the advantage that is capable of distinguishing between dry and wet snow. One of the

* The research works described herein have been realized within the project INTESOR (SAE-20081013), which is partially funded by CDTI (Centro para el Desarrollo Tecnolgico Industrial) of the Spanish Ministry of Science and Innovation.

E. Hüllermeier, R. Kruse, and F. Hoffmann (Eds.): IPMU 2010, Part II, CCIS 81, pp. 221–230, 2010.

disadvantages of this modality is due to the presence of SAR layover shadowing [8]. These phenomena are provoked by the orography, i.e. in the part of a mountain that is in the opposite site as the origin of the SAR backscatter. Both disadvantages are expected to be overcome through the simultaneous usage empowered by their fusion, which is also known as merging or aggregation. The fusion approach has been attempted previously [5], which uses a Bayesian approach. However no numerical results are given in that work. No other works have been found in the literature dealing with the fusion of optical and SAR modalities for snow detection.

The fusion stage is implemented herein through different fusion operators, which can be analyzed within the soft data fusion framework [14]. These operators generalize the most used fusion operators used in real world applications. This is achieved by the introduction of different parameters that improve the flexibility of the operators in the fusion implementation. We attain the optimal parameterization of several operators and the comparison of their performance in the detection of snow. We make use of operators introduced recently. Few works described their employment in real world applications. To the best of our knowledge this particularly applies to the field of remote sensing, where either harder operators , e.g. average, or general purpose pattern recognition techniques, e.g. kernel machines, are used therefore ignoring the advances in the research topic dealing with aggregation operators.

2 Framework for Snow Detection

The processing framework being evaluated herein first takes into account the fuzzification of the input remote sensing modalities. Thence the fusion of the resulting membership functions is attained. The fusion stage can be implemented with two different architectures. We detail the overall framework in the following paragraphs.

The detection of snow on the optical modality is based on the so-called NDSI index, which presents a large value when snow is supposed to be present in the image. For further information on this index the reader is referred to [6][13]. This index presents a real value in the interval $[-1, 1]$. Snow is usually characterized by values over a threshold of 0.4. Since the fusion operators, which will be used in the framework, are defined in the positive interval $[0, 1]$, a previous mapping has to be applied on the raw NDSI. This is achieved in the framework described herein by applying a monotonic increasing fuzzy trapezoidal membership function. The ramp is centered in this case in 0.4. The initial and origin points of the ramp have been respectively placed in 0.1 and 0.7. The result of the formerly mentioned procedure is a fuzzy NDSI index, defined in the real interval [0,1]. The effect of such a fuzzification on the image domain can be observed in Fig. 1.

Usually snow is detected on the SAR modality by applying an expert system, which generates two binary maps with the presence of wet and dry snow[1]. This

[1] One of the possible categorizations of snow is this taking into account wet and dry snow. Summarizing it can be stated that wet snow is about to melt, whereas dry snow not.

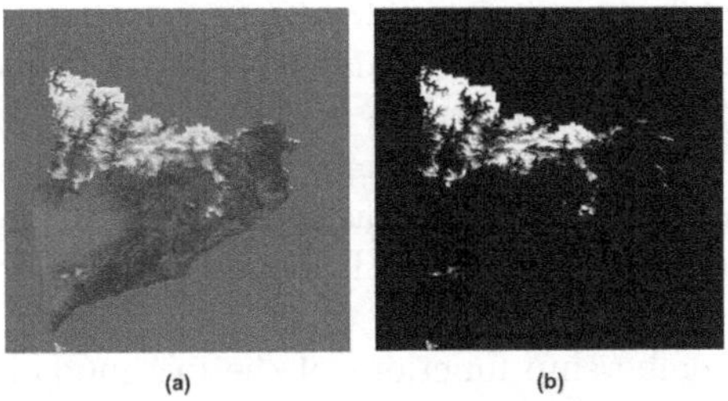

Fig. 1. NDSI (a) and fuzzy NDSI (b). This last one results from fuzzifying the former one with a monotonic increasing trapezoidal membership function centered in 0.4. Both images have been normalized for display in the interval [0, 255].

system takes into account the result of a template matching procedure with a SAR reference image with no snow, which is given in form of a matching ratio, the temperature, and the height at each pixel. The overall system is referred in [10][16]. The results of the expert system are depicted in Figs. 2a-b.

However the binary maps generated by the formerly mentioned expert system can not be used in a framework with soft data fusion. This is due to the definition scope of the fusion operators in the real interval [0, 1]. Therefore the original expert system has been substituted by a fuzzy expert system that tries to map the structure of the original rule system into the fuzzy domain. The original expert system takes into account three different rules based on the application of crisp AND and OR operators, together with some operations that take image neighborhood into account. The transformation into a fuzzy rule system has been done by substituting the AND and OR binary operators through respectively MIN and MAX ones. Moreover the neighborhood operations respect the original expert system structure. The goal of the resulting fuzzy rule system is thence the generation of the fuzzy membership functions corresponding to dry and wet snow presence (see Figs. 2c-d). A complete description of this fuzzy rule system can be addressed on demand at [15].

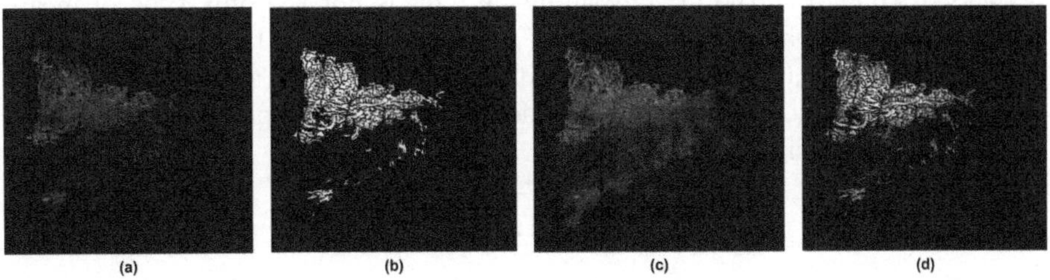

Fig. 2. Wet (a) and dry (b) snow detected by the system currently used in Starlab's service based on SAR. Both images are binary. Fuzzy membership functions for wet (c) and dry (d) snow used in the fusion system presented herein.

The fusion procedure is only possible after a previous co-registration of the two modalities both in the temporal and the spatial domains. This operation attains setting up a common reference in these two domains. In the temporal domain this can be achieved by taking into account images of the corresponding satellites acquired in similar dates. In case of the spatial one, this is achieved by affine transformations and interpolation of one of the modalities in the other's measurement grid.

Once the fuzzy membership functions of the two modalities has been generated, they go through a fusion stage. Two different membership functions are fused in this stage, what we denote as bimodal fusion. Moreover the bimodal fusion can be undertaken as well with two different architectures. The wet and dry snow membership functions can be separately fused with the fuzzy NDSI membership function. On the other hand, both SAR membership functions can be previously fused and the result itself fused with the fuzzy NDSI. We use the maximum operator for fusing the two SAR membership functions. This selection is motivated by the fact that we would like to analyze in this second architecture the potentiality of a fuzzy SAR detection without taking into account if dry or wet snow are detected. The performance of these two different architectures has been analyzed.

3 Bi-modal Fusion Operators for Snow Detection

The fusion stage is mainly formed by a fusion operator, which undertakes the fusion of the incoming variables. In the results presented herein we have compared the performance of five different fusion operators, namely: Power Mean, Yager S-norm, Weighted Sum, Uninorm based on the application of the Yager norms and the Arithmetic Mean, and Ordered Weighted Averaging (OWA). We briefly describe this operators in the following for the sake of completeness.

3.1 Power or Generalized Mean

The mean is one of the most well-know fusion operators. It is used in statistics for finding the central location of a probability distribution. This particular mean is denoted though as the arithmetic mean. This is not the only type of mean. There are other mean operators like the geometric mean or the harmonic mean.

One interesting aspect in this context is the existence of a parametric generalization of all these expressions [2], which is known as the power or generalized mean, with expression

$$z = \left(\frac{1}{n}\sum_{i=1}^{n} x_i^m\right)^{1/m}, \tag{1}$$

whose value depends on the real-valued parameter m, e.g. for m=1 results in the arithmetic mean and for m=2 is denoted as the quadratic mean. As it can be observed the quadratic mean is similar to the Euclidean distance but for a scale factor that makes the operator result remain in the unit hypercube.

3.2 Yager S-Norm

T- and S-norms, whose fundamentals were introduced in [9], are aggregation operators related with the concept of statistical metrical spaces [17]. Here the goal was to join different statistical estimations into a unique value by means of an operator. Later on though T- and S-norms were taken within fuzzy sets as means of aggregating fuzzy membership functions. There are many T- and S-norms [7], from which the Yager S-norm has been selected herein after a preliminary study taking the diversity of operators to be analyzed into consideration. This S-norm presents the following expression:

$$z = \min\{1, (x_1^p + x_2^p)^{1/p}\}, \tag{2}$$

where $p \in [0, \infty]$.

3.3 Weighted Sum

The weighted sum is an operator used in different application domains, e.g. descriptive statistics, neural networks. It presents the same structure as the mean but with the particularity that weights can be established on the values being operated

$$z = \sum_{i=1}^{n} w_i x_i, \tag{3}$$

where the sum of the weights is usually normalized to sum up 1 (this ensures as well working in the unit hypercube). It is worth mentioning that the work in [5] makes use of a somehow equivalent operator.

3.4 Uninorm Based on Yager Norms

Uni-norms were introduced in [19]. Uni-norms generalize T- and S-norms by introducing an arbitrary neutral element denoted as e [12] defined in $[0, 1]$ such that $U(x, e) = x$. There exist a mathematical expression to map T- and S-norms into Uni-norms. The mapping $U \to T, S$ holds for the unit hypercube, whereas $T, S \to U$, only for the spaces $[0, e]^2$ and $[e, 1]^2$. In the other subspaces the uninorm shows a compensating behavior, i.e. the result value is between minimum and maximum. There is a particular type of uni-norms denoted as representable among which we can find the operators used in well-known fusion paradigms of expert systems, like MYCIN and PROSPECTOR [3]. The work in [11] presents the concept of absorbing norm, which in some sense is dual to this of uni-norm.

One can see uni-norms and absorbing-norms as two different ways of combining T- and S-norms. Thus in the uni-norms the subspace $[0, e] \times [0, e]$ is occupied by a T-norm, whereas $[e, 1] \times [e, 1]$ by a S-norm. In the remaining two spaces there is a compensatory operator, although this is not a condition of the operator. The only condition is that the resulting operator must be commutative and associative. In the case of absorbing norms with absorbing elements a, the space

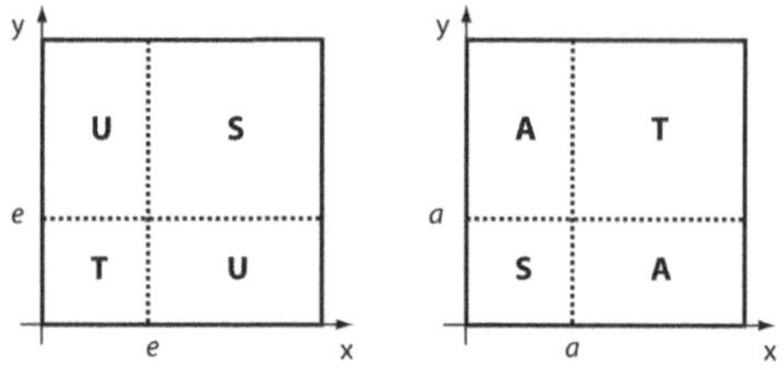

Fig. 3. Schematic description of uni- (left) and absorbing-norms (right) behavior w.r.t. T- and S-norms behavior (T, S). The different behavior is respectively bounded by the value of the neutral element e and the absorbing element a. Areas of compensation behavior are respectively indicated by U and A, i.e. $\min() \leq U(), A() \leq \max()$.

$[0, a] \times [0, a]$ is occupied by a S-norm, and $[a, 1] \times [a, 1]$ by a T-norm. For the remaining spaces applies the same as in the case of the uni-norms. These facts can be represented in graphical form with the diagrams depicted in Fig. 3.

Since the T-norm with the largest value is the minimum and the S-norm with the smallest one is the maximum [1], no the T- and S-norm can be placed in the U and A quadrants. Moreover these two quadrants have to be filled by compromise operators like means or min/max itself. In the works described herein we have selected a uninorm based on the Yager T- and S-norms, and on the arithmetic mean in the U-quadrant.

3.5 Ordered Weighted Averaging

A generalization of the average, where the weighting is established upon ordinal data, was proposed in [18] and denoted as Ordered Weighted Averaging (OWA). OWA operator presents the following expression:

$$z = \sum_{i=1}^{n} w_{(i)} x_{(i)}, \tag{4}$$

where $w_{(i)}$ are the weights of the operator. The bracketed subindices state for a sorting operation that is applied on x_i before aggregating their values, e.g. (1) state for the larger x_i, (n) for the lowest one. As it can be observed the weights are hence applied on the sorting result. This results in a unique weighting set, but that is applied to different channels on each canonical region of the unit hypercube [14].

4 Fusion Results and Performance Evaluation

In this section we analyze the results obtained by applying the different fusion operators described in the former section. For this purpose we first present the evaluation methodology. Then the presentation of two different types of results is attained. First, the results obtained by the parameterization of the fusion operators. Here the goal is to find the set of parameters that make a particular fusion operator work in an optimal manner. Lastly we perform a direct comparison of the performance achieved among the different operators when being optimally parameterized.

4.1 Evaluation Methodology

One way of automate measuring the performance in classification/detection systems takes into account the comparison of the obtained results with a validated ground truth (GT). In case of the snow detection we do not count with in-situ measurements of snow presence. Therefore the ground truth has to be generated by manually labeling some images of the region. It was decided to generate the ground truth based on the RGB channels of the optical satellite images. The ground truth was generated by manually thresholding the RGB image in order to detect all areas perceived on the image as snow. This perception is based on both the characteristic color hue of snow and geographical a priori expectancy of snow. However directly thresholding RGB images can make cloud areas appear as snow.

Once the ground truth images have been generated, we can use them for computing the performance. A test set formed by approximately 80 million measurement points for performance evaluation. We can consider the snow detection as a binary classification problem. Therefore it is appropriate to use the so-called Receiver Operating Curves, which relates True Positive Rate (TPR) to the False Positive Rate (FPR) with respect to the value of a particular parameter, usually the detection threshold, in a binary classification. ROCs are an accepted method for performance evaluation in pattern recognition and data mining [4]. They can be used for visual inspection of performance. The closer to the point of TPR=1, FPR=0, the better. If a numerical comparison is needed the Area Under the Curve can be used [4]. This is the numerical integral of the ROC with respect to the FPR. In an AUC comparison, the larger, the better.

4.2 Optimal Parameterization of Operators

Once the measure of the system performance has been established, we attain the elucidation of the procedure for the parameter search. The vectorial implementation of the fusion operators allows the computational efficient extensive search of optimal parameters. We denote extensive search as the procedure whereby all possible parameter values are employed in the computation of the fusion result. Once this is achieved, the ROC and the corresponding AUC of the fusion result are computed. The parameter whose application delivers a maximal AUC is selected as the optimal parameter value. This can be achieved both numerically and by visual inspection of the ROC set.

4.3 Operator Comparison

After optimally parameterizing the operators, we can proceed to their comparison in terms of detection performance. As described in former sections the fusion operator can be applied on three different bimodal fusion operations. Therefore

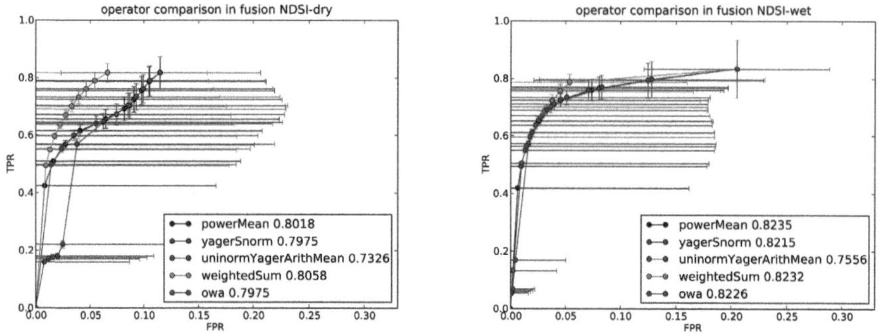

Fig. 4. Average ROC and AUC (in the legend) over the different test days when fusing the fuzzy NDSI and fuzzy dry snow (left), and fuzzy NDSI and fuzzy wet snow (right) memberships through different operators. The operators are respectively parameterized with their optimal parameter set. Depicted FPR range $[0, 0.33]$, so FPR variances appear correspondingly larger.

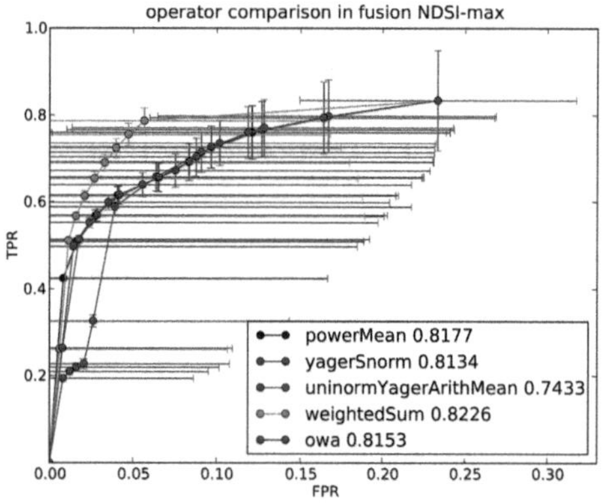

Fig. 5. Average ROC and AUC (in the legend) over the different test days when fusing the fuzzy NDSI and fuzzy SAR snow memberships through different operators. The fuzzy SAR membership is computed through the application of a maximum operator on the fuzzy dry and wet snow membership functions (see motivation in text). The operators are respectively parameterized with their optimal parameter set. Depicted FPR range $[0, 0.33]$, so FPR variances appear correspondingly larger.

we have undertaken the comparison of the operators optimally parameterized for each of these operations. For this comparison we use again the visualization of the ROC curves together with the numerical evaluation of the AUC. Moreover we have add the visualization of the variances both in the TPR and the FPR axes to the ROC curves, which has been denoted in [4] as threshold averaging for summarizing ROCs. We depict in the following figures (see Figs. 4, and 5) the performance obtained for the fusion operators being evaluated when applied in the three different bimodal operations.

5 Conclusions and Future Work

As it can be observed in the results (see Figs. 4, and 5), the evaluated operators deliver a very similar performance. This does not apply for the employed uninorm, which delivers the worse results in both analyzed architectures in terms of AUC. Moreover there are no significant differences between both architectures.

One interesting aspect relates to the flexibility of the operators. All outperforming ones present a larger flexibility than the weighted sum. This can be observed on hand of the larger range of values that their ROC include. This feature allows a better tuning of the detection performance to the operational requirements, which constitutes an important advantage for its implementation.

One important point we have to improve is this of the GT estimation. Since it is established based on optical spectral bands (RGB), its value tends to favor the NDSI in front of the SAR modalities. With this in mind we have to extremely tune the parameters of the SAR fuzzy expert system and find other ways of extracting this ground truth, e.g. through in-situ measurements. The test set has to be extended as well, in order to evaluate seasonal variations of the system performance in order to achieve its operational deployment. Finally an interesting aspect to be analyzed in the future is the comparison of performance when using bimodal fusion and when the operators aggregate all three information channels. For associative operators, like the T- and S-norm, this should not make any difference, but possibly for the remaining ones.

References

1. Bloch, I.: Information combination operators for data fusion: a comparative review with classification. IEEE Transactions on Systems, Man and Cybernetics, Part A 26(1), 52–67 (1996)
2. Bullen, P.: Handbook of Means and their Inequalities. Kluwer, Dordrecht (2003)
3. De Baets, B., Fodor, J.: Van melle's combining function in mycin is a representable uninorm: an alternative proof. Fuzzy Sets Syst. 104(1), 133–136 (1999)
4. Fawcett, T.: An introduction to roc analysis. Pattern Recogn. Lett. 27(8), 861–874 (2006), http://dx.doi.org/10.1016/j.patrec.2005.10.010
5. Haefner, H., Piesbergen, J.: High alpine snow cover monitoring osing ers-1 sar and landsat tm data. In: Baumgartner, M., Schultz, G.A., Johnson, A.I. (eds.) Remote Sensing and Gographie Information Systems for Design and Operation of Water Resources Systems (Proceedings of Rabat Symposium S3) April 1997, vol. 242. IAHS Publ. (1997)
6. Hall, D.K., Riggs, G.A., Salomonson, V.V.: Development of methods for mapping global snow cover using moderate resolution imaging spectroradiometer data. Remote Sensing of Environment 54(2), 127–140 (1995)
7. Klement, E.P., Mesiar, R., Pap, E.: Triangular Norms. Trends in Logic, vol. 8, Ist edn. Springer, Heidelberg (July 2000)
8. Kropatsch, W.G., Strobl, D.: The generation of sar layover and shadow maps from digital elevation models. IEEE Transactions on Geoscience and Remote Sensing 28(1), 98–107 (2002)
9. Menger, K.: Statistical metrics. Proceedings of the National Academy of Sciences of the United States of America 28(12), 535–537 (1942)

10. Nagler, T., Rott, H.: Retrieval of wet snow by means of multitemporal sar data. IEEE Transactions on Geoscience and Remote Sensing 38(2), 754–765 (2002)
11. Rudas, I.J.: New types of aggregation operators in intelligent systems: absorbing norms and evolutionary operators. In: Proceedings of IEEE International Symposium on Industrial Electronics, ISIE 2001, vol. 1, pp. 404–412 (2001)
12. Rudas, I.J., Fodor, J.: Information aggregation in intelligent systems using generalized operators. International Journal of Computers, Communications & Control 1(1), 47–57 (2006)
13. Salomonson, V., Appel, I.: Estimating fractional snow cover from modis using the normalized difference snow index. Remote Sens. Envir. 89(3), 351–360 (2004)
14. Soria-Frisch, A.: Soft Data Fusion for Computer Vision. Ph.D. thesis, TU Berlin (2005)
15. Soria-Frisch, A.: Optical-SAR fusion based snow detection. Tech. Rep. TN00186, Starlab Barcelona (2009)
16. Storvold, R., Malnes, E.: Snow covered area retrieval using envisat asar wideswath in mountainous areas. In: Proc. IEEE International Geoscience and Remote Sensing Symposium, IGARSS 2004, vol. 3, pp. 1845–1848 (2004)
17. Thorp, E.: Best possible triangle inequalities for statistical metric spaces. Proceedings of the American Mathematical Society 11(5), 734–740 (1960)
18. Yager, R.R.: On ordered weighted averaging aggregation operators in multicriteria decisionmaking. IEEE Trans. Syst. Man Cybern. 18(1), 183–190 (1988)
19. Yager, R.R., Rybalov, A.: Uninorm aggregation operators. Fuzzy Sets Syst. 80(1), 111–120 (1996)

Color Recognition Enhancement by Fuzzy Merging

Vincent Bombardier, Emmanuel Schmitt, and Patrick Charpentier

Centre de Recherche en Automatique de Nancy (CRAN), CNRS UMR n°7039
Faculté des Sciences, Bd des Aiguillettes – BP 239
54506 Vandoeuvre-lès-Nancy Cedex, France
{vincent.bombardier,emmanuel.schmitt,patrick.charpentier}
@cran.uhp-nancy.fr

Abstract. This paper deals with color matching in a wood quality control problem. The main difficulty consists in the recognition of gradual color in an industrial context. The wood, which is a natural material, implies a subjective processing to make the controls. The current methods do not take into account the human aspect of the process. An improvement consists in integrating the imprecision of this subjectivity by using the concept of Fuzzy Sensor. Such a sensor has been developed and done with a Fuzzy Rule Classifier which is quite efficient with imprecise data. Then, in multi-face color matching case, the color recognition is enhanced by merging the outputs of the sensors used together. A specific fuzzy merging operator is proposed to use and compared with more classical merging methods. The obtained results show the efficiency of the proposed enhancement.

Keywords: Image Processing, Color Matching, Pattern Recognition, Fuzzy Sensor, Fuzzy Merging.

1 Introduction

The problem of matching wood pieces is an important issue. It affects many activities of the timber industry (veneer, paneling, manufacture of squares, etc ...). Since we try to get a product, a result of the assembly of several component pieces, with a visual consistency, this problem is found.

In this paper, a new method is proposed for the matching of red oak pieces to produce squares which will be assembled to become stair handrails (fig.1). The wood pieces have to be assembled according to their color of appearance. The aim is mainly to provide a wood piece that seems to be homogeneous and massive.

The colors which must be identified are intrinsically fuzzy (impact of the wooden fiber in the wooden color notion). Thus the colour descriptors calculated on the images are uncertain and imprecise. Moreover, the user of the system expresses his needs under the guise of linguistic terms. The output classes are subjective and non-disjointed. The perception of colors is then gradual. For example, there are no strict bounds between a "red" wood and a "light red" wood. In this case, the difference is given by the wood expert and not by the vision expert. Even if the contrast is not very good in this picture, the real hue of the wood board is not evident for a non-specialist In this way, the fuzzy logic concept stands out as the most flexible technique [1].

E. Hüllermeier, R. Kruse, and F. Hoffmann (Eds.): IPMU 2010, Part II, CCIS 81, pp. 231–240, 2010.

Another difficulty takes place in small number of samples that it can be obtained for learning step. Thus, the industrialist cannot provide a consequent and homogeneous data set because of the rarity of particular wood color. So, we proposed to use for the color recognition step a Fuzzy Rule Classifier, which is well-adapted in this industrial context [2].

In the following sections, the color identification vision system is first detailed. Then some backgrounds on the Fuzzy Linguistic Rule Classifier are introduced. Finally the merging method and its application in an industrial context are presented.

2 Proposed Matching Process

This study concerns the development of a matching system for wooden boards according to their colorimetric aspect [3]. This recognition is carried out in real time on the industrial production line. These lines may reach speeds up to 400 meters of board length per minute. After the color identification step, done by the vision system, color information is sent to an optimization step. Then each board is sent to a sorting line or to a cutting line. The cutting line aims to split the boards into uniformly colored piece of wood. The sorting line aims to group pieces of wood into specific classes, whose number and definition are given by the final customer. The boundary classes are very subjective in both cases.

The originality of the process concerns the color sorting which is realized only on the wooden board edges (board thickness). Indeed, the machining of handrails requires a uniform color in a large thickness (Fig. 1). To obtain this large thickness, three boards are glued by their face. So, the final product makes illusion of a product carved in an uncut wood piece. However, operators use the two wide faces to take their single and global decision and it is necessary to make the same classification only by taking into account the edge decisions.

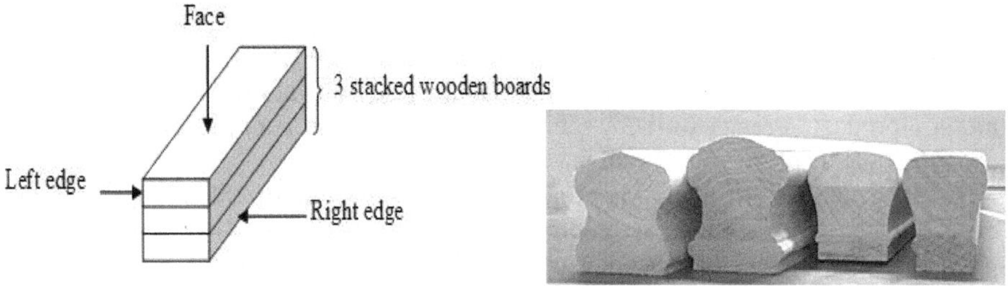

Fig. 1. Schematic representation of the final products considered in the study

The used acquisition system is made up of one type of linear sensors: CCD color cameras. This CCD sensor provides the red, green and blue components of the signal. The signals are sampled at the rate of 1500 lines per second. Each line is composed of 900 pixels. Fig. 2 illustrates the acquisition system with the processing parts in the case of two color cameras.

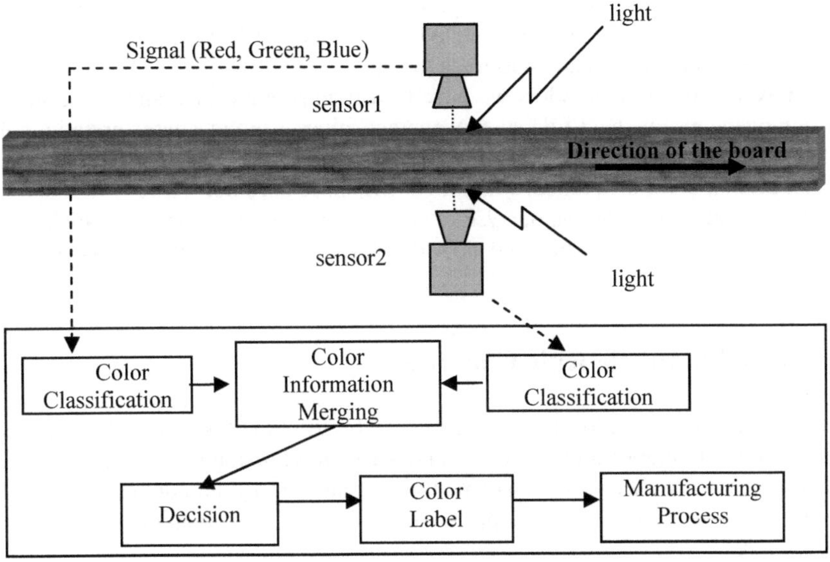

Fig. 2. Acquisition system and processing parts

Several factors can have an impact on the measurements provided by these sensors. Among these parameters, it can be listed the ageing of the acquisition system, the ambient temperature of use, and the precision of the wooden board convoying, i.e. the precision on the distance sensor/board. By integrating correction models concerning the evoked parameters, the imprecision in the measurements will be reduced [2]. That is why we decide to develop our system under the form of an appearance fuzzy sensor [4].

Two aspects are essential to characterizing color: the reference color space and the characteristic vector.

One of the most common color spaces denoted RGB, organizes the color information of an image into its red, green, and blue components. However, the International Commission on Illumination (CIE) does not recommend its use because the color components are not independent of one another [5]. Other popular color spaces include the Lab and HSV (Hue, Saturation, Value (intensity)) spaces. Many studies on color space selection have been conducted elsewhere, i.e. [6] [7]. After conducting several internal tests on various sets of wood samples, we decided to work in the Lab space because it provides the best color discrimination in term of recognition rates. We have also done this choice in relation with an objective criterion funded on Δ_{cielab} recommended distance [8]. This could perhaps be explained because this colorimetric reference space represents colors in the same order than humans do and the color class definitions are given by customers.

In the same way, it is necessary to characterize a color with a set of parameters which are extracted from the image. This set, called "characteristic vector" characterizes color in a simpler way. We choose one of the simplest attribute responding to the calculation time criterion: the average of L, a, b values in the Region of Interest (ROI). The size of 300 lines was selected after a study of ROI size impact in processing time between 50 lines and 450 lines [8].

It must be noticed that colors we have to recognize are closed and are supposed to be homogeneous in the ROI, so the average gives a good characterization of the R.O.I. CIELab space has metric color difference sensitivity and is very convenient to measure small color difference, while the RGB space does not [30].

The second part of the vision system is the color identification step which uses for input the characteristic vector and provides for output the label of processed ROI's board.

3 Fuzzy Linguistic Rule Classifier

For the color identification in wood context, the used methods are often based on k nearest neighbor algorithm [9] [10] or on distance minimization algorithms [11] [12].

However, these methods are not really well-adapted to the applicative context as described in section 1 (color subjectivity in human classification, gradual output classes ...).

The Fuzzy Rule Classifier (FRC) [13], based on a fuzzy linguistic rule mechanism, is more convenient for our industrial context. Indeed, it has got a good generalisation capacity as it is shown in several comparisons done with other classifiers such as bayesian classifier, K-Nearest Neighbour, Neural Network or Support Vector Machine [14]. It is able to take into account the graduality of the output classes.

This fuzzy recognition method is a supervised mechanism divided into three parts as shown in Fig. 3: Input fuzzification, Fuzzy rule generation and Model adjustment.

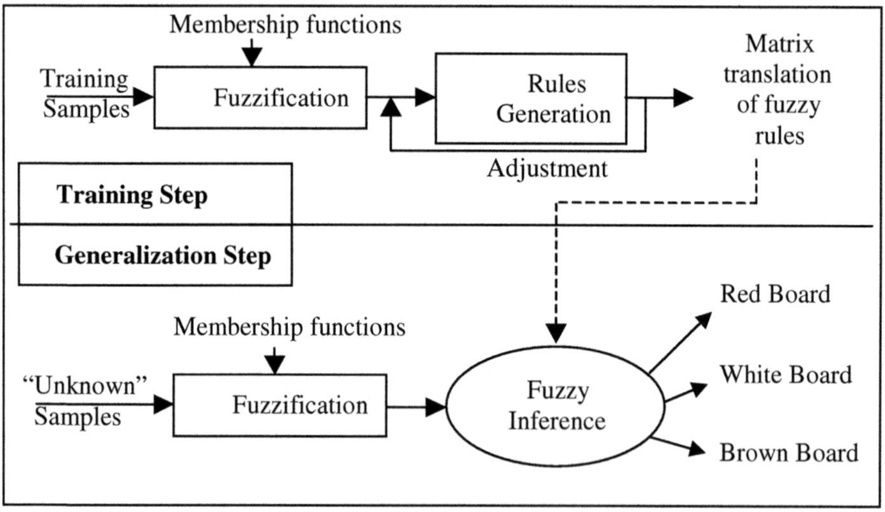

Fig. 3. Overall description of the Fuzzy Rule Classifier

3.1 Input Fuzzification

The fuzzification step aims to translate numerical variables into linguistic variables. A linguistic variable [15] is defined by a triple value (V, X, Tv) where:

- V is a variable (area, size, etc.) defined on a set of reference X
- X is the universe of discourse (field of variation of V)
- Tv is the chosen vocabulary to describe in a symbolic way the values of V (small, big, etc.).

The set Tv = {A1, A2, ...}, finite or infinite, contains normalized fuzzy subsets of X which are usable to characterize V. Each fuzzy subset, Ai, is defined by the membership degree μAi(x). This fuzzification step defines the decomposition number of the considered variable to provide the fuzzy rule premises.

For example the L color composant could have a Weak, Medium or High value. The symbolic vocabulary associated with this variable is then Tv = {Weak, Medium, High}. So, this variable will be split into three terms and characterized by a vector composed of three membership degrees: $[\mu Weak(x), \mu Medium(x), \mu High(x)]^T$.

The different terms are chosen in relation to the expert vocabulary. The number of terms used to qualify a linguistic variable is, generally, empirically defined. But, the industrial user, who is not an expert in pattern recognition, often chooses a regular distribution of terms, generally having more terms than are needed. Whenever the number of terms increases, so does the number of rules and, thus, the overall complexity of the entire system. Classical automatic methods are based on Genetic Algorithm [16] or Clustering [17]. The main drawback of these methods is that they need lots of samples to be efficient.

The chosen fuzzification method is based on the study of the output class typicality [18]. From the Typicality measure T(V), the correlation (Corr) and the cross-correlation (Xcorr), coefficients are computed for each output class. Then, from the ratio Corr/Xcorr, which characterizes inter-class similarities, the number of fuzzification terms is determined. Their positions are obtained by calculating the mean value of the samples belonging to the considered output classes [14].

3.2 Fuzzy Rule Generation

This second step allows the defining of "If… Then…" fuzzy rules. For instance:

« IF L is (High) AND a is (Big) AND b is (Medium) THEN the Color IS (Light Red) »

Each rule describes the perceived color, related to the system. Such rules can be classified into two categories: conjunctive rules and implicative rules. These two categories are regrouped respectively. On the one hand, there are the possibility rules and the anti-gradual rules and, on the other hand, the certitude rules and the gradual rules. The conjunctive rules are derived from the data analysis field where reasoning mechanisms are led by the data whereas implicative rules are most utilized in the cognitive sciences field where reasoning is led by knowledge [19].

For this application, conjunctive reasoning mechanisms have logically been selected. Each rule is activated in parallel and a disjunction operator combines the intermediate results. Moreover, this inference mechanism assures the consistency of the rule base [20]. If no information is processed, that is, the input space is not covered by the rule set; the output gives an "unknown defect" class. The two main models using these rules are the Larsen's model and the Mamdani's model. The Sugeno's model is not suitable in this case because the aim is not to achieve numerical output values.

The chosen classifier is based on Ishibuchi's algorithm which provides an automatic rule generation step [21]. There are many methods, which automatically obtain fuzzy rules according to data sets such as a genetic algorithm [22], but the Ishibushi's algorithm is quite simple and gives better results [14]. Moreover, its inference mechanism follows the Larsen's model, which is better than the Mamdani's model, because the Product is more adapted than the Minimum for the manipulation of several premises [23]. In fact, it allows non-linear splitting of variable input space. The iterative version of the Ishibushi's algorithm [24] is used here. It allows to adjust the input space splitting by supporting the rule of having the maximum response.

3.3 Model Adjustment

The adjustment represents the iterative part of the algorithm. The following mechanism allows for the adjustment of the decomposition of representative space according to the achieved results. From the training patterns, the algorithm generates the first model and a CF confident coefficient is calculated from the truth degree of each rule.

If the classification rate is below ą threshold, defined by the user, the iterative part is performed to adjust this rate by modifying the CF value. This coefficient is increased if the sample confirms the rule and decreased in the contrary case

The output of the fuzzy classifier is a fuzzy vector whose components indicate the possibility that such board belongs to such color class. It can be noticed that these possibility degrees are not complementary.

Usually, the final decision is taken at this level by using a disjunction operator as max. The board is allocated to the color class corresponding of the maximal possibility degree. However, in this industrial context, two color sensors are used to give the board final color. So, it will be better not to take the decision sensor by sensor but to keep the possibility degree and to decide by merging both fuzzy information provided by each sensor. The next section details this merging step.

4 Multi-face Matching by Fuzzy Merging

4.1 Fuzzy Merging Operator Used

Then, at this level, both delivered information must be merged to provide a single decision to classify the board. To define the final color thanks to the collected information, it is currently used a Symbolic Merging performed only from the symbolic terms provided by each colorimetric fuzzy sensor [25] [26]. The *AND* operator is applied to merge the data. Thus, a color class is allocated to a final product, only if the two sides have been recognized as the same color (*IF* Red on side 1 *AND IF* Red on side 2 *THEN* color is RED). All the other cases are considered as a rejection class. This Symbolic Merging is very restrictive.

In this case, the fuzzy outputs of the system are not used. By using the possibility degree provided by the each sensor, the uncertainty of the results can be taken into account. For instance, the results of two fuzzy sensors can be considered. The first sensor S_1 gives the color class Red with a membership degree value equal to 0.8 and the class Brown with a membership degree value equal to 0.3. The second sensor S_2 gives the color class Red with a membership degree value equal to 0.5 and the class Brown with a membership degree value equal to 0.6. If the maximum value for each sensor is considered, the merging result allocates the rejection class. However, the output of sensor S2 seems not to bet sure because the difference between the two possibility degrees is reduced. Thus, if the fuzzy output is considered by using a merging operator, these possibility degrees can be taken into account.

In order to totally exploit the fuzzy outputs through their possibility degrees, it will be better to use a suitable merging operator [27]. In our application case, a wise operator is proposed to be used [28].The presented operator has been developed in CRAN research team and have been soon applied in similar industrial vision context.

It is defined from two linguistic variables x and y.

$$F(x, y) = \min\left[1, \frac{\min(x, y)}{1 - \min(x, y)}\right] \quad (1)$$

where F(x,y) is the wise operator

x and y are the linguistic variables.

From the calculation of the expression (1) for all possibility degrees of each color classes, the merging results are obtained for the color class, which allows to check these relations (2).

$$\begin{cases} \text{if } x \le y \text{ then } x \le F(x, y) \le y \\ \text{if } y \le x \text{ then } y \le F(x, y) \le x \end{cases} \quad (2)$$

In the case where possibility degrees of both linguistic variables are superior to 0.5, the relation (3) is checked.

$$F(x, y) \ge \max(x, y) \quad (3)$$

In this way, by using a fuzzy merging of both single-face fuzzy results, the sorting rates in the rejection class can be reduced as it can be seen in the following section. A numerical example of using this fuzzy merging operator is given in [2].

4.2 Results

The whole tests have been made on "red oak " boards. This wood specie is divided for this application case into 6 "customer" color classes: Dark Brown, Brown, Light Brown, Dark Red, Red and Light Red.

In order to check the choice of the merging operator, the final recognition rates obtained after merging are then compared. Two databases used have been provided by the industrialist, one corresponding to the acquisition data of the Left Sensor, and the second relating to the Right Sensor. The learning database, used for generating the rule model is composed with 316 data samples. The database used in the generalization step hold 627 samples.

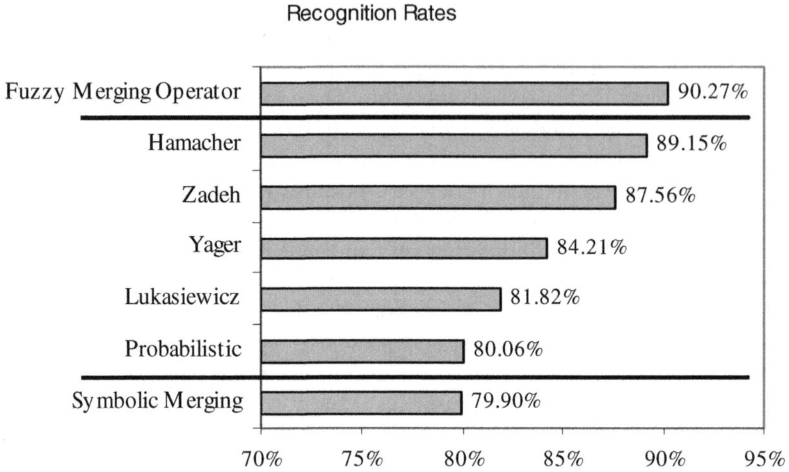

Fig. 4. Comparison of the recognition rates provided by merging both right and left sensors

The recognition rates provided by the Fuzzy Rule Classifier for both Left and Light sensors are respectively 85.65% and 86.12% with generalization data set. A comparison with other classifiers on this study case can be found in [2]. These recognition rates are satisfactory but using a Symbolic Merging has defined in section 4.1, the final recognition rate decreases to 79.9% as shown in Fig 4. Others comparisons are given in this figure, with usual fuzzy merging operators [27].

To do this comparison, the intermediate results provided by each T-norm are aggregated using Max operator. Thus, the outputs delivered by probabilistic and Zadeh T-norm respectively correspond to the implementation of Larsen and Mamdani models whose rules are the ones used in the Symbolic Merging.

The results show that using a fuzzy merging operator really increases the final recognition rate and then the global classification. Moreover, the proposed F operator gives an even better result. Indeed, the final recognition rate reaches 90%.

In a specific case, where the customer tolerates the matching of similar faces in terms of luminance (ie matching a Red face with a Light Red face), the final classification rate increases to 95%.

5 Concluding Remarks

In this paper, a wood color classification has been presented. The color perception is a very subjective notion and it is strongly linked to the wood species or to its final use. This industrial application supports the development of an original Colorimetric Fuzzy Sensor [2]. However, the classification can even more be enhanced in the specific application matching case. So, we have proposed to do this matching with a fuzzy merging operator rather than using a Symbolic Merging. These results show the good behaviour of the proposed method and especially of the presented fuzzy merging method (F operator).

The main evolution of the proposed system concerns the expansion to other notions than color. The wooden board appearance is not due to the only concept of hue. Texture and wood grain are also to be taken into account. That is why we would like to develop an appearance fuzzy sensor putting together several wood appearance attributes.

Then, further investigations aim to reduce the number of generated rules in order to improve the interpretability of the rule set given by the Fuzzy Rule Classifier. For that, it could be consider the use a tree version of it [29] where the configuration of each fuzzy Inference System is led by expertise and expert knowledge integration.

Another investigation way concerns the integration of Fuzzy Information delivered by such a sensor in the global Information System of the plant.

References

1. Zadeh, L.A.: Fuzzy sets. Information and control 8, 338–353 (1965)
2. Bombardier, V., Schmitt, E., Charpentier, P.: A fuzzy sensor for color matching vision system. Measurement 42, 189–201 (2009)
3. Sangwine, S.J., Horne, R.E.N.: The colour image Handbook. Chapman and Hall, Boca Raton (1998)
4. Benoit, E., Foulloy, L.: High functionalities for intelligent sensors, application to fuzzy colour sensor. Measurement 30, 161–170 (2001)
5. International Commission on Illumination. Colorimetry, 2nd edn., Publication CIE 15.2 (1986), http://www.cie.co.at/
6. Burd, N.C., Dorey, A.P.: Intelligent transducers. Journal of Microcomputer Applications 7, 87–97 (1984)
7. Leon, K., Mery, D., Pedreschi, F., Leon, F.: Color measurement in L*a*b* units from RGB digital images. Food research international 39, 1084–1091 (2006)
8. Schmitt, E.: Contribution au Système d'Information d'un Produit Bois. Appariement automatique de pièces de bois selon des critères de couleur et de texture, Ph.D. Thesis, Henri Poincaré University, Nancy, France (2007)
9. Kline, D.E., Surak, C., Araman, P.A.: Automated hardwood lumber grading a multiple sensor machine vision technology. Computers and electronic in agriculture 41, 139–155 (2003)
10. Maenpaa, T., Viertola, J., Pietikainen, M.: Optimising Colour and Texture Features for Real-time Visual Inspection. Pattern Analysis Applications 6, 169–175 (2003)
11. Lu, Q.: A Real-Time System for Color Sorting Edge-Glued Panel Parts, PhD Thesis, Faculty of the Virginia Polytechnic Institute and State University (1997)
12. Daul, C., Rosch, R., Claus, B.: Building a color classification system for textured and hue homogeneous surfaces: system calibration and algorithm. Machine Vision and Applications 12, 137–148 (2000)
13. Schmitt, E., Mazaud, C., Bombardier, V., Lhoste, P.: A Fuzzy Reasoning Classification Method for Pattern Recognition. In: Proc. 15th Int. Conf. on Fuzzy Systems, FUZZIEEE 2006, Vancouver, Canada, pp. 5998–6005 (2006)
14. Schmitt, E., Bombardier, V., Charpentier, P.: Self-Fuzzification Method according to Typicality Correlation for Classification on Tiny Data Sets. In: IEEE Conference on Fuzzy Systems, London, UK, pp. 1072–1077 (2007)
15. Zadeh, L.A.: The concept of linguistic variable and its application to approximate reasoning. Information sciences 8, 199–249 (1975)

16. Cordon, O., Gomide, F., Herrera, F., Hoffmann, F., Magdalena, L.: Ten years of genetic fuzzy systems: current framework and new trends. Fuzzy Sets and Systems 141, 5–31 (2004)
17. De Carvalho, F.A.T.: Fuzzy c-means clustering methods for symbolic interval data. Pattern Recognition Letters 28, 423–437 (2007)
18. Forest, J., Rifqi, M., Bouchon-Meunier, B.: Class Segmentation to Improve Fuzzy Prototype Construction: Visualization and Characterization of Non Homogeneous Classes. In: IEEE World Congress on Computational Intelligence, Vancouver, Canada, pp. 555–559 (2006)
19. Dubois, D., Prade, H.: What are Fuzzy rules and how to use them. Fuzzy Sets and Systems 84, 169–185 (1996)
20. Dubois, D., Prade, H., Ughetto, L.: Checking the coherence and redundancy of fuzzy knowledge bases. IEEE Trans. Fuzzy Systems 5, 398–417 (1997)
21. Ishibuchi, H., Nozaki, K., Tanaka, H.: Distributed representation of fuzzy rules and its application to pattern classification. Fuzzy Sets and Systems 52, 21–32 (1992)
22. Alcala, R., Alcala-Fdez, J., Herrera, F., Otero, J.: Genetic learning of accurate and compact fuzzy rule based systems based on the 2-tuples linguistic representation. Int. Journal of Approximate reasoning 44, 45–64 (2007)
23. Berthold, M.R.: Mixed fuzzy rule formation. Int. Journal. of Fuzzy Sets and Systems 32, 67–84 (2003)
24. Ishibuchi, H., Nozaki, K., Tanaka, H.: A Simple but powerful heuristic method for generating fuzzy rules from numeric data. Fuzzy sets and systems 86, 251–270 (1997)
25. Mauris, G., Benoit, E., Foulloy, L.: Fuzzy Linguistic Methods for the Aggregation of Complementary Sensor Information. Aggregation and Fusion of Imperfect Information, 215–230 (1998)
26. Fagin, R.: Combining Fuzzy Information from Multiple Systems. Jour. of Computer and System Sciences 57, 83–99 (1999)
27. Dubois, D., Prade, H.: On the use of aggregation operations in information fusion processes. Fuzzy Sets and Systems 142, 143–161 (2004)
28. Perez Oramas, O.: Contribution à une méthodologie d'intégration de connaissances pour le traitement d'images. Application à la détection de contours par règles linguistiques floues, PhD Thesis, Université Henri Poincaré de Nancy (2000)
29. Bombardier, V., Mazaud, C., Lhoste, P., Vogrig, R.: Contribution of Fuzzy Reasoning Method to knowledge Integration in a wood defect Recognition System. Computers in Industry Journal 58, 355–366 (2007)
30. Cheng, H.D., Li, J.: Fuzzy homogeneity and scale-space approach to color image segmentation. Pattern Recognition 36, 1545–1562 (2003)

Towards a New Generation of Indicators for Consensus Reaching Support Using Type-2 Fuzzy Sets

Witold Pedrycz[1], Janusz Kacprzyk[1,2], and Sławomir Zadrożny[1,3]

[1] Systems Research Institute, Polish Academy of Sciences, Warsaw, Poland
[2] Warsaw School of Information Technology (WIT), Warsaw, Poland
[3] Technical University of Radom, Radom, Poland
{pedrycz,kacprzyk,zadrozny}@ibspan.waw.pl

Abstract. In this paper, we consider group decision making setting and propose a novel concept of indicators which are meant to guide the discussion in the group that should lead to consensus. Preferences of the group members are assumed to be fuzzy sets of options. The proposed indicators help analyze the structure of preferences in the group. Their derivation is expressed as an optimization problem. The preferences of a member of the group are finally represented as type-2 fuzzy sets (and interval-valued fuzzy sets, in particular). It is shown that this higher order construct plays a pivotal role in the quantification of variability present in the preferences of the group members. We introduce a constructive way of estimation of type-2 membership functions by invoking a principle of justifiable granularity.

Keywords: decision-making under fuzziness, consensus reaching support, type-2 fuzzy sets, information granularity.

1 Introduction

This paper concerns a problem related to decision making in a group of agents, either human or software type, that may be viewed as being concerned with both group decision making or decision making in multiagent system. The basic setting is as follows. We have a group of options (alternatives, decisions, etc.) and a group of individuals (decision makers, experts, agents, etc.). The individuals provide their testimonies concerning their opinions on options in question. We assume these testimonies (preferences) to be expressed as a fuzzy decision meant as a fuzzy set in the set of options.

In the group decision making model adopted, these preferences are just a starting point for a discussion within the group. It is expected that thanks to argumentation, exchange of information etc. the members of the group change their opinion and, ultimately, arrive at consensus. Some means are needed to provide the group with relevant information concerning both the decision problem considered and the current state of the discussion (cf. our recently proposed model of the overall consensus reaching process [1]). Here we focus on the latter aspect and propose some new tools to analyze preferences of the individuals vs. the preferences of the other members.

Our objective is three-fold. We propose how to quantify the consistency of the individual's preferences by solving an optimization task. Then we show how original

E. Hüllermeier, R. Kruse, and F. Hoffmann (Eds.): IPMU 2010, Part II, CCIS 81, pp. 241–250, 2010.

preferences, expressed by the membership function of a fuzzy set of options, may be augmented by incorporating variability which is conveniently captured via type-2 fuzzy sets. By doing so, we highlight the usefulness of higher order fuzzy set constructs and motivate their further usage. Notice that while the conceptual underpinnings of type-2 fuzzy sets were quite commonly reported in the literature, cf. [2,3,4], very little attention has been paid to their origin and algorithmic aspects of these constructs including an estimation of the underlying membership functions.

2 Problem Formulation and the Optimization Procedure

We assume that we have a set of n options, $O = \{o_1, o_2, \ldots, o_n\}$, and a set of c individuals, $E=\{e_1, e_2, \ldots, e_c\}$ who provide their testimonies as fuzzy decisions, i.e. fuzzy sets in the set of options O. Let us denote the fuzzy decision of individual e_i as F_i, and thus the whole set of fuzzy decisions of all members of the group as $\mathcal{F} = \{F_1, F_2, \ldots, F_c\}$.

There is also a person, called a *moderator*, who is responsible for running the session with the individuals, aimed at arriving at a *consensual* fuzzy decision. Thus, the individuals get involved in a discussion to present arguments for their preferences, learn arguments of the others, get some new information on the decision problem, etc.; cf. our recent paper on a comprehensive model for consensual group decision making [1]. The session is finished as soon as the individuals reach consensus or exceed a time limit for the session.

In such a setting it is impractical to mean consensus as the full agreement, of all individuals on all options which is usually not needed and, moreover, such a binary concept of consensus does not help in guiding the session. Thus, some flexible definitions of consensus were proposed, notably by Kacprzyk and Fedrizzi [5,6], for the case of fuzzy preference relations. Similar concepts were proposed by Kacprzyk and Zadrozny [7,8,1] for different (fuzzy) representations of preferences too.

A measure of consensus in the group is a primary indicator for the moderator and individual group members to get oriented how far they are from consensus and if the discussion proceeds in a proper direction. However, some other indicators are also needed to provide more detailed information on the state of the discussion and positions of the particular individuals. Such indicators and the whole systems for supporting discussion guidance have been proposed; cf., e.g., [9,10,8], but mainly for the preference relations representing opinions of the individuals. Though there is a rich literature on fuzzy sets aggregation within group decision making (social choice, etc.), cf., e.g., [11,12,13,14,15,16,17,18,19], the problem of consensus reaching support in this setting is not addressed properly. Here we attempt to fill this gap.

In our approach, we introduce a new indicator, denoted $Q(e_i)$, which is computed for each individual $e_i \in E$ and is meant to convey information on the consistency of an individual's e_i preferences with respect to preferences of the rest of the group. It is defined as follows. Consider for a given individual $e_i \in E$ a vector $W(e_i)$ such that:

$$W(e_i) = [w_1(e_i), \ldots, w_c(e_i)], \qquad W(e_i) \in [0,1]^c, \qquad \sum_{j \neq i}^{c} w_j(e_i) = 1 \tag{1}$$

The interpretation of the weight vector $W(e_i)$ is: the higher the value of the weight $w_j(e_i)$ is, the more consistent (relatively) the corresponding fuzzy decision (F_j) is with the fuzzy decision of individual e_i, i.e., with F_i. We also assume that $w_i(e_i)=1$, and the interpretation of this assumption is straightforward. These weights are *implicit* in their nature (so their numeric values are not explicitly given) and will be revealed and quantified through the following optimization process, which is used to compute the value of the $Q(e_i)$ indicator at the same time. Namely:

$$Q(e_i) = \min_{W(e_i)} \sum_{j=1}^{c} w_j^m(e_i)\,||F_j - F_i|| \qquad (2)$$

subject to (1)

where $m > 1$ is a parameter; $||.||$ denotes a distance between two fuzzy decisions, which may be given in various ways. Note that if there exists j such that $||F_j - F_i|| = 0$, then the solution to (2) is trivial. Thus, in what follows we assume that if such j does exist then $||F_j - F_i||$ is replaced with a small number ε, which is parameter of the method. By using the Lagrange multipliers, the weight values are expressed as follows

$$w_j(e_i) = \frac{1}{\sum\limits_{\substack{k=1\\k\neq i}}^{c}\left(\frac{||F_j - F_i||}{||F_k - F_i||}\right)^{1/(m-1)}} \qquad \forall j \neq i \qquad (3)$$

Thus, each individual e_i knowing "his or her" vector $W(e_i)$ knows how his preferences, expressed by his fuzzy decision, are consistent with the preferences of other members of the group: the higher values of the components of the vector the more consistent (relatively) the preferences are.

Also, an individual e_{i_0} such that $Q(e_{i_0}) = \min_{e_i \in E} Q(e_i)$ may be meant as holding the preferences most representative for the group; we denote his or her fuzzy decision F_{i_0} as G in what follows. However, G does not reflect the diversity of fuzzy decisions exhibited by individual parties. It would be beneficial to quantify the variability within individual fuzzy decisions in $\mathcal{F}$ and articulate it within G.

This brings us to the concept of fuzzy sets of higher order, and more specifically interval-valued fuzzy sets. More descriptively, we could envision that G has been expanded to a more abstract conceptual entity which is capable of quantifying the underlying diversity of individual fuzzy decisions. The crux of the process is schematically visualized in Fig. 1. Note that one of elements of $\mathcal{F}$ has been selected and then augmented to the next level of abstraction, by enhancing it with information on that diversity of individual fuzzy decisions; this effect is stressed by showing G in a different way than F_j, as a cloudly entity. This analysis is meaningful for each of the individuals (their fuzzy decisions), not only for the one corresponding to G.

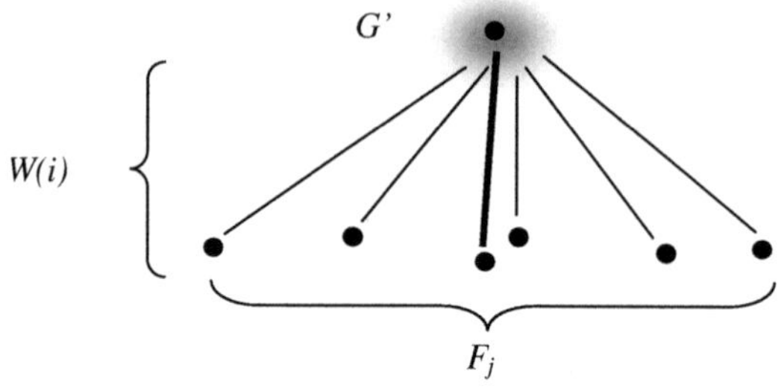

Fig. 1. An individual fuzzy decision G confronted with the whole collection of individual fuzzy decisions and represented at a higher level of abstraction as an augmented (accounting for the variability of the whole collection) fuzzy decision G'

The quantification of diversity of fuzzy decisions is performed by interval-valued values of membership function of G which in its essence elevates its level of abstraction to an interval-valued fuzzy set. The quantification of this effect is realized by invoking a so-called *principle of justifiable granularity* due to Pedrycz [20,21] whose essence and computational aspects are covered in the next section.

The optimization criterion (1) can be augmented by assigning weights to options. This may be done by replacing the distance function $\|\cdot\|$ by its weighted counterpart. Consider the following distance

$$\|F_i - F_j\|_Z = \sum_{k=1}^{n} z_k^r |\mu_{F_i}(o_k) - \mu_{F_j}(o_k)| \tag{4}$$

where $Z = [z_1, z_2, \ldots, z_n]^T \in [0,1]^n$ is a weight vector associated with the set of options O; $\mu_{F_i}(\cdot)$ denotes the membership function of fuzzy set F_i (fuzzy decision of individual e_i); $r > 1$ is a parameter. The use of the Hamming distance between two fuzzy sets is motivated by its robustness (always a desirable property of the method) though we should bear in mind that the absolute values it contains may sometimes be inappropriate.

The optimization problem (2) is reformulated as follows

$$Q(e_i) = \min_{W(i),\, Z(i)} \sum_{j=1}^{c} w_j^m(e_i) \|F_j - F_i\|_{Z(e_i)} \tag{5}$$

subject to (1) and: $z_k(e_i) \in [0,1]$ $\quad \sum_{k=1}^{n} z_k(e_i) = 1$

The constraints in the problem may be again handled via the Lagrange multipliers, to obtain an unconstrained optimization problem formulation and to solve it using alternate iterative optimization of the weight vectors $W(e_i)$ and $Z(e_i)$. The interpretation of the weights vector $Z(e_i)$ is: the higher the value of $z_k(e_i)$ is, the more consistent is the fuzzy decision F_i with fuzzy decisions of other group members, with respect to option o_k. Thus, now each individual e_i knows who in the group holds similar preferences, which is indicated by vector $W(e_i)$, as well as in regards to which options are his or her preferences most consistent with preferences of other members of the group, what is indicated by the vector $Z(e_i)$.

An interesting and useful generalization of the model discussed so far can be proposed. The indicator Q (cf. (2)) may be considered for a subset of individuals instead for one individual. Denote the index set of the members of such a subset by I. Then the indicator is defined via the following minimization problem:

$$Q(\{e_i\}_{i\in I}) = \min_{W} \sum_{i\in I} \sum_{j=1}^{c} w_j^m(e_i) \| F_j - F_i \| \tag{6}$$

$$\text{subject to } \forall i\in I, \forall j \;\; w_j^m(e_i) \in [0,1] \quad \text{and} \quad \forall i \in I, \sum_{j\notin I} w_j^m(e_i) = 1$$

where the weights form a matrix $W = [w_j(e_i)]$ and each column of W is a vector of weights showing consistency of preferences (fuzzy decision) of individual e_i with the preferences of the rest of the group.

In order to obtain the weights we again use the Lagrange multipliers and obtain:

$$w_j(e_i) = \frac{1}{\sum_{\substack{k=1\\k\notin I}}^{c} \left(\frac{\|F_j - F_i\|}{\|F_k - F_i\|} \right)^{1/(m-1)}} \quad \text{for } j \notin I, i \in I \tag{7}$$

and we assume $w_j(e_i) = 1$ for $i,j \in I$.

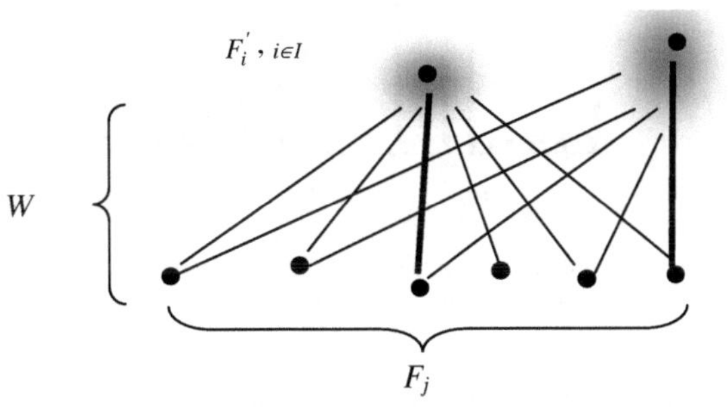

Fig. 2. A subset of individual fuzzy decisions $\{F_i\}_{i\in I}$ confronted with the whole collection of individual fuzzy decisions and represented at the higher level of abstraction as a set of augmented fuzzy decisions

As before the value of the indicator $Q(\{e_i\}_{i\in I})$ may be accompanied by the quantification of diversity of the preferences of the group when confronted with the preferences of the subgroup determined by the index set I. This may be best presented as in Fig. 2 meant similarly as Fig. 1. In the next section we discuss in detail how this quantification is formed.

3 The Principle of Justifiable Granularity

The idea behind the *principle of justifiable granularity*, proposed by Pedrycz [20,21], is to quantify the variability in a set of available membership degrees as some information granule such as an interval or a fuzzy set. In our problem, these are membership degrees of an option o_k to fuzzy decisions of the particular individuals, i.e., $\mu_{F_1}(o_k)$, $\mu_{F_2}(o_k)$, ..., $\mu_{F_c}(o_k)$; denote them for brevity as u_1, u_2, ..., u_c. Now assume that we are looking for some additional information for/about individual e_l. Denote by u_0 the membership degree $\mu_{F_l}(o_k)$, i.e., the membership degree of the option under consideration o_k to the fuzzy set representing the fuzzy decision of individual e_l. We compute the value of the indicator $Q(e_l)$ according to (2)-(3) and thus obtain the weights vector $W(e_l)$. Now, with each u_i we may associate the weight $w_i(e_l)$ and thus get a set of pairs (u_i, $w_i(e_l)$), – remember that by definition $w_l(e_l)$=1, and thus we have a pair (u_0, 1) among them. Given this set of pairs, cf. Fig. 3, we are interested in representing the weights $w_i(e_l)$ by spanning an interval [u_-, u_+] around u_0 so that it realizes an intuitively appealing procedure: includes as many membership degrees u_i of individuals with preferences compatible with e_l (i.e., with high $w_i(e_l)$) as possible, and excludes as many membership degrees u_j of individuals with preferences incompatible with e_l (i.e., with low $w_j(e_l)$) as possible. In this sense we form an interval as a suitable information granule capturing the diversity in the pairs (u_i, $w_i(e_l)$); cf. Fig. 3.

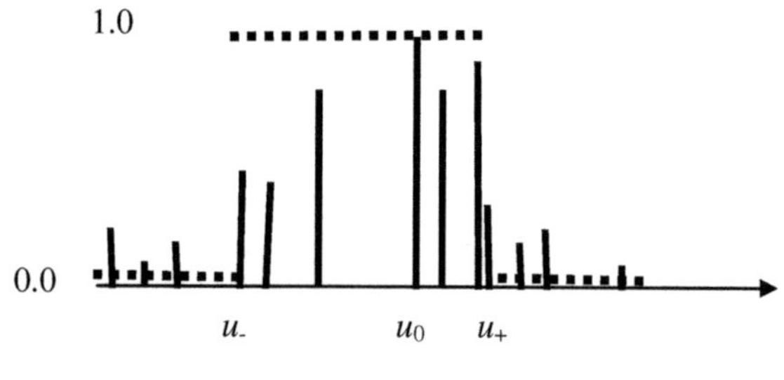

Fig. 3. Computing the interval representation of numeric values through the principle of justifiable granularity by elevating and suppressing the weights

The formal rule of constructing the interval is as follows. Such an interval [u_-, u_+] is chosen, which minimizes the joint cost of including membership degrees of incompatible individuals and excluding membership degrees of compatible individuals, what may be quantified as follows for an individual e_i:

if $u_i \in [u_-, u_+]$ then elevate $w_i(e_l)$ to 1

if $u_i \notin [u_-, u_+]$ then reduce $w_i(e_l)$ to 0

Thus, the bounds of [u_-, $\underline{u_+}$] are subject to optimization under the criterion that the total changes to the weights (being equal either to 1- $w_i(e_l)$, when a weight is elevated, or $w_i(e_l)$, when a weight is reduced) are made as small as possible. The values of u_- and u_+ are thus chosen via the following minimization problem:

$$\min_{u_-,u_+ \in R:\, u_- \le u_+} \left\{ \sum_{u_i \in [u_-,u_+]} [1 - w_i(e_l)] + \sum_{u_i \notin [u_-,u_+]} w_i(e_l) \right\} \tag{8}$$

The above performance index minimized in (8) is guided by the changes in the weight vector $W(e_l)$. We can consider another variant that is based on the area criterion. Instead of changes in the weights themselves, we consider the minimization of the changes in the area affected by the elevation or suppression of the corresponding weights. The essence of this technique is illustrated in Fig. 4. The corresponding minimization problem is expressed as:

$$\min_{u_-,u_+ \in R:\, u_- \le u_+} |A_1 + A_2 - A_3 - A_4| \tag{9}$$

with the areas shown in the same figure. The resulting approach reflects the minimization of area changes.

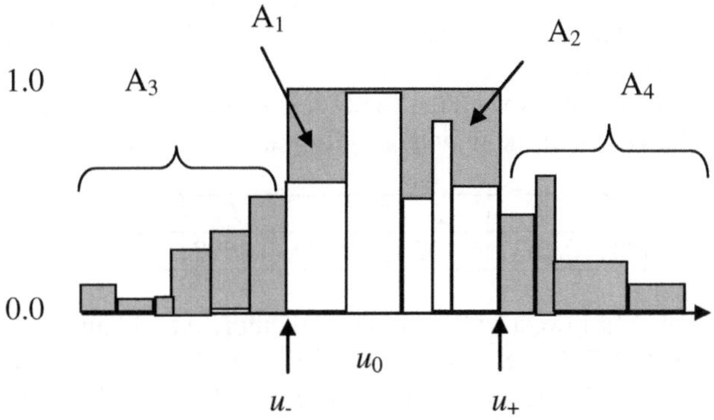

Fig. 4. A pictorial representation of area changes – the adjustment of positions of u_- and u_+ so that the changes in the area become minimized. The area affected (where the weights are modified) is shown in grey color.

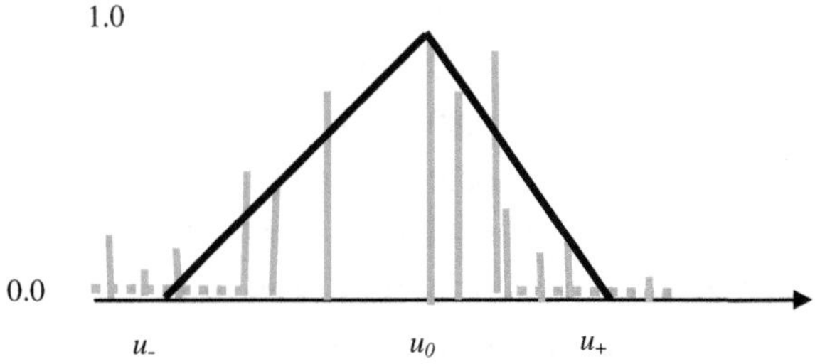

Fig. 5. Triangular fuzzy set representing individual membership grades; the cutoff points u_- and u_+ are optimized by running standard linear regression

The information granule can obviously be expressed as some fuzzy set too. In particular, the triangular fuzzy sets can be very convenient and intuitively appealing a tool because they have a clear semantics. As shown in Fig. 5, the modal value of the membership function is then u_0.

The optimized fuzzy set is spanned over [0, 1] with the slopes of the membership functions optimized individually for the data positioned below and above u_0. The standard linear regression applied here returns the parameters u_- and u_+ in the membership function.

The result obtained through the use of the principle of justifiable granularity being either an interval or some type of a fuzzy set defined over the unit interval gives rise to type-2 fuzzy sets. In the first case we form interval-valued fuzzy sets with membership intervals given by $[u_-, u_+]$. In the second case, we end up having triangular fuzzy sets defined in the unit interval.

4 A Numerical Example

The family of fuzzy decisions $F_1, F_2, \ldots, F_9$ defined in the space of 11 options is shown in Table 1. The particular rows correspond to $F_1, F_2, \ldots, F_9$, and the columns to the particular options

The distance function between the fuzzy sets is implemented via the Hamming distance. Below the values of the indicator Q for particular individuals are presented:

F_1	F_2	F_3	F_4	F_5	F_6	F_7	F_8	F_9
0.4525	0.4834	0.4837	0.4524	0.4779	0.4592	0.4680	0.4985	0.5304

For F_4, which has lowest value of this indicator, we obtain the following weight vector: $W(e_4)$ = [0.087 0.108 0.183 1.00 0.076 0.094 0.129 0.090 0.228], which indicates that F_9 and F_3 are the most consistent with F_4, while F_5 and F_8 exhibit the least consistency with F_4.

Table 1. Example of 9 individual fuzzy decisions defined in an 11-dimensional space of options; the consecutive rows of the matrix concern the individual fuzzy decisions

$$\begin{bmatrix} 0.90 & 0.90 & 0.80 & 0.50 & 0.50 & 0.50 & 0.78 & 0.10 & 1.00 & 0.40 & 0.20 \\ 0.80 & 0.70 & 0.40 & 0.90 & 0.60 & 0.80 & 0.20 & 0.70 & 1.00 & 0.20 & 0.22 \\ 0.70 & 0.70 & 0.30 & 0.50 & 0.80 & 1.00 & 0.70 & 0.34 & 0.32 & 0.30 & 0.61 \\ 0.60 & 0.60 & 0.85 & 0.50 & 0.80 & 0.90 & 0.40 & 0.40 & 0.50 & 0.90 & 0.87 \\ 0.50 & 0.20 & 0.20 & 0.85 & 0.90 & 0.90 & 1.00 & 0.12 & 0.80 & 0.40 & 0.33 \\ 0.40 & 0.00 & 0.10 & 0.05 & 0.90 & 0.20 & 0.60 & 0.51 & 0.90 & 0.95 & 0.70 \\ 0.30 & 0.20 & 0.20 & 0.60 & 0.20 & 0.20 & 0.70 & 0.57 & 1.00 & 0.69 & 1.00 \\ 0.20 & 0.70 & 0.70 & 0.10 & 0.61 & 0.00 & 0.00 & 0.11 & 0.80 & 0.00 & 1.00 \\ 0.00 & 0.70 & 0.84 & 0.30 & 0.40 & 0.40 & 0.30 & 0.95 & 1.00 & 0.15 & 0.00 \end{bmatrix}$$

Let us now consider the indicator Q for two element subset of E. In particular, let us analyze the subset $\{F_2, F_4\}$. The weights matrix, cf. (6), takes the following form (for convenience we show its two rows separately):

$W_1 = [0.142\ \ 1.00\ \ 0.076\ \ 0.118\ \ 0.142\ \ 0.142\ \ 0.117\ \ 0.142\ \ 0.117]$

$W_2 = [0.114\ \ 0.153\ \ 0.157\ \ 1.00\ \ 0.114\ \ 0.143\ \ 0.100\ \ 0.114\ \ 0.100]$

and we can see that, in the case of W_1, the most consistent with F_2 are F_1, F_5, F_6 and F_8ʾ , while in the case of W_2 the most consistent with F_4 are F_3 and F_2.

5 Conclusions

Group decision making via reaching consensus plays an important role in many decision making scenarios. The task of arriving at consensus, even meant – as we assume here – in a flexible way may be a real challenge. Consensus has to be obtained during discussion which is expected to incur some revisions of individual preferences. Thus the role of a group decision support system in such a setting is to support the group members with any kind of information that may help to clarify preferences, focus discussion on most important aspects of the decision problem, etc. In this paper we propose a set of new indicators that may be of use.

Moreover, since we have multiple decision makers who present their testimonies as fuzzy decisions, the aggregation of their individual fuzzy decisions is a crucial task and may be a subject of an additional analysis as it is an important output of the session but this issue is not yet explicitly considered here.

Acknowledgments

Support from the Ministry of Science and Higher Education under Grant N N519 404734 is gratefully acknowledged.

References

1. Kacprzyk, J., Zadrożny, S.: Soft computing and Web intelligence for supporting consensus reaching. Soft Computing 14(8), 833–846 (2010)
2. Türksen, I.B.: Type 2 representation and reasoning for CWW. Fuzzy Sets and Systems 127, 17–36 (2002)
3. Mendel, J.M.: On answering the question, Where do I start in order to solve a new problem involving interval type-2 fuzzy sets? Information Sciences 179(19), 3418–3431 (2009)
4. Zhou, S.M., Chiclana, F., John, R.I., Garibaldi, J.M.: Type-1 OWA operators for aggregating uncertain information with uncertain weights induced by type-2 linguistic quantifiers. Fuzzy Sets and Systems 159(24), 3281–3296 (2008)
5. Kacprzyk, J., Fedrizzi, M.: A 'soft' measure of consensus in the setting of partial (fuzzy) preferences. European Journal of Operational Research 34, 316–325 (1988)
6. Kacprzyk, J., Fedrizzi, M.: A 'human-consistent' degree of consensus based on fuzzy login with linguistic quantifiers. Mathematical Social Sciences 18(3), 275–290 (1989)

7. Zadrożny, S., Kacprzyk, J.: An internet-based group decision and consensus reaching support system. In: Yu, X., Kacprzyk, J. (eds.) Applied Decision Support with Soft Computing, pp. 263–275. Springer, Heidelberg (2003)
8. Zadrożny, S.: An approach to the consensus reaching support in fuzzy environment. In: Kacprzyk, J., Nurmi, H., Fedrizzi, M. (eds.) Consensus under Fuzziness, pp. 83–109. Kluwer, Boston (1996)
9. Herrera-Viedma, E., Martinez, L., Mata, F., Chiclana, F.: A consensus support system model for group decision-making problems with multi-granular linguistic preference relations. IEEE Trans. on Fuzzy Systems 13(5), 644–658 (2005)
10. Herrera-Viedma, E., Mata, F., Martınez, L., Perez, L.G.: An adaptive module for the consensus reaching process in group decision making problems. In: Torra, V., Narukawa, Y., Miyamoto, S. (eds.) MDAI 2005. LNCS (LNAI), vol. 3558, pp. 89–98. Springer, Heidelberg (2005)
11. Cholewa, W.: Aggregation of fuzzy opinions - an axiomatic approach. Fuzzy Sets and Systems 17, 249–258 (1985)
12. Montero de Juan, F.J.: Aggregation of fuzzy opinion in a non-homogeneous group. Fuzzy Sets and Systems 25, 15–20 (1987)
13. Dubois, D., Koning, J.-L.: Social choice axioms for fuzzy set aggregation. Fuzzy Sets and Systems 43, 257–274 (1991)
14. Fung, L.W., Fu, K.S.: An axiomatic approach to rational decision making in a fuzzy environment. In: Zadeh, L., Fu, K.S., Tanaka, T., Shimura, M. (eds.) Fuzzy Sets and Their Applications to Cognitive and Decision Processes, pp. 227–256. Academic Press, New York (1975)
15. Dubois, D., Prade, H.: A review of fuzzy set aggregation connectives. Information Sciences 36, 85–121 (1985)
16. Choudhury, A.K., Shankar, R., Tiwari, M.K.: Consensus-based intelligent group decision-making model for the selection of advanced technology. Decision Support Systems 42(3), 1776–1799 (2006)
17. Herrera, F., Martínez, L., Sánchez, P.J.: Managing non-homogeneous information in group decision making. European Journal of Operational Research 166(1), 115–132 (2005)
18. Tiryaki, F., Ahlatcioglu, B.: Fuzzy portfolio selection using fuzzy analytic hierarchy process. Information Sciences 179(1-2), 53–69 (2009)
19. Tsabadze, T.: A method for fuzzy aggregation based on group expert evaluations. Fuzzy Sets and Systems 157(10), 1346–1361 (2006)
20. Bargiela, A., Pedrycz, W.: Granular Computing. An Introduction. Springer, Heidelberg (2002)
21. Pedrycz, W.: Granular computing in multi-agent systems. In: Wang, G., Li, T., Grzymala-Busse, J.W., Miao, D., Skowron, A., Yao, Y. (eds.) RSKT 2008. LNCS (LNAI), vol. 5009, pp. 3–17. Springer, Heidelberg (2008)

Multiagent Decision Making, Fuzzy Prevision, and Consensus

Antonio Maturo[1] and Aldo G.S. Ventre[2]

[1] Department of Social Sciences, University of Chieti-Pescara, via dei Vestini, 31, 66013, Chieti, Italy
amaturo@unich.it, antmatya@yahoo.it
[2] Department of Culture of the Project and Benecon Center, Second University of Napoli, via San Lorenzo, 1, 81031, Aversa, Italy
aldo.ventre@unina2.it, aldoventre@yahoo.it

Abstract. Some multi-objective and multi-person decision making models are introduced. Objectives and alternatives are considered as elements of two algebras of events, and weights and scores are assumed to be values of crisp or fuzzy measures. Then the problem of verifying coherence is considered and the aggregation of scores is performed with respect to a suitable t-conorm. Moreover decision makers are identified with points of a metric space, and consensus in a group is obtained if the pairwise distances of the elements of the group do not exceed a fixed threshold. Procedures for enhancing, or achieving consensus are considered. Finally, further multi-objective and multi-person decision making models are introduced based on the concept of prevision. Precisely objectives and alternatives are assumed to be random numbers and weights and scores crisp or fuzzy previsions.

Keywords: Multiagent decision making, fuzzy measure, coherence, fuzzy prevision, metric spaces, consensus.

1 Introduction

Procedures for activating, enhancing and achieving consensus are widely dealt with in literature (see, e.g., [1], [2], [3], [4], [5], [6]). The assessment of scores is often performed by applying the Analytic Hierarchy Process [7], [8]. In Sec. 2 we will remark that normalization conditions utilized in AHP include the implicit assumptions that the objectives behave as incompatible and exhaustive events; furthermore similar assumptions hold also for the alternatives. Moreover the weights of the objectives define a finitely additive probability distribution, the scores of the alternatives with respect to an objective form a conditional probability distribution, and the weighed scores of the alternatives with respect to the objectives define a probability distribution on the pairs (alternative, objective).

We here wish to overcome these implicit assumptions and exhibit a more general and flexible Group Decision Model, in which:

E. Hüllermeier, R. Kruse, and F. Hoffmann (Eds.): IPMU 2010, Part II, CCIS 81, pp. 251–260, 2010.

- objectives or alternatives can be compatible or not exhaustive events;
- the scores are not necessarily probabilities, but, in general, they are suitable and coherent fuzzy measures (see, e.g., [9], [10], [11], [12]).

In the last Sec. a different point of view is introduced. Objectives (and/or alternatives) are seen as finite de Finetti random numbers [13]. We recall that the concept of de Finetti random number generalizes that of event; then it is a logical, non-probabilistic, concept. Indeed, an event E can be considered as a real function in $\{E, E^c\}$ with values on {0, 1}, while a de Finetti random number is any real function defined in a partition of the certain event. The extension of the concept of coherent probability to random numbers is that of coherent prevision [13].

We present also an extension of the coherent fuzzy measure to de Finetti random numbers, by considering the fuzzy prevision, introduced in some papers of ours ([14], [15], [16]).

2 Multiagent and Multiobjective Crisp Decision Making Models

The classical multiobjective decision making model is based on a triplet (A, O, S), where $A = \{A_1, A_2, \ldots A_m\}$ is the set of the *alternatives*, $O = \{O_1, O_2, \ldots O_n\}$ is the set of the *objectives*, and S is the *matrix of scores*, whose rows are the alternatives, and columns the objectives. The element s_{ij} of S is a nonnegative real number that measures to what extent the alternative A_i satisfies the objective O_j. All the scores s_{ij} belong to the same scale, usually the interval [0, 1]. The classical procedure to obtain the global score of the alternative A_i is as follows:

- a weight w_j is assigned to every objective O_j, where w_j is a nonnegative real number;
- the global score $s(A_i)$ of the alternative A_i is the sum, on the index j, of the products $w_j s_{ij}$, i.e.,

$$s(A_i) = w_1 s_{i1} + w_2 s_{i2} + \ldots + w_n s_{in}. \tag{1}$$

The preferred alternative is that having the maximum global score. Many Authors, especially when adopting AHP procedure [7], assume the normality conditions:

$$w_1 + w_2 + \ldots + w_n = 1, \tag{2}$$

$$\forall j \in \{1, 2, \ldots, n\},\ s_{1j} + s_{2j} + \ldots + s_{mj} = 1. \tag{3}$$

We here propose that objectives and alternatives play the role of the events in probability. So the objectives are considered as subsets of a universal set U, whose elements are called *micro-objectives*, and the alternatives are assumed to be subsets of another universal set V, whose elements are called *micro-alternatives*. These hypotheses imply that two objectives (resp. two alternatives) may have some elements in common or be disjoint. Moreover, the union of the objectives (resp. alternatives) can be equal to U (resp. V), but can also be strictly contained in U (resp. V).

In such a framework, the weight w_j of the objective O_j is the value $m_O(O_i)$ of a finitely additive measure m_O on the algebra a(O) generated by O and such that $m_O(U)=1$, i.e., m_O must be a de Finetti coherent probability on O. The condition (2) is the equivalent of the "implicit assumption" that the objectives are considered "pairwise disjoint and such that their union is U"; briefly, to say this *à la* de Finetti, they are assumed to be "a partition of the certain event U".

Similarly, for every fixed objective O_j, the score s_{ij}, $i \in \{1, 2, \dots, m\}$ is the value $\mu(A_i/O_j)$ of a conditional finitely additive measure μ on A×O. If the alternatives are assumed to be "pairwise disjoint and such that their union is V", i.e., they are "a partition of the certain event V", then the conditions (**3**) are the de Finetti coherence conditions for probability conditional assessments [13], [17], [18].

In this line of thinking the function m: $(A_i, O_j) \in A \times O \to w_j s_{ij} \in [0, 1]$ is a coherent probability assessment on A×O. The conditions (**2**) and (**3**) are the coherence conditions on m in the hypothesis that A×O is a partition of the certain event V×U.

If the hypotheses that A and O are partitions of the certain event are not verified, coherence conditions different from (**2**) and (**3**) are obtained, dependent on the logical conditions among the events (A_i, O_j). These conditions reduce to the existence of solutions of a suitable linear system, whose equations have the numbers $w_j s_{ij}$ as known terms.

If m is a coherent probability assessment on A×O, let m* be the extension of m to the algebra a(A×O) generated by A×O. For every $i \in \{1, 2, \dots, m\}$, the global value $s(A_i)$ of the alternative A_i is the probability of the union, respect to j, of the events (A_i, O_j). If these events are pairwise disjoint, then the value $s(A_i)$ is given by formula (**1**), otherwise we have to consider a different formula, dependent on the logical relations among the events O_j (see, e.g., [13], [17], [18], [19], [20], [21]).

3 Multiagent and Multiobjective Fuzzy Decision Making Models

Let us recall, from [9], [10], [11], [12], [22], some definitions and results.

Definition 1. A *t-conorm* ⊕ is a binary operation on the interval [0, 1] non decreasing in each argument, associative, commutative and having 0 as neutral element. A t-conorm is said to be *Archimedean* if it is continuous and, for every x in (0, 1), $x \oplus x > x$. An Archimedean t-conorm is said to be *strict* if it is strictly increasing in the open square $(0, 1)^2$.

The following representation theorem [22] holds.

Theorem 1. A binary operation ⊕ on [0, 1] is an Archimedean t-conorm if and only if there exists a strictly increasing and continuous function g: $[0, 1] \to [0, +\infty]$, with g(0)=0, and such that

$$x \oplus y = g^{(-1)}(g(x) + g(y)). \tag{4}$$

Function $g^{(-1)}$ denotes the pseudo-inverse of g, i.e.:

$$g^{(-1)}(x) = g^{-1}(\min(x, g(1))). \tag{5}$$

Moreover ⊕ is strict if and only if g(1) =+∞. The function g, called an *additive generator* of ⊕, is unique up to a positive constant factor.

Example 1. For $\lambda > -1$, x, y∈[0, 1], the Sugeno t-conorm [9] is defined as:

$$x \oplus_\lambda y = \min(x + y + \lambda\, x\, y, 1). \tag{6}$$

It is a non strict Archimedean t-conorm with additive generator

$$g_\lambda(x) = (\log(1 + \lambda\, x))/\lambda. \tag{7}$$

In particular for $\lambda = 0$, (6) reduces to the *bounded sum*:

$$x \oplus_0 y = \min(x + y, 1). \tag{8}$$

with additive generator $g_0(x) = x$, and, as $\lambda \to -1$ Sugeno t-conorm reduces to the *algebraic sum*:

$$x \oplus_{-1} y = \min(x + y - x\, y, 1), \tag{9}$$

that is a strict Archimedean t-conorm with additive generator $g_{-1}(x) = -\log(1-x)$.

Definition 2. Let X be a universal set and F a family of subsets of X containing Ø, X. A set function m: F → [0, 1] with m(Ø) = 0 and m(X) = 1, is said to be:

(a) a *normalized measure* (or *simple fuzzy measure*) on (X, F) if:

$$\forall A, B \in F,\ A \subseteq B \Rightarrow m(A) \le m(B) \quad \text{(monotonicity)} \tag{10}$$

(b) a *decomposable measure* on (X, F) with respect to a t-conorm ⊕, or ⊕-*decomposable measure* if:

$$\forall A, B \in F,\ A \cup B \in F,\ A \cap B = \emptyset \Rightarrow m(A \cup B) = m(A) \oplus m(B) \quad (\oplus\text{ - additivity}) \tag{11}$$

Definition 3. We say that a ⊕-decomposable measure m on (X, F) is *coherent* if there exists an extension m* of m to the algebra a(F) generated by F.

Remark 1. The ⊕ - decomposable measures generalize the finitely additive probabilities considered in [13], [17], [18]. Indeed a finitely additive probability is a ⊕ - decomposable measure m, with ⊕ the Sugeno t-conorm with $\lambda = 0$, satisfying the further property:

$$\forall A, B \in a(F),\ A \cap B = \emptyset \Rightarrow m(A) + m(B) \le 1. \tag{12}$$

Moreover definition 3 introduces an extension of the de Finetti coherence to the ⊕-decomposable measures.

Let us also remark that every decomposable measure is also a simple fuzzy measure. Then decomposable measure seems to be the most general reasonable extension of finitely additive probability.

From now on we assume that every considered t-conorm $\oplus$ is nonstrict Archimedean with additive generator g. The condition (**12**) leads us to introduce the following definition.

Definition 4. We say that a $\oplus$-decomposable measure m on (X, F) is *g-bounded* if:

$$\forall A, B \in a(F),\ A \cap B = \emptyset \ \Rightarrow\ g(m(A)) + g(m(B)) \leq g(1). \tag{13}$$

Remark 2. From (**13**) a finitely additive probability can be characterized as a *g-bounded* Sugeno t-conorm with $\lambda = 0$.

Condition (**13**) gives rise to the possibility to extend the normality conditions in a fuzzy context.

3.1 Decision Making with Objectives and Alternatives That Are Partitions of the Certain Event

Like in Sec. 2, we assume that objectives and alternatives play the role of the events in probability. So the objectives are considered as subsets of a universal set U, whose elements are called *micro-objectives*, and the alternatives are assumed to be subsets of another universal set V, whose elements are called *micro-alternatives*.

We consider now the case that objectives and alternatives are partitions of the certain event. Then two objectives (resp. two alternatives) are disjoint. Moreover, the union of the objectives (resp. alternatives) is equal to U (resp. V).

In the general context of $\oplus$ - decomposable measures we can assume the weight w_j of the objective O_j is the value $m_O(O_j)$ of a $\oplus$ - decomposable and g-bounded measure m_O on the algebra a(O) generated by O. The coherence condition is an extension of (**2**), precisely, from Weber's classification theorem (see, e.g., [11], [19], [20], [21]), it is given by formula:

$$g(w_1) + g(w_2) + \ldots + g(w_n) = g(1). \tag{14}$$

Similarly, for every fixed objective O_j, every score s_{ij}, $i \in \{1, 2, \ldots, m\}$ is the value $\mu(A_i/O_j)$ of a conditional $\oplus$ - decomposable and g-bounded measure μ on A×O. If the alternatives are assumed to be pairwise disjoint and such that their union is V, i.e., they are a partition of the certain event V, then the coherence conditions are extension of (**3**) and are given by the formula:

$$\forall j \in \{1, 2, \ldots, n\},\ g(s_{1j}) + g(s_{2j}) + \ldots + g(s_{mj}) = g(1). \tag{15}$$

In such a framework, for every alternative A_i, the global score of A_i is given by

$$s(A_i) = w_1 \otimes s_{i1} \oplus w_2 \otimes s_{i2} \oplus \ldots \oplus w_n \otimes s_{in}, \tag{16}$$

where $\otimes$ is a suitable t-norm, e.g. the *conjugate* t-norm [11].

3.2 Decision Making with Objectives and Alternatives Not Necessarily Partitions of the Certain Event

If the hypotheses that A and O are partitions of the certain event are not verified, we have coherence conditions different from (**14**) and (**15**), and dependent on the logical relations among the events A_i or O_j.

Let C_t, t = 1, 2, ... , s be the set of atoms of the objectives and let $a_{jt} = 1$, if $C_t \subseteq O_j$ and $a_{jt} = 0$, if $C_t \subseteq O_j^c$. The assessment of weights w_j, $0 \leq w_j < 1$ over the events O_j, is *coherent* w. r. to a ⊕-decomposable and g-bounded measure m, with additive generator g, if there is a solution $x = (x_1, x_2, \ldots, x_s) \in [0, 1]^s$ of the following system:

$$a_{j1}g(x_1) + a_{j2}g(x_2) + \ldots + a_{js}g(x_s) = g(w_j),\ j = 1,\ldots, n, \tag{17}$$

with the condition

$$g(x_1) + g(x_2) + \ldots + g(x_s) = g(1). \tag{18}$$

Analogous coherence conditions hold related to the coherence of the assessment of scores s_{ij}, i = 1, 2, … , m, of the alternatives with respect to the objective O_j.

Let K_r, r = 1, 2, ... , h, be the set of atoms of the alternatives and $b_{ir} = 1$, if $K_r \subseteq A_i$ and $b_{ir} = 0$, if $K_r \subseteq A_i^c$. The assessment of scores s_{ij} of the alternatives with respect to the objective O_j, $0 \leq s_{ij} < 1$, is *coherent* w. r. to a ⊕-decomposable and g-bounded measure m, with additive generator g, if there is a solution $z^j = (z_1^j, z_2^j, \ldots, z_h^j) \in [0, 1]^h$ of the following system:

$$b_{i1}g(z_1^j) + b_{i2}g(z_2^j) + \ldots + b_{ih}g(z_h^j) = g(s_{ij}),\ i = 1,\ldots, m, \tag{19}$$

with the condition

$$g(z_1^j) + g(z_2^j) + \ldots + g(z_h^j) = g(1). \tag{20}$$

If the coherence conditions are satisfied, then the global score of the alternative A_i is given by the formula:

$$s(A_i) = d_1\ (x_1 \otimes s_{i1}) \oplus d_2\ (x_2 \otimes s_{i2}) \oplus \ldots \oplus d_s\ (x_s \otimes s_{is}), \tag{21}$$

where $d_t = 1$ if the atom C_t is contained in at least an objective, and $d_t = 0$ otherwise.

In general the system (**17**) has several solutions, and, for every atom C_t, there is an interval $[a_t, b_t]$ such that a_t and b_t are, respectively, the minimum and the maximum value of x_t such that there exists a solution $x = (x_1, x_2, \ldots, x_s)$ of the system (**17**).

Then there is uncertainty about the values x_t of the formula (**21**) and we can think that every number x_t must be replaced with a suitable triangular fuzzy number x_t* having the interval $[a_t, b_t]$ as support. Zadeh extension based operations can be replaced with alternative fuzzy operations preserving the triangular shape (see, e.g., [23], [24], [25], [26], [27]). Then the global score of the alternative A_i is the triangular fuzzy number s*(A_i) given by:

$$s^*(A_i) = d_1\ (x_1^* \otimes s_{i1}) \oplus d_2\ (x_2^* \otimes s_{i2}) \oplus \ldots \oplus d_s\ (x_s^* \otimes s_{is}). \tag{22}$$

4 Consensus Reaching

In the assumed model (Sec.2), where objectives and alternatives are seen as events in probability, the scores of the alternatives are probabilities. In a multiobjective multi-agent decision making context, when, e. g., a committee is charged of making a decision of social relevance, a particular ranking of the alternatives is determined by each agent. In the literature (see, e. g., [1], [2], [3], [5], [6], [8], we refer throughout the present Sec.) dealing with such a context, an alternative ranking procedure widely used is AHP [7].

In order to reach a collective satisfying decision, the members of the committee (or a majority of them) have to agree about the decision. Usually a collective decision is laboriously built; indeed it is the fruit of compromises and negotiations, and possibly the action of a chairman external to the committee.

What is needed a collective decision to be accepted inside the committee and recognized by a social group, is the consensus, or a good degree of consensus, reached among the members of the decision making group.

Like discussions of the real life, the debates in a committee produce changes in the points of view, or the positions, of the decision makers. As a result, any two positions may move closer or farther when the debate develops. A geometrical modelling (see refs above) for such a situation looks at each "position" as a point of a Euclidean space whose coordinates are alternative scores assessed by each decision maker, the difference of the positions of two agents is measured by the distance between representative points, the dynamics, going closer and farther, is monitored by some convergence to or divergence from a suitable ideal central point. The set of the points, that represent the rankings of the decision makers, form a cloud that, during the debate, is going deformed.

When consensus is reached a suitable subset of points (a majority) in the cloud concentrates into a spherical neighbourhood. Of course, the ranking of the alternatives and, as a consequence, the coordinates of the points, depend on the adopted ranking procedure. If the linear operations, proper of the AHP procedure, i. e. the usual multiplication and addition, are replaced by a triangular norm and the conjugate conorm, the cloud of points follows a new trajectory, and possibly reaching consensus is subject to a different lot.

5 Prevision Based Decision Making Models

A different and more general point of view consists in considering objectives and alternatives are bounded de Finetti random numbers. Then every objective (resp. alternative) is characterized by a function X: $\Pi_X \rightarrow R$, where Π_X is a partition of the universal set U (resp. V) of micro-objectives (resp. micro-alternatives) and a real number X(E) is associated to every element E of the partition. The number X(E) represents the utility of E in the objective (resp. alternative).

In such a framework the weights w_j associated to the objectives can be interpreted as the *previsions* of the objectives.

We recall [13], [14], [15], [16] that a *prevision* P on a set S of de Finetti bounded random numbers is a function P: S $\rightarrow$ R such that:

(P1) for every $X \in S$, $\inf X \leq P(X) \leq \sup X$;
(P2) for every $X, Y \in S$, $X + Y \in S \Rightarrow P(X + Y) = P(X) + P(Y)$;
(P3) for every $a \in R$, $X \in S$, $a X \in R \Rightarrow P(a X) = a P(X)$.

A prevision P on S is said to be *coherent* if there exists an extension of P to the vector space of the linear combinations of elements of S.

If for every element X of S, the range of X is contained in {0, 1}, then every X can be identified with the event $X^{-1}(1)$, the union of the events $E \in \Pi_X$ such that $X(E) = 1$, and prevision reduces to finitely additive probability. Coherence conditions (P1), (P2), (P3) reduce to the ones of de Finetti coherent probability.

Then, if the objectives are bounded de Finetti random numbers, the coherence conditions (**2**) or (**14**) are replaced by coherence conditions of the prevision on the set of objectives, i.e., by the existence of an extension of the assessment $(w_1, w_2, \ldots, w_n)$ to the vector space generated by the objectives.

In an analogous way, for every objective O_j, the scores s_{ij}, $i = 1, 2, \ldots, m$, are an assessment of (conditional) prevision on the alternatives. Then coherence conditions (**3**) or (**15**) are replaced by coherence prevision conditions on the alternatives, i.e., for every objective O_j, by the existence of an extension of the assessment $(s_{1j}, s_{2j}, \ldots, s_{mj})$ of (conditional) scores of alternatives to the vector space generated by the alternatives.

In this framework, for every alternative A_i, we can assume the formula (**1**) gives the global score $s(A_i)$ of A_i, with a new meaning in terms of prevision. Precisely the number $s(A_i)$ is the sum, with respect to j, of the previsions s_{ij} of the random numbers associated to the pairs (alternative A_i, objective O_j) multiplied by the scalars w_j.

As an extension of the concepts of *simple fuzzy measure* and *decomposable measures*, referring to events, we can introduce those of *simple fuzzy prevision* and *decomposable prevision* applied to bounded de Finetti random numbers [14], [15], [16].

Let S be a set of bounded random numbers with universe U containing the null function $0: U \rightarrow 0$, and the unity function $1: U \rightarrow 1$.

We define *simple fuzzy prevision* on S every function $P: S \rightarrow R$ such that:
(SFP1) for every $X \in S$, $\inf X \leq P(X) \leq \sup X$;
(SFP2) for every $X, Y \in S$, $X \leq Y \Rightarrow P(X) \leq P(Y)$.

Let P be a simple fuzzy prevision on S. If J is an interval containing the range of P and $\oplus$ is an operation on J we define P as a *$\oplus$-decomposable prevision* on (U, S) if:

$$\forall X, Y \in S, X + Y \in S \Rightarrow P(A + B) = P(A) \oplus P(B) \quad (\oplus\text{ - additivity}) \tag{23}$$

A $\oplus$-decomposable prevision P on S is said to be *coherent* if there exists an extension of P to the vector space of the linear combinations of elements of S.

The operation $\oplus$ is defined as an extension of t-conorm and can be defined an extension of the concept of additive generator [16]. By utilizing a suitable $\oplus$-decomposable prevision, we can assume the global score of every alternative A_i is given by the formula:

$$s(A_i) = w_1 s_{i1} \oplus w_2 s_{i2} \oplus \ldots \oplus w_n s_{in}. \tag{24}$$

The global score $s(A_i)$ of the alternative A_i is the $\oplus$-sum, on the index j, of the $\oplus$-decomposable previsions s_{ij} of the random numbers associated to the pairs (alternative A_i, objective O_j) multiplied by the scalars w_j.

Of course, in formula (**24**) the multiplication can be replaced by a suitable operations ⊗, defined as an extension of the concept of t-norm.

6 Conclusions

The main motivation that leads us to consider previsions is the representation of an objective (resp. alternative) as a set of micro-objectives (resp. micro-alternatives) implies that the utility of the objective w.r.to a micro-objective is either complete or null. But, in general, the utility of an objective O_j w.r.to a micro-objective is partial, and the function X_j: U → [0, 1] that associates the utility of O_j to every micro-objective ω w.r.to ω is a de Finetti random number.

For instance the objective "environmental respect" has different importance for the micro-objectives "immediate scholastic performance" or "perspective of health after 10 years". Analogously the alternative "choosing motorway" has different utilities for the micro-alternatives "choosing food in the travel" and "choosing car for travel".

Then the interpretation of objectives as de Finetti random numbers, and weights and scores as previsions, seems to arise in a natural way from an in-depth analysis of the decision making problem. The prevision measures the weight (resp. score) of an objective and is a global and intuitive summary of the weights (resp. scores) of micro-objectives (usually unknown, guessed, not calculated) and their utility respect to the objective. The coherence conditions about previsions assure their mathematical and logical consistence for applications.

References

1. Ehrenberg, D., Eklund, P., Fedrizzi, M., Ventre, A.G.S.: Consensus in distributed soft environments. Reports in Computer Science and Mathematics, Ser. A (88) (1989)
2. Carlsson, C., Ehrenberg, D., Eklund, P., Fedrizzi, M., Gustafsson, P., Lindholm, P., Merkurieva, G., Riissanen, T., Ventre, A.G.S.: Consensus in distributed soft environments. European J. Operational Research 61, 165–185 (1992)
3. Eklund, P., Rusinowska, A., De Swart, H.: Consensus reaching in committees. European Journal of Operational Research 178, 185–193 (2007)
4. Herrera-Viedma, E., Alonso, S., Chiclana, F., Herrera, F.: A Consensus Model for Group Decision Making with Incomplete Fuzzy Preference Relations. IEEE Transactions on Systems Fuzzy Systems 15(5), 863–877 (2007)
5. Maturo, A., Ventre, A.G.S.: Models for Consensus in Multiperson Decision Making. In: NAFIPS 2008 Conference Proceedings, New York, USA. IEEE Press, Los Alamitos (2008)
6. Maturo, A., Ventre, A.G.S.: Aggregation and consensus in multiobjective and multiperson decision making. International Journal of Uncertainty, Fuzziness and Knowledge-Based Systems 17(4), 491–499 (2009)
7. Saaty, T.L.: The Analytic Hierarchy Process. McGraw-Hill, New York (1980)
8. Maturo, A., Ventre, A.G.S.: An Application of the Analytic Hierarchy Process to Enhancing Consensus in Multiagent Decision Making. In: ISAHP 2009, Proceeding of the International Symposium on the Analytic Hierarchy Process for Multicriteria Decision Making, July 29- August 1, paper 48, pp. 1–12. University of Pittsburg, Pittsburgh (2009)

9. Sugeno, M.: Theory of fuzzy integral and its applications, Ph.D. Thesis, Tokyo (1974)
10. Banon, G.: Distinction between several subsets of fuzzy measures. Int. J. Fuzzy Sets and Systems 5, 291–305 (1981)
11. Weber, S.: Decomposable measures and integrals for Archimedean t-conorms. J. Math. Anal. Appl. 101(1), 114–138 (1984)
12. Berres, M.: Lambda additive measure spaces. Int. J. Fuzzy Sets and Systems 27, 159–169 (1988)
13. de Finetti, B.: Theory of Probability. J. Wiley, New York (1974)
14. Maturo, A., Tofan, I., Ventre, A.G.S.: Fuzzy Games and Coherent Fuzzy Previsions. Fuzzy Systems and A.I. Reports and Letters 10, 109–116 (2004)
15. Maturo, A., Ventre, A.G.S.: On Some Extensions of the de Finetti Coherent Prevision in a Fuzzy Ambit. Journal of Basic Science 4(1), 95–103 (2008)
16. Maturo, A., Ventre, A.G.S.: Fuzzy Previsions and Applications to Social Sciences. In: Kroupa, T., Vejnarová, J. (eds.) Proceedings of the 8th Workshop on Uncertainty Processing (Wupes 2009), Liblice, Czech Rep, September 19-23, pp. 167–175 (2009)
17. Coletti, G., Scozzafava, R.: Probabilistic Logic in a Coherent Setting. Kluver Academic Publishers, Dordrecht (2002)
18. Dubins, L.E.: Finitely additive conditional probabilities, conglomerability, and disintegrations. The Annals of Probability 3, 89–99 (1975)
19. Maturo, A., Squillante, M., Ventre, A.G.S.: Consistency for assessments of uncertainty evaluations in non-additive settings. In: Amenta, P., D'Ambra, L., Squillante, M., Ventre, A.G.S. (eds.) Metodi, modelli e tecnologie dell'informazione a supporto delle decisioni, pp. 75–88. Franco Angeli, Milano (2006)
20. Maturo, A., Squillante, M., Ventre, A.G.S.: Consistency for nonadditive measures: analytical and algebraic methods. In: Reusch, B. (ed.) Computational Intelligence, Theory and Applications, pp. 29–40. Springer, Berlin (2006)
21. Maturo, A., Squillante, M., Ventre, A.G.S.: Decision Making, fuzzy Measures, and hyperstructures. Advances and Applications in Statistical Sciences (to appear)
22. Ling, C.H.: Representation of associative functions. Publ. Math. Debrecen 12, 189–212 (1965)
23. Zadeh, L.: The concept of a linguistic variable and its application to approximate reasoning. Inf. Sci. 8, Part I:199–249, Part 2: 301–357 (1975)
24. Zadeh, L.: The concept of a linguistic variable and its applications to approximate reasoning. Part III. Inf Sci. 9, 43–80 (1975)
25. Dubois, D., Prade, H.: Fuzzy numbers: An overview. In: Bedzek, J.C. (ed.) Analysis of fuzzy information, vol. 2, pp. 3–39. CRC-Press, Boca Raton (1988)
26. Yager, R.: A characterization of the extension principle. Fuzzy Sets Syst. 18, 205–217 (1986)
27. Maturo, A.: Alternative Fuzzy Operations and Applications to Social Sciences. International Journal of Intelligent Systems 24, 1243–1264 (2009)

A Categorical Approach to the Extension of Social Choice Functions

Patrik Eklund[1], Mario Fedrizzi[2], and Hannu Nurmi[3]

[1] Department of Computing Science, Umeå University, Sweden
peklund@cs.umu.se
[2] Department of Computer and Management Sciences, University of Trento, Italy
mario.fedrizzi@unitn.it
[3] Department of Political Science, University of Turku, Finland
hnurmi@utu.fi

Abstract. Are we interested in choice functions or function for choice? Was it *my choice* or did *I choose*? In the end it is all about sorts and operators, terms as given by the term monad over the appropriate category, and variable substitutions as morphisms in the Kleisli category of that particular term monad.

Keywords: Choice function, monad, Kleisli category, substitution.

1 Introduction

The theory of choice under uncertainty has been considered for a long time, starting from the monumental work of von Neumann and Morgenstern [19], one of the success stories of economic and social sciences. The theory rested on solid axiomatic foundations, formally based on the expected utility model of preferences over random prospects, and it stood ready to provide the theoretical framework for newly emerging paradigm of information revolution in economics and social sciences.

Even though there is a substantial body of evidence that decision makers systematically violate the basic tenets of expected utility theory, see e.g. [1,12], nevertheless the major impact of the effort of von Neumann and Morgenstern was that they settled the foundations for a new mathematical methodology that marked a turning point in the so called welfare economics and its mathematical framework of social choice theory.

The problem of modelling social choices involving conflicting interests and concerns has been explored for a long time, but social choice theory as a systemic discipline was born around the time of the French Revolution. As a matter of fact, the formal discipline of social choice was pioneered by French mathematicians Borda [3] and de Condorcet [6], who addressed, in rather mathematical terms, voting problems and related procedures.

It's widely agreed today that the most important advance in the theory of social choice during the past century was Arrow's discovery [2] that a few appealing criteria for social ranking methods are mutually incompatible. The crucial

E. Hüllermeier, R. Kruse, and F. Hoffmann (Eds.): IPMU 2010, Part II, CCIS 81, pp. 261–270, 2010.

technical advance in Arrow's approach that led to the impossibility theorem was the consideration of a variety of individual preference profiles that might arise in a choice process involving three or more individuals. Arrow's impossibility theorem generated a huge amount of research work in response including many other impossibility results [17], and also led, as Sen [23] pointed out, to the "diagnosis of a deep vulnerability in the subject that overshadowed Arrow's immensely important constructive program of developing a systematic social choice theory that could actually work".

After the introduction of the concept of fuzzy binary relation by Zadeh [24], the first applications of fuzzy sets to social choice appeared rather shortly and the concept of fuzzy rationality was introduced by Montero de Juan [15], as a mean to escape from impossibility theorems. So far, much literature now exists on many important results of the fuzzy counterpart of traditional social choice theory and the interested reader can find selected overviews in Kacprzyk and Fedrizzi [13], Carlsson et al [4], and Nurmi [21].

The paper is organized as follows. Section 2 describes relevant parts of the classical approach to social choice. In Section 3 we present a 'relations as mappings to powersets' approach to social choice. In Section 4 we discuss the pragmatics of choice underlying underlying the classical approaches and formalisms for choice functions. Section 5 then describes the categorical tools, monads and Kleisli categories, needed for describing functions for choice. Section 6 concludes the paper.

2 Classic Approach to Social Choice

The starting point for describing choice is a mapping

$$f : X_1 \times X_n \to Y$$

where agents $i \in \{1, \dots, n\}$ are choosing or arranging elements in sets X_i. The social choice related to all $x_i \in X_i, i = 1, \dots, n$ is then represented by $f(x_1, \dots, x_n)$ as an element in the set Y.

Usually, $X_1 = \dots = X_n = Y$, and the social choice function

$$f : X \times \dots \times X \to X \tag{1}$$

is e.g. said to respect unanimity if $f(x, \dots, x) = x$, i.e. if all the individual choices coincide and the resulting social choice indeed is that coincidation.

Concerning individual choice there is clear distinction between *I choose* and *my choice*, the former being the mechanism of choosing, sometimes considered to include the result of *me choosing*, and the latter being only the result of choosing. For social choice, f is then correspondingly the mechanism for *we choose*, or *somebody or something chooses on behalf of us*, and $f(x_1, \dots, x_n)$ is *our choice*. Rationality of choice [2] is then obviously in respective mechanisms for individual as well as social choices.

Traditional mathematical modelling of social choice deals with aggregation of preference or preference values. More specifically, it focuses on amalgamating individual opinions into a collective one. Formally, social choice rules have been construed as functions or correspondences. The earliest view, adopted by Arrow [2], is to study social welfare functions, the arguments of which are named components of social states. These are rules that map n-tuples of individual preferences (orderings [2]) into a collective preferences:

$$f : (X^m)^n \to X^m$$

Note that the individual preferences are m-tuples, i.e. maps n-tuples of m-tuples into m-tuples. The underlying assumption is that X is an ordering $(X, \preceq)$ with some suitable properties.

In the case of $\preceq$ being a total order, sometimes called connected relation, for any $(x_1, \ldots, x_m) \in X^m$, there is always a permutation $(x'_1, \ldots, x'_m)$ of $(x_1, \ldots, x_m)$ such that $x'_1 \succeq \cdots \succeq x'_m$. Equivalently, for all $x_1, x_2 \in X$, we always have either $x_1 \preceq x_2$ or $x_1 \succeq x_2$ (or both in case $x_1 = x_2$).

For an individual preference $x^{(i)} = (x_1^{(i)}, \ldots, x_m^{(i)})$, i.e. $x_1^{(i)} \succeq \cdots \succeq x_m^{(i)}$, the order of $x_k^{(i)}$ in the array $x^{(i)}$ then indicates the preference value of individual i for alternative k.

Note that the preference value in this case is an ordinal value and not a scale value. Sometimes choice functions are written

$$f : (R^m)^n \to R^m$$

i.e. using the real line, or some suitable closed interval within the real line, for the preference (scale) values, and $x_k^{(i)}$ is then interpreted as the real value assigned by the individual for the alternative. In the scale case, the alternative is thus mapped to a real value, whereas in the ordinal case, the symbol for the ordinal itself is used within the ordering. Using closed interval scales means using total orders when scale values are viewed as ordinal values in the total order of that closed interval. In this paper we prefer the ordinal view, which indeed does not exclude also using scale values. A scale value would then be an additional specification by the individual with respect to the alternatives. The literature is somewhat confusing from this point of view as we seldom see a distinction between a *decision-maker choosing* and a *decision-maker's choice*.

Example 1. Consider three individuals ι_1, ι_2, and ι_3 providing preferences for three alternatives A, B, and C. Let the individual preference relations or preference profiles be the following:

ι_1	ι_2	ι_3
A	B	B
C	C	C
B	A	A

i.e. ι_1 possesses the 3-tuple $x^{(1)} = (A, C, B)$, ι_2 the 3-tuple $x^{(2)} = (B, C, A)$, and ι_3 the 3-tuple $x^{(3)} = (B, C, A)$. Further, let the social choice be $f(x^{(1)}, x^{(2)}, x^{(3)}) = x^{(1)}$.

3 A 'Relations as Mappings to Powersets' Approach to Social Choice

Computing with preferences is less transparent with orderings built into the set X of alternatives. The ordering, as a relation, is made more explicit when viewing a (binary) relation ρ on X, i.e. $\rho \subseteq X \times X$ as its corresponding mapping $\rho : X \to PX$, where PX is the powerset of X. We then start with an unordered set of alternatives X, i.e. a set X of unrelated elements. The choice function f can be extended to the powersets of its domain and range, namely,

$$Pf : P[(X^m)^n] \to P[X^m]$$

Further, as

$$(P[X^m])^n \subseteq P[(X^m)^n]$$

we may also consider the well-defined restriction

$$Pf_{|S} : (P[X^m])^n \to P[X^m],$$

where $S = (P[X^m])^n$ is the set of all relations over X^m.

A social welfare function in this 'relations as mappings to powersets' approach is then

$$\varphi : (P[X^m])^n \to P[X^m],$$

In other words, social welfare functions, in this more transparent view, map individual preference orderings, or profiles, into collective preference orderings, and indeed not just mapping the m-tuples. In Arrow's original work [2], the mapping f is the social welfare mapping, including the assumption of underlying orderings.

A further consideration is weakening the constraint that the outcome of the social choice also is a preference. We may indeed want to restrict to the case where a unordered set of alternatives is the outcome of the social choice function. In this case we are dealing with mappings of the form

$$\phi : (P[X^m])^n \to PX$$

Thus, a social choice function specifies, for each subset of alternatives and a preference profile, a set of "chosen" or "best" alternatives. We are interested also in social decision functions

$$\psi : (P[X^m])^n \to X$$

i.e., a social decision function, a.k.a. resolute social choice function, assigning to each preference profile and subset of alternatives a unique alternative.

Example 2. In the profile of the above example the function

$$\phi(x^{(1)}, x^{(2)}, x^{(3)}) = \begin{cases} \{C\} & \text{if everyone ranks C first} \\ \{A, B\} & \text{otherwise} \end{cases}$$

specifies $\{A, B\}$ as the social choice. This is non-resolute, while the following is a resolute, i.e. social decision function:

$$\psi(x^{(1)}, x^{(2)}, x^{(3)}) = \begin{cases} \{C\} \text{ if person 3 ranks C first} \\ \{A\} \text{ otherwise} \end{cases}$$

The welfare and choice functions used in the above above examples are not intuitively reasonable or fair. In the first instance the welfare function always results in the collective preference relation that coincides with that of the first individual. It is thus an example of a dictatorial social welfare function. The example of a social choice function, in turn, seems biased in favor of some alternatives with respect to others. The example of social decision function seems to treat both individuals and alternatives in a biased way. Obviously, there is more to social choice than to define a correspondence or mapping from individual opinions to collective ones.

The most celebrated result in the social choice theory is undoubtedly Arrow's impossibility theorem which essentially proves the incompatibility of several choice desiderata. Its dramatic effect thus hinges on how desirable are those desiderata. It will be recalled that Arrow's desiderata are:

- unrestricted domain: the function is defined for any preference profile,
- Pareto condition: if everyone prefers alternative x to alternative y, then this is also the social preference,
- independence of irrelevant alternatives (the social preference between any two alternatives depends only on the individual preferences between those alternatives), and
- non-dictatorship: no individual alone determines the social preference relation irrespective of others.

The impossibility theorem states that no social welfare function satisfies all these desiderata[2,22]. It is noteworthy that apart from non-dictatorship the conditions are trivially satisfied by individual preference relations. So, it can be argued that the thinking underlying Arrow's approach is that social preferences are structurally similar to those of individuals.

Arrow's theorem deals with social welfare functions. Another classic result, viz. the Gibbard-Satterthwaite theorem, focuses on social decision functions [10,16]. It is also an incompatibility result. It states that all reasonably unbiased social decision functions are either manipulable or dictatorial. By reasonably unbiased function we mean one that is neutral (no alternative is discriminated for or against), anonymous (no individual is discriminated for or against) and non-trivial (for any alternative, one can construct a preference profile so that this alternative is the social choice). Manipulability of a social decision function means that there is a profile where an individual may benefit from misrepresenting his/her preferences, given the other individuals' votes. In somewhat different terminology, manipulability of a social decision function means that there is a profile so that sincere revelation of preferences by all individuals does not lead to a Nash equilibrium.

Example 3. Consider the following:

ι_1	ι_2	ι_3
A	B	C
B	C	A
C	A	B

Suppose that the choice function is the amendment procedure whereby the alternatives are voted upon in pairs so that the majority winner of the first contest faces the remaining alternative in the second comparison. The majority winner of the latter is then declared the overall winner. Assume that the agenda is: (1) A vs. B, and (2) the winner of (1) vs. C. With all voters voting according to their preferences, the overall winner is C. Suppose now that individual 1 misrepresent his preference by voting for B is the first vote. Ceteris paribus, this would lead to B becoming the overall winner. Hence the original outcome C is not a Nash equilibrium and, hence, the amendment procedure is manipulable.

The two theorems above perhaps the best known, but by no means the only ones in social choice theory (see e.g. [17]). Historically, the theorems are a relatively recent newcomer in the voting theory field. More down-to-earth approaches focus on specific desiderata and their absence or presence in the existing or proposed voting systems. Often the absence of an intuitively obvious desideratum is expressed in the form of a paradox. Perhaps the best one of these is Condorcet's paradox or the phenomenon of cyclic majorities. An instance of this paradox can be seen in the preceding example. The majority preference relation formed on the basis of paired comparison is obviously cyclic: $A \succeq B \succeq C \succeq A \ldots$. This means that whichever alternative is chosen, the majority of individuals would prefer some other alternative to it. Thus, a majority of voters is frustrated whatever the outcome. And yet, it is the majority rule that determines the winner at each stage of voting.

Another well-known paradox is related to plurality (one-person-one-vote) system. It is known as Borda's paradox.[1]

Example 4. Consider the following:

4 voters	3 voters	2 voters
A	B	C
B	C	B
C	A	A

With one-person-one-vote alternative A is likely to win. Yet, in pairwise majority comparisons it would be defeated by all other alternatives.

[1] Marquis de Condorcet and Chevalier de Borda were 18'th century member of the French Academy of Sciences. Their main contributions to the theory of voting can be found in [18].

What makes these two settings paradoxical is the unexpected or surprising outcome: the system does not seem to be working in the intended manner. Majority voting is expected to yield an unambiguous winner as it does when only two alternatives are at hand. The alternative voted for by a plurality of voters is expected to be the best also in pairwise comparisons. Similar paradoxical observations are the no-show paradox, additional support (monotonicity) paradox, inconsistency and various aggregation paradoxes (referendum and multiple elections paradoxes) [9,20].

4 Operator-Based Choice

Having made the important distinction between choice and mechanism for choice, we will at this point briefly mention some formalism involving signatures and their algebras, i.e. more clearly show where we will be syntactic and where we are semantic in our further discussion. A signature $\Sigma = (S, \Omega)$ consists of sorts, or types, in S, and operators in Ω. More precisely, Ω is a family of sets $(\Omega_n)_{n \leq k}$, where n is the arity of the operators in Ω_n. An operator $\omega \in \Omega_n$ is syntactically written as

$$\omega : s_1 \times \ldots \times s_n \to s$$

where $s_1, \ldots, s_n, s \in S$. Operators in Ω_0 are constants. Given a set of variables we may construct the set of all terms over the signature. This set is usually denoted $\mathsf{T}_\Omega X$, and its elements are denoted $(n, \omega, (t_i)_{i \leq n})$, $\omega \in \Omega_n$, $t_i \in \mathsf{T}_\Omega X, i = 1, \ldots, n$, or $\omega(t_1, \ldots, t_n)$.

In this algebraic formalism, ω is the mechanism, e.g. of choosing, and $\omega(t_1, \ldots, t_n)$ a result, or choice. Note that both the operator ω as well as the term $\omega(t_1, \ldots, t_n)$ are syntactic representations of mechanisms for choosing and choices. Algebras of signatures provide the semantics. Each sort $s \in S$ then has a semantic domain $A(s)$, a set representing all possible values for the sort. The semantics of ω is then a mapping

$$A(\omega) : A(s_1) \times \ldots \times A(s_n) \to A(s).$$

Note the distinction between $\times$ being syntactic and $\times$ semantically being the cartesian product[2].

For the choice function (1) we should then note that it is indeed its semantic representation which has an underlying choice operator in its signature.

Generally speaking, signatures are the basis for terms, which in turn provide the building blocks for sentences in a logic. Sentences do not suffice, as a logic additional needs a satisfaction relation $\models$, based on the algebraic models of the signature, and an entailment relation $\vdash$, where its power lies in the axioms of the logic and inference rules for entailment.

Coming back to rationality, it has been said ([14]) that behaviour is based on custom more than rationality. Thus we may intuitively say that custom is based

[2] The cartesian product of sets is the categorical product of objects in the category of sets.

on particular algebras acting as models and used in $\models$, whereas rationality is based on representable sentences motored by $\vdash$.

Extensions of choice including preferences is now either involving just the results or also the mechanisms. Using just the results means extending X to the set X^m of all m-tuples of elements in X. This is then based on the assumption that there are underlying relations (mechanisms) providing particular permutations. The choice function including preferences is then φ. However, this indeed hides the mechanisms of individual choice, and therefore the operators. Including preferences as mechanisms means representing the preference relation in a more general manner. In Section 5 we view relations as substitutions involving powersets and this opens up for more general views of relations.

5 Monads, Kleisli Categories and Substitutions

A monad (or triple, or algebraic theory) over a category $\mathtt{C}$ is written as $\mathbf{F} = (\mathsf{F}, \eta, \mu)$, where $\mathsf{F} : \mathtt{C} \to \mathtt{C}$ is a (covariant) functor, and $\eta : \mathsf{id} \to \mathsf{F}$ and $\mu : \mathsf{F} \circ \mathsf{F} \to \mathsf{F}$ are natural transformations for which $\mu \circ \mathsf{F}\mu = \mu \circ \mu\mathsf{F}$ and $\mu \circ \mathsf{F}\eta = \mu \circ \eta\mathsf{F} = id_\Phi$ hold.

A Kleisli category $\mathtt{C}_\mathbf{F}$ for a monad $\mathbf{F}$ over a category $\mathtt{C}$ is given with objects in $\mathtt{C}_\mathbf{F}$ being the same as in $\mathtt{C}$, and morphisms being defined as $Hom_{\mathtt{C}_\mathbf{F}}(X, Y) = Hom_\mathtt{C}(X, \mathsf{F}Y)$. Morphisms $f \colon X \rightharpoondown Y$ in $\mathtt{C}_\mathbf{F}$ are thus morphisms $f \colon X \to \mathsf{F}Y$ in $\mathtt{C}$, with $\eta_X \colon X \to \mathsf{F}X$ being the identity morphism. Composition of morphisms in $\mathtt{C}_\mathbf{F}$ is defined as

$$(X \xrightharpoondown{f} Y) \diamond (Y \xrightharpoondown{g} Z) = X \overset{\mu_Z \circ \mathsf{F}g \circ f}{\longrightarrow} \mathsf{F}Z.$$

Morphisms in $\mathtt{C}_\mathbf{F}$ are general *variable substitutions.*

Let $\mathtt{Set}$ be the category of sets and mappings, and $\mathsf{Set}(L)$, where L is a completely distributive lattice, be the category with objects being pairs (A, α) where $\alpha \colon A \to L$ and morphisms $(A, \alpha) \xrightarrow{f} (B, \beta)$ are mappings $f \colon A \to B$ such that $\beta(f(a)) \geq \alpha(a)$ for all $a \in A$. Note that Set is not isomorphic to $\mathsf{Set}(2)$, where $2 = \{0, 1\}$.

In the usual covariant powerset monad (P, η, μ), over $\mathtt{Set}$, we have $\mathsf{P}X$ being the powerset of X, $\eta_X(x) = \{x\}$ and $\mu_X(\mathcal{B}) = \bigcup \mathcal{B}$. The category of 'sets and relations', i.e. where objects are sets and morphisms $f : X \to Y$ are ordinary relations $f \subseteq X \times Y$ with composition of morphisms being relational composition, is isomorphic to the Kleisli category $\mathtt{Set}_\mathbf{P}$. Relations $R \subseteq X \times Y$ correspond exactly to substitutions $\rho_R : X \to \mathsf{P}Y$., i.e. elements of $Hom_{\mathtt{C}_\mathbf{P}}(X, Y)$.

For the construction of the term monads over $\mathsf{Set}(L)$ and Set, respectively, i.e. paradigms for non-classical variable substitutions with terms, see [11,7,8].

Example 5. Let $\mathtt{NAT} = (S_\mathtt{NAT}, \Omega_\mathtt{NAT})$ be the signature of natural numbers (or the signature for the "algebra of natural numbers"), i.e. $S_\mathtt{NAT} = \{nat\}$ and $\Omega_\mathtt{NAT} = \{\mathtt{0} : \to \mathtt{nat}, \mathtt{succ} : \mathtt{nat} \to \mathtt{nat}\}$.

In Set, for $\Omega = \Omega_\mathtt{NAT}$ we have $\Omega_0 = \{\mathtt{0} : \to \mathtt{nat}\}$ and $\Omega_1 = \{\mathtt{succ} : \mathtt{nat} \to \mathtt{nat}\}$. Further, $(\Omega_0)_\mathtt{Set} \times id^0 A = \Omega_0$ and $(\Omega_1)_\mathtt{Set} \times id^1 A = \{(1, \mathtt{succ}, a) \mid a \in A\}$.

For $\mathbf{T}_\Omega$ we have $T^0_\Omega A = A$ and $T^1_\Omega A = \coprod\{\Omega_0, (\Omega_1)_{\mathtt{Set}} \times id^1 A\} = \{(0, 0, ())\} \cup \{(1, \mathtt{succ}, a) \mid a \in A\}$. From this we then continue to $T^2_\Omega A = (\coprod_{n\leq 1}((\Omega_n)_{\mathtt{Set}} \times id^n) \circ \bigvee_{\kappa<2} T^\kappa_\Omega A = \{(0, 0, ())\} \cup \{(1, \mathtt{succ}, t) \mid t \in T^0_\Omega A \cup T^1_\Omega A\}$, and $T^\iota_\Omega A$ is unfolded similarly.

Correspondingly, in $\mathsf{Set}(L)$, we have $(\Omega_0, \vartheta_0)_{\mathtt{Set}(L)} \times id^0(A, \alpha) = (\Omega_0, \vartheta_0)$ and $(\Omega_1, \vartheta_1)_{\mathtt{Set}(L)} \times id^1(A, \alpha) = (\Omega_1, \vartheta_1) \times (A, \alpha) = (\Omega_1 \times A, \vartheta_1 \times \alpha)$, where $(\vartheta_1 \times \alpha)(\mathtt{succ}, a) = \vartheta_1(\mathtt{succ}) \wedge \alpha(a)$. For $\mathbf{T}_{(\Omega,\vartheta)}$ we then correspondingly have $T^0_{(\Omega,\vartheta)}(A, \alpha) = (A, \alpha) = (T^0_\Omega A, \beta^0)$, where $\beta^0 = \alpha$, and

$$\begin{aligned} T^1_{(\Omega,\vartheta)}(A,\alpha) &= (\coprod_{n\leq 1}((\Omega_n, \vartheta_n)_{\mathtt{Set}(L)} \times id^n) \circ \bigvee_{\kappa<1} T^\kappa_{(\Omega,\vartheta)}(A,\alpha) \\ &= \coprod\{(\Omega_0, \vartheta_0), (\Omega_1 \times A, \vartheta_1 \times \alpha)\} \\ &= (T^1_\Omega A, \beta^1), \end{aligned}$$

where $\beta^1((0, 0, ())) = \vartheta_0(0)$ and $\beta^1((1, \mathtt{succ}, a)) = \vartheta_1(\mathtt{succ}) \wedge \alpha(a)$. Similarly, from this we continue to

$$\begin{aligned} T^2_{(\Omega,\vartheta)}(A,\alpha) &= (\coprod_{n\leq 1}((\Omega_n, \vartheta_n)_{\mathtt{Set}(L)} \times id^n) \circ \bigvee_{\kappa<2} T^\kappa_{(\Omega,\vartheta)}(A,\alpha) \\ &= \coprod\{(\Omega_0, \vartheta_0), (\Omega_1, \vartheta_1) \times \bigvee_{\kappa<2} T^\kappa_{(\Omega,\vartheta)}(A,\alpha)\} \\ &= \coprod\{(\Omega_0, \vartheta_0), (\Omega_1, \vartheta_1) \times [T^0_{(\Omega,\vartheta)}(A,\alpha) \vee T^1_{(\Omega,\vartheta)}(A,\alpha)]\}, \end{aligned}$$

and $T^\iota_{(\Omega,\vartheta)}(A, \alpha)$ unfolds accordingly.

Setting up the generalized functions for choice is now as follows. Let $\mathtt{C}$ be the Kleisli category $\mathtt{Set}(L)_{\mathbf{T}_{(\Omega,\vartheta)}}$. Further, let $\mathcal{X} = Hom_{\mathtt{C}}(X, Y)$ be the corresponding set of substitutions capturing the notion of *me choosing*. The generalized function of choice is then

$$\varphi : (P[\mathcal{X}^m])^n \to P[\mathcal{X}^m],$$

Note the obvious next step to generalize P, as a monad, to any monad representing generalized relations in the best of ways. Monadicity thus resides both in *me choosing* as well is in *my choosing*, thus providing an overall monadic extension of social choice functions.

6 Conclusions

The categorical approach to generalization of the choice functions has several advantages. The underlying categories, carrying the uncertainties are exposed more clearly. Distinctions between *choosing* and *choice* is made more explicit. Further work in these directions will introduce more example, more pragmatics, in the area of applied social choice.

References

1. Allais, M.: Le comportment de l'homme rationel devant le risque, critique des postulates et axioms de l'ecole americaine. Econometrica 21, 446–503 (1953)
2. Arrow, K.J.: Social Choice and Individual Values. Wiley, New York (1951)
3. Borda, J.C.: Memoir sur les elections au scrutin, Histoire de l'Academie Royale des Sciences, Paris (1781)
4. Carlsson, C., Fedrizzi, M., Fullér, R.: Fuzzy Logic in Management. Kluwer Academic Publishers, Boston (2004)
5. Chichilnisky, G.: Social choice and the topology of spaces of preferences. Adv. Math. 37, 165–176 (1980)
6. de Condorcet, M.: Essai sur l'application de l'analyse a des decisions rendues a la pluralité des voix. L'Imprimerie Royale, Paris (1785)
7. Eklund, P., Gähler, W.: Fuzzy filter functors and convergence, Applications of category theory to fuzzy subsets. In: Rodabaugh, S.E., et al. (eds.) Theory and Decision Library B, pp. 109–136. Kluwer, Dordrecht (1992)
8. Eklund, P., Galán, M.A., Kortelainen, J., Stout, L.N.: Paradigms for non-classical substitutions. In: Proc. 39th IEEE International Symposium on Multiple-Valued Logic (ISMVL 2009), Naha, Okinawa (Japan), May 21-23, pp. 77–79 (2009)
9. Fishburn, P.C., Brams, S.J.: Paradoxes of preferential voting. Mathematics Magazine 56, 201–214 (1983)
10. Gibbard, A.: Manipulation of voting schemes: a general result. Econometrica 41, 587–601 (1973)
11. Gähler, W.: Monads and convergence. In: Proc. Conference Generalized Functions, Convergences Structures, and Their Applications, Dubrovnik (Yugoslavia), pp. 29–46. Plenum Press, NewYork (1988)
12. Kahneman, D., Tversky, A.: Prospect theory: An analysis of decision under risk. Econometrica 47, 263–291 (1979)
13. Kacprzyk, J., Fedrizzi, M.: Multiperson Decision Making Models Using Fuzzy Sets and Possibility Theory. Kluwer Academic Publishers, Boston (1990)
14. Mill, J.S.: Principles of Political Economy, 7th edn. Longmans, Green and Co., London (1909); 1st edn. (1948)
15. Montero de Juan, J.: Arrow's theorem under fuzzy rationality. Behavioral Science 32, 267–273 (1987)
16. Satterthwaite, M.A.: Strategy-Proofness and Arrow's Conditions. Journal of Economic Theory 10, 187–217 (1975)
17. Kelly, J.S.: Arrow Impossibility Theorems. Academic Press, New York (1978)
18. McLean, I., Urken, A.B. (eds.): Classics of Social Choice. The University of Michigan Press, Ann Arbor (1995)
19. von Neumann, J., Morgenstern, O.: Theory of Games and Economic Behavior, 3rd edn. Princeton University Press, Princeton (1953)
20. Nurmi, H.: Voting Paradoxes and How to Deal with Them, Berlin, Heidelberg, New York (1999)
21. Nurmi, H.: Fuzzy social choice: A selective overview. Soft Computing 12, 281–288 (2008)
22. Sen, A.K.: Collective Choice and Social Welfare. Oliver & Boyd, Edinburgh (1970)
23. Sen, A.: The possibility of social choice. American Economic Review 89, 349–378 (1999)
24. Zadeh, L.A.: Fuzzy sets. Information and Control 8, 338–353 (1965)

Signatures for Assessment, Diagnosis and Decision-Making in Ageing

Patrik Eklund

Umeå University, Department of Computing Science, SE-90187 Umeå, Sweden
peklund@cs.umu.se

Abstract. Computerized decision-making in social and health care is traditionally focused on representation and implementation of know-how and guidelines. Less attention has been paid to underlying data structures and formalizations of ontologies. In this paper we show how underlying signatures of a logic can be based on monads over suitable categories. Furthermore, we argue in favour of using application domain specific logic, and even cross-functional logic as enabled by general logics. Our examples are drawn from decision-making with assessment scales and consensus guidelines in social and health care of older people.

Keywords: Ageing, general logic, monad, substitution.

1 Introduction

Computerized decision-making in social and health care is traditionally focused on representation and implementation of know-how and guidelines. Data structures and ontologies are developed by several organizations and even industry federations like `openEHR`[1]. These, however, view ontologies not as part of underlying logics for decision-making, but rather as standards and terminologies including skeletons and frameworks of informal logic structures. Thus, ontologies are enforced to concentrate on generating informal logics, then relying on formalizations like 'description logic' [2] to host inference mechanisms in those logics. Still, data structures and signatures are not considered enough and in required detail.

Furthermore, *sharing* of information is not enough, we have to *communicate* it. And it's not just about information, it's about knowledge and decision-making. Communication of decision is a most challenging area as communication means moving information, rules and decisions represented in one functional and professional domain to another. This requires transformation of logics, not just sharing of information. We could also say that sharing is a database view, whereas transformations are morphisms in the categorical framework of general logics [20].

[1] ... the principal challenge for health ICT is to represent the semantics of the sector ... requires a knowledge-oriented computing framework that includes ontologies, terminology and a semantically enabled health computing platform in which complex meaning can be represented and shared ... the main problem in health is the lack of shareable and computable information ... [24].

E. Hüllermeier, R. Kruse, and F. Hoffmann (Eds.): IPMU 2010, Part II, CCIS 81, pp. 271–279, 2010.

Programming in logic boils down to manipulation of terms and substitution with terms. Classical terms won't suffice. An ontology building upon classical terms, trying to enhance missing parts in the underlying structures by being clever about inference, becomes logically sterile and basically useless in formal frameworks.

We also need to make a distinction between imprecise or vague information, and being formal and accurate in reasoning with vague values. Furthermore, a value may be vague as produced by a crisp operation, or a value is vague since the underlying operation is vague. 'This is a *possible* dementia' claimed by an expert has a different vagueness as compared to the same expressed by a novice. Therefore we may argue that the vague value should be recorded together with the vagueness value of the operation producing that value. From formal point of view this is all about underlying categories and monads, and in this paper we will indeed show how the signatures reside in term monads over chosen categories. Our approach is thus monadic, and we consider monads over suitable categories. Our application domain is elderly care, in particular from the viewpoint of assessments in old age psychiatry and early diagnosis of cognitive disorder.

2 Social and Health Care Assessment of Older Persons

The elderly population continues to grow, and the speed of growth is a challenge for excellence and timeliness with respect to sustainable developments within management of ageing. Communities and regions need to deliver innovative procedures for evaluation and quality assurance, thereby also providing financial as well as social benefits and cohesion of solutions for management of ageing. This in the end is enabled by enhanced processes for observation and monitoring that facilitate earlier diagnosis and intervention. Consolidation of social intelligence within the paradigm shift from individual care to team-based care also contributes significantly to the modernizing of social models. Team orientation and its success then also becomes dependent of innovation with respect to information sharing and knowledge transformation.

Early diagnosis of dementia[2] [5] is of utmost importance, and assessment scales in old age psychiatry [6] play significant roles. Further, guideline ([23]) adherence must be guaranteed, and our approach is to ensure strictness in ground category based logic formalization of the underlying information and reasoning machineries [7].

Assessment scales cover areas and conditions such as ADL (activities of daily living), dementia, depression, delirium, substance-based, nutrition and pain.

[2] Dementia is not a disease but rather a syndrome due to disease of the brain, which is progressive and includes disturbances e.g. in memory, orientation, comprehension, learning capacity, language, and judgement. Impairments of cognitive functions appears together with deterioration in emotional control, social behaviour, and motivation. This syndrome occurs typically in Alzheimer's disease and in cerebrovascular diseases.

Typical scales in ADL are the basic and instrumental ADL scales BADL and IADL. In dementia we distinguish between cognitive and non-cognitive scales. The Mini-Mental State Examination (MMSE [16]) is the most typical scale for assessment involving cognitive decline. The Clinical Dementia Rating (CDR) is useful when assessing ADL disability and dementia progression in one scale. Non-cognitive scales include e.g. the Neuropsychiatric Inventory (NPI) which covers assessment involving behavioural and psychological symptoms of dementia. Further, the Geriatric Depression Scale GDS [22] is the most widely used scale for observation of depressions. Dementia weakens ability, but it is important also to note how e.g. aphasia and depression contributes to that weakening.

Elderly care management mostly involves decisions concerning levels of services. Older persons are supported in their own homes or in various forms of residential living and nursing homes. Decision-making is also about transitions between service levels. Basically, there is a queue for residential living, nursing homes and wards. Decision-making can be objective only with broad and detailed information enabled by assessment scale data. Certainly, other information and considerations are included, but assessment scales form the backbone of this decision-making. The overall information management process is a clear Observe-Assess-Decide, where observation typically are made within home care, assessments by regional home care adminstration offices, and decisions are made by multi-professional groups.

Municipal and regional home care organization consists of groups of home care teams by geographic subdivision. Daily administration communicates with home care teams, residential services, and health care providers. The office maintains assessment scale data, and blends these with patient record information. Further, the administration office deals with referrals, and planning for decision-making and multi-professional board meetings, including geriatricians, elderly care managers, adminstration office managers, home care area coordinators, and selected team leaders.

Observations of assessment scale data requires to make an important distinction between *I observe* and *my observation*, the former being the process of observation, and the latter being only the result of the observation. Uncertainty resides both in process as well as result.

From signature point of view, process is operation and result is the outcome of operation. Early work on fuzzy sets and fuzzification focused only on fuzzification of truth values and membership, and the underlying language including signatures and sentences were not considered for fuzzification. In [14] we initiated developments on generalized fuzzy logic by considering generalized notions of terms including a substitution theory for generalized terms. The composed monad $\boldsymbol{L}_{id} \bullet \boldsymbol{T}_{\Omega}$ became a natural candidate for generalized terms. A further direction towards more substitution paradigms was taken in [13] where lattice-valuedness is moved to the underlying category now being $\mathsf{Set}(L)$, i.e. the Goguen category of fuzzy sets [17]. Previous work on decision support in elderly care used only generalized substitutions over Set. The distinction between *I observe* and *my observation* then clearly suggest using lattice-valuedness as represented within the underlying category Set(L). We will see how attribute values, appearing in assessment scales are canonically represented within the lattice-valued ground category.

3 Observation Is Process and Data over Set(L)

The term monad over Set(L) is a strict categorical methodology constructed as follows [18,10,14,13]. Let $id^0(A,\alpha) = (\{\varnothing\}, \top)$ and $id^n(A,\alpha) = (id^n A, id^n(\alpha)$, where $id^n(\alpha)(a_1,\ldots,a_n) = \bigwedge_{i=1}^n \alpha(a_i)$. A constant functor $(A,\alpha)_{\mathsf{Set}(L)}$ assigns any (X,ξ) to (A,α) and all morphisms $f\colon (X,\xi) \longrightarrow (Y,\upsilon)$ to the identity morphism $id_{(A,\alpha)}$.

Let k be a cardinal number and $\{(\Omega_n,\vartheta_n) \mid n \leq k\}$ be a family of L-sets. We have

$$\coprod_{n\leq k} (\Omega_n,\vartheta_n)_{\mathsf{Set}(L)} \times id^n(X,\xi) = \Big(\bigcup_{n\leq k} \{n\} \times \Omega_n \times id^n X, \beta\Big), \tag{1}$$

where $\beta(n,\omega,(x_i)_{i\leq n}) = \vartheta_n(\omega) \wedge id^n(\xi)((x_i)_{i\leq n})$, $\omega \in \Omega_n$ and $(x_i)_{i\leq n} \in X^n$.

Consider $(\Omega,\vartheta) = \coprod_{n\leq k}(\Omega_n,\vartheta_n)$ as a lattice-valued operator domain, i.e. $\vartheta_n : \Omega_n \longrightarrow L$. We use transfinite induction:

$$T^0_{(\Omega,\vartheta)} = id$$

$$T^1_{(\Omega,\vartheta)}(X,\xi) = \Big(\bigcup_{n\leq k} \{n\} \times \Omega_n \times id^n X, \beta\Big)$$

$$T^\iota_{(\Omega,\vartheta)}(X,\xi) = \coprod_{n\leq k} (\Omega_n,\vartheta_n)_{\mathsf{Set}(L)} \times id^n \bigvee_{\kappa<\iota} T^\kappa_{(\Omega,\vartheta)}(X,\xi)$$

for each ordinal $\iota > 1$. Finally, let

$$T_{(\Omega,\vartheta)}(X,\xi) = \bigvee_{\iota<\bar{k}} T^\iota_{(\Omega,\vartheta)}(X,\xi)$$

where $\bar{k}$ is the least cardinal greater than k and $\aleph_0$.

This makes $T_{(\Omega,\vartheta)}$ a functor and it is further extendable to a monad $\mathbf{T}_{(\Omega,\vartheta)}$ [13].

The process representing *I observe* is then a $\omega \in \Omega_n$, and $\vartheta(\omega)$ is the uncertainty attached to this particular process of observation. Correspondingly, *my observation* is more related to the value $(n,\omega,(t_i)_{i\leq n})$ that is the result when operating with ω on terms $t_i, i \leq n$. Note that this value is frequently written $\omega(t_1,\ldots,t_n)$. The notation $(n,\omega,(t_i)_{i\leq n})$ is, however, suitable in the purely categorical construction of $\mathbf{T}_{(\Omega,\vartheta)}$.

4 An Example

In this section we provide an example for an older person having indications of both depression as well as dementia. We will discuss the situation, on the one hand, from information and assessment scale point of view, and, on the other hand, from patient and care giver point of view.

4.1 Information and Assessment

For clarity, and in order to demonstrate our signature approach, we will restrict to using only `GDS-4` as a depression scale, and `MMSE` as a dementia scale. We will make some use also of the **`Hackinski`** score for vascular dementia.

The geriatric depression scale `GDS` [22] includes 30 questions, where the older persons investigated are asked to reply either `yes` or `no`. The main indication of `GDS` is to rate depression in elderly people. The scale does not require skill of a trained interviewer. `GDS` correlates with diagnostic criteria (DSM-IV guidelines [23]) symptoms for depression. `GDS` is weaker on monitoring change, as compared e.g. to the **`Hamilton`** scale. In `GDS`, the interval for mild depression is 11-20 positive answers (some positive answers are `no`'s). A 15-item version, `GDS-15`, correlates significantly with the full `GDS` scale. Even a 4-item version, `GDS-4`, can be used successfully in screening.

In this section we use the `GDS-4` scale, which includes the following questions:

1. *Are you basically satisfied with your life?* (`NO/yes`)
2. *Do you feel that your life is empty?* (`no/YES`)
3. *Are you afraid that something bad is going to happen to you?* (`no/YES`)
4. *Do you feel happy most of the time?* (`NO/yes`)

The `MMSE` test [16] has a maximal score of 30, and includes e.g. two questions on orientation, one on time, another on place. The time orientation question is the following:

> *What is the year, time of the year, date, month, weekday?* (1 point each, max 5 points)

There is a useful, even if weak, correlation between `MMSE` scores and severities of dementia. A score of 18-23 points indicates a mild dementia, 12-17 points a moderate dementia, and 0-11 a severe dementia. Roughly speaking, older persons with mild dementia can still manage in their own homes, whereas moderate and moderate to severe dementias typically imply residential living and nursing homes. Severe dementia patients are treated in nursing homes and hospital wards.

Note how a reply, e.g. to `GDS-4` question 1, is viewed as $\mathtt{NO} \in \Omega_0$ and $\vartheta(\mathtt{NO}) \in L$.

4.2 A Patient Oriented View

Consider the following dialogues between a patient and care givers. 'John', a librarian emeritus, is the older person having ADL problems and beginning to show symptoms of depression cognitive decline. 'Patrick' is a home care social worker, accidentally also having been trained as a logician. 'Cindy' is a primary health care physician and originally a trained pediatrician.

Patrick making a house call in John's home:

Patrick: *Let me help you with that rollator.*
John: *Yes .. Auhh! .. My knee always reminds me.*
Patrick: *Life treats you ..*
John: *No longer in the best of ways.*
Patrick: *Okey, there you go. Can you move?*
John: *I'll try ..*

John visiting Cindy at the health care centre to update his medications:

Cindy: *Okey, these are now your new pills. Be sure always to take them on time.*
John: *Thank you.*
Cindy: *Anything else we can do for you?*
John: *No, thank you, I'm fine.*
Cindy: *Otherwise life treating you well?*
John: *Yes, yes, ... everything is ok.*

Concerning question 1 in GDS-4, even if not directly presented to John, Patrick would say that John clearly replied `NO`, whereas Cindy does not observe anything else but `yes`. Thus $\vartheta^{Patrick}(\texttt{NO}) = \texttt{present} \in L$ and $\vartheta^{Cindy}(\texttt{NO}) = \texttt{absent} \in L$. According to the chart, Patrick has observed a possible, or even close to

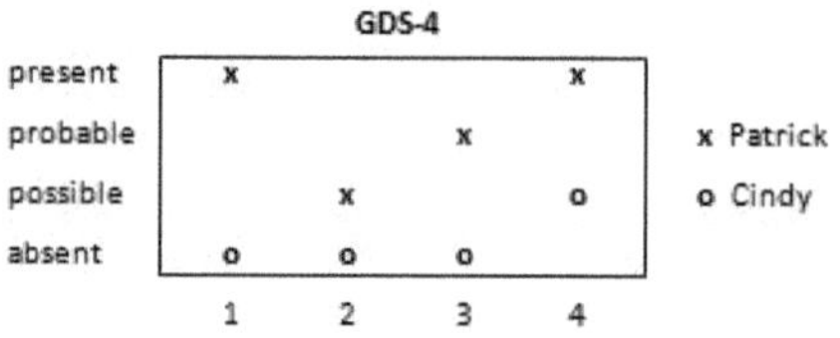

probable depression, which should be further investigated. Cindy makes no such observations.

Sharing of information is now transformation of data in logics. The reasoning above clearly invites to saying that Patrick's and Cindy's logic, $\vdash^{Patrick}$ and $\vdash^{Cindy}$, respectively, may not coincide. This is where general logics [20] comes into the picture, and there are on-going research on monadic extensions of institutions, which aims at enabling this type of transformation in logics.

The following rule base is typical for a patient situation, where a municipality multiprofessional group needs to decide on services assigned to the patient:

```
{
  presently in supported home care;
  GDS-4 by social worker shows possible depression;
  recent acceleration of cognitive decline;
  MMSE less than 23, observation from last year;
```

```
  stroke three years ago;
  pharmacologic treatment for hypertension;
  ADL levels are ...
  Social ability is ...
  Environmental issues measure ...
}
|-
{
  continue with supported home care;
  update service levels;
  referral:
    GDS-15, and consider testing for Major Depressive Episode;
    new MMSE test;
    CLOCK test;
}
```

The reason for `CLOCK` test is that stroke, hypertension treatment and depression, together with emotional incontinence and a seemingly abrupt onset of cognitive failure, in the `Hackinski` score, also called `Ischaemic Score`, is on the limit to indicate vascular dementia. The `CLOCK` test is a quick method to receive further indications, before possibly proceeding with a brain scan that reveal focal signs needed for accurate diagnosis.

Note also that the inference mechanism $\vdash$ in this logic needs to 'share' information as provided by $\vdash^{Patrick}$, $\vdash^{Cindy}$ and others.

In a municipality view, resources are usually limited. In rural areas, the situation may be even worse, as seen from the following table.

Year / Population	2010 / 10 000		2020 / 10 000	
	Customers	Caregivers	Customers	Caregivers
Population	10 000		10 000	
Home care	300	30	400	30
Nursing home	25	12	25	12

The population will not change, and the average age increases. This means that tax income does not increase, and this in turn means that caregiver resources or number of beds cannot be increased. As the average age increases, cognitive and other disorders increase together with severity degrees also adding to the situation. Thus, in 2020, the rural municipality is faced with having the same total amount of care giving resource for home care and the nursing home, but the severity of the disorders is then higher. From decision-making point of view, the queue to receiving municipality support is quite different from what it was 2010, and objectivity in decision-making is put under stress.

Formal assessment scale based investigation are more and more expensive, and therefore observations in a broad sense, as provided by all professional groups involved in care giving, must contribute with information into decision-making processes. In these situations, arithmetics won't do. We need logic.

5 Conclusion

Our signatures acting in respective term monads clearly invites to further research on how assessment scales and consensus guidelines for diagnosis integrate within underlying categories and logics.

References

1. Adámek, J., Herrlich, H., Strecker, G.: Abstract and concrete categories. Wiley-Interscience, New York (1990)
2. Baader, F., Calvanese, D., McGuinness, D., Nardi, D., Patel-Schneider, P. (eds.): The Description Logic Handbook: Theory, Implementation, and Applications. Cambridge University Press, Cambridge (2003)
3. Barr, M., Wells, C.: Toposes, Triples and Theories. Springer, Heidelberg (1985)
4. Beck, J.: Distributive laws. In: Seminars on Triples and Categorical Homology Theory, 1966/67. Lecture Notes in Mathematics, vol. 80, pp. 119–140. Springer, Heidelberg (1969)
5. Burns, A., O'Brien, J., Ames, D.: Dementia, 3rd edn. Hodder Arnold (2005)
6. Burns, A., Lawlor, B., Craig, S.: Assessment Scales in Old Age Psychiatry, 2nd edn. Martin Dunitz (2004)
7. Eklund, P.: Assessment scales and consensus guidelines encoded in formal logic. Journal of Nutrition, Health and Aging (19th IAGG World Congress of Gerontology and Geriatrics, Paris) (2009)
8. Eklund, P.: Non-classical logic for elderly care management. In: Magdalena, L., Ojeda-Aciego, M., Verdegay, J.L. (eds.) Proceedings of IPMU 2008, 21th Int. Conf. Information Processing and Management of Uncertainty in Knowledge-based Systems, Torremolinos (Malaga), June 22-27, pp. 646–651 (2008)
9. Eklund, P.: General logics and management of ageing. In: Third 2008 International Conference on Convergence and Hybrid Information Technology (ICCIT 2008), pp. 424–429. IEEE Computer Society, Los Alamitos (2008)
10. Eklund, P., Gähler, W.: Fuzzy Filter Functors and Convergence. Applications of category theory to fuzzy subsets. In: Rodabaugh, S.E., Klement, E.P., Höhle, U. (eds.) Theory and Decision Library B, pp. 109–136. Kluwer, Dordrecht (1992)
11. Eklund, P., Gähler, W.: Completions and Compactifications by Means of Monads. In: Lowen, R., Roubens, M. (eds.) Fuzzy Logic, State of the Art, pp. 39–56. Kluwer, Dordrecht (1993)
12. Eklund, P., Gähler, W.: Partially ordered monads and powerset Kleene algebras. In: Proc. 10th Information Processing and Management of Uncertainty in Knowledge Based Systems Conference, IPMU 2004 (2004)
13. Eklund, P., Galán, M.A., Kortelainen, J., Stout, L.N.: Paradigms for non-classical substitutions. In: Proc. 39th IEEE International Symposium on Multiple-Valued Logic (ISMVL 2009), Naha, Okinawa, (Japan), May 21-23, pp. 77–79 (2009)
14. Eklund, P., Galán, M.A., Ojeda-Aciego, M., Valverde, A.: Set functors and generalised terms. In: Proc. 8th Information Processing and Management of Uncertainty in Knowledge-Based Systems Conference (IPMU 2000), pp. 1595–1599 (2000)
15. Eklund, P., Helgesson, R., Lindgren, H.: Towards refinement of clinical evidence using general logics. In: Rutkowski, L., Tadeusiewicz, R., Zadeh, L.A., Zurada, J.M. (eds.) ICAISC 2008. LNCS (LNAI), vol. 5097, pp. 1029–1040. Springer, Heidelberg (2008)

16. Folstein, M., Folstein, S., McHugh, P.: Mini Mental State - A practical method for grading the cognitive state on patients for the clinician. Journal of Psychiatry Research 12, 189–198 (1975)
17. Goguen, J.A.: L-fuzzy sets. J. Math. Anal. Appl. 18, 145–174 (1967)
18. Gähler, W.: Monads and convergence. In: Proc. Conference Generalized Functions, Convergences Structures, and Their Applications, Dubrovnik (Yugoslavia), pp. 29–46. Plenum Press, New York (1988)
19. Lawvere, F.W.: Functorial Semantics of Algebraic Theories. Dissertation, Columbia University (1963)
20. Meseguer, J.: General logics. In: Ebbinghaus, H.-D., et al. (eds.) Logic Colloquium 1987, pp. 275–329. Elsevier, North-Holland (1989)
21. Richardson, G.D.: A Stone-Čech compactification for limit spaces. Proc. Amer. Math. Soc. 25, 403–404 (1970)
22. Yesavage, J.: Development and validation of geriatric depression screening scale: A preliminary report. Journal of Psychiatry 17, 37–49 (1983)
23. American Psychiatric Association, Diagnostic and statistical manual of mental disorders, Fourth Edition (DSM-IV-TR), Text Revisions, American Psychiatric Association (2000)
24. openEHR, `http://www.openehr.org`

A Default Risk Model in a Fuzzy Framework

Hiroshi Inoue and Masatoshi Miyake

School of Management, Tokyo University of Science
Kuki, Saitama 346-8512 Japan

Abstract. A default risk model is provided by using option pricing theory in a fuzzy framework in consideration of a simple company comprised of a single type of the debt that is free from profit payment and a single type of capital that is liberated from dividend. The model is based on the assumption that asset value of a company is the sum of total market value of stock and debt value, considering a situation where the asset value becomes below the debt value is default. For constructing the default risk model, a new variable is defined to derive a formula to evaluate the probability of default. Thus, an EDP (estimated default probability) model with the first and second moment is proposed and since debt value is fluctuated as asset price some bounds are established to somehow admit the fluctuation of the debt values, employing fuzzy number in the total market value of asset value.

Keywords: Default risk model, Total market value, Stock price, Debt value, Moments, Fuzzy numbers, Black-Scholes-Merton model.

1 Introduction

On September 15 of 2008, Lehman Brothers, one of the largest investment banks, declared bankruptcy, having failed to find an investor, a buyer, or government guarantees. The failure of Lehman and the government's determination not to rescue the firm sharply raised investor fears. Since then this world has suffered from a financial crisis. Severity of cumulative debts has been deepened and risk control of increasing derivative transactions has become a problematic point. Thus, it becomes very important not only for financial institutions but also for general companies in possession of stocks or bonds to evaluate management abilities, credit power of companies in a fair and simple manner. Especially with the financial institutions including banks, soundness in management is required in performing international business endeavors and in making transactions competitive enough to cope with foreign financial institutions in a field of international competitions. In controlling such a kind of a credit risks, an estimation method of the default probability for the clients with whom credit should be provided becomes important. Proper estimation of the default probability becomes very important information when observation is made with proper pricing of the bond values to which credit risks are applied or with adequate profit brought about by market transactions, resulting in foundation to evaluate these factors.

As a representative method of structural approach taken up in this study, Merton [2], who stipulates default as a situation where asset values become below debt values

E. Hüllermeier, R. Kruse, and F. Hoffmann (Eds.): IPMU 2010, Part II, CCIS 81, pp. 280–288, 2010.

as default, modeled default risks of the bonds by an option pricing theory. Black and Cox[3] developed a model that default is caused at the time when the value of the profit of the company reaches a certain low threshold value by relaxing the assumption based on a framework of the Merton model [2] as a foundation. However, Merton or Black-Scholes model [1,2](B-S-M model) may not be appropriate to evaluate default probability since the EDP value is occasionally underestimated for heightening of volatility usually caused by critical affairs of the company and increase of a ratio of the leverage. The model assumes the feature that default occurs only when a company runs through all its assets. On the other hand, it is not straightforward to estimate asset value in obtaining EDP since the asset value itself cannot be directly observed in the market and the conventional methods allow to using nonlinear equation with stock price data, giving complicated calculations.

Miyake and Inoue [9] propose a new methodology in place of these to estimate asset value by taking advantage of moments. Levy[6] uses moments for Asian option pricing and Inoue et al. [7] derive weighted Asian rate option, then it is applied to strike type by Miyake and Inoue [8]. Miyake and Inoue assume that asset value as the sum of the total market value of current stock and debt value , and propose a default model based on moments to estimate its probability, showing its adequacy with the application of Japanese companies.

While the total market value is easily found with multiplying stock value daily observed in the market by the number of issued total stocks, for total market value of debt is difficult to be observed in the market. In evaluation EDP of companies, Ronn and Verma [10] estimate debt value from the data of stock value, and Ando and Marushige [12] use book value of the debt. However, the debt value as well as stock price actually shows fluctuation whose trend is not regular or systematic.

Thus, in this study, we propose a methodology to estimate default probability with the first and second moments in fuzzy approach in which debt value in repayment time may be described with fuzzy numbers. The work allows us to be able to cope with different circumstances we may face in the vicinity of default point.

2 Default Risk Model of B-S-M

A default risk model is provided by using an option pricing theory by the B-S-M model in consideration of a simple company comprised of a single type of the debt that is free from profit payment and a single type of capital that is liberated from dividend. This model is based on the assumption that a situation where asset value becomes below the debt is default. Then the assumption below is made with the EDP estimation by means of the B-S-M model. The asset value of a company follows a stochastic process below,

$$dA_\tau = \mu_A A_\tau d\tau + \sigma_A A_\tau dW_\tau \tag{1}$$

where μ_A is the expected profit ratio of the asset value, σ_A is the volatility of the asset value and dW_τ is standard Brownian motion. From (1) the asset value A_τ of the $\tau(0 \le \tau \le T)$ time point can be expressed for A_τ.

$$A_\tau = A_0 \exp((\mu_A - \sigma_A^2/2)\tau + \sigma_A W_\tau), \tag{2}$$

where $W_\tau \sim N(0,\tau)$. Denoting the debt value at the time point τ by D_τ, the default event can be expressed as $\{A_\tau \le D_\tau\}$. The default probability EDP on the Black-Scholes-Merton model is given below.

$$\begin{aligned} EDP &= P(\ln A_\tau \le \ln D_\tau) \\ &= 1 - N\left(\frac{\ln(A_0/D_\tau) + (\mu_A - \sigma_A^2/2)\tau}{\sigma_A\sqrt{\tau}}\right) \end{aligned} \tag{3}$$

where $N(\cdot)$ is an accumulated probability density function of the standard normal distribution.

In this study, we make the following assumptions for the four parameters as shown below. 1) Let the asset value A_0 be the sum of the total market value of current stock E_0 and debt value D_0 2) The debt value cannot be observed in the market, and let it be understood that D_τ is book value of the debt 3) Let the expected profit ratio of the asset value be the risk-free interest rate.

Denote volatility of the total market value of current stock by σ_E. On the assumption of 1), the expression below is obtained from Itô 's lemma because the volatility σ_E is observable on the market. That is,

$$\sigma_E E_0 = N(x)\sigma_A A_0 \tag{4}$$

where

$$x = \frac{\ln(A_0/D_\tau) + (r + \sigma_A^2/2)\tau}{\sigma_A\sqrt{\tau}}$$

3 Default Risk Model with the First and Second Moments

We consider the first and second moments of the asset value and by defining a new variable we may take the following procedure.

a) The first and second moments such as a mean value and variance of the sum of the total value of stock price and debt value together are derived.
b) Then, a new variable X following geometric Brownian motion, where fluctuation in the model evaluation period coincides with the moments of the sum of the total value of stock price and debt value, is assumed.
c) After obtaining the first and second moments of the new variable and by letting these moments be equal to those obtained in a) we can find the expected profit ratio and volatility with respect to the new variable X .

According to Merton[2], the asset value is expressed as the sum of the total market value of stock price and debt value. While the total market value is easily found by multiplying stock value daily observed in the market by the number of issued total stocks, for total market value of debt is difficult to be observed in the market. On the assumption that total market value of current stock price follows geometric Brownian motion, the total value of current price of stock E_τ can be expressed.

$$E_\tau = E_0 \exp((\mu_E - \sigma_E^2 / 2)\tau + \sigma_E W_\tau) \tag{5}$$

where μ_E, σ_E are the expected profit ratio and volatility of total market values. Miyake and Inoue [8] estimate EDP by using book value of the debt as substitute for debt value for valuing debt. As mentioned above the total market value of stock price is expressed as

$$A_\tau = E_\tau + D_\tau \tag{6}$$

The first and second moments of A_τ are obtained below

$$\begin{aligned} E[A_\tau] &= E_0 e^{\mu_E \tau} + D_\tau \\ E[A_\tau^2] &= E_0^2 e^{(2\mu_E + \sigma_E^2)\tau} + 2E_0 D_\tau e^{\mu_E \tau} + D_\tau^2 \end{aligned} \tag{7}$$

Next, assume a new variable X follows stochastic process with Brownian motion whose expected profit ratio and volatility may be obtained in (9) and (10).

$$dX_\tau = \mu_x X_\tau d\tau + \sigma_x X_\tau dW_\tau \tag{8}$$

Since we assume that the fluctuation of the new variable during the evaluation period of the variable coincides with the first and second moments of the total value of stock price and debt value, letting the first and second moments be equivalent to those obtained above, the drift ratio and volatility can be obtained as below,

$$\mu_x = \frac{1}{\tau} \ln \left(\frac{E_0 e^{\mu_E \tau} + D_\tau}{X_0} \right) \tag{9}$$

and

$$\sigma_x = \sqrt{\frac{1}{\tau} \ln \left(\frac{E_0^2 e^{(2\mu_E + \sigma_E^2)\tau} + 2E_0 D_\tau e^{\mu_E \tau} + D_\tau^2}{X_0^2} \right) - 2\mu_x} \tag{10}$$

By defining default as an event that asset value X_τ becomes below debt value D_τ, default probability is obtained below.

$$\begin{aligned} EDP &= P(X_\tau < D_\tau) \\ &= 1 - N\left(\frac{\ln(X_0 / D_\tau) + (\mu_x - \sigma_x^2 / 2)\tau}{\sigma_x \sqrt{\tau}} \right) \end{aligned} \tag{11}$$

Remark1. Comparison of B-S-M model and our model with moments.
Look at EDP transition of the proposed model and conventional model for Mical which is a retail business company and became a default company on September 1st of 2001. Table 1 shows that a considerable amount of dissociation is seen with the EDP of both models in the vicinity where default was made.

Table 1

!	2001/08/29	2001/09/06	2001/09/14	2001/09/17	2001/09/25
Proposed model	23.059%	25.163%	26.835%	45.632%	94.092%
B-S-M model	20.536%	22.121%	22.694%	34.997%	40.558%

The default point comes up to 26% with the proposed model, and this becomes 94.1% in October 2001 to be increased to 98.0% in December 2001. Contrary to the above, the EDP of the default point reaches 22.7% in case of the conventional model. On October 2001, the EDP becomes 39.2% to be decreased to 21.7% in December 2001. Thus contradictory behavior is noticed as seen for example in decrease of the EDP after the default. For more other examples see Miyake and Inoue [9]. Fig.1 shows that though both the models indicate similar movement with EDP the proposed model remarkably improves EDP in the vicinity of default point, representing the results fitted and accepted for more actual phenomenon.

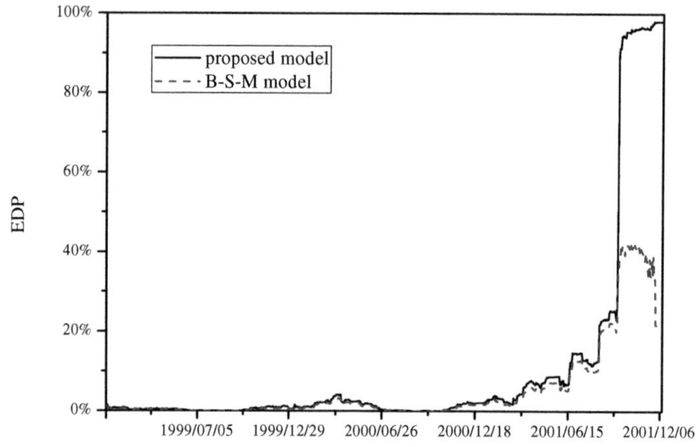

Fig. 1. EDP for B-S-M model and proposed model

4 Default Risk Model with Fuzzy Nature

So far, we have tried to estimate the default probability of a company with replacing debt value by book value of debt. However, note that the debt value as well as the stock price may be fluctuated during the evaluation period. Then, it is expected for some bounds to somehow admit the fluctuation of the debt values, incorporating fuzzy nature in the total market value of asset value.

For debt value D_τ at time τ, we consider triangular fuzzy number $\tilde{D}_\tau = ((\tilde{D}_\tau)_L, (\tilde{D}_\tau)_C, (\tilde{D}_\tau)_R)$. The membership function of the triangular fuzzy number $\tilde{D}_\tau$ is defined below.

$$\varphi_{\tilde{D}_\tau}(x)=\begin{cases} 0, & x<(\tilde{D}_\tau)_L \\ (x-(\tilde{D}_\tau)_L)/((\tilde{D}_\tau)_C-(\tilde{D}_\tau)_L), & (\tilde{D}_\tau)_L<x<(\tilde{D}_\tau)_C \\ ((\tilde{D}_\tau)_U-x)/((\tilde{D}_\tau)_U-(\tilde{D}_\tau)_C), & (\tilde{D}_\tau)_C<x<(\tilde{D}_\tau)_U \\ 0, & x>(\tilde{D}_\tau)_U \end{cases} \tag{12}$$

The α-level set of $\tilde{D}_\tau$ is then, for any $\alpha\in[0,1]$

$$(\tilde{D}_\tau)_\alpha=[(\tilde{D}_\tau)_\alpha^L,(\tilde{D}_\tau)_\alpha^U] \tag{13}$$

That is,

$$(\tilde{D}_\tau)_\alpha^L=(1-\alpha)(\tilde{D}_\tau)_L+\alpha(\tilde{D}_\tau)_C,\ (\tilde{D}_\tau)_\alpha^U=(1-\alpha)(\tilde{D}_\tau)_R+\alpha(\tilde{D}_\tau)_C$$

Thus, the asset value of the company firm is described.

$$X_\tau^*=E_\tau+E[\tilde{D}_\tau] \tag{14}$$

where $E[(\tilde{D}_\tau)_\alpha]$ indicates the expectation of the triangular fuzzy number. According to Carlsson and Fuller[13] $E[(\tilde{D}_\tau)_\alpha]$ is expressed as below

$$E[\tilde{D}_\tau]=(\tilde{D}_\tau)_C+\frac{(\tilde{D}_\tau)_R+(\tilde{D}_\tau)_L-2(\tilde{D}_\tau)_C}{6} \tag{15}$$

Thus, the expected profit ratio of the firm with fuzzy nature is obtained by (9),

$$\mu_{x^*}=\frac{1}{\tau}\ln\left(\frac{E_0e^{\mu_E\tau}+E[\tilde{D}_\tau]}{X_0^*}\right) \tag{16}$$

and the volatility becomes, by (10),

$$\sigma_{x^*}=\sqrt{\frac{1}{\tau}\ln\left(\frac{E_0^2e^{(2\mu_E+\sigma_E^2)\tau}+2E_0E[\tilde{D}_\tau]e^{\mu_E\tau}+E[\tilde{D}_\tau]^2}{(X_0^*)^2}\right)-2\mu_{x^*}} \tag{17}$$

Next, assume the following stochastic process which has equivalent expected profit ratio and volatility as (16),(17) and follows geometric Brownian motion

$$dX_\tau^*=\mu_{x^*}X_\tau^*d\tau+\sigma_{x^*}X_\tau^*dW_\tau \tag{18}$$

By considering (18) as asset value and defining default as event that asset value becomes below the fuzzy number of debt value $\tilde{D}_\tau$ at time τ the default risk model with fuzzy nature is obtained below

$$\begin{aligned}(\tilde{EDP})_\alpha&=P(X_\tau^*\in(\tilde{D}_\tau)_\alpha)\\&=[(\tilde{EDP})_\alpha^L,(\tilde{EDP})_\alpha^U]\end{aligned} \tag{19}$$

where $(\tilde{D}_\tau)_\alpha$ is an interval with level as in (13). The left end point $(EDP)_\alpha^L$ and right end point $(EDP)^U$ become

$$(\tilde{EDP})_\alpha^L = 1 - N(\tilde{d}_\alpha^U), \quad (\tilde{EDP})_\alpha^U = 1 - N(\tilde{d}_\alpha^L) \tag{20}$$

where $\tilde{d}_\alpha^L, \tilde{d}_\alpha^U$ are

$$\tilde{d}_\alpha^L = \frac{\ln(X_0^* / (\tilde{D}_\tau)_\alpha^U) + (\mu_{x^*} - \sigma_{x^*}^2 / 2)\tau}{\sigma_{x^*}\sqrt{\tau}}, \quad \tilde{d}_\alpha^U = \frac{\ln(X_0^* / (\tilde{D}_\tau)_\alpha^L) + (\mu_{x^*} - \sigma_{x^*}^2 / 2)\tau}{\sigma_{x^*}\sqrt{\tau}}$$

5 Simulation Results with Different αValues

Figure 2,3 and 4 show intervals with respect to EDP obtained in previous section for α =0.6, 0.8, 1.0. With different α values transition of EDP is observed for a default company Mical illustrated in Remark1 of section 3. For α =0.6 or 0.8 it does not seem to suggest any pattern which may lead to be associated with the figure for α =1.0.

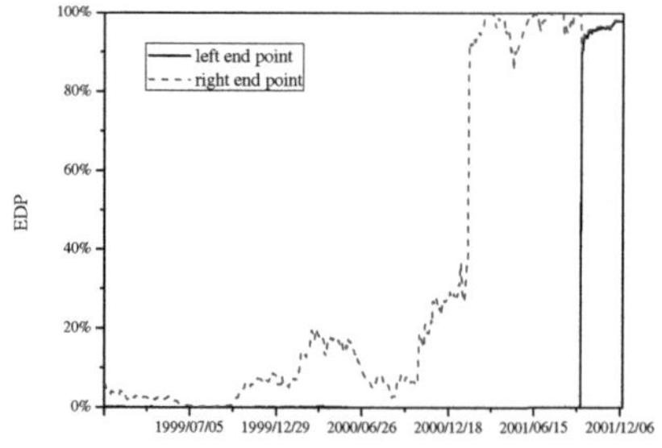

Fig. 2. EDP for α=0.6

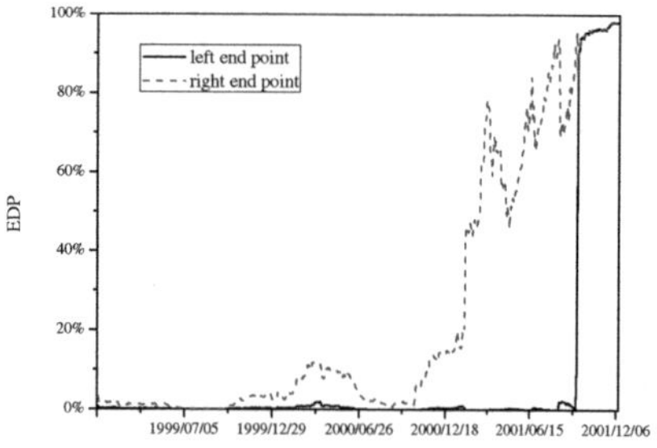

Fig. 3. EDP for α=0.8

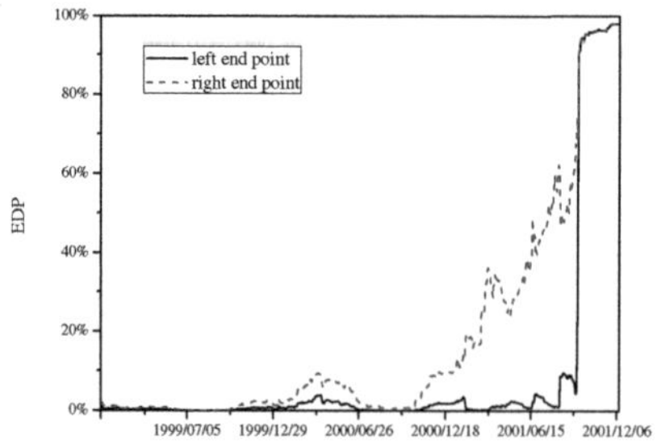

Fig. 4. EDP for α=0.9

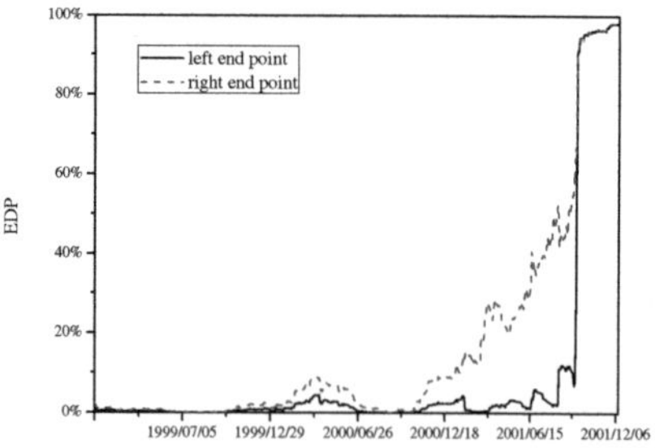

Fig. 5. EDP for α=0.92

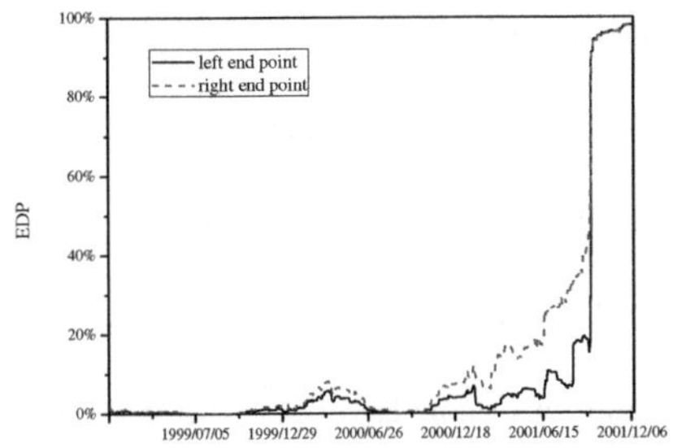

Fig. 6. EDP for α=0.96

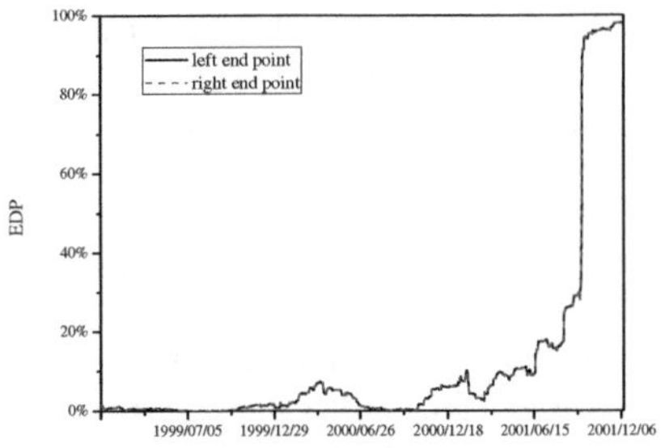

Fig.7. EDP for α=1.0

We note that there are two types of debt, one is current liabilities and the other fixed liabilities. In this example, for fuzzy number $\tilde{D}_\tau$ the left end point $(\tilde{D}_\tau)_L$ is calculated as current liabilities +0.5×fixed liabilities and the right end point $(\tilde{D}_\tau)_R$ as current liabilities×1.5 + fixed liabilities. The implication is interpreted as follows, for lower limit the liabilities don't increase and the fixed liabilities usually remain more than one year before its repayment time, which allows less fluctuation. On the contrary, the liabilities allow to increase for some reason during the evaluation period. For example, at =0.6 the difference between lower limit and upper limit spreads to 58.33% on June 15 of 2001, but at =0.8the difference makes decreased to 25.48%, and finally it becomes 0 at =1.0.

6 Concluding Remarks

In this study, a default risk model was presented in which the first and second moments are incorporated in the model to be sure of more preciseness so that the estimated probability for default gets higher than the conventional model by Merton [2]. The reason exists in that companies approaching to default situation show higher values of leverage D/A, and the volatility of both the model plays a crucial role. Thus, the respective volatility σ_x and σ_A in EDP formulas dominates each of the whole expression of normal distribution and σ_x^2 is larger than σ_x^2 so that the above-mentioned may be understood.

However, the method is not sufficient in the light of the fluctuation of debt value and the debt value in the model remains unchanged, and hence is needed to admit the fluctuation in a sense of fuzzy approach. The additional work allows us to be able to cope with different circumstances we may face in the vicinity of default point. Thus, the debt value which may be expected at maturity time is treated with fuzzy feature, providing allowable intervals for the estimated probability of default with optimistic and pessimistic views.

References

1. Black, F., Scholes, M.: The Pricing of Options and Corporate Liabilities. J. Political Economy 81(3), 637–654 (1973)
2. Merton, R.C.: On the Pricing of Corporate Debt: The Risk Structure of Interest Rates. J. Finance 29, 449–470 (1974)
3. Black, F., Cox, J.C.: Valuing Corporate Securities: Some Effects of Bond Indenture Provisions. J. Finance 31, 351–367 (1976)
4. Longstaff, F.A., Schwartz, A.E.S.: A Simple Approach to Valuing Risky Fixed and Floating Rate Debt. J. Finance 50, 789–821 (1995)
5. Turnbull, S., Wakeman, L.: A Quick Algorithm for Pricing European Average. J. Financial Quantitative Analysis, 377–389, September 26 (1991)
6. Levy, E.: Pricing European Average Rate Currency Options. J. International Money and Finance 11, 474–491 (1992)
7. Inoue, H., Miyake, M., Takahashi, S., Yu, M.: Option Pricing for which Payoff Depends on Weighted Sums of Prices. In: Proc. of Hawaii International Conference on Statistics, Mathematics and Related Fields (2007)
8. Miyake, M., Inoue, H.: Note on Weighted Average Strike Asian Options. In: Proc. IPMU 2008 (Information Processing and Management of Uncertainty in Knowledge-Based Systems), pp. 601–607 (2008)
9. Miyake, M., Inoue, H.: A Default Probability Estimation Model: An Application to Japanese Companies. Journal of Uncertain Systems, 3(3), 210–220 (2009)
10. Ronn, E.I., Verma, A.K.: Pricing Risk-Adjusted Deposit Insurance: An Option-Based Model. J. Finance 41, 871–895 (1989)
11. Boness, A.J.: Elements of a Theory of Stock-option Value. J. Political Economy 72(2), 163–175 (1964)
12. Ando, T., Marushige, K.: Estimation Method of Default Probability with Knock-out Option Approach~Comparative Analysis with European Option Approach~ IMES Discussion Paper (in Japanese) No. 2001-J-4 (2001)
13. Carlsson, C., Fuller, R.: On possibilistic mean value and variance of fuzzy numbers. Fuzzy Sets and Systems 122, 315–326 (2001)
14. Carlsson, C., Fuller, R.: A fuzzy approach to real option valuation. Fuzzy Sets and Systems 139, 297–312 (2003)

On a Fuzzy Weights Representation for Inner Dependence AHP

Shin-ichi Ohnishi[1], Takahiro Yamanoi[1], and Hideyuki Imai[2]

[1] Faculty of Engineering, Hokkai-Gakuen University,
0640926 Sapporo, Japan
{ohnishi,yamanoi}@eli.hokkai-s-u.ac.jp
[2] Graduate School of Information Science and Technology, Hokkaido University,
0600814 Sapporo, Japan
imai@ist.hokudai.ac.jp

Abstract. The AHP (Analytic Hierarchy Process) has been widely used in decision making. Inner dependence method AHP is one technique for the case in which criteria do not have enough independency. However using original AHP or inner dependence method, the data and results often lose their reliability because the comparison matrix does not always have sufficient consistency. In these cases, fuzzy representation for weighting criteria and alternatives using results from a sensitivity analysis is useful. In this paper, we first present weights of criteria of normal AHP by means of fuzzy sets, then modified fuzzy weights is calculated. Overall weights of alternatives can also be calculated by employing some assumptions. The results show how inner dependence AHP has fuzziness when the comparison matrix is not sufficiently consistent and each criterion has not enough independency.

Keywords: Decision making, AHP (Analytic Hierarchy Process), Fuzzy sets, Inner Dependence, Sensitivity analysis.

1 Introduction

AHP (Analytic Hierarchy Process) was proposed by Saaty T.L. in 1977 [1], [2]. The method has been popular and widely used in the field of decision making, since it can reflect humans feelings naturally.

Usually normal AHP must assume independency among criteria, although it is difficult to choose enough independent criteria. Inner dependence method AHP[10] is one technique of solving this kind of problem even in case of criteria have dependency.

However, the comparison matrix often does not have enough consistency when AHP or Inner dependence method is used since, for instance, a problem may contain too many criteria for decision making. In these cases, we consider that answers from decision-makers (i.e. components of the comparison matrix) have ambiguity or fuzziness. For resolving this type of problem, fuzzy reciprocal components have been proposed as components of the data matrix in some research [12]. In this paper, we consider that weights should also have ambiguity or fuzziness. Therefore, it is necessary to represent these weights by use of fuzzy sets.

E. Hüllermeier, R. Kruse, and F. Hoffmann (Eds.): IPMU 2010, Part II, CCIS 81, pp. 289–297, 2010.

Sensitivity analysis is applied to Inner dependence AHP to analyze the amount the components of a pairwise comparison matrix influences the weights and consistency of a matrix. This makes it possible to show the magnitude of the fuzziness in the weights.

In previous researches, we proposed a new representation for weights of criteria and alternatives in normal AHP [7][8][11] and inner dependence method[13]. In this paper, a refined representation of weights of inner dependence method is proposed. It is represented as L-R fuzzy numbers by using the results from the sensitivity analysis and fuzzy operations. Then we can show a representation of fuzziness as a result of inner dependence when a comparison matrix does not have enough consistency.

2 Inner Dependence AHP

2.1 Process of Normal AHP

(Process 1) Representation of structure by a hierarchy. The problem under consideration can be represented in a hierarchical structure. The highest level of the hierarchy consists of a unique element that is the overall objective. At the lower levels, there are multiple criterion (i.e. elements within a single level) with relationships among elements of the adjacent higher level to be considered. The criterion are evaluated using subjective judgments of a decision maker. Elements that lie at the upper level are called parent elements while those that lie at lower level are called child elements. Alternative elements are put at the lowest level of the hierarchy

(Process 2) Paired comparison between elements at each level. A pairwise comparison matrix A is created from a decision maker's answers. Let n be the number of elements at a certain level. The upper triangular components of the comparison matrix a_{ij} ($i< j = 1,\ldots,n$) are 9, 8, .. , 2, 1, 1/2, …, or 1/9. These denote intensities of importance from activity i to j. The lower triangular components a_{ji} are described with reciprocal numbers as follows

$$a_{ji} = 1/a_{ij} \tag{1}$$

in addition, for diagonal elements, let $a_{ii} = 1$. The lower triangular components and diagonal elements are occasionally omitted from the written equation as they are evident if upper triangular components are shown. The decision maker should make $n(n-1)/2$ paired comparisons at a level with n elements.

(Process 3) Calculations of weight at each level. The weights of the elements, which represent grade of importance among each element, are calculated from the pairwise comparison matrix. The eigenvector that corresponds to a positive eigenvalue of the matrix is used in calculations throughout in this paper.

(Process 4) Priority of an alternative by a composition of weights. The composite weight can be calculated from the weights of one level lower. With repetition, the weights of the alternative, which are the priorities of the alternatives with respect to the overall objective, are finally found.

2.2 Consistency

Since components of the comparison matrix are obtained by comparisons between two elements, coherent consistency is not guaranteed. In AHP, the consistency of the comparison matrix A is measured by the following consistency index (C.I.)

$$\text{C.I.} = \frac{\lambda_A - n}{n - 1}, \tag{2}$$

where n is the order of matrix A, and λ_A is its maximum eigenvalue.

It should be noted that $\text{C.I.} \geq 0$ holds. And if the value of C.I. becomes smaller, then the degree of consistency becomes higher, and vice versa. The comparison matrix is consistent if the following inequality holds.

$$\text{C.I.} \leq 0.1$$

Also consistency ratio (C.R.) is defined as

$$\text{C.R.} = \frac{\text{C.I.}}{M},$$

where M is random consistency value. However we only employ C.I., since we mainly use 4 or 5-dimensional data whose random consistency value is not far from 1.

2.3 Inner Dependence Method

Usually normal AHP must assume independency among criteria, although it is difficult to choose enough independent criteria. Inner dependence method AHP[10] is one technique of solving this kind of problem even in case of criteria have dependency.

In the method, using a dependency matrix $F=\{ f_{ij} \}$, we can calculate real weights $\boldsymbol{w}^{(n)}$ as follows,

$$\boldsymbol{w}^{(n)}=F\boldsymbol{w} \tag{3}$$

where $\boldsymbol{w}$ is weights from independent criterion, i.e. normal weights of normal AHP and F is calculated as eigen value of influenced matrix.

3 Sensitivity Analysis

When AHP is used, the comparison matrix is often inconsistent or large differences among the overall weights of the alternatives do not appear. Thus, it is very important to investigate how the components of a pairwise comparison matrix influence the consistency or weights. Sensitivity analysis is used to analyze how results are influenced when certain variables change. Therefore, it is necessary to establish a sensitivity analysis of AHP.

In our research, a previously proposed method [7] is used to evaluate the fluctuation of the consistency index and weights when a comparison matrix is perturbed. This method is useful as it does not change the structure of the data.

Evaluating the consistency index and the weights of a perturbed comparison matrix are performed as follows.

(1) Perturbations $\varepsilon a_{ij}d_{ij}$ are imparted to component a_{ij} of a comparison matrix, and the fluctuation of the consistency index and the weight are expressed by the power series of ε.
(2) Fluctuations of the consistency index and the weights are represented by the linear combination of d_{ij}.
(3) By the coefficient of d_{ij}, it can be shown that how the component of the comparison matrix gives influence on the consistency index and the weight.

Since the pairwise comparison matrix A is a positive square matrix, the following Perron- Frobenius theorem [4] holds.

Theorem 1 (Perron – Frobenius). *For a positive square matrix A, the following holds true.*

1. *Matrix A has a positive eigenvalue. If λ_A is the largest eigenvalue then λ_A is a simple root. The positive eigenvector $\mathbf{w}$, corresponding to λ_A, exists. λ_A is called the Frobenius root of A.*
2. *Any positive eigenvectors of A are the constant multiples of $\mathbf{w}$.*
3. *The absolute value of the eigenvalues of A, except for λ_A, is smaller than λ_A.*
4. *The Frobenius root of the transposed matrix A' is equivalent to the Frobenius root of A.*

This theorem ensures the existence of a weight vector in a pairwise comparison matrix.

From Theorem 1, the following theorem regarding a perturbed comparison matrix holds true [7].

Theorem 2. *Let $A = (a_{ij})$, $i,j = 1,\ldots,n$ be a comparison matrix and let $A(\varepsilon) = A+\varepsilon D_A$, $D_A=(a_{ij}d_{ij})$ be a matrix that has been perturbed. Moreover, let λ_A be the Frobenius root of A with $\mathbf{w}_1$ being the corresponding eigenvector. Let $\mathbf{w}_2$ be the eigenvector corresponding to the Frobenius root of transposed matrix A', then, the Frobenius root $\lambda(\varepsilon)$ of $A(\varepsilon)$ and the corresponding eigenvector $\mathbf{w}_1(\varepsilon)$ can be expressed as follows*

$$\lambda(\varepsilon) = \lambda_A + \varepsilon\lambda^{(1)} + o(\varepsilon), \tag{4}$$

$$\mathbf{w}_1(\varepsilon) = \mathbf{w}_1 + \varepsilon \mathbf{w}^{(1)} + \mathbf{o}(\varepsilon), \tag{5}$$

where

$$\lambda^{(1)} = \frac{\mathbf{w}_2' D_A \mathbf{w}_1}{\mathbf{w}_2' \mathbf{w}_1}, \tag{6}$$

$\boldsymbol{w}^{(1)}$ *is an n-dimension vector that satisfies*

$$(A-\lambda_A I)\boldsymbol{w}^{(1)} = -(D_A - \lambda^{(1)} I)\boldsymbol{w}_1, \tag{7}$$

where $\boldsymbol{o}(\varepsilon)$ *denotes an n-dimension vector in which all components are* $o(\varepsilon)$.

Proof of this theorem can be found in Ohnishi's paper [7].

3.1 Sensitivity Analysis for Consistency Index

Regarding a fluctuation of the consistency index, the following corollary can be obtained from Theorem 2.

Corollary 1. *Using an appropriate* g_{ij}, *we can represent the consistency index* C.I.(ε) *of the perturbed comparison matrix as follows*

$$\text{C.I.}(\varepsilon) = \text{C.I.} + \varepsilon \sum_i^n \sum_j^n g_{ij} d_{ij} + o(\varepsilon). \tag{8}$$

(Proof)

From the definition of the consistency index (3) and (4),

$$\text{C.I.}(\varepsilon) = \text{C.I.} + \varepsilon \frac{\lambda^{(1)}}{n-1} + o(\varepsilon).$$

Let $\boldsymbol{w}_1=(w_{1i})$ and $\boldsymbol{w}_2=(w_{2i})$ from (6). $\lambda^{(1)}$ is can now be represented as

$$\lambda^{(1)} = \frac{1}{\boldsymbol{w}_2 \, {}^{\prime} \boldsymbol{w}_1} \sum_i^n \sum_j^n w_{2i} a_{ij} w_{1j} d_{ij},$$

therefore, the second part of the right side is expressed by a linear combination of d_{ij}. (Q.E.D)

p_{ij} in equation (8) in Corollary 1 shows the influence of comparison matrix components on the consistency.

On the other hand, since the comparison matrix $A(\varepsilon) = (a_{ij}(\varepsilon))$ is reciprocal, then $a_{ji}(\varepsilon) = 1/a_{ij}(\varepsilon)$ and becomes

$$a_{ji} + \varepsilon a_{ji} d_{ji} = \frac{1}{a_{ij}} - \varepsilon \frac{d_{ij}}{a_{ij}} + o(\varepsilon). \tag{9}$$

Here, since $a_{ji} = 1/a_{ij}$,

$$d_{ji} = -d_{ij} \tag{10}$$

is obtained. The impact on the consistency can be easily shown by use of this property.

3.2 Sensitivity Analysis for Weights

With regards to the fluctuation in weighs, the following corollary can also be obtained from Theorem 2.

Corollary 2. *Using an appropriate* $h_{ij}^{(k)}$, *we can represent the fluctuation* $\mathbf{w}^{(1)}=(w_k^{(1)})$ *of the weight (i.e. the eigenvector corresponding to the Frobenius root) as follows*

$$w_k^{(1)} = \sum_i^n \sum_j^n h_{ij}^{(k)} d_{ij} \tag{11}$$

(Proof)

The k-th row component of the right side of (7) in Theorem 2 is represented as

$$\sum_i^n \sum_j^n \{ \frac{w_{1k}\, w_{2i}\, a_{ij}\, w_{1j}}{\mathbf{w}_2{}' \, \mathbf{w}_1} - \delta(i,k) a_{ij} w_{1j} \} d_{ij},$$

and is expressed by a linear combination of d_{ij}. Here,$\delta(i,k)$ is Kronecker's symbol

$$\delta(i,k) = \begin{cases} 1 & (i = k), \\ 0 & (i \neq k). \end{cases}$$

In contrast, since λ_A is a simple root, Rank(A-$\lambda_A I$) = n-1. Accordingly, the weight vector is normalized as

$$\sum_k^n (w_k + \varepsilon w_k^{(1)}) = \sum_k^n w_k = 1,$$

then the condition is as follows.

$$\sum_k^n w_k^{(1)} = 0. \tag{12}$$

By using an elementary transformation to formula (7) in the condition above, we also can represent $w_k^{(1)}$ by linear combinations of d_{ij}. (Q.E.D)

As seen in equation (5) in Theorem 2, the component that has a great influence on weight $\mathbf{w}_1(\varepsilon)$ is the component which has the greatest influence on $\mathbf{w}^{(1)}$. $q_{ij}^{(k)}$ in equation (11) from Corollary 2 shows how the influence by the components of a comparison matrix on the weights can be calculated.

The influence can also be shown easily by use of equation (10).

4 A Weights Representation

The comparison matrix often has poor consistency (i.e. 0.1<C.I.<0.2) because it encompasses several criterion. In these cases, the components of a comparison matrix are considered to have fuzziness since they result from the fuzzy judgment of humans. Therefore, weights should be treated as fuzzy numbers.

4.1 L-R Fuzzy Numbers

To represent fuzziness of weight w_{1k}, an L-R fuzzy number is used.

L-R fuzzy number

$$M = (m, \alpha, \beta)_{LR}$$

is defined as fuzzy sets whose membership function is as follows.

$$\mu_M(y) = \begin{cases} R\left(\dfrac{y-m}{\beta}\right) & (y > m), \\ L\left(\dfrac{m-y}{\alpha}\right) & (y \leq m). \end{cases}$$

where $L(y)$ and $R(y)$ are shape function which satisfies

(1) $L(y) = L(-y)$,

(2) $L(0) = 1$,

(3) $L(y)$ is a non increasing function

4.2 Fuzzy Weights of Criteria

From the fluctuation of the consistency index, the multiple coefficient $g_{ij}h_{ij}^{(k)}$ in Corollary 1 and 3 is considered as the influence on a_{ij}.

Since g_{ij} is always positive, if the coefficient $h_{ij}^{(k)}$ is positive, the real weight of criterion k is considered to be larger than w_{1k}. Conversely, if $h_{ij}^{(k)}$ is negative, the real weight of activity k is considered to be smaller. Therefore, the sign of $h_{ij}^{(k)}$ represents the direction of the fuzzy number spread. The absolute value $g_{ij}|h_{ij}^{(k)}|$ represents the size of the influence.

On the other hand, if C.I. becomes bigger, then the judgment becomes more fuzzy.

Consequently, multiple C.I. $g_{ij}|h_{ij}^{(k)}|$ can be regarded as a spread of a fuzzy weight $\tilde{w}_k$ concerned with a_{ij}.

Definition 1 (fuzzy weight). *Let $w^{(n)}_k$ be a crisp weight of criterion k of inner dependence model, and g_{ij} $|h_{ij}^{(k)}|$ denote the coefficients found in* Corollary 1 and 3. *If* 0.1<C.I.<0.2, *then a fuzzy weight $\tilde{w}_k$ is defined by*

$$\tilde{w}_k = (w_k, \alpha_k, \beta_k)_{LR} \tag{13}$$

where

$$\alpha_k = \text{C.I.}\sum_i^n \sum_j^n s(-, h_{kij}) g_{ij} \mid h_{kij} \mid, \tag{14}$$

$$\beta_k = \text{C.I.}\sum_i^n \sum_j^n s(+, h_{kij}) g_{ij} \mid h_{kij} \mid, \tag{15}$$

$$s(+,h) = \begin{cases} 1, (h \geq 0) \\ 0.(h < 0) \end{cases}, \quad s(-,h) = \begin{cases} 1, (h < 0) \\ 0.(h \geq 0) \end{cases}$$

4.3 Fuzzy Weights for Inner Dependence Method

For inner dependence method, we can calculate modified fuzzy weights using a dependency matrix $F=\{ f_{ij} \}$ as follows,

$$\tilde{w}_k^{(n)} = (w_k^{(n)}, \alpha_k^{(n)}, \beta_k^{(n)})_{LR} \tag{16}$$

where $w_k^{(n)}, \alpha_k^{(n)}, \beta_k^{(n)}$ are calculated by fuzzy multiple operations and equation(3) and definition 1.

Then fuzzy weights of alternatives are also calculated with local crisp weights of alternatives with respect to certain criterion. However, the results from the operation of fuzzy numbers are frequently too ambiguous to interpret.

Fuzzy weights of criteria are normalized thus their sum is 1, therefore we can avoid much ambiguity since this condition has been considered [9].

In general, operating with some constraints is difficult but can be accomplished if every fuzzy membership function is linear. Especially for every normal triangular function with a core u_i, the constraint $\sum_i^n u_i = 1$ holds, and the order of singleton coefficients is assumed. Thus, the upper and lower limit of α -cut sets of linear sum can be easily calculated.

Let $f_t(x_k)$ be a crisp local weight of alternative t with respect to criterion k, and in this paper, assume $0 \leq f_t(x_1) \leq f_t(x_2) \leq \cdots \leq f_t(x_n)$. Then, the overall weight of an alternative t is also the L-R fuzzy number and is represented as follows.

$$\tilde{v}_t = (v_t, l_t, r_t)_{LR}$$

where

$$v_t = \sum_k^n w_k^{(n)} f_t(x_k),$$

$$l_t = v_t - \inf \operatorname{supp}(\tilde{v}_t), \quad r_t = \sup \operatorname{supp}(\tilde{v}_t) - v_t$$

In the above equations, inf supp, sup supp are lower and upper limits of support sets and are calculated as follows.

$$\inf \operatorname{supp}(\tilde{v}_t) =$$

$$\max_j \left[\sum_{i=1}^{j-1} (w_i^{(n)} + \beta_i^{(n)}) f_t(x_i) + \sum_{i=j+1}^{n} (w_i^{(n)} - \alpha_i^{(n)}) f_t(x_i) \right.$$

$$\left. + \left\{ 1 - \sum_{i=1}^{j-1} (w_i^{(n)} + \beta_i^{(n)}) - \sum_{i=j+1}^{n} (w_i^{(n)} - \alpha_i^{(n)}) \right\} f_t(x_j) \right]$$

$$\sup \operatorname{supp}(\tilde{v}_t) =$$

$$\min_j \left[\sum_{i=1}^{j-1} (w_i^{(n)} - \alpha_i^{(n)}) f_t(x_i) + \sum_{i=j+1}^{n} (w_i^{(n)} + \beta_i^{(n)}) f_t(x_i) \right.$$

$$\left. + \left\{ 1 - \sum_{i=1}^{j-1} (w_i^{(n)} - \alpha_i^{(n)}) - \sum_{i=j+1}^{n} (w_i^{(n)} + \beta_i^{(n)}) \right\} f_t(x_j) \right]$$

5 Conclusions

We proposed a refined representation for the inner dependence overall weights of alternatives by use of fuzzy sets and the result of a sensitivity analysis for cases in which dependency of criteria exist. Our approach shows how to represent weights, as well as how the result of AHP has fuzziness, when the data are not enough consistent. Also we can reduce ambiguity in the representation compared to previous representation in which normal fuzzy operations is used.

References

1. Saaty, T.L.: A scaling method for priorities in hierarchical structures. J. Math. Psy. 15(3), 234–281 (1977)
2. Saaty, T.L.: The Analytic Hierarchy Process. McGraw-Hill, New York (1980)
3. Saaty, T.L.: Scaling the membership function. European J. of O.R. 25, 320–329 (1986)
4. Saito, M.: An Introduction to Linear Algebra. Tokyo University Press (1966)
5. Tanaka, Y.: Recent advance in sensitivity analysis in multivariate statistical methods. J. Japanese Soc. Comp. Stat. 7(1), 1–25 (1994)
6. Tone, K.: The Game Feeling Decision Making. Nikka-giren Press, Tokyo (1986)
7. Ohnishi, S., Imai, H., Kawaguchi, M.: Evaluation of a Stability on Weights of Fuzzy Analytic Hierarchy Process using a sensitivity analysis. J. Japan Soc. for Fuzzy Theory and Sys. 9(1), 140–147 (1997)
8. Ohnishi, S., Imai, H., Yamanoi, T.: Weights Representation of Analytic Hierarchy Process by use of Sensitivity Analysis. In: IPMU 2000 Proceedings (2000)
9. Dubois, D., Prade, H.: Possibility Theory An Approach to Computerized Processing of Uncertainty. Plenum Press, New York (1988)
10. Saaty, T.L.: Inner and Outer Dependence in AHP. University of Pittsburgh (1991)
11. Ohnishi, S., Yamanoi, T., Imai, H.: A Fuzzy Representation for Weights of Alternatives in AHP. New Dimensions in Fuzzy Logic and Related Technologies II, 311–316 (2007)
12. Ohnishi, S., Dubois, D., Prade, H., Yamanoi, T.: A Fuzzy Constraint-based Approach to the Analytic Hierarchy Process. Uncertainty and Intelligent Information Systems, 217–228 (2008)
13. Ohnishi, S., Yamanoi, T., Imai, H.: A Fuzzy Weight Representation for Inner Dependence Method AHP, 2009 IFSA World congress/ EUSFLAT conference(IFSA-EUSFLAT 2009), 1612—1617 (2009)

Different Models with Fuzzy Random Variables in Single-Stage Decision Problems*

Luis J. Rodríguez-Muñiz and Miguel López-Díaz

Department of Statistics and Operations Research, University of Oviedo
c/ Calvo Sotelo s/n, E33071 Oviedo, Spain
{luisj,mld}@uniovi.es

Abstract. In this paper we examine two different models using fuzzy random variables as the tool for dealing with single-stage decision problems with imprecise assessments of utilities. Both of them are oriented to prove the equivalence between normal and extensive forms of Bayesian analysis. The first model uses Fubini-type techniques to obtain the result whereas the second does not construct a product space and the result is obtained by different techniques. Addition of fuzzy-valued sample information is also considered.

Keywords: Bayesian decision analysis, Fuzzy random variable, Iterated expectation, Uncertainty modeling.

1 Introduction

In a single-stage decision problem imprecision is frequent in the assessment of utilities. An exact numerical quantification of the consequences of decision maker's choice uses to be hard to measure in a real scale, therefore it is a common way to deal with that uncertainty to model the utilities by a fuzzy utility function. And the way to construct that fuzzy utility function is based on a proper definition of a fuzzy random variable.

In this paper we suggest a method to perform a Bayesian analysis within that class of problems involving fuzzy random variables. And it is based in the procedure followed in the real-valued case: exchanging the order of iterated integrals for guaranteeing the equivalence between normal and extensive forms of Bayesian analysis. Therefore, we can perform this exchange by Fubini-type theorems, when applicable, or by alternative methods when product measurability is not attained, even not defined (see [10]).

Several studies have been developed before to evaluate imprecise utilities, see for instance [29,28,7,8,9,11,1,3,13,16,2,22,27].

Now we are combining some results from [11], [12], [13], [24] and [27] to produce a general method that gathers imprecise assessments both in random experiments and in utilities.

* Authors acknowledge financial support by Grant MTM2008-01519 from Ministry of Science and Innovation, Government of Spain.

E. Hüllermeier, R. Kruse, and F. Hoffmann (Eds.): IPMU 2010, Part II, CCIS 81, pp. 298–305, 2010.

2 Notation and Preliminaries

By $\mathcal{K}_c$ we will denote the class of nonempty compact convex subsets of $\mathbb{R}$, endowed with a semilinear structure by means of the Minkowski addition and the product by a scalar. Also, we will consider the *Hausdorff metric* on $\mathcal{K}_c$ (see [6]). On a measurable space $(\Omega, \mathcal{A})$ we can define $S : \Omega \to \mathcal{K}_c$ a random set as a $\mathcal{A}|\mathcal{B}_{d_H}$-measurable mapping ([14]).

A random set S is said to be *integrably bounded with respect to measure* μ, if $\|S\| \in L^1(\Omega, \mathcal{A}, \mu)$, where $\|S\|(\omega) = \sup_{x \in S(\omega)} \|x\|$. The *integral*, or *expected value* in case of μ being a probability, of S, is given by the *Kudo-Aumann integral* and it will be denoted either by $\int_\Omega S(\omega)\, d\mu(\omega)$ ([14]).

By $\mathcal{F}_c$ we will denote the class of *fuzzy sets* $U : \mathbb{R} \to [0,1]$ such that $U_\alpha \in \mathcal{K}_c$ for all $\alpha \in [0,1]$, being $U_\alpha = \{x \in \mathbb{R} : U(x) \geq \alpha\}$ for $\alpha \in (0,1]$, and $U_0 = \mathrm{cl}\{x \in \mathbb{R} : U(x) > 0\}$.

The class $\mathcal{F}_c$ can be endowed with a semilinear structure, defining addition and product by a scalar by means of Zadeh's extension principle ([30,19]). On $\mathcal{F}_c$ we consider the d_∞ metric ([19]). The *magnitude* of $U \in \mathcal{F}_c$ is given by $\|U\| = d_\infty(U, \mathbf{1}_{\{0\}}) = d_H(U_0, \{0\})$.

Being $(\Omega, \mathcal{A})$ a measurable space, a mapping $X : \Omega \to \mathcal{F}_c$ is said to be a *random upper semicontinuous function* (r.u.s.f. for short) if $X_\alpha : \Omega \to \mathcal{K}_c$ with $X_\alpha(\omega) = (X(\omega))_\alpha$ for all $\omega \in \Omega$, is a random set for all $\alpha \in [0,1]$ ([21,4]).

A r.u.s.f. X is said to be *integrably bounded* with respect to a measure $\mu : \mathcal{A} \to \mathbb{R}$, if the mapping $\|X\| \in L^1(\Omega, \mathcal{A}, \mu)$, where $\|X\| : \Omega \to \mathbb{R}$ is given by $\|X\|(\omega) = \|X(\omega)\|$ for all $\omega \in \Omega$.

If X is an integrably bounded r.u.s.f. its *integral*, denoted by $\int_\Omega X(\omega)\, d\mu(\omega)$ or $E(X|\mu)$, can be defined ([21]) as the unique set in $\mathcal{F}_c$ such that $E(X|\mu)_\alpha = E(X_\alpha|\mu)$ for every $\alpha \in [0,1]$. When $\Omega = [a,b]$, we will use also the notation $\int_a^b X(\omega)\, d\mu(\omega)$.

If μ is a probability measure, an r.u.s.f. is also referred to as a *fuzzy random variable* (f.r.v., for short) and its integral as *the (fuzzy) expected value of* X.

Let $\mathcal{D}_j$ denote the smallest σ-field on $\mathcal{F}_c$ such that all the projections $j_\alpha : \mathcal{F}_c \to \mathcal{K}_c$ (with $j_\alpha(A) = A_\alpha$ for all $A \in \mathcal{F}_c$) are $\mathcal{D}_j|\mathcal{B}_{d_H}$-measurable. In [4] it is proved that given a measurable space $(\Omega, \mathcal{A})$, $X : \Omega \to \mathcal{F}_c$ is a fuzzy random variable if, and only, if, X is $\mathcal{A}|\mathcal{D}_j$-measurable. Hence, if X is a f.r.v. associated with the space $(\Omega, \mathcal{A}, P)$, then X induces a probability distribution on $(\mathcal{F}_c, \mathcal{D}_j)$, which will be denoted by ξ^X and it is given by $\xi^X(B) = P(\{\omega \in \Omega : X(\omega) \in B\})$ for all $B \in \mathcal{D}_j$. Along this paper, for EX we will use the alternative notations $\int_\Omega X(\omega)\, dP(\omega)$ and $\int_{\mathcal{F}_c} A\, d\xi^X(A)$.

If $\Omega \subset \mathbb{R}^k$ with $k \in \mathbb{N}$, $\mathcal{B}_\Omega$ will denote the Borel σ-field on Ω. Given $(\Omega, \mathcal{B}_\Omega)$ and $m_1, m_2 : \mathcal{B}_\Omega \to [0, \infty]$ two σ-finite measures, $m_1 \ll m_2$ will indicate that m_1 is absolutely continuous with respect to m_2, and $\frac{dm_1}{dm_2}$ will denote a Radon-Nikodym derivative of m_1 with respect to m_2. If it is supposed that there exists a continuous Radon-Nikodym derivative, then $\frac{dm_1}{dm_2}$ will denote this particular function.

When it is necessary to rank fuzzy sets, we will use the criterion introduced in [5]. Thus, $U \in \mathcal{F}_c$ will be considered to be *greater than or equal to* $W \in \mathcal{F}_c$ *in the* λ, μ*-average sense* (and we will denote it by $U \geq_{\lambda,\mu} W$) if $V_\mu^\lambda(U) \geq V_\mu^\lambda(W)$, where $\lambda \in [0,1]$ represents a kind of degree of optimism/pessimism and μ is a measure on $[0,1]$ (see [18] for more details).

3 Single-Stage Decision Problem

We will use the following notation: Θ is the state space and $\mathcal{E}_\Theta$ is a σ-field on Θ, and $\mathbf{A}$ is the action space. First, we introduce the concept of fuzzy utility function.

Definition 1. *A mapping* $U : \Theta \times \mathbf{A} \to \mathcal{F}_c$ *is said to be a fuzzy utility function on* $\Theta \times \mathbf{A}$ *if*

i) for every $a \in \mathbf{A}$*, the projection* $U_a : \Theta \to \mathcal{F}_c$ *is an r.u.s.f. on* $(\Theta, \mathcal{E}_\Theta)$*,*
ii) for every pair $a_1, a_2 \in \mathbf{A}$*,* a_1 *will be considered preferred or indifferent to* a_2 *with respect to a probability distribution* ξ *on* $(\Theta, \mathcal{E}_\Theta)$*, if* $E(U_{a_1}|\xi) \geq_{\lambda,\mu} E(U_{a_2}|\xi)$ *(for fixed* $\lambda \in [0,1]$ *and measure* μ*).*

The decision problem with fuzzy utilities will be denoted by $(\Theta, \mathbf{A}, U)$.

A Bayesian context is considered, therefore we assume the existence of a probability distribution π on $(\Theta, \mathcal{E}_\Theta)$, *the prior distribution.* Thus, the "value" of the decision problem will be the fuzzy value $E(U_{a^\pi}|\pi)$, where a^π is a *prior Bayes action* in the λ, μ-average sense, this is, $a^\pi \in \mathbf{A}$ verifies $E(U_{a^\pi}|\pi) \geq_{\lambda,\mu} E(U_a|\pi)$ for all $a \in \mathbf{A}$.

As in the case of real-valued utilities, it is useful for increasing the expected utility to incorporate *sample information.* Let $\mathbf{X}$ be a statistical experiment characterized by a probability space $(\mathbb{X}, \mathcal{E}_\mathbb{X}, P_\theta)$, where $\theta \in \Theta$, $\mathcal{E}_\mathbb{X}$ is a σ-field on $\mathbb{X}$ and the experimental distribution P_θ depends on the true unknown state θ. We will denote by P the *marginal* (also called *predictive*) distribution of the experiment.

Once we have obtained an experimental outcome, $\mathbf{X} = x$, the fuzzy expected utility associated with an action $a \in \mathbf{A}$ is given by $E(U_a|\pi_x)$, where π_x is the *posterior distribution* of θ given $\mathbf{X} = x$, obtained on the basis of Bayes' formula. Therefore, a *posterior Bayes action* is any $a^{\pi_x} \in \mathbf{A}$ such that $E(U_{a^{\pi_x}}|\pi_x) \geq_{\lambda,\mu} E(U_a|\pi_x)$ for every $a \in \mathbf{A}$.

An important remark should be done here. Usually, $\mathbb{X} \subset \mathbb{R}^k$ and the Borel σ-field on $\mathbb{X}$, $\mathcal{B}_\mathbb{X}$, is taken as $\mathcal{E}_\mathbb{X}$. But, depending on the context, we will be able to incorporate not only real-valued experimental information (as in [12]) but also fuzzy-valued information, by using the probability distribution induced by an experimental fuzzy random variable (as in [24]).

A *decision rule* is a way to make a decision based on the sample information. More formally:

Definition 2. *Let* $(\mathbb{X}, \mathcal{B}_\mathbb{X}, P_\theta)$ *be the probability space of a statistical experiment* $\mathbf{X}$ *associated with the decision problem* $(\Theta, \mathbf{A}, U)$*. A decision rule is a mapping* $d : \mathbb{X} \to \mathbf{A}$*.*

When considering the *normal Bayesian analysis*, we should find a *Bayes decision rule*, that is, a rule d_B such that

$$\int_{\Theta}\left(\int_{\mathbb{X}} U(\theta, d_B(x))\, dP_\theta(x)\right) d\pi(\theta) \geq_{\lambda,\mu} \int_{\Theta}\left(\int_{\mathbb{X}} U(\theta, d(x))\, dP_\theta(x)\right) d\pi(\theta)$$

for every decision rule d. In this case, the "value" of the problem is

$$\int_{\Theta}\left(\int_{\mathbb{X}} U(\theta, d_B(x))\, dP_\theta(x)\right) d\pi(\theta). \tag{1}$$

On the other hand, we can consider the *extensive Bayesian analysis*. We should obtain for each sample information x a posterior Bayes action a^{π_x}, and consider the decision rule which associates with each x an action a^{π_x}. In this analysis, the "value" of the experiment $\mathbf{X}$ is quantified by the fuzzy expected terminal utility, defined as follows:

Definition 3. *Given* $(\Theta, \mathbf{A}, U)$ *a decision problem and* $\mathbf{X} = (\mathbb{X}, \mathcal{B}_{\mathbb{X}}, P_\theta)$*, an associated experiment, the fuzzy expected terminal utility of* $\mathbf{X}$ *is given by*

$$U_t(\mathbf{X}) = \int_{\mathbb{X}}\left(\int_{\Theta} U(\theta, a^{\pi_x})\, d\pi_x(\theta)\right) dP(x). \tag{2}$$

Hence, the coherence Bayesian analysis passes through obtaining equivalent results when performing each one of the forms, normal and extensive. That is: Are (1) and (2) equal?

4 Fubini-Type Model

When we are considering a product measurable space structure, in which not only the cartesian product set is constructed but also the product σ-field, we can use, under certain conditions, Fubini-type results. We are examining now the conditions to guarantee the equivalence between the two forms of the Bayesian analysis, that is, when (1) and (2) are equal, in the λ, μ-average sense (see [12,17]).

Theorem 1. *Let* $(\Theta, \mathbf{A}, U)$ *be a decision problem, let* $\Theta \subset \mathbb{R}$ *and let* π *be a prior probability on* $(\Theta, \mathcal{E}_\Theta)$ *such that the probability space is complete. Let* $\mathbf{X} = (\mathbb{X}, \mathcal{E}_{\mathbb{X}}, P_\theta)$ *be an associated experiment, and let* P *be the marginal distribution.*
If, for every decision rule $d : \mathbb{X} \to \mathbf{A}$ *it holds that:*

i) the mapping $U(\cdot, d(\cdot)) : \Theta \times \mathbb{X} \to \mathcal{F}_c$ *is an integrably bounded r.u.s.f. with respect to* $(\Theta \times \mathbb{X}, \mathcal{E}_\Theta \otimes \mathcal{E}_{\mathbb{X}}, \Pi)$ *(being* $\mathcal{E}_\Theta \otimes \mathcal{E}_{\mathbb{X}}$ *the product* σ*-algebra and* Π *the joint probability distribution),*
ii) there exists $f \in L^1(\Pi)$ *such that* $\|U(\theta, d(x))\| \leq f(\theta, x)$*, for every* $(\theta, x) \in \Theta \times \mathbb{X}$*,*
iii) for every $\theta \in \Theta$*, it holds that projection* $f_\theta \in L^1(P_\theta)$ *and expectation* $\int_{\mathbb{X}} f_\theta(x) dP(x) \in L^1(P)$*,*

iv) for every $x \in \mathbb{X}$, it holds that projection $f_x \in L^1(\pi_x)$ and expectation $\int_\Theta f_x(\theta)d\pi_x(\theta) \in L^1(\pi)$

then,

$$\int_{\mathbb{X}} \left(\int_\Theta U(\theta, d(x))\, d\pi_x(\theta) \right) dP(x) = \int_\Theta \left(\int_{\mathbb{X}} U(\theta, d(x))\, dP_\theta(x) \right) d\pi(\theta),$$

whatever the decision rule d.

Once we have established the conditions for exchangeability of integrals, we can state the following result abut equivalence of Bayesian analyses.

Theorem 2. *Assume the conditions of Theorem 1. Let us consider the mapping which associates with each sample $x \in \mathbb{X}$ a posterior Bayes action a^{π_x}. If this mapping satisfies the definition of decision rule, then it is a Bayes decision rule. Moreover, $U_t(\mathbf{X})$ is equal, in the λ, μ-average sense, to the fuzzy expected utility associated with any Bayes decision rule, this is*

$$U_t(\mathbf{X}) =_{\lambda,\mu} \int_{\mathbb{X}} \left(\int_\Theta U(\theta, d_B(x))\, d\pi_x(\theta) \right) dP(x)$$

Remark 1. At this point, we have to remark that these results hold when the experiment is modeled by a fuzzy random variable X, since in that case $\mathbb{X} \subseteq \mathcal{F}_c$ and we can consider the σ-field induced by X, as described in Section 2, that is $(\mathcal{F}_c, \mathcal{D}_j)$ and the marginal distribution P will be the induced probability ξ^X. Since no other constraints have been imposed on the experimental sample space, we can use Theorems 1 and 2 also when adding fuzzy information to our single-stage decision problem (in [24,25] we can see how to incorporate that type of information in graphical tools for decision problems).

5 Non Fubini-Type Model

In this section we analyze how to proceed in Bayesian analysis when no product space can be constructed or, even if it is defined, conditions about product measurability or integrability are not attained.

In this case we have to require different conditions. First main difference with respect to model in Section 4 is that, in this case, we have to ask Θ, the state space, to be an interval of $\mathbb{R}$,. This condition does not use to be very constraining in most of usual contexts, moreover if we consider Bayesian decision analysis in Statistical Inference problems. The reason why we have to impose this condition is the way we obtain the exchange in the order of integration: not based on a product space, but just considering separated integrals. $\mathcal{B}_\Theta$ is the Borel σ-field on Θ, with m the Borel measure and $\mathbf{A}$ is the action space. When considering decision rules in this framework we also have to modify conditions in Theorem 1 in order to fulfill properties established in [26]. Mainly, we are asking for different integrable bounds and measurability of sections. Second difference is that we need now the sample space, $\mathbb{X}$ to be a real subset. We will later provide details about the reason of this constraint.

Theorem 3. *Let $(\Theta, \mathbf{A}, U)$ be a decision problem, let $\Theta \subset \mathbb{R}$ and let π be a prior probability on $(\Theta, \mathcal{B}_\Theta)$ such that $\pi \ll m$ with a continuous Radon-Nikodym derivative. Let $\mathbf{X} = (\mathbb{X}, \mathcal{B}_\mathbb{X}, P_\theta)$ be an associated experiment, and let P be the marginal distribution. For every $\theta \in \Theta$, suppose that $P_\theta \ll P$ and there exists a continuous Radon-Nikodym derivative. For every $x \in \mathbb{X}$, let π_x be the posterior distribution on $(\Theta, \mathcal{B}_\Theta)$ such that $\pi_x \ll m$ with a continuous Radon-Nikodym derivative. If, for every decision rule d it holds that:*

i) for every $\theta \in \Theta$, $U(\theta, d()) : \mathbb{X} \to \mathcal{F}_c$ is an integrably bounded r.u.s.f. with respect to P_θ,
ii) for every $x \in \mathbb{X}$, $U(, d(x)) : \Theta \to \mathcal{F}_c$ is an integrably bounded r.u.s.f. with respect to π_x, moreover, it is continuous a.s. $[P]$,
iii) there exists $h_1 \in L^1(\mathbb{X}, \mathcal{B}_\mathbb{X}, P)$ such that $\left\|U(\theta, d(x))\frac{d\pi_x}{dm}(\theta)\right\| \le h_1(x)$ a.s. $[P]$ for every $\theta \in \Theta$, and the mapping $x \mapsto U(\theta, d(x))\frac{d\pi_x}{dm}(\theta)$ is continuous a.e. $[m]$,
iv) there exists $g \in L^1(\Omega, \mathcal{B}_\Omega, m)$ such that for every $x \in \mathbb{X}$, it holds that $\|U(\theta, d(x))\frac{d\pi_x}{dm}(\theta)\| \le g(\theta)$ a.e. $[m]$ for every $x \in \mathbb{X}$,
v) the mapping $\theta \mapsto U(\theta, d(x))\frac{dP_\theta}{dP}(x)$ is continuous on Θ a.s. $[P]$,
vi) there exists $h_2 \in L^1(\mathbb{X}, \mathcal{B}_\mathbb{X}, P)$ such that $\left\|U(\theta, d(x))\frac{dP_\theta}{dP}(x)\right\| \le h_2(x)$ a.s. $[P]$ for every $\theta \in \Theta$,
vii) there exists $g' \in L^1(\mathbb{X}, \mathcal{B}_\mathbb{X}, P)$ with $\int_\Theta \|U(\theta, d(x))\| \, d\pi_x(\theta) \le g'(x)$.

If for every $\theta \in \Theta$, it holds that $\frac{d\pi_x}{dm}(\theta) = \frac{dP_\theta}{dP}(x)\frac{d\pi}{dm}(\theta)$ a.s. $[P]$, then

$$\int_\mathbb{X}\left(\int_\Theta U(\theta, d(x))\, d\pi_x(\theta)\right) dP(x) = \int_\Theta\left(\int_\mathbb{X} U(\theta, d(x))\, dP_\theta(x)\right) d\pi(\theta)$$

whatever the decision rule $d : \mathbb{X} \to \mathbf{A}$ may be.

As a consequence, the following key result, which states the equivalence between the normal and extensive forms of Bayesian analysis, is obtained.

Theorem 4. *Assume the conditions of Theorem 3. Let us consider the mapping which associates with each sample $x \in \mathbb{X}$ a posterior Bayes action a^{π_x}. If this mapping satisfies the definition of decision rule, then it is a Bayes decision rule. Moreover, $U_t(\mathbf{X})$ is equal, in the λ, μ-average sense, to the fuzzy expected utility associated with any Bayes decision rule, this is*

$$U_t(\mathbf{X}) =_{\lambda,\mu} \int_\mathbb{X}\left(\int_\Theta U(\theta, d_B(x))\, d\pi_x(\theta)\right) dP(x)$$

Remark 2. A natural question arises now: Can we add fuzzy-valued experimental information in this model developed in Section 5? The answer is no, at least by now. The requirement of being $\mathbb{X} \subseteq \mathbb{R}$ is derived from the use of fuzzy-valued version of the fundamental theorem of calculus for Hukuhara derivative (see [15,20,23]) to obtain the result about exchangeability of integrals. Therefore, to obtain such a result with a fuzzy-valued sample parameter would need a proper definition of differential for a fuzzy-valued function with a fuzzy-valued parameter and, thus, similar results to fundamental theorem of calculus in those cases.

References

1. Billot, A.: An existence theorem for fuzzy utility functions: A new elementary proof. Fuzzy Sets and Systems 74, 271–276 (1995)
2. Bordley, F.: Reformulating decision theory using fuzzy set theory and Shafer's theory of evidenc. Fuzzy Sets and Systems 139, 243–266 (2003)
3. Chen, C.B., Klein, C.M.: A simple approach to ranking a group of aggregated fuzzy utilities. IEEE Transactions on Systems, Man, and Cybernetics 27, 26–35 (1997)
4. Colubi, A., Domínguez-Menchero, J.S., López-Díaz, M., Ralescu, D.A.: A $D_E[0,1]$ representation of random upper semicontinuous functions. Proceedings of the American Mathematical Society 130, 3237–3242 (2002)
5. De Campos, L.M., González, A.: A subjective approach for ranking fuzzy numbers. Fuzzy Sets and Systems 29, 145–153 (1989)
6. Debreu, G.: Integration of correspondences. In: Proc. Fifth Berkeley Sympos. Math. Statist. and Probability, 1965/66. Contributions to Probability Theory, Part 1, vol. II, pp. 351–372. University of California Press, Berkeley (1967)
7. Dubois, D., Prade, H.: Additions of interactive fuzzy numbers. IEEE Transactions on Automatic Control 26, 926–936 (1981)
8. Dubois, D., Prade, H.: The use of fuzzy numbers in decision analysis. In: Fuzzy Information and Decision Processes, pp. 309–321. North-Holland, Amsterdam (1982)
9. Dubois, D., Prade, H.: Possibility Theory. In: An Approach to Computerized Processing of Uncertainty. Plenum Press, New York (1988)
10. Friedman, H.: A consistent Fubini-Tonelli theorem for nonmeasurable functions. Illinois Journal of Mathematics 24, 390–395 (1980)
11. Gil, M.A., Jain, P.: Comparison of experiments in statistical decision problems with fuzzy utilities. IEEE Transactions on Systems, Man, and Cybernetics 22, 662–670 (1992)
12. Gil, M.A., López-Díaz, M.: Fundamentals and Bayesian analyses of decision problems with fuzzy-valued utilities. International Journal of Approximate Reasoning 15, 203–224 (1996)
13. Gil, M.A., López-Díaz, M., Rodríguez-Muñiz, L.J.: An improvement of a comparison of experiments in statistical decision problems with fuzzy utilities. IEEE Transactions on Systems, Man, and Cybernetics 28, 856–864 (1998)
14. Hiai, F., Umegaki, H.: Integrals, conditional expectations and martingales of multivalued functions. Journal of Multivariate Analysis 7, 149–182 (1977)
15. Hukuhara, M.: Intégration des applications mesurables dont la valeur est un compact convexe. Funkcialaj Ekvacioj 10, 205–223 (1967)
16. Krätschmer, V.: Coherent lower previsions and Choquet integrals. Fuzzy Sets and Systems 138, 469–484 (2003)
17. López-Díaz, M., Gil, M.A.: Reversing the order of integration in iterated expectations of fuzzy random variables, and statistical applications. Journal of Statistical Planning and Inference 74, 11–29 (1998)
18. López-Díaz, M., Gil, M.A.: The λ-average value and the fuzzy expectation of a fuzzy random variable. Fuzzy Sets and Systems 99, 347–352 (1998)
19. Puri, M.L., Ralescu, D.A.: Différentielle d'une fonction floue. Comptes Rendus de l'Académie des Sciences. Série I. Mathématique 293, 237–239 (1981)
20. Puri, M.L., Ralescu, D.A.: Differentials of fuzzy functions. Journal of Mathematical Analysis and Applications 91, 552–558 (1983)
21. Puri, M.L., Ralescu, D.A.: Fuzzy random variables. Journal of Mathematical Analysis and Applications 114, 409–422 (1986)

22. Rébillé, Y.: Decision making over necessity measures through the Choquet integral criterion. Fuzzy Sets and Systems 157, 3025–3039 (2006)
23. Rodríguez-Muñiz, L.J., López-Díaz, M.: Hukuhara derivative of the fuzzy expected value. Fuzzy Sets and Systems 138, 593–600 (2003)
24. Rodríguez-Muñiz, L.J., López-Díaz, M., Gil, M.A.: Equivalence between normal and extensive forms of Bayesian analysis in statistical decision problems with imprecise utilities. European Journal of Operational Research 167, 444–460 (2005)
25. Rodríguez-Muñiz, L.J., López-Díaz, M.: Influence diagrams with super value nodes involving imprecise information. European Journal of Operational Research 179, 203–219 (2007)
26. Rodríguez-Muñiz, L.J., López-Díaz, M.: On the exchange of iterated expectations of random upper semicontinuous functions. Statistics and Probability Letters 77, 1628–1635 (2007)
27. Rodríguez-Muñiz, L.J., López-Díaz, M.: A new framework for the Bayesian analysis of single-stage decision problems with imprecise utilities. Fuzzy Sets and Systems 159, 3271–3280 (2008)
28. Tong, R.M., Bonissone, P.P.: A linguistic approach to decision making with fuzzy sets. IEEE Transactions on Systems, Man, and Cybernetics 10, 716–723 (1980)
29. Watson, S.R., Weiss, J.J., Donnell, M.L.: Fuzzy decision analysis. IEEE Transactions on Systems, Man, and Cybernetics 9, 1–9 (1979)
30. Zadeh, L.A.: The concept of a linguistic variable and its application to approximate reasoning. Parts I, II and III. Information Science 8, 199-249; 8; 301–357; 9; 43–80 (1975)

A Neuro-Fuzzy Decision Support System for Selection of Small Scale Business

Rajendra Akerkar[1] and Priti Srinivas Sajja[2]

[1] Senior Researcher, Vestlandsforsking, Norway
rak@vestforsk.no
[2] Associate Professor, Sardar Patel University, India
priti_sajja@yahoo.com

Abstract. Artificial Neural Network (ANN) and Fuzzy Logic (FL) are two important and useful technologies having their strengths and weaknesses. The combination of fuzzy logic and neural networks constitutes a powerful means for intelligent system development and offers dual advantages of the technologies. This article describes four approaches of neuro-fuzzy systems with their broad design and also presents general structure of a business advisory system using hybrid neuro-fuzzy approach. The system utilizes ANN that considers basic parameters and data from the environment for selection of a small-scale business in the given area and generates rules accordingly. Finally, the article presents sample rules extracted from the neuro-fuzzy system, screens for the interface design and parameters for implementation.

Keywords: Decision Support, Rural Development, Business Advisory System, Implementation of Parameters, Neuro-Fuzzy Systems, Rule Extraction.

1 Introduction

Both Artificial Neural Network (ANN) and Fuzzy Logic (FL) are powerful design techniques with their strengths and weaknesses. The combination of fuzzy logic and neural networks constitutes a powerful means for designing classical control scheme, analytical and diagnostic systems, image processing, forecasting, and prediction etc. and hence achieves dual advantages of both the technologies.

Fuzzy Logic was formulated by Zadeh [1] as multi-valued logic between 0 and 1 which was developed as a response to the fact that human routinely categorizes items into classes that have no specific boundaries. FL is widely applicable especially in system controls that are very complex, uncertain and cannot be modeled precisely even under various assumptions and approximations. Such systems provide framework for approximate reasoning and allow qualitative knowledge about a problem to be implemented into systems using fuzzy rules. This makes system more user friendly, readable, easy to maintain and effectively reduces the complexity by reducing number of rules. The key benefit of FL is simple "if-then" relations to describe systems behavior, which leads to simpler solution in less design time. However, the designer has to derive the "if-then" rules from the data sets manually, if neural network technique is not utilized; which requires a major effort with large data sets.

E. Hüllermeier, R. Kruse, and F. Hoffmann (Eds.): IPMU 2010, Part II, CCIS 81, pp. 306–315, 2010.

On other hand, the ANN works on implicit representation of knowledge and promises a solution when data sets contain knowledge about the system to be designed, as it can train itself from the data sets. As neural net solutions remain a "black box", it is difficult to interpret or manually change it. Further, it is difficult to document knowledge and provide explanation and reasoning to justify the solution as knowledge is not represented explicitly in the network. In addition to these, lack of an easy way to verify and optimize a neural net solution is probably the major limitation of the ANN technique.

ANN can learn from large data sets while FL solutions are simple, easy to verify and optimize. Clever combinations of these two technologies deliver best of both worlds. Neuro-fuzzy system is combination of ANN and FL technologies in such a way that ANN techniques are used to determine parameters of FL systems. Basic objective of such system is to improve one system by means of another. A more important aspect of the hybridization is that the system should always be interpretable in terms of fuzzy "if-then-else" rules [2].

In Section 2 of this paper four major approaches of modeling neuro-fuzzy systems are discussed. Section 3 discusses structure and rule extraction of pure hybrid neuro-fuzzy system. Small scale industries play an important role in development of nation, particularly when the large degree of population belongs to rural areas. From the data available at the corporate sector and domain heuristic collection, a neuro-fuzzy system can be used to decide a suitable business effectively. The structure and design of an advisory system using modified hybrid model of neuro-fuzzy system is described in Section 4.

2 Approaches of Neuro-Fuzzy Computing

There are four major approaches of the hybrid neuro-fuzzy computing namely (i) fuzzy neural network model; (ii) concurrent neuro-fuzzy model; (iii) co-operative neuro-fuzzy model; and (iv) hybrid neuro-fuzzy model.

The first model, Fuzzy Neural Network is combination of enhanced learning capabilities of neural network by fuzzy system. This approach is also useful in creating neural network that operates on fuzzy inputs. This approach proposed by Sajja [3], [4] offers dual advantages of FL and ANN technologies with utmost simplicity and ease of implementation in a distributed way. Both the FL and ANN components can be developed and tested independently and integrated further. This approach is also useful in creating neural network that operates on fuzzy inputs [4], [5], [6], [7].

The second model, Co-operative Neuro-Fuzzy uses the neural net to decide and enhance parameters of fuzzy system. Once parameters and rules are defined, the fuzzy system solves problem more efficiently [2], [5].

In the third model also, which is known as Concurrent Neuro-Fuzzy model, ANN determines the FL systems parameters as mentioned in the second model. The only difference is that, this process is continuous. This is useful for dynamic situations where the interface variables can be directly measured by FL system. According to Abraham [5], this model highlights opportunity of use of ANN as pre-processor and/or interface of FL systems.

Hybrid neuro-fuzzy model supports special neural network with fuzzy parameters. Here, the neural network in multi layer architecture can extract fuzzy rules. This is an alternative way of development of the fuzzy system using distributed approach in layers. The multipurpose models like ANFIS [8], NEFCLASS [9], Fuzzy Rule Net [10] etc. have been developed using this approach. Many researchers such as Rutkowski and Cpalka [11] and Sajja [12] have designed flexible neuro-fuzzy applications using the same approach.

Fig. 1 demonstrates broad framework of the above approaches.

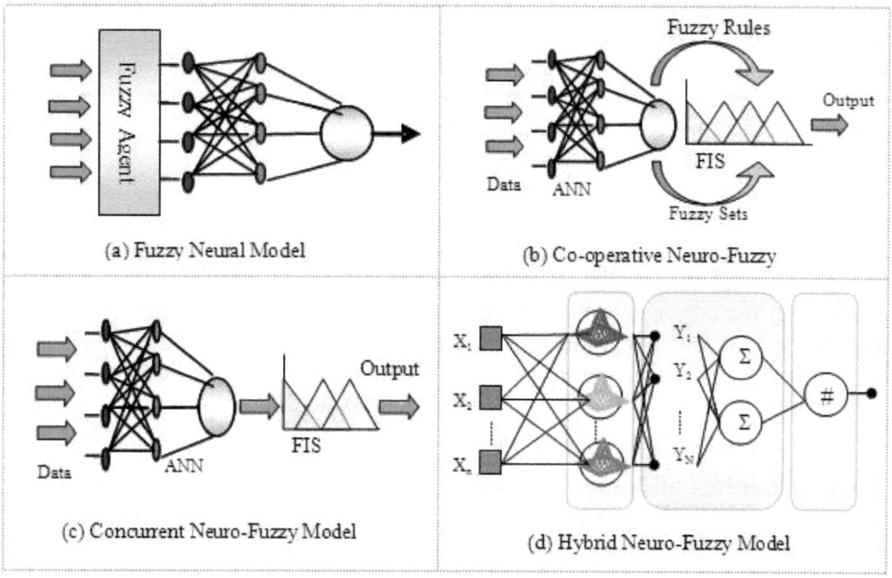

Fig. 1. Approaches of neuro-fuzzy computing

Hybrid Neuro-fuzzy systems have wide scope and ample abilities in different fields. Jang [13] applied fuzzy neural net ANFIS to the control of the inverted pendulum as well as for modeling of nonlinear functions. Takagi [14] proposed a NN structural method based on approximate reasoning rules and applied it to the pattern recognition problem. State of the art modeling techniques for neuro-fuzzy system have been thoroughly discussed by Ajit Abraham [5]. A survey of neuro-fuzzy rule generation is presented by Mitra and Hayashi [15]. Recently a new document page segmentation method using a neuro-fuzzy methodology is proposed by Laura Caponetti, et al [16]. Siu-Yeung Cho, Chai Quek, Shao-Xiong Seah, and Chin-Hui Chong [17] presented a neuro-fuzzy network for visual based traffic monitoring system. A course selection and advisory system is designed by Sajja [18] using neuro-fuzzy approach. For systems where there is availability of ample data sets but production of generalized rule is tedious, hybrid neuro-fuzzy system is effective solution. Application discussed in section 4 is suitable for hybrid neuro-fuzzy approach and hence it is decided to use the same. Section 3 discusses the hybrid approach in detail.

3 General Structure and Rule Extraction in Hybrid Neuro-Fuzzy System

The neuro-fuzzy system has the ability of self-learning and it can auto generate rules from data historians other than from expert's prior knowledge [19], [20]. A neuro-fuzzy system can be trained to develop "if-then" fuzzy rules and determine membership functions for input and output variables of the system [21]. The aim of the rule extraction is to get rules in the following form from input–output data historians:

$$R^{(k)} : \text{If } x_1 \text{ is } A_1,\ x_2 \text{ is } A_2\ , \ldots.,\ x_m \text{ is } A_m \text{ then } Y \text{ is } Y_k$$

Where $R^{(K)}$ represents the kth rule; x_1, x_2, x_m are input variables; Y is the output variable; A_1, A_2, A_m are input fuzzy sets; and Y_K is corresponding output fuzzy sets.

Such neuro-fuzzy systems can be initialized with the input data set given [19], [20]. For the dataset containing n dimensions with m sample sets, the jth sample can be represented as{x1j, x2j, x3j, x4j,...Xnj, Yj}, (j=1,2,...,m).

The sample data set is used to generate the initial decision table as below.

U	x_1	x_2	$x_{...}$	x_i	x_n	y
1	x_{11}	X_{21}	$x_{...1}$	x_{i1}	x_{n1}	y_1
...	$x_{1...}$	$X_{2...}$	$x_{......}$	x_i ...	$x_{n...}$	$y_{...}$
m	x_{1m}	X_{2m}	$x_{...m}$	x_{im}	x_{nm}	y_m

The above initial decision table can be used to initialize the equivalent neural network using the confidence function at r^{th} rule is $\mu_r = \prod \mu A_k (X_{ij})$.

Structure of the neuro-fuzzy system [21], [22] is described in detail in Fig. 2.

Layer 1 shown in the Fig. 2 is input layer. Each neuron in this layer transmits normalized values from the environment to the next layer.

Layer 2 is fuzzification layer. Neurons in this layer represent fuzzy sets used in the antecedents of fuzzy rules. A fuzzification neuron receives a normalized input from the previous layer and determines the degree to which this input belongs to the neuron's fuzzy set. The activation function of a membership neuron is set to the function that specifies the neuron's fuzzy set.

Layer 3 is fuzzy rule layer. Each neuron represents a fuzzy rule. A fuzzy rule neuron receives inputs from the fuzzification neurons of layer 2 that represent fuzzy sets in the rule antecedents. For instance, neuron R1, which corresponds to Rule 1, receives inputs from neurons A1 and B1. In a neuro-fuzzy system, intersection can be implemented by the product operator or by Minimum operator.

Layer 4 is output membership layer. Neurons in this layer represent fuzzy sets used in the consequent of fuzzy rules. An output membership neuron combines all its inputs by using the fuzzy operation union. The probabilistic OR or Maximum operator can implement this operation.

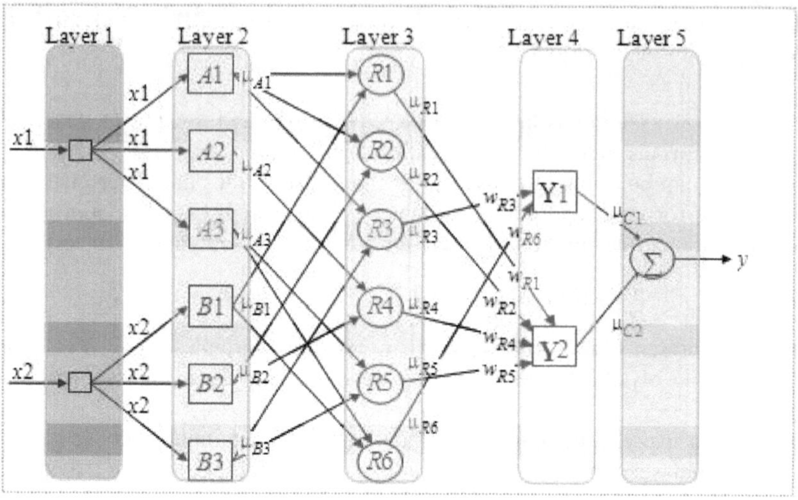

Fig. 2. Hybrid neuro-fuzzy system

Layer 5 is defuzzification layer. Each neuron in this layer represents a single output of the neuro-fuzzy system. It takes the output fuzzy sets clipped by the respective integrated firing strengths and combines them into a single fuzzy set. Neuro-fuzzy systems can apply standard defuzzification methods, including techniques like centroid and sum-product composition.

4 Application

A significant percentage of world's population lives in rural area. One way to accelerate economic development is to create employment in the rural sectors, which needs information/advice about rural industries. In developing countries like India, small-scale industries are playing a special role in employment creation with low capital investment. These kind of small industries are labor intensive, require less capital investment and they provide employment to the target audience. One of the major requirements for the field is the support in decision making to select appropriate business.

Since human experts and their knowledge are limited commodities in the field, such automated system can help in proper selection of business and motivate users for the business by providing necessary guidance.

Several systems have been developed to automate advisory system using different technologies such as multi layer knowledge-based system considering the layers of information and knowledge related to the application of business advisory [23, 24]. However, such systems are single PC based legacy systems that offer comparatively less degree of self learning (in comparison with the proposed approach). Most of the advisory systems are for big business activities like product innovation, enterprise wide resource planning, supply chain management and small process like payroll and

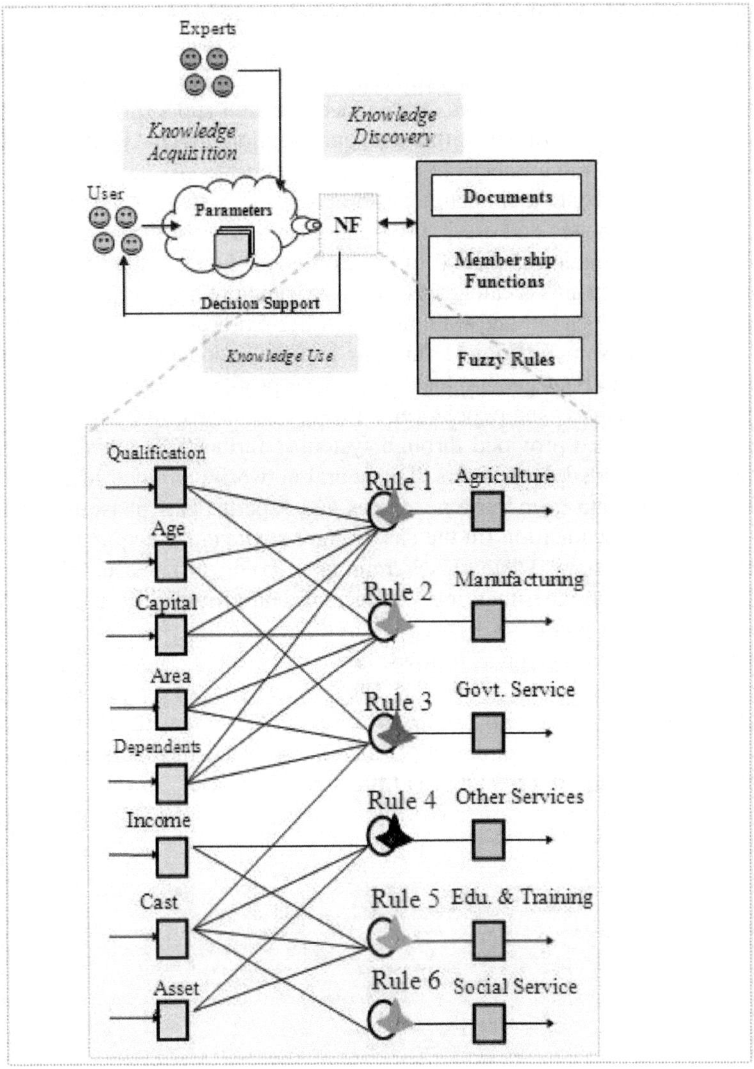

Fig. 3. Structure of the system

attendance management. Advisory and counseling systems also being used in student course selection [12] and determination of student's aptitude using theory of multiple intelligence [25].

The small scale business is normally supported through office automation packages and accounting software. Full fledged advisory system may be planned to accelerate employment using the fuzzy approach. The neural network utilizes available data for learning trends and providing effective decisions. Users' data and parameters are collected through fuzzy linguistic parameters and explain with proper justification. Lot of history data are available at different level such as population information from government, information from corporate & NGOs sector, districts, blocks, villages, etc. This leads to the development of an automated support that learns from the data and generates rules to provide decision support. On the basis of above objectives, an electronic adviser using hybrid neuro-fuzzy approach is presented for small and cottage industries. This also provides an intended decision support. For such assistance and advisory, different parameters like age, qualification, income, and interest of user, environment of area, and number of dependent are considered.

Fig. 3 represents basic architecture of the proposed system. As denoted in Fig. 3, environmental parameters and user's information are collected through fuzzy linguistic interface. This information is passed to the base neuro-fuzzy system which categories the users into broad business categories such as agriculture, small scale manufacturing, service providers, education and training. The rules extracted are stored in the system along with membership function definitions and other documents. Other documents contain information about geographical and social profiles of the areas under consideration, possible business and population.

The business advice provided through system is further fine tuned by fuzzy rules and details are provided to the users. The neural network is trained in offline manner with the available data from various sources and experts. This phase may be considered as knowledge acquisition (in the case when knowledge is acquired from experts) or knowledge discovery (when ANN training is going on) phase. The discovered knowledge is used in experimenting with actual data provided by users, which falls under knowledge use phase.

Example rules can be given as follows:

If qualification(-0.76), area(-1.6) and dependents(1.9) then opt for more education.

If qualification(1.6), capital(-1.0), area(-2.9) and dependents(0) then opt for job.

Fig. 4 describes initial membership function, adjusted membership functions for the input variable *Qualification* and final fuzzy rules obtained.

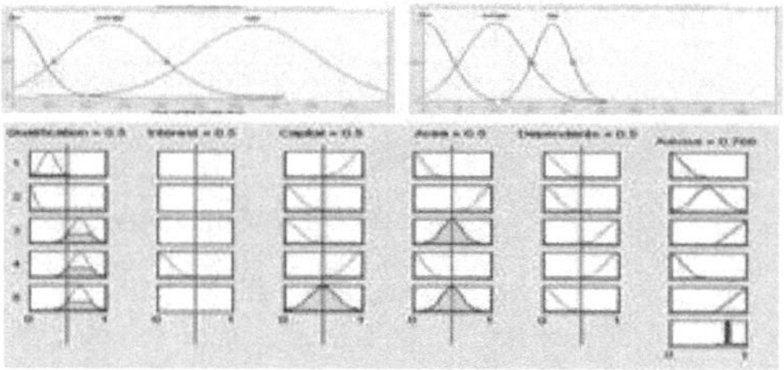

Fig. 4. Initial membership functions (top left), adjusted membership functions (top right) for the input variable *Qualification* and final fuzzy rules obtained

Parameter settings [26] for the neuro-fuzzy model are given below. (The software NEFCLASS-J developed by Nauck and Kruse [9] can be used for implementation.)

Parameters	Details
Training data file	Advisory.dat
Number and type of fuzzy sets	Multiple Gaussian

Aggregation function	Maximum
Size of rule base	Automatically determined
Learning rule	Best per class
Fuzzy sets constraints	(i) Keep relative order (ii) Always overlap
Learning rate	0.1
Stop Control	Maximum number of epochs=100 Minimum number of epochs=0 Number of epochs after optimum=10 Admissible classification errors=0

Fig. 5 represents interface design showing input and output screens for the advisory system.

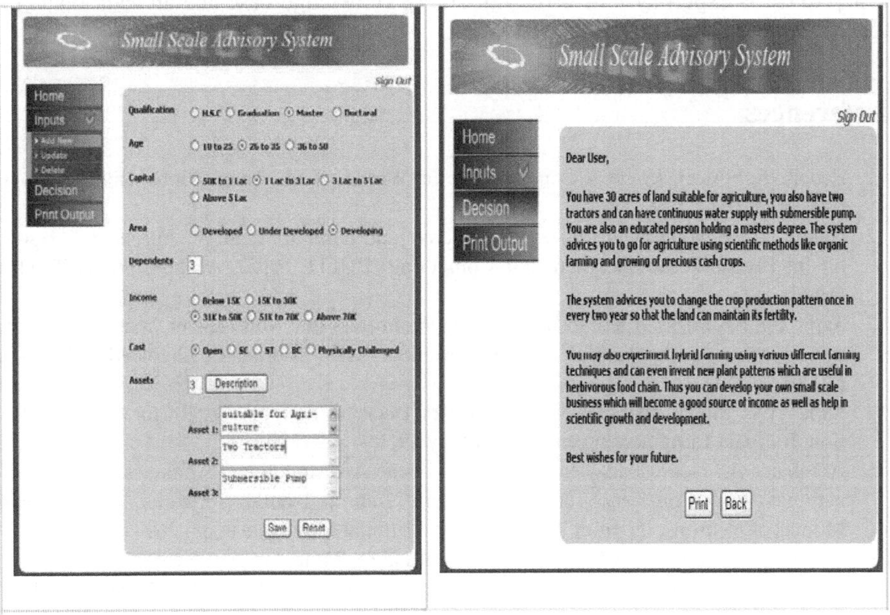

Fig. 5. Interface design

5 Conclusion

ANN is a useful technique and has received a lot of attention because of its ability to learn from large data sets. However, a major drawback associated with the use of neural network for decision-making is the lack of explanation capability and explicit representation of knowledge. While they can achieve a high predictive accuracy rate, the reasoning behind how they reach their decisions is not readily available. On other hand FL technique is simple, flexible and has ability to provide explanation explicitly. Combining both the techniques serves dual advantages like self-learning, flexibility, and explicit explanation.

The proposed advisory system learns from the data available at multiple sources and generates rules to provide decision support along with justification of the advice. Simultaneously, there is a critical requirement of interactive user interface to increase effectiveness and scope of the advisory system. FL based system solves the problem efficiently. So business advisory system is thought as one of the suitable candidates to demonstrate the approach.

The proposed advisory system is meant for the users seeking advisory for selection of possible small-scale business. The system can be uploaded on net by Non-Government Organizations (NGOs) and by associations/trusts serving citizens of rural areas; and can be freely distributed at district or village level for rural uplift. A friendly interface with such systems increases scope and effectiveness in a considerable fashion. Such advisory system can be more generalized and can be considered as a step towards expert system shell with empty knowledge base and an additional editor to document domain knowledge.

References

1. Zadeh, L.: Fuzzy Logic - Computing with Words. IEEE Transactions on Fuzzy Systems 4(2), 103–111 (1996)
2. Nauck, D.: Beyond Neuro-Fuzzy: Perspective and Directions. In: 3rd European Congress on Intelligent Techniques and Soft Computing (EUFIT 1995), Aachen, pp. 1159–1164 (1995)
3. Sajja, P.S.: A Fuzzy Agent to Input Vague Parameters into Multi-Layer Connectionist Expert System: An Application for Stock Market. ADIT Journal of Engineering 3, 30–32 (2006)
4. Sajja, P.S.: Fuzzy Artificial Neural Network Decision Support System for Course Selection. Journal of Engineering and Technology 19, 99–102 (2006)
5. Abraham, A.: Neuro-Fuzzy Systems: State-of-the-Art Modeling Techniques. In: 6th International Work-Conference on Artificial and Natural Neural Networks: Connectionist Models of Neurons, Learning Processes and Artificial Intelligence, pp. 269–276 (2001)
6. Narazaki, H., Ralescu, A.L.: A Synthesis Method for Multi-Layered Neural Network using Fuzzy Sets. In: IJCAI 1991: Workshop on Logic in Artificial Intelligence, Sydney, pp. 54–66 (1991)
7. Halgamuge, S.K., Glesner, M.: Neural Networks in Designing Fuzzy Systems for Real Word Applications. Fuzzy Sets and Systems 65, 1–12 (1994)
8. Jang, J.S.: Neuro-Fuzzy Modeling: Architectures, Analyses and Applications, Ph.D Thesis, University of California, Berkeley (1992)
9. Nauck, D., Kruse, R.: NEFCLASS, A Neuro-Fuzzy Approach for the Classification of Data. In: ACM Symposium on Applied Computing, Nashville, pp. 461–465 (1995)
10. Tschichold-Gurman, N.: Generation and Improvement of Fuzzy Classifiers with Incremental Learning using Fuzzy Rulenet. In: ACM Symposium on Applied Computing, Nashville, pp. 466–470 (1995)

11. Rutkowski, L., Cpałka, K.: Flexible Neuro-Fuzzy Systems. IEEE Transactions on Neural Networks 14(1), 554–574 (2003)
12. Sajja, P.S.: Type-2 Fuzzy user Interface for Artificial Neural Network-Based Decision Support System for Course Selection. International Journal of Computing and ICT Research 2(2), 96–102 (2008)
13. Jang, J.S.: ANFIS: Adaptive-Network-Based Fuzzy Inference System. IEEE Transactions on Systems, Man and Cybernetics 23(3), 665–685 (1993)
14. Takagi, H.: Fusion Technology of Fuzzy Theory and Neural Networks: Survey and Future Directions. In: International Conference on Fuzzy Logic & Neural Networks, Iizuka, Japan, pp. 13–26 (1990)
15. Mitra, S., Hayashi, Y.: Neuro-Fuzzy Rule Generation: Survey in Soft Computing Framework. IEEE Transactions on Neural Networks 11(3), 748–768 (2000)
16. Caponetti, L., Castiello, C., Górecki, P.: Document Page Segmentation using Neuro-Fuzzy Approach. Applied Soft Computing 8(1), 118–126 (2008)
17. Cho, S., Quek, C., Seah, S., Chong, C.: HebbR^2-Taffic: A Novel Application of Neuro-Fuzzy Network for Visual Based Traffic Monitoring System. Expert System Application 36(3), 6343–6356 (2009)
18. Sajja, P.S.: Type-2 Fuzzy Interface for Artificial Neural Network. In: Anbumani, K., Nedunchezhian, R. (eds.) Soft Computing Applications in Database Technology: Techniques and Issues. IGI Global Book Publishing, Hershey (2010)
19. Jian, X.L., Hui, H.S.: Developing Soft Sensor using Hybrid Soft Computing Methodology: A Neuro-Fuzzy System Based on Rough Set Theory and Genetic Algorithms. Soft Computing 10(1), 54–60 (2006)
20. Jun, L., Ding, L., Hua, Y., Sheng, W., Xia, H.: A New Strategy for Optimizing the Parameters Updating Algorithm of Fuzzy Neural Controller. Soft Computing - A Fusion of Foundations, Methodologies and Applications 10(1), 61–67 (2006)
21. Negnevitsky, M.: Artificial Intelligence: A Guide to Intelligent Systems. Pearson Education Limited, England (2002)
22. Kasabov, N.: NeuCom - A Neuro-Computing Decision Support Environment (2007), `http://www.aut.ac.nz/research/research_institutes/kedri/research_centres/centrefor_data_mining_and_decision_support_systems/neucom.htm`
23. Sajja, P.S.: Knowledge-Based Systems for Socio-Economic Rural Development. Ph.D Thesis, Sardar Patel University, India (2000)
24. Sajja, P.S.: Multi-Layer Connectionist Model of Expert System for an Advisory System. In: National Level Seminar - Tech. Symposia on IT Futura, Anand, India (2006)
25. Mankad, K.B., Sajja, P.S.: A Design of Encoding Strategy and Fitness Function for Genetic-Fuzzy System for Classification of Student Skills. In: International Conference on Signals, Systems and Automation (ICSSA 2009), Vallabh Vidyanagar, India (2009)
26. Dixon, B.: Prediction of Ground Water Vulnerability using an Integrated GIS-Based Neuro-Fuzzy Techniques. Journal of Spatial Hydrology 4(2), 1–38 (2004)

Bond Management: An Application to the European Market

José Manuel Brotons

Department of Economic and Financial Studies, Miguel Hernandez University,
Avda. de la universidad, s/n, 03202, Elche, Spain
jm.brotons@umh.es

Abstract. Active bond management strategies rely on expectations of interest rate movements or changes in yield-spread relationships. However, the variation of the duration increases the risk of a portfolio, that why the decision maker will have to chose the combination of expected return (mid-point of the fuzzy number) and risk (width of the fuzzy number) which provides the higher utility. The construction of a fuzzy return risk map will allow the DM to know the over risk and the over return as regards immunization strategy for each duration and for each risk aversion of the DM. Finally, we present an application to the European market in which DM will have to forecast the future interest rate and will have to compare for each considered duration, the final yield of the portfolio with the expected one in order to check the validity of the model.

Keywords: Bond management, utility function, European market, Euribor, risk.

1 Introduction

Asset-Liability Management can be traced back to Redington [1] who suggested that the assets and liabilities in actuarial valuation should be treated parallelly. Such authors as Gerber [2] point out that the use of stochastic models is not suitable for the prediction of long term interest rates. They consider it is much more realistic to use discount rates based on fuzzy numbers (FN), since the only data available is that provided by experts.

Terceño *et al.* [3], Brotons [4] and Brotons and Terceño [5] apply immunization strategy in a fuzzy environment. Vercher *et al.* [6] use Sengupta's methodology [7] to optimize a fuzzy portfolio under downside risk measures. Consequently, the main goal of the present study is i) to design active management strategies in a fuzzy environment to determine the return and the risk of a portfolio and ii) to construct return risk maps in a fuzzy environment, using fuzzy utility functions that allow DMs to choose the most appropriate combination of risk and return by modifying portfolio duration.

The structure of the paper is as follows. Section 2 introduces the risk and return of a portfolio using active and passive bond management. The next section presents the construction of a risk and return map of active management. Section 4 illustrates the approach with an application to the European market. Finally, section 5 exposes the most relevant conclusions.

E. Hüllermeier, R. Kruse, and F. Hoffmann (Eds.): IPMU 2010, Part II, CCIS 81, pp. 316–323, 2010.

2 Risk and Return in Active Management

Let us assume that at time t_0 we have a portfolio. For an Investor Planning Horizon (IHP) of $T-t_0$ years, and assuming that the DM has n bonds[1] with maturities t_1-t_0 ... t_n-t_0, the proportion of the investment in each kind of bond has to fulfil the set of equations $\sum_{s=1}^{n} x_s = 1$; $\sum_{s=1}^{n} x_s (t_s - t_0) = T - t_0$. The DM will have to buy b_s type s, $s=1...n$. The cost of the portfolio at t_0 is P_0^F (the prices of the bonds are crisp numbers, P_0^s, and the interest rate is a crisp number, a particular realization of the unknown fuzzy number), being $P_0^F = \sum_{s=1}^{n} b_s P_0^s$ and the value at the IPH $P_{IPH}^F(i) = \sum_{s=1}^{n} b_s P_0^s (1+i)^{T-t_0}$.

According to the dynamic immunization theorem [8], if there is a variation in the interest rates from i to $\tilde{i}_s = (i_s, l_s, r_s)$ with probability p_s, $s=1,...,n$, the value of the portfolio at the IPH moment will be at least $P_{IPH}^F(i)$. The value of the portfolio is $\tilde{P}_{IPH}^F = \left(P_{IPH_C}^F, l_{P_{IPH}}^F, r_{P_{IPH}}^F\right)$[2]. Note that if there is no change in interest rates the accumulated portfolio value remains at $P_{IPH}^F(i)$. However, if interest rates either increase or decrease, the portfolio value increases (due to bond convexity), without any change in the portfolio composition. Now, we define the portfolio return as,

$$\tilde{h}_{IPH} = \left(h_{IPH_C}, l_{h_{IPH}}, r_{h_{IPH}}\right) = \left(\begin{array}{l} \left(\dfrac{P_{IPH_C}^F}{P_0^F}\right)^{\frac{1}{T-t_0}} - 1, \left(\dfrac{P_{IPH_C}^F}{P_0^F}\right)^{\frac{1}{T-t_0}} - \left(\dfrac{P_{IPH_C}^F + r_{P_{IPH}}^F}{P_0^F}\right)^{\frac{1}{T-t_0}}, \\ \left(\dfrac{P_{IPH_C}^F - l_{P_{IPH}}^F}{P_0^F}\right)^{\frac{1}{T-t_0}} - \left(\dfrac{P_{IPH_C}^F}{P_0^F}\right)^{\frac{1}{T-t_0}} \end{array} \right) \quad (1)$$

Consequently, the midpoint of the portfolio return is $m(\tilde{h}_{IPH}) = h_{IPH_C} + \frac{1}{2}\left(r_{h_{IPH}} - l_{h_{IPH}}\right)$ and the width $w(\tilde{h}_{IPH}) = l_{h_{IPH}} + r_{h_{IPH}}$. However, the width for an immunized portfolio is nearly zero.

Assuming, without loss of generality, that the DM expects a reduction in the interest rate, the duration must be increased if the return is to be higher, but the risk will be greater. For a duration equal to D^*, with $D^* > D$, the proportion of the assets must be $\left(x_1^*, ..., x_n^*\right)$, and the number of each kind of bonds that a DM needs to buy is

[1] We assume zero coupon bonds for simplicity, because duration is equal to maturity.

[2] Interest rate is a fuzzy number; however, bonds are bought at a particular price that involves a particular crisp interest rate. That is why we consider a crisp interest if there is no change, and a fuzzy interest rate for any unknown change.

$\left(b_1^*,...,b_n^*\right)=\left(P_0^F x_1^* / P_0^1,...,P_0^F x_n^* / P_0^n\right)$. Hence, the value of the portfolio at the IPH is $\tilde{P}_{IPH,D^*,\tilde{i}}^F=\left(P_{IPH_C}^F\left(D^*,\tilde{i}^*\right),l_{P_{IPH}}^F\left(D^*,\tilde{i}^*\right),r_{P_{IPH}}^F\left(D^*,\tilde{i}^*\right)\right)$, a function of D^* and $\tilde{i}^*=\sum_{s=1}^n p_s\left(i_s,l_s,r_s\right)$.

The return for the new interest rate $\tilde{i}^*=\sum_{s=1}^n p_s\left(i_s,l_s,r_s\right)=\left(i_C^*,l_i^*,r_i^*\right)$ is:

$$\tilde{h}_{IPH,D^*}\left(\tilde{i}\right)=\left(h_{IPH,D^*_C},l_{h_{IPH,D^*}},r_{h_{IPH,D^*}}\right)=$$
$$=\left(\begin{array}{l}\left(\dfrac{P_{IPH_C}^F\left(D^*\right)}{P_0^F}\right)^{\frac{1}{T-t_0}}-1,\left(\dfrac{P_{IPH_C}^F\left(D^*\right)}{P_0^F}\right)^{\frac{1}{T-t_0}}-\left(\dfrac{P_{IPH_C}^F\left(D^*\right)+r_{P_{IPH}}^F\left(D^*\right)}{P_0^F}\right)^{\frac{1}{T-t_0}},\\ \left(\dfrac{P_{IPH_C}^F\left(D^*\right)-l_{P_{IPH}}^F\left(D^*\right)}{P_0^F}\right)^{\frac{1}{T-t_0}}-\left(\dfrac{P_{IPH_C}^F\left(D^*\right)}{P_0^F}\right)^{\frac{1}{T-t_0}}\end{array}\right) \tag{2}$$

The midpoint of the portfolio return,

$$m\left(\tilde{h}_{IPH}\left(D^*\right)\right)=h_{IPH_C}\left(D^*\right)+\frac{1}{2}\left(r_{h_{IPH}}\left(D^*\right)-l_{h_{IPH}}\left(D^*\right)\right) \tag{3}$$

and the width,

$$w\left(\tilde{h}_{IPH}\left(D^*\right)\right)=l_{h_{IPH}}\left(D^*\right)+r_{h_{IPH}}\left(D^*\right) \tag{4}$$

However, our goal is to compare these results with those of an immunized portfolio. The over return $m^*\left(\tilde{h}_{IPH}\left(D^*\right)\right)$ is defined as:

$$\begin{aligned}m^*\left(\tilde{h}_{IPH}\left(D^*\right)\right)&=m\left(\tilde{h}_{IPH}\left(D^*\right)\right)-m\left(\tilde{h}_{IPH}\right)=\\&=h_{IPH_C}\left(D^*\right)+\frac{1}{2}\left(r_{h_{IPH}}\left(D^*\right)-l_{h_{IPH}}\left(D^*\right)\right)-h_{IPH_C}-\frac{1}{2}\left(r_{h_{IPH}}-l_{h_{IPH}}\right)\end{aligned} \tag{5}$$

and the over risk is:

$$w^*\left(\tilde{h}_{IPH}\left(D^*\right)\right)=w\left(\tilde{h}_{IPH}\left(D^*\right)\right)-w\left(\tilde{h}_{IPH}\right)=l_{h_{IPH}}\left(D^*\right)+r_{h_{IPH}}\left(D^*\right)-l_{h_{IPH}}-r_{h_{IPH}} \tag{6}$$

We can construct the fuzzy risk return map taking for each duration, the over risk corresponding to each over return. It is possible to check that the over risk increases in the same way that the over return increases. So, we will have one relation over risk – over return for each duration.

3 Risk and Return Map of Active Management

Once the fuzzy risk map has been constructed, a fuzzy utility function must be defined so that the duration that provides the DM with the best utility can be chosen.

However, the membership function or the probability distribution in an inexact environment is often not specified. In the crisp environment utility functions have been widely studied (see for example [9]). For an approximation to the possibilistic risk premium associated with a fuzzy number, see Carlsson *et al.* [10], Karimi *et al.* [11] and Georgescu [12].

The return of the portfolio at t is not a TFN, and sometimes it is very difficult to define it, especially for the construction of fuzzy utility functions. For this purpose, Sengupta's methodology [7] has been proposed, so the midpoint and the half-width of the interval-valued expectations return at IPH have been taken as the return and the risk of the portfolio. In our opinion, it is the easiest and most powerful way of constructing fuzzy utilities.

The following assumptions have been commonly assumed about the utility function u: more money is better than less (u strictly increases with higher return), and risk aversion, we assume that u is risk averse, and decreases when the risk increases.

The starting point for the utility function definition is an immunized portfolio, the duration of which is equal to that of the IPH, which has a guaranteed return and will increase if there is any variation in interest rate. The initial value of the initial width of the interval used as the return risk is near zero. If the DM wants to increase the portfolio return, he will have to vary the portfolio duration and, consequently, the risk will increase, because the initial risk is almost zero due to the immunization. Only a duration that enhances the portfolio return will be chosen, because any variation will increase the risk. Our main objective is to compare the result of passive and active bond management.

According to active bond management, for an expected increase (decrease) in the interest rate, the DM will have to reduce (increase) the portfolio duration. Nevertheless, this rule presents some problems, as for example, what should a DM do if he expects the interest rates to change from i_0 to $\tilde{i}_s = \left(i_{s_C}, l_{i_s}, r_{i_s}\right)$ with probability p_s for $s = 1,\ldots,n$ and $\sum_{s=1}^{n} p_s = 1$. And, how much will he have to decrease (increase) the portfolio duration to maximize the DM utility?

Denoting the return at the IPH, $m\left(\tilde{h}_{IPH}(D)\right)$, with $m(D)$ and the risk, $w\left(\tilde{h}_{IPH}(D)\right)$, with $w(D)$, in active bond management, several hypotheses can be analyzed: i) increase in risk (width of the possible interval) and decrease in the return (central point of the interval): $m\left(D^*\right) < m(D)$ and $w\left(D^*\right) > w(D)$. Active management is rejected; ii) decrease in the risk and return: this is unlikely because the width of it is almost zero; iii) increase in the return and decrease in the risk, as in the situation above, this is very unlikely, and iv) increase in risk and return. This hypothesis must be analysed. According to these situations, the utility function must be constructed assuming that:

- For $m\left(D^*\right) < m(D)$, DM must consider only immunization.
- For $m\left(D^*\right) - w\left(D^*\right) > m(D) - w(D)$, DM must consider only active bond management.

- For others durations, the utility membership function must be constructed as $\frac{m(D^*)-w(D^*)-m(D)+w(D)}{w(D^*)-w(D)}$. Higher values of this quotient indicates a best consideration for active bond management.

Consequently, the membership function of the utility function for active bond management is the following one, in which the DM will have to choose the D^* that provides the higher utility. For instance, if utility for D^* is zero, the DM will have to choose D, if utility is one, will have to choose D^*, for other situations, DM will have to consider the value of the utility, the lower values provide the best results for immunization.

$$\mu_{GA}(D^*)=\begin{cases} 0 & m(D^*)<m(D) \\ \frac{m(D^*)-w(D^*)-m(D)+w(D)}{w(D^*)-w(D)} & m(D)-w(D)<m(D^*)-w(D^*)<m(D) \\ 1 & m(D^*)-w(D^*)>m(D)-w(D) \end{cases} \quad (7)$$

To define the DM's risk aversion, the following concept must be introduced

$$\varphi_{GA}(D^*)=(\mu_{GA}(D^*))^p, \mu_{GA}(D^*)\in[0,1], p\in\left[\frac{1}{F},F\right] \quad (8)$$

where F is the (non-fuzzy) level of the pessimism parameter for any DM. When the value of $p=1/F$, the DM is said to be an absolutely pessimistic person who takes no risk so, most of the time, he will prefer passive management.

4 Empirical Application

We shall illustrate the proposed methodology with an application to the European market. We are going to consider the period from January 2007 to December 2009, an IPH of 10 months, and two zero coupon bonds with maturities 6 and 12 months. 10 months Euribor rate will be used as the market interest rate at each day of the three years.

Only two situations will be assumed, the first one that we will consider the pessimistic one with a probability 0,40 (obviously, the optimistic one with a probability 0,6) with an interest rate of 80% of the one considered as the market interest, with left and right hand of 75% and 85% of it respectively. On the other hand, for the optimistic assumption we will consider as central point 120% of the market interest, with left and right hand of 115% and 125% of it respectively (optimistic and pessimistic assumption are symmetrical, however, any other hypothesis can be considered). Fig. 1 shows the nine month Euribor rate. On the other hand, Fig. 2. Expected return for a 9 month duration. shows the expected return according the former assumptions for the whole period.

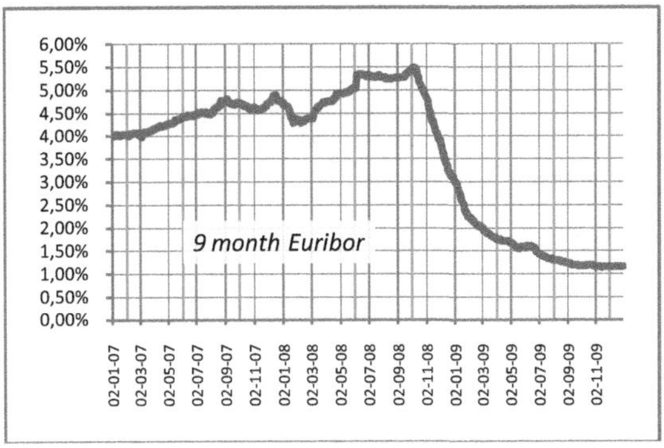

Fig. 1. 9 month Euribor rate

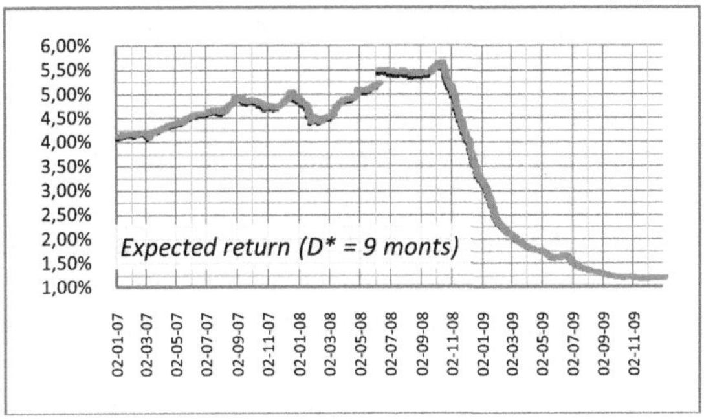

Fig. 2. Expected return for a 9 month duration

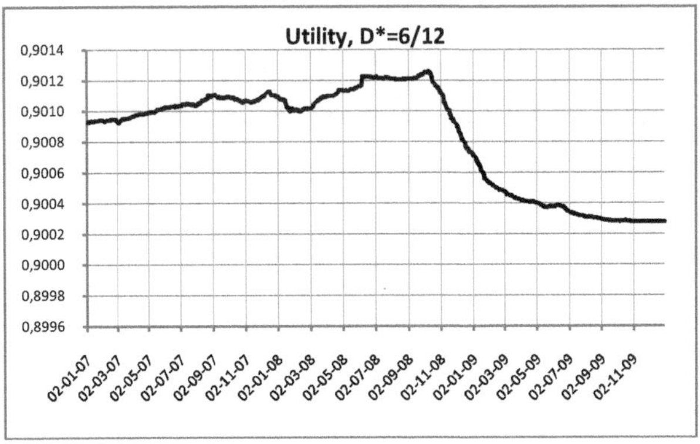

Fig. 3. Utility for a 6 month duration

According to (7) we get the utility function for each duration (from 6 months to 12). Fig. 3 and Fig. 4 show the utility for active bond management, for duration 6 and 10 months respectively. Higher values of utility show more predisposition of the DM to use active bond management.

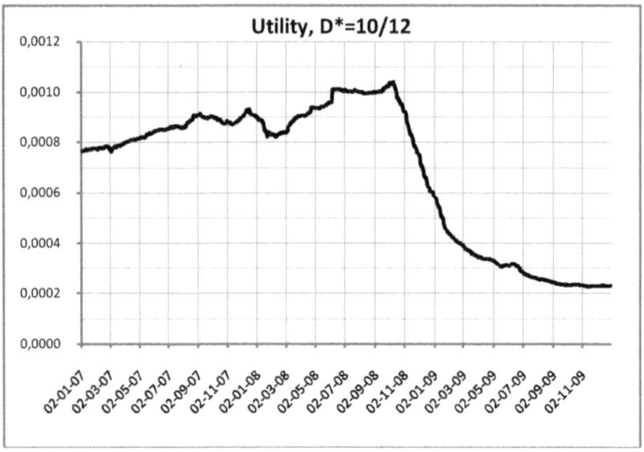

Fig. 4. Utility for a 10 month duration

As we expect a reduction in the interest rate with a probability of 0,4, the best result is obtained for durations of 6 and 7 months. The final step is to check if we get higher returns for those durations. For this purpose, we have assumed that the short time bone has a maturity of 6 months, and we reinvest it at the 4 month Euribor rate existing 6 months later. On the other hand, the long term bond will be sold according to the 2 month Euribor rate existing 10 months later. According to these premises,

Fig. 5 shows the excess of the real return for each duration from the expected return (mid point). The better results are obtained durations of 6, 7 and 8 months (those who get the better utility function).

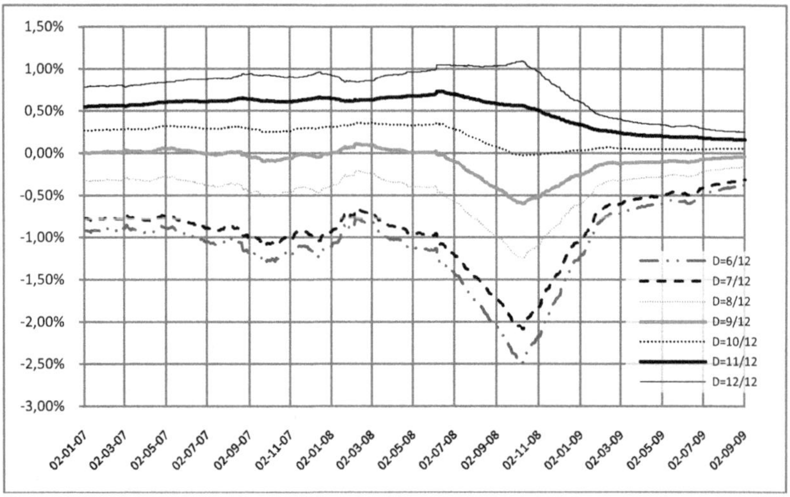

Fig. 5. Excess of the real return for each duration from the expected return (mid point)

5 Conclusions

Knowledge about future interest rates and its probabilities is very uncertain. Fuzzy methodology allows us to work in the field of uncertainty. Most of the time, the only thing that the DM knows about the future interest is the interval within which it can vary. Although fuzzy methodology is very convenient, full knowledge of the membership functions is unlikely. This is why in this paper we deal with the midpoint and half-width of the fuzzy numbers as a measure of their return and risk, respectively.

Several hypotheses, with their probabilities, can be considered, not like a single point, but as an interval in which the rate of interest can change with the same probability.

The proposed utility function allows us to choose the best duration for each interest rate forecast, assuming the use of fuzzy random variables. In this case, the membership function shows us the degree of fulfilment of the DM utility.

Finally, the application to the European market (for short periods) shows the evolution of the utility function in the considered period, and the result of each decision in yields terms.

References

1. Redington, F.M.: Review of the Principles of Life Office Valuations. J. Inst. Actu. 78(350-3), 286–340 (1952)
2. Gerber, H.U.: Life Insurance Mathematics. Springer, Heidelberg (1995)
3. Terceño, A., Brotons, J.M., Fernandez, A.: Immunization strategy in a fuzzy environment. Fuzzy Econ. Rev. XII(2), 95–116 (2007)
4. Brotons, J.M.: Return Risk Map in a fuzzy environment. In: Vanhoof, K., Ruan, D., Li, T., Wets, G. (eds.) Intelligent Decision Making Systems. World Scientific Proceedings Series on Computer Engineering and Information Science, vol. 2, pp. 106–111 (2009)
5. Brotons, J.M., Terceño, A.: Risk premium in the Spanish Market: an empirical study. J. Econ. Computation Econ. Cybernetics Stud. Res. 1, 81–100 (2010)
6. Vercher, E., Bermudez, J.D., Segura, J.V.: Fuzzy portfolio optimization under down site risk measures. Fuzzy Sets Syst. 158, 769–782 (2007)
7. Sengupta, A., Pal, T.K.: Theory and Methodology on comparing interval numbers. Eur. J. Oper. Res. 127, 28–43 (2000)
8. Khang, C.: A dynamic global immunization strategy in the world of multiple interest rate changes: A dynamic immunization and minmax theorem. J. Financ. Quant. Anal. 18, 355–363 (1983)
9. Bell, D.E.: Risk, Return, and Utility. Management Sci. 41(1), 23–30 (1995)
10. Carlsson, C., Fullér, R., Majlender, P.: A possibilistic approach to selecting portfolios with highest utility score. Fuzzy Sets Syst. 131, 13–21 (2002)
11. Karimi, I., Hüllermeier, E.: Risk Assessment system of natural hazards: A new approach based on fuzzy probability. Fuzzy Sets Syst. 158, 987–999 (2007)
12. Georgescu, I.: Possibilistic. Risk aversion. Fuzzy Sets Syst. 160, 2608–2619 (2009)

Estimating the Brazilian Central Bank's Reaction Function by Fuzzy Inference System

Ivette Luna, Leandro Maciel, Rodrigo Lanna F. da Silveira, and Rosangela Ballini

Department of Economic Theory
Institute of Economics, UNICAMP
Sao Paulo, Brazil 13083–857
{ivette,ballini,rodrigolanna}@eco.unicamp.br,
leandro_maciell@hotmail.com

Abstract. A modelling strategy based on the application of fuzzy inference system is shown to provide a powerful and efficient method for the identification of non-linear and linear economic relationships. The procedure is particularly suitable for the estimation of ill-defined systems in which there is considerable uncertainty about the nature and range of key input variables. In addition, no prior knowledge is required about the form of the underlying relationships. Trend, cyclical and irregular components of the model can all be processed in a single pass. The potential benefits of the fuzzy logic approach are illustrated using a model to explain regime changes in Brazilian nominal interest rates. The results suggest that the relationships in the model are basically non-linear.

Keywords: Fuzzy Inference System; Economic modelling; Uncertainty; Non-linear estimation.

1 Introduction

The application of dynamic linear estimation procedures has greatly increased our empirical understanding of the economic system, but there still remains considerable uncertainty about the form and time-stability of many of the key functional relationships.

Part of the problem is that many of the underlying relationships in the economic system may be highly non-linear, and the application of linear estimation methods may lead to significant mistake for precification of both the structure and dynamics of the system. A second major problem arises from the fact that many of the theoretical concepts underlying empirical models are actually quite vague and there is considerable uncertainty about the precise meaning and range of key input variables.

One way to handle problems of the kind discussed above, particularly those connected with uncertainty and imprecision about input values and theoretical relationships, is to apply the framework of fuzzy inference systems. Fuzzy inference systems have been successfully applied in fields such as automatic control, data classification, decision analysis, expert systems, and time series forecasting [1].

However, how to select a suitable number of fuzzy rules for the model structure is still an open problem, which is normally handled via trial and error. Different structures are built, adjusted and tested. The one with better performance is chosen as the more adequate.

E. Hüllermeier, R. Kruse, and F. Hoffmann (Eds.): IPMU 2010, Part II, CCIS 81, pp. 324–333, 2010.

In this paper, we suggest a fuzzy inference system (FIS) for the estimation of the Brazilian Central Bank's reaction function. The FIS is a simplified version of the proposal in in [7]. In spite of the same model structure between the FIS and the C-FSM proposed in [7], they have a different learning algorithm. The C-FSM is a constructive model, which is always initialized with two fuzzy rules, and initial model parameters adjusted via the traditional EM algorithm. Indeed, its structure is varies during the offline learning processes, since the constructive algorithm considers adding and pruning conditions and operators, in order to determine an adequate model structure for an specific problem.

Here, FIS structure is defined in two phases. In the first phase, an initial rule based system composed by a set of fuzzy rules is generated using a Subtractive Clustering algorithm (SC), originally proposed in [2]. In a second phase, the model is re-adjusted using the Expectation Maximization algorithm, where all the model parameters are adjusted considering as a start point results obtained with the SC algorithm.

After this introduction, the paper proceeds as follows. Section 2 presents the fuzzy inference system and the learning method proposed. Section 3 presents the application and simulation results. Finally, some conclusions are presented in Section 4.

2 Fuzzy Inference System - FIS

This section introduces the structure of the fuzzy inference system and the learning algorithm for model structure and parameters update.

2.1 Model Structure

Let $\mathbf{x}^k = [x_1^k,\ x_2^k,\ \ldots,\ x_p^k] \in \mathbb{R}^p$ denotes the input vector at instant k, $k \in \mathbb{Z}_0^+$; $\hat{y}^k \in \mathbb{R}$ is the output model, for the correspondent input $\mathbf{x}^k$. The input space represented by $\mathbf{x}^k \in \mathbb{R}^p$, is partitioned into M sub-regions, and each of these is represented by a fuzzy rule; $k = 0, 1, 2, \ldots$ is the time index (Figure 1). The antecedents of each fuzzy **If-Then** rule (R_i) are represented by their respective centers $\mathbf{c}_i \in \mathbb{R}^p$ and covariance matrices $\mathbf{V}_i|_{p \times p}$. The consequents are represented by local linear models, with output y_i, $i = 1, \ldots, M$ defined by:

$$y_i^k = \phi^k \times \theta_i{}^T \tag{1}$$

where $\phi^k = [1\ x_1^k\ x_2^k\ \ldots x_p^k]$; $\theta_\mathrm{i} = [\theta_{i0}\ \theta_{i1}\ \ldots\ \theta_{ip}]$ is the coefficient vector of the local linear model for the $i-th$ rule.

Each input pattern has a membership degree associated with each region of the input space partition. This is calculated through membership functions $g_i(\mathbf{x}^k)$ that vary according to centers and covariance matrices related to the fuzzy partition, and are computed by:

$$g_i(\mathbf{x}^k) = g_i^k = \frac{\alpha_i \cdot P[\, i \mid \mathbf{x}^k\,]}{\sum_{q=1}^{M} \alpha_q \cdot P[\, q \mid \mathbf{x}^k\,]} \tag{2}$$

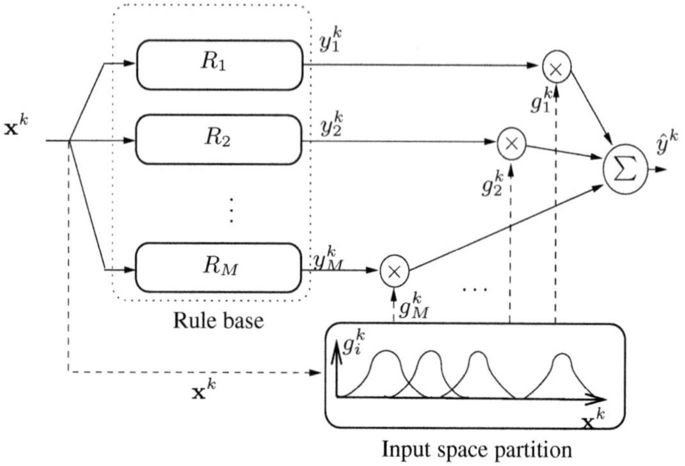

Fig. 1. A general FIS structure

where α_i are positive coefficients satisfying $\sum_{i=1}^{M} \alpha_i = 1$ and $P[\, i \mid \mathbf{x}^k]$ is defined according to

$$P[\, i \mid \mathbf{x}^k \,] = \frac{1}{(2\pi)^{p/2} \det(\mathbf{V}_i)^{1/2}} \exp\left\{ -\frac{1}{2}(\mathbf{x}^k - \mathbf{c}_i)\mathbf{V}_i^{-1}(\mathbf{x}^k - \mathbf{c}_i)^T \right\} \tag{3}$$

where $\det(\cdot)$ is the determinant function. The model output $y(k) = \hat{y}^k$, which represents the predicted value for future time instant k is calculated by means of a non-linear weighted averaging of local outputs y_i^k and its respective membership degrees g_i^k, i.e.

$$\hat{y}(\mathbf{x}^k) = \hat{y}^k = \sum_{i=1}^{M} g_i^k \, y_i^k \tag{4}$$

2.2 Learning Algorithm

First, an initial structure composed by fuzzy rules is defined, and its parameters are adjusted via the traditional Expectation Maximization (EM) algorithm, originally proposed in [5] for mixture of experts models.

Model structure is initialized using the unsupervised clustering algorithm called the Subtractive Clustering Algorithm (SC), proposed in [2]. This algorithm provide a set of M clusters from an specific training data set presented to the algorithm. Patterns processed by the SC algorithm are composed by the input-output patterns used in a second stage for model optimization.

These groups are associated to a set of fuzzy rules codified in the FIS structure. Therefore, after the number of fuzzy rules is defined, we proceed to initialize the model parameters, for $i = 1, \ldots, M$, according to the following criteria:

- $\mathbf{c}_i^0 = \psi_i^0|_{1\ldots p}$, where $\psi_i^0|_{1\ldots p}$ is composed by the first p components of the $i-$th center found by the SC algorithm;

- $\sigma_i^0 = 1.0$;
- $\theta_i^0 = [\psi_i^0|_{p+1}\, 0\, \ldots\, 0]_{1\times p+1}$, where $\psi_i^0|_{p+1}$ is the $p+1-$th component of the $i-$th center found by the SC algorithm;
- $\mathbf{V}_i^0 = 10^{-4}\mathbf{I}$, where $\mathbf{I}$ is a $p \times p$ identity matrix;
- $\alpha_i^0 = 1/M$.

After this initialization, model parameters are re-adjusted based on the traditional offline EM algorithm, following an iterative sequence of EM steps, given incomplete data y^k. It means that, a complete data is composed by the output variable y^k and a missing data. The goal of the EM algorithm is to find a set of model parameters, which will maximize the log-likelihood $\mathcal{L}$, of the observed values of y^k at each M step of the learning process. This objective function is defined by

$$\mathcal{L}(D,\, \mathbf{\Omega}) = \sum_{k=1}^{N} \ln\left(\sum_{i=1}^{M} g_i(\mathbf{x}^k,\, \mathbf{C}) \times P(y^k \mid \mathbf{x}^k,\, \theta_i)\right) \tag{5}$$

where $D = \{\mathbf{x}^k, y^k | k = 1, \ldots, N\}$, $\mathbf{\Omega}$ contains all model parameters and $\mathbf{C}$ contains just the antecedents parameters (centers and covariance matrices). However, for maximizing $\mathcal{L}(D,\, \mathbf{\Omega})$, it is necessary to estimate the missing data h_i^k (E step). This missing data, according to mixture of experts theory, is known as the posterior probability of $\mathbf{x}^k$ belong to the active region of the $i-$th local model.

When the EM algorithm is adapted for adjusting fuzzy systems, h_i^k may also be interpreted as a posterior estimate of membership functions defined by Eq. (2). So, h_i^k is calculated as

$$h_i^k = \frac{\alpha_i P(i \mid \mathbf{x}^k) P(y^k \mid \mathbf{x}^k,\, \theta_i)}{\sum_{q=1}^{M} \alpha_q P(q \mid \mathbf{x}^k) P(y^k \mid \mathbf{x}^k,\, \theta_q)} \tag{6}$$

for $i = 1, \ldots, M$. These estimates are called as "posterior", because these are calculated assuming $y^k,\ k = 1, \ldots, N$ as known. Moreover, conditional probability $P(y^k|\mathbf{x}^k, \theta_i)$ is defined by:

$$P(y^k \mid \mathbf{x}^k,\, \theta_i) = \frac{1}{\sqrt{2\pi\sigma_i^2}} \exp\left(-\frac{[y^k - y_i^k]^2}{2\sigma_i^2}\right) \tag{7}$$

with σ_i^2 estimated as:

$$\sigma_i^2 = \left(\sum_{k=1}^{N} h_i^k [y^k - y_i^k]^2\right) / \sum_{k=1}^{N} h_i^k \tag{8}$$

Hence, the EM algorithm for determining FIS parameters can be summarized as:

1. E step: Estimate h_i^k via Eq. (6);
2. M step: Maximize Eq. (5) and update model parameters, with optimal values calculated as:

$$\alpha_i = \frac{1}{N}\sum_{k=1}^{N} h_i^k \tag{9}$$

$$\mathbf{c}_i = \left(\sum_{k=1}^{N} h_i^k \mathbf{x}^k\right) / \sum_{k=1}^{N} h_i^k \tag{10}$$

$$\mathbf{V}_i = \left(\sum_{k=1}^{N} h_i^k (\mathbf{x}^k - \mathbf{c}_i)'(\mathbf{x}^k - \mathbf{c}_i)\right) / \sum_{k=1}^{N} h_i^k \tag{11}$$

for $i = 1, \ldots, M$, where M is the size of the fuzzy rule base, N is the number of input-output patterns at the training set. For all these equations, $\mathbf{V}_i$ was considered as a positive diagonal matrix, as an alternative to simplify the problem and avoid infeasible solutions. An optimal solution for θ_i is derived solving the following equation:

$$\sum_{k=1}^{N} \frac{h_i^k}{\sigma_i^2} \left(y^k - \phi^k \times \theta_i\right) \cdot \phi^k = 0 \tag{12}$$

where σ_i is the standard deviation for each local output y_i, $i = 1, \ldots, M$, with σ_i^2 defined by Eq.(8). After parameters adjustment, calculate the new value for $\mathcal{L}(D, \mathbf{\Omega})$.

3. If convergence is achieved, then stop the process, else return to step 1.

There are some differences to consider if he FIS structure is compared to a basic one given by the SC algorithm. First, The FIS structure has coefficients α_i as parameters, which are not directly initialized by the SC algorithm. Secondly, consequents assumed by the SC algorithm are singletons, whereas the FIS considers local linear models (which are a function of the input vector). Therefore, even though the SC algorithm can be used directly for modeling purposes, it is still necessary a global optimization, considering all the FIS parameters, which is performed using the EM algorithm in this paper.

3 An Application to the Central Bank's Reaction Function

In order to develop an econometric model of the Takagi and Sugeno-type, we need to determine the underlying structure of the fuzzy system and its parameters. The model is identified by a fuzzy modelling method, using economic theory in combination with a set of input - output data. The first step in structure identification is to choose the relevant explanatory variables. As with any estimation procedure, this is the point at which a careful consideration of economic theory is important, to ensure that only relevant inputs to the model are used. In the present context, it is important because a fuzzy model will always attempt to match the chosen inputs and outputs.

Probably, the most well known reaction function is the Taylor Rule, proposed in [10], by which the Central Bank uses the nominal interest rate to minimize the total variance of inflation and output. Taking the Taylor Rule, this section illustrates how a fuzzy inference system of the first-order Takagi-Sugeno type can be used to model the relationship between the nominal interest rate, inflation and output. Our principal objective is to demonstrate the potential benefits of the method, particularly its power in handling non-linear relationships. For this reason, we consider a relatively simple version of the relationship which contains only three input variables.

We study whether there is evidence that the Brazilian nominal interest rate followed a nonlinear process between March 2002 and September 2008. A brief description of the data set used is made below.

3.1 The Data

The data sources are Banco Central do Brasil, IBGE (Instituto Brasileiro de Geografia e Estatística), and IPEA (Instituto de Pesquisa Econômica Aplicada). The nominal interest rate used is the annualized end-of-period Taxa Selic[1], controlled by the Central Bank. Output is measured by monthly industrial production and the output gap is measured as the residual from a Hodrick-Prescott filter [4] applied to the monthly index of industrial production. The inflation rate is calculated by a monthly wholesale index (IPCA). The index is computed between the 21st day of the previous month and 20th day of the reference month [8].

We use the end-of-period interest rate in order to avoid endogeneity problems. The end-of-period interest rate is the rate of the last day of the month. In that case, it is clear that inflation could be considered pre-determined. Moreover, it also reasonable to assume that the nominal interest rate of the last day of the month will not affect the output of the same month.

Another important point to discuss is whether or not the series considered in this paper have a non-stationary behavior. Although the nominal interest rate is a variable controlled by the central bank and the hypothesis of a unit-root seems not to be a reasonable one, the usual unit-root tests did not reject the null hypothesis of a unit-root. In [8], it is argued that this may happen because of the convergent behavior of the series during the period analyzed in the paper and the relative small number of observations (80). All the other series were considered non-stationary by the usual tests. Hence, the models were constructed taking the series in first difference.

3.2 Estimating Results

In this section, a fuzzy inference system and a linear model will be estimated and then compared. Given the inherent flexibility of the fuzzy modelling procedure, before we present estimates of the nominal interest rate equation, it is appropriate to say a few words about the criteria for model selection. In this paper we applied standard model selection procedure BIC [9], which indicated the following inputs: first

[1] The Selic (Special Settlement and Custody System) rate is an overnight rate expressed per year, which is obtained by an average weighted, taking the operations total in one-day with federal public bond. These transactions are made between Brazilian Central Bank and authorized financial institutions. Selic rate is the basic rate used as reference by the monetary policy.

difference of the nominal interest rate i and output gap $\tilde{y}$ in $t-1$, first difference of the deviation of the inflation rate π with respect to the target π^*, $(\pi - \pi^*)$, in t and $t-1$[2].

As our objective in this paper is not to estimate the 'best' model as such, but rather to illustrate the potential power of fuzzy modelling and how the structure of the model can be adjusted to achieve the desired degree of generality. For this reason, our approach is to use the full sample to estimate a series of models of increasing generality and then to examine how the forms of the underlying relationships vary across the estimated models. This provides a useful way of determining the robustness of the identified relationships. The performance of each model is shown with respect to the Root Mean Square Error (RMSE) and the pattern of the associated residuals. We also show plots of the actual and model outputs.

3.3 A Linear Estimate

As a point of reference, we begin by briefly presenting the estimates of a single-equation linear model. This is the equivalent of estimating a fuzzy model in which the entire data set is effectively encompassed by a single cluster, so that the inputs are each described by a single membership function and the behavior of system is shown by a single rule, represented by a simple linear equation.

We estimate a linear model where the error are normally and independently distributed. As the usual unit-root tests [3] did not reject the null hypothesis of a unit-root, we consider the first difference of the historical database. We found the following results:

$$\Delta i_t = \underset{(0.063)}{0.720}\ \Delta i_{t-1} - \underset{(0.024)}{0.042}\ \Delta \tilde{y}_{t-1} + \underset{(0.083)}{0.177}\ \Delta \Pi_t + \underset{(0.087)}{0.178}\ \Delta \Pi_{t-1}$$

where $\Pi = (\pi - \pi^*)$ and the values between parentheses bellow the estimates are the standard errors. We note that all coefficients are statistically significant at 10% and have the desired signs. The RMSE is 0.327 and Figures 2 (a) and (b) show, respectively, the estimated interest rates and deviations for the period considered.

In Figures 2 (a) and (b) the very large positive errors in periods usually associated to currency crises (specially during 2002 and early 2003) suggest that a nonlinear model may be more adequate to represent the Brazilian nominal interest rate. Furthermore, there are some evidence that the model may not be correctly specified, indicated by Jarque-Bera test [6] where the hypothesis of normally distributed residuals is strongly rejected with significance level 5%.

3.4 Estimates of the Fuzzy Model

In this section we present the results for a fuzzy inference models. We have already said that the initial step in the identification of the fuzzy model is to use the subtractive

[2] Inflation targeting (IT) is a monetary policy strategy, formulated by Central Banks, that makes public a target for annual inflation rate (π^*). The Brazilian Central Bank adopted the IT in May of 1999. The inclusion of the difference $(\pi - \pi^*)$ can be explained by the fact that any deviation of the inflation rate with the respective target will produce interest rate changes, conducted by Central Bank, to correct this situation.

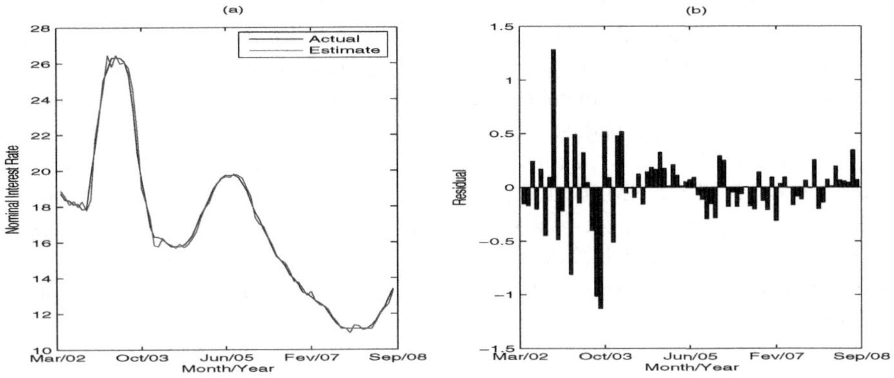

Fig. 2. Nominal interest rate: (a) Actual and Estimated by a linear model; (b) Residual plot for linear model

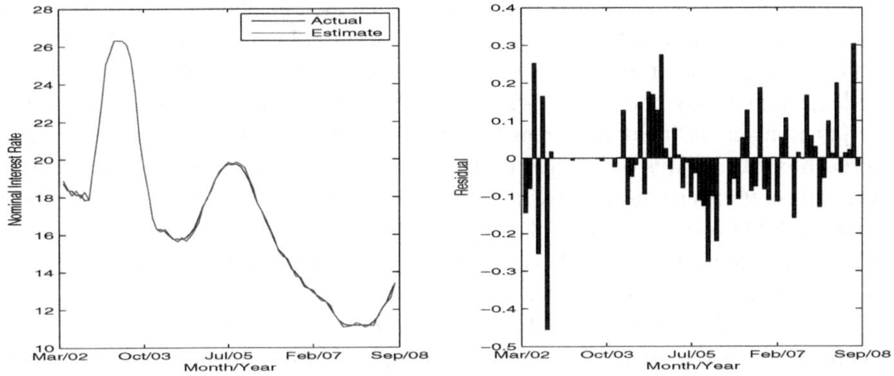

Fig. 3. Nominal interest rate: (a) Actual and Estimated by a fuzzy model; (b) Residual plot for fuzzy model

clustering method to determine the number of fuzzy rules and the rule premise membership functions. For the purpose of this exercise, we used the Gaussian form for the membership functions and chose a cluster radius igual to 0.15 and was just sufficient to generate a model with three membership functions. Moreover the index are fixed in $T = 12, \alpha_{min} = 0.001$ and $f_{forget} = 0.9$.

After this initialization, model parameters are re-adjusted based on the EM algorithm, following an iterative sequence of EM steps.

The underlying structure of the inference fuzzy model can be represented as

$$\hat{y}(\mathbf{x}^t) = \sum_{i=1}^{3} g_i^t \, y_i^t$$

where $\hat{y}$ denotes the first difference of the nominal interest rate estimated, $\mathbf{x}^t$ is the input vector at instant t, g_i^t is the membership degree and y_i^t is the output of the local linear models. The relative weights of the rules g_i^t are determined by the positions of the inputs in their respective membership functions and the parameters of the equations are estimated via the EM method. The RMSE of this model is igual 0.121.

The increased flexibility of this model leads to an improvement in performance, judged in terms of the RMSE, and plots of the actual and estimated outputs shown in Figure 3(a) confirm the increased explanatory power of the model. The associated residual plot for the model, shown in Figure 3(b), indicates that the error structure is much closer to white noise, although there is still some discernible pattern in the residuals.

4 Conclusions

In this paper, we have demonstrated that a modelling procedure based on the application of fuzzy logic has considerable potential as a complement to traditional linear and non-linear estimation methods. The method involves the identification and estimation of a series of rules, described by local linear relationships, which are weighted according to the position of input observations in their respective fuzzy membership functions. The behaviour of any linear or non-linear system is then approximated via a weighted interpolation across the local regions of the model. We have seen that the fuzzy logic approach is particularly well-suited to the estimation of ill-defined systems in which there is theoretical and quantitative uncertainty about the nature and range of key input variables. Additional strengths of the approach are that it requires no prior knowledge of the functional form of the underlying relationships, and is also robust with respect to outlying observations or noisy data.

To illustrate the potential benefits of the fuzzy modelling procedure, we used it in conjunction with cluster identification techniques to estimate the Brazilian Central Bank's reaction function to determine the interest rate. We showed how the flexibility of the model, and its ability to track any underlying non-linearity, can readily be increased by expanding the number of operational rules in the system.

Acknowledgement

The last author thank the Brazilian National Research Council, CNPq, for grants 302407/2008-1.

References

1. Angelov, P., Filev, D.P., Kasabov, N.: Evolving Intelligent Systems: Methodology and Applications. Wiley-IEEE Press (March 2010)
2. Chiu, S.: A cluster estimation method with extension to fuzzy model identification. In: Proceedings of The IEEE International Conference on Fuzzy Systems, June 1994, vol. 2, pp. 1240–1245 (1994)

3. Dickey, D.A., Fuller, W.: Distribution of the estimators for autoregressive time series with a unit root. Journal of the American Statistical Association 74, 427–431 (1979)
4. Hodrick, R., Prescott, E.: Postwar U.S. Business Cycles: An Empirical Investigation. Journal of Money, Credit and Banking 29, 1–16 (1997)
5. Jacobs, R., Jordan, M., Nowlan, S., Hinton, G.: Adaptive Mixture of Local Experts. Neural Computation 3(1), 79–87 (1991)
6. Jarque, C.M., Bera, A.K.: Efficient tests for normality, homoscedasticity and serial independence of regression residuals. Economics Letters 6(3), 255–259 (1980)
7. Luna, I., Soares, S., Ballini, R.: A Constructive-Fuzzy System Modeling for Time Series Forecasting. In: Proceedings of The International Joint Conference on Neural Networks (2007)
8. Salgado, M.J.S., Garcia, M.G.P., Medeiros, M.C.: Monetary Policy During Brazil's Real Plan: Estimating The Central Bank's Reaction Function. Technical Report 442, Department of Economics, Pontifical Catholic University of Rio de Janeiro (August 2004)
9. Schwarz, G.: Estimating the dimension of a model. Annals of Statistics 6(2), 461–468 (1978)
10. Taylor, J.: Discretion Versus Policy Rules in Practice. In: Carnegie-Rochester Conference Series on Public Policy, vol. 39, pp. 195–214 (1993)

Do Uncertainty and Fuzziness Present Themselves (and Behave) in the Same Way in Hard and Human Sciences?

Settimo Termini

Dipartimento di Matematica Università di Palermo,
Via Archirafi, 34 - 90123 Palermo, Italy
European Center for Soft Computing, Mieres (Asturias), Spain
termini@math.unipa.it

Abstract. In the present paper the question whether uncertainty and fuzziness present themselves and behave in the same way (or not) in hard and human sciences will be briefly discussed. This problem came out from the attempt to answer the question asked by Lotfi Zadeh on the (apparent) strangeness of a very limited use of fuzzy sets in human sciences.

Keywords: Uncertainty, fuzziness, hard sciences, human sciences, two cultures, use of formal methods in human sciences.

1 Introduction

The problem which is addressed in the present paper is the way in which uncertainty, and in particular its specific facet represented by fuzziness, presents itself in different disciplines, looking – in particular – for possible meaningful differences when human sciences or hard sciences are, respectively, taken into account. As a matter of fact, I have been induced to consider this general problem starting from one observation done by Lotfi Zadeh who, rightly, judged strange the fact that Fuzzy Sets Theory has been so less used in Human Sciences.

An attempt will be done, in these pages, to propose some hypotheses for explaining this fact. If a phenomenon is persistent for many decades, in fact, it cannot be simply ascribed to the chance but it is reasonable to think that unusual, deep (or anyway, unknown) reasons have worked – in the underground – for producing it.

In the rest of this Introduction, I shall, first, briefly focus the setting, the general context, in which these considerations should be done; secondly, I will state in a synthetic way a related difficult problem, and, finally, some reference to a few related questions will be done. In Section 2, starting from Zadeh's question, I shall argue in favour of the thesis that what impeded a rapid and widely diffused use of fuzzy sets theory results and methodologies in Human Sciences, was not only a lack of correct communication but resides in deep and difficult questions connected with the fact that the valuation of *precision* and *rigor* is different in different disciplines.

Let me add that – although no single formula is present in the paper - all the considerations and remarks which are present here spring out from a rethinking of

E. Hüllermeier, R. Kruse, and F. Hoffmann (Eds.): IPMU 2010, Part II, CCIS 81, pp. 334–343, 2010.

technical results, in the light of the problems discussed. A clarification of conceptual questions – besides being of great intrinsic interest – plays a crucial role also in the development of purely technical results in *new* directions. I am, in fact, firmly convinced that conceptual clarifications subsequent to, and emerging from epistemological analyses help the focusing of *innovative* paths of investigation. Also from this point of view – besides their intrinsic value and interest – both the scholarly investigation of Rudolf Seising [1] and his edited volume [2] are very important references which have been very useful for clarifying a few aspects along the writing of the present paper.

1.1 The Context

There is a long history regarding the mutual relationship existing between the (so-called) humanities and the (so-called) hard sciences. The past year witnessed the completion of just half a century from the publication of a precious booklet on "The Two Cultures" by the English physicist and novelist Charles Percy Snow [3], dealing with the problem of the breakdown of communication between the two main forms of culture of our times, the sciences and the humanities, and whose appearance provoked – 50 years ago – intense debates and strong discussions.

It is not clear whether this debate can ever reach a definite conclusion at this very general level whilst it is relatively clear that the problems present themselves with different nuances and specificities according to the particular fields of inquiry and disciplines considered and the way in which it is presumed or asked they should interact. It seems that also the general, and specific tradition of a given Country can play a no negligible role.

For instance, in Italy, where – in the past century – there has been a strong contraposition between the "two cultures" the situation (the context) should be considered different from the one described by Snow in which more than a strong contraposition there was a very profound lack of communication. It is also interesting to observe that, remaining always in Italy – in contrast to the just mentioned contraposition that dominated many decades of the twentieth century – it seems possible also to trace an old and long lasting tradition in Italian Literature that stresses a dialogue between scientists and humanities [4]. This tradition has the same roots of the scientific revolution of the seventeenth century and is also connected to the founding father of Italian literature. According to this school of thought, there is, in fact, a line connecting Dante, Galileo, Leopardi, Italo Calvino, Primo Levi, all authors in whose writings not only there is no opposition between the "two cultures" but an innovative literary language is a tool, a powerful tool, for expressing new scientific results or a new (scientific) Weltanschaung. All this, in turn, contribute to establishing new clear relationships between Science and Society. In this setting also the way in which science is communicated and the audience which is explicitly chosen, or seen, as the privileged target plays a central role [5].

This is the general context in which the questions asked in the paper should be seen, although to reach not too vague conclusions it is important to examine specific questions and problems. Many of the questions and proposals done along the years by Zadeh are utmost original and innovative so that they have required also to look anew at some epistemological aspects for a satisfactory assessment of the considered problems, before trying to tackle (and solve) them. This applies also to his question of the limited use in Human Sciences of results and techniques of Fuzzy Sets and Soft Computing (from now on, FS&SC), a question which, then, must not be considered as an occasional comment, as it could – *prima facie* – appear.

1.2 A Crucial (Difficult) Problem

The general motivations previously outlined indicate that for affording concretely the question posed by the title of this paper, we must look carefully at the meaningful features of every specific discipline and, also, to the context in which a certain question is asked or a given concept is used. So we cross and are obliged to face here the old (and also very difficult besides being, under many aspects, still mysterious[1]) *problem of interdisciplinarity* (see, for instance, [6], [7]). Everyone who has crossed interdisciplinary problems knows that the best way for obtaining good results is to be very cautious and prudent in using concepts (and tools) outside the domain in which they were initially conceived and developed. This fact generates also some "family resemblance" between different disciplines at least from the epistemological point of view (see [8]). In the case considered in the present paper we have to be particularly careful since we are trying to establish bridges between very distant domains.

1.3 Some Related Questions

Some related questions which will not be treated here, in the present preliminary attempt to focus the problem, are the following ones. First, one must remember that many problems (technical, epistemological, psychological, etc) posed by the use of formal methods in human sciences are still crucial, unsettled and strongly debated. Secondly, the process leading from an informal notion as used in everyday language to its regimentation inside scientific theories (but also in specialized languages) is to be carefully taken into account. This second problem has been superbly discussed by Rudolf Carnap who introduced the two notions of *explicandum* and *explicatum* for explaining aspects of this process. The help that Carnap's analysis can provide also for studying in a general way the relationships between FS&SC and Human Sciences has been briefly indicated in [3]. In the present paper, for reason of space, these questions – however crucial they are – will not be taken into account.

2 Zadeh's Question and a Tentative Answer

In the present section starting from Zadeh's question, I shall focus a few general points aiming at a preliminary chartering of the territory. More complete analyses of the topics briefly surveyed here will involve – as already stressed – a careful consideration of many other aspects (at least, the ones mentioned in subsection 1.3, above).

---

[1] Of course, those aspects of interdisciplinarity still remaining *mysterious* are not at all mysterious but are (simply!) those emerging from the extreme complexity of forcing different disciplines to interact while asking them to preserve their proper specificities, methodologies, levels of rigor, etc., (and, then , aspects very difficult to treat). In our case we must also play a specific tribute to the constraints imposed by the novelty of the theory (FS&SC) on one side, and the refractoriness to be imbedded into (more or less) formal methods on the other side (Human Sciences).

2.1 Zadeh's Question

Rudolf Seising, proposing and presenting a special session on *"Uncertainty, Vagueness and Fuzziness in Humanities and Social Sciences"* rightly remembered that Lotfi Zadeh has been struck by the fact that the theory of fuzzy sets has not been very diffused and strongly used in human sciences and reports a quotation from a 1994 interview:

"I expected people in the social sciences, economics, psychology, philosophy, linguistics, politics, sociology, religion and numerous other areas to pick up on it. It's been somewhat of a mystery to me why even to this day, so few social scientists have discovered how useful it could be."[2]

This is the question I want to address in this paper, trying to find out some reasons of this fact and so contributing to reduce the mistery to which Zadeh refers. In the following pages I shall present a tentative answer to Zadeh's question or, better, an hypothesis regarding the reasons why – forty five years after the appearance of the founding paper of the theory – there has not been a wide use of Zadeh's theories in Human and Social Sciences. I shall not provide an answer, however tentative, to the question asked, but I shall formulate a few hypotheses regarding what can be involved in the situation. I am, in fact, convinced that Zadeh has not simply asked a question but has touched a crucial point of the interaction between human sciences and hard sciences which cannot be solved simply by asking a question and providing an answer. So, if these hypotheses grasp some truths we shall remain with a lot of additional work to be done (as always happens in science, a (possible) solution of a specific point opens a lot of new questions).

2.2 A Remark and a Few Tentative Hypotheses

The remark has to do with the epistemological (and ontological) features of both. Although many people are probably convinced that there exist strong epistemological similarities between FS&SC and Human Sciences, it is more difficult to find analyses and descriptions of these similarities. (I leave completely out of the considerations done in this paper the problem of possible similarities of ontological type, but see [10] and [11]).

Let us consider the problem of *rigour*. One of the crucial points in my view is that in both fields we are looking for rigour in the same way and this way is in many aspects different from the one in which the rigour is looked for in Hard Sciences. The kind of rigour that is meaningful and useful in both FS&SC and Human Sciences, in fact, is different from the one that displays his benefic effects in hard sciences.

In all the fields of investigation one is looking – among many things – also for rigour. Let us consider the case of Hard Sciences. Roughly, what we require from a good theory is that it – at least – should model the chosen piece of reality and be able to forecast the output of experiments. For this second aspect, one is looking then for a stronger and stronger correspondence between the numerical output computed by the

[2] "Lotfi A. Zadeh. Creator of Fuzzy Logic", Interview by Betty Blair, 1994 *Azerbaijan International* Winter 1994 (2.4).

theory and the numerical output measured by the experimental apparatus. This is not certainly, in general, the case in Human Sciences.

Let us make a *digression* to clarify this point. The exacteness of a writer or of a resercher in philology is not based on the number of decimal ciphers after the dot that a given theory produces for some meaningful parameters (and which allows a comparison with the measurements done in corresponding experiments). Human Sciences are also looking for exactness but a sort of different one, which is difficult also to define if we have in mind the numerical-previsional model, typical of Hard Sciences. We could, perhaps, say that is the exactness of having grasped in a meaningful way some of the central aspect of a given problem. For instance, in a literary text of very good quality we see that we cannot change the words without loosing something, we cannot change in some cases even the simple *order* of the words. The text produced by an outstanding writer is exactly what was needed to express something. This is certainly true for poetry, but also for any literary text of very high quality[3].

In a certain sense the same is true also in FS&SC. It suffices to remember the very frequent remarks done by Zadeh on the fact that humans very efficiently act without doing any "measurement" (and numerical computations). Also in the case of FS&SC, then, we are moving in a universe in which we have to do with an exactness which is different from the one of *measurements* and *numerical* precision.

Let us now go back to the problem of a possible dialogue between FS&SC and Human Sciences (and the fact that it has not been very strong in the past decades). The (tentative) hypothesis which I want to propose is that one of the causes resides just on the fact that both FS&SC and Human Sciences share the same methodological and epistemological attitude towards the problem exactness and rigour: both look – as already observed – for a sort of exactness not based on numerical precision.

But why an epistemological similarity could be the cause of a difficulty in the interaction of the two fields? My answer is twofold. From one side, I think that this *same similarity* is difficult to master. We are accustomed to think that the interaction with formal, hard, sciences is a way for introducing *their kind* of precision (a quantitative one) into a (still) imprecise field. From the other side, I think that the same fact that the kind of precision which is possible to introduce can be different from “numerical precision” obliges to reflect to the kind of operation one is trying to do.

Of course, the problems proper to each of FS&SC and Human Sciences *are* different (and different also from those appearing in the traditional approaches of hard sciences) notwithstanding the epistemological similarities mentioned above. To pick up the differences, acknowledging the epistemological similarities is, in my view, a crucial passage. Regarding their methodological similarity, let us limit, here, to take into account only their not considering *numerical precision* as crucial. The challege is to evaluate which kind of advantage can emerge for these two worlds if the traditional passage through the caudin forks of numerical measurements and computation is not the crucial point. FS&SC provide a very flexible language in which we can creatively

---

[3] And this is in fact one of the central points and difficulties of the translation of a text in different languages. Commenting the difficulties faced in early Cybernetics by the mechanical, cybernetic, translation from one natural language to another one, it was observed that some of the difficulties are caused by the lack of a formal (mechanical) theory of meaning. But for having a good mechanical translation of literary tests there are other things still missing, among which a mechanical theory of "exactness", which is difficult also to envisage.

pick up locally meaningful tools which can efficiently clarify some specific aspects of the considered problems. But there is no formal machinery that automatically solves the problems. If we do not have clear this point in mind it is easy to have difficulty in contrasting at the root the objections that the formalism and the language of fuzzy sets theory can complicate and not simplify the description and understanding of some pieces of reality. Just, for giving an example, the *full* machinery of many valued logic *is* more complex of the one of classical logic. It is difficult to have a non occasional interaction, if we are not able to convince people working in human sciences that: a) we can have advantages even without numerical evalutions, and b) by introducing "degrees" we are not opening the Pandora's box of the full machinery of a more complex formalism.

Considerations regarding methodological and epistemological similarities between FS&SC in Human Sciences can help, then, to understand aspects of the relationships between these fields of inquiry. The natural question that arises at this point is whether an examination also of their respective ontological basic assumptions could provide relevant information. But, as was said above, I leave this as a completely untouched problem here.

3 A Few Conceptual Corollaries

Let us list in a very rough and rudimentary way same of the consequences that in my view immediately emerge from the previous attempt at analyzing the question.

A. *Precision* is not an exclusive feature of hard sciences. However the way in which this concept is needed and is used in Human Sciences and in Hard Sciences *is* different.

B. We must observe that the same is true also when we consider the problem in different scientific disciplines. Also when mathematics and physics are taken, for instance, into account, we can easily observe differences. Dirac's *delta function*, for instance, was considered in no way acceptable by mathematicians until Schwartz developed the Theory of Distributions, while physicists – still being completely aware and sharing all the mathematical objections to the contradictions conveyed by this notion – used it by strictly controlling the way in which it was used. The same could be argued for other notions which have been used in Theoretical Physics, i.e., Renormalization, Feynman's diagrams.

C. The notion of *precision* that can be fruitfully used in Fuzzy Sets and Soft Computing has features belonging partly to Human Sciences and partly to Hard Sciences. This fact makes particularly interesting and flexible the tools provided by FS & SC but, at the same time, makes more complex their use in new domains since it requires their intelligent adaptation to the considered problems and not a "mechanical application".

D. The "mechanical application", however, is exactly what each of the two partners in an interdisciplinary collaboration, usually, expect each from the other. Let us consider disciplines an imaginary example. Let A and B two disciplines involved, A-scientist sees a difficulty for the solution of a certain problem arising from features and questions typical of discipline B and asks to his colleague B-scientist for their solution, presuming that he can provide on the spot a (routine) answer (what above, was called a *mechanical* solution). However, for non-trivial problems, usually B-scientist is unable to provide a mechanical solution, since, for instance, the posed problem either can be particularly complex or outside the mainstream of B discipline. The expected mechanical interaction does not happen, since it is not feasible. However, an interaction is possible. What A and B – representing with these letters both the disciplines and the scientists – can do is to cooperate for a creative solution and not for a routine mechanical application of already existing tools.

E. The problem appears particularly delicate when Human Sciences and formal techniques are involved for reasons which I am tempted to define of *sociological* nature. I am, in fact, convinced that the epistemological problems are indeed strong, but not crucial in creating *difficulties*. These can emerge more from both the attitudes and the expectations of the people involved. We could, for instance, meet categories of the following types. People that so strongly believe in the autonomy of their discipline that they do not think that other disciplines can provide any help. People so strongly trusting to the power of the tools coming from hard sciences that they expect these techniques may work very well without any *creative adaptation* to the considered problem.

F. The difficulties of point E) above are magnified if the tools to be used cannot, by their same nature, be mechanically applied and – in the light of point C above – this is often the case for tools borrowed from FS & SC.

One is tempted to ask whether one could use Carnap's procedure to see whether one is using – in different disciplines – different explicata of the intuitive notions of rigour and precision. This question, besides being very slippery, presents the additional difficulty of not forgetting the fact that rigour and precision are also – in some sense, metatheoretic notions and, in the case of interdisciplinary investigations, intertheoretic notions. Another type of *corollaries* has to do with the fact that all these considerations impinge strongly (also) with technical work and specific investigations. In the following I shall list a few items just for clarifying my point, leaving a more detailed analysis to future occasions.

Vagueness. It has been often asked the question of the nature of vagueness and whether it should be eliminated from logic or from scientific discussions. Alternatively, whether it can (should) be formalized and whether fuzzy sets can provide an adequate or possible formalization (see [12]). The literature on this topic is overwhelming and I shall refer here to a few papers of mine only for providing information on my point of view on the problem [13], [14]. Let me observe that a preliminary investigation as the one outlined here can be useful to tackle correctly the problem, since it is completely different to move in context in which we can (and prefer) eliminate vagueness and another one in which we stress "The Value of Vagueness in everyday Communication" (see [15]). So Uncertainty and Fuzziness can present themselves (and behave) in a different way not only in Hard and Human Sciences but also in in problems and contexts belonging to very near fields.

Logical principles. General logical principles have been considered for long periods as part of philosophical tradition before the mathematical turn of logic after (Boole and) Frege and Russell. The problems and results connected to the development of fuzzy sets has provided the occasion of revisiting them in a enlarged context. It is clear that an evaluation of the paths which is possible to follow as well as of the type and level of precision cannot but follow a conceptual clarification of the context in which we move. Enric Trillas, has attempted to revisiting the role played by "logical principles" in many papers (see, for instance, [16]). It is interesting to read these papers having in mind more remote (innovative) investigations, for instance the ones of Jan Lukasiewicz [17] to appreciate how much (and in which directions) the development of Fuzzy Sets Theory has changed the general context, allowing to ask new questions (or asking them in a different way). This is a typical field in which same points appear under different view and we can also see the different role played by – apparently – the same things behaving now as a theorem, now as a principle.

Measuring fuzziness, varying information and all that. The same need of the specification of a context emerge for an analysis of the possible use of measures of fuzziness in different fields as well as of the construction of a (dynamic theory) of information [18], [19], [20]. Possible problems arising from a mechanical (uncritical) application of information theory to problems of "visual arts" have been stressed by Rudolf Arnheim [21] already in 1971. The extended corpus of the results of the theory of Fuzzy Sets allows to state the problem in the correct way if we consider both the problem and the context and determine correspondingly, then, the type of *precision* we must look for. For a few preliminary results in this direction see [22].

Fuzziness as a key for approaching problems in new ways. The same idea of fuzziness needs the contemporary use of epistemological analyses, formal developments and applications. See the subsequent developments and refinements of the same ideas – as they appear, for instance, in [23], [24] and [25] – to understand that Zadeh is really provoking to go out of the established, conventional, scientific tradition. His idea of "computing with words", for instance, really breaks with the *basic notion* of computation; the idea of "manipulating perceptions" goes nearer to Husserl's ideas than to the Galileian tradition. In developing – in a more or less ortodox way – these fruitful suggestions we cannot but use all the tools (formal, epistemological, mathematical, conceptual, linguistic) we have at hand.

4 Conclusions

In conclusion we can say that not only *Uncertainty and Fuzziness present themselves (and behave) in different ways in Hard and Human Sciences* but they present themselves in different ways also when we consider more near and more specific disciplines. Usually we are unaware of this behavior since we automatically select our tools and confine their use to what is suitable for the resolution of our specific problem. When we do not apply the relevant distinctions we obtain contradictory conclusions like the ones stressed by Arnheim. This, however, happens only occasionally. General conceptual analyses, I think, allow to definitely conclude that – if we consider the large dicothomy between Hard and Soft Sciences – the Theory of Fuzzy Sets interestingly manifests selected aspects and features of both. This allows us to provide at least one partial answer to Zadeh's question: their interaction has been very limited not only due to a generic lack of communication (and of mutual knowledge), but also (or mainly) since a true, profound and stable interaction requires the clarification of what each of the two actors is asking (and – reciprocally – is able to give) to the other one. The interaction while appearing, in principle, very natural and, apparently, straightforward, in practice, poses – in fact – very subtle conceptual problems (different from the ones arising in a "more traditional" interdisciplinary interaction between *near* fields or *very distant* fields). It is exactly for this reason, that it is not a trivial thing to realize. An innovative application, for instance, cannot be obtained through a routine application of fuzzy techniques to problems of human sciences done in a mechanical way, but can only emerge from a specific interaction on the selected problems and questions. The fact that in both human sciences and fuzzy sets the kind of precision requested and required is not of *numerical* type but of a more *linguistic* nature has a twofold consequence. It can facilitate – at the beginning – the dialogue and the interaction; however, good, innovative results can *only* spring out from a truly creative, interdisciplinary work which takes into account the specific and special features of the considered topics.

Acknowledgments. I want to deeply and friendly thank Enric Trillas, who in Naples, last year, during the Workshop *"Memoria e progetto"* provoked me with many challenging intellectual problems. A sincere thank goes also to Rudi Seising who kindly invited me to participate to the *"Seminar on Soft Computing in Humanities and Social Sciences"*, last September, an occasion which added unvaluable intellectual stimuli to Enric's provocations. Finally I want to sincerely thank the referees for their careful reading of the paper and many useful suggestions.

References

1. Seising, R.: The Fuzzification of Systems. Springer, Heidelberg (2007)
2. Seising, R. (ed.): Views on Fuzzy Sets and Systems from Different Perspectives. Springer, Heidelberg (2009)
3. Snow, C.P.: The Two Cultures and the Scientific Revolution. Cambridge University Press, Cambridge (1959)
4. Greco, P.: L'astro narrante. Springer, Milan (2009)
5. Greco, P.: L'idea pericolosa di Galileo. Storia della comunicazione della scienza nel Seicento. UTET, Turin (2009)
6. Termini, S.: Imagination and Rigor: their interaction along the way to measuring fuzziness and doing other strange things. In: Termini, S. (ed.) Imagination and Rigor, pp. 157–176. Springer, Milan (2006)
7. Termini, S.: Remarks on the development of Cybernetics. Scientiae Matematicae Japonicae 64(2), 461–468 (2006)
8. Tamburrini, G., Termini, S.: Do Cybernetics, System Science and Fuzzy Sets share some epistemological problems? I. An analysis of Cybernetics. In: Proc. of the 26th Annual Meeting Society for General Systems Research, Washington, D.C., January 5-9, pp. 460–464 (1982)
9. Termini, S.: Explicandum vs Explicatum and Soft Computing. In: Seising, R., Sanz, V. (eds.) Proceedings of the I. International Seminar on Soft Computing in Humanities and Social Sciences. Springer, Heidelberg (2010) (to appear)
10. Termini, S.: The formalization of vague concepts and the traditional conceptual framework of mathematics. In: Proc. VII International Congress of Logic, Methodology and Philosophy of Science, Salzburg, vol. 3, section 6, pp. 258–261 (1983)

11. Tamburrini, G., Termini, S.: Some Foundational Problems in the Formalization of Vagueness. In: Gupta, M.M., Sanchez, E. (eds.) Fuzzy Information and Decision Processes, pp. 161–166. North-Holland, Amsterdam (1982)
12. Terricabras, J.M., Trillas, E.: Some remarks on vague predicates. Theoria 10, 1–12 (1988)
13. Termini, S.: Aspects of vagueness and some epistemological problems related to their formalization. In: Skala, H.J., Termini, S., Trillas, E. (eds.) Aspects of Vagueness, pp. 205–230. D. Reidel (1984)
14. Termini, S.: Vagueness in Scientific Theories. In: Singh, M.G. (ed.) Encyclopedia of Systems and Control, pp. 4993–4996. Pergamon Press, Oxford (1988)
15. Kluck, N.: Some Notes on the Value of Vagueness in Everyday Communication. In: Hüllermeier, E., Kruse, R., Hoffmann, F. (eds.) IPMU 2010, Part I. CCIS, vol. 80, pp. 65–84. Springer, Heidelberg (2010)
16. Trillas, E.: Non contradiction, Excluded middle and Fuzzy sets. In: Di Gesù, V., Pal, S.K., Petrosino, A. (eds.) Fuzzy Logic and Applications. LNCS (LNAI), vol. 5571, pp. 1–11. Springer, Heidelberg (2009)
17. Lukasiewicz, J.: Philosophical remarks on many-valued systems of propositional logic. In: Borkowski, L. (ed.) J. Lukasiewicz Selected Works, pp. 153–178. North-Holland, Amsterdam (1970)
18. De Luca, A., Termini, S.: A definition of a non probabilistic entropy in the setting of fuzzy sets theory. Information and Control 20, 301–312 (1972)
19. De Luca, A., Termini, S.: Entropy and energy measures of a fuzzy set. In: Gupta, M.M., Ragade, R.K., Yager, R.R. (eds.) Advances in Fuzzy Set Theory and Applications, pp. 321–338. North-Holland, Amsterdam (1979)
20. De Luca, A., Termini, S.: Entropy Measures in the Theory of Fuzzy Sets. In: Singh, M.G. (ed.) Encyclopedia of Systems and Control, pp. 1467–1473. Pergamon Press, Oxford (1988)
21. Arnheim, R.: Entropy and Art: An Essay on Disorder and Order. The University of California Press, Berkeley (1971)
22. Termini, S.: On some vagaries of vagueness and information. Annals of Mathematics and Artificial Intelligence 35, 343–355 (2002)
23. Zadeh, L.A.: Fuzzy sets. Information and Control 8, 338–353 (1965)
24. Zadeh, L. A.: Foreword. In: Dubois, D., Prade, H. (eds.): Fundamentals of Fuzzy Sets. Kluwer Academic Publishers, Dordrecht (2000)
25. Zadeh, L.A.: From Computing with Numbers to Computing with Words—from Manipulation of Measurements to Manipulation of Perceptions. Int. J. Appl. Math. Comput. Sci. 12, 307–324 (2002)

Some Notes on the Value of Vagueness in Everyday Communication

Nora Kluck

Department of Philosophy, RWTH Aachen University
Eilfschornsteinstr. 16, D-52062 Aachen, Germany
Nora.Kluck@rwth-aachen.de

Abstract. From a logical point of view, vagueness is a problem. It even was called a "philosopher's nightmare"; it is often only regarded as one of the logical shortcomings of natural languages. But nevertheless, vagueness also has some advantages, especially regarding everyday communication. It already begins in first language acquisition: Here are no sharp boundaries of predicates learned. Thanks to vagueness, we do not have to precisify predicates ad infinitum, we can communicate successfully and efficiently. Vague predicates also take into account the limitations of human perception.

Keywords: Vagueness, Communication, Semantics, Pragmatics, Philosophy of Language.

1 Introduction: Vagueness – A Logical Problem

If 10,000 grains of wheat constitute a heap, 9,999 also do: If you take away one grain, it does not make a difference for the application of the predicate "heap". But if you reiterate that step 9998 times, you only will have one grain left – and you still would have to call it a heap, because one grain does not make a difference at any of the previous steps. But one grain of wheat is clearly *not* a heap: the conclusion, though derived from true premises, is false: Here we have a paradox, the so-called *sorites-paradox* (from Greek σωρός, soros = heap).

This paradox is known since antiquity and can be traced back to Eubulides of Miletus (4th century BC). Since then, philosophers thought about blurred boundaries of predicate-extensions. The problem always is the same: Where is the borderline for the correct application of a predicate? Which number of grains constitutes a heap, and at which number does the heap turn into a non-heap? There is no sharp borderline for the application of this predicate, and this phenomenon occurs for many predicates of natural languages – e.g. up to which height is somebody "small", how many hairs may a "bald man" have on his head? (The latter is known as the *falakros puzzle*, which also comes from Eubulides.) This sort of predicates are called *vague*.

This kind of vagueness is different from *underspecification*, which is also often called vagueness. For example the sentence "This conference takes place somewhere in Germany" is vague in this sense, but it *can* be precisified. Underspecification is not addressed in this paper, although it also is useful in communication; think of election

E. Hüllermeier, R. Kruse, and F. Hoffmann (Eds.): IPMU 2010, Part II, CCIS 81, pp. 344–349, 2010.

campaigns or international negotiations. The kind of vagueness I want to discuss here is the one which gives rise to the above mentioned sorites-paradox, also called *sorites-vagueness*.

Sorites-vagueness is pervasive in natural language; borderline cases occur everywhere. Ernesto Napoli called vagueness a "philosopher's nightmare" ([1], 115), because it gives rise to serious logical problems: The law of excluded middle (that every proposition is either true or false) does not seem to hold. If I say "This is a heap", pointing at an amount of grains which is a borderline case of "heap", it is not clear whether the proposition I expressed is true or false.

The current debate about vagueness in philosophy started in 1923, when Bertrand Russell published his seminal paper "Vagueness" ([2]). Since then, philosophers developed many divergent theories to handle the logical problems caused by vagueness. Some introduced more truth-values (*degree theories*; *three-valued-logic*; *fuzzy logic*: Tye [3], Machina [4], Sainsbury [5], Edgington [6]), some worked with admissible precisifications and super-truth (*supervaluationism*: Mehlberg [7], Fine [8], Keefe [9]), some denied the existence of ordinary things because of the lack of sharp boundaries (*nihilsm*: Unger [10]), or they postulated the existence of unknowable sharp borderlines (*epistemic theory*: Cargile [11], Campbell [12], Sorensen [13], Williamson [14]). Others treat vagueness as context-dependency (*contextualism*: Kamp [15], Bosch [16], Raffman [17], Fara [18], Shapiro [19]). For an overview of these theories, see e.g. [14] or [20].

All these theories have advantages and shortcomings which cannot be discussed here. But they have one important feature in common: They regard vagueness as a problem. From a logical point of view, it clearly is. But vagueness also has advantages, which deserve a closer look.

Natural languages contain a lot of vague predicates. This suggests the assumption that languages would have developed differently if vagueness were only disadvantageous and unfavourable. In everyday communication, natural languages containing vague predicates work perfectly well. From time to time, a misunderstanding might occur because of vagueness, but in these cases language provides tools to handle it: We can ask our interlocutor what he has meant or ask him to be more precise. But in the very most cases, vagueness does not lead to problems in communication.

Vagueness is no obstacle for communication; communication even works better with vague than with precise predicates. This claim is often found in linguistic and philosophical literature, but there it is not told why this is the case. I want to sketch some of the advantages of vagueness in everyday communication in the following sections; I will continue working on this topic in the context of my dissertation project.

2 Vagueness in First Language Acquisition

Let us first have a look at language acquisition. Here the communication with vague predicates already begins, because predicates are not acquired with sharp boundaries.

For early word learning, ostensive definitions are important. So children acquire a new word in a special case, perhaps the word "dog" while the mother is pointing at the family's dog. From this special case children conclude how to name other objects which are similar to the first one. This first one is the prototype around which the extension of the predicate grows (see [21], 273).

At the age of 1 to 2½ years, children overextend predicates because of the similarity of objects: They call e.g. everything "moon" which is yellow, round or has the form of the crescent moon, e.g. a lemon slice (see [21], 266). Overextension seems to be a successful communicative strategy (see [22], 35): It allows the child to refer to objects, even if the precise fitting word is still unknown. Instead of saying nothing, the next appropriate word is used. Bloom describes it like that: "It is almost as if the child were reasoning, 'I know about dogs, that thing is not a dog, I don't know what to call it, but it is like a dog!'" ([23], 79). The more words the child knows, the less overextension is needed to achieve the communicative goals.

The process of first language acquisition sheds light on a second aspect: The predicate-extension is not learned with sharp boundaries. The child does not learn the meaning of the word "heap" by counting the grains or the meaning of "red" by measuring the wave-length of light. It perhaps learns that a "mug" and a "cup" are different things, but it does not acquire an exact borderline between them (for the difference between "mug" and "cup" see [24]). So there always will be borderline cases, even after first language acquisition.

3 Limitations of the Human Perception

If our predicates were not vague, we often could not apply them, because our perception and our discriminative abilities are limited. Without counting, we simply cannot know whether a heap of wheat contains one grain more than another; our eyes are not suitable for perceiving the exact number of grains.

Waismann points out: "Suppose a pattern-book were shown to me, and I was later asked whether this was the colour I had seen, perhaps I would not be able to decide. [...] Notice that, in this case, it is quite natural to use a vague term (‚light colour') to express the indeterminacy of the impression. If language was such that each and every word was particular and each colour word had a definite, clearly defined meaning, we should find we could not use it." ([25], 21). Our impressions are indeterminate, so it is easier to express them with indeterminate predicates than with precise ones.

If there were a sharp borderline between heaps and non-heaps, we could not see with the naked eye whether something is a heap or not, and so we could not apply the predicate "heap" correctly without additional tools or without spending a lot oft time counting grains.

So vague expressions enable us to use words in an economic way: As one grain makes no difference for the application of the word "heap", vagueness relieves us from counting them.

4 Where to Stop Precisification?

If we wanted to get rid of vague predicates in everyday communication, we had to try to precisify them. But where to stop precisification? Why stopping at the precisification-level of a grain? There even are borderline cases of "grain": What about a damaged grain? Here the problem of borderline cases rises again. And why not get further and count molecules or atoms?

The problem will not be solved by counting grains – so it seems better not even to start counting, because it only would be a waste of time.

5 Communication Success, Efficiency and Flexibility

Vague predicates are crucial for reaching our communicative goals; thanks to their flexibility, we can communicate successfully and more efficiently compared to a formal, non-vague language.

5.1 Communication Success

Thanks to vague predicates, communication succeeds in cases where it would fail with precise predicates. My interlocutor knows what I mean, even if I talk about borderline cases. Perhaps we would not both have called the thing I am pointing at a "heap", but the other person knows which object I refer to, because there is no sharp boundary which excludes the application of the predicate "heap" for this set of grains. The communication process succeeds, and that is what matters for everyday communication, even if it gives rise to logical problems (for my interlocutor, the grains do not constitute a heap, but he understands when I call them a heap; nevertheless, from a logical point of view, it cannot be a heap and a non-heap at the same time).

Wittgenstein points out that we can work perfectly well with vague predicates: "But is it senseless to say: 'Stand roughly there'?" ([26], §71). In most cases we do not mind whether the word is vague, e.g. in the case of the word "game": "What still counts as a game and what no longer does? Can you give the boundary? No. You can *draw* one; for none has so far been drawn. (But that never troubled you before when you used the word 'game')." ([26], §68)

As long as communication succeeds, there is no reason to call for more precision. But if it does not succeed any more, we have to think about more precise predicates, as Quine emphasizes: "When sentences whose truth values hinge on the penumbra of a vague word do gain importance, they cause pressure for a new verbal convention or changed trend of usage that resolves the vagueness in its relevant portion. We may prudently let vagueness persist until such pressure arises, since meanwhile we are in an inferior position for judging which reforms might make for the most useful conceptual scheme."([27], 128). We can precisify predicates; either in the given situation to avoid misunderstandings or as speech community, as Quine suggests. But in most cases precisification is not necessary. In fact, too much precision would not only be unnecessary, but even an obstacle, as Schaff points out ([28], 90).

5.2 Flexibility

"Vagueness is a precondition of the flexibility of ordinary language", says Williamson ([14], 70): The lack of sharp borderlines makes vague predicates more *flexible* than precise ones, because they can be applied in more situations.

Frege compares natural languages to the human hand, while formal, non-vague languages are like specialized tools: "We build ourselves artificial hands-tools for special purposes which function more exactly than the hand is capable of doing. And how is this exactness possible? Through the very rigidity and inflexibility of the parts, the lack of which makes the hand so dexterous." ([29], 158). Formal, non-vague languages serve better for the purposes of logic and mathematics, like a tool which is made for a special task – but only for that one. Natural languages are not suitable for these special tasks, but serve better for everyday communication, because they are more flexible.

5.3 Efficiency

Communication with vague predicates is not just successful, but it is also *efficient*: It is successful in a small amount of time. If I had to count grains before I could determine whether "heap" applies or not, it simply would take too long. Some predicates would require special tools to decide about their application: To decide e.g. about colours, I would have to measure the wave-length; that is impossible without technical devices, as was pointed out in section 3. For most communicative goals it is better to use a word quickly which does not fit exactly than to use an exactly fitting word after a long time of deliberating.

Vague predicates can be used without measuring, counting etc. So in everyday situations, we can communicate efficiently in a limited amount of time. Therefore it is advantageous that the difference of one grain does not determine the application of a term like "heap".

6 Concluding Remarks

Vagueness clearly has some advantages in everyday communication. Of course, for special purposes we have to precisify our language and even occasionally have to count electrons. But for everyday use, natural languages with vague predicates serve perfectly well and yield their own advantages. With them, we can achieve our communicative goals in a small amount of time without enhancing our perception with technical devices. So we can use vague languages more flexible than non-vague ones.

Vagueness still gives rise to logical problems, but natural languages get along with these problems. They are the price to pay for communicative success, efficiency and flexibility in everyday communication.

References

1. Napoli, E.: Is Vagueness a Logical Enigma? Erkenntnis 23, 115–121 (1985)
2. Russell, B.: Vagueness. The Australasian Journal of Philosophy 1(2), 84–92 (1923)
3. Tye, M.: Sorites Paradoxes and the Semantics of Vagueness. Philosophical Perspectives 8, 189–206 (1994)
4. Machina, K.: Truth, Belief, and vagueness. Journal of Philosophical Logic 5, 47–78 (1976)
5. Sainsbury, R.M.: Degrees of Belief and Degrees of Truth. Philosophical Papers 15, 97–106 (1986)
6. Edgington, D.: Vagueness by degrees. In: Keefe, R., Smith, P. (eds.) Vagueness. A Reader, pp. 294–316. The MIT Press, Cambridge (1999)
7. Mehlberg, H.: The Reach of Science. University of Toronto Press, Toronto (1958)
8. Fine, K.: Vagueness, truth and logic. Synthese 30, 265–300 (1975)

9. Keefe, R.: Theories of vagueness. Cambridge University Press, Cambridge (2000)
10. Unger, P.: There are no ordinary things. Synthese 41, 117–154 (1979); Reprinted in: Graff, D., Williamson, T. (eds.): Vagueness. The International Research Library of Philosophy, pp. 3–40. Ashgate/Dartmouth, Aldershot (2002)
11. Cargile, J.: The Sorites Paradox. British Journal for the Philosophy of Science 20, 193–202 (1969)
12. Campbell, R.: The Sorites Paradox. Philosophical Studies 26, 175–191 (1974)
13. Sorensen, R.A.: Vagueness and contradiction. Clarendon Press, Oxford (2001)
14. Williamson, T.: Vagueness. Routledge, London (1994)
15. Kamp, H.: The Paradox of the Heap. In: Mönnich, U. (ed.) Aspects of philosophical logic. Some logical forays into central notions of linguistics and philosophy, pp. 225–277. Daniel Reidel Publishing Company, Dordrecht (1981)
16. Bosch, P.: 'Vagueness' is Context-Dependence. A Solution to the Sorites-Paradox. In: Ballmer, T.T., Pinkal, M. (eds.) Approaching Vagueness, pp. 189–210. North Holland, Amsterdam (1983)
17. Raffman, D.: Vagueness and Context-relativity. Philosophical Studies 81, 175–192 (1996)
18. Fara, D.: Graff: Shifting Sands: An Interest-Relative Theory of Vagueness. Philosophical Topics 28, 45–81 (2000); Originally published under the name "Delia Graff"
19. Shapiro, S.: Vagueness in Context. Clarendon Press, Oxford (2006)
20. Keefe, R., Smith, P.: Introduction: theories of vagueness. In: Keefe, R., Smith, P. (eds.) Vagueness. A Reader, pp. 2–57. The MIT Press, Cambridge (1999)
21. Bowerman, M.: The Acquisition of Word Meaning. An Investigation in some Current Conflicts. In: Waterson, N., Snow, C. (eds.) The Development of Communication, pp. 263–287. John Wiley & Sons, Chichester (1978)
22. Clark, E.: What's in a word? On the child's acquisition of semantics in his first language. In: Moore, T.E. (ed.) Cognitive development and the acquisition of language, pp. 66–110. Academic Press, New York (1973)
23. Bloom, L.M.: One word at a time: The use of single word utterances before syntax. Mouton, The Hague (1973)
24. Labov, W.: The boundaries of words an their meanings. In: Bailey, C.J.N., Shuy, R.W. (eds.) New ways of analyzing variation in English, pp. 340–373. Georgetown University Press, Washington (1973)
25. Waismann, F.: Language strata. In: Flew, A. (ed.) Logic and Language. Second series, pp. 11–31. Basil Blackwell, Oxford (1953)
26. Wittgenstein, L.: Philosophical Investigations. The German text with a revised English translation. 3rd edn. Blackwell, Oxford (2001)
27. Quine, W.V.O.: Word and Object. The MIT Press, Cambridge (1960)
28. Schaff, A.: Unscharfe Ausdrücke und die Grenzen ihrer Präzisierung. In his Essays über die Philosophie der Sprache, pp. 65–94. Europäische Verlagsanstalt et al., Frankfurt am Main et al. (1968)
29. Frege, G.: Frege: On the Scientific Justification of a Concept-Script. Translated by J.M. Bartlett. Mind 73(290), 155–160 (1964); Originally published as: Über die wissenschaftliche Berechtigung einer Begriffsschrift. In: Zeitschrift für Philosophie und philosophische Kritik NF, vol. 81, pp. 48–56 (1882)

On Zadeh's "The Birth and Evolution of Fuzzy Logic"

Yücel Yüksel

Department of Philosophy, Faculty of Letters, Istanbul University,
34459 Vezneciler, Istanbul, Turkey
yyuksel@istanbul.edu.tr

Abstract. Lotfi A. Zadeh, in his article entitled "The Birth and Evolution of Fuzzy Logic" discusses R.E. Kalman's and W. Kahan's strong criticisms of fuzzy logic and presents his answers to these criticisms. The main subject of this debate consisted of a discussion of the criticisms targeting at the concept of the linguistic variable and, consequently, at fuzzy logic. The main aim of my paper is to expose the methods of the positive natural sciences which generally form the basis of these criticisms, to analyze and to evaluate scientific and epistemological theories of science historians and philosophers of science such as T.S. Kuhn and K.R. Popper, and to try to show the importance of fuzzy logic in this context. There is a huge amount of work to be done within the framework of the philosophy and sociology of science in order to provide some hints about the future of fuzzy logic, therefore it must be stated that this paper can only be considered as a modest contribution to this vast field.

Keywords: fuzzy logic, philosophy of science, sociology of science.

1 Introduction

Lotfi A. Zadeh, who is the pioneer of fuzzy logic, in his article which is entitled "The Birth And Evolution Of Fuzzy Logic" and which is based on a lecture presented on the occasion of the award of the 1989 Honda Prize in Japan [15], discusses R.E. Kalman's [16] and W. Kahan's [17] strong criticisms of fuzzy logic and presents his answers to these criticisms. Especially the debate which took place between Zadeh and Kalman at "Man and Computer Conference" in Bordeaux, France, 1972 and Zadeh's opinions about the debate constitute the basis for this article.

2 Preciseness of Modern Sciences and Fuzziness

For Zadeh, Kalman is a scientist who strictly adheres to the Cartesian tradition in science. In his article Zadeh firstly speaks about the Cartesian tradition and Lord Kelvin[1], who is one of its foremost spokesmen. In order to expose clearly the opposition between Kalman and himself he makes the following statement [15]:

[1] Lord Kelvin, in the chapters III and IV of his work entitled Elements of Natural Philosophy, gives detailed explanations on the importance of experience and devices of measurement in the analysis of nature and on how to use these devices, which are necessary for the precise measurement of time, space, force and mass in relation to observation (see [13]).

E. Hüllermeier, R. Kruse, and F. Hoffmann (Eds.): IPMU 2010, Part II, CCIS 81, pp. 350–355, 2010.

The Cartesian tradition of respect for what is quantitative and precise, and disdain for what is qualitative and imprecise is too deep-seated to be abandoned without a fight. The basic tenet of this tradition was stated succinctly by Lord Kelvin. …He wrote, "In physical science a first essential step in the direction of learning any subject is to find principles of numerical reckoning and practicable methods for measuring some quality connected with it. (p. 95)

In order to show clearly the content of the debate between Zadeh on one side and Kalman and Kahan as well as the established inclination to consider science almost as divine on the other, I will, on the following pages, need to give a couple of long quotations one after the other. These long quotations function as evidences of the fact that prominent scholars may from time to time be far from being as objective as they claim to be in their scientific work and consequently, it will be, in my opinion, the best attitude to include them in this study without adding any comments.

Zadeh cites Kalman's serious criticisms of fuzzy logic [15]:

…No doubt Professor Zadeh's enthusiasm for fuzziness has been reinforced by the prevailing political climate in the U.S.—one of unprecedented permissiveness. "Fuzzification" is a kind of scientific permissiveness; …I cannot conceive of "fuzzification" as a viable alternative for the scientific method; I even believe that it is healthier to adhere to Hilbert's naive optimism, "Wir wollen wissen: wir werden wissen".

It is very unfair for Professor Zadeh to present trivial examples …and then imply (though not formally claim) that his vaguely outlined methodology can have an impact on deep scientific problems. In any case, if the "fuzzification" approach is going to solve any difficult problems, this is yet to be seen. (pp. 96-97)

As a reply to these criticisms Zadeh says that [15]:

To view Professor Kalman's rather emotional reaction to my presentation in a proper perspective, I should like to observe that, up to a certain point in time, Professor Kalman and I have been traveling along the same road, by which I mean that both of us believed in the power of mathematics, in the eventual triumph of logic and precision over vagueness. But then I made a right turn, or maybe even started turning backwards, whereas Professor Kalman has stayed on the same road. Thus, today, I no longer believe, as Professor Kalman does, that the solution to the kind of problems referred to in my talk lies within the conceptual framework of classical mathematics. In taking this position, I realize, of course, that I am challenging scientific dogma. …Now, when one attacks dogma, one must be prepared to become the object of counterattack on the part of those who believe in the status quo. …Nevertheless, I believe that, in time, the concepts that I have presented will be accepted and employed in a wide variety of areas. …Indeed, in retrospect, the somewhat unconventional ideas suggested by me may well be viewed as self-evident to the point of triviality. (p. 97)

Kalman continues his criticisms in the following way [15]:

Professor Zadeh misrepresents my position if he means to say that I view scientific research solely in terms rigidly precise or even "classical" mathematical models.

...This is not to argue that only rigorously rational methods (of conventional science) should be used. But if one proposes to deprecate this tool (which, when properly understood and used, has given us many striking successes), he should at least provide some hard evidence of what can be gained thereby. Professor Zadeh's fears of unjust criticism can be mitigated by recalling that the alchemists were not prosecuted for their beliefs but because they failed to produce gold. (pp. 97-98)

Zadeh also refers to Kahan's criticisms which are similar to those of as Kalman [15]:

Fuzzy theory is wrong, wrong, and pernicious" says William Kahan, a professor of computer sciences and mathematics at Cal whose Evans Hall office is a few doors from Zadeh's. "I can not think of any problem that could not be solved better by ordinary logic. ...What Zadeh is saying is the same sort of thing as, "Technology got us into this mess and now it can't get us out". Well, technology did not get us into this mess. Greed and weakness and ambivalence got us into this mess. What we need is more logical thinking, not less. The danger of fuzzy theory is that it will encourage the sort of imprecise thinking that has brought us so much trouble. (p. 98)

As is also understood from Kalman's, Kahan's and Zadeh's opinions, there is a wide difference between the Cartesian tradition which goes after the absolute precision in science and fuzzy logic which is based upon vagueness and subjectivity. Modern science which is improved by the Cartesian tradition as Zadeh states or in more general terms the method and mentality of positivistic natural sciences[2] have had very great achievements in determining and explaining phenomena by all means. Inevitably these achievements have toughened the mentality of positivistic natural sciences and paved the way for the spread of its influences in every respect. Thus it is not surprising that these very skeptical, assertive and probably deliberately inhospitable criticisms toward fuzzy logic appear in such a "safe" scientific environment. I think that another reason of these hostile criticisms is that Kalman, Kahan and other scientists who share the same ideas with them were unaware of the strong discussions which were taking place almost concurrently with the birth of fuzzy logic, especially between T.S. Kuhn and K.R. Popper as two prominent critical thinkers and also others like P.K. Feyerabend, I. Lakatos about the philosophy and sociology of science. It looks almost impossible to talk about these very important discussions (e.g. about paradigm notion or some questions such as "is there scientific revolution?") in a detailed way in this short and modest paper. Hence at first my aim is to touch briefly upon Kuhn's theory about scientific constitution and progression by the help of his book entitled *The Structure of Scientific Revolutions* [4] and to make a comparison between "normal science" as one of his basic notions and the scientific environment which is followed and defended by Kalman, Kahan and other similar scientists or thinkers.

According to Kuhn scientific communities always tend to describe the place where scientists stand now as the best stage and consequently any current conclusion is considered to be a result of a series of positive and continuous scientific activities and

[2] What we mean here by "the method of positivistic natural sciences" is mathematical physics which was developed by Galileo and Descartes and which reached its highest point with Newton. The language of this method, in other words, the language of the natural sciences which combine experiences with precise mathematical formulas, has in time become the language of the modern science (for details see [3], [2], [1], [7] and also [12]).

cumulative increase of knowledge. Yet Kuhn thought that such a perception is only an illusion and very dangerous. Because this illusion is a basis for an immorality which can turn into a scientific ideology and as part of a scientific ideology scientific communities can believe that only they have the truest scientific route. In his *The Structure of Scientific Revolutions* Kuhn's main aim is to claim that it is possible to introduce a quite different scientific process and philosophical conclusions in the same history of science [5]. At this point, we can take a look at Kuhn's views on this matter [4]:

Normal science, the activity in which most scientists inevitably spend almost all their time, is predicated on the assumption that the scientific community knows what the world is like. Much of the success of the enterprise derives from the community's willingness to defend that assumption, if necessary at considerable cost. Normal science, for example, often suppresses fundamental novelties because they are necessarily subversive of its basic commitments. (p. 5)

In this context, it is possible to see the extreme loyalty of Kalman, Kahan and others to two-valued logic or ordinary logic as a sign of their ideological scientific approach.

Kuhn believes that every conceptual system has a distinctive semantics. If we have to interpret a novelty which arises from a different semantics then we have to explore the semantics which is different from our semantics. Otherwise to interpret this novelty within our semantics is unthinkable. By means of this idea Kuhn asserts that it is impossible to make progress by arguing opposite views in the same scientific tradition. To prefer between opposite views is to change our belief system rather than a technical changeover [5].

In Kuhn's words [4],

...If, therefore, these epistemological counter instances are to constitute more than a minor irritant, that will be because they help to permit the emergence of a new and different analysis of science within which they are no longer a source of trouble. Furthermore, if a typical pattern, which we shall later observe in scientific revolutions, is applicable here, these anomalies will then no longer seem to be simply facts. From within a new theory of scientific knowledge, they may instead seem very much like tautologies, statements of situations that could not conceivably have been otherwise. (p. 78)

It is interesting that Kuhn's definition of the sociological and psychological processes experienced in the transition from the usual state defined by scientific communities as normal science to an alternative scientific system as well as his definition of that newly emergent situation summarizes the processes of the theory of fuzzy logic as well.

The strongest criticism against Kuhn's above mentioned views came from Popper. He, in his article entitled "Normal Sciences and Its Dangers" [10], states that he is disturbed by Kuhn's emphasis on sociological and psychological elements in the development of science and blames Kuhn of falling into logical relativism [5].

Due to the fact that Popper puts logic to the centre in his theory of the development of science, we encounter two significant problems in terms of philosophy of science. The first of these is the problem of "induction" which Popper believes totally lost its validity with David Hume. And the other one is the principle of "falsifiability" [11] which Popper offers as an alternative central method for science [5]. For fuzzy logic those two problems are the subjects of two different fields of study. However, to make a brief explanation, it will not be wrong to state that today fuzzy logic has the capacity to analyze within its own systematic both the "probability theory" (see [6], [9]) which the scientists developed in order to make "induction" more effective [5] and the "falsifiability" that Popper based on two-valued logic. For fuzzy logic, which considers the truth of any proposition as a gradual situation, Popper's principle of falsifiability is one that can be re-interpreted[3].

Taking also into consideration the special significance that Kuhn attaches to sociology and psychology in scientific developments, it can be expected that fuzzy logic, which is based on the vagueness and subjectivity which are the materials of sociology and psychology, can, in addition to its achievements in the technical area against the two-valued logic, also contributes considerably to the way scientific developments can be defined logically as Popper would prefer it.

3 Conclusion

Although the current paradigm which is based on two-valued logic still exists and is influential, in my opinion, fuzzy logic has the potential to be the essence or at least the starter of a potential change of paradigm. My argument should never be interpreted as a radical assertion to negate the current paradigm altogether. I am only suggesting that it has to be taken seriously into consideration and evaluated that the problems arising from the fact that the concept of uncertainty has not been taken seriously enough in the ways the current paradigm has been put forward by many scholars and that it should be accepted that these two frames of thought can actually co-exist and be employed simultaneously without negating one another in very effective and functional ways within various adequate contexts.

I would like to finish my presentation with a quotation from Kuhn, which, I think, gives clues about the development of fuzzy logic [4]:

At the start a new candidate for paradigm may have few supporters, and on occasions the supporters' motives may be suspect. Nevertheless, if they are competent, they will improve it, explore its possibilities, and show what it would be like to belong to the community guided by it. And as that goes on, if the paradigm is one destined to win its fight, the number and strength of the persuasive arguments in its favor will increase. More scientists will then be converted, and the exploration of the new paradigm will go on. Gradually the number of experiments, instruments, articles, and books based upon the paradigm will multiply. Still more men, convinced of the new view's fruitfulness, will adopt the new mode of practicing normal science, until at last only a few elderly hold-outs remain. And even they, we cannot say, are wrong. Though the historian can always find men -Priestley, for instance- who were unreasonable to resist for as long as they did, he will not find a point at which resistance becomes illogical or unscientific. At most he may wish to say that the man who continues to resist after his whole profession has been converted has ipso facto ceased to be a scientist. (p. 159)

---

[3] For detailed information on Probability theory and approximate truth see [14], [8].

References

1. Descartes, R.: The World and Other Writings. Trans. and Ed. by Stephen Gaukroger. Cambridge University Press, UK (1998)
2. Galilei, G.: Dialogues Concerning Two New Sciences. Trans. by Henry Crew & Alfonso De Salvio. General Publishing Company Ltd., Canada (1954)
3. Galilei, G.: Dialogue Concerning the Two Chief World Systems. Trans. by Stillman Drake. University of California Press, USA (1967)
4. Kuhn, T.S.: The Structure of Scientific Revolutions. International Encyclopedia of Unified Science. 2nd edn., USA (enlarged). The University of Chicago Press (1970)
5. Kuyas, N.: Çevirmenin Sunusu. In: Yapısı, B.D. (ed.) The Structure of Scientific Revolutions. Trans. by Nilüfer Kuyas, pp. 7–32. Alan Yayıncılık, Istanbul (1991)
6. Montero, J.: Fuzzy Logic and Science. In: Seising, R. (ed.) Views on Fuzzy Sets and Systems from Different Perspectives: Philosophy and Logic, Criticisms and Applications, pp. 70–73. Springer, Heidelberg (2009)
7. Newton, I.: The Principia: The Mathematical Principles of Natural Philosophy. In: Bernard Cohen, I., Whitman, A.M. (eds.) University of California Press, USA (1999)
8. Niskanen, V.A.: Soft Computing Methods in Human Sciences. Springer, Germany (2004)
9. Nurmi, H.: Probability and Fuzziness – Echoes from 30 Years Back. In: Seising, R. (ed.) Views on Fuzzy Sets and Systems from Different Perspectives: Philosophy and Logic, Criticisms and Applications, pp. 163–170. Springer, Heidelberg (2009)
10. Popper, K.R.: Normal Science and its Dangers, Criticism and The Growth of Knowledge. In: Lakatos, I., Musgrave, A. (eds.), pp. 51–58. Cambridge University Press, UK (1970)
11. Popper, K.R.: The Logic of Scientific Discovery, 2nd edn., Routledge, pp. 57–73 (2002)
12. Seising, R.: Fuzzy Sets and Systems and Philosophy of Science. In: Seising, R. (ed.) Views on Fuzzy Sets and Systems from Different Perspectives: Philosophy and Logic, Criticisms and Applications, pp. 1–4. Springer, Heidelberg (2009)
13. Thomson Sir, W., Tait, P.G.: Elements of Natural Philosophy, pp. 106–129. MacMillan and Co. Publishers to the University of Oxford, London (1873)
14. Zadeh, L.A.: Fuzzy Logic and Approximate Reasoning. In: Klir, G.J., Yuan, B. (eds.) Fuzzy Sets, Fuzzy Logic, and Fuzzy Systems: Selected Papers by Lotfi Asker Zadeh, World Scientific Publishing Co. Pte. Ltd., Singapore (1996)
15. Zadeh, L.A.: The Birth and Evolution of Fuzzy Logic. International Journal of General Systems 17, 95–105 (1990)
16. http://www.ieeeghn.org/wiki/index.php/Rudolf_E._Kalman
17. http://www.eecs.berkeley.edu/~wkahan/

Complexity and Fuzziness in 20th Century Science and Technology

Rudolf Seising

European Centre for Soft Computing
Edificio Científico-Tecnológico. 3ª Planta
C Gonzalo Gutiérrez Quirós S/N
33600 Mieres, Asturias, Spain
Rudolf.Seising@softcomputing.es

Abstract. This historical and philosophical paper shows the parallel views of Warren Weaver and Lotfi Zadeh in the 1950s and 1960, respectively, on mathematics in science and technology and their calls for "new mathematics" to solve a new class of problems.

Keywords: Complexity, fuzziness, science, technology, history, philosophy.

1 Introduction

This historical and philosophical paper deals with important changes in science and technology in the 20th century which we associate with the concepts of *complexity* and *fuzziness*. Here, *complexity* refers to the big change in science – roughly spoken – from physics to the life sciences that started in the thirties of the 20th century. The initiation of this change is deeply connected with the work of the American mathematician and science administrator Warren Weaver (1894-1978).

Weaver was a very active and exciting man of science in a very long period of the 20th century. In 1948/49 he did not only publish the most well-known introductory and popularizing paper "The Mathematics of Communication" [1] on Shannon's "Mathematical Theory of Communication" [2], [5] and the very influential memorandum "Translation" [3] on the possible use of computers to translating natural languages. But he also wrote the article "Science and Complexity" [4] where he identified a class of scientific problems which science has as yet little explored or conquered". Weaver argued that these problems can neither be reduced to a simple formula nor they be solved with methods of probability theory. To solve such problems he pinned his hope on the power of digital computers and on interdisciplinary collaborating "mixed teams".

The second concept that we discuss in this paper is the concept of *fuzziness* which is well-known in the scientific community attending to IPMU-conferences for many years. *Fuzziness* indicates a change in science, too: the addition of scientific methods and techniques that are different from usual (classical) mathematics: the methods of fuzzy sets and systems which are rooting in Lotfi Zadeh's call for a non-probabilistic

E. Hüllermeier, R. Kruse, and F. Hoffmann (Eds.): IPMU 2010, Part II, CCIS 81, pp. 356–365, 2010.

and non-statistical mathematical theory in 1962.[1] It is understood that Zadeh kept sets of problems at the back of his mind, that are very similar to Weaver's newly-discovered scientific problems, when he described problems and applications of *System theory* and its relations to network theory, control theory, and information theory in the paper "From Circuit Theory to System Theory" [7]. He pointed out that "largely within the past two decades, by the great progress in our understanding of the behaviour of both inanimate and animate systems—progress which resulted on the one hand from a vast expansion in the scientific and technological activities directed toward the development of highly complex systems for such purposes as automatic control, pattern recognition, data-processing, communication, and machine computation, and, on the other hand, by attempts at quantitative analyses of the extremely complex animate and man-machine systems which are encountered in biology, neurophysiology, econometrics, operations research and other fields" [7]. In this paper he wrote: "In fact, there is a fairly wide gap between what might be regarded as "animate" system theorists and "inanimate" system theorists at the present time, and it is not at all certain that this gap will be narrowed, much less closed, in the near future. There are some who feel that this gap reflects the fundamental inadequacy of the conventional mathematics – the mathematics of precisely-defined points, functions, sets, probability measures, etc. – for coping with the analysis of biological systems, and that to deal effectively with such systems, which are generally orders of magnitude more complex than man-made systems, we need a radically different kind of mathematics, the mathematics of fuzzy or cloudy quantities which are not describable in terms of probability distributions. Indeed, the need for such mathematics is becoming increasingly apparent even in the realm of inanimate systems, for in most practical cases the a priori data as well as the criteria by which the performance of a man-made system is judged are far from being precisely specified or having accurately-known probability distributions" [7].

2 Zadeh's Fuzzy Sets and Systems

In 1962 Zadeh called for "fuzzy mathematics" without exact knowing, what kind of theory he would create later on in his first journal article on this subject, in 1965, where he introduced the new mathematical entities – "fuzzy sets" – as classes or sets that "are not classes or sets in the usual sense of these terms, since they do not dichotomize all objects into those that belong to the class and those that do not". In fuzzy sets "there may be a continuous infinity of grades of membership, with the grade of membership of an object x in a fuzzy set A represented by a number $f_A(x)$ in the interval [0,1]" [8]. In the same year the *Symposium on System Theory* took place at the Polytechnic Institute in Brooklyn where Zadeh presented "A New View on System Theory". A shortened version of the paper delivered at this symposium appeared in the proceedings under the title "Fuzzy Sets and Systems" and Zadeh defined for the first time the concept of a "fuzzy system" as a system S where "(input) $u(t)$, output $y(t)$, or state $s(t)$ of S or any combination of them ranges over fuzzy sets. ([8], p. 33).

[1] For the history of the theory of Fuzzy Sets see: [6].

3 Intelligent Systems – Humanistic Systems

In the 1950s, computers became popular as "electronic brains" or "thinking machines", and a after the lounging of *Artificial Intelligence* (AI) in 1965 this new research program spread to many scientific and technological communities throughout the world. AI-history includes a number of successes, but to date it has lagged behind expectations. AI became a field of research aimed at developing computers and computer programs that act "intelligently" even though no human being controls these systems. AI methods became methods of computing with numbers and finding exact solutions. On the other hand, humans are able to resolve such tasks very well, as Zadeh mentioned very often over the last decades, beginning in 1950 when he served as a moderator at a debate on digital computers at Columbia University between Claude E. Shannon, Edmund C. Berkeley, the author of the book *Giant Brains or Machines That Think* published in 1949 [9], and Francis J. Murray, a mathematician and consultant to IBM. In the same year the British mathematician Alan M. Turing published his famous article "Computing Machinery and Intelligence" [10] in the journal *Mind*. "Can machines think?" was the question and he proposed the well-known imitation game, now called the Turing test, to decide whether a computer or a program could think like a human being or not. Unaware of Turing's philosophical article, Zadeh wrote the paper "Thinking Machines – A New Field in Electrical Engineering", which appeared in the student journal *The Columbia Engineering Quarterly* in New York City in 1950 [11] (Fig. 1). He asked, "How will 'electronic brains' or 'thinking machines' affect our way of living?" and "What is the role played by electrical engineers in the design of these devices?" ([11], p. 12.]

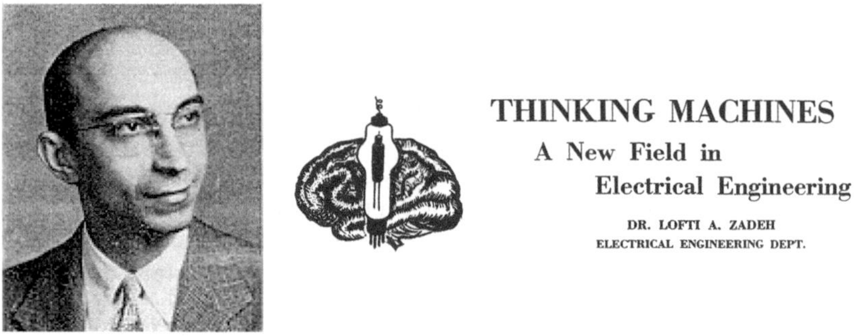

Fig. 1. Left: Lotfi A.Zadeh in the 1950s; right: an illustration from Zadeh's article [11]

In conclusion, Zadeh stated that "thinking machines" do not think as humans do. From the mid-1980s he focused on "Making Computers Think like People". [12] For this purpose, the machine's ability "to compute with numbers" was supplemented by an additional ability that was similar to human thinking.

Zadeh was and is inspired by the "remarkable human capability to perform a wide variety of physical and mental tasks without any measurements and any computations". In many papers he has given everyday examples of such tasks: parking a car, playing golf, deciphering sloppy handwriting, and summarizing a story. Underlying this, is the human ability to reason with perceptions – "perceptions of time, distance, speed, force, direction, shape, intent, likelihood, truth and other attributes of physical and mental objects." ([13], p. 903).

In the 1970s, Zadeh distinguished between mechanic (or inanimate or man-made) systems at one hand and humanistic systems at the other hand and he saw the following state of the art in computer technology: „Unquestionably, computers have proved to be highly effective in dealing with mechanistic systems, that is, with inanimate systems whose behavior is governed by the laws of mechanics, physics, chemistry and electromagnetism. Unfortunately, the same cannot be said about humanistic systems, which – so far at least – have proved to be rather impervious to mathematical analysis and computer simulation." In a footnote he explained that a "humanistic system" be "a system whose behaviour is strongly influenced by human judgement, perception or emotions. Examples of humanistic systems are: economic systems, political systems, legal systems, educational systems, etc. A single individual and his thought processes may also be viewed as a humanistic system." ([14], p. 200) In the main text he argued then, "that the use of computers has not shed much light on the basic issues arising in philosophy, literature, law, politics, sociology and other human-oriented fields. Nor have computers added significantly to our understanding of human thought processes—excepting, perhaps, some examples to the contrary that can be drawn from artificial intelligence and related fields." ([14], p. 200)

Computers have been very successful in mechanic systems but they could not be that successful humanistic systems in the field of non-exact sciences. Zadeh argued that that this is the case because oft his so-called Principle of Incompatibility that he established in 1973 for the concepts of exactness and complexity: "The closer one looks at a 'real world' problem, the fuzzier becomes its solution." [15]2 With this principle there is a difference between system analysis and simulations that base on precise number computing at one hand and analysis and simulations of humanistic systems at the other hand. Zadeh conjectured that precise quantitative analysis of the behaviour of humanistic systems are not meaningful for "real-world societal, political, economic, and other types of problems which involve humans either as individuals or in groups." ([15], p. 28)

4 Weaver's Midcentury Expectations on Science and Technology

The "Age of intelligent systems" was initiated in the middle of the 20th century when many of the scientific-technological achievements that were developed in research projects during the Second World War became generally known by the public. At that time Warren Weaver wrote three important papers:

- "The Mathematics of Communication" [1] – a re-interpretation of the article "A Mathematical Theory of Communication" [2] by the electronic engineer and mathematician Claude Elwood Shannon (1916-2001) for broader scientific audiences. Later, Weaver modified and accentuated this text with the

[2] More explicitly, Zadeh wrote: „Stated informally, the essence of this principle is, that as the complexity of a system increases, our ability to make precise and yet significant statements about it's behaviour diminishes until a threshold is reached beyond which precision and significance (or relevance) become almost mutually exclusive characteristics." [15].

new title "Recent Contributions to the Mathematical Theory of Communication" [16] that was published together with Shannon's paper in the book *The Mathematical Theory of Communication*. [5]

- "Translation" – a memorandum that circulated to some twenty or thirty acquaintances, which was to stimulate the beginnings of research on machine translation in the United States. In 1955, this text appeared in a Collection of essays on *Machine translation of Language*, see [3].
- "Science and Complexity" [4] – an article that based upon material for Weaver's introductory contribution to a series of radio talks, presenting aspects of modern science by 97 scientists, given as intermission programs during broadcasts of the New York Philharmonic-Symphonies. Weaver edited the written contributions in the book *The Scientists Speak* [17] and one year later "Science and Complexity", which arose from the book's first chapter, was published in the *American Scientist* [4].

In the first paper of this list Weaver argued that Shannon's "Mathematical theory of communication" did not even touch upon any of the semantic and effectiveness or pragmatic problems, but that the concepts of information and communication therefore must not be identified with the "meaning" of the symbols. But then he wrote "The theory goes further. Though ostensibly applicable only to problems at the technical level, it is helpful and suggestive at the levels of semantics and effectiveness as well." [1]

In the second paper, Weaver brooded whether it is unthinkable to design digital computers which would translate documents between natural human languages, Weaver speculated "that the way to translate from Chinese to Arabic, or from Russian to Portuguese, is not to attempt the direct route [...]. Perhaps the way is to descend, from each language, down to the common base of human communication – the real but as yet undiscovered universal language – and – then re-emerge by whatever particular route is convenient." [3]

In the third paper Weaver identified a "region" of problems "which science has as yet [1947/1948] little explored or conquered". These problems, he wrote, can neither be reduced to a simple formula nor can they be solved with methods of probability theory. To solve such problems he pinned his hope on the power of computers and on interdisciplinary collaborating "mixed teams". [4]

Weaver's midcentury expectations on the progress in science and technology seem to be anticipating important topics in the field of Soft Computing (SC) and Computational Intelligence (CI): vague, fuzzy or approximate reasoning, the meaning of concepts, and "to descend from each language, down to the common base of human communication—the real but as yet undiscovered universal language—". [3] This seems similar to Zadeh's concept of "precisiated natural language" [18] – and obviously Zadeh's thinking induced a big change in science and technology in the 20th century. However, there is no direct relation between the work of Weaver and Zadeh[3] but these aspects make it worth to study Weavers writings in this context.

---

[3] In a personal message Zadeh answered to the author's question whether he was familiar with Weaver's papers in the 1940s and 1950s that he did not read the papers [4, 5]. He also wrote: "It may well be the case that most people near the center [of the "world of information theory and communication" in that time] did not appreciate what he had to say. In a sense, he may have been ahead of his time." [19]

5 Weaver's "Science and Complexity"

In the introductory paragraph of "Science and Complexity" Weaver asked: "How can we get a view of the function that science should have in the developing future of man? How can we appreciate what science really is and, equally important, what science is not? It is, of course, possible to discuss the nature of science in general philosophical terms. For some purposes such a discussion is important and necessary, but for the present a more direct approach is desirable." Weaver then overviewed the "three and a half centuries" of modern science and he took "a broad view that tries to see the main features, and omits minor details." [4]

Regarding the history of sciences, Weaver said "that the seventeenth, eighteenth, and nineteenth centuries formed the period in which physical sciences learned variables, which brought us the telephone and the radio, the automobile and the airplane, the phonograph and the moving pictures, the turbine and the Diesel engine, and the modern hydroelectric power plant." [4] Compared to that, he assessed the development of life sciences else wise: "The concurrent progress in biology and medicine was also impressive, but that was of a different character. The significant problems of living organisms are seldom those in which one can rigidly maintain constant all but two variables. Living things are more likely to present situations in which a half-dozen or even several dozen quantities are all varying simultaneously, and in subtly interconnected ways. Often they present situations in which the essentially important quantities are either non-quantitative, or have at any rate eluded identification or measurement up to the moment. Thus biological and medical problems often involve the consideration of a most complexly organized whole."4

In summary, Weaver distinguished here between "problems of simplicity" that "physical science before 1900 was largely concerned with", and another type of problems that "life sciences, in which these problems of simplicity are not so often significant", are concerned with. The life sciences "had not yet become highly quantitative or analytical in character", Weaver stated in the late 1940s. Then, he enlarged on the new developed approach of probability and statistics in the area of exact sciences at around 1900: "Rather then study problems which involved two variables or at most three or four, some imaginative minds went to the other extreme, and said. »Let us develop analytical methods which can deal with two billion variables.« That is to say, the physical scientists, with the mathematician often in the vanguard, developed powerful techniques of probability theory and statistical mechanics to deal with what may be problems of disorganized complexity", a phrase that "calls for explanation" as he wrote, and he entertained this as follows: A problem of disorganized complexity "is a problem in which the number of variables is very large, and one in which each of the many variables has a behavior which is individually erratic, or perhaps totally unknown. However, in spite of this helter-skelter, or unknown, behavior of all the individual variables, the system as a whole possesses certain orderly and analyzable average properties."4

Weaver emphasized that probability theory and statistical techniques "are not restricted to situations where the scientific theory of the individual events is very well known" but he also attached importance to the fact that they can also "be applied to situations [...] where the individual event is as shrouded in mystery as is the chain of complicated and unpredictable events associated with the accidental death of a healthy man." He stressed "the more fundamental use which science makes of these new techniques. The motions of the atoms which form all matter, as well as the motions of the stars which form the universe, come under the range of these new techniques. The fundamental laws of heredity are analyzed by them. The laws of thermodynamics, which describe basic and inevitable tendencies of all physical systems, are derived from statistical considerations. The entire structure of modern physics, our present concept of the nature of the physical universe, and of the accessible experimental facts concerning it, rest on these statistical concepts. Indeed, the whole question of evidence and the way in which knowledge can be inferred from evidence are now recognized to depend on these same statistical ideas, so that probability notions are essential to any theory of knowledge itself."4

But there is more to this paper than that! In this article at the end of the 1940's Weaver mentioned – may be for the first time at all – a trichotomy of scientific problems: In addition to, and in-between, the "problems of simplicity" and the "problems of disorganized complexity" he identified another kind of scientific problems: "One is tempted to oversimplify, and say that scientific methodology went from one extreme to the other—from two variables to an astronomical number—and left untouched a great middle region. The importance of this middle region, moreover, does not depend primarily on the fact that the number of variables involved is moderate—large compared to two, but small compared to the number of atoms in a pinch of salt. The problems in this middle region, in fact, will often involve a considerable number of variables. The really important characteristic problems of this middle region, which science has as yet little explored or conquered, lies in the fact that these problems, as contrasted with the disorganized situations which statistics can cope, show the essential feature of organization. In fact, one can refer to this group of problems as those of organized complexity."4 (Fig. 2) He listed examples of such problems:

- What makes an evening primrose open when it does?
- Why does salt water fail to satisfy thirst?
- Why can one particular genetic strain of microorganism synthesize within its minute body certain organic compounds that another strain of the same organism cannot manufacture?
- Why is one chemical substance a poison when another, whose molecules have just the same atoms but assembled into a mirror-image pattern, is completely harmless?
- Why does the amount of manganese in the diet affect the maternal instinct of an animal?
- What is the description of aging in biochemical terms?
- What meaning is to be assigned to the question: Is a virus a living organism?
- What is a gene, and how does the original genetic constitution of a living organism express itself in the developed characteristics of the adult?
- Do complex protein molecules "know how" to reduplicate their pattern, and is this an essential clue to the problem of reproduction of living creatures?

Although these problems are complex, they are not problems "to which statistical methods hold the key" but they are "problems which involve dealing simultaneously with a sizable number of factors which are interrelated into an organic whole". All these are not problems of disorganized complexity but, "in the language here proposed, problems of organized complexity."[4] Weaver specified some more of these questions:

- On what does the prize of wheat depend?
- How can currency be wisely and effectively stabilized?
- To what extend is it safe to depend on the free interplay of such economic forces as supply and demand?
- To what extend must systems of economic control be employed to prevent the wide swings from prosperity to depression?
- How can one explain the behavior of pattern of a group of persons such as a labor union, or a group of manufacturers, or a racial minority?
- With a given total of national resources that can be brought to bear, what tactics and strategy will most promptly win a war, or better: what sacrifices of present selfish interest will most effectively contribute to a stable, decent, and peaceful world?

With regard to these problems Weaver stressed that the involved variables are "all interrelated in a complicated, but nevertheless not in helter-skelter, fashion" that these complex systems have "parts in close interrelations", and that "something more is needed than the mathematics of averages."[4]

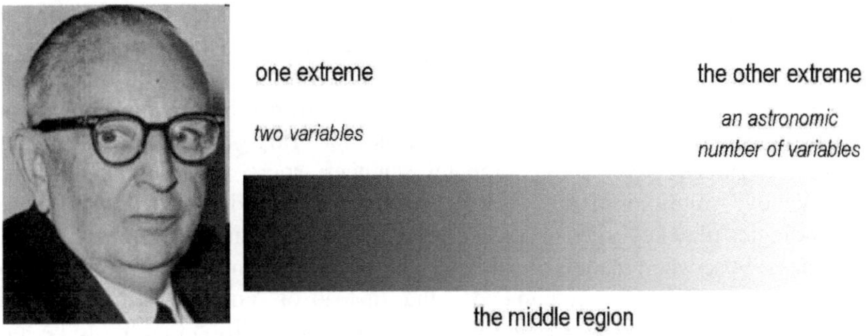

Fig. 2. Left: Warren Weaver; right: Weaver's trichotomy of scientific problems

"These problems—and a wide range of similar problems in the biological, medical, psychological, economic, and political sciences—are just too complicated to yield to the old nineteenth-century techniques …" and "these new problems, moreover, cannot be handled with the statistical techniques so effective in describing average behaviour in problems of disorganized complexity." "These new problems – and the future of the world depends of many of them, requires science to make a third great advance, an advantage that must be even greater than the nineteenth-century conquest of problems of simplicity or the twentieth-century victory over problems of disorganized complexity. Science must, over the next 50 years, learn to deal with these problems of organized complexity."4

In my judgment science performed this task in fact with some new concepts and theories, which have – of course – their roots in earlier decades or centuries, but have got developed in the second half of the 20th century, e.g. self-organization, synergetic, chaos theory, fractals, and the technologies of SC with the central theory of fuzzy sets and systems!

6 Outlook

As we have seen in this paper, the methodology of fuzzy sets and systems have been used to solving problems of humanistic problems, societal, political, economic, and other types of problems. Many of these problems are problems of "organized complexity" in the sense of Weaver's classification.

From his experience in the World War II, Weaver found among the "wartime development of new types of electronic computers" a second wartime advance, the "mixed-team" approach of operational analysis: "Although mathematicians, physicists, and engineers were essential, the best of the groups also contained physiologists, biochemists, psychologists, and a variety of representatives of other fields of the biochemical and social sciences. Among the outstanding members of English mixed teams, for example, were an endocrinologist and an X-ray crystallographer. Under the pressure of war, these mixed teams pooled their resources and focused all their different insights on the common problems. It as found, in spite of the modern tendencies toward intense scientific specialization, that members of such diverse groups could work together and could form a unit which was much greater than the mere sum of its parts. It was shown that these groups could tackle certain problems of organized complexity, and get useful answers." [4] Not only in wartimes but also in times of peace Weaver considered possible that mixed teams that bridge the gaps between natural sciences, engineering sciences, computer sciences, social sciences and humanities could achieve solutions of the world's problems. Continuing this thinking Zadeh's methodologies play key roles in science and technology of the 21st century.

In the 1990s, when Zadeh pleaded for the establishment of the research field of Soft Computing (SC), he recommended that instead of "an element of competition" between the complementary methodologies of SC "the coalition that has to be formed has to be much wider: it has to bridge the gap between the different communities in various fields of science and technology and it has to bridge the gap between science and humanities and social sciences! SC is a suitable candidate to meet these demands because it opens the fields to the humanities.

Acknowledgments. Work leading to this paper was partially supported by the *Foundation for the Advancement of Soft Computing*, Mieres (Asturias) Spain.

References

1. Weaver, W.: The Mathematics of Communication. Scientific American 181, 11–15 (1948)
2. Shannon, C.E.: A Mathematical Theory of Communication. The Bell System Technical Journal 27, 379–423, 623–656 (1948) (Also in Ref. 3)

3. Weaver, W.: Translation. In: Locke, W.N., Booth, A.D. (eds.) Machine translation of Languages: fourteen essays, pp. 15–23. Technology Press of the MIT, John Wiley & Sons, Inc., Cambridge, New York (1955)
4. Weaver, W.: Science and Complexity. American Scientist 36, 536–544 (1948)
5. Shannon, C.E., Weaver, W.: The Mathematical Theory of Communication. Univ. of Illinois Press, Urbana (1949)
6. Seising, R.: The Fuzzification of Systems. The Genesis of Fuzzy Set Theory and Its Initial Applications. Developments up to the 1970s. Studies in Fuzziness and Soft Computing, vol. 216, Springer (1970)
7. Zadeh, L.A.: From Circuit Theory to System Theory. In: Proc. of the IRE, vol. 50, pp. 856–865 (1962)
8. Zadeh, L.A.: Fuzzy Sets and Systems. In: Fox, J. (ed.) System Theory. Microwave Res. Inst. Symp. Ser. XV, pp. 29–37. Polytech. Pr., Brooklyn (1965)
9. Berkeley, E.C.: Giant Brains or Machines that Think. John Wiley & Sons, Chapman & Hall, New York, London (1949)
10. Turing, A.M.: Computing machinery and intelligence, Mind LIX (236), pp. 433–460 (October 1950)
11. Zadeh, L.A.: Thinking machines – a new field in electrical engineering. Columbia Engineering Quarterly, 12–13, 30-31 (January 1950)
12. Zadeh, L.A.: Making Computers Think like People. IEEE Spectrum 8, 26–32 (1984)
13. Zadeh, L.A.: The Birth and Evolution of Fuzzy Logic – A Personal Perspective. Journal of Japan Society for Fuzzy Theory and Systems 11(6), 891–905 (1999)
14. Zadeh, L.A.: The Concept of a Linguistic Variable and its Application to Approximate Reasoning–I. Information Science 8, 199–249 (1975)
15. Zadeh, L.A.: Outline of a new approach to the analysis of complex systems and decision processes. IEEE Trans. SMC SMC-3(1), 28–44 (1973)
16. Weaver, W.: Recent Contributions to the Mathematical Theory of Communication, in Ref. 3
17. Weaver, W.: The Scientists Speak. Boni & Gaer Inc. (1947)
18. Zadeh, L.A.: Precisiated Natural Language (PNL). AI Magazine 25(3), 74–92 (2004)
19. E-mail: L. A. Zadeh to R. Seising, May 23 (2009)

Educational Software of Fuzzy Logic and Control

José Galindo and Enrique León-González

E.U. Politécnica, Universidad de Málaga
jgg@lcc.uma.es, c47761g@copitima.com

Abstract. The so-called Educational Software for the Introduction to Fuzzy Control (SDICD in Spanish acronym) has been developed in the University of Málaga, Spain, with the purpose of extending fuzzy logic and fuzzy control to any interested person. This software is more than an electronic book. It includes a good set of chapters, animated graphics, some interactive examples, and auto-evaluation tests, among other capabilities. In the interactive examples, the user may change the parameters and evaluate the different results.

Keywords: Educational Software, Fuzzy Logic Book, Fuzzy Control Book, Fuzzy Controlled Greenhouse, Interactive Fuzzy Control Example, Fuzzy Control Simulation.

1 Introduction

SDICD (Spanish acronym of *Software Didáctico para la Introducción al Control Difuso*) is an Educational Software for the Introduction to Fuzzy Control, and consequently, to Fuzzy Logic [6][7]. Basic concepts of these items are explained in an educational form, easy to understand for persons with a knowledge level of a first year university student. This software is complementary to SCD (*Software de Control Difuso*) [3][4], a software for simulating systems with fuzzy control, highly customizable with regards to fuzzy rules, fuzzy t-norms and other fuzzy control options. SCD is a practical application and, then, the user requires basic and theoretical knowledge. This knowledge is included in the educational software, which is presented here.

The application is designed like an electronic book with the purpose of facilitating users or students to learn the principles of fuzzy logic in general [6] and of fuzzy control in particular [7], in such a way that he/she can learn these concepts in a progressive way, and trying to simplify the mathematical concepts, definitions, and principles, using a graphical and interactive system, in order to achieve an easy understanding of the key concepts of fuzzy logic and, so, to introduce us in a fuzzy controller.

Fuzzy logic and fuzzy control are subjects included in the curriculum of some qualifications, university degrees and PhD, such as computer science and industrial engineering. As we show below, the list of topics of our approach is comprehensive enough to cover most of university syllabus in this field, and in other case, it can be useful, at least, as a good introduction.

The mobility between the different sections in this our electronic book is very intuitive, using indexes, links and the different buttons of this application. At all times, user may look up the references and the meanings of the important terms, which are

E. Hüllermeier, R. Kruse, and F. Hoffmann (Eds.): IPMU 2010, Part II, CCIS 81, pp. 366–375, 2010.

found while the user is studying. User can use an auto-evaluation tool in order to measure the acquired knowledge in an objective and fair way.

One of the pillars is the help of two graphical examples, developed to provide a practical tool to see, try and understand the theoretical concepts. These examples are interactive, i.e. user may modify the inputs and the characteristics of both fuzzy controllers, and then user may check how each modification affects the system. Both interactive examples allow to understand the development and inner operation of two fuzzy controllers, and even to simulate fuzzy logic based systems.

We want to emphasize two complementary tools, very useful if they are used with SDICD: Xfuzzy software [9], which allows to design and to verify complex fuzzy systems (http://www2.imse-cnm.csic.es/Xfuzzy), and also the electronic book FLEB [1], which utilizes Xfuzzy.

Section 2 explains the programming environment of this application. Section 3 summarizes briefly the main characteristics of SDICD. Finally, some conclusions and future lines are shown.

2 Brief Programming Information

The application has been developed using Microsoft Visual Basic 6.0® [2], because we wanted to get a Windows application with a familiar interface to most users.

With regards to the functional viewpoint, the application has been designed to allow the user to navigate in a dynamic, intuitive and structured way. In this sense, there are indexes and links systems, which allow the user to move easily from one part to other one, using sections and subsections.

The software uses a wide file system, which are loaded when each one is needed. In this file system we can find all texts, graphics and formulae, scientific references, the glossary, the evaluation system (test), and the user control system, which are registered in SDICD. In this software we can find an application so-called Tools Editor, which allows an administrator user (without limited access) to modify the glossary, tests and references.

3 Features

We summarize in eight subsections the main characteristics of the SDICD software.

3.1 List of Chapters

SDICD is an application like an electronic book. We can find 151 pages or screen slides. The book includes seven chapters, following some previous works like [7]:

- **Chapter 1: Fuzzy Logic.** This section introduces the user to the *fuzzy logic* basic concepts and advantages, with regards to the *classic logic,* and without any formulae (highlighting how the imprecise data are an important part of our usual life). It includes a basic historic evolution, clarifies how and when to use fuzzy logic, and some of the main applications are mentioned.

- **Chapter 2: Fuzzy Sets Theory.** It explains basic and mathematical definitions, such as *fuzzy set*, *membership functions*, different types of these functions, and different methods to obtain these functions (their mathematical definition).
- **Chapter 3: Operations and Concepts with Fuzzy Sets.** This chapter is divided in two sections. First of all, it defines the main concepts to characterize fuzzy sets (support, kernel, height, etc.). After, it develops several operations for fuzzy sets, like *unary operations* (normalization, etc.), *union*, *intersection* and complement or *negation* operations, including the main *t-norms* and *s-norms* (we can see a good compendium in [6]). Finally, this chapter explains some of the most interesting operations to *compare* fuzzy sets, including equality indexes, distance measurements and, of course, the possibility and necessity measures.
- **Chapter 4: Fuzzy Relations and Fuzzy Numbers.** With regards to *fuzzy relations*, the explanations are concentrated on the operations of extension and cylindrical projection. About *fuzzy numbers*, the chapter introduces the user to the theoretical definitions. After, the *extension principle* is widely explained because it is a fundamental concept in the fuzzy logic.
- **Chapter 5: Basic Concepts about Control.** It gives the user fundamental notions about the different techniques and theories about *control*. This chapter emphasizes the importance of the human experience and how fuzzy control tries to simulate it.
- **Chapter 6: Fuzzy Controllers.** This is the most important and longest chapter. Here, the different parts of a fuzzy controller are broken down. Special attention is paid to the *fuzzyfication* module, the fuzzy knowledge base, the inference engine and the final and optional *defuzzyfication* module.
- **Chapter 7: Tuning Methods and Types of Fuzzy Controllers.** In this last chapter, the fuzzy controllers are classified with regards different features, such as the tuning methods and the operation mode.

3.2 Control of Users

There are two user types, the *administrator* and the *student* or standard user. The administrator user will be able to execute some application tools limited for the other users, such as, the query of the marking in the test of all registered users, and the Tools Editor. Of course, this user is exempt from doing the tests.

The control of users is the module to register or delete users, to allow an user to continue his/her study, and to query the marks of each test/user.

3.3 Menus

SDICD has four main menus: File, Tools, Examples and Help. The File menu has options to save and open user sessions, print, Acrobat Reader path (configuration) and Exit.

The Tools menu is useful to see the references and the glossary (ordered definitions of key terms), or to do the test related to each chapter.

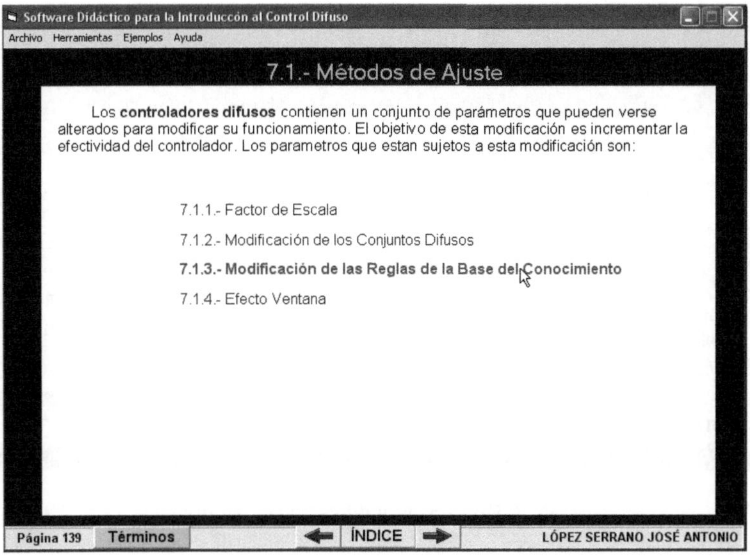

Fig. 1. Mobility in a normal page of SDICD

The Examples menu includes the two interesting examples, useful for understanding fuzzy control. These examples are explained in subsection 3.8.

The Help menu includes the typical About option, the User Handbook and the PDF Glossary.

3.4 Mobility

Running this software, we can see some fixed buttons which are useful to move the attention to different pages of the book. Figure 1 shows, in the bottom part, the button for the Table of Contents (*Índice* in Spanish), and on both sides two arrows useful for turning page backward and forward.

On the other hand, using the keyboard we can always navigate in the application. Besides, all pages include links, useful to access to interesting and related pages.

3.5 Glossary and References

In each page, it is easy to see the definitions of the key terms and the related references. We can see the key terms in blue and the references in orange. The most simple method to do that is using the buttons *Terms* and *References* (Figures 1 and 2), located in the state bar, just next to the page number and the left arrow button (to go to the previous page). These buttons only appear when there is any term and/or reference in the current page.

When the user clicks one of these buttons, a list appears with all the terms or references in the current page (Figure 2). Then, the user can choose one key term or reference in order to see a windows with the information related to the selected term or reference. This window will disappear pressing the Ok button.

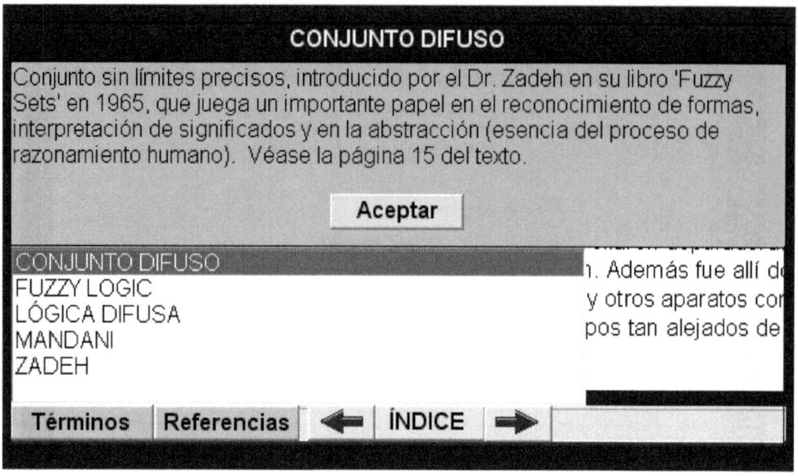

Fig. 2. Key Terms and References Buttons

3.6 Test

SDICD has now more than 140 questions distributed among the seven chapters. The user can execute this evaluation tool using the Tools menu, option Test.

Besides, when one student reaches the last page of each chapter, this student finds the option of doing the test of that chapter. Each test includes ten random questions and order of questions and possible answers are also random.

When any user answers all questions, the system shows the mark. Then the user could revise the test and see the chosen answers and the correct and incorrect ones.

When the user does all the test of all the chapters, SDICD gives the user the possibility of doing a general test with twenty questions of all the chapters.

3.7 Figures and Dynamic Representation

Throughout the pages of SDICD we can find hundreds of graphics, figures and formulas. SDICD offers to the students the possibility of some clear and concise explanations about the figures. You need to site the mouse on the figure for this. Then, a yellowish rectangle appears with the explanations about that figure, such as we can see on Figure 3.

Some figures and formulas have a *Representation* button just next to them (Figure 4). This button is useful to help understand this issue. In this case, SDICD shows the final image. Thus, when we press the *Representation* button, we see an animation that clarifies the process to reach the final image. Figure 4 shows an image representing on the left the compatibility measure between fuzzy sets B and A, on the right. The computation of this compatibility measure is not intuitive, and any explanation could be useful. Then, the *Representation* button shows, step by step, how to obtain the compatibility measure between B and A.

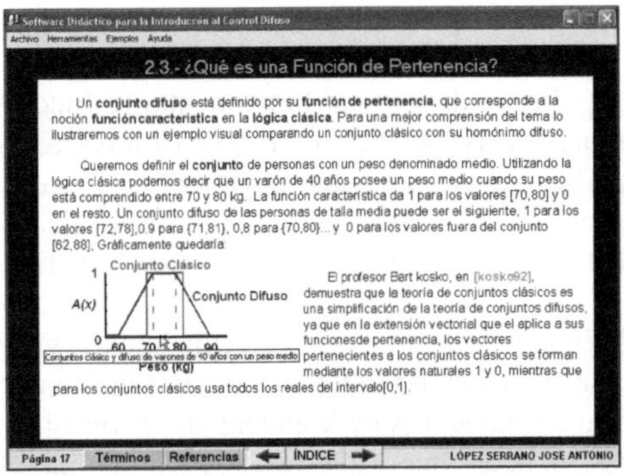

Fig. 3. One page of SDICD with an image

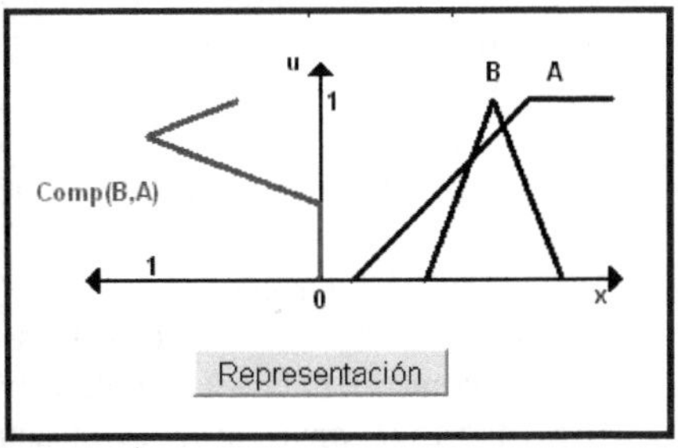

Fig. 4. Image with the Representation button

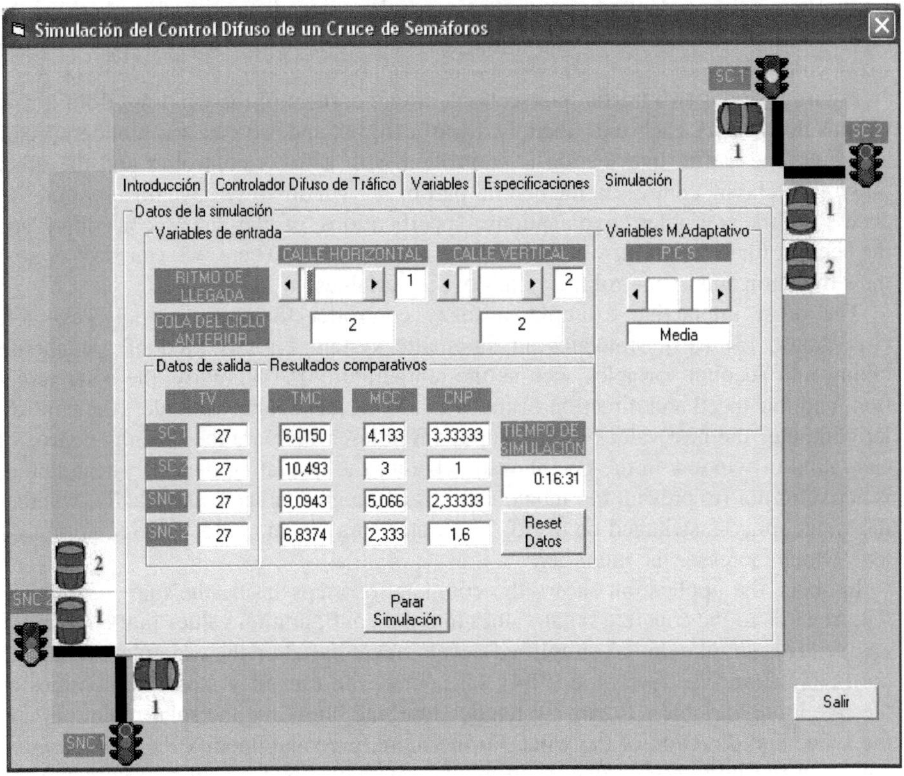

Fig. 5. Fuzzy and Non-Fuzzy Control Simulations in a Traffic Intersection

3.8 Interactive Examples

SDICD includes two interactive examples. We think they are fundamental for understanding fuzzy control. These examples are the Simulation of Fuzzy Control for a Traffic Intersection (crossroads with traffic lights), and the Simulation of Fuzzy Control for an Industrial Vegetable Greenhouse [4][5]. The student is allowed to modify the input values in the input variables and the controller configuration, for both examples.

In the Traffic Intersection example (Figure 5) we can compare using statistics the performance of the traffic intersection controller using both a fuzzy controller and a fixed time controller. This example is based on the works by Mandani and Pappis [8].

Here, the four input variables measures, in each one of the two streets, the rate of vehicle arrivals, and the number of vehicles waiting in the tail. The output value is the time of each state for all traffic lights in the intersection. This example allows to understand how to design a fuzzy controller starting from some previous requirements. Furthermore, the student can compare the results using different values both in the input variables and in the configuration parameters in the adaptive module of this fuzzy controller.

Figure 5 shows two traffic intersections in two corners of the window. Both traffic intersections have, each one, their two traffic lights and we can see numbered cars arriving. One traffic intersection is controlled with a fuzzy controller and the other one uses a classic control. In the central part of the window, we have different tabs to access to the variables, explanations, specifications of the fuzzy controller, and the section for simulation, which is shown in that figure. There, we can start or stop the simulation and to control the values of the variables.

The other interactive example, a fuzzy controller for an Industrial Vegetable Greenhouse [3][4][5], simulates an automatic system for this kind of greenhouse taking into account variables such as the temperature, the humidity, the solar radiation, and the speed and direction of the wind. With these input variables, the controller computes the best values in the output variables, which are the opening degree of each kind of window in the greenhouse, whether the insulating material (heat shield) is active or not (to prevent too much heat inside the greenhouse), and finally, whether the water spray is switched on or off. The water spray is an indirect refrigeration system, which increases the humidity.

In short, the application shows the computation steps inside the fuzzy controller, with regards to the concrete input values and the configuration values in the inference engine. This simulation is a simplified development based on the example included in the Fuzzy Controller Software [3][4]. Of course, the user may modify the values of the five input variables (Figure 6): temperature, the humidity, the solar radiation, and the speed and direction of the wind. Furthermore, user may modify the configuration values in the inference engine, such as, the operator AND function to compute the activation degree of each rule, the implication function, or the defuzzyfication function. This software allows the user to see the computed values in each step.

In the inputs tab (Figure 7) the fuzzyfication module in a fuzzy controller is simulated. For each input variable, we get the fuzzy value according to the precision assigned to the corresponding sensor. Graphic representation of each value is very useful to follow the fuzzy controller process.

In the activation degree tab, we can see, for each rule, the antecedents, consequents, and the activation degree according to the input values and the operator AND function chosen by the user.

In the fuzzy implication tab (Figure 8), the user can see the fuzzy sets generated in each rule. In this step, the system takes into account the activated consequents in each rule, the activation degree in each rule, and the implication function chosen by the user.

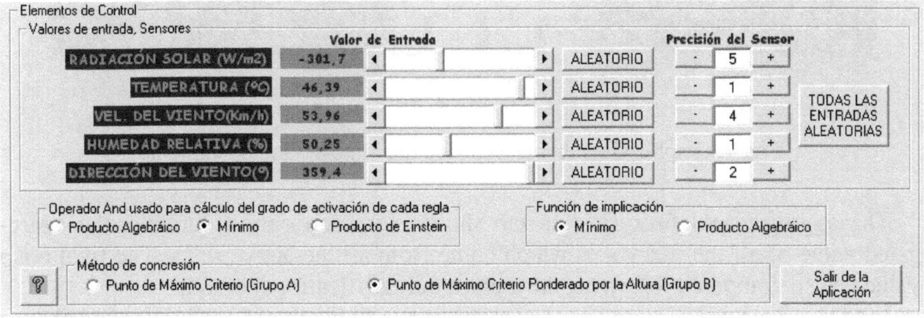

Fig. 6. Input Variables in the Greenhouse Example

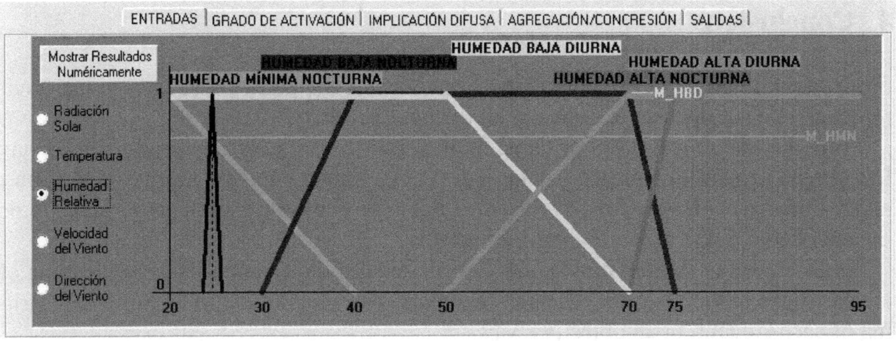

Fig. 7. Fuzzy Sets of the Fuzzyficated Inputs

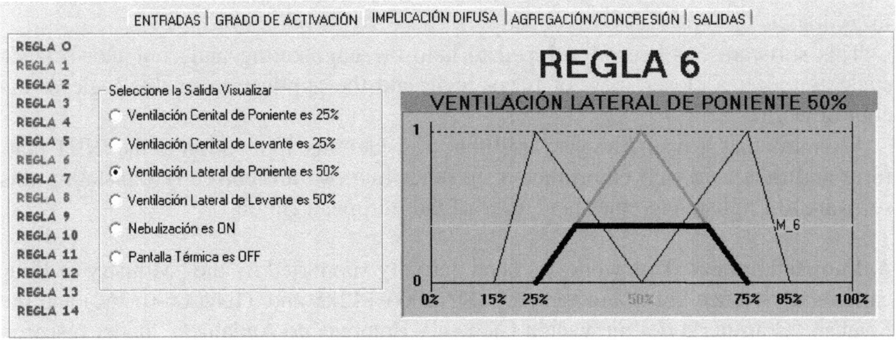

Fig. 8. Fuzzy Implication

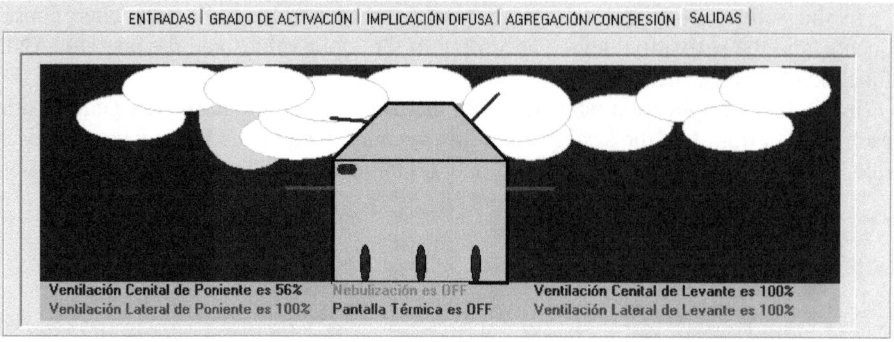

Fig. 9. Graphic Greenhouse Showing the State of Output Variables

The aggregation/defuzzyfication tab shows, for each output variable, the aggregated value of all the fuzzy sets which come from all the active rules. The final crisp value is also computed following the chosen defuzzyfication method.

In the outputs tab (Figure 9) we can observe, graphical and easily, the greenhouse estate in each moment.

4 Conclusions

From a functional point of view, SDICD has been developed with an easy to use environment, and so, the user accesses to all the options easily.

From the theoretical point of view, SDICD tries to use easy explanations for many fuzzy logic principles and fuzzy control concepts. Mathematical concepts, definitions, applications and many ideas have been explained to introduce students to fuzzy logic in general and to the inner of a fuzzy controller.

At all times the student may quickly query scientific references and the meaning of any key term in the glossary. The auto-evaluation system is used to measure the learning of each student, in an objective way.

An strong and very useful point is the inclusion of two interactive examples, explained above, which allow the user to see and understand the development of two easy fuzzy controllers, and then to simulate these Rule Based Systems (RBS) with fuzzy logic.

This software has been developed to help the engineering and computer science students to get a closer view of fuzzy logic and the application of this logic in the control world.

We think that it would be very useful to translate this application to the .NET platform and then to a web environment. In other line, we are currently translating this software to English and other languages of the European Union.

Acknowledgments. This work has been partially supported by the "Ministry of Education and Science" of Spain (projects TIN2006-14285 and TIN2006-07262) and the Spanish "Consejería de Innovación Ciencia y Empresa de Andalucía" under research project TIC-1570.

References

1. Bermúdez, A., Barriga, A., Baturone, I., Sánchez-Solano, S.: FLEB: A Fuzzy Logic E-Book. In: Proc. European Symposium on Intelligent Technologies, Hybrid Systems and their implementation on Smart Adaptive Systems, Tenerife, pp. 549–554 (2001)
2. Sierra, C., Francisco, J.: Enciclopedia de Microsoft Visual 6. Editorial Ra-Ma (1999)
3. Rodríguez, E., Calixto: Software para Control Difuso de todo tipo de Sistemas (SCD): Aplicación al Control de Invernaderos Industriales. Proyecto Fin de Carrera, Ingeniería Técnica Industrial, Universidad de Málaga (2003)
4. Escobar, C., Galindo, J.: Software Genérico de Control Difuso: Aplicación en Agricultura Industrial. In: XII Congreso Español sobre Tecnologías y Lógica Fuzzy (ESTYLF 2004), Jaén, Spain, pp. 551–556 (2004)
5. Escobar, C., Galindo, J.: Fuzzy Control in Agriculture: Simulation Software. In: Marín, J., Koncar, V. (eds.) Industrial Simulation Conference 2004 (ISC 2004), Málaga, Spain, June 2004, pp. 45–49 (2004), `http://www.lcc.uma.es`
6. Galindo, J.: Introduction and Trends to Fuzzy Logic and Fuzzy Databases. In: Handbook of Research on Fuzzy Information Processing in Databases, vol. I, pp. 1–33. Information Science Reference, Hershey (2008)
7. Galindo, J.: Curso Introductorio de Conjuntos y Sistemas Difusos (Lógica Difusa y Aplicaciones), `http://www.lcc.uma.es/~ppgg/FSS` (accedido 2010)
8. Mandani, E., Pappis, C.: A Fuzzy Logic Controller for a Traffic Intersection. IEEE Trans. on Systems, Man and Cybernetics SMC-7, 707–717 (1977)
9. Moreno-Velo, F.J., Baturone, I., Sánchez-Solano, S., Barriga, A.: Rapid Design of Fuzzy Systems with XFUZZY. In: Proc. IEEE Int. Conference on Fuzzy Systems, St. Louis, pp. 342–347 (2003)

A Fuzzy Distance between Two Fuzzy Numbers

Saeid Abbasbandy* and Saeide Hajighasemi

Department of Mathematics, Science and Research Branch
Islamic Azad University, Tehran, 14515/775, Iran
Tel.: +98(912)1305326 (S. Abbasbandy)
abbasbandy@yahoo.com

Abstract. In this paper by using Hausdorff distance as a maximum distance between two fuzzy numbers, a new fuzzy distance is introduced between two fuzzy numbers. Several examples are used to show preference of the proposed fuzzy distance to others.

1 Introduction

The methods of measuring of distance between fuzzy numbers have became important due to the significant applications in diverse fields like remote sensing, data mining, pattern recognition and multivariate data analysis and so on. Several distance measures for precise numbers are well established in the literature.

Several researchers focused on computing the distance between fuzzy numbers [1,2,3,6,8,9]. Usually the distance methods basically compute crisp distance values for fuzzy numbers. Naturally a logical question occurs to us: if the numbers themselves are not known exactly, how can the distance between them be an exact value? In view of this, Voxman [9] first introduced a fuzzy distance for fuzzy numbers. Therefore a distance measure for fuzzy numbers is that the distance between two uncertain numbers should also be an uncertain number, logically.

Section 2 describes the basic notation and definitions of fuzzy numbers, support and α-cut of fuzzy numbers. Also the fuzzy distance of Voxman is described in Section 2.1. A new distance measure between fuzzy numbers is defined in Section 3 and a fuzzy distance measure in Section 4. Ambiguity and fuzziness of fuzzy distance measure are investigated in Section 4.1. Finally, conclusions are drawn in Section 5.

2 Preliminaries

A fuzzy set on a set X is a function $\mu : X \to [0,1]$. The support of μ, supp μ is the closure of the set $\{x \in X \mid \mu(x) > 0\}$.

Definition 1. *[9] A fuzzy number is a fuzzy set* $\mu : \mathbb{R} \to [0,1]$ *on* $\mathbb{R}$ *satisfying*

(i) μ *is upper semi-continuous;*

* Corresponding author.

E. Hüllermeier, R. Kruse, and F. Hoffmann (Eds.): IPMU 2010, Part II, CCIS 81, pp. 376–382, 2010.

(ii) supp μ *is a closed and bounded interval;*
(iii) *if* supp $\mu = [a, b]$, *then there exist* $c, d,\ a \le c \le d \le b$, *such that* μ *is increasing on the interval* $[a, c]$, *equal to 1 on the interval* $[c, d]$ *and decreasing on the interval* $[d, b]$.

We let F denote the family of all fuzzy numbers. If $\mu \in F$, then for each α, $0 < \alpha \le 1$, the α-cut of μ, is defined by

$$\mu_\alpha = \{x \in X \mid \mu(x) \ge \alpha\}.$$

The α-cut representation of μ is the pair of functions, $(L(\alpha), R(\alpha))$, defined by

$$L(\alpha) = \begin{cases} \inf\{x \mid x \in \mu_\alpha\} & \text{if } \alpha > 0, \\ \inf\{x \mid x \in \text{supp } \mu\} & \text{if } \alpha = 0, \end{cases}$$

and

$$R(\alpha) = \begin{cases} \sup\{x \mid x \in \mu_\alpha\} & \text{if } \alpha > 0, \\ \sup\{x \mid x \in \text{supp } \mu\} & \text{if } \alpha = 0. \end{cases}$$

If μ is a fuzzy number then the compliment of μ, μ_c, is the fuzzy set defined by $\mu_c(x) = 1 - \mu(x)$.

If K is the set of compact subsets of $\mathbb{R}^2$, and A and B are two subsets of $\mathbb{R}^2$ then the Hausdorff metric $H : K \times K \to [0, \infty)$ is defined by [9]

$$H(A, B) = \max\{\sup_{b \in B} d_E(b, A), \sup_{a \in A} d_E(a, B)\},$$

where d_E is the usual Euclidean metric for $\mathbb{R}^2$.

Definition 2. *The metric* d_∞ *on* $F \times F$ *is defined by*

$$d_\infty(\mu, \nu) = \sup_{\alpha \in [0,1]} \{H(\mu_\alpha, \nu_\alpha)\}.$$

Definition 3. *The* μ *is a triangular fuzzy number, and We write* $\mu = (x_0, \sigma, \beta)$, *with defuzzifier* x_0, *and left fuzziness* $\sigma > 0$ *and right fuzziness* $\beta > 0$ *is a fuzzy set where the membership function is as*

$$\mu(x) = \begin{cases} \frac{1}{\sigma}(x - x_0 + \sigma), & x_0 - \sigma < x \le x_0, \\ \frac{1}{\beta}(x_0 - x + \beta), & x_0 \le x < x_0 + \beta, \\ 0, & \text{otherwise}. \end{cases}$$

2.1 Fuzzy Distance Given by Voxman

Here, we briefly describe the fuzzy distance measure by Voxman [9]. The fuzzy distance function on F, $\Delta : F \times F \rightarrow F$, is define by

$$\Delta(\mu,\nu)(z) = \sup_{|x-y|=z} \min\{\ \mu(x), \nu(y)\}.$$

For each pair of fuzzy numbers μ and ν, let $\Delta_{\mu\nu}$ denote the fuzzy number $\Delta(\mu,\nu)$. If the α-cut representations of μ and ν are $(A_1^L(\alpha), A_1^R(\alpha))$ and $(A_2^L(\alpha), A_2^R(\alpha))$, respectively, then the α-cut representation of $\Delta_{\mu\nu}$, $(L(\alpha), R(\alpha))$, is given by

$$L(\alpha)=\begin{cases} \max\ \{A_2^L(\alpha)-A_1^R(\alpha), 0\} & \text{if } \frac{1}{2}(A_1^L(1)+A_1^R(1)) \leq \frac{1}{2}(A_2^L(1)+A_2^R(1)), \\ \max\ \{A_1^L(\alpha)-A_2^R(\alpha), 0\} & \text{if } \frac{1}{2}(A_2^L(1)+A_2^R(1)) \leq \frac{1}{2}(A_1^L(1)+A_1^R(1)), \end{cases}$$

and

$$R(\alpha) = \max\ \{A_1^R(\alpha) - A_2^L(\alpha), A_2^R(\alpha) - A_1^L(\alpha)\}.$$

3 A New Distance between Two Fuzzy Numbers

Let μ and ν be two arbitrary fuzzy numbers with α-cut representations $(A_1^L(\alpha), A_1^R(\alpha))$ and $(A_2^L(\alpha), A_2^R(\alpha))$, respectively. The distance between μ and ν is defined as

$$d(\mu,\nu) = \Bigg| \int_0^{\frac{1}{2}} [(1-\alpha)(A_1^R(\alpha) - A_2^R(\alpha)) + \alpha(A_1^L(\alpha) - A_2^L(\alpha))]d\alpha \qquad (1)$$

$$+ \int_{\frac{1}{2}}^1 [\alpha(A_1^R(\alpha) - A_2^R(\alpha)) + (1-\alpha)(A_1^L(\alpha) - A_2^L(\alpha))]d\alpha \Bigg| .$$

In other words, right dominance has preference to the left dominance.

Theorem 1. *For fuzzy numbers μ, ν and ω, we have*

(i) $d(\mu,\nu) \geq 0$ *and* $d(\mu,\mu) = 0$*;*
(ii) $d(\mu,\nu) = d(\nu,\mu)$*;*
(iii) $d(\mu,\nu) \leq d(\mu,\omega) + d(\omega,\nu)$.

Proof. We consider only *(iii)*. Suppose μ and ν have α-cut representations as before, and ω has α-cut representation $(A_3^L(\alpha), A_3^R(\alpha))$. By (1), we have

$$d(\mu,\nu) = \Bigg| \int_0^{\frac{1}{2}} [(1-\alpha)\big(A_1^R(\alpha) - A_3^R(\alpha) + A_3^R(\alpha) - A_2^R(\alpha)\big)$$

$$+\alpha\big(A_1^L(\alpha) - A_3^L(\alpha) + A_3^L(\alpha) - A_2^L(\alpha)\big)]d\alpha$$

$$+\int_{\frac{1}{2}}^{1}[\alpha\big(A_1^R(\alpha)-A_3^R(\alpha)+A_3^R(\alpha)-A_2^R(\alpha)\big)$$
$$+(1-\alpha)\big(A_1^L(\alpha)-A_3^L(\alpha)+A_3^L(\alpha)A_2^L(\alpha)\big)]d\alpha\Big|$$
$$\leq\left|\int_0^{\frac{1}{2}}[(1-\alpha)(A_1^R(\alpha)-A_3^R(\alpha))+\alpha(A_1^L(\alpha)-A_3^L(\alpha))]d\alpha\right.$$
$$\left.+\int_{\frac{1}{2}}^{1}[\alpha(A_1^R(\alpha)-A_3^R(\alpha)+(1-\alpha)(A_1^L(\alpha)-A_3^L(\alpha))]d\alpha\right|$$
$$+\left|\int_0^{\frac{1}{2}}[(1-\alpha)(A_3^R(\alpha)-A_2^R(\alpha))+\alpha(A_3^L(\alpha)-A_2^L(\alpha))]d\alpha\right.$$
$$\left.+\int_{\frac{1}{2}}^{1}[\alpha(A_3^R(\alpha)-A_2^R(\alpha)+(1-\alpha)(A_3^L(\alpha)-A_2^L(\alpha))]d\alpha\right|=d(\mu,\omega)+d(\omega,\nu).$$

□

Since we introduce this distance by dominance, similarity Hausdorff distance we can be proved these properties

(i) $d(u+w,v+w)=d(u,v)$ for every $u,v,w\in F$,
(ii) $d(u+v,\tilde{0})\leq d(u,\tilde{0})+d(v,\tilde{0})$ for every $u,v\in F$,
(iii) $d(\lambda u,\lambda v)=|\lambda|d(u,v)$ for every $u,v\in F$ and $\lambda\in\mathbb{R}$,
(iv) $d(u+v,w+z)\leq d(u,w)+d(v,z)$ for u,v,w, and $z\in F$.

Theorem 2. *For two fuzzy numbers μ and ν, We have $d(\mu,\nu)\leq\ d_\infty(\mu,\nu)$.*

Proof. By definition $d(.,.)$ we have,

$$d(\mu,\nu)=\left|\int_0^{\frac{1}{2}}(1-\alpha)(A_1^R(\alpha)-A_2^R(\alpha))d\alpha+\int_0^{\frac{1}{2}}\alpha(A_1^L(\alpha)-A_2^L(\alpha))d\alpha\right.$$
$$\left.+\int_{\frac{1}{2}}^{1}\alpha(A_1^R(\alpha)-A_2^R(\alpha))d\alpha+\int_{\frac{1}{2}}^{1}(1-\alpha)(A_1^L(\alpha)-A_2^L(\alpha))]d\alpha\right|.$$

By assumption $d_\infty(\mu,\nu)=M$, we have $\left|A_1^R(\alpha)-A_2^R(\alpha)\right|\leq M$ and $\left|A_1^L(\alpha)-A_2^L(\alpha)\right|\leq M$ and mean value theorem for integrals, We obtain

$$d(\mu,\nu)\leq M\int_0^{\frac{1}{2}}(1-\alpha)d\alpha+M\int_0^{\frac{1}{2}}\alpha d\alpha$$
$$+M\int_{\frac{1}{2}}^{1}\alpha d\alpha+M\int_{\frac{1}{2}}^{1}(1-\alpha)d\alpha$$
$$=M\int_0^1(1-\alpha)d\alpha+M\int_0^1(\alpha)d\alpha=M$$

Therefore

$$d(\mu,\nu)\leq d_\infty(\mu,\nu).$$

□

Table 1. Comparison of d and d_∞

μ	ν	$d(\mu,\nu)$	$d_\infty(\mu,\nu)$
(4,3,1)	(0,1,2)	$\frac{27}{8}$	4
(3,2,2)	(4,3,1)	0.5	1
(2,1,1)	(4,1,1)	2	2
(4,1,1)	(6,2,2)	2.25	3
(2,1,4)	(3,2,2)	0.125	1
(2,1,1)	(6,1,1)	4	4
(3,2,2)	(3,1,1)	0.25	1

See Table 1 for comparison between Hausdorff distance and d distance for some triangular fuzzy numbers. We can see that $d(\mu,\nu) \le d_\infty(\mu,\nu)$ in all examples.

4 New Fuzzy Distance between Two Fuzzy Numbers

Let two fuzzy numbers μ and ν, with α-cut representation $(A_1^L(\alpha), A_1^R(\alpha))$ and $(A_2^L(\alpha), A_2^R(\alpha))$, respectively, are given. By $d(.,.)$ and $d_\infty(.,.)$, we can introduce the fuzzy distance by a symmetric triangular fuzzy number as follows:

$$\widetilde{d}(\mu,\nu) = \left(\frac{d(\mu,\nu)+d_\infty(\mu,\nu)}{2}, \frac{d_\infty(\mu,\nu)-d(\mu,\nu)}{2}, \frac{d_\infty(\mu,\nu)-d(\mu,\nu)}{2}\right), \quad (2)$$

with α-cut representation $(\lambda_\alpha(\mu,\nu), \rho_\alpha(\mu,\nu))$.

The proposed fuzzy distance (2) satisfies fuzzy distance properties followed in Kaleva and Seikkala [7].

Theorem 3. *For fuzzy numbers μ, ν and ω, we have*

(i) $\widetilde{d}(\mu,\nu) = \widetilde{0}$ *if only if* $\mu = \nu$;
(ii) $\widetilde{d}(\mu,\nu) = \widetilde{d}(\nu,\mu)$;
(iii) $\lambda_\alpha(\mu,\nu) \le \lambda_\alpha(\mu,\omega) + \lambda_\alpha(\omega,\nu)$ *and* $\rho_\alpha(\mu,\nu) \le \rho_\alpha(\mu,\omega) + \rho_\alpha(\omega,\nu)$.

Proof. (*i*) By definition of fuzzy zero, $\widetilde{0}(x) = \begin{cases} 1, & x = 0, \\ 0, & x \ne 0, \end{cases}$, from assumption $\widetilde{d}(\mu,\nu) = \widetilde{0}$, we obtain $d(\mu,\nu) + d_\infty(\mu,\nu) = 0$. Since $d(\mu,\nu)$ and $d_\infty(\mu,\nu)$ are positive numbers, we have $d(\mu,\nu) = d_\infty(\mu,\nu) = 0$ and hence $\mu = \nu$. Also, converse is obvious.

(*ii*) By properties of $d(.,.)$ and $d_\infty(.,.)$, it is obvious.

(*iii*) By definition of $\lambda_\alpha(\mu,\nu)$, we have

$$\lambda_\alpha(\mu,\nu) = d(\mu,\nu) + \alpha\Big(\frac{d_\infty(\mu,\nu)-d(\mu,\nu)}{2}\Big) = \Big(1-\frac{\alpha}{2}\Big)d(\mu,\nu) + \frac{\alpha}{2}d_\infty(\mu,\nu)$$

$$\leq \left(1-\frac{\alpha}{2}\right)\left(d(\mu,\omega)+d(\omega,\nu)\right)+\frac{\alpha}{2}\left(d_\infty(\mu,\omega)+d_\infty(\omega,\nu)\right) = \lambda_\alpha(\mu,\omega)+\lambda_\alpha(\omega,\nu),$$

because $(1-\frac{\alpha}{2}) > 0$. For $\rho(\mu,\nu)$, we have the similar proof. □

4.1 Ambiguity and Fuzziness of a Fuzzy Number

Delgado et al. [4,5] have extensively studied two attributes of fuzzy numbers, ambiguity and fuzziness. Ambiguity may be seen as a 'global spread' of the membership function, whereas the fuzziness involve a comparison between the fuzzy set and its complement. These concepts are defined as follow :

$$A(\mu) = \int_0^1 S(\alpha)[R(\alpha) - L(\alpha)]d\alpha,$$

$$F(\mu) = \int_0^1 S(\alpha)[q-p]d\alpha - \left\{ \int_{\frac{1}{2}}^1 S(\alpha)[L_c(\alpha) - p]d\alpha + \int_{\frac{1}{2}}^1 S(\alpha)[R(\alpha) - L(\alpha)]d\alpha \right.$$
$$+ \int_{\frac{1}{2}}^1 S(\alpha)[q - R_c(\alpha)]d\alpha + \int_0^{\frac{1}{2}} S(\alpha)[L(\alpha) - p]d\alpha + \int_0^{\frac{1}{2}} S(\alpha)[R_c(\alpha) - L_c(\alpha)]d\alpha$$
$$\left. + \int_0^{\frac{1}{2}} S(\alpha)[q - R(\alpha)]d\alpha \right\},$$

where supp $\mu = [p,q]$ and $(L(\alpha), R(\alpha))$ be the α-cut representations of μ. Also μ_c be the complement of μ with α-cut representations $(L_c(\alpha), R_c(\alpha))$. The function $S : [0,1] \to [0,1]$ is an increasing function and $S(0) = 0$ and $S(1) = 1$, [9]. We say that S is a regular reducing function if $\int_0^1 S(\alpha)d\alpha = \frac{1}{2}$. A routine calculation shows for $S(\alpha) = \alpha$, we have

$$F(\mu) = \int_0^{\frac{1}{2}} [R(\alpha) - L(\alpha)]d\alpha + \int_{\frac{1}{2}}^1 [L(\alpha) - R(\alpha)]d\alpha.$$

Table 2. Comparison of ambiguity and fuzziness

μ	ν	$A(\tilde{d}(\mu,\nu))$	$F(\tilde{d}(\mu,\nu))$	$A(\Delta(\mu,\nu))$	$F(\Delta(\mu,\nu))$
(3,2,2)	(4,3,1)	$\frac{1}{12}$	$\frac{1}{8}$	$\frac{68}{75}$	$\frac{17}{20}$
(2,1,1)	(4,1,1)	0	0	$\frac{2}{3}$	1
(4,1,1)	(6,2,2)	$\frac{1}{8}$	$\frac{3}{16}$	$\frac{53}{54}$	$\frac{4}{3}$
(2,1,4)	(3,2,2)	$\frac{7}{48}$	$\frac{7}{32}$	$\frac{203}{216}$	1
(2,1,1)	(6,1,1)	0	0	$\frac{2}{3}$	1
(3,2,2)	(3,1,1)	$\frac{1}{8}$	$\frac{3}{16}$	$\frac{1}{2}$	$\frac{3}{4}$

Table 2 shows that the ambiguity and the fuzziness of $\widetilde{d}(\mu, \nu)$ are less than of the ambiguity and fuzziness of $\Delta(\mu, \nu)$, which is defined by Voxman [9], for some examples. We can see that, when the support of μ and ν are disjoint, then $\widetilde{d}(\mu, \nu) = d_\infty(\mu, \nu)$ and in this case, $A(\widetilde{d}(\mu, \nu)) = F(\widetilde{d}(\mu, \nu)) = 0$.

5 Conclusions

Here, a new distance measure has been introduced for computing crisp distances for fuzzy numbers. It is reasonable, the distance between two uncertain numbers should also be an uncertain number. Voxman first introduced the concept of fuzzy distance for fuzzy numbers. In this paper, we introduce another fuzzy distance measure between two fuzzy numbers. However, the method proposed in this paper compute a fuzzy distance value with less ambiguity and fuzziness as compared to that of Voxman's method, which has been shown by some examples.

Acknowledgements

The authors would like to thank the anonymous referees for their constructive suggestions and comments.

References

1. Abbasbandy, S., Hajjari, T.: A new approach for ranking of trapezoidal fuzzy numbers. Comput. Math. Appl. 57, 413–419 (2009)
2. Chakraborty, C., Chakraborty, D.: A theoretical development on a fuzzy distance measure for fuzzy numbers. Math. Comput. Modeling 43, 254–261 (2006)
3. Cheng, C.H.: A new approach for ranking fuzzy numbers by distance method. Fuzzy Sets and Systems 95, 307–317 (1998)
4. Delgado, M., Vila, M.A., Voxman, W.: On a canonical representation of fuzzy numbers. Fuzzy Sets and Systems 93, 125–135 (1998)
5. Delgado, M., Vila, M.A., Voxman, W.: A fuzziness measure for fuzzy numbers: Applications. Fuzzy Sets and Systems 94, 205–216 (1998)
6. Grzegorzewski, P.: Distances between intuitionistic fuzzy sets and/or interval-valued fuzzy sets based on the Hausdorff metric. Fuzzy Sets and Systems 148, 319–328 (2004)
7. Kaleva, O., Seikkala, S.: On fuzzy metric spaces. Fuzzy Sets and Systems 12, 215–229 (1984)
8. Tran, L., Duckstein, L.: Comparison of fuzzy numbers using a fuzzy distance measure. Fuzzy Sets and Systems 130, 331–341 (2002)
9. Voxman, W.: Some remarks on distances between fuzzy numbers. Fuzzy Sets and Systems 100, 353–365 (1998)

On the Jaccard Index with Degree of Optimism in Ranking Fuzzy Numbers

Nazirah Ramli[1] and Daud Mohamad[2]

[1] Department of Mathematics and Statistics,
Faculty of Computer and Mathematical Sciences, Universiti Teknologi MARA
Pahang, 26400, Bandar Jengka, Pahang, Malaysia
nazirahr@pahang.uitm.edu.my
[2] Department of Mathematics, Faculty of Computer and Mathematical Sciences,
Universiti Teknologi MARA, 40450 Shah Alam Selangor, Malaysia
daud2010salam.uitm.edu.my

Abstract. Ranking of fuzzy numbers plays an important role in practical use and has become a prerequisite procedure for decision-making problems in fuzzy environment. Jaccard index similarity measure has been introduced in ranking the fuzzy numbers where fuzzy maximum, fuzzy minimum, fuzzy evidence and fuzzy total evidence are used in determining the ranking. However, the fuzzy total evidence is obtained by using the mean aggregation which can only represent the neutral decision maker's perspective. In this paper, the degree of optimism concept which represents all types of decision maker's perspectives is applied in calculating the fuzzy total evidence. Thus, the proposed method is capable to rank fuzzy numbers based on optimistic, pessimistic and neutral decision maker's perspective. Some properties which can simplify the ranking procedure are also presented.

Keywords: degree of optimism; fuzzy total evidence; Jaccard index; ranking fuzzy numbers.

1 Introduction

In fuzzy environment, the ranking of fuzzy numbers is an important procedure for decision-making and generally becomes one of the main issues in fuzzy theory. Various techniques of ranking fuzzy numbers have been developed such as distance index by Cheng [1], signed distance by Yao and Wu [2] and Abbasbandy and Asady [3], area index by Chu and Tsao [4], index based on standard deviation by Chen and Chen [5], score value by Chen and Chen [6], distance minimization by Asady and Zendehnam [7] and centroid index by Wang and Lee [8]. These methods range from the trivial to the complex, including one fuzzy number attribute to many fuzzy number attributes. The similarity measure concept using Jaccard index has also been proposed in ranking fuzzy numbers. This method was first introduced by Setnes and Cross [9] where the agreement between each pair of fuzzy numbers in similarity manner is evaluated. The mean aggregation

E. Hüllermeier, R. Kruse, and F. Hoffmann (Eds.): IPMU 2010, Part II, CCIS 81, pp. 383–391, 2010.

is applied in obtaining the fuzzy total evidence which then is used in determining the ranking of the fuzzy numbers.

The development of ranking fuzzy numbers using similarity measure with Jaccard index is limited in the literature except for some discussion on its properties by Cross and Setnes [10] and [11] and Cross and Sudkamp [12]. In 2009, Ramli and Mohamad [13] applied the function principle approach to the Jaccard index in determining the fuzzy maximum and fuzzy minimum which upgrades the capability of the index in ranking to both normal and non-normal fuzzy sets in a simpler manner.

However, the mean aggregation used in the Jaccard index can only represent the neutral decision maker's perspective and as the ranking of fuzzy numbers is commonly implemented in the decision-making problems, it is crucial to consider all types of decision maker's perspectives. In this paper, the degree of optimism concept which represents all types of decision maker's perspectives is applied in calculating the fuzzy total evidence. The properties of the proposed method which can simplify the ranking procedure are also presented.

2 Fuzzy Numbers

In this section, we briefly review the definition of fuzzy numbers.

A fuzzy number is a fuzzy subset in the universe discourse that is both convex and normal. The membership function of a fuzzy number A can be defined as

$$f_A(x) = \begin{cases} f_A^L(x) & , a \le x \le b \\ 1 & , b \le x \le c \\ f_A^R(x) & , c \le x \le d \\ 0 & , \text{otherwise} \end{cases}$$

where
f_A^L is the left membership function that is increasing and $f_A^L : [a,b] \to [0,1]$. f_A^R is the right membership function that is decreasing and $f_A^R : [c,d] \to [0,1]$. If f_A^L and f_A^R are linear and continuous, then A is a trapezoidal fuzzy number denoted as (a,b,c,d). Triangular fuzzy numbers which are special cases of trapezoidal fuzzy numbers with $b{=}c$ are denoted as (a,b,d).

3 A Review on Fuzzy Jaccard Ranking Index

Based on the psychological ratio model of similarity from Tversky [14], which is defined as

$$S_{\alpha,\beta}(X,Y) = \frac{f(X \cap Y)}{f(X \cap Y) + \alpha f(X \cap \bar{Y}) + \beta f(Y \cap \bar{X})},$$

various index of similarity measures have been proposed which depend on the values of α and β. Typically, the function f is taken to be the cardinality function.

For $\alpha = \beta = 1$, the psychological ratio model of similarity becomes the Jaccard index similarity measure which is defined as

$$S_{1,1}(X,Y) = \frac{f(X \cap Y)}{f(X \cup Y)}.$$

The objects X and Y described by the features are replaced with fuzzy sets A and B which are described by the membership functions. The fuzzy Jaccard index similarity measure is defined as

$$S_J(A,B) = \frac{|A \cap B|}{|A \cup B|}$$

where $|A|$ denotes the cardinality of fuzzy set A, $\cap$ and $\cup$ can be replaced by t-norm and s-norm respectively.

The fuzzy Jaccard ranking procedure by Setnes and Cross [9] is presented as follows:

Step 1: For each pair of fuzzy numbers A_i and A_j where $i, j = 1, 2, \ldots, n$, find the fuzzy minimum and fuzzy maximum between A_i and A_j.

Step 2: Calculate the evidences of $E(A_i \geq A_j)$, $E(A_j \leq A_i)$, $E(A_j \geq A_i)$ and $E(A_i \leq A_j)$ which are defined based on fuzzy Jaccard index as

$$\begin{aligned} E(A_i \geq A_j) &= S_J(MAX(A_i, A_j), A_i), \\ E(A_j \leq A_i) &= S_J(MIN(A_i, A_j), A_j), \\ E(A_j \geq A_i) &= S_J(MAX(A_i, A_j), A_j), \\ E(A_i \leq A_j) &= S_J(MIN(A_i, A_j), A_i). \end{aligned}$$

To simplify, C_{ij} and c_{ji} are used to represent $E(A_i \geq A_j)$ and $E(A_j \leq A_i)$, respectively. Likewise, C_{ji} and c_{ij} are used to denote $E(A_j \geq A_i)$ and $E(A_i \leq A_j)$ respectively.

Step 3: Calculate the total evidences $E_{total}(A_i \geq A_j)$ and $E_{total}(A_j \geq A_i)$ which are defined based on the mean aggregation concept as

$$E_{total}(A_i \geq A_j) = \frac{C_{ij} + c_{ji}}{2}$$

and

$$E_{total}(A_j \geq A_i) = \frac{C_{ji} + c_{ij}}{2}.$$

To simplify, $E_{\geq}(i,j)$ and $E_{\geq}(j,i)$ are used to represent $E_{total}(A_i \geq A_j)$ and $E_{total}(A_j \geq A_i)$, respectively.

Step 4: For two fuzzy numbers, compare the total evidences in **Step 3** which will result the ranking of the two fuzzy numbers A_i and A_j as follows:
i. $A_i \succ A_j$ if and only if $E_{\geq}(i,j) > E_{\geq}(j,i)$.

ii. $A_i \prec A_j$ if and only if $E_{\geq}(i,j) < E_{\geq}(j,i)$.
iii. $A_i \approx A_j$ if and only if $E_{\geq}(i,j) = E_{\geq}(j,i)$.

Step 5: For n fuzzy numbers, develop $n \times n$ binary ranking relation $R_{>}(i,j)$, which is defined as

$$R_{>}(i,j) = \begin{cases} 1 \text{ , } E_{\geq}(i,j) > E_{\geq}(j,i) \\ 0 \quad\quad \text{, otherwise.} \end{cases}$$

Step 6: Develop a column vector $[O_i]$ where O_i is the total element of each row of $R_{>}(i,j)$ and is defined as $O_i = \sum_{j=1}^{n} R_{>}(i,j)$ for $j = 1, 2, \ldots, n$.

Step 7: The total ordering of the fuzzy numbers A_i corresponds to the order of the elements $[O_i]$ in the column vector $[O_i]$.

4 Fuzzy Jaccard Ranking Index with Degree of Optimism

We propose fuzzy Jaccard ranking index with Hurwicz optimism-pessimism criterion as follows:

Steps 1-2: These steps are similar with fuzzy Jaccard ranking index.

Step 3: Calculate the total evidences $E_{total}(A_i \geq A_j)$ and $E_{total}(A_j \geq A_i)$ which are defined based on the degree of optimism concept as

$$E_{total}(A_i \geq A_j) = \beta C_{ij} + (1-\beta)c_{ji}$$

and

$$E_{total}(A_j \geq A_i) = \beta C_{ji} + (1-\beta)c_{ij}$$

where $\beta \in [0,1]$ represents the degree of optimism. Conventionally, $\beta = 0$, $\beta = 0.5$ and $\beta = 1$ represent very pessimistic, neutral and very optimistic decision maker's perspective, respectively. $E_{\geq}(i,j)$ and $E_{\geq}(j,i)$ are used to replace $E_{total}(A_i \geq A_j)$ and $E_{total}(A_j \geq A_i)$, respectively.

Steps 4-7: These steps are similar with fuzzy Jaccard ranking index.

Lemma 1. *For two fuzzy numbers A_i and A_j with $C_{ij} - c_{ji} - C_{ji} + c_{ij} \neq 0$ and $\beta_{ij} = \frac{c_{ij}-c_{ji}}{C_{ij}-c_{ji}-C_{ji}+c_{ij}}$ where C_{ij}, c_{ji}, C_{ji} and c_{ij} denote the evidences $E(A_i \geq A_j)$, $E(A_j \leq A_i)$, $E(A_j \geq A_i)$ and $E(A_i \leq A_j)$ respectively, the results of Jaccard ranking index with degree of optimism β are,*

1. *$A_i \approx A_j$ if and only if $\beta = \beta_{ij}$.*
2. *$A_i \succ A_j$ if and only if*
   (a) $\beta > \beta_{ij}$ with $C_{ij} - c_{ji} - C_{ji} + c_{ij} > 0$.
   (b) $\beta < \beta_{ij}$ with $C_{ij} - c_{ji} - C_{ji} + c_{ij} < 0$.
3. *$A_i \prec A_j$ if and only if*
   (a) $\beta < \beta_{ij}$ with $C_{ij} - c_{ji} - C_{ji} + c_{ij} > 0$.
   (b) $\beta > \beta_{ij}$ with $C_{ij} - c_{ji} - C_{ji} + c_{ij} < 0$.

Proof. Let A_i and A_j be two fuzzy numbers with $C_{ij} - c_{ji} - C_{ji} + c_{ij} \neq 0$ and $\beta_{ij} = \frac{c_{ij}-c_{ji}}{C_{ij}-c_{ji}-C_{ji}+c_{ij}}$.

1. Let $A_i \approx A_j$, then $E_{\geq}(i,j) = E_{\geq}(j,i)$.
   Thus, $\beta C_{ij} + (1-\beta)c_{ji} = \beta C_{ji} + (1-\beta)c_{ij}$, and upon simplification, we obtain $\beta = \frac{c_{ij}-c_{ji}}{C_{ij}-c_{ji}-C_{ji}+c_{ij}} = \beta_{ij}$.
   Therefore, if $A_i \approx A_j$,
   then $\beta = \frac{c_{ij}-c_{ji}}{C_{ij}-c_{ji}-C_{ji}+c_{ij}} = \beta_{ij}$ with $C_{ij} - c_{ji} - C_{ji} + c_{ij} \neq 0$.
   Similarly, if $\beta = \beta_{ij} = \frac{c_{ij}-c_{ji}}{C_{ij}-c_{ji}-C_{ji}+c_{ij}}$ with $C_{ij} - c_{ji} - C_{ji} + c_{ij} \neq 0$, then $\beta(C_{ij} - c_{ji} - C_{ji} + c_{ij}) = c_{ij} - c_{ji}$, and rearranging the equation will give $\beta C_{ij} + (1-\beta)c_{ji} = \beta C_{ji} + (1-\beta)c_{ij}$.
   Thus, $E_{\geq}(i,j) = E_{\geq}(j,i)$ or $A_i \approx A_j$.
   Hence, if $\beta = \beta_{ij} = \frac{c_{ij}-c_{ji}}{C_{ij}-c_{ji}-C_{ji}+c_{ij}}$ with $C_{ij} - c_{ji} - C_{ji} + c_{ij} \neq 0$, then $A_i \approx A_j$.
   This proves that, $A_i \approx A_j$, if and only if
   $\beta = \beta_{ij} = \frac{c_{ij}-c_{ji}}{C_{ij}-c_{ji}-C_{ji}+c_{ij}}$ and $C_{ij} - c_{ji} - C_{ji} + c_{ij} \neq 0$.
2. Let $A_i \succ A_j$, then $E_{\geq}(i,j) > E_{\geq}(j,i)$.
   Thus, $\beta C_{ij} + (1-\beta)c_{ji} > \beta C_{ji} + (1-\beta)c_{ij}$, and upon simplification, we obtain $\beta(C_{ij} - c_{ji} - C_{ji} + c_{ij}) > c_{ij} - c_{ji}$.
   For $C_{ij} - c_{ji} - C_{ji} + c_{ij} > 0$, then $\beta > \frac{c_{ij}-c_{ji}}{C_{ij}-c_{ji}-C_{ji}+c_{ij}} = \beta_{ij}$.
   While for $C_{ij} - c_{ji} - C_{ji} + c_{ij} < 0$, then $\beta < \frac{c_{ij}-c_{ji}}{C_{ij}-c_{ji}-C_{ji}+c_{ij}} = \beta_{ij}$.
   Therefore, if $A_i \succ A_j$, then
   (a) $\beta > \frac{c_{ij}-c_{ji}}{C_{ij}-c_{ji}-C_{ji}+c_{ij}} = \beta_{ij}$ with $C_{ij} - c_{ji} - C_{ji} + c_{ij} > 0$.
   (b) $\beta < \frac{c_{ij}-c_{ji}}{C_{ij}-c_{ji}-C_{ji}+c_{ij}} = \beta_{ij}$ with $C_{ij} - c_{ji} - C_{ji} + c_{ij} < 0$.
   Similarly, if $\beta > \frac{c_{ij}-c_{ji}}{C_{ij}-c_{ji}-C_{ji}+c_{ij}} = \beta_{ij}$ with $C_{ij} - c_{ji} - C_{ji} + c_{ij} > 0$, then $\beta(C_{ij} - c_{ji} - C_{ji} + c_{ij}) > c_{ij} - c_{ji}$, and rearranging the inequality will give $\beta C_{ij} + (1-\beta)c_{ji} > \beta C_{ji} + (1-\beta)c_{ij}$.
   Thus, $E_{\geq}(i,j) > E_{\geq}(j,i)$ or $A_i \succ A_j$.
   Hence, if $\beta > \frac{c_{ij}-c_{ji}}{C_{ij}-c_{ji}-C_{ji}+c_{ij}} = \beta_{ij}$ with $C_{ij} - c_{ji} - C_{ji} + c_{ij} > 0$, then $A_i \succ A_j$.
   Similarly, we can prove that if $\beta < \frac{c_{ij}-c_{ji}}{C_{ij}-c_{ji}-C_{ji}+c_{ij}} = \beta_{ij}$
   with $C_{ij} - c_{ji} - C_{ji} + c_{ij} < 0$, then $A_i \succ A_j$.
   This proves that $A_i \succ A_j$ if and only if
   (a) $\beta > \beta_{ij}$ with $C_{ij} - c_{ji} - C_{ji} + c_{ij} > 0$.
   (b) $\beta < \beta_{ij}$ with $C_{ij} - c_{ji} - C_{ji} + c_{ij} < 0$.
3. In similar manner, we can also prove that $A_i \prec A_j$ if and only if
   (a) $\beta < \beta_{ij}$ with $C_{ij} - c_{ji} - C_{ji} + c_{ij} > 0$.
   (b) $\beta > \beta_{ij}$ with $C_{ij} - c_{ji} - C_{ji} + c_{ij} < 0$.

Lemma 2. *For two fuzzy numbers A_i and A_j with $C_{ij} - c_{ji} - C_{ji} + c_{ij} = 0$ where C_{ij}, c_{ji}, C_{ji} and c_{ij} denote the evidences $E(A_i \geq A_j)$, $E(A_j \leq A_i)$, $E(A_j \geq A_i)$*

and $E(A_i \le A_j)$ respectively, the results of Jaccard ranking index with degree of optimism β are,

1. *If $c_{ij} - c_{ji} > 0$, then for all $\beta \in [0,1]$, $A_i \prec A_j$.*
2. *If $c_{ij} - c_{ji} < 0$, then for all $\beta \in [0,1]$, $A_i \succ A_j$.*
3. *If $c_{ij} - c_{ji} = 0$, then for all $\beta \in [0,1]$, $A_i \approx A_j$.*

Proof. Let A_i and A_j be two fuzzy numbers with $C_{ij} - c_{ji} - C_{ji} + c_{ij} = 0$.

1. Let $c_{ij} - c_{ji} > 0$.
   Thus, for all $\beta \in [0,1]$, $\beta(C_{ij} - c_{ji} - C_{ji} + c_{ij}) = 0 < c_{ij} - c_{ji}$ or $\beta(C_{ij} - c_{ji} - C_{ji} + c_{ij}) < c_{ij} - c_{ji}$, and rearranging the inequality will give $\beta C_{ij} + (1-\beta)c_{ji} < \beta C_{ji} + (1-\beta)c_{ij}$.
   $E_{\ge}(i,j) < E_{\ge}(j,i)$ or $A_i \prec A_j$.
   Therefore, if $C_{ij} - c_{ji} - C_{ji} + c_{ij} = 0$ and $c_{ij} - c_{ji} > 0$, then $A_i \prec A_j$ for all $\beta \in [0,1]$.
2. Let $c_{ij} - c_{ji} < 0$.
   Thus, for all $\beta \in [0,1]$, $\beta(C_{ij} - c_{ji} - C_{ji} + c_{ij}) = 0 > c_{ij} - c_{ji}$ or $\beta(C_{ij} - c_{ji} - C_{ji} + c_{ij}) > c_{ij} - c_{ji}$, and rearranging the inequality will give $\beta C_{ij} + (1-\beta)c_{ji} > \beta C_{ji} + (1-\beta)c_{ij}$.
   $E_{\ge}(i,j) > E_{\ge}(j,i)$ or $A_i \succ A_j$.
   Therefore, if $C_{ij} - c_{ji} - C_{ji} + c_{ij} = 0$ and $c_{ij} - c_{ji} < 0$, then $A_i \succ A_j$ for all $\beta \in [0,1]$.
3. In similar manner, we can prove that if
   $C_{ij} - c_{ji} - C_{ji} + c_{ij} = 0$ and $c_{ij} - c_{ji} = 0$, then $A_i \approx A_j$ for all $\beta \in [0,1]$.

5 Implementation

In this section, eight sets of numerical examples are presented to illustrate the validity and advantages of the Jaccard with degree of optimism ranking properties. Tables 1 and 2 show the ranking results for Sets 1-4 and Sets 6-8 respectively.

Set 1 : $A_1 = (0.2, 0.5, 0.8)$, $A_2 = (0.4, 0.5, 0.6)$.
Set 2 : $A_1 = (0.2, 0.5, 1)$, $A_2 = (0.155, 0.645, 0.8)$.
Set 3 : $A_1 = (0.1, 0.2, 0.4, 0.5)$, $A_2 = (0.2, 0.3, 0.4)$.
Set 4 : $A_1 = (0, 3, 4)$, $A_2 = (1.5, 2, 4.5)$.
Set 5 : $A_1 = (3, 5, 9)$, $A_2 = (1.5, 6.5, 8)$.
Set 6 : $A_1 = (5, 6, 13)$, $A_2 = (2, 6.5, 9, 12.5)$.
Set 7 : $A_1 = (1, 7, 10, 12)$, $A_2 = (2, 6.5, 9, 12.5)$.
Set 8 : $A_1 = (1, 2, 5)$, $A_2 = [1, 2, 2, 4]$ with membership function,

$$f_{A_2}(x) = \begin{cases} \sqrt{1-(x-2)^2} & ,1 \le x \le 2 \\ \sqrt{1-\frac{1}{4}(x-2)^2} & ,2 \le x \le 4 \\ 0 & ,\texttt{otherwise.} \end{cases}$$

Table 1. Comparative Results for Sets 1-4

	Set 1	Set 2	Set 3	Set 4
Abbasbandy & Asady [3]	$A_1 \approx A_2$	$A_1 \prec A_2$	$A_1 \approx A_2$	$A_1 \approx A_2$
Chu & Tsao [4]	$A_1 \approx A_2$	$A_1 \prec A_2$	$A_1 \approx A_2$	$A_1 \approx A_2$
Chen & Chen [6]	$A_1 \prec A_2$	$A_1 \prec A_2$	$A_1 \prec A_2$	$A_1 \prec A_2$
Asady & Zendehnam [7]	$A_1 \approx A_2$	$A_1 \prec A_2$	$A_1 \approx A_2$	$A_1 \approx A_2$
Wang & Lee [8]	$A_1 \approx A_2$	$A_1 \succ A_2$	$A_1 \approx A_2$	$A_1 \prec A_2$
Jaccard Index [9]	$A_1 \approx A_2$	$A_1 \prec A_2$	$A_1 \approx A_2$	$A_1 \prec A_2$
Proposed Method	$A_1 \prec A_2, \beta \in [0, 0.5)$ $A_1 \approx A_2, \beta = 0.5$ $A_1 \succ A_2, \beta \in (0.5, 1]$	$A_1 \prec A_2, \beta \in [0, 0.61)$ $A_1 \approx A_2, \beta = 0.61$ $A_1 \succ A_2, \beta \in (0.61, 1]$	$A_1 \prec A_2, \beta \in [0, 0.5)$ $A_1 \approx A_2, \beta = 0.5$ $A_1 \succ A_2, \beta \in (0.5, 1]$	$A_1 \prec A_2$, $\beta \in [0, 1]$

6 Discussion

In Table 1, we have the following results: In Sets 1 and 3, for Abbasbandy and Asady's [3], Chu and Tsao's [4], Asady and Zendehnam's [7], Wang and Lee's [8] and the Jaccard index [9], the ranking order is $A_1 \approx A_2$. This is the shortcoming of [3], [4], [7], [8] and [9] that cannot discriminate the ranking between two different fuzzy numbers. However, the proposed index ranks the fuzzy numbers based on decision makers' perspective where for pessimistic $A_1 \prec A_2$, neutral $A_1 \approx A_2$ and optimistic decision makers $A_1 \succ A_2$. For Set 4, [3], [4] and [7] also cannot discriminate the ranking between A_1 and A_2. The proposed index has $A_1 \prec A_2$ for all types of decision makers' perspective which is consistent with [6], [8] and the Jaccard index [9]. In Set 2, the proposed index with pessimistic and neutral decision makers ranks $A_1 \prec A_2$ which is similar to the previous indices except [8]. However, the optimistic decision makers give three different ranking results, $A_1 \prec A_2$ for $\beta \in (0.5, 0.61)$, $A_1 \approx A_2$ for $\beta = 0.61$ and $A_1 \succ A_2$ for $\beta \in (0.61, 1]$. This indicates that the equal ranking result does not necessarily occur for neutral decision makers.

Based on Table 2, we have the following results: In Set 5, the proposed index with pessimistic decision makers have three types of ranking results while neutral and optimistic decision makers rank $A_1 \prec A_2$. For Sets 6 and 7, [3] and [7]

Table 2. Comparative Results for Sets 5-8

	Set 5	Set 6	Set 7	Set 8
Abbasbandy & Asady [3]	$A_1 \prec A_2$	$A_1 \approx A_2$	$A_1 \approx A_2$	*
Chu & Tsao [4]	$A_1 \succ A_2$	$A_1 \prec A_2$	$A_1 \succ A_2$	$A_1 \succ A_2$
Chen & Chen [6]	$A_1 \succ A_2$	$A_1 \prec A_2$	$A_1 \succ A_2$	*
Asady & Zendehnam [7]	$A_1 \prec A_2$	$A_1 \approx A_2$	$A_1 \approx A_2$	*
Wang & Lee [8]	$A_1 \succ A_2$	$A_1 \succ A_2$	$A_1 \prec A_2$	$A_1 \succ A_2$
Jaccard Index [9]	$A_1 \prec A_2$	$A_1 \succ A_2$	$A_1 \approx A_2$	$A_1 \succ A_2$
Proposed Method	$A_1 \succ A_2, \beta \in [0, 0.49)$ $A_1 \approx A_2, \beta = 0.49$ $A_1 \prec A_2, \beta \in (0.49, 1]$	$A_1 \succ A_2, \beta \in [0, 0.54)$ $A_1 \approx A_2, \beta = 0.54$ $A_1 \prec A_2, \beta \in (0.54, 1]$	$A_1 \prec A_2, \beta \in [0, 0.5)$ $A_1 \approx A_2, \beta = 0.5$ $A_1 \succ A_2, \beta \in (0.5, 1]$	$A_1 \succ A_2,$ $\beta \in [0, 1]$

'*': the method cannot rank general fuzzy numbers

cannot discriminate the ranking between the fuzzy numbers while the proposed index ranks the fuzzy numbers based on decision makers perspective. In Set 8, [3], [6] and [7] cannot rank the general fuzzy number A_2. The results for other methods are $A_1 \succ A_2$ and similar with all types of decision makers of the proposed index.

7 Conclusion

This paper proposes the degree of optimism concept in determining the fuzzy total evidence which is capable to rank fuzzy numbers based on all types of decision maker's perspective. The proposed index can rank fuzzy numbers effectively for cases where some of the previous ranking methods failed such as [3], [4], [6], [7], [8] and [9]. Some properties which are based on the values of fuzzy evidences C_{ij}, c_{ji}, C_{ji} and c_{ij} are developed. These properties can simplify the lengthy procedure as we only have to calculate from Step 1 to Step 2 to get the ranking results, rather than calculating from Step 1 to Step 4. Thus, it reduces the computational procedure and is practically applicable to solve the ranking problems in fuzzy environment.

References

1. Cheng, C.H.: A New Approach for Ranking Fuzzy Numbers by Distance Method. Fuzzy Sets and Systems 95, 307–317 (1998)
2. Yao, J.S., Wu, K.: Ranking Fuzzy Numbers based on Decomposition Principle and Signed Distance. Fuzzy Sets and Systems 116, 275–288 (2000)
3. Abbasbandy, S., Asady, B.: Ranking of Fuzzy Numbers by Sign Distance. Information Sciences 176, 2405–2416 (2006)
4. Chu, T.C., Tsao, C.T.: Ranking Fuzzy Numbers with an Area between the Centroid Point and Original Point. Computers and Mathematics with Applications 43, 111–117 (2002)
5. Chen, S.J., Chen, S.M.: A New Method for Handling Multicriteria Fuzzy Decision Making Problems using FN-IOWA Operators. Cybernatics and Systems 34, 109–137 (2003)
6. Chen, S.J., Chen, S.M.: Fuzzy Risk Analysis based on the Ranking of Generalized Trapezoidal Fuzzy Numbers. Applied Intelligence 26, 1–11 (2007)
7. Asady, B., Zendehnam, A.: Ranking Fuzzy Numbers by Distance Minimization. Applied Mathematical Modelling 31, 2589–2598 (2007)
8. Wang, Y.J., Lee, H.S.: The Revised Method of Ranking Fuzzy Numbers with an Area between the Centroid and Original Points. Computers and Mathematics with Applications 55, 2033–2042 (2008)
9. Setnes, M., Cross, V.: Compatibility based Ranking of Fuzzy Numbers. In: 1997 Conference of North American Fuzzy Information Processing Society (NAFIPS), Syracuse, New York, pp. 305–310 (1997)
10. Cross, V., Setnes, M.: A Generalized Model for Ranking Fuzzy Sets. In: 7th IEEE World Congress on Computational Intelligence, Anchorage, Alaska, pp. 773–778 (1998)
11. Cross, V., Setnes, M.: A Study of Set Theoretic Measures for Use with the Generalized Compatibility-based Ranking Method. In: 1998 Conference of North American Fuzzy Information Processing Society (NAFIPS), Pensacola, FL, pp. 124–129 (1998)
12. Cross, V., Sudkamp, T.: Similarity and Compatibility in Fuzzy Set Theory: Assessment and Applications. Physica-Verlag, New York (2002)
13. Ramli, N., Mohamad, D.: A Function Principle Approach to Jaccard Ranking Fuzzy Numbers. In: Abraham, A., Muda, A.K., Herman, N.S., Shamsuddin, S.M., Huoy, C.H. (eds.) SOCPAR 2009. Proceedings of International Conference of Soft Computing and Pattern Recognition, pp. 324–328. IEEE, Inc., Malacca (2009)
14. Tversky, A.: Features of Similarity. Psychological Review 84, 327–352 (1977)

Negation Functions in the Set of Discrete Fuzzy Numbers

Jaume Casasnovas and J. Vicente Riera

Department of Mathematics and Computer Science, University of Balearic Islands, 07122 Palma de Mallorca, Spain
{jaume.casasnovas,jvicente.riera}@uib.es

Abstract. The aim of this paper is to build a strong negation $\mathcal{N}$ on the bounded distributive lattice, $\mathcal{A}_1^L$, of discrete fuzzy numbers whose support is a subset of consecutive natural numbers of the finite chain $L = \{0, 1, \cdots, m\}$, from the only negation on L. Moreover, we obtain the $\mathcal{N}$-dual t-norm(t-conorm) of a $\mathcal{T}(\mathcal{S})$ t-norm(t-conorm) on $\mathcal{A}_1^L$.

1 Introduction

Negations and strong negations on the unit interval and their applications in classic fuzzy set theory were studied and characterized by many authors [9,11,17]. In the sixties, Schweizer and Sklar [16] defined a t-conorm S from t-norm T and a strong negation n on $[0,1]$ on the following way: $S(x,y) = 1 - T(n(x), n(y))$. Therefore, considering the standard negation $n(x) = 1 - x$ as complement of x in the unit interval, the previous expression explains the name t-conorm. Another interesting use of negation functions can be found in fuzzy logic, where a generalization of the classic implication "$p \to q = \neg p \vee q$" called S-implication is defined, obtained from a strong negation and a S-conorm. In this sense, the contributions on intuitionistic fuzzy connectives [2,10] are very interesting, especially the study of intuitionistic fuzzy negators and of intuitionistic fuzzy implicators obtained from this fuzzy negators. On discrete settings [15], we wish to point out that there is not any strong negation on the chain $L = \{0 < \cdots < +\infty\}$ but on the finite chain $L = \{0 < \cdots < m\}$ there exists a unique strong negation given by $n(x) = m - x$ for all $x \in L$.

Voxman [18] introduced the concept of discrete fuzzy number such as a fuzzy subset on $\mathbb{R}$ with discrete support and analogous properties to a fuzzy number. It is well known that arithmetic and lattice operations between fuzzy numbers are defined using the Zadeh's extension principle [14]. But, in general, for discrete fuzzy numbers this method fails [3,4,5,19]. We have studied this drawback [3,4,5] and we have obtained new closed operations in the set of discrete fuzzy numbers. In particular, we showed [6] that $\mathcal{A}_1$, the set of discrete fuzzy numbers whose support is a sequence of consecutive natural numbers, is a distributive lattice. In this lattice, we considered a partial order, obtained in a usual way, from the lattice operations of this set. So, from this partial order, we investigated [7] the extension of monotone operations defined on a discrete setting to a

E. Hüllermeier, R. Kruse, and F. Hoffmann (Eds.): IPMU 2010, Part II, CCIS 81, pp. 392–401, 2010.

closed binary operation of discrete fuzzy numbers. In this same article, we also investigated different properties such as the monotonicity, commutativity and associativity. The objective of the present article is on the one hand, to construct negation functions on the bounded distributive lattice of discrete fuzzy numbers whose support is a subset of consecutive natural numbers of the finite chain $L = \{0, 1, \cdots, m\}$ obtained from the unique strong negation L and, on the other hand, to define the dual t-conorm of a t-norm on $\mathcal{A}_1^L$.

2 Preliminaries

2.1 Triangular Norms and Conorms on Partially Ordered Sets

Let $(P; \leq)$ be a non-trivial bounded partially ordered set (poset) with "e" and "m" as minimum and maximum elements respectively.

Definition 1. *[1]A* triangular norm *(briefly t-norm) on P is a binary operation $T : P \times P \rightarrow P$ such that for all $x, y, z \in P$ the following axioms are satisfied:*

1. $T(x, y) = T(y, x)$ *(commutativity)*
2. $T(T(x, y), z) = T(x, T(y, z))$ *(associativity)*
3. $T(x, y) \leq T(x', y')$ *whenever* $x \leq x', y \leq y'$ *(monotonicity)*
4. $T(x, m) = x$ *(boundary condition)*

Definition 2. *A* triangular conorm *(t-conorm for short) on P is a binary operation $S : P \times P \rightarrow P$ which, for all $x, y, z \in P$ satisfies (1), (2), (3) and (4'): $S(x, e) = x$, as boundary condition.*

2.2 Triangular Norms and Conorms on Discrete Settings

Let L be the totally ordered set $L = \{0, 1, \ldots, m\} \subset \mathbb{N}$. A t-norm(t-conorm) defined on L will be called a discrete t-norm(t-conorm).

Definition 3. *[12,15] A t-norm(t-conorm) $T(S) : L \times L \rightarrow L$ is said to be* smooth *if it satisfies $T(S)(x + 1, y) - T(S)(x, y) \leq 1$ and $T(S)(x, y + 1) - T(S)(x, y) \leq 1$.*

Definition 4. *[15] A t-norm(t-conorm) $T : L \times L \rightarrow L$ is said to be* divisible *if it satisfies: For all $x, y \in L$ with $x \leq y$, there is $z \in L$ such that $x = T(y, z)(y = S(x, z))$.*

Proposition 1. *[15] Given a t-norm(t-conorm) $T(S)$ on L, it is equivalent:*

1. *$T(S)$ is smooth*
2. *$T(S)$ is divisible*

2.3 Discrete Fuzzy Numbers

By a fuzzy subset of $\mathbb{R}$, we mean a function $A : \mathbb{R} \to [0,1]$. For each fuzzy subset A, let $A^\alpha = \{x \in \mathbb{R} : A(x) \geq \alpha\}$ for any $\alpha \in (0,1]$ be its α-level set (or α-cut). By $supp(A)$, we mean the support of A, i.e. the set $\{x \in \mathbb{R} : A(x) > 0\}$. By A^0, we mean the closure of $supp(A)$.

Definition 5. *[18] A fuzzy subset A of $\mathbb{R}$ with membership mapping $A : \mathbb{R} \to [0,1]$ is called* discrete fuzzy number *if its support is finite, i.e., there exist real numbers $x_1, ..., x_n \in \mathbb{R}$ with $x_1 < x_2 < ... < x_n$ such that $supp(A) = \{x_1, ..., x_n\}$, and there are natural numbers s, t with $1 \leq s \leq t \leq n$ such that:*

1. *$A(x_i)$=1 for any natural number i with $s \leq i \leq t$ (*core*)*
2. *$A(x_i) \leq A(x_j)$ for each natural number i, j with $1 \leq i \leq j \leq s$*
3. *$A(x_i) \geq A(x_j)$ for each natural number i, j with $t \leq i \leq j \leq n$*

Remark 1. If the fuzzy subset A is a discrete fuzzy number then the support of A coincides with its closure, i.e. $supp(A) = A^0$.

From now on, we will denote the set of discrete fuzzy numbers by DFN and the abbreviation *dfn* will denote a discrete fuzzy number.

Theorem 1. *[19] (Representation of discrete fuzzy numbers) Let A be a discrete fuzzy number. Then the following statements (1)-(4) hold:*

1. *A^α is a nonempty finite subset of $\mathbb{R}$, for any $\alpha \in [0,1]$*
2. *$A^{\alpha_2} \subset A^{\alpha_1}$ for any $\alpha_1, \alpha_2 \in [0,1]$ with $0 \leq \alpha_1 \leq \alpha_2 \leq 1$*
3. *For any $\alpha_1, \alpha_2 \in [0,1]$ with $0 \leq \alpha_1 \leq \alpha_2 \leq 1$, if $x \in A^{\alpha_1} - A^{\alpha_2}$ we have $x < y$ for all $y \in A^{\alpha_2}$, or $x > y$ for all $y \in A^{\alpha_2}$*
4. *For any $\alpha_0 \in (0,1]$, there exist some real numbers α_0' with $0 < \alpha_0' < \alpha_0$ such that $A^{\alpha_0'} = A^{\alpha_0}$ (i.e. $A^\alpha = A^{\alpha_0}$ for any $\alpha \in [\alpha_0', \alpha_0]$).*

Theorem 2. *[19] Conversely, if for any $\alpha \in [0,1]$, there exists $A^\alpha \subset \mathbb{R}$ satisfying analogous conditions to the (1)-(4) of Theorem 1, then there exists a unique $A \in DFN$ such that its α-cuts are exactly the sets A^α for any $\alpha \in [0,1]$.*

2.4 Maximum and Minimum of Discrete Fuzzy Numbers

Let A, B be two dfn and $A^\alpha = \{x_1^\alpha, \cdots, x_p^\alpha\}$, $B^\alpha = \{y_1^\alpha, \cdots, y_k^\alpha\}$ their α-cuts respectively.

In [5], for each $\alpha \in [0,1]$, we consider the following sets,

$$min_w(A,B)^\alpha = \{z \in supp(A) \bigwedge supp(B) | \min(x_1^\alpha, y_1^\alpha) \leq z \leq \min(x_p^\alpha, y_k^\alpha)\} \text{ and}$$

$$max_w(A,B)^\alpha = \{z \in supp(A) \bigvee supp(B) | \max(x_1^\alpha, y_1^\alpha) \leq z \leq \max(x_p^\alpha, y_k^\alpha)\}$$

where $supp(A) \bigwedge supp(B) = \{z = min(x,y) | x \in supp(A), y \in supp(B)\}$ and $supp(A) \bigvee supp(B) = \{z = max(x,y) | x \in supp(A), y \in supp(B)\}$.

Proposition 2. *[5] There exist two unique discrete fuzzy numbers, that we will denote by $min_w(A,B)$ and $max_w(A,B)$, such that they have the above defined sets $min_w(A,B)^\alpha$ and $max_w(A,B)^\alpha$ as α-cuts respectively.*

The following result is not true, in general, for the set of discrete fuzzy numbers[6].

Proposition 3. *[6] The triplet ($\mathcal{A}_1$,min_w,max_w) is a distributive lattice, where $\mathcal{A}_1$ denotes the set of discrete fuzzy numbers whose support is a sequence of consecutive natural numbers.*

Remark 2. [6] Using these operations, we can define a partial order on $\mathcal{A}_1$ on the usual way:
$A \preceq B$ if and only if $min_w(A,B) = A$, or equivalently, $A \preceq B$ if and only if $max_w(A,B) = B$ for any $A, B \in \mathcal{A}_1$. Equivalently, we can also define the partial ordering in terms of α-cuts:
$A \preceq B$ if and only if $\min(A^\alpha, B^\alpha) = A^\alpha$
$A \preceq B$ if and only if $\max(A^\alpha, B^\alpha) = B^\alpha$

2.5 Discrete Fuzzy Numbers Obtained by Extending Discrete t-Norms(t-Conorms) Defined on a Finite Chain

Let us consider a discrete t-norm(t-conorm) $T(S)$ on the finite chain $L = \{0, 1, \cdots, m\} \subset \mathbb{N}$. Let $\mathcal{D}_L$ be the subset of the discrete fuzzy numbers $\mathcal{D}_L = \{A \in DFN \text{ such that } supp(A) \subseteq L\}$ and $A, B \in \mathcal{D}_L$. If X and Y are subsets of L, then the subset $\{T(x,y)|x \in \mathsf{X}, y \in \mathsf{Y}\} \subseteq L$ will be denoted by $T(\mathsf{X},\mathsf{Y})$. Analogously, $S(\mathsf{X},\mathsf{Y}) = \{S(x,y)|x \in \mathsf{X}, y \in \mathsf{Y}\}$.

So, if we consider the α-cut sets, $A^\alpha = \{x_1^\alpha, ..., x_p^\alpha\}$, $B^\alpha = \{y_1^\alpha, ..., y_k^\alpha\}$, for A and B respectively then $T(A^\alpha, B^\alpha) = \{T(x,y)|x \in A^\alpha, y \in B^\alpha\}$ and $S(A^\alpha, B^\alpha) = \{S(x,y)|x \in A^\alpha, y \in B^\alpha\}$ for each $\alpha \in [0,1]$, where A^0 and B^0 denote $supp(A)$ and $supp(B)$ respectively.

Definition 6. *[7]For each $\alpha \in [0,1]$, let us consider the sets*

$$C^\alpha = \{z \in T(supp(A), supp(B))| \min T(A^\alpha, B^\alpha) \le z \le \max T(A^\alpha, B^\alpha)\}$$

$$D^\alpha = \{z \in S(supp(A), supp(B))| \min S(A^\alpha, B^\alpha) \le z \le \max S(A^\alpha, B^\alpha)\}$$

Remark 3. [7]From the monotonicity of the t-norm(t-conorm) $T(S)$,

$$C^\alpha = \{z \in T(supp(A), supp(B))|T(x_1^\alpha, y_1^\alpha) \le z \le T(x_p^\alpha, y_k^\alpha)\}$$

$$D^\alpha = \{z \in S(supp(A), supp(B))|S(x_1^\alpha, y_1^\alpha) \le z \le S(x_p^\alpha, y_k^\alpha)\}$$

For $\alpha = 0$ then $C^0 = T(supp(A), supp(B))$ and $D^0 = S(supp(A), supp(B))$.

Theorem 3. *[7] There exists a unique discrete fuzzy number that will be denoted by $\mathcal{T}(A,B)(\mathcal{S}(A,B))$, such that its α-cuts sets $\mathcal{T}(A,B)^\alpha(\mathcal{S}(A,B)^\alpha)$ are exactly the sets $C^\alpha(D^\alpha)$ for each $\alpha \in [0,1]$ and $\mathcal{T}(A,B)(z) = sup\{\alpha \in [0,1] : z \in C^\alpha\}(\mathcal{S}(A,B)(z) = sup\{\alpha \in [0,1] : z \in D^\alpha\})$.*

Remark 4. [7] From the previous theorem, if $T(S)$ is a discrete t-norm(t-conorm) on L, we see that it is possible to define a binary operation on $\mathcal{D}_L = \{A \in DFN | supp(A) \subseteq L\}$,

$$\begin{aligned} \mathcal{T}(\mathcal{S}) : \mathcal{D}_L \times \mathcal{D}_L &\longrightarrow \mathcal{D}_L \\ (A, B) &\longmapsto \mathcal{T}(A, B)(\mathcal{S}(A, B)) \end{aligned}$$

that will be called *the extension of the t-norm T(t-conorm S) to $\mathcal{D}_L$*. Moreover, $\mathcal{T}$ and $\mathcal{S}$ are commutative and associative binary operations. Also, if we restrict these operations on the subset $\{A \in \mathcal{A}_1 \mid supp(A) \subseteq L = \{0, 1, \cdots, m\}\} \subseteq \mathcal{D}_L$ we showed that $\mathcal{T}$ and $\mathcal{S}$ are increasing operations as well.

2.6 Negation on Bounded Lattices

Definition 7. *[11] A negation on a bounded lattice $\mathcal{L} = (\mathcal{L}, \vee, \wedge, 0, 1)$ is a mapping $n : \mathcal{L} \to \mathcal{L}$ such that*

i) $x \leq y$ *implies* $n(x) \geq n(y)$
ii) $n^2(x) \geq x$ *for all* $x \in \mathcal{L}$*(being* $n^2(x) = n(n(x))$*)*
iii) $n(1) = 0$

If $n^2 = Id$ then n will be called strong negation and in the other cases, n will be called weak negation.

Remark 5. Let us notice the following facts:

1. [15] If the lattice is a bounded chain and n is a strong negation then n is a strictly decreasing bijection with $n(0) = 1$ and $n(1) = 0$.
2. [15] If the lattice is the bounded finite chain $L = \{0, 1, \cdots, m\}$ then there is only one strong negation n which is given by $n(x) = m - x$ for all $x \in L$.
3. [13] If we consider a negation n on the closed interval $[0, 1]$, then the associated negation on the set of closed intervals on $[0, 1]$ is defined by $N : \mathcal{I}([0, 1]) \to \mathcal{I}([0, 1])$ where $N([a, b]) = [n(b), n(a)]$.
4. Analogously to item 3, it is possible to consider a strong negation on the set of closed intervals of the finite chain $L = \{0, 1, \cdots, m\}$ from a strong negation $\widehat{n}$ on L, as follows: $N : \mathcal{I}(L) \to \mathcal{I}(L)$ where $N([a, b]) = [\widehat{n}(b), \widehat{n}(a)]$.

3 Distributive Bounded Lattices on $\mathcal{A}_1$

According to proposition 3, we know that $\mathcal{A}_1$ constitutes a partially ordered set which is a lattice. Now, using this fact, we want to see that the set $\mathcal{A}_1^L$ is a bounded distributive lattice with the operations $\min_w$ and $\max_w$, considered in proposition 2, as lattice operations.

Proposition 4. *If $A, B \in \mathcal{A}_1^L$ then $\min_w(A, B)$ and $\max_w(A, B)$ belong to the set $\mathcal{A}_1^L$.*

Proof. According to proposition 3, if $A, B \in \mathcal{A}_1^L \subset \mathcal{A}_1$ then the discrete fuzzy numbers $\max_w(A,B)$ and $\min_w(A,B) \in \mathcal{A}_1$. On the other hand, it is easy to see that the sets $supp(A) \bigwedge supp(B)$ and $supp(A) \bigvee supp(B)$ are subsets of L. So, $\min_w(A,B)^\alpha$ and $\max_w(A,B)^\alpha$ are subsets of L for each $\alpha \in [0,1]$. Hence, the discrete fuzzy numbers $\min_w(A,B)$ and $\max_w(A,B)$ belong to the set $\mathcal{A}_1^L$. □

Theorem 4. *The triplet* $(\mathcal{A}_1^L, \min_w, \max_w)$ *is a bounded distributive lattice.*

Proof. The distributive lattice structure stems from propositions 2, 3 and 4. Moreover, it is straightforward to see that the natural number m, which is the maximum of the chain L, as a discrete fuzzy number (i.e. it is the discrete fuzzy number M such that it has only the natural number m as support) is the greatest element of the distributive lattice $\mathcal{A}_1^L$. Analogously, the natural number 0, which is the minimum of the chain L, as a discrete fuzzy number (i.e. it is the discrete fuzzy number O such that it has only the natural number 0 as a support) is the least element of the distributive lattice $\mathcal{A}_1^L$. □

Theorem 5. *[8] Let $T(S)$ be a divisible t-norm(t-conorm) on L and let*

$$\begin{aligned} \mathcal{T}(\mathcal{S}) : \mathcal{A}_1^L \times \mathcal{A}_1^L &\rightarrow \mathcal{A}_1^L \\ (A,B) &\longmapsto \mathcal{T}(\mathcal{S})(A,B) \end{aligned}$$

be the extension of t-norm(t-conorm) $T(S)$ to $\mathcal{A}_1^L$,where $\mathcal{T}(A,B)$ and $\mathcal{S}(A,B)$ are defined according to theorem 3. Then, $\mathcal{T}(\mathcal{S})$ is a t-norm(t-conorm) on the bounded set $\mathcal{A}_1^L$.

4 Negations on $(\mathcal{A}_1^L, \min_w, \max_w)$

From now on, the α-cuts of a discrete fuzzy number $A \in \mathcal{A}_1^L$ will be denoted by $A^\alpha = \{x_1^\alpha, \cdots, x_p^\alpha\}$ or equivalently by $[x_1^\alpha, x_p^\alpha]$ where $[x_1^\alpha, x_p^\alpha] = \{z \in \mathbb{N} \mid x_1^\alpha \leq z \leq x_p^\alpha\}$ for each $\alpha \in [0,1]$. Moreover, if X is a subset of consecutive natural numbers where x_1, x_p denote the maximum and the minimum of X, then we will denote X as the closed interval $[x_1, x_p] = \{z \in \mathbb{N} \mid x_1 \leq z \leq x_p\}$.

Lemma 1. *Let n be the strong negation on L. If $\mathsf{X} \subseteq L$ is a subset of consecutive natural numbers then the set $N(\mathsf{X}) = \{z = n(x) \mid x \in \mathsf{X}\}$ is a subset of consecutive natural numbers as well.*

Proof. As n is a strictly decreasing bijection on L then $N(\mathsf{X}) = N([x_1, x_p]) =$ (from remark 5)$[n(x_p), n(x_1)] = \{z \in \mathbb{N} | n(x_p) \leq z \leq n(x_1)\}$. □

Remark 6. We know that if $A \in \mathcal{A}_1^L$ then its α-cuts A^α (for all $\alpha \in [0,1]$) are sets of consecutive natural numbers, where A^0 denotes the support of A. Then from lemma 1, $N(A^\alpha)$ are sets of consecutive natural numbers too.

Proposition 5. *Let us consider $A \in \mathcal{A}_1^L$ being $A^\alpha = [x_1^\alpha, x_p^\alpha]$ its α-cuts for each $\alpha \in [0,1]$. Moreover, for each $\alpha \in [0,1]$ let us consider the sets $\mathcal{N}(A)^\alpha = \{z \in N(supp(A)) | \min(N(A^\alpha)) \leq z \leq \max(N(A^\alpha))\}$. Then there exists a unique discrete fuzzy number, that will be denoted by $\mathcal{N}(A)$, such that it has the sets $\mathcal{N}(A)^\alpha$ as α-cuts.*

Proof. We know that if $A \in \mathcal{A}_1^L$ then its α-cuts are sets of consecutive natural numbers for each $\alpha \in [0,1]$. So, from remark 6, the sets $N(A^\alpha)$ for each $\alpha \in [0,1]$ as well. Moreover, from the monotonicity of the strong negation n and according to remark 6, $\mathcal{N}(A)^\alpha = \{z \in N(supp(A)) | \min(N(A^\alpha)) \leq z \leq \max(N(A^\alpha))\} = \{z \in N(supp(A)) | n(x_p^\alpha) \leq z \leq n(x_1^\alpha)\} = [n(x_p^\alpha), n(x_1^\alpha)]$. Now we show that the set $\mathcal{N}(A)^\alpha$ fulfills for each $\alpha \in [0,1]$ the conditions 1-4 of theorem 1 and then, if we apply the theorem 2 then the proposition holds. Indeed,

1. $\mathcal{N}(A)^\alpha$ is a nonempty finite set, because A^α is a nonempty finite set (the discrete fuzzy numbers are normal fuzzy subsets) and $N(supp(A))$ is a finite set.
2. We wish to see that the relation $\mathcal{N}(A)^\beta \subseteq \mathcal{N}(A)^\alpha$ for any $\alpha, \beta \in [0,1]$ with $0 \leq \alpha \leq \beta \leq 1$ holds.
   Because if $A \in \mathcal{A}_1^L$ and
   $$A^\alpha = \{x_1^\alpha, ..., x_p^\alpha\}, A^\beta = \{x_1^\beta, ..., x_r^\beta\},$$
   then
   $$A^\beta \subseteq A^\alpha \text{ implies } x_1^\alpha \leq x_1^\beta \text{ and } x_r^\beta \leq x_p^\alpha \tag{1}$$
   Moreover, as n is a strong negation on L and from the relation (1) then we obtain:
   $$n(x_1^\beta) \leq n(x_1^\alpha)$$
   $$n(x_r^\beta) \geq n(x_p^\alpha)$$
   And combining the previous conditions,
   $$n(x_p^\alpha) \leq n(x_r^\beta) \leq n(x_1^\beta) \leq n(x_1^\alpha)$$
   Therefore, $\mathcal{N}(A)^\beta \subseteq \mathcal{N}(A)^\alpha$.
3. If $x \in \mathcal{N}(A)^\alpha$ hence $x \in \mathcal{N}(supp(A))$ and x does not belong to $\mathcal{N}(A)^\beta$, then either $x < n(x_r^\beta)$, which is the minimum of $\mathcal{N}(A)^\beta$, or $x > n(x_1^\beta)$, which is the maximum of $\mathcal{N}(A)^\beta$.
4. As $A \in \mathcal{A}_1^L$, then from theorem 1(of representation of discrete fuzzy numbers), for each $\alpha \in (0,1]$ there exists a real number α' with $0 < \alpha' < \alpha$ such that for each $r \in [\alpha', \alpha]$, $A^\alpha = A^r$. Then
   $$\min(A^r) = \min(A^\alpha) \text{ and } \max(A^r) = \max(A^\alpha)$$
   for each $r \in [\alpha', \alpha]$. Therefore
   $$\min(N(A^r)) = \min(N(A^\alpha))$$
   $$\max(N(A^r)) = \max(N(A^\alpha)) \text{ for each } r \in [\alpha', \alpha]$$
   Hence,
   $$\begin{aligned}\mathcal{N}(A)^\alpha &= \{z \in N(supp(A)) | \min(N(A^\alpha)) \leq z \leq \max(N(A^\alpha))\} \\ &= \{z \in N(supp(A)) | \min(N(A^r)) \leq z \leq \max(N(A^r))\} \\ &= \mathcal{N}(A)^r \text{ for each } r \in [\alpha', \alpha].\end{aligned}$$
   □

Example 1. Let us consider the finite chain $L = \{0,1,2,3,4,5,6,7\}$ and the discrete fuzzy number $A \in \mathcal{A}_1^L$, $A = \{0.3/1, 0.5/2, 0.7/3, 1/4, 0.8/5\}$. Then $\mathcal{N}(A) = \{0.8/2, 1/3, 0.7/4, 0.5/5, 0.3/6\}$.

Proposition 6. *Let us consider the strong negation n on the finite chain $L = \{0, 1, \cdots, m\}$. The mapping*

$$\begin{array}{rl} \mathcal{N} : \mathcal{A}_1^L & \longrightarrow \mathcal{A}_1^L \\ A & \mapsto \mathcal{N}(A) \end{array}$$

where $\mathcal{N}(A)$ is the discrete fuzzy number such that it has as support the sets $[n(x_p^\alpha), n(x_1^\alpha)]$ for each $\alpha \in [0,1]$, being $[x_1^\alpha, x_p^\alpha]$ the α-cuts of A, is a strong negation on the bounded distributive lattice $\mathcal{A}_1^L = (\mathcal{A}_1^L, \min_w, \max_w)$.

Proof. It is obvious from the previous proposition 5, that $\mathcal{N}(A) \in \mathcal{A}_1^L$ because $A \in \mathcal{A}_1^L$ and n is a strong negation on L. Now, we wish to show that $\mathcal{N}$ is a nonincreasing and involutive mapping. For this reason, let us consider $A, B \in \mathcal{A}_1^L$ being $A^\alpha = [x_1^\alpha, x_p^\alpha]$ and $B^\alpha = [y_1^\alpha, y_k^\alpha]$ for each $\alpha \in [0,1]$, their α-cut sets for A and B respectively, with $A \preceq B$. By hypothesis, as $A \preceq B$ from remark 2, this condition implies that $[x_1^\alpha, x_p^\alpha] \leq [y_1^\alpha, y_k^\alpha]$ for each $\alpha \in [0,1]$, i.e. $x_1^\alpha \leq y_1^\alpha$ and $x_p^\alpha \leq y_k^\alpha$ for each $\alpha \in [0,1]$. Now, from remark 5, as n is a strong negation then the inequality $[n(y_k^\alpha), n(y_1^\alpha)] \leq [n(x_p^\alpha), n(x_1^\alpha)]$ holds for each $\alpha \in [0,1]$, i.e. $\mathcal{N}(B) \preceq \mathcal{N}(A)$.

Finally, the involution property of the mapping $\mathcal{N}$ follows from remark 5, because n is a strong negation on L. □

Proposition 7. *Let us consider the strong negation n on the finite chain $L = \{0, \cdots, m\}$ and a divisible discrete t-norm(t-conorm) $T(S)$ on L. If $A, B \in \mathcal{A}_1^L$, the following statements*

i) $\mathcal{S}(\mathcal{N}(A), B) \in \mathcal{A}_1^L$
ii) $\mathcal{S}(\mathcal{N}(A), \mathcal{T}(A, B)) \in \mathcal{A}_1^L$
iii) $\mathcal{S}(\mathcal{T}(\mathcal{N}(A), \mathcal{N}(B)), B) \in \mathcal{A}_1^L$

hold, where $\mathcal{T}(\mathcal{S})$ denote the extension of the t-norm(t-conorm) $T(S)$ to $\mathcal{A}_1^L$ and $\mathcal{N}$ denotes the strong negation considered in proposition 6.

Proof. It is straightforward from theorem 5 and proposition 6. □

5 t-Norms and t-Conorms on $\mathcal{A}_1^L$ Obtained from a Negation

It is well known[15] that if T a t-norm on the finite chain L and n is the strong negation on L then $S_n(x, y) = n(T(n(x), n(y)))$ is a t-conorm on L. And reciprocally, if S is a t-conorm on L and n is the strong negation on L then $T_n(x, y) = n(S(n(x), n(y)))$ is a t-norm on L. A similar result can be obtained in the bounded distributive lattice $\mathcal{A}_1^L$.

Proposition 8. *Let T, S be a divisible t-norm and t-conorm on L respectively. And, let $\mathcal{T}, \mathcal{S}$ be their extensions on $\mathcal{A}_1^L$. If $A, B \in \mathcal{A}_1^L$ then the following statements*

i) $\mathcal{N}(\mathcal{T}(\mathcal{N}(A), \mathcal{N}(B))) \in \mathcal{A}_1^L$
ii) $\mathcal{N}(\mathcal{S}(\mathcal{N}(A), \mathcal{N}(B))) \in \mathcal{A}_1^L$

hold, where $\mathcal{N}$ denotes the strong negation obtained in proposition 6.

Proof. It is straightforward from theorem 5 and proposition 6 because $\mathcal{T}, \mathcal{S}$ and $\mathcal{N}$ are closed operations on $\mathcal{A}_1^L$ for all pair $A, B \in \mathcal{A}_1^L$.

Remark 7. Let $A, B \in \mathcal{A}_1^L$ be with support the sets $supp(A) = \{x_1, \cdots, x_n\}$ and $supp(B) = \{y_1, \cdots, y_q\}$. Let $A^\alpha = \{x_1^\alpha, ..., x_p^\alpha\}$, $B^\alpha = \{y_1^\alpha, ..., y_k^\alpha\}$ be the α-cuts for $A, B \in \mathcal{A}_1^L$ respectively. Then for each $\alpha \in [0, 1]$,

$$\mathcal{N}(\mathcal{T}(\mathcal{N}(A), \mathcal{N}(B))^\alpha = \{z \in N(supp(\mathcal{T}(\mathcal{N}(A), \mathcal{N}(B)))) \text{ such that}$$
$$\min(N(\mathcal{T}(\mathcal{N}(A), \mathcal{N}(B))^\alpha)) \leq z \leq \max(N(\mathcal{T}(\mathcal{N}(A), \mathcal{N}(B))^\alpha))\} =$$
$$\{z \in [nT(n(x_1), n(y_1)), nT(n(x_n), n(y_q))] \text{ such that}$$
$$nT(n(x_1^\alpha), n(y_1^\alpha)) \leq z \leq nT(n(x_p^\alpha), n(y_k^\alpha))\} =$$

(If $\overline{S}$ is the dual t-conorm of T, then we know [15] that $nT(n(x), n(y))) = \overline{S}(x, y)$)

$$\{z \in [\overline{S}(x_1, y_1), \overline{S}(x_n, y_q)] \text{ such that } \overline{S}(x_1^\alpha, y_1^\alpha) \leq z \leq \overline{S}(x_p^\alpha, y_k^\alpha)\} = \overline{\mathcal{S}}(A, B)^\alpha$$

where $\overline{\mathcal{S}}$ denotes the extension of $\overline{S}$ on $\mathcal{A}_1^L$. Analogously,

$$\mathcal{N}(\mathcal{S}(\mathcal{N}(A), \mathcal{N}(B))^\alpha) =$$
$$\{z \in [\overline{T}(x_1, y_1), \overline{T}(x_n, y_q)] \text{ such that } \overline{T}(x_1^\alpha, y_1^\alpha) \leq z \leq \overline{T}(x_p^\alpha, y_k^\alpha)\} = \overline{\mathcal{T}}(A, B)^\alpha$$

where $\overline{\mathcal{T}}$ denotes the extension of $\overline{T}$ on $\mathcal{A}_1^L$.

Theorem 6. *Let S be a divisible t-conorm on L and let $\mathcal{S}$ be its extension on $\mathcal{A}_1^L$. Let n be, the strong negation on L. The binary operation*

$$\begin{aligned} \mathcal{T}_\mathcal{N} : \mathcal{A}_1^L \times \mathcal{A}_1^L &\to \mathcal{A}_1^L \\ (A, B) &\longmapsto \mathcal{T}_\mathcal{N}(A, B) \end{aligned}$$

where $\mathcal{T}_\mathcal{N}(A, B) = \mathcal{N}(\mathcal{S}(\mathcal{N}(A), \mathcal{N}(B)))$is a t-norm on the bounded set $\mathcal{A}_1^L$,which will be called the dual t-norm *of $\mathcal{S}$ w.r.t the strong negation $\mathcal{N}$. Analogously, if T is a divisible t-norm on L then the binary operation*

$$\begin{aligned} \mathcal{S}_\mathcal{N} : \mathcal{A}_1^L \times \mathcal{A}_1^L &\to \mathcal{A}_1^L \\ (A, B) &\longmapsto \mathcal{S}_\mathcal{N}(A, B) \end{aligned}$$

where $\mathcal{S}_\mathcal{N}(A, B) = \mathcal{N}(\mathcal{T}(\mathcal{N}(A), \mathcal{N}(B)))$is a t-conorm on the bounded set $\mathcal{A}_1^L$, which will be called the dual t-conorm *of $\mathcal{T}$ w.r.t the strong negation $\mathcal{N}$.*

Proof. From proposition 8, $\mathcal{A}_1^L$ is closed under the binary operation $\mathcal{T}_\mathcal{N}$. Now, according to remark 7, the binary operation $\mathcal{T}_\mathcal{N}$ is a t-conorm on $\mathcal{A}_1^L$.

Analogously we can see that $\mathcal{S}_\mathcal{N}$ is a t-conorm on $\mathcal{A}_1^L$. □

Acknowledgments. We would like to express our thanks to anonymous reviewers who have contributed to improve this article. This work has been partially supported by the MTM2009-10962 project grant.

References

1. De Baets, B., Mesiar, R.: Triangular norms on product lattices. Fuzzy Sets and Systems 104, 61–75 (1999)
2. Bustince, H., Kacprzyk, J., Mohedano, V.: Intiutionistic Fuzzy Sets. Application to Intuitionistic Fuzzy Complementation. Fuzzy Sets and Systems 114, 485–504 (2000)
3. Casasnovas, J., Riera, J.V.: On the addition of discrete fuzzy numbers. WSEAS Transactions on Mathematics, 549–554 (2006)
4. Casasnovas, J., Riera, J.V.: Discrete fuzzy numbers defined on a subset of natural numbers. In: Castillo, O., Melin, P., Montiel Ross, O., Sepúlveda Cruz, R., Pedrycz, W., Kacprzyk, J. (eds.) Theoretical Advances and Applications of Fuzzy Logic and Soft Computing: Advances in Soft Computing, vol. 42, pp. 573–582. Springer, Heidelberg (2007)
5. Casasnovas, J., Riera, J.V.: Maximum and minimum of discrete fuzzy numbers. In: Angulo, C., Godo, L. (eds.) Frontiers in Artificial Intelligence and Applications: artificial intelligence research and development, vol. 163, pp. 273–280. IOS Press, Amsterdam (2007)
6. Casasnovas, J., Riera, J.V.: Lattice properties of discrete fuzzy numbers under extended min and max. In: Proceedings IFSA-EUSFLAT, Lisbon, pp. 647–652 (2009)
7. Casasnovas, J., Riera, J.V.: Extension of discrete t-norms and t-conorms to discrete fuzzy numbers. In: Fifth international summer school on aggregation operators (AGOP 2009), Palma de Mallorca, pp. 77–82 (2009)
8. Casasnovas, J., Riera, J.V.: Triangulars norms and conorms on the set of discrete fuzzy numbers. Accepted IPMU (2010)
9. Esteva, F., Domingo, X.: Sobre funciones de negación en [0,1]. Sthocastica 4(2), 144–166 (1980)
10. Deschrijver, G., Cornelis, C., Kerre, E.E.: Implication in intuitionistic fuzzy and interval-valued fuzzy set theory: construction, classification, application. Internat. J. Approx. Reason 35(1), 55–95 (2004)
11. Esteva, F.: Negaciones en la teoría de conjuntos difusos. Sthocastica V, 33–44 (1981)
12. Fodor, J.C.: Smooth associative operations on finite ordinals scales. IEEE Trans. on Fuzzy Systems 8, 791–795 (2000)
13. Jenei, S.: A more efficient method for defining fuzzy connectives. Fuzzy Sets and Systems 90, 25–35 (1997)
14. Klir, G., Bo, Y.: Fuzzy sets and fuzzy logic. Theory and applications. Prentice Hall, Englewood Cliffs (1995)
15. Mayor, G., Torrens, J.: Triangular norms on discrete settings. In: Klement, E.P., Mesiar, R. (eds.) Logical, Algebraic, Analytic, and Probabilistic Aspects of Triangular Norms, pp. 189–230. Elsevier, Amsterdam (2005)
16. Schweizer, B., Sklar, A.: Associative functions and statistical triangle inequalities. Publ. Math. Debrecen 8, 169–186 (1961)
17. Trillas, E.: Sobre funciones de negación en la teoría de los subconjuntos borrosos. Stochastica III-1, 47–59 (1979)
18. Voxman, W.: Canonical representations of discrete fuzzy numbers. Fuzzy Sets and Systems 54, 457–466 (2001)
19. Wang, G., Wu, C., Zhao, C.: Representation and Operations of discrete fuzzy numbers. Southeast Asian Bulletin of Mathematics 28, 1003–1010 (2005)

Trapezoidal Approximation of Fuzzy Numbers Based on Sample Data

Przemysław Grzegorzewski[1,2]

[1] Systems Research Institute, Polish Academy of Sciences
ul. Newelska 6, 01-447 Warsaw, Poland
[2] Faculty of Mathematics and Information Science, Warsaw University of Technology
Plac Politechniki 1, 00-661 Warsaw, Poland
pgrzeg@ibspan.waw.pl, pgrzeg@mini.pw.edu.pl
http://www.ibspan.waw.pl/~pgrzeg

Abstract. The idea of the membership functions construction form a data sample is suggested. The proposed method is based on the trapezoidal approximation of fuzzy numbers.

Keywords: fuzzy numbers, membership function, trapezoidal approximation.

1 Introduction

Fuzzy set theory provides tools for dealing with imprecise measurements or concepts expressed in natural language. These tools enable us not only to represent these imprecise or vague objects but also to manipulate them in may of ways and for various purposes.

The lack of precision is mathematically expressed by the use of membership functions which describe fuzzy sets. A membership function may be perceived as a generalization of characteristic function that assumes not only binary values of 1 and 0 corresponding to membership or nonmembership, respectively, but admits also intermediate values for partial membership or gradual possibility.

Since fuzzy set represent imprecise objects, their membership functions may differ from person to person even under the same circumstances.

Although the problem of constructing membership functions that capture adequately the meanings of imprecise terms employed in a particular application is not a problem of fuzzy theory per se but belongs to much more general area of knowledge acquisition, it is also very important for further tasks performed within the framework of fuzzy set theory like information processing and data management including effective data representation and necessary calculations.

Numerous methods for constructing membership functions have been described in the literature. All these methods may be classified into direct or indirect methods both further classified to methods that involve one expert or require multiple experts (see [11]).

If the universe of discourse X is discrete an expert is expected to assign to each given element $x \in X$ its membership grade $\mu_A(x)$ that, according to his or

E. Hüllermeier, R. Kruse, and F. Hoffmann (Eds.): IPMU 2010, Part II, CCIS 81, pp. 402–411, 2010.

her opinion, best captures the meaning of the linguistic term represented by the fuzzy set A.

However, if $X = \mathbb{R}$ (or any other continuous universe of discourse) the problem of constructing membership functions can be solved by either defining the membership function completely in terms of a justifiable mathematical formula or exemplifying it for some selected elements, which often can be treated as sample data.

All these approaches when membership grades are assigned directly by a single expert or aggregated for the opinion poll, based on rankings or deduced from some information available, are called subjective. There exist also, so-called, objective approaches when membership degrees are derived with help of mathematical statistics or assigned according to some rules derived from control theory methods. Other objective methods utilize neural networks as a part of a neuro-fuzzy modelling system or genetic/evolutionary algorithms (initially chosen parameters are changed by applying special optimization techniques).

In this paper we consider just a situation when a sample of n data points described by a set of ordered pairs $\{(x_i, a_i) : i = 1, \ldots, N\}$ is given, where $a_i \in [0, 1]$ denotes a grade of membership of $x_i \in \mathbb{R}$ in a fuzzy set A for each $i = 1, \ldots, N$. Further on a data set $\{(x_i, a_i) : i = 1, \ldots, N\}$ is used for constructing a membership function μ_A of A. Traditionally an appropriate curve-fitting method is applied. This method requires a suitable class of functions (triangular, trapezoidal, S-shaped, bell-shaped, etc.) chosen with respect to the opinion of an expert, based on some theory, previous experience, or experimental comparison with other classes.

Below we propose another approach to membership function construction from sample data designed and specialized for fuzzy numbers. The origin of our method goes back to approximations of fuzzy numbers, especially to trapezoidal approximation.

2 Fuzzy Numbers and Trapezoidal Approximations

Let A denote a fuzzy number, i.e. such fuzzy subset A of the real line $\mathbb{R}$ with membership function $\mu_A : \mathbb{R} \to [0, 1]$ which is (see [4]): normal (i.e. there exist an element x_0 such that $\mu_A(x_0) = 1$), fuzzy convex (i.e. $\mu_A(\lambda x_1 + (1 - \lambda)x_2) \geq \mu_A(x_1) \wedge \mu_A(x_2)$, $\forall x_1, x_2 \in \mathbb{R}$, $\forall \lambda \in [0, 1]$), μ_A is upper semicontinuous, supp$A$ is bounded, where supp$A = cl(\{x \in \mathbb{R} : \mu_A(x) > 0\})$, and cl is the closure operator. A space of all fuzzy numbers will be denoted by $\mathbb{F}(\mathbb{R})$.

Moreover, let $A_\alpha = \{x \in \mathbb{R} : \mu_A(x) \geq \alpha\}$, $\alpha \in (0, 1]$, denote an α-cut of a fuzzy number A. As it is known, every α-cut of a fuzzy number is a closed interval, i.e. $A_\alpha = [A_L(\alpha), A_U(\alpha)]$, where $A_L(\alpha) = \inf\{x \in \mathbb{R} : \mu_A(x) \geq \alpha\}$ and $A_U(\alpha) = \sup\{x \in \mathbb{R} : \mu_A(x) \geq \alpha\}$.

For two arbitrary fuzzy numbers A and B with α-cuts $[A_L(\alpha), A_U(\alpha)]$ and $[B_L(\alpha), B_U(\alpha)]$, respectively, the quantity

$$d(A,B) = \left(\int_0^1 [A_L(\alpha) - B_L(\alpha)]^2 d\alpha + \int_0^1 [A_U(\alpha) - B_U(\alpha)]^2 d\alpha \right)^{1/2} \tag{1}$$

is the distance between A and B (for more details we refer the reader to [5]).

It is obvious that the results of our calculations on fuzzy numbers strongly depend on the shape of the membership functions of these numbers. In particular, less regular membership functions lead to more complicated calculations. Additionally, fuzzy numbers with simpler shape of membership functions often have more intuitive and more natural interpretation. This is the reason that approximation methods for simplifying original membership functions fuzzy numbers are of interest.

A sufficiently effective simplification of a membership function can be reached by the piecewise linear curves leading to triangle, trapezoidal or orthogonal membership curves. These three mentioned shapes are particular cases of the so-called trapezoidal membership function defined as

$$\mu(x) = \begin{cases} 0 & \text{if} \quad x < t_1, \\ \frac{x-t_1}{t_2-t_1} & \text{if} \quad t_1 \le x < t_2, \\ 1 & \text{if} \quad t_2 \le x \le t_3, \\ \frac{t_4-x}{t_4-t_3} & \text{if} \quad t_3 < x \le t_4, \\ 0 & \text{if} \quad t_4 < x, \end{cases} \tag{2}$$

where $t_1, t_2, t_3, t_4 \in \mathbb{R}$ and $t_1 \le t_2 \le t_3 \le t_4$. A family of all trapezoidal fuzzy numbers will be denoted by $\mathbb{F}^T(\mathbb{R})$.

By (2) any trapezoidal fuzzy number is completely described by four real numbers $t_1 \le t_2 \le t_3 \le t_4$ that are borders of its support and core. Naturally, it is much easier to process and manage such simpler objects. And this is just the main reason that so many researchers are interested in trapezoidal approximations (see, e.g. [1,2,3,6,7,8,9,10,13]).

The matter of such a trapezoidal approximation consists in finding an appropriate approximation operator $T : \mathbb{F}(\mathbb{R}) \to \mathbb{F}^T(\mathbb{R})$ which produces a trapezoidal fuzzy number $T(A)$ closest to given original fuzzy number A with respect to distance (1), i.e.

$$d(A,T(A)) = \left(\int_0^1 [A_L(\alpha) - T(A)_L(\alpha)]^2 d\alpha + \int_0^1 [A_U(\alpha) - T(A)_U(\alpha)]^2 d\alpha \right)^{1/2} . \tag{3}$$

Since the membership function of $T(A) \in \mathbb{F}^T(\mathbb{R})$ is given by (2) hence the α-cuts of $T(A)$ have a following form $(T(A))_\alpha = [t_1 + \alpha(t_2 - t_1), t_4 - \alpha(t_4 - t_3)]$ so the equation (3) reduces to

$$d(A,T(A)) = \left(\int_0^1 [t_1 + \alpha(t_2 - t_1) - A_L(\alpha)]^2 d\alpha \right. \tag{4}$$
$$\left. + \int_0^1 [t_4 - \alpha(t_4 - t_3) - A_U(\alpha)]^2 d\alpha \right)^{1/2} .$$

Additionally we may demand some requirements the operator T should fulfil which warrant that our approximation would possess some desired properties, like preservation of some fixed parameters or relations, continuity, etc.

It seems that the idea of the trapezoidal approximation indicated above might be also fruitfully applied in membership function determination from sample data.

3 Constructing Trapezoidal Membership Functions

First of all let us realize that trapezoidal fuzzy numbers are quite sufficient in most of practical situations because of their simplicity in calculations, interpretation and computer implementation (see, e.g., [12]). These arguments are also valid when we consider the problem of constructing a membership function. It is simply reasonable to leave some room for deviations in estimating membership functions since small variations in membership functions must not, as a result of our calculations, turn into major differences. Therefore, further on we will restrict our attention to the determination of a trapezoidal fuzzy number from sample data.

Suppose that our sample data given by $S = \{(x_i, a_i) : i = 1, \ldots, N\}$ are perceptions of the unknown fuzzy number A. In other words, the true shape of the membership function μ_A describing A is not known and the only information about μ_A is delivered by N sample points. Nevertheless we still want to approximate this fuzzy number A by the nearest trapezoidal fuzzy number $T(A)$, as it has been discussed in previous section. Of course, the minimization of the distance $d(A, T(A))$ given by (3) cannot be performed because of the ignorance of the membership function. But as a matter of fact we may try to find $T(A)$ solving a slightly modified optimization problem, i.e. through the minimization of $F(\widehat{A}(S), T(A))$, where $\widehat{A}(S)$ is a counterpart of A based on the data set S, while F is a discrete version of the distance (4). Before defining F we have to introduce some natural assumptions related to the fact that we consider not arbitrary fuzzy sets but fuzzy numbers.

Definition 1. *A data set $S = \{(x_i, a_i) : x_i \in \mathbb{R}, 0 \leq a_i \leq 1, i = 1, \ldots, N\}$, where $N \geq 3$, is called* ***proper*** *if it contains at least three elements $(x_i, 0), (x_j, 1)$ and $(x_k, 0)$ such that $x_i < x_j < x_k$.*

Thus the data set is proper if and only if it contains both points with full membership and points with full nonmembership grades. For any proper data set we can find following four values:

$$w_2 = \min_{i=1,\ldots,N} \{x_i : a_i = 1\}, \tag{5}$$

$$w_3 = \max_{i=1,\ldots,N} \{x_i : a_i = 1\}, \tag{6}$$

$$w_1 = \max_{i=1,\ldots,N} \{x_i : x_i < w_2, a_i = 0\}, \tag{7}$$

$$w_4 = \min_{i=1,\ldots,N} \{x_i : x_i > w_3, a_i = 0\}. \tag{8}$$

Let us also adopt a following definition that characterize these data sets which could be nicely approximated by fuzzy numbers.

Definition 2. *A data set* $S = \{(x_i, a_i) : x_i \in \mathbb{R}, 0 \le a_i \le 1, i = 1, \ldots, N\}$ *is called* ***regular*** *if it is proper and the following conditions are satisfied:*
(a) $a_i = 1$ *if and only if* $w_2 \le x_i \le w_3$
(b) $a_i = 0$ *if and only if* $x_i \le w_1$ *or* $x_i \ge w_4$.

Such assumptions as required for a data set to be regular are not too restrictive in the case of fuzzy numbers since according to the definition for each fuzzy number A one can indicate both points that surely belong to A and points that surely do not belong to A (which is guaranteed by the normality and the bounded support of a fuzzy number, respectively). However, it seems that the trapezoidal approximation would be also justified for slightly relax criteria of so-called ϵ-regularity.

Definition 3. *A data set* $S = \{(x_i, a_i) : x_i \in \mathbb{R}, 0 \le a_i \le 1, i = 1, \ldots, N\}$ *is called* ***ϵ-regular*** *if it is proper and there exist such* $\epsilon \in (0, \frac{1}{2})$ *that the following conditions are satisfied:*
(a) $a_i \in [1-\epsilon, 1]$ *for* $w_2 \le x_i \le w_3$
(b) $a_i \in [0, \epsilon]$ *for* $x_i \le w_1$ *or* $x_i \ge w_4$.

Each regular data set is, of course, ϵ-regular data set with $\epsilon = 0$.

For further calculations we need two subsets S^* and S^{**} of the initial data set S, including these observations which are crucial for estimating the left arm and the right arm of a fuzzy number, respectively. Thus we have

$$S^* = \{(x_i, a_i) \in S : w_1 \le x_i \le w_2\}, \tag{9}$$
$$S^{**} = \{(x_i, a_i) \in S : w_3 \le x_i \le w_4\}. \tag{10}$$

Let $I = \{i : (x_i, a_i) \in S^*\}$ denote the set of the indices of all those observations that belong to S^* while $J = \{i : (x_i, a_i) \in S^{**}\}$ denote the set of the indices of all those observations that belong to S^{**}. Let us also assume that $\#S^* = n$ and $\#S^{**} = m$, where $\#$ stands for the cardinality of a set. It is easily seen that if our data set S is proper then both S^* and S^{**} are not empty. For each proper data set $n \ge 2$ and $m \ge 2$.

Now we are ready for defining function F announced above. Namely, let

$$F(t_1, t_2, t_3, t_4) = \sum_{i \in I} [t_1 + a_i(t_2 - t_1) - x_i]^2 + \sum_{j \in J} [t_4 - a_j(t_4 - t_3) - x_j]^2. \tag{11}$$

It is easily seen that (11) is a natural discrete counterpart of the square of the distance (4) where instead of integration we have summation over those α-cuts which in the data set. This function, considered as a function of parameters t_1, t_2, t_3, t_4, may represent a loss that happen when A (not known exactly but represented by data set S) is approximated by a trapezoidal fuzzy number T characterized by four real numbers t_1, t_2, t_3, t_4. Therefore, to obtain a good trapezoidal approximation we have to minimize this loss (the distance). However, if we like a satisfying approximation we should also remember about some additional constraints on some parameters. Finally we may describe our optimization problem as follows:

$$F(t_1, t_2, t_3, t_4) \longrightarrow \min \tag{12}$$

$$\text{subject to} \quad w_1 \leq t_1 \leq t_2 \leq w_2, \tag{13}$$

$$w_3 \leq t_3 \leq t_4 \leq w_4. \tag{14}$$

Please note, that the constraints (13) and (14) correspond to natural expectation that having regular data set the core of the approximation will contain all the data points from the sample that surely belong to the concept described by the fuzzy number and simultaneously the data points from the sample that surely does not belong to that concept will be left outside the support, i.e. $[w_2, w_3] \subseteq \text{core}T(A)$ and $\text{supp}T(A) \subseteq [w_1, w_4]$. For any regular or ϵ-regular data set we also have $\{x_i : a_i = 1\} \in \text{core}T(A)$ and $\{x_i : a_i = 0\} \notin \text{supp}T(A)$.

The requested trapezoidal fuzzy number $T(A) = T(t_1, t_2, t_3, t_4)$ which minimizes the loss function (12) with respect to constraints (13) and (14) will be called the *the trapezoidal fuzzy number nearest to the data set S*.

4 Main Result

In this section we present the solution of the optimization problem stated in the previous section and show the sketch of the proof.

Theorem 1. *Let $S = \{(x_i, a_i) : x_i \in \mathbb{R}, 0 \leq a_i \leq 1, i = 1, \ldots, N\}$ denote at least ϵ-regular sample data set. Then the left arm of the trapezoidal fuzzy number $T(A) = T(t_1, t_2, t_3, t_4)$ nearest to S is defined by t_1 and t_2 given as follows:*

(a) if

$$w_1 + \frac{\sum\limits_{i \in I} (x_i - w_1)(1 - a_i)}{\sum\limits_{i \in I} a_i(1 - a_i)} < w_2 < w_1 + \frac{\sum\limits_{i \in I} (x_i - w_1) a_i}{\sum\limits_{i \in I} a_i^2} \tag{15}$$

then

$$t_1 = w_1 \tag{16}$$

$$t_2 = w_2 \tag{17}$$

(b) if

$$w_1 > \frac{(\sum\limits_{i \in I} x_i)(\sum\limits_{i \in I} a_i^2) - (\sum\limits_{i \in I} x_i a_i)(\sum\limits_{i \in I} a_i)}{n \sum\limits_{i \in I} a_i^2 - (\sum\limits_{i \in I} a_i)^2} \tag{18}$$

then

$$t_1 = w_1 \tag{19}$$

$$t_2 = w_1 + \frac{\sum\limits_{i \in I} (x_i - w_1) a_i}{\sum\limits_{i \in I} a_i^2} \tag{20}$$

(c) if

$$w_2 < \frac{(\sum_{i\in I} x_i a_i)(n - \sum_{i\in I} a_i) - (\sum_{i\in I} x_i) \sum_{i\in I} a_i(1-a_i)}{n \sum_{i\in I} a_i^2 - (\sum_{i\in I} a_i)^2} \tag{21}$$

then

$$t_1 = w_2 - \frac{\sum_{i\in I} (w_2 - x_i)(1-a_i)}{\sum_{i\in I} (1-a_i)^2} \tag{22}$$

$$t_2 = w_2 \tag{23}$$

(d) otherwise

$$t_1 = \frac{(\sum_{i\in I} x_i)(\sum_{i\in I} a_i^2) - (\sum_{i\in I} x_i a_i)(\sum_{i\in I} a_i)}{n \sum_{i\in I} a_i^2 - (\sum_{i\in I} a_i)^2} \tag{24}$$

$$t_2 = \frac{(\sum_{i\in I} x_i a_i)(n - \sum_{i\in I} a_i) - (\sum_{i\in I} x_i) \sum_{i\in I} a_i(1-a_i)}{n \sum_{i\in I} a_i^2 - (\sum_{i\in I} a_i)^2}. \tag{25}$$

The right arm of that trapezoidal fuzzy number $T(A) = T(t_1, t_2, t_3, t_4)$ nearest to S is described by t_3 and t_4 given as follows:

(a) if

$$w_4 - \frac{\sum_{j\in J} (w_4 - x_j) a_j}{\sum_{j\in J} a_j^2} < w_3 < w_4 - \frac{\sum_{j\in J} (w_4 - x_j)(1-a_j)}{\sum_{j\in J} a_j(1-a_j)} \tag{26}$$

then

$$t_3 = w_3 \tag{27}$$

$$t_4 = w_4 \tag{28}$$

(b) if

$$w_4 < \frac{(\sum_{j\in J} x_j)(\sum_{j\in J} a_j^2) - (\sum_{j\in J} x_j a_j)(\sum_{j\in J} a_j)}{n \sum_{j\in J} a_j^2 - (\sum_{j\in J} a_j)^2} \tag{29}$$

then

$$t_3 = w_4 - \frac{\sum_{j\in J} (w_4 - x_j) a_j}{\sum_{j\in J} a_j^2} \tag{30}$$

$$t_4 = w_4 \tag{31}$$

(c) if

$$w_3 > \frac{(\sum_{j\in J} x_j a_j)(m - \sum_{j\in J} a_j) - (\sum_{j\in J} x_j) \sum_{j\in J} a_j(1-a_j)}{m \sum_{j\in J} a_j^2 - (\sum_{j\in J} a_j)^2} \tag{32}$$

then

$$t_3 = w_3 \tag{33}$$

$$t_4 = w_3 + \frac{\sum_{j\in J} (x_j - w_3)(1-a_j)}{\sum_{j\in J} (1-a_j)^2} \tag{34}$$

(d) otherwise

$$t_3 = \frac{(\sum_{j\in J} x_j a_j)(m - \sum_{j\in J} a_j) - (\sum_{j\in J} x_j) \sum_{j\in J} a_j(1-a_j)}{m \sum_{j\in J} a_j^2 - (\sum_{j\in J} a_j)^2} \tag{35}$$

$$t_4 = \frac{(\sum_{j\in J} x_j)(\sum_{j\in J} a_j^2) - (\sum_{j\in J} x_j a_j)(\sum_{j\in J} a_j)}{m \sum_{j\in J} a_j^2 - (\sum_{j\in J} a_j)^2}. \tag{36}$$

Proof: One may easily observe that F is a sum of the two nonnegative summands with separated arguments and the constraints are separated too. Hence we may substitute the original optimization problem into two simpler ones, corresponding to the left and the right arm of a fuzzy number, respectively. Thus we have

$$F^*(t_1, t_2) = \sum_{i\in I} [t_1 + a_i(t_2 - t_1) - x_i]^2 \longrightarrow \min \tag{37}$$

subject to

$$\mathbf{g}^*(t_1, t_2) = [w_1 - t_1, t_1 - t_2, t_2 - w_2] \le 0 \tag{38}$$

and

$$F^{**}(t_3, t_4) = \sum_{j\in J} [t_4 - a_j(t_4 - t_3) - x_j]^2 \longrightarrow \min \tag{39}$$

subject to

$$\mathbf{g}^{**}(t_3, t_4) = [w_3 - t_3, t_3 - t_4, t_4 - w_4] \le 0. \tag{40}$$

Let us start from the first problem (37)-(38). By the Karush-Kuhn-Tucker theorem, if (t_1^*, t_2^*) is a local minimizer for the problem of minimizing F^* subject to $\mathbf{g}^*(t_1, t_2) \le 0$ then there exist the Karush-Kuhn-Tucker multiplier $\eta = (\eta_1, \eta_2, \eta_3)$ such that

$$\begin{aligned} DF^*(t_1^*, t_2^*) + \eta^T D\mathbf{g}^*(t_1^*, t_2^*) &= \mathbf{0}^T, \\ \eta^T \mathbf{g}^*(t_1^*, t_2^*) &= 0, \\ \eta &\ge 0. \end{aligned}$$

After some calculations we get

$$DF^*(t_1,t_2)=\left[2\sum_{i\in I}[t_1+a_i(t_2-t_1)-x_i](1-a_i), 2\sum_{i\in I}[t_1+a_i(t_2-t_1)-x_i]a_i\right],$$

$$D\mathbf{g}^*(t_1,t_2)=\begin{bmatrix}-1 & 0\\ 1 & -1\\ 0 & 1\end{bmatrix}.$$

Therefore, we can rewrite the Karush-Kuhn-Tucker conditions in a following way

$$(n-2\sum_{i\in I}a_i+\sum_{i\in I}a_i^2)t_1+(\sum_{i\in I}a_i-\sum_{i\in I}a_i^2)t_2-\sum_{i\in I}x_i+\sum_{i\in I}x_ia_i-\frac{\eta_1-\eta_2}{2}=0,$$

$$(\sum_{i\in I}a_i-\sum_{i\in I}a_i^2)t_1+t_2\sum_{i\in I}a_i^2-\sum_{i\in I}x_ia_i-\frac{\eta_2-\eta_3}{2}=0,$$

$$\eta_1(w_1-t_1)=0,$$

$$\eta_2(t_1-t_2)=0,$$

$$\eta_3(t_2-w_2)=0,$$

$$\eta_1\geq 0,$$

$$\eta_2\geq 0,$$

$$\eta_3\geq 0.$$

To find points that satisfy the above conditions, we should consider eight possible situations related to vector η. Some careful but tedious calculations show that if $\eta_1>0$, $\eta_2=0$, $\eta_3>0$ then as a solution we obtain (16)-(17). Situation $\eta_1>0$, $\eta_2=\eta_3=0$ leads to (19)-(20) while for $\eta_1=\eta_2=0$, $\eta_3>0$ we get (22)-(23). Finally, $\eta_1=\eta_2=\eta_3=0$ produces (24)-(25). In all other cases, i.e. $\eta_1=\eta_3=0$, $\eta_2>0$; $\eta_1>0$, $\eta_2>0$, $\eta_3=0$; $\eta_1=0$, $\eta_2>0$, $\eta_3>0$ and $\eta_1>0$, $\eta_2>0$, $\eta_3>0$ there are no solutions.

Now we have to verify that all our solutions $\mathbf{t}=(t_1,t_2)$ satisfy the second-order sufficient conditions. For this we form a matrix $\Psi(\mathbf{t},\eta)=D^2F^*(\mathbf{t})+\eta D^2\mathbf{g}^*(\mathbf{t})$. One check easily that for all our solutions $\mathbf{t}$ we have $\mathbf{y}^T\Psi(\mathbf{t},\eta)\mathbf{y}>0$ for all vectors $\mathbf{y}$ in the tangent space to the surface defined by active constraints, i.e. $\{\mathbf{y}: D^2\mathbf{g}^*(\mathbf{t})\mathbf{y}=0\}$. Therefore, we conclude that we have received four different solutions t_1^*, t_2^* for the problem of minimizing F^* subject to $\mathbf{g}^*(t_1,t_2)\leq 0$.

Nearly identical reasoning leads to four different solutions t_3^*, t_4^* for the problem of minimizing F^{**} subject to some constraints described above. This completes the proof.

5 Conclusions

In this paper we have suggested a new method for constructing membership function based on sample data. The general idea of the proposed approach goes back to the trapezoidal approximation of fuzzy numbers.

We have shown, that depending on a data set we obtain one of the four possible left arms and also one of the four possible right arms of the final trapezoidal fuzzy number. It does not exclude that as a result we get a triangular fuzzy number (what can happen if $t_2 = t_3$).

Although in the paper we have utilized a single data set that correspond to situation typical for a single expert, our approach might be also generalized for the multiple experts problem. In that case we have to aggregate information delivered by several data sets, say $S_1, \ldots, S_k$, and then follow the steps shown in the paper.

It is worth mentioning that the suggested approach could be applied not only for the classical fuzzy numbers but also for one-sided fuzzy numbers (see [5]) which are sometimes of interest (e.g. in possibility theory).

References

1. Abbasbandy, S., Asady, B.: The nearest approximation of a fuzzy quantity in parametric form. Applied Mathematics and Computation 172, 624–632 (2006)
2. Abbasbandy, S., Amirfakhrian, M.: The nearest trapezoidal form of a generalized LR fuzzy number. International Journal of Approximate Reasoning 43, 166–178 (2006)
3. Ban, A.: Approximation of fuzzy numbers by trapezoidal fuzzy numbers preserving the expected interval. Fuzzy Sets and Systems 159, 1327–1344 (2008)
4. Dubois, D., Prade, H.: Operations on fuzzy numbers. Int. J. Syst. Sci. 9, 613–626 (1978)
5. Grzegorzewski, P.: Metrics and orders in space of fuzzy numbers. Fuzzy Sets and Systems 97, 83–94 (1998)
6. Grzegorzewski, P.: Trapezoidal approximations of fuzzy numbers preserving the expected interval - algorithms and properties. Fuzzy Sets and Systems 159, 1354–1364 (2008)
7. Grzegorzewski, P.: New algorithms for trapezoidal approximation of fuzzy numbers preserving the expected interval. In: Magdalena, L., Ojeda-Aciego, M., Verdegay, J.L. (eds.) Proceedings of the Twelfth International Conference on Information Processing and Management of Uncertainty in Knowledge-Based Systems, IPMU 2008, Spain, Torremolinos, Málaga, pp. 117–123 (2008)
8. Grzegorzewski, P.: Algorithms for trapezoidal approximations of fuzzy numbers preserving the expected interval. In: Bouchon-Meunier, B., Magdalena, L., Ojeda-Aciego, M., Verdegay, J.-L., Yager, R.R. (eds.) Foundations of Reasoning under Uncertainty, pp. 85–98. Springer, Heidelberg (2010)
9. Grzegorzewski, P., Mrówka, E.: Trapezoidal approximations of fuzzy numbers. Fuzzy Sets and Systems 153, 115–135 (2005)
10. Grzegorzewski, P., Mrówka, E.: Trapezoidal approximations of fuzzy numbers - revisited. Fuzzy Sets and Systems 158, 757–768 (2007)
11. Klir, G.J., Yuan, B.: Fuzzy Sets and Fuzzy Logic. Theory and Applications. Prentice Hall, Englewood Cliffs (1995)
12. Pedrycz, W.: Why triangular membership functions? Fuzzy Sets and Systems 64, 21–30 (1994)
13. Yeh, C.T.: Trapezoidal and triangular approximations preserving the expected interval. Fuzzy Sets and Systems 159, 1345–1353 (2008)

Multiple Products and Implications in Interval-Valued Fuzzy Set Theory

Glad Deschrijver

Fuzziness and Uncertainty Modelling Research Unit,
Department of Applied Mathematics and Computer Science,
Ghent University, B–9000 Gent, Belgium
Glad.Deschrijver@UGent.be
http://www.fuzzy.ugent.be

Abstract. When interval-valued fuzzy sets are used to deal with uncertainty, using a single t-norm to model conjunction and a single implication leads to counterintuitive results. Therefore it is necessary to look beyond the traditional structures such as residuated lattices, and to investigate whether these structures can be extended using more than one product and implication. In this paper we will investigate under which conditions a number of properties that are valid in a residuated lattice are still valid when different products and implications are used.

1 Introduction

Fuzzy set theory is a valuable tool for problems that have to deal with imprecision or vagueness. However, it is not so appropriate to deal with situations in which the membership degree is uncertain. Interval-valued fuzzy set theory [1,2] is an extension of fuzzy set theory in which to each element of the universe a closed subinterval of the unit interval is assigned which approximates the unknown membership degree. Another extension of fuzzy set theory is intuitionistic fuzzy set theory introduced by Atanassov [3]. In Atanassov's intuitionistic fuzzy set theory together with the membership degree a degree of non-membership is given; this allows to model information both in favour and in disfavour of the inclusion of an element in a set. In [4] it is shown that Atanassov's intuitionistic fuzzy set theory is mathematically equivalent to interval-valued fuzzy set theory and that both are equivalent to L-fuzzy set theory in the sense of Goguen [5] w.r.t. a special lattice $\mathcal{L}^I$.

Triangular norms (t-norms for short) are often classified based on the properties they satisfy (see e.g. [6]). From a logical point of view, a particularly useful property for a t-norm T on a bounded lattice $(L, \leq)$ is the residuation principle, i.e. the existence of an implication I_T satisfying $T(x, y) \leq z$ iff $x \leq I_T(y, z)$ for all x, y and z in L; the corresponding structure is a residuated lattice. In [7,8,9,10] an extension of residuated lattices in interval-valued fuzzy set theory called interval-valued residuated lattices are investigated; these are residuated lattices on the set of closed intervals of a bounded lattice such that the set of trivial intervals (intervals with only one element) is closed under the product

E. Hüllermeier, R. Kruse, and F. Hoffmann (Eds.): IPMU 2010, Part II, CCIS 81, pp. 412–419, 2010.

and implication. In [9] the interval-valued residuated lattices based on the unit interval are completely characterized in terms of residuated lattices on the unit interval.

In the above mentioned structures there is only one product and only one implication. In practice however, several products and implications may be needed. Being totally uncertain about $x \in A$ (represented by $A(x) = [0, 1]$ in interval-valued fuzzy set theory) and also about $x \in B$ does not necessarily imply that we have total uncertainty about $x \in A \cap B$. For example, in information retrieval, when searching documents that contain e.g. "fuzzy subgroup" and "Christ", then the user (or the system) might be completely uncertain whether those terms occur in the document, but he is almost sure that the terms cannot occur together, so the membership degree of x in $A \cap B$ should be close to $[0, 0]$ (no membership). On the other hand, when searching for documents that contain "fuzzy" and "subgroup", then a membership degree of x in $A \cap B$ close to $[0, 1]$ (total uncertainty) is more appropriate, since one of the terms does not exclude the other. So in the same application it might be necessary to model conjunction by two different t-norms. The aim of this paper is to make the first steps in constructing a structure in which several t-norms and implications are available. Therefore, we will investigate under which conditions the properties that are valid in a residuated lattice are still valid when different products and implications are used.

2 Preliminary Definitions

Definition 1. *We define $\mathcal{L}^I = (L^I, \leq_{L^I})$, where*

- $L^I = \{[x_1, x_2] \mid (x_1, x_2) \in [0, 1]^2$ *and* $x_1 \leq x_2\}$,
- $[x_1, x_2] \leq_{L^I} [y_1, y_2]$ *iff* $x_1 \leq y_1$ *and* $x_2 \leq y_2$, *for all* $[x_1, x_2]$, $[y_1, y_2]$ *in* L^I.

Similarly as Lemma 2.1 in [4] it can be shown that $\mathcal{L}^I$ is a complete lattice.

Definition 2. [1,2] *An interval-valued fuzzy set on U is a mapping $A : U \to L^I$.*

Definition 3. [3] *An Atanassov's intuitionistic fuzzy set on U is a set*

$$A = \{(u, \mu_A(u), \nu_A(u)) \mid u \in U\}, \tag{1}$$

where $\mu_A(u) \in [0, 1]$ denotes the membership degree and $\nu_A(u) \in [0, 1]$ the non-membership degree of u in A and where for all $u \in U$, $\mu_A(u) + \nu_A(u) \leq 1$.

An Atanassov's intuitionistic fuzzy set A on U can be represented by the $\mathcal{L}^I$-fuzzy set A given by

$$\begin{aligned} A : U &\to L^I : \\ u &\mapsto [\mu_A(u), 1 - \nu_A(u)], \end{aligned} \tag{2}$$

In Figure 1 the set L^I is shown. Note that each $x = [x_1, x_2] \in L^I$ is represented by the point $(x_1, x_2) \in \mathbb{R}^2$.

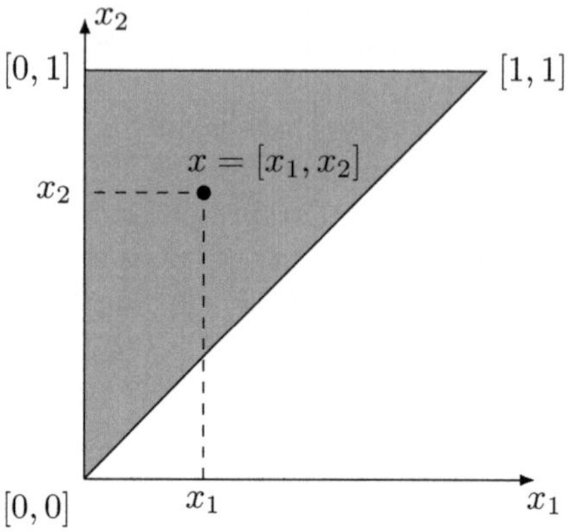

Fig. 1. The grey area is L^I

In the sequel, if $x \in L^I$, then we denote its bounds by x_1 and x_2, i.e. $x = [x_1, x_2]$. The smallest and the largest element of $\mathcal{L}^I$ are given by $0_{\mathcal{L}^I} = [0, 0]$ and $1_{\mathcal{L}^I} = [1, 1]$. Note that, for x, y in L^I, $x <_{L^I} y$ is equivalent to $x \leq_{L^I} y$ and $x \neq y$, i.e. either $x_1 < y_1$ and $x_2 \leq y_2$, or $x_1 \leq y_1$ and $x_2 < y_2$. We define for further usage the set $D = \{[x_1, x_1] \mid x_1 \in [0, 1]\}$.

Definition 4. *A t-norm on $\mathcal{L}^I$ is a commutative, associative, increasing mapping $\mathcal{T} : (L^I)^2 \to L^I$ which satisfies $\mathcal{T}(1_{\mathcal{L}^I}, x) = x$, for all $x \in L^I$.*

A t-conorm on $\mathcal{L}^I$ is a commutative, associative, increasing mapping $\mathcal{S}$: $(L^I)^2 \to L^I$ which satisfies $\mathcal{S}(0_{\mathcal{L}^I}, x) = x$, for all $x \in L^I$.

An implication on $\mathcal{L}^I$ is a mapping $\mathcal{I} : (L^I)^2 \to L^I$ which is decreasing in its first argument, increasing in its second argument and for which $\mathcal{I}(0_{\mathcal{L}^I}, 0_{\mathcal{L}^I}) = \mathcal{I}(0_{\mathcal{L}^I}, 1_{\mathcal{L}^I}) = \mathcal{I}(1_{\mathcal{L}^I}, 1_{\mathcal{L}^I}) = 1_{\mathcal{L}^I}$ and $\mathcal{I}(1_{\mathcal{L}^I}, 0_{\mathcal{L}^I}) = 0_{\mathcal{L}^I}$.

Definition 5. *A residuated lattice is a structure $\mathcal{L} = (L, \wedge, \vee, *, \Rightarrow, 0, 1)$ in which $\wedge$, $\vee$, $*$ and $\Rightarrow$ are binary operators on the set L and*

- *$(L, \wedge, \vee)$ is a bounded lattice with 0 as smallest and 1 as greatest element,*
- *$*$ is commutative and associative, with 1 as neutral element, and*
- *$x * y \leq z$ iff $x \leq y \Rightarrow z$ for all x, y and z in L (residuation principle).*

Let $\mathcal{T}$ be a t-norm on $\mathcal{L}^I$ which satisfies the residuation principle. Then the implication $\mathcal{I}_\mathcal{T}$ given by $\mathcal{I}_\mathcal{T}(x, y) = \sup\{z \mid z \in L^I \text{ and } \mathcal{T}(x, z) \leq y\}$ is called the residual implication of $\mathcal{T}$.

Theorem 1. [9] *Let $\mathcal{T}$ be a t-norm on $\mathcal{L}^I$. The structure $(L^I, \inf, \sup, \mathcal{T}, \mathcal{I}_\mathcal{T}, 0_{\mathcal{L}^I}, 1_{\mathcal{L}^I})$ is a residuated lattice with $\mathcal{T}(D, D) \subseteq D$ and $\mathcal{I}_\mathcal{T}(D, D) \subseteq D$ if and only if there exists a t-norm T on $([0, 1], \leq)$ which satisfies the residuation principle and a $t \in [0, 1]$ such that $\mathcal{T} = \mathcal{T}_{T,t}$, where for all x and y in L^I:*

$$\mathcal{T}_{T,t}(x, y) = [T(x_1, y_1), \max(T(t, T(x_2, y_2)), T(x_1, y_2), T(x_2, y_1))]. \quad (3)$$

In this case, the residual implication $\mathcal{I}_{\mathcal{T}}$ of $\mathcal{T}$ is given by, for all x and y in L^I,

$$\mathcal{I}_{\mathcal{T}_{T,t}}(x,y) = [\min(I_T(x_1,y_1), I_T(x_2,y_2)), \min(I_T(x_1,y_2), I_T(x_2, I_T(t,y_2)))], \tag{4}$$

where I_T is the residual implication of T.

Proposition 1. *Let T_1 and T_2 be t-norms on $([0,1],\leq)$ and $(t_1,t_2) \in [0,1]^2$. Then*

$$\mathcal{T}_{T_1,t_1} \leq_{L^I} \mathcal{T}_{T_2,t_2} \quad \textit{iff} \quad T_1 \leq T_2 \textit{ and } t_1 \leq t_2 \,. \tag{5}$$

Proposition 2. *Let T_1 and T_2 be t-norms on $([0,1],\leq)$ and $(t_1,t_2) \in [0,1]^2$. Then*

$$\mathcal{I}_{\mathcal{T}_{T_1,t_1}} \leq_{L^I} \mathcal{I}_{\mathcal{T}_{T_2,t_2}} \quad \textit{iff} \quad I_{T_1} \leq I_{T_2} \textit{ and } t_1 \geq t_2 \,. \tag{6}$$

3 Multiple Products and Implications

In this section we will assume that $(L, \wedge, \vee, *_k, \Rightarrow_k, 0, 1)_{k\in K}$ is a family of residuated lattices. We will check whether the properties in the following proposition still hold when we use different operators $*_i$ and $\Rightarrow_j$, with i and j in K.

Proposition 3. [11,12,13] *Let $(L, \wedge, \vee, *, \Rightarrow, 0, 1)$ be a residuated lattice. Then for all x, y, z, u in L it holds:*

1. $x \leq y$ *iff* $x \Rightarrow y = 1$,
2. $1 \Rightarrow x = x$,
3. $x * y \leq x \wedge y$,
4. $y \leq x \Rightarrow y$,
5. $x * (x \Rightarrow y) \leq x \wedge y$,
6. $x \vee y \leq (x \Rightarrow y) \Rightarrow y$,
7. $(x \Rightarrow y) * z \leq x \Rightarrow (y * z)$,
8. $(x \Rightarrow y) \Rightarrow ((y \Rightarrow z) \Rightarrow (x \Rightarrow z)) = 1$,
9. $(y \Rightarrow z) \Rightarrow ((x \Rightarrow y) \Rightarrow (x \Rightarrow z)) = 1$,
10. $((x * y) \Rightarrow z) = (x \Rightarrow (y \Rightarrow z)) = (y \Rightarrow (x \Rightarrow z))$,
11. $(x \Rightarrow y) \Rightarrow ((z \Rightarrow u) \Rightarrow ((y \Rightarrow z) \Rightarrow (x \Rightarrow u))) = 1$,
12. $(x \vee y) \Rightarrow x = y \Rightarrow x$,
13. $x \Rightarrow (x \wedge y) = x \Rightarrow y$,
14. $x \Rightarrow y \leq (x \wedge z) \Rightarrow (y \wedge z)$.

For the ease of notations we will assume that $1, 2, \ldots$ are indices in K and we will use $*_1, \Rightarrow_2, \ldots$ instead of the more general notations $*_{i_1}, \Rightarrow_{i_2}, \ldots$

Proposition 4. *Let $(L, \wedge, \vee, *_k, \Rightarrow_k, 0, 1)_{k\in K}$ be a family of residuated lattices. Assume that $\{1,2\} \subseteq K$. Then for all x, y and z in L it holds that*

$$x *_1 (x \Rightarrow_2 y) \leq x \wedge y \tag{7}$$

if and only if $\Rightarrow_2 \leq \Rightarrow_1$.

Proposition 5. *Let $(L, \wedge, \vee, *_k, \Rightarrow_k, 0, 1)_{k \in K}$ be a family of residuated lattices. Assume that $\{1, 2\} \subseteq K$. Then for all x, y and z in L it holds that*

$$x \vee y \leq (x \Rightarrow_1 y) \Rightarrow_2 y \tag{8}$$

if and only if $\Rightarrow_1 \leq \Rightarrow_2$.

Proposition 6. *Let $(L, \wedge, \vee, *_k, \Rightarrow_k, 0, 1)_{k \in K}$ be a family of residuated lattices. Assume that $\{1, 2, 3, 4\} \subseteq K$.*

1. *If for all x, y and z in L it holds that*

$$(x \Rightarrow_1 y) *_2 z \leq x \Rightarrow_3 (y *_4 z), \tag{9}$$

   then $\Rightarrow_1 \leq \Rightarrow_3$, $*_2 \leq *_4$ and $*_3 \leq *_4$.
2. *Conversely, if $\Rightarrow_1 \leq \Rightarrow_3$, $*_2 \leq *_3$ and $*_3 \leq *_4$, then (9) holds for all x, y and z in L.*

Proposition 7. *Let $(L, \wedge, \vee, *_k, \Rightarrow_k, 0, 1)_{k \in K}$ be a family of residuated lattices. Assume that $\{1, 2, \ldots, 5\} \subseteq K$.*

1. *If for all x, y and z in L it holds that*

$$(x \Rightarrow_1 y) \Rightarrow_2 ((y \Rightarrow_3 z) \Rightarrow_4 (x \Rightarrow_5 z)) = 1, \tag{10}$$

   then $\Rightarrow_1 \leq \Rightarrow_5$, $\Rightarrow_3 \leq \Rightarrow_4$ and $\Rightarrow_3 \leq \Rightarrow_5$.
2. *Conversely, if $\Rightarrow_1 \leq \Rightarrow_5$, $\Rightarrow_3 \leq \Rightarrow_5$ and $*_4 \leq *_5$, then (10) holds for all x, y and z in L.*

Proposition 8. *Let $(L, \wedge, \vee, *_k, \Rightarrow_k, 0, 1)_{k \in K}$ be a family of residuated lattices. Assume that $\{1, 2, \ldots, 5\} \subseteq K$.*

1. *If for all x, y and z in L it holds that*

$$(y \Rightarrow_1 z) \Rightarrow_2 ((x \Rightarrow_3 y) \Rightarrow_4 (x \Rightarrow_5 z)) = 1, \tag{11}$$

   then $\Rightarrow_1 \leq \Rightarrow_4$, $\Rightarrow_1 \leq \Rightarrow_5$ and $\Rightarrow_3 \leq \Rightarrow_5$.
2. *Conversely, if $\Rightarrow_1 \leq \Rightarrow_5$, $\Rightarrow_3 \leq \Rightarrow_5$ and $*_4 \leq *_5$, then (11) holds for all x, y and z in L.*

Proposition 9. *Let $(L, \wedge, \vee, *_k, \Rightarrow_k, 0, 1)_{k \in K}$ be a family of residuated lattices. Assume that $\{1, 2, 3, 4\} \subseteq K$.*

1. *If for all x, y and z in L it holds that*

$$(x *_1 y) \Rightarrow_2 z \leq x \Rightarrow_3 (y \Rightarrow_4 z), \tag{12}$$

   then $\Rightarrow_2 \leq \Rightarrow_3$, $\Rightarrow_2 \leq \Rightarrow_4$ and $*_4 \leq *_1$.
2. *Conversely, if either $\Rightarrow_2 \leq \Rightarrow_3$, $\Rightarrow_2 \leq \Rightarrow_4$ and $*_2 \leq *_1$, or $\Rightarrow_2 \leq \Rightarrow_4$, $\Rightarrow_4 \leq \Rightarrow_3$ and $*_4 \leq *_1$, then (12) holds for all x, y and z in L.*

Proposition 10. *Let $(L, \wedge, \vee, *_k, \Rightarrow_k, 0, 1)_{k \in K}$ be a family of residuated lattices. Assume that $\{1, 2, 3, 4\} \subseteq K$.*

1. *If for all x, y and z in L it holds that*
$$(x *_1 y) \Rightarrow_2 z \geq x \Rightarrow_3 (y \Rightarrow_4 z), \tag{13}$$
then $\Rightarrow_2 \geq \Rightarrow_3$, $\Rightarrow_2 \geq \Rightarrow_4$ and $*_4 \geq *_1$.
2. *Conversely, if either $\Rightarrow_2 \geq \Rightarrow_3$, $\Rightarrow_2 \geq \Rightarrow_4$ and $*_2 \geq *_1$, or $\Rightarrow_2 \geq \Rightarrow_4$, $\Rightarrow_4 \geq \Rightarrow_3$ and $*_4 \geq *_1$, then (13) holds for all x, y and z in L.*

Corollary 1. *Let $(L, \wedge, \vee, *_k, \Rightarrow_k, 0, 1)_{k \in K}$ be a family of residuated lattices. Assume that $\{1, 2, 3, 4\} \subseteq K$.*

1. *If for all x, y and z in L it holds that*
$$(x *_1 y) \Rightarrow_2 z = x \Rightarrow_3 (y \Rightarrow_4 z), \tag{14}$$
then $*_4 = *_1$ and $\Rightarrow_1 = \Rightarrow_2 = \Rightarrow_3 = \Rightarrow_4$.
2. *Conversely, if $*_2 = *_1$ and $\Rightarrow_1 = \Rightarrow_2 = \Rightarrow_3 = \Rightarrow_4$, then (13) holds for all x, y and z in L.*

Proposition 11. *Let $(L, \wedge, \vee, *_k, \Rightarrow_k, 0, 1)_{k \in K}$ be a family of residuated lattices. Assume that $\{1, 2, \ldots, 7\} \subseteq K$.*

1. *If for all x, y, z and u in L it holds that*
$$(x \Rightarrow_1 y) \Rightarrow_2 ((z \Rightarrow_3 u) \Rightarrow_4 ((y \Rightarrow_5 z) \Rightarrow_6 (x \Rightarrow_7 u))) = 1, \tag{15}$$
then $\Rightarrow_3 \leq \Rightarrow_4$, $\Rightarrow_3 \leq \Rightarrow_6$, $\Rightarrow_1 \leq \Rightarrow_7$, $\Rightarrow_3 \leq \Rightarrow_7$ and $\Rightarrow_5 \leq \Rightarrow_7$.
2. *Conversely, if $\Rightarrow_1 \leq \Rightarrow_7$, $\Rightarrow_3 \leq \Rightarrow_7$, $\Rightarrow_5 \leq \Rightarrow_7$, $*_4 \leq *_7$ and $*_6 \leq *_7$, then (15) holds for all x, y, z and u in L.*

Proposition 12. *Let $(L, \wedge, \vee, *_k, \Rightarrow_k, 0, 1)_{k \in K}$ be a family of residuated lattices. Assume that $\{1, 2\} \subseteq K$. Then for all x and y in L it holds that*
$$(x \vee y) \Rightarrow_1 x \leq y \Rightarrow_2 x \tag{16}$$
if and only if $\Rightarrow_1 \leq \Rightarrow_2$ (and similarly if we replace in both places $\leq$ by $\geq$).

Proposition 13. *Let $(L, \wedge, \vee, *_k, \Rightarrow_k, 0, 1)_{k \in K}$ be a family of residuated lattices. Assume that $\{1, 2\} \subseteq K$. Then for all x and y in L it holds that*
$$x \Rightarrow_1 (x \wedge y) \leq x \Rightarrow_2 y \tag{17}$$
if and only if $\Rightarrow_1 \leq \Rightarrow_2$ (and similarly if we replace in both places $\leq$ by $\geq$).

Proposition 14. *Let $(L, \wedge, \vee, *_k, \Rightarrow_k, 0, 1)_{k \in K}$ be a family of residuated lattices. Assume that $\{1, 2\} \subseteq K$. Then for all x, y and z in L it holds that*
$$x \Rightarrow_1 y \leq (x \wedge z) \Rightarrow_2 (y \wedge z) \tag{18}$$
if and only if $\Rightarrow_1 \leq \Rightarrow_2$.

Note that the above properties are in particular valid for the case that L is the set L^I, $*_k = \mathcal{T}_{T_k,t_k}$ and $\Rightarrow_k = \mathcal{I}_{\mathcal{T}_{T_k,t_k}}$. In this case we can even obtain conditions in terms of T_k and t_k, using Proposition 1 and 2. For example, from Proposition 4 the following property can be deduced.

Proposition 15. *Let* $(L^I, \inf, \sup, \mathcal{T}_{T_k,t_k}, \mathcal{I}_{\mathcal{T}_{T_k,t_k}}, 0_{\mathcal{L}^I}, 1_{\mathcal{L}^I})_{k\in K}$ *be a family of residuated lattices. Assume that* $\{1,2\} \subseteq K$*. Then for all* x, y *and* z *in* L^I *it holds that*

$$\mathcal{T}_{T_1,t_1}(x, \mathcal{I}_{\mathcal{T}_{T_2,t_2}}(x,y)) \le \inf(x,y) \tag{19}$$

if and only if $I_{T_2} \le I_{T_1}$ *and* $t_2 \ge t_1$*.*

Using the fact that Proposition 4 also holds for residuated lattices based on T_k, we obtain the following.

Corollary 2. *Let* $(L^I, \inf, \sup, \mathcal{T}_{T_k,t_k}, \mathcal{I}_{\mathcal{T}_{T_k,t_k}}, 0_{\mathcal{L}^I}, 1_{\mathcal{L}^I})_{k\in K}$ *be a family of residuated lattices. Assume that* $\{1,2\} \subseteq K$*. Then for all* x, y *and* z *in* L^I *it holds that*

$$\mathcal{T}_{T_1,t_1}(x, \mathcal{I}_{\mathcal{T}_{T_2,t_2}}(x,y)) \le \inf(x,y) \tag{20}$$

if and only if $t_2 \ge t_1$ *and for all* x_1, y_1 *and* z_1 *in* $[0,1]$ *it holds that*

$$T_1(x, I_{T_2}(x,y)) \le \min(x,y). \tag{21}$$

4 Conclusion

Starting from the observation that in interval-valued fuzzy set theory using only one t-norm and implication is not sufficient in order to obtain useful results, we started investigating whether we can extend traditional residuated lattices with several products and implications. Therefore we considered a list of properties that are valid in residuated lattices and we investigated under which conditions these properties still hold when we use different products and implications. In the future we will investigate other properties (also involving the negation generated by the implication), and we will investigate whether notions such as BL-algebras, MV-algebras can be extended with multiple products and implications.

References

1. Sambuc, R.: Fonctions Φ-floues. Application à l'aide au diagnostic en pathologie thyroidienne. PhD thesis, Université de Marseille, France (1975)
2. Gorzałczany, M.B.: A method of inference in approximate reasoning based on interval-valued fuzzy sets. Fuzzy Sets and Systems 21(1), 1–17 (1987)
3. Atanassov, K.T.: Intuitionistic fuzzy sets. Physica-Verlag, Heidelberg (1999)
4. Deschrijver, G., Kerre, E.E.: On the relationship between some extensions of fuzzy set theory. Fuzzy Sets and Systems 133(2), 227–235 (2003)
5. Goguen, J.A.: L-fuzzy sets. Journal of Mathematical Analysis and Applications 18(1), 145–174 (1967)

6. Klement, E.P., Mesiar, R., Pap, E.: Triangular norms. Kluwer Academic Publishers, Dordrecht (2000)
7. Van Gasse, B., Cornelis, C., Deschrijver, G., Kerre, E.E.: On the properties of a generalized class of t-norms in interval-valued fuzzy logics. New Mathematics and Natural Computation 2(1), 29–41 (2006)
8. Van Gasse, B., Cornelis, C., Deschrijver, G., Kerre, E.E.: Triangle algebras: a formal logic approach to interval-valued residuated lattices. Fuzzy Sets and Systems 159(9), 1042–1060 (2008)
9. Van Gasse, B., Cornelis, C., Deschrijver, G., Kerre, E.E.: A characterization of interval-valued residuated lattices. International Journal of Approximate Reasoning 49(2), 478–487 (2008)
10. Van Gasse, B., Cornelis, C., Deschrijver, G., Kerre, E.E.: The pseudo-linear semantics of interval-valued fuzzy logics. Information Sciences 179(6), 717–728 (2009)
11. Hájek, P.: Metamathematics of fuzzy logic. Kluwer Academic Publishers, Dordrecht (1998)
12. Höhle, U.: Commutative, residuated l-monoids. In: Höhle, U., Klement, E.P. (eds.) Non-classical logics and their applications to fuzzy subsets: a handbook of the mathematical foundations of fuzzy set theory, pp. 53–106. Kluwer Academic Publishers, Dordrecht (1995)
13. Turunen, E.: Mathematics behind fuzzy logic. Physica-Verlag, Heidelberg (1999)

Fuzzy Ontology and Information Granulation: An Approach to Knowledge Mobilisation

Christer Carlsson[1], Matteo Brunelli[2,3], and Jozsef Mezei[2]

[1] IAMSR, Åbo Akademi University,
Joukahainengatan 3-5A, Åbo, FI-20520, Finland
christer.carlsson@abo.fi
[2] Turku Centre for Computer Science, Åbo Akademi University,
Joukahainengatan 3-5A, Åbo, FI-20520, Finland
{matteo.brunelli, josef.mezei}@abo.fi
[3] Department of Computer and Management Sciences, University of Trento,
Via Inama 5, I-38122 Trento, Italy

Abstract. In order to find a common conceptual framework to address the problems of representation and management of imprecise and vague knowledge in a semantic web we have analysed the role played by information granulation in human cognition. Fuzzy granulation underlies the basic concepts of linguistic variable and fuzzy if-then rules, which play a major role in the applications of fuzzy logic to the representation and treatment of imprecision and vagueness. Fuzzy information granulation is central to fuzzy logic because it is central to concept formation in human reasoning and to the design of intelligent systems, and therefore is central to the modelling of fuzzy ontologies.

Keywords: Fuzzy ontology, information granulation, fuzzy numbers.

1 Introduction

A key challenge for research in knowledge management is to find ways to extend Semantic Web to include soft semantics, which would make it possible to represent imprecise knowledge and information. A possible way to work out the extension is to use fuzzy sets and approximate reasoning schemes to create a fuzzy ontology structure. *Knowledge mobilisation* is an enhancement of knowledge management methods and technology and represents the next step in using information technology in management processes which is commonly believed to improve the quality of planning, problem solving and decision making.

Ontologies are proper formalisms for the representation of knowledge in modern information systems and represent one of the most frequently used knowledge representation models. Ontology offers several advantages over other formalisms: (i) information sharing among different people (or software agents) because the primitive elements are anchored in meanings relevant to a context; (ii) knowledge reuse because the primitive elements can be (re)used in different combinations;

E. Hüllermeier, R. Kruse, and F. Hoffmann (Eds.): IPMU 2010, Part II, CCIS 81, pp. 420–429, 2010.

(iii) the definition of the primitive elements offers separation between declarative and procedural knowledge, and (iv) the definition of elements in a context enables the acquisition and analysis of knowledge.

The most common reason for constructing and implementing ontology is to get a basis for sharing knowledge. A second common reason is knowledge reuse. Ontology is especially designed to save time and effort in the knowledge acquisition processes by offering ways to use parts of previously built knowledge models (or even reusing the models as such), to combine knowledge models to form new constructs and to adapt general models to specific domains.

As soon as we have developed ontology we have a formal specification and model of the domain, which we can use to validate and verify the knowledge we believe we have about the domain. The validation can be done formally through reasoning schemes that produce testable conclusions and results, or it can be carried out as an empirical validation process. Validated knowledge is normally included in a repository and can be reused and built on to form new, enhanced knowledge.

We are developing a fuzzy ontology as the core of knowledge mobilisation and building on fuzzy description logic [fDL] as we will have key descriptions that will be imprecise and fuzzy set theory is more suitable than crisp ontologies to handle uncertainty and vagueness. A good example of the application of fuzziness for knowledge mobilization can be found in [11]. Fuzzy DLs extend classical DLs with *concepts* that are fuzzy sets of *individuals* and *roles* that are represented by binary relations. The interpretation of *concepts* and *roles* in an fDL is carried out in such a way that: (i) an element of a domain may belong to a *concept* with a grade in $[0, 1]$; (ii) a pair of elements of a domain is part of a *role* with a grade in $[0, 1]$. The semantics of the constructors used in the fDL is built around fuzzy sets; e.g. the concept *intersection* is given by a t-norm. Axioms are extended to the fuzzy case. The basis of our model is the fuzzy extension of the $\mathcal{SHOIN}(\mathtt{D})$, the corresponding Description Logic of the ontology description language OWL DL [12].

We can define a fuzzy inclusion relation between two fuzzy concepts with a terminological axiom for concepts. Approximate reasoning (AR) schemes use translation rules to represent common linguistic statements in terms of propositions in everyday language which can represent imprecise data, information and knowledge and are often accepted intuitively, which builds fast, robust and effective conclusions which are useful for a number of practical purposes.

We are pursuing a quest to build a wine ontology that can help turn amateurs into wine connoisseurs. One of the tasks of the wine ontology is to define similarity among the wines using keyword taxonomy and considering an existing database of wine descriptions to make it useful for consumers who are not wine experts. The database of wines is built from the opinions and judgments of wine experts. We then build AR schemes to allow the (amateur) wine drinkers to subjectively assess the expert judgments in a series of wine drinking scenarios (cf. formal dinner, business dinner, dinner with family, dinner with friends, pick nick, candle light dinner) and to subjectively determine the importance of the different properties and attributes of the wines. In this way we want to identify wines which would be best suited for different wine drinking scenarios.

In this paper we have addressed one of the problems we have encountered when developing the wine ontology: what if the *best suited* wine is not available, what would then be the *most similar* wine to replace the original choice? For this purpose we use fuzzy information granulation to represent the attributes of wines with fuzzy keyword taxonomy and to find the most similar wine in terms of the keywords. The rest of the paper is structured as follows: in section 2 we have worked out a simple ontology using fuzzy logic in order to introduce the conceptual framework; in section 3 we introduce our approach to a fuzzy ontology to demonstrate the idea of an application with the wine ontology and section 4 gives some conclusions.

2 A Simple Ontology Using Fuzzy Logic

The most widely accepted definition of ontology is probably that given by Gruber [7]: 'An ontology is an explicit specification of a conceptualization'. Ontologies have become increasingly popular and explored since the beginning of the 1990s and Parry has given a definition for fuzzy ontology that proposes to represent it as a fuzzy relation [9,10]. So far, ontologies, especially in their fuzzy version, have shown their utility in many fields as, for instance, information retrieval [10] and summarization [8]. We want to propose a new application such that, given an object/keyword k_i or a logical combination of objects $k_{i1}, \dots, k_{iq}$, our ontology-based model returns the object k_j which is the best substitute of object k_i or of the logical combination of objects $k_{i1}, \dots, k_{iq}$.

In this paper we consider a simple case and do not distinguish between categories of keywords and types of relationship. We only assume that there is a finite non-empty set of keywords $K = \{k_1, \dots, k_n\}$ and that a complete ontology of keywords can be represented by means of a relation $R \subseteq K \times K$. *Fuzzy ontology* can be represented by means of a relation associated with a membership function instead of a characteristic function. A membership function [16] is a mapping

$$\mu_R : K \times K \to L \tag{1}$$

where L is a lattice (or, more generally, a partially ordered set). It is important to stress that the real unit interval is simply a special case [6]. In fact, for instance, a set of linguistic labels, e.g. $\{bad, good, excellent\}$, is a lattice. Let us also notice that our definition of an ontology fits with that given by Parry [10] with the only exception that we do not ask for any normalization of the membership function of the fuzzy relation. Nevertheless, for the sake of simplicity, if not differently stated, we assume $[0, 1]$ to be the codomain of μ_R. Then, the degree of relation between k_i and k_j is defined as follows

$$\mu_R(k_i, k_j) = \begin{cases} 1, & \text{if } k_i \text{ and } k_j \text{ are definitely related} \\ \gamma \in]0,1[, & \text{if } k_i \text{ and } k_j \text{ are to some extent related} \\ 0, & \text{if } k_i \text{ and } k_j \text{ are not related} \end{cases}$$

Clearly, the semantic underlying the relation must be the same, i.e. the meaning of R does not change over the domain $K \times K$, but this does not necessarily imply that the relation must be symmetric. For instance, if we mean $\mu_R(k_i, k_j)$ to be the degree to which k_i includes k_j, then we might have $\mu_R(k_i, k_j) \neq \mu_R(k_j, k_i)$ for some i, j.

Values of the relation can be estimated in different ways. We are going to list three possible methods.

Computational approach. Following this approach, provided that we already have a sufficiently large knowledge base, the value of $R(k_i, k_j)$ can be estimated. Using this approach enables us to *learn* the information from the system.

Example 1. Given a list of objects, wines in this example, there can be a database from which we can find attributes for the objects. Therefore, we can have a normalized table as it follows

Table 1. Excerpt from a wine database

	acid	expensive	···	alcoholic
Cabernet X	0.1	0.2	···	0.5
Chardonnay Y	0.3	0.5	···	0.6
···	···	···	···	···
Shiraz Z	0.5	0.4	···	0.3

which can be written in matrix form as

$$\mathbf{M} = \begin{pmatrix} 0.1 & 0.2 & \cdots & 0.5 \\ 0.3 & 0.5 & \cdots & 0.6 \\ \cdots & \cdots & \cdots & \cdots \\ 0.5 & 0.4 & \cdots & 0.3 \end{pmatrix}. \tag{2}$$

If we denote the i-th row of $\mathbf{M}$ as $\mathbf{m}_i$, it is then possible to estimate entries of R using some distance function, for example the normalized norm of the difference between the two vectors

$$r_{ij} = \frac{1}{\sqrt{n}} \, ||\mathbf{m}_i - \mathbf{m}_j|| \tag{3}$$

Subjective approach. According to the subjective approach, values of R are estimated subjectively according to the perception of some experts. A good point in this method is that it allows subjectivity and it is possible to build a group decision making model whenever a group of experts is asked to give the values.

Imprecision based approach. This is mainly an extension of the second approach. Values are not given under the form of real numbers but under some other forms, e.g. linguistic labels, fuzzy numbers, intervals. This theory is based on the theory of uncertain probabilities. Uncertainty is added to the model.

Let us observe how these three methods are interchangeable in estimating the values of the relation. Namely, we can employ different methods for estimating different $\mu_R(k_i, k_j)$. Also in the case of a wine ontology, all the three methods seem to be coherent.

The relation R can be represented by means of a matrix $\mathbf{R} = (r_{ij})_{n\times n}$, with $r_{ij} := \mu_R(k_i, k_j)$. In its more extended form it is

$$\mathbf{R} = (r_{ij})_{n\times n} = \begin{pmatrix} r_{11} & \cdots & r_{1n} \\ \vdots & \ddots & \vdots \\ r_{n1} & \cdots & r_{nn} \end{pmatrix} \tag{4}$$

Example 2. Let us conclude this section with a toy example of a fuzzy ontology that we will return to later on in this paper. If we assume that we have four keywords, then the following could be a possible outcome when we compare the similarities of keywords and want to find the best substitute.

$$\mathbf{R} = (r_{ij})_{4\times 4} = \begin{pmatrix} 1 & minor & 0.5 & moderate \\ minor & 1 & 0.2 & [0.2, 0.3] \\ 0.4 & [0.3, 0.5] & 1 & [0.6, 0.8] \\ 0.3 & [0.1, 0.4] & significant & 1 \end{pmatrix} \tag{5}$$

3 An Approach to a Fuzzy Ontology

Our proposal is to use a fuzzy ontology, which in our case is represented by the fuzzy relation R, which estimates degrees of substitution between keywords. More precisely, the semantic underlying the relation is to which extent, in a given context, a keyword (the first component of a pair) can be suitably used in place of another (the second component).

3.1 The Ordering of the Fuzzy Numbers

Thus, our aim is to find the most suitable keyword to replace a given keyword given a logical combination of some other keywords. In the first step we have a matrix $\mathbf{R}$ as in the previously proposed example. Then we want to represent these numbers, intervals or linguistic labels such that a common framework, e.g. a proper metric, can be applied to all the entries of $\mathbf{R}$. To do this, we need to work with a numerical matrix. First of all, let us give the following definition of fuzzy number

Definition 1 (Fuzzy number [4]). *A fuzzy number is a convex fuzzy set on the real line* $\mathbb{R}$ *such that*

a) $\exists x_0 \in \mathbb{R}, \mu_A(x_0) = 1$
b) μ_A *is piecewise continuous*

Furthermore, we call $\mathcal{F}$ *the family of all fuzzy numbers.*

It is well-known, and straightforward to check, that real numbers and intervals can be seen as special cases of fuzzy numbers. Moreover, fuzzy numbers are a representation of possibility distributions [17]. Let us now explain how to translate a relation R into a fuzzy number based relation $\tilde{R}$ which we will hereafter represent by means of the matrix $\tilde{\mathbf{R}} = (\tilde{r}_{ij})_{n\times n} \in \mathcal{F}^{n\times n}$.

Real numbers and intervals: As mentioned, real numbers and intervals are both special types of fuzzy numbers and therefore they can be left unchanged.

Linguistic labels: Given a set D including all the possible linguistic descriptions for the degree of relation, if we consider that each membership function is associated to a unique fuzzy number, then we can define a mapping

$$\phi : D \to \mathcal{F} \tag{6}$$

For the example, we may use four linguistic labels, i.e. *minor*, *moderate*, *significant*, *total*, associated with the following fuzzy numbers:

$$\mu_{minor}(t) = \begin{cases} 1 & \text{if } 0 \le t \le 0.25 \\ \dfrac{0.5 - t}{0.25} & \text{if } 0.25 \le t \le 0.5 \\ 0 & \text{otherwise} \end{cases}$$

$$\mu_{moderate}(t) = \begin{cases} \dfrac{t}{0.25} & \text{if } 0 \le t \le 0.25 \\ 1 & \text{if } 0.25 \le t \le 0.5 \\ \dfrac{0.75 - t}{0.25} & \text{if } 0.5 \le t \le 0.75 \\ 0 & \text{otherwise} \end{cases}$$

$$\mu_{significant}(t) = \begin{cases} \dfrac{t - 0.25}{0.25} & \text{if } 0.25 \le t \le 0.5 \\ 1 & \text{if } 0.5 \le t \le 0.75 \\ \dfrac{1 - t}{0.25} & \text{if } 0.75 \le t \le 1 \\ 0 & \text{otherwise} \end{cases}$$

$$\mu_{total}(t) = \begin{cases} \dfrac{t - 0.5}{0.25} & \text{if } 0.5 \le t \le 0.75 \\ 1 & \text{if } 0.75 \le t \le 1 \\ 0 & \text{otherwise} \end{cases}$$

Example 3. Using this procedure we can transform the example from section 1 into the following form:

$$\tilde{\mathbf{R}} = \begin{pmatrix} (1,1,1,1) & (0,0,0.25,0.5) & (0.5,0.5,0.5,0.5) & (0,0.25,0.5,0.75) \\ (0,0,0.25,0.5) & (1,1,1,1) & (0.2,0.2,0.2,0.2) & (0.2,0.2,0.3,0.3 \\ (0.4,0.4,0.4,0.4) & (0.3,0.3,0.5,0.5) & (1,1,1,1) & (0.6,0.6,0.8,0.8) \\ (0.3,0.3,0.3,0.3) & (0.1,0.1,0.4,0.4) & (0.25,0.5,0.75,1) & (1,1,1,1) \end{pmatrix} \tag{7}$$

where each entry, i.e. fuzzy number, is described by means of a quadruple (a, b, c, d) where $[a, d]$ is the support and $[b, c]$ is the kernel. Furthermore, we assume that the membership function is piecewise linear.

In the next step, by means of the obtained matrix $\tilde{\mathbf{R}}$ we are going to order these fuzzy numbers. First, we consider only the simplest case where the query contains only one keyword. We check the fuzzy numbers in the row corresponding to the keyword, excluding the entry on the main diagonal, and the result is the keyword with the highest ordering value. From [13] we choose the following ordering relation.

Adamo's approach [1]. Given λ in $(0, 1]$, we simply evaluate the fuzzy quantity based on the utmost right point of the λ-cut. Therefore, the ordering index is

$$AD_\lambda(A) = a_\lambda^+.$$

if $B = (a, b, c, d)$ is a trapezoidal fuzzy number, then $AD_\lambda(B) = d - \lambda(d - c)$

Adamo's approach is very appealing because (i) it is the only method satisfying the seven axioms proposed in [13] and (ii) it works, contrarily to some other approaches (see [14]), with fuzzy numbers such that $a = b = c = d$. In the following section we examine how to deal with queries which involve more than one keyword.

3.2 Logical Combinations

If we have a logical combination of keywords, we can translate the problem into the language of optimization. In the simplest case above, we need to maximize the value of the ordering relation over a chosen row.

The *negation* means that we have to minimize over the values. Given an index set $I \subset \{1, \ldots, n\}$ with the corresponding keywords, we want to find the keyword which is the furthest from a chosen keyword i. Then we have to find for which j the following value achieves the minimum:

$$\bigwedge_{j \in I} \tilde{r}_{ij} \tag{8}$$

The *conjunction* of some keywords means that we want to have a high substitution degree for all of them. Evaluating according to the ordering relation we obtain ordered m-tuples for every other keyword and we have to find the one with the highest possible value in both coordinates at the same time. In other words we would like to find the Pareto-optimal points over this discrete set of two-dimensional points. We need to solve the following problem

$$\arg\max_{i \notin I} \min_{j \in I} \tilde{r}_{ij} \tag{9}$$

In the case of *disjunction*, we would like to find keywords with a high value at least for one of the two chosen keywords, that is, we would like to find the weakly

Pareto-optimal points over that set. In logical terms, we want to know for which i the value of

$$\bigvee_{i \notin I} \bigvee_{j \in I} \tilde{r}_{ij} \tag{10}$$

is maximized. We can find a solution solving

$$\arg\max_{i \notin I} \max_{j \in I} \tilde{r}_{ij} \tag{11}$$

Example 4. Let us show one example of a query. Let us assume that we want to find the most similar wine to k_a *and* k_b and we clearly want to exclude the trivial solutions k_a and k_b. Logically, we want to find i such that the conjunction is maximized. As seen before, this is a max-min problem, and we are interested in the argument (index) which maximizes it. Namely, we want to solve the problem (9). There may be multiple solutions. In this case there is no problem as they can all be used at the same time.

If we continue our example with the matrix $\tilde{\mathbf{R}}$, we can calculate Adamo's ordering index for every fuzzy number. We will use the value $\frac{1}{2}$ for λ.

- The index for a fuzzy number obtained from a real number is the same, the number itself for every λ.
  So $AD_{0.5}(\tilde{r}_{11}) = 1, AD_{0.5}(\tilde{r}_{13}) = 0.5, AD_{0.5}(\tilde{r}_{22}) = 1, AD_{0.5}(\tilde{r}_{23}) = 0.2, AD_{0.5}(\tilde{r}_{31}) = 0.4, AD_{0.5}(\tilde{r}_{33}) = 1, AD_{0.5}(\tilde{r}_{41}) = 0.3, AD_{0.5}(\tilde{r}_{44}) = 1.$
- The index for a fuzzy number obtained from a real interval is the same, the right endpoint for every λ.
  So $AD_{0.5}(\tilde{r}_{24}) = 0.3, AD_{0.5}(\tilde{r}_{32}) = 0.5, AD_{0.5}(\tilde{r}_{34}) = 0.8, AD_{0.5}(\tilde{r}_{42}) = 0.4.$
- We can calculate the index for a trapezoidal fuzzy number from the formula:
  So $AD_{0.5}(\tilde{r}_{12}) = 0.375, AD_{0.5}(\tilde{r}_{14}) = 0.625, AD_{0.5}(\tilde{r}_{21}) = 0.375, AD_{0.5}(\tilde{r}_{43}) = 0.875.$

With $r^\star_{ij} := AD_{0.5}(\tilde{r}_{ij})$ we obtain the following matrix:

$$\mathbf{R}^\star = (r^\star_{ij})_{n \times n} = \begin{pmatrix} 1 & 0.375 & 0.5 & 0.625 \\ 0.375 & 1 & 0.2 & 0.3 \\ 0.4 & 0.5 & 1 & 0.8 \\ 0.3 & 0.4 & 0.875 & 1 \end{pmatrix} \tag{12}$$

If we want to find the most similar keyword to k_1, then from the first row we can see, that this is k_4. If we want to determine the most similar keyword to k_1 and k_2, then we have to compare the $(0.5, 0.2)$ and the $(0.625, 0.3)$ two-dimensional vectors with respect to the Pareto domination rule, and we see that the solution is k_4. It is possible to obtain multiple solutions.If we want to determine the most similar keyword to k_2 or k_3, then we can see that k_1 and k_4 are both weakly Pareto-optimal solutions.

3.3 The Wine Ontology

We are testing the first constructs of the fuzzy ontology with intuitive everyday examples [2,3]. Let us consider two ontology classes: *wine* and *food*. The subclasses of *wine* and *food* are {*red*, *rose*, *white*} and {*salad*, *beef*, *chicken*, *fish*, *pork*, *game*, *soup*, *dessert*} respectively. We define different properties of wines, and create keywords for them such as *location*, *price*, *acidity* and *alcohol* level, and introduce the corresponding linguistic labels, such as *high*, *medium* and *low* for each slot. For each linguistic label of a profile, we build a membership function, which we assume to be piecewise defined and linear.

The first problem we have to take into account is that the different types of attributes are sometimes not comparable. The first step is that we should consider not only one but several matrices corresponding to different properties. This case is more complex, but we can use the Pareto-type approach. A general query consists of several attributes which are not necessarily from the same category, rather each of them belongs to different groups. Formally we have an observation in the form of a list:

$$L = l_{1,1}, l_{1,2}, \ldots, l_{n,m_n}$$

The first number in the index determines from which category the keyword is coming, in general we have n; the second value in the index, m_i, tells us the number of items from a particular category; from the ith we have m_i. We can represent the query as an $(m_1 + ... + m_n)$-tuple. Usually we have one keyword from each category and we do not necessarily consider each one of them. In the first step we determine similar keywords with respect to the categories, and we can compose them to obtain a set of tuples. Then we can determine the Pareto or weakly Pareto optimal vectors.

4 Conclusions

Ontology is especially designed to save time and effort in the knowledge acquisition processes by offering ways to use parts of previously built knowledge models (or even reusing the models as such) to form new constructs and to adapt general models to specific domains.

We are developing a fuzzy ontology as the core of knowledge mobilisation and building on fuzzy description logic [fDL] as we will have key descriptions that will be imprecise. We have showed that we can use fuzzy keyword ontology for imprecise queries. In this paper we have addressed one particular problem in more detail - "what if the *best suited* wine is not available, what would then be the *most similar* wine to re-place the original choice"? For this purpose we have used fuzzy information granulation to represent the attributes of wines with fuzzy keyword taxonomy and to find the *most similar* wine in terms of the keywords.

In the future extensions of this paper we will consider the general case, where we will include a more complex keyword taxonomy in the ontology and a family of relationships, and describe how we can handle this with the modifications of the presented method.

References

1. Adamo, J.M.: Fuzzy decision trees. Fuzzy sets and systems 4, 207–219 (1980)
2. Calegari, S., Ciucci, D.: Integrating Fuzzy Logic in Ontologies. In: Proceedings of the 8th International Conference on Enterprise Information Systems, pp. 66–73 (2006)
3. Calegari, S., Ciucci, D.: Granular computing applied to ontologies. International Journal of Approximate Reasoning 51(4), 391–409 (2010)
4. Dubois, D., Prade, H.: Operations on fuzzy numbers. International Journal of Systems Science 9(6), 613–626 (1978)
5. Dubois, D., Prade, H.: Fuzzy Sets and Systems: Theory and Applications. Academic Press, New York (1980)
6. Goguen, J.A.: *L*-Fuzzy Sets. Journal of mathematical analysis and applications 18, 145–174 (1967)
7. Gruber, T.: A translation approach to portable ontology specification. Knowledge Acquisition 5(2), 199–220 (1993)
8. Lee, C.-S., Jian, Z.-W., Huang, L.-K.: A fuzzy ontology and its application to news summarization. IEEE Transactions on Systems, Man and Cybernetics, Part B 35(5), 859–880 (2005)
9. Parry, D.: Fuzzification of a standard ontology to encourage reuse. In: 2004 IEEE International Conference on Information Reuse and Integration, Las Vegas, USA, pp. 582–587 (2004)
10. Parry, D.: Fuzzy ontologies for information retrieval on the WWW. In: Bouchon-Meunier, B., Gutierrez Rios, J., Magdalena, L., Yager, R.R. (eds.) Fuzzy Logic and the semantic Web. Capturing Intelligence Series. Elsevier, Amsterdam (2006)
11. Romero, J.G.: Knowledge Mobilization: Architectures, Models and Applications. PhD Thesis, University of Granada (2008)
12. Straccia, U.: A fuzzy description logic for the Semantic Web. In: Sanchez, E. (ed.) Fuzzy Logic and the Semantic Web. Capturing Intelligence Series, ch. 4, vol. 1, pp. 73–90. Elsevier, Amsterdam (2006)
13. Wang, X., Kerre, E.E.: Reasonable properties for the ordering of fuzzy quantities (I). Fuzzy Sets and Systems 118, 375–385 (2001)
14. Yager, R.R.: Ranking fuzzy subsets over the unit interval. In: Proc. of the IEEE Conference on Decision and Control, pp. 1435–1437 (1978)
15. Yager, R.R., Filev, D.P.: Induced ordered weighted averaging operators. IEEE Transaction on Systems, Man and Cybernetics Part A 29, 141–150 (1999)
16. Zadeh, L.A.: Fuzzy Sets. Information and Control 8, 338–353 (1965)
17. Zadeh, L.A.: Fuzzy sets as a basis for a theory of possibility. Fuzzy Sets and Systems 1, 3–28 (1978)

Adjoint Pairs on Interval-Valued Fuzzy Sets

Jesús Medina*

Department of Mathematics
University of Cádiz
jesus.medina@uca.es

Abstract. In this paper the authors present the definition of interval operator associated to two general increasing operators, on the set of subintervals of $[0,1]$, and how its residuated implication must be defined, if the initial operator have adjoint implications. These results are necessary in several frameworks where mechanisms for reasoning under uncertainty are needed, such as decision and risk analysis, engineering design, and scheduling. We will show three framework where the interval values are used and, hence, where the results presented here can be useful.

1 Introduction

The methods of using intervals, either symbolic (inf, sup) or numerical $[a,b]$, to describe uncertain information have been adopted in several mechanisms, and they are useful in applications such as decision and risk analysis, engineering design, and scheduling. Intervals are also used in some frameworks for generalized logic programming such as the *hybrid probabilistic logic programs* [8] and the *probabilistic deductive databases* [16], where an interval $[a,b]$ is given as the probability, which cannot be a fixed value in $[0,1]$ but we need a rank of values. Just note that, in the latter framework, one could write rules like:

$$\text{paper\_accepted} \xleftarrow{\langle [0.7,0.95],[0.03,0.2]\rangle} \text{good\_work}, \text{good\_referees}$$

where we have a complex confidence value containing two probability intervals, one for the case where *paper_accepted* is true and other for the case where *paper_accepted* is false (there can exist some lack of information, or undefinedness, in these intervals).

The use of intervals is very common in the Atanassov's intuitionistic fuzzy sets environment [2,3], indeed, both intuitionistic fuzzy sets and interval-valued fuzzy sets are equivalent to L-fuzzy sets in the sense of Goguen [12] w.r.t. a special sublattice of $[0,1] \times [0,1]$, see [10]. They handle two values which can be translated to the extremes of an interval. Hence, in order to calculate some results in this framework, it is interesting how can the operators be defined, if we want to consider t-norms on $[0,1]$ or some other kind of operator more general.

* Partially supported by the Spanish Science Ministry under grant TIN2009-14562-C05-03 and by Junta de Andalucía under grant P09-FQM-5233.

E. Hüllermeier, R. Kruse, and F. Hoffmann (Eds.): IPMU 2010, Part II, CCIS 81, pp. 430–439, 2010.

The multi-adjoint logic programming is a general logical frameworks recently introduced, which is receiving considerable attention [15,21]. The multi-adjoint framework originated as a generalisation of several non-classical logic programming frameworks whose semantic structure is the multi-adjoint lattice, in which a lattice is considered together with several conjunctors and implications making up adjoint pairs. The particular details of the different approaches were abstracted away, retaining only the minimal mathematical requirements guaranteeing operability. In particular, conjunctors were required to be neither commutative nor associative.

In [17] the authors implement the procedural semantics given by the immediate consequence operator using neural networks. Although this is restricted to ordinal sums defined on $[0, 1]$.

Later, in [19] the authors use the above neural network to handle interval in a parallel way, but the possible inconsistence results are not studied. Thus, the results given in this paper will improve this paper.

Thus, the usefulness of this approach for real applications, such as the given ones from the probabilistic frameworks, where the probabilities are given as subintervals, and Atanassov's intuitionistic fuzzy sets, is clear.

The plan of this paper is the following: in Section 2 we recall the basics t-norm and how we can define operators on subinterval either from this or from other more general operators; later, in Section 3 we introduce the natural residuated implications from the general operators defined above; in Section 4 some frameworks in which we can use the results given in the previous sections are given; the paper ends with some conclusions and prospects for future work.

2 From Conjunctors on $[0, 1]$ to $\mathcal{C}[0, 1]$

The more usual generalization in $[0, 1]$ of the classical conjunctor is the t-norm. For example, the t-norm Gödel, product and Łukasiewicz are defined for each $x, y \in [0, 1]$ as:

$$\&_G(x, y) = \min\{x, y\} \qquad \&_P(x, y) = x \cdot y \qquad \&_L(x, y) = \max\{0, x + y - 1\}$$

The natural extension of these conjunctors to another defined in the set of subintervals, $\mathcal{C}[0, 1]$, is the mapping $\&_T \colon \mathcal{C}[0, 1] \times \mathcal{C}[0, 1] \to \mathcal{C}[0, 1]$, defined, for each $[a, b], [c, d] \subseteq [0, 1]$, as:

$$\&_{\mathbf{T}}([a, b], [c, d]) = [\&_T(a, c), \&_T(b, d)]$$

It is easy to check that, if $\&_T$ is a t-norm then $\&_{\mathbf{T}}$ is another one. But, if we use two different t-norms $\&_{T_1}$, $\&_{T_2}$, in order to define the operator $\&_{\mathbf{T_{12}}} \colon \mathcal{C}[0, 1] \times \mathcal{C}[0, 1] \to \mathcal{C}[0, 1]$ as

$$\&_{\mathbf{T_{12}}}([a, b], [c, d]) = [\&_{T_1}(a, c), \&_{T_2}(b, d)] \tag{1}$$

for all $[a, b], [c, d] \subseteq [0, 1]$; it is possible that this will not be well defined, that is, we could not obtain an interval: $\&_{T_1}(a, c) \not\leq \&_{T_2}(b, d)$. For example, if $\&_{T_1}$ and

$\&_{T_2}$ are the Gödel and Łukasiwicz t-norms respectively, we have that:

$$\&_{\mathbf{T_{12}}}([0.3, 0.4], [0.5, 0.6]) = [\min\{0.3, 0.5\}, \max\{0, 0.4 + 0.6 - 1\}] = [0.3, 0]$$

which is not a subinterval of $[0, 1]$.

As consequence, in general, we cannot use two t-norms on $[0, 1]$ to define another one on $\mathcal{C}[0, 1]$, because we may obtain an inconsistence.

Therefore, given two t-norms $\&_{T_1}$ and $\&_{T_2}$, such that $\&_{T_1} \leq \&_{T_2}$ pairwise, we have that $\&_{\mathbf{T_{12}}}$, defined in Equation 1, is also a t-norm.

In order to avoid the inconsistence about the order between the t-norms, we could introduce the following definitions: $\&_{T_1}$ and $\&_{T_2}$ as:

$$\&_{\mathbf{T_{12}}}([a, b], [c, d]) = [\min\{\&_{T_1}(a, c), \&_{T_2}(b, d)\}, \&_{T_2}(b, d)]$$

or $\&_{\mathbf{T_{12}}}([a, b], [c, d]) = [\&_{T_1}(a, c), \max\{\&_{T_1}(a, c), \&_{T_2}(b, d)\}]$, but these are not t-norms, because, although the left side is less or equal than the right one, $\&_{\mathbf{T_{12}}}$ could not be associative, as we show in the following example:

Example 1. Let $\&_S$ be the operator defined as:

$$\&_S(a, b) = \begin{cases} 0.7 \cdot \&_L(\frac{a}{0.7}, \frac{b}{0.7}) & \text{if } a, b \in [0, 0.7] \\ \min(x, y) & \text{otherwise} \end{cases}$$

which is a t-norm, exactly an ordinal sum [13], and where $\&_L$ is the Łukasiewicz implication. Then, we have that the operator $\&_{SP}$ defined as:

$$\&_{\mathbf{SP}}([a, b], [c, d]) = [\min\{\&_S(a, c), \&_P(b, d)\}, \&_P(b, d)]$$

where $\&_P$ is the product t-norm, is not associative. Let us to consider the intervals $[0.1, 0.4], [0.7, 0.7], [0.8, 0.8]$, then

$$\begin{aligned} \&_{\mathbf{SP}}((\&_{\mathbf{SP}}[0.1, 0.4], [0.7, 0.7]), [0.8, 0.8]) &= \&_{\mathbf{SP}}([\min\{0.1, 0.28\}, 0.28], [0.8, 0.8]) \\ &= [0.143, 0.224] \\ \&_{\mathbf{SP}}([0.1, 0.4], (\&_{\mathbf{SP}}([0.7, 0.7], [0.8, 0.8])) &= \&_{\mathbf{SP}}([0.1, 0.4], [\min\{0.7, 0.56\}, 0.56]) \\ &= [0.0, 0.224] \end{aligned}$$

which prove that $\&_{\mathbf{SP}}$ is not associative.

If we consider the other definition that consider the maximum, we obtain the same problem, therefore the natural definition of an t-norm in $\mathcal{C}[0, 1]$ from two t-norms in $[0, 1]$ is the definition given in [14]. Now, we will rewrite it for two increasing operator which not need be t-norms:

Definition 1. *Given an order $\preceq$ in $\mathcal{C}[0, 1]$, where $(\mathcal{C}[0, 1], \preceq)$ is a lattice, and two increasing operators $\&_1, \&_2 \colon [0, 1] \times [0, 1] \to [0, 1]$, where $\&_1$ is pointwise less or equal than $\&_2$, the operator $\&_{12}$ defined, for each $[a, b], [c, d]$ in $\mathcal{C}[0, 1]$, as:*

$$\&_{12}([a, b], [c, d]) = [\&_1(a, c), \&_2(b, d)]$$

is called the interval operator associated to $\&_1$ and $\&_2$.

Some particular kinds of this operator are very used in several mechanisms for reasoning under uncertainty, and they have many applications, such as decision and risk analysis, engineering design, and scheduling. If $\&_1$ and $\&_2$ are t-norms then $\&_{12}$ is another t-norm [14]. Moreover, we can remark that if $\&_1$ and $\&_2$ are commutativity and associativity then $\&_{12}$ is a probabilistic composition function used in the probabilistic strategies introduced in [8], and that all these functions are well founded.

Examples of this kind of functions introduced by Dekhtyar and Subrahmanian are:

$$\&_{inc}([a,b],[c,d]) = [\&_P(a,c), \&_P(b,d)] \quad \textit{Conjunctive independence}$$
$$\&_{igc}([a,b],[c,d]) = [\&_L(a,c), \&_G(b,d)] \quad \textit{Conjunctive ignorance}$$
$$\&_{pcc}([a,b],[c,d]) = [\&_G(a,c), \&_G(b,d)] \quad \textit{Conjunctive positive correlation}$$

Note that, in the conjunctive ignorance, two different t-norms are used: the Łukasiewicz and Gödel t-norms. However, we cannot define a p-strategy from an interval operator associated to the Gödel and Łukasiewicz t-norms, respectively, as we showed at the beginning of this section.

Another important remark is that we may consider any order on $\mathcal{C}[0,1]$. The order relations more usual defined on the set of subintervals of $[0,1]$ (see [11]), are:

- The *truth order*, defined for each $[a,b],[c,d] \subseteq [0,1]$ as

$$[a,b] \leq_t [c,d] \quad \text{if and only if} \quad a \leq c, b \leq d$$

- The *knowledge order*, defined for each $[a,b],[c,d] \subseteq [0,1]$ as

$$[a,b] \leq_k [c,d] \quad \text{if and only if} \quad a \leq c, d \leq b$$

Hence, $(\mathcal{C}[0,1], \leq_t)$ and $(\mathcal{C}[0,1], \leq_k)$ are lattices where the supremum and infimum are defined, for each $[a,b],[c,d] \in \mathcal{C}[0,1]$, as following:

$$\begin{aligned}
\sup_t\{[a,b],[c,d]\} &= [\sup\{a,c\}, \sup\{b,d\}];\\
\inf_t\{[a,b],[c,d]\} &= [\inf\{a,c\}, \inf\{b,d\}];\\
\sup_k\{[a,b],[c,d]\} &= [\sup\{a,c\}, \inf\{b,d\}];\\
\inf_k\{[a,b],[c,d]\} &= [\inf\{a,c\}, \sup\{b,d\}]
\end{aligned}$$

3 Adjoint Implications

In logic programming it is very usual to consider t-norms and its residuated implications [13]. Here, we will consider general increasing operators defined on $\mathcal{C}[0,1]$ and we will present the residuated implication as usual [7,13,20] and we will introduce the corresponding definition of the residuated implication associated to an interval operator defined from two operators, if these last operators have residuated implications.

In general, given a lattice $(L, \preceq)$ and an isotonic operator $\&: L \times L \longrightarrow L$ in both arguments, we can define another one $\leftarrow: L \times L \longrightarrow L$ increasing in the first argument and decreasing in the second, as follows:

$$z \leftarrow_{\&} y = \sup\{x \in L \mid x \,\&\, y \preceq z\} \tag{2}$$

which, if it satisfy that the supremum is a maximum, that is, if $(z \leftarrow y) \,\&\, y \preceq z$, it is called the *residuum* or *adjoint implication* of $\&$. The pair $(\&, \leftarrow)$ is called *adjoint pair* and the residuum verify some interesting properties as:

- It is a generalization of the classical implication.
- It is monotone increasing in the first argument and decreasing in the second argument.
- The adjoint property: $x \preceq (z \leftarrow y)$ iff $(x \,\&\, y) \preceq z$, where $x, y, z \in L$.

Example 2. The residuum of the Gödel, product and Łukasiewicz t-norms, for each $z, y \in [0, 1]$, are defined, respectively, as:

$$z \leftarrow_G y = \begin{cases} 1 & \text{if } y \leq z \\ z & \text{otherwise} \end{cases} \qquad \begin{aligned} z \leftarrow_P y &= \min(1, z/y) \\ z \leftarrow_L y &= \min\{z - y + 1, 1\} \end{aligned}$$

The aim of this section is to study if an interval operator defined on $\mathcal{C}[0, 1]$ has a residuum and how we can obtain it. In the usual unit interval lattice $([0, 1], \leq)$, if we consider two adjoint pairs $(\&_i, \leftarrow_1)$, $(\&_j, \leftarrow_2)$, then we have that:

$$[z_1, z_2] {\leftarrow_{\&_{ij}}} [y_1, y_2] \neq [z_1 \leftarrow_i y_1, z_2 \leftarrow_j y_2]$$

where $[x_1, x_2], [y_1, y_2], [z_1, z_2] \in \mathcal{C}[0, 1]$ and $\&_{ij}$ is the interval operator associated to $\&_i$ and $\&_j$, as the following example shows:

Example 3. Let $(\&_P, \leftarrow_P)$, $(\&_G, \leftarrow_G)$ be the product and Gödel adjoint pairs. The interval conjunctor associated to $\&_P$, $\&_G$ is defined, for each $[a, b], [c, d] \in \mathcal{C}[0, 1]$ as $[a, b] \,\&_{PG} [c, d] = [a \,\&_P\, c, b \,\&_G\, d]$, but the operator $\leftarrow_{PG}: \mathcal{C}[0, 1] \times \mathcal{C}[0, 1] \to \mathcal{C}[0, 1]$ defined as

$$[a, b] \leftarrow_{PG} [c, d] = [a \leftarrow_P c, b \leftarrow_G d]$$

It is not well defined because $[a \leftarrow_P c, b \leftarrow_G d]$ could not be an interval since $\leftarrow_G$ is pointwise less or equal than $\leftarrow_P$. Therefore, $\leftarrow_{PG}$ is not the adjoint implication of $\&_{PG}$.

We can check that the adjoint implication of $\&_{PG}$, is equal to

$$\begin{cases} [z_1 \leftarrow_P y_1, z_2 \leftarrow_G y_2] & \text{if } z_1 \leftarrow_P y_1 \leq z_2 \leftarrow_G y_2 \\ [z_1 \leftarrow_G y_1, z_2 \leftarrow_G y_2] & \text{otherwise} \end{cases}$$

As extension of this result, on a non t-norms environment, the following theorem shows how we may obtain the adjoint implication[1] of $\&_{ij}$ from the adjoint implications of $\&_i$ and $\&_j$. This result is an extension of the given one in [1,4] and also a shorter proof it is presented.

[1] In order to not confuse the subindex of the conjunctors with the extremes of the intervals, we will write i, j as the subindex of the conjunctors $\&$.

Theorem 1. *Given the lattice of subintervals $(\mathcal{C}([0,1]), \preceq)$, two adjoint pairs $(\&_i, \leftarrow_i), (\&_j, \leftarrow_j)$, and an interval operator $\&_{ij}$ associated to $\&_i$ and $\&_j$; we have that the adjoint implication $\leftarrow_{ij} : \mathcal{C}[0,1] \times \mathcal{C}[0,1] \to \mathcal{C}[0,1]$, defined as in Eq. 2, verify that:*

$$[z_1, z_2] \leftarrow_{ij} [y_1, y_2] = [\inf\{z_1 \leftarrow_i y_1, z_2 \leftarrow_j y_2\}, z_2 \leftarrow_j y_2]$$

for all $[y_1, y_2], [z_1, z_2] \in \mathcal{C}[0,1]$.

Proof. Clearly, it is well defined and we only need to prove the equality. For that, we will consider the truth order and we will prove two inequalities. The first one is the following:

$$[z_1, z_2] \leftarrow_{ij} [y_1, y_2] \preceq [z_1, z_2]\} \preceq_t [\inf\{z_1 \leftarrow_i y_1, z_2 \leftarrow_j y_2\}, z_2 \leftarrow_j y_2]$$

Let $[y_1, y_2]$, $[z_1, z_2] \in \mathcal{C}[0,1]$ be, given $[x_1, x_2]$ such that $[x_1, x_2] \&_{ij} [y_1, y_2] \preceq [z_1, z_2]$, by the definition of $\&_{ij}$, we have that $x_1 \&_i y_1 \leq z_1$ and $x_2 \&_j y_2 \leq z_2$, therefore

$$x_1 \leq \sup\{x \mid x \&_i y_1 \leq z_1\} = z_1 \leftarrow_i y_1$$
$$x_2 \leq \sup\{x \mid x \&_j y_2 \leq z_2\} = z_2 \leftarrow_j y_2$$

and, as $x_1 \leq x_2$, since $[x_1, x_2] \in \mathcal{C}[0,1]$, we obtain that

$$[x_1, x_2] \preceq_t [\inf\{z_1 \leftarrow_i y_1, z_2 \leftarrow_j y_2\}, z_2 \leftarrow_j y_2]$$

for all $[x_1, x_2] \in \mathcal{C}[0,1]$.

Now, by the definition of supremum, we prove the first inequality:

$$\begin{aligned}[z_1, z_2] \leftarrow_{ij} [y_1, y_2] &= \sup\{[x_1, x_2] \in [0,1] \mid [x_1, x_2] \&_{ij} [y_1, y_2] \leq [z_1, z_2]\} \\ &\preceq_t [\inf\{z_1 \leftarrow_i y_1, z_2 \leftarrow_j y_2\}, z_2 \leftarrow_j y_2]\end{aligned}$$

where $[x_1, x_2], [y_1, y_2], [z_1, z_2] \in \mathcal{C}[0,1]$.

The other inequality that we need to prove is:

$$[\inf\{z_1 \leftarrow_i y_1, z_2 \leftarrow_j y_2\}, z_2 \leftarrow_j y_2] \preceq_t [z_1, z_2] \leftarrow_{ij} [y_1, y_2]$$

Applying the adjoint property, the above inequality is equivalent to:

$$[\inf\{z_1 \leftarrow_i y_1, z_2 \leftarrow_j y_2\}, z_2 \leftarrow_j y_2] \&_{ij} [y_1, y_2] \preceq_t [z_1, z_2]$$

which is equivalent to prove the inequalities:

$$\inf\{z_1 \leftarrow_i y_1, z_2 \leftarrow_j y_2\} \&_i y_1 \leq z_1$$
$$(z_2 \leftarrow_j y_2) \&_j y_2 \leq z_2$$

The second inequality is clear from the adjoint property. The first one is given by the inequality $(z_1 \leftarrow_1 y_1) \&_1 y_1 \leq z_1$ and the monotonicity of $\&_1$:

$$\begin{aligned}\inf\{z_1 \leftarrow_i y_1, z_2 \leftarrow_j y_2\} \&_i y_1 &\leq (z_1 \leftarrow_i y_1) \&_i y_1 \\ &\leq z_1\end{aligned}$$

Therefore, an interval operator $\&_{ij}$, associated to two increasing operators $\&_i$, $\&_j$, has an adjoint implication if the operators have adjoint implications. Moreover, the resultant adjoint implication $\leftarrow_{ij}$ must be equal to:

$$[\inf\{z_1 \leftarrow_i y_1, z_2 \leftarrow_j y_2\}, z_2 \leftarrow_j y_2]$$

for all $[y_1, y_2], [z_1, z_2] \in \mathcal{C}[0,1]$. Hence, if $\leftarrow_{ij}$ is defined as above, then $(\&_{ij}, \leftarrow_{ij})$ is an adjoint pair.

4 The Interval-Valued Operators Are Needed

As we comment above, the mappings $f \colon E \to (\mathcal{C}[0,1], \leq)$, called *interval-valued fuzzy set* (IVFS), are common in several mechanisms for reasoning under uncertainty, as well as they have many applications, such as decision and risk analysis, engineering design, and scheduling.

4.1 Intuitionistic Fuzzy Set

We can apply the results to Atanassov's intuitionistic fuzzy sets [2,3], because it is well known that an IVFS can be represented by an Intuitionistic Fuzzy Set (IFS) and each IFS can be represented by an IFS. For example, in [10], the authors show that both intuitionistic fuzzy sets and interval-valued fuzzy sets are equivalent to L-fuzzy sets in the sense of Goguen [12] w.r.t. a special sublattice of $[0,1] \times [0,1]$.

Definition 2. *Given a set E, an* intuitionistic fuzzy set *is the set*

$$\{\langle x, \mu(x), \nu(x)\rangle \mid x \in E\}$$

where $\mu, \nu \colon E \to [0,1]$ and $\mu(x) + \nu(x) \leq 1$ for all $x \in E$.

To understand this concept we consider an example given in [3]. Let E be the set of all countries with elective governments. Assume that we know for every country $x \in E$ the percentage of the electorate that have voted for the corresponding government. Denote it by $M(x)$ and let $\mu(x) = \dfrac{M(x)}{100}$ (degree of membership, validity, etc.). We define $\nu(x)$ (degree of non-membership, uncertainty, etc.) as the number of votes given to parties of persons outside the government.

Clearly, we can see that given a IFS, $\{\langle x, \mu(x), \nu(x)\rangle \mid x \in E\}$, we obtain the interval $[\mu(x), 1-\nu(x)]$ for each $x \in E$, therefore we can define a IVFS. Also, all operators used with IFS can be transformed for the IVFS case.

In the last years IFS were applied in different areas, such that in the process of decision making, medical diagnosis, in decision making in medicine, chemistry, credit risk assessment, etc. Also, some people have described the key process variable and corrective actions of the waste water treatment plant with biosorption using theory of IFSs.

4.2 Probabilistic Deductive Databases

Also, for dealing with probabilities is necessary to resort to intervals of probability. This is particularly clear in the Probabilistic Deductive Databases system [16], where we can express rules like:

$$\left(\text{paper\_accepted} \xleftarrow{\langle[0.7,0.95],[0.03,0.2]\rangle} \text{good\_work, good\_referees}; ind, pc\right)$$

Here we have a complex confidence value containing two probability intervals, one for the case where *paper_accepted* is true and other for the case where *paper_accepted* is false (there can exist some lack of information, or undefinedness, in these intervals). The label *ind* indicates that *good_work* is assumed to be independent from *good_referees*, while *pc* specified the way how the results of the several rules for *paper_accepted* are (disjunctively) combined.

4.3 Neural Networks

Another importante observation is about the neural net introduced in [18] and improved later in [17] in order to admit ordinal sums. A neural net based implementation of propositional $[0,1]$-valued multi-adjoint logic programming [20] with the advantage that, at least potentially, we can calculate in parallel the answer for any query, was presented.

If we want extend this implementation to $\mathcal{C}[0,1]$, we could duplicate the multi-adjoint logic program considering the left extreme of the intervals as truth values of the facts and rules in the first program and on the other truth values are the right extreme, for example:

Example 4. Let $\mathbb{P}$ be the following multi-adjoint logic program with support of truth values $(\mathcal{C}[0,1], \leq_t)$:

$$\langle A \leftarrow \&_{LG}(\&_{GL}(B,C),D),[1,1]\rangle \qquad \langle B \leftarrow [0.6,0.7]\rangle$$
$$\langle C \leftarrow [0.6,0.7]\rangle \qquad \langle D \leftarrow [0.3,0.5]\rangle$$

where A, B, C are propositional symbols, $\leftarrow$ is any implication in $\mathcal{C}[0,1]$ and $\&_{GL}$ and $\&_{LG}$ are the corresponding operator associated to the t-norms $\&_G$ and $\&_L$. We can consider the following programs $\mathbb{P}_1$ and $\mathbb{P}_2$:

$$\langle A \leftarrow \&_{LG}(\&_{GL}(B,C),D),1\rangle \qquad \langle A \leftarrow \&_{LG}(\&_{GL}(B,C),D),1\rangle$$
$$\langle B \leftarrow 0.6\rangle \qquad \langle B \leftarrow 0.7\rangle$$
$$\langle C \leftarrow 0.6\rangle \qquad \langle C \leftarrow 0.7\rangle$$
$$\langle D \leftarrow 0.3\rangle \qquad \langle D \leftarrow 0.5\rangle$$

After running the nets associated to $\mathbb{P}_1$ and $\mathbb{P}_2$ respectively, we obtain, for the propositional symbol A, the values 0 and 0.4 respectively, however the interval $[0,0.4]$ is not the truth value of the minimal model for A because we obtain an inconsistence when we calculate the value from the subformula $\&_{GL}(B,C)$, that is:

$$\&_{GL}([0.6,0.7],[0.6,0.7]) = [0.6,0.4] \notin \mathcal{C}[0,1]$$

Therefore, it is necessary to implement another neural net to consider programs with truth values in the set of subintervals of $[0, 1]$, as we remark in this paper.

5 Conclusions and Future Work

We are presented the definition of interval operator associated to two general increasing operators and the main result, that we are introduced, is about the adjoint implications of this kind of operators. Concretely, if the initial operator has an adjoint implication then the interval operator there exists and its definition is fixed. This result is necessary in several frameworks, in general ones, as in the Atanassov's intuitionistic fuzzy sets or in probabilistic frameworks, where the probabilities are given as subintervals, such that in the probabilistic deductive databases; or in a more particular application, in the interval-valued neural network implementation of the multi-adjoint $T_{\mathbb{P}}$ operator.

As future work, we aim to generalize the results given in this paper to subintervals of lattices and compare them with the given ones in [5,6,9]. Moreover, we want to study the definition of adjoint implication given, if we consider t-norms, in order to compare with some papers develops on this environment, such as [22].

References

1. Alcalde, C., Burusco, A., Fuentes-González, R.: A constructive method for the definition of interval-valued fuzzy implication operators. Fuzzy Sets and Systems 153, 211–227 (2005)
2. Atanassov, K.: Intuitionistic fuzzy sets. Fuzzy Sets and Systems 20(1), 87–96 (1986)
3. Atanassov, K.: Intuitionistic fuzzy sets. Springer/Physica-Verlag, Berlin (1999)
4. Burusco, A., Alcalde, C., Fuentes-González, R.: A characterization for residuated implications on i[0, 1]. Mathware & Soft Computing 12, 155–167 (2005)
5. Cornelis, C., Deschrijver, G., Kerre, E.E.: Implication in intuitionistic fuzzy and interval-valued fuzzy set theory: construction, classification, application. Internat. J. Approx. Reason. 35(1), 55–95 (2004)
6. Cornelis, C., Deschrijver, G., Kerre, E.E.: On the representation of intuitionistic fuzzy t-norms and t-conorms. IEEE Trans. Fuzzy Systems 12(1), 45–61 (2004)
7. Damásio, C.V., Pereira, L.M.: Monotonic and residuated logic programs. In: Benferhat, S., Besnard, P. (eds.) ECSQARU 2001. LNCS (LNAI), vol. 2143, pp. 748–759. Springer, Heidelberg (2001)
8. Dekhtyar, A., Subrahmanian, V.S.: Hybrid probabilistic programs. Journal of Logic Programming 43, 187–250 (2000)
9. Deschrijver, G.: Implication in intuitionistic fuzzy and interval-valued fuzzy set theory: construction, classification, application. Fuzzy Sets and Systems 159, 1597–1618 (2008)
10. Deschrijver, G., Kerre, E.E.: On the relationship between some extensions of fuzzy set theory. Fuzzy Sets and Systems 133, 227–235 (2003)
11. Fitting, M.C.: Bilattices and the semantics of logic programming. Journal of Logic Programming 11, 91–116 (1991)
12. Goguen, J.A.: L-fuzzy sets. Journal of Mathematical Analysis and Applications 18, 145–174 (1967)

13. Hájek, P.: Metamathematics of Fuzzy Logic. Trends in Logic. Kluwer Academic Publishers, Dordrecht (1998)
14. Jenei, S.: A more efficient method for defining fuzzy connectives. Fuzzy Sets and Systems 90(1), 25–35 (1997)
15. Julian, P., Moreno, G., Penabad, J.: On fuzzy unfolding: A multi-adjoint approach. Fuzzy Sets and Systems 154(1), 16–33 (2005)
16. Lakshmanan, L.V.S., Sadri, F.: On a theory of probabilistic deductive databases. Theory and Practice of Logic Programming 1(1), 5–42 (2001)
17. Medina, J., Mérida-Casermeiro, E., Ojeda-Aciego, M.: Descomposing ordinal sums in neural multi-adjoint logic programs. In: Lemaître, C., Reyes, C.A., González, J.A. (eds.) Advances in Artificial Intelligence. LNCS (LNAI), vol. 3315, pp. 717–726. Springer, Heidelberg (2004)
18. Medina, J., Mérida-Casermeiro, E., Ojeda-Aciego, M.: A neural implementation of multi-adjoint logic programming. Journal of Applied Logic 2(3), 301–324 (2004)
19. Medina, J., Mérida-Casermeiro, E., Ojeda-Aciego, M.: Interval-valued neural multi-adjoint logic programs. In: Mira, J., Álvarez, J.R. (eds.) Mechanisms Symbols, and Models Underlying Cognition. LNCS, vol. 3561, pp. 521–530. Springer, Heidelberg (2005)
20. Medina, J., Ojeda-Aciego, M., Vojtáš, P.: Multi-adjoint Logic Programming with Continuous Semantics. In: Eiter, T., Faber, W., Truszczyński, M. (eds.) LPNMR 2001. LNCS (LNAI), vol. 2173, pp. 351–364. Springer, Heidelberg (2001)
21. Medina, J., Ojeda-Aciego, M., Vojtáš, P.: Similarity-based unification: a multi-adjoint approach. Fuzzy Sets and Systems 146, 43–62 (2004)
22. Mesiar, R., Mesiarová, A.: Residual implications and left-continuous t-norms which are ordinal sums of semigroups. Fuzzy Sets and systems 143, 47–57 (2004)

Optimistic Arithmetic Operators for Fuzzy and Gradual Intervals - Part I: Interval Approach

Reda Boukezzoula and Sylvie Galichet

LISTIC, Laboratoire d'Informatique, Systems, Traitement de l'Information et de la Connaissance – Université de Savoie
BP. 80439 – 74944 Annecy le vieux Cedex, France
{Reda.boukezzoula,Sylvie.galichet}@univ-savoie.fr

Abstract. This two part paper proposes new and exact arithmetic operations for intervals and their extension to fuzzy and gradual ones. Indeed, it is well known that the practical use of interval and fuzzy arithmetic operators gives results more imprecise than necessary or in some cases, even incorrect. This problem is due to the overestimation effect induced by computing interval arithmetic operations. In this part, the *M*idpoint-*R*adius (*MR*) representation is considered to define new exact and optimistic subtraction and division operators. These operators are extended to fuzzy and gradual intervals in the Part II.

Keywords: Exact and optimistic Interval Arithmetic, Midpoint-Radius space.

1 Introduction

Nowadays, interval arithmetic [4], [8], [9], [10], [14] is a frequently encountered problem, especially in decision making, control theory, fuzzy systems and approximate reasoning [1], [2], [3], [11].

Usually, intervals are used in the Endpoints form (*EP*) where intervals are represented by their lower and upper bounds. However, the main drawback of this representation resides in the fact that the uncertainty (the Radius) is not directly exhibited in the interval representation [5], [6]. Moreover, for many practical situations, it is more natural to define intervals in the Midpoint-Radius (*MR*) space. Indeed, in this representation the "nominal" best value (the Midpoint) of an entity is considered together with an upper bound of its measurement error (the Radius) [5], [6], [14], [15], [16]. In this paper, the Midpoint-Radius (*MR*) representation is adopted.

It is well known that standard interval arithmetic operations give a guaranteed enclosure of the solution, which is usually too large or more imprecise. The origin of the problem is the memoryless nature of interval arithmetic operations [8], [9], [14]. It follows that these interval operators may produce some counterintuitive results where addition and subtraction (multiplication and division) are not inverse operations. Thus, a way around this problem must be searched for outside standard arithmetic operations. In this context and as mentioned in [12], [13], the exact inversion of addition and multiplication operations is fundamental in interval and fuzzy arithmetic analysis, with applications to the solving of fuzzy equations, fuzzy integral, differential equations and

E. Hüllermeier, R. Kruse, and F. Hoffmann (Eds.): IPMU 2010, Part II, CCIS 81, pp. 440–450, 2010.

interval regression. So, in order to overcome these problems and to obtain exact inverse operators, a new subtraction and division operators are proposed according to the *MR* representation. The idea explored consists in characterizing the quantity of uncertainties contained in the handled variables, for which exact inversion can be obtained.

This paper part is structured in the following way. In section 2, some concepts of intervals are introduced. Section 3 is devoted to the presentation of the new interval arithmetic operators. Section 4 is dedicated to the overestimation errors induced by the conventional operators with regard to the new proposed ones. Concluding remarks are finally given in Section 5.

2 Relevant Concepts and Notations

2.1 Interval Concepts

A real interval a is a continuous bounded subset of $\mathbb{R}$ defined by the set of elements lying between its endpoints (lower and upper limits) a^- and a^+. Given an interval a, its Midpoint M_a, its Radius R_a and its Width W_a are defined by (see Fig. 1):

$$M_a = (a^- + a^+)/2\ ,\ R_a = (a^+ - a^-)/2\ ,\ R_a \geq 0 \text{ and } W_a = (a^+ - a^-) = 2.R_a \quad (1)$$

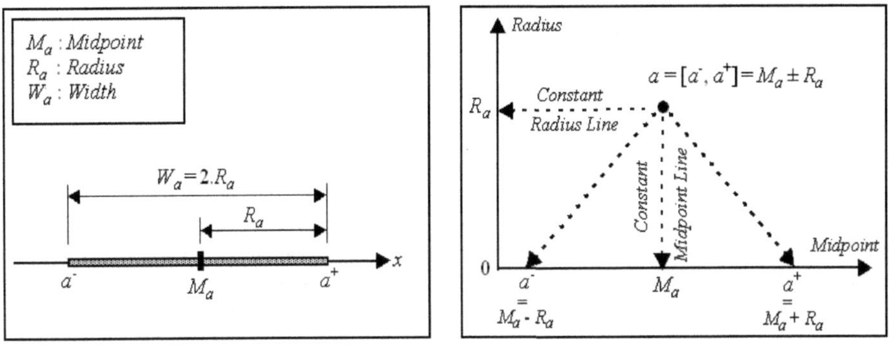

a. Endpoints Interval Representation   b. *MR* Interval Representation

Fig. 1. *EP* and *MR* Representations

In this paper, to avoid confusion, the equality between real numbers is denoted by = and the equality between intervals is represented by ⊜. An interval a can be characterized by its endpoints a^- and a^+ or by its Midpoint M_a and Radius R_a. Indeed, in the Endpoints (*EP*) representation, the interval a is denoted a ⊜$[a^-, a^+]$. The same interval can be represented by the pair (M_a, R_a), where $R_a \geq 0$ in the Midpoint-Radius (*MR*) space. In this case the interval a is denoted a⊜(M_a, R_a). The relation between the *EP* and *MR* notations is straightforward, i.e.

$$a^- = M_a - R_a \text{ and } a^+ = M_a + R_a \quad (2)$$

The equality between two intervals a and b is interpreted as follows:

$$a \ominus b \Leftrightarrow \begin{cases} M_a = M_b \text{ and } R_a = R_b & \text{for } MR \text{ notation} \\ a^- = b^- \text{ and } a^+ = b^+ & \text{for } EP \text{ notation} \end{cases} \quad (3)$$

Given an interval a, two particular cases can be distinguished:

- If $R_a = 0 \Leftrightarrow (a^- = a^+) \Rightarrow a \ominus (M_a, 0)$: a is a thin (point) interval.
- If $M_a = 0 \Leftrightarrow (a^+ = -a^-) \Rightarrow a \ominus (0, R_a)$: a is a zero-symmetric interval.

According to the MR representation, the relative extent of an interval denoted RX_a [5][6] is defined by:

$$RX_a = R_a / M_a \text{ ; for } M_a \neq 0 \quad (4)$$

As illustrated on the polar graph (see Fig. 2), the RX definition leads to:

$$|RX_a| \geq 1, \text{ if } : 0 \in a \text{ ; and} : |RX_a| < 1, \text{ if } : 0 \notin a \quad (5)$$

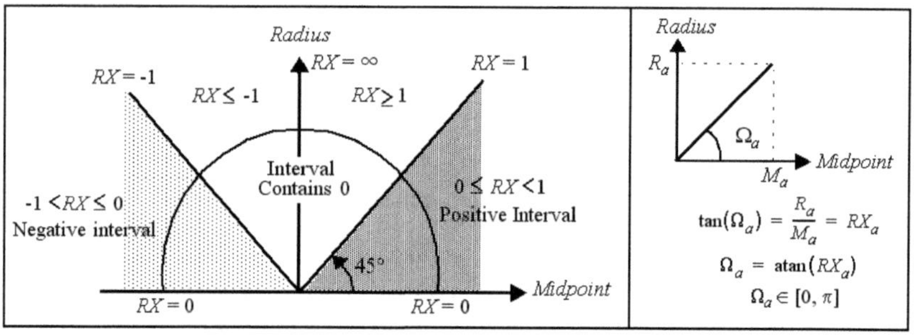

Fig. 2. The Polar Graph of the *RX* function

In the sequel, the set of all intervals on $\mathbb{R}$ is denoted $\mathbb{I}(\mathbb{R})$. In the same way, the set of all intervals which do not contain zero is denoted:

$$\mathbb{I}^*(\mathbb{R}) = \{a \in \mathbb{I}(\mathbb{R})\ R_a < |M_a|\} = \{a \in \mathbb{I}(\mathbb{R})\ |RX_a| < 1\}$$

2.2 Standard Interval Arithmetic

The standard interval arithmetic operations can be defined in the MR space by the following expressions [7]:

- *Addition*: for $a \in \mathbb{I}(\mathbb{R})$ and $b \in \mathbb{I}(\mathbb{R})$, the addition is straightforward, i.e.

$$a \oplus b \ominus (M_a, R_a) \oplus (M_b, R_b) \ominus (M_a + M_b, R_a + R_b) \quad (6)$$

- *Subtraction*: for $a \in\in \mathbb{I}(\mathbb{R})$ and $b \in \mathbb{I}(\mathbb{R})$, this operation is defined by:

$$a \ominus b \ominus (M_a, R_a) \ominus (M_b, R_b) \ominus (M_a - M_b, R_a + R_b) \quad (7)$$

- *Multiplication by a scalar*: for $a \in \mathbb{I}(\mathbb{R})$ and for any scalar $\sigma \in \mathbb{R}$

$$\sigma \circledast a \ominus \sigma \circledast (M_a, R_a) \ominus (\sigma . M_a, |\sigma| R_a) \quad (8)$$

- *Multiplication*: for $a \in \mathbb{I}(\mathbb{R})$ and $b \in \mathbb{I}(\mathbb{R})$, the multiplication operation:

$$a \otimes b \ominus (M_a, R_a) \otimes (M_b, R_b) \tag{9}$$

can be simplified according to the values of the interval operands:

1. If $|RX_a| \leq 1$ and $|RX_b| \leq 1$

$$a \otimes b \ominus (M_a M_b + sign(M_a M_b).R_a R_b,\ |M_a| R_b + |M_b| R_a) \tag{10}$$

2. If $|RX_a| > 1 \geq |RX_b|$ or $|RX_a| \geq |RX_b| > 1$

$$a \otimes b \ominus \theta_b \circledast (M_a, R_a) \ominus (\theta_b.M_a, |\theta_b| R_a);\ \text{where : } \theta_b = sign(M_b).(|M_b| + R_b) \tag{11}$$

3. If $|RX_b| > 1 \geq |RX_a|$ or $|RX_b| \geq |RX_a| > 1$

$$a \otimes b \ominus \theta_a \circledast (M_b, R_b) \ominus (\theta_a.M_b, |\theta_a| R_b);\ \text{where: } \theta_a = sign(M_a).(|M_a| + R_a) \tag{12}$$

- *Inversion*: for $a \in \mathbb{I}^*(\mathbb{R})$, this operation can be defined by:

$$\oslash a \ominus \delta_a \circledast a\ ;\ \text{ where: } \delta_a = (M_a^2 - R_a^2)^{-1} \tag{13}$$

The inverse (reciprocal) of an interval is reduced to a multiplication by a scalar.

- *Division*: for $a \in \mathbb{I}(\mathbb{R})$ and $b \in \mathbb{I}^*(\mathbb{R})$, this operation is defined as a multiplication by an inverse:

$$a \oslash b \ominus (M_a, R_a) \oslash (M_b, R_b) \ominus a \otimes (\oslash b) \ominus a \otimes (\delta_b \circledast b) \ominus \delta_b \circledast (a \otimes b) \tag{14}$$

The division is reduced to a multiplication operator weighted by a scalar.

3 Optimistic and Exact Inverse Operators

According to the standard interval arithmetic it can be stated that $b \oplus (a \ominus b) \not\ominus a$ and $b \otimes (a \oslash b) \not\ominus a$. Moreover, as $a \ominus a \not\ominus 0$ and $a \oslash a \not\ominus 1$, it is obvious that the used operators produce counterintuitive results. In this case, it follows that the x solution of the equation $a \oplus x \ominus d$ is not, as we would expect, $x \ominus d \ominus a$. The same annoyance appears when solving the equation $a \otimes x \ominus e$ whose solution is not given by $x \ominus e \oslash a$ as expected. Indeed, the usual interval operators give results more imprecise than necessary. This problem is related to the lack of inverses in the calculus of interval quantities. In this context, as the addition and subtraction (resp. multiplication and division) are not reciprocal operations, it is not possible to solve inverse problems exactly using these operators. Thus, a way around this problem must be searched for outside standard arithmetic operations. In the context of optimistic interval computing, we propose to use new subtraction and division operators, denoted respectively $\boxminus$ and $\boxdiv$ which are exact inverse of the addition $\oplus$ and multiplication $\otimes$ operators.

3.1 The Proposed Operator $\boxminus$

A. ***Proposition*** **1:** for $a \in \mathbb{I}(\mathbb{R})$ and $b \in \mathbb{I}(\mathbb{R})$, the exact operator $\boxminus$ is defined by:

$$a \boxminus b ⊜ (M_a, R_a) \boxminus (M_b, R_b) ⊜ (M_a\text{-}M_b, R_a\text{-}R_b) ⊜ (M_\Phi, R_\Phi) ⊜ \Phi \quad (15)$$

- ***Proof***

This proof is straightforward. Indeed, the operator $\boxminus$ is the exact inverse of $\oplus$ if:

$$b \oplus (a \boxminus b) ⊜ (a \boxminus b) \oplus b ⊜ a \quad (16)$$

By substituting (15) in (16), it follows:

$$(M_b, R_b) \oplus (M_\Phi, R_\Phi) ⊜ (M_b, R_b) \oplus (M_a - M_b, R_a - R_b) ⊜ (M_a, R_a) ⊜ a$$

***B. Proposition* 2:** For $a \in \mathbb{I}(\mathbb{R})$ and $b \in \mathbb{I}(\mathbb{R})$, the result produced by the exact difference operator $\boxminus$ is an interval if and only if:

$$R_a \geq R_b \quad (17)$$

In other words, the interval a must be broader than b.

- ***Proof***

According to equation (15), it is clear that $a \boxminus b$ can always be computed. However, the difference operation is only valid when the obtained result Φ is an interval, which means that the operator $\boxminus$ produces an interval if Φ satisfies the following condition:

$$R_\Phi \geq 0 \Rightarrow R_a - R_b \geq 0 \Rightarrow R_a \geq R_b \quad (18)$$

3.2 The Proposed Operator $\boxdiv$

***A. Proposition* 3:** for $a \in \mathbb{I}(\mathbb{R})$ and $b \in \mathbb{I}^*(\mathbb{R})$, the exact division operator $\boxdiv$ is given by:

$$a \boxdiv b ⊜ (M_a, R_a) \boxdiv (M_b, R_b) ⊜ (M_\Phi, R_\Phi) ⊜ \Phi; \text{ where:} \quad (19)$$

$$M_\Phi = \begin{cases} M_a/\theta_b : \text{if } |RX_a| > 1 \\ M_a/(M_b + \psi_\Phi . R_b . sign(M_a)) : \text{if } |RX_a| \leq 1 \end{cases} \quad (20)$$

$$\text{and: } R_\Phi = \begin{cases} R_a/|\theta_b| : \text{if } |RX_a| > 1 \\ \psi_\Phi . M_\Phi : \text{if } |RX_a| \leq 1 \end{cases};$$

$$\text{with: } \theta_b = sign(M_b).(|M_b| + R_b) \text{ and } \psi_\Phi = \frac{|RX_a| - |RX_b|}{sign(M_a.M_b) - RX_a.RX_b}$$

It follows that:

$$RX_\Phi = \begin{cases} sign(M_b).RX_a; \text{if} : |RX_a| > 1 \\ \psi_\Phi ; \text{if } |RX_a| \leq 1 \end{cases}$$

- ***Proof***

The operator $\boxdiv$ is the exact inverse of $\otimes$ if the following equation is verified:

$$b \otimes (a \boxdiv b) ⊜ (a \boxdiv b) \otimes b ⊜ a \quad (21)$$

As $b \in \mathbb{I}^*(\mathbb{R})$ $(0 \notin b)$, it can be stated that $|RX_b|<1$. In this case, by substitution of the expression of Φ given by (20) in equation (21) and according to the relative extent of the interval a, two cases can be distinguished:

1. Case 1: If $|RX_a|>1 \Rightarrow |RX_\Phi|>1$, according to equation (11), the expression of equation (21) is reduced to:

$$b \otimes (a \boxdiv b) \ominus b \otimes \Phi = \theta_b \circledast (M_\Phi, R_\Phi) = (\theta_b.M_\Phi, |\theta_b|.R_\Phi) \ominus (M_a, R_a)$$

which leads to $M_\Phi = M_a/\theta_b$ and $R_\Phi = R_a/|\theta_b|$.

2. Case 2: If $|RX_a|\leq 1 \Rightarrow |RX_\Phi|\leq 1$, according to equation (10), the expression of equation (21) can be written as:

$$b \otimes \Phi \ominus (M_b.M_\Phi + sign(M_b.M_\Phi).R_b.R_\Phi,\ |M_b|R_\Phi + |M_\Phi|R_b) \ominus (M_a, R_a)$$

So, according to the interval equality definition (see equation (3)), it follows:

$$\begin{cases} |M_b|R_\Phi + |M_\Phi|R_b = R_a \\ M_b M_\Phi + sign(M_b M_\Phi) R_b R_\Phi = M_a \end{cases}$$

In this case, as $sign(M_\Phi) = sign(M_b M_a) \Rightarrow sign(M_b M_\Phi) = sign(M_a)$, it follows:

$$\frac{|M_b|.|M_\Phi|.(|RX_\Phi| + |RX_b|)}{|M_b||M_\Phi|.(sign(M_a) + RX_b.RX_\Phi)} = RX_a \Rightarrow \frac{|\psi_\Phi| + |RX_b|}{sign(M_a) + RX_b.\psi_\Phi} = RX_a$$

Thus,

$$|\psi_\Phi| = \frac{|RX_a| - |RX_b|}{(1 - |RX_a|.|RX_b|)} \quad \text{or:} \quad \psi_\Phi = \frac{|RX_a| - |RX_b|}{(sign(M_a.M_b) - RX_a.RX_b)} \Rightarrow R_\Phi = \psi_\Phi.M_\Phi$$

By substitution of $R_\Phi = \psi_\Phi.M_\Phi$ it follows that $M_\Phi = M_a/(M_b + \psi_\Phi.R_b.sign(M_a))$.

***B. Proposition* 4:** For $a \in \mathbb{I}(\mathbb{R})$ and $b \in \mathbb{I}^*(\mathbb{R})$, the result produced by the exact division operator $\boxdiv$ is an interval if and only if:

$$|RX_a| \geq |RX_b| \tag{22}$$

In other words, the interval a must be more extended than b.

- ***Proof***

As $b \in \mathbb{I}^*(\mathbb{R})$ $(0 \notin b)$, it is obvious that $|RX_b|<1$. In this context, according to the interval a, two cases can be distinguished

1. Case 1: If $|RX_a|>1 \Rightarrow |RX_\Phi|>1$:

In this case, according to the definition of Φ (see equation (20)), it is obvious that R_Φ is always positive and the solution Φ represents always an interval.

2. Case 2 : If $|RX_a|\leq 1 \Rightarrow |RX_\Phi|\leq 1$:

The result Φ is an interval if and only if: $R_\Phi \geq 0 \Leftrightarrow \psi_\Phi.M_\Phi \geq 0$. In this case, it is obvious that M_Φ and ψ_Φ must have the same *sign*. So, two cases are considered:

a. If $M_\Phi > 0 \Rightarrow sign(M_\Phi) = sign(M_a M_b) = 1 \Rightarrow sign(M_a M_b) - RX_a.RX_b > 0$.

In this case, $\psi_\Phi \geq 0$ is verified if and only if: $|RX_a| - |RX_b| \geq 0 \Rightarrow |RX_a| \geq |RX_b|$.

b. If $M_\Phi < 0 \Rightarrow sign(M_a M_b) = -1 \Rightarrow sign(M_a M_b) - RX_a.RX_b < 0$

In this case, $\psi_\Phi < 0$ is verified if and only if: $|RX_a| - |RX_b| \geq 0 \Rightarrow |RX_a| \geq |RX_b|$

4 Overestimation between Interval Operators

From practical point of view it is important to be able to calculate and estimate the overestimation error between the conventional and the new proposed operators.

According to the definition of the *RX* function (see equation (4)), it can be stated that the latter characterise the relative extent of an interval with respect to its midpoint position, i.e., the relative degree of uncertainty of the number approximated by the interval. So, the *RX* function is used in this paper since it is natural, in the *MR* representation, for quantifying the degree of uncertainty in intervals computing.

Let us suppose that z and p are respectively resulted intervals from two distinct interval operators O_z and O_p. Let us also define:

$$\Delta_{RX} = |RX_z| - |RX_p|; \text{ for: } M_z \neq M_p$$

as an indicator for the uncertainty quantification error between the intervals z and p. We call the interval z more extended (or more uncertain) than the interval p when $\Delta_{RX} > 0$, i.e., $|RX_z| > |RX_p|$. In this case, the indicator Δ_{RX} can be interpreted as an overestimation error between the operator O_z and O_p. In the opposite case, i.e., $\Delta_{RX} < 0$ the interval z is less extended than p. Obviously, intervals are always more extended (more uncertain) than real numbers. When $\Delta_{RX} = 0$ ($|RX_z| = |RX_p|$), the intervals z and p have the same extent (are equally uncertain). For all zero-symmetric intervals, we assume that the *RX* function is undefined.

It can be stated that if the intervals z and p are centred on the same midpoint, i.e., $M_z = M_p$ or zero-symmetric ones, the overestimation between intervals can be quantified and interpreted by the radius difference, i.e.,

$$\Delta_R = R_z - R_p$$

A. Overestimation error *between* $\ominus$ *and* $\boxminus$

According to the definition of the operators $\ominus$ and $\boxminus$ (see equation (7) and (15)), it can be stated that the resulted intervals $s_1 \ominus a \ominus b$ and $s_2 \ominus a \boxminus b$ are centred on the same midpoint value. As illustrated in Fig. 3, it can be observed that the conventional operator $\ominus$ is more uncertain than the new difference $\boxminus$. In this case, the overestimation error between the two operators is given by the following equation:

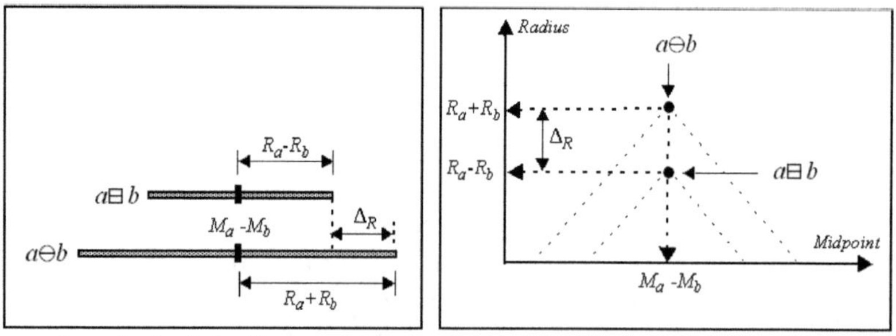

a. Over-estimation in the *EP* representation b. Over-estimation in the *MR* representation

Fig. 3. Overestimation error between $\ominus$ and $\boxminus$

$$\Delta_R = R_{s_1} - R_{s_2} = R_a + R_b - (R_a - R_b) = 2.R_b \tag{23}$$

B. Overestimation error between $\oslash$ *and* $\boxdiv$

In this case, as $0 \notin b$ ($|RX_b|<1$) it is obvious that $M_b \neq 0$. So, according to the position of the midpoint of the interval a, three different cases are to be distinguished:

- If $|RX_a|>1$ and $M_a = 0$:

In this case, the conventional division operator $\oslash$ can be simplified to:

$$a \oslash b \circeq \delta_b \circledast (a \otimes b) \circeq \delta_b \circledast (0, |\theta_b| R_a) \circeq (0, |\theta_b||\delta_b| R_a)$$

As $|RX_b|<1 \Rightarrow \delta_b = (M_b^2 - R_b^2)^{-1} > 0$, the following result is obtained:

$$a \oslash b \circeq (0, |\theta_b|.\delta_b.R_a) \tag{24}$$

In the same time, according to the definition of the operator $\boxdiv$ it can be written:

$$a \boxdiv b \circeq (0, R_a/|\theta_b|) \tag{25}$$

According to equations (24) and (25), the resulted interval $d_1 \circeq a \oslash b$ and $d_2 \circeq a \boxdiv b$ are zero-symmetrical intervals. As illustrated in Fig. 4, it can be deduced that the operator $\oslash$ is more uncertain than the exact division one. In this case, the overestimation error is given by the following equation:

$$\Delta_R = R_{d_1} - R_{d_2} = |\theta_b|.\delta_b.R_a - R_a/|\theta_b| = 2.R_a R_b.\delta_b \tag{26}$$

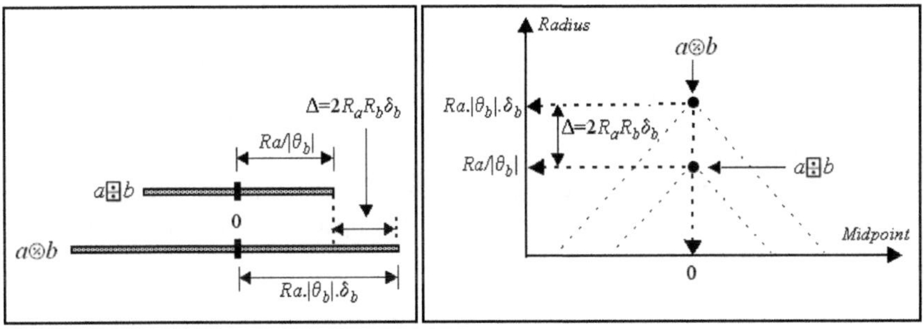

a. Over-estimation in the *EP* representation b. Over-estimation in the *MR* representation

Fig. 4. Overestimation error between $\oslash$ and $\boxdiv$ (case: $|RX_a|>1$ and $M_a = 0$)

- If $|RX_a|>1$ and $M_a \neq 0$:

In this case, the division operation $\oslash$ gives the following result:

$$a \oslash b \circeq \delta_b \circledast (a \otimes b) \circeq \delta_b \circledast (\theta_b M_a, |\theta_b| R_a) \circeq (\delta_b.\theta_b.M_a, |\theta_b|.\delta_b.R_a) \tag{27}$$

From the definition of the operator $\boxdiv$, it follows:

$$a \boxdiv b \circeq (M_a/\theta_b, R_a/|\theta_b|) \tag{28}$$

From equations (27) and (28) it can be stated the intervals $d_1 \circeq a \oslash b$ and $d_2 \circeq a \boxdiv b$ are not centered on the same midpoint. In this case, the following result is obtained:

$$\Delta_{RX} = |RX_{d_1}| - |RX_{d_2}| = RX_a.sign(M_b) - RX_a.sign(M_b) = 0$$

So, the two operators have the same extent. In other words, the two operations produce the same relative uncertainty. In this case, let us determine the Midpoint and Radius translations between the operators (see Fig.5):

$$|\Delta_M| = |\delta_b.\theta_b.R_a - M_a/\theta_b| = |2.M_a R_b.\delta_b.sign(M_b)| = 2|M_a|R_b.\delta_b$$

$$\text{and : } \Delta_R = |\theta_b|.\delta_b.R_a - R_a/|\theta_b| = 2.R_a R_b.\delta_b \geq 0$$

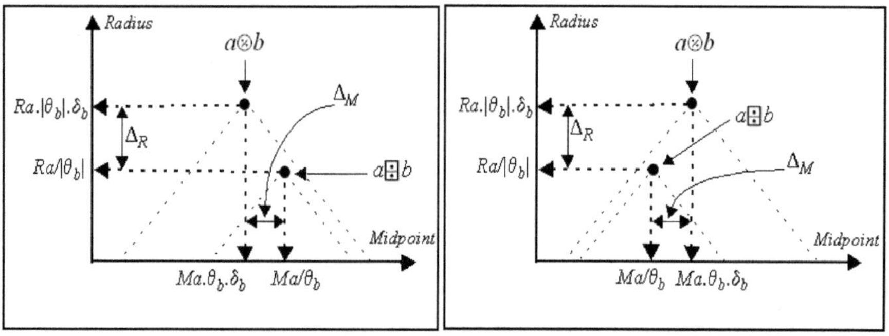

a. Over-estimation in the MR representation $sign(M_a) \neq sign(M_b)$

b. Over-estimation in the MR representation $sign(M_a) = sign(M_b)$

Fig. 5. Overestimation error between $\oslash$ and $\boxed{\div}$ (case: $|RX_a|>1$ and $M_a \neq 0$)

According to Fig. 5, it can be stated that even if the operators $\oslash$ and $\boxed{\div}$ have the same extent, the conventional operator $\oslash$ is broader than the proposed one.

- If $|RX_a| \leq 1$:

In this case, like the previous case, the intervals $a\oslash b$ and $a\boxed{\div}b$ are not centered on the same midpoint. According to the definition of the operator $\boxed{\div}$, the following equation can be determined:

$$|RX_{d_2}| = |RX_\Phi| = (|RX_a| - |RX_b|)/(1 - |RX_a|.|RX_b|)$$

According to the definition of the operator $\otimes$, the following equation is obtained:

$$|RX_{a\otimes b}| = (|RX_a| + |RX_b|)/(1 + |RX_a|.|RX_b|)$$

As, the division operator $\oslash$ is reduced to a multiplication operator weighted by a scalar, it can be deduced that $|RX_{a\otimes b}| = |RX_{d_1}|$.
In this case, the overestimation error between the two operators is given by:

$$\Delta_{RX} = |RX_{d_1}| - |RX_{d_2}| = 2|RX_b|(1 - RX_a^2)/(1 - RX_a^2.RX_b^2) \geq 0 \qquad (29)$$

The overestimation error illustration when a and b are positive is given in Fig. 6.

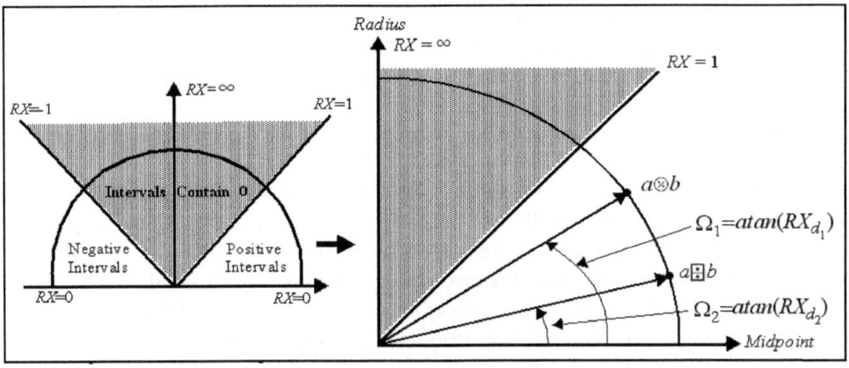

Fig. 6. Overestimation error between ⊗and ⊡ (case $|RX_a| \leq 1$)

In this case:

$$RX_{d_1} = (RX_a + RX_b)/(1 + RX_a.RX_b) = \tan(\Omega_1) \Rightarrow \Omega_1 = \mathrm{a}\tan(RX_{d_1})$$

$$RX_{d_2} = (RX_a - RX_b)/(1 - RX_a.RX_b) = \tan(\Omega_2) \Rightarrow \Omega_2 = \mathrm{a}\tan(RX_{d_2})$$

It is important to note here that the proposed approach has an evident methodological value. Let us only point out that it provides an exact quantification and a visual illustration of the overestimation errors between the conventional operators and the exact inverse proposed ones. Moreover, it can be proven that the inclusion property is ensured, i.e., $a \boxminus b \subseteq a \ominus b$ and $a ⊡ b \subseteq a \oslash b$.

5 Conclusion

This paper presents a new methodology for the implementation of the subtraction and the division operators between intervals with their existence conditions. Based on these new operators, an optimistic counterpart of the usual interval arithmetic has been proposed. The proposed operators are exact inverses of the addition and multiplication operators and can solve the overestimation problem well known in interval arithmetic. This paper deals with subtraction and division operations, of course, other operators can be defined and illustrated in a similar manner.

References

1. Boukezzoula, R., Foulloy, L., Galichet, S.: Inverse Controller Design for Interval Fuzzy Systems. IEEE Transactions On Fuzzy Systems 14(1), 111–124 (2006)
2. Ceberio, M., Kreinovich, V., Chopra, S., Longpré, L., Nguyen, H.T., Ludäscher, B., Baral, C.: Interval-type and affine arithmetic-type techniques for handling uncertainty in expert systems. Journal of Computational and Applied Mathematics 199, 403–410 (2007)
3. Galichet, S., Boukezzoula, R.: Optimistic Fuzzy Weighted Average. In: Int. Fuzzy Systems Association World Congress (IFSA/EUSFLAT), Lisbon, Portugal, pp. 1851–1856 (2009)

4. Kaufmann, A., Gupta, M.M.: Introduction to fuzzy arithmetic: Theory and Applications. Van Nostrand Reinhold Company Inc., New York (1991)
5. Kulpa, Z.: Diagrammatic representation for interval arithmetic. Linear Algebra and its Applications 324(1-3), 55–80 (2001)
6. Kulpa, Z., Markov, S.: On the inclusion properties of interval multiplication: a diagrammatic study. BIT Numerical Mathematics 43, 791–810 (2003)
7. Markov, S.: Computation of Algebraic Solutions to Interval Systems via Systems of Coordinates. In: Kraemer, W., Wolff von Gudenberg, J. (eds.) Scientific Computing, Validated Numerics, Interval Methods, pp. 103–114. Kluwer, Dordrecht (2001)
8. Moore, R.E.: Interval Analysis. Prentice-Hall, Englewood Cliffs (1966)
9. Moore, R.E.: Methods and applications of interval analysis. SIAM, Philadelphia (1979)
10. Moore, R., Lodwick, W.: Interval analysis and fuzzy set theory. Fuzzy Sets and Systems 135(1), 5–9 (2003)
11. Rauh, A., Kletting, M., Aschemann, H., Hofer, E.P.: Reduction of overestimation in interval arithmetic simulation of biological wastewater treatment processes. Journal of Computational and Applied Mathematics 199(2), 207–212 (2007)
12. Stefanini, L., Bede, B.: Generalization of Hukuhara differentiability of interval-valued functions and interval differential equations. Nonlinear Analysis 71, 1311–1328 (2009)
13. Stefanini, L.: A generalization of Hukuhara difference and division for interval and fuzzy arithmetic. Fuzzy sets and systems (2009), doi:10.1016/j.fss.2009.06.009
14. Sunaga, T.: Theory of an Interval Algebra and its application to Numerical Analysis. RAAG Memories 2, 547–564 (1958)
15. Warmus, M.: Calculus of Appoximations. Bulletin Acad. Polon. Science, C1. III IV, 253–259 (1956)
16. Warmus, M.: Approximations and inequalities in the calculus of approximations: classification of approximate numbers. Bulletin Acad. Polon. Science, Ser. Math. Astr. et Phys. IX, 241–245 (1961)

Optimistic Arithmetic Operators for Fuzzy and Gradual Intervals - Part II: Fuzzy and Gradual Interval Approach

Reda Boukezzoula and Sylvie Galichet

LISTIC, Laboratoire d'Informatique, Systems, Traitement de l'Information et de la Connaissance – Université de Savoie
BP. 80439 – 74944 1nnecy le vieux Cedex, France
{Reda.boukezzoula,Sylvie.galichet}@univ-savoie.fr

Abstract. This part aims at extending the proposed interval operators detailed in the Part I to fuzzy and gradual intervals. Recently, gradual numbers have been introduced as a means of extending standard interval computation methods to fuzzy and gradual intervals. In this paper, we combine the concepts of gradual numbers and the *M*idpoint-*R*adius (*MR*) representation to extend the interval proposed operators to fuzzy and gradual intervals. The effectiveness of the proposed operators is illustrated by examples.

Keywords: Fuzzy and Gradual Intervals, Exact Operators, Midpoint-Radius Representation.

1 Introduction

It is well known that the usual arithmetic operations on real numbers can be extended to the ones defined on fuzzy numbers [5], [15], [16], [21] by means of Zadeh's extension principle [22], [23]. A direct implementation of this principle is computationally expensive due to the requirement of solving a nonlinear programming problem [4], [18]. To overcome this deficiency, many researchers have investigated this problem by observing the fuzzy numbers as a collection of α-levels [1], [9], [10], [13]. In this context, fuzzy arithmetic is performed using conventional interval arithmetic according to the α-cut representation.

Recently, Fortin *et al.* introduced the concept of gradual numbers [8] which provides a new outlook on fuzzy intervals. Indeed, a gradual number is defined by an assignment function able to represent the essence of graduality. In this case, a fuzzy interval can be represented as a pair of gradual numbers (lower and upper bounds), which allows the extension of standard interval computation methods to fuzzy intervals. Such bounds are called profiles (left and right profiles) [8]. In this case, the interval arithmetic operations defined for intervals can be directly extended to fuzzy intervals [2], [3], [7], [8]. As detailed in [8], only monotonic gradual numbers are useful to define fuzzy intervals. However, some fuzzy interval computations can lead to non-monotonic gradual numbers which are not fuzzy subsets and cannot be represented by fuzzy intervals since the interval boundaries are not monotonic. In order to overcome this problem, the notion of an interval of gradual numbers or gradual

E. Hüllermeier, R. Kruse, and F. Hoffmann (Eds.): IPMU 2010, Part II, CCIS 81, pp. 451–460, 2010.

interval for short is adopted in this paper [8]. In this case, no monotonicity assumption is imposed on the gradual interval boundaries. In this framework, a fuzzy interval can be viewed as a particular case of a gradual interval where the interval boundaries are monotonic. In this second Part, the interval arithmetic operations defined for intervals in the first Part are directly extended to the gradual and fuzzy ones. Indeed, according to the gradual concept, it can be stated that a gradual interval can be viewed as a conventional interval in a space of functions (interval bounds functions) [6][8].

This paper part is structured in the following way. In section 2, some concepts of fuzzy and gradual intervals are introduced. Section 3 is devoted to the extension of the interval operators to gradual intervals. Section 4 is dedicated to illustrative examples. Concluding remarks are finally given in Section 5.

2 Relevant Concepts and Notations

2.1 Fuzzy and Gradual Intervals

An interval a can be represented by a membership function μ_a which takes the value 1 over the interval and 0 anywhere else (see Fig. 1.a). Indeed, this representation supposes that all possible values of the interval a belong to it with the same membership degree [14], [17]. In this case, an interval can be viewed as a Boolean uncertainty representation. So, a value in the interval is possible and a value outside it is impossible. As mentioned in [6], [8], [12], the idea of fuzziness is to move from the Boolean context to a gradual one. Hence, fuzziness makes the boundaries of the interval softer and thus making the uncertainty gradual (the transition from an impossible value to an entirely possible one is gradual). In order to represent the essence of graduality, the concept of gradual numbers has been recently proposed [6], [8]. Indeed, a gradual number is defined by an assignment function from $(0,1]\rightarrow\mathbb{R}$. In other words, it is simply a number which is parameterized by $\lambda\in(0,1]$.

According to the concept of gradual numbers, a gradual interval $A(\lambda)$ can de described by an ordered pair of two gradual numbers $A^-(\lambda)$ and $A^+(\lambda)$, where $A^-(\lambda)$ is a gradual lower bound and $A^+(\lambda)$ is a gradual upper bound of $A(\lambda)$ (see Fig. 1.b). In this context, no monotonicity assumption is imposed on the gradual interval boundaries. It is obvious that for an unimodal interval A we have $A^-(1)= A^+(1)= A(1)$ (see Fig. 1). It is important to note here that if the boundaries of conventional intervals are real numbers (points), the boundaries of gradual intervals are functions. Thus, in the same way that the interval a is denoted $[a^-, a^+]$ in the End-Points (*EP*) representation, the gradual interval A will be denoted $[A^-(\lambda), A^+(\lambda)]$ in its End-Functions (*EF*) space. Moreover, a unimodal gradual interval defines a unimodal fuzzy interval only if $A^-(\lambda)$ and $A^+(\lambda)$ must satisfy the following properties: $A^-(\lambda)$ is an increasing function and $A^+(\lambda)$ is a decreasing function. So, a fuzzy interval can be viewed as a particular case of a gradual interval. In this paper, we will additionally assume that $A^-(\lambda)$ and $A^+(\lambda)$ are continuous and their domains are extended to interval [0,1] ($A^-(0)$ and $A^+(0)$ are defined). In consequence, the corresponding membership function of a fuzzy interval is continuous and has a compact support. For more details on gradual numbers and their relationships with fuzzy intervals, we refer the reader to [6], [8], [12].

In this framework, a unimodal fuzzy interval A with a support $A(0)⊜[A^{-}(0), A^{+}(0)]$ and a kernel $A(1)$, is defined by:

For $\lambda \in [0,1]$:

$$A(\lambda) ⊜ [A^{-}(\lambda), A^{+}(\lambda)]\text{; where: }\begin{cases} A^{-}(\lambda) = Inf\{x / \mu_A(x) \geq \lambda; x \geq A^{-}(0)\} \\ A^{+}(\lambda) = Sup\{x / \mu_A(x) \geq \lambda; x \leq A^{+}(0)\} \end{cases} \quad (1)$$

According to the *MR* representation, the gradual interval $A(\lambda)$ is given by:

$$A(\lambda) ⊜ (M_{A(\lambda)}, R_{A(\lambda)}) \text{ ; with : } R_{A(\lambda)} \geq 0 \quad (2)$$

The relation between the *EF* and the *MR* representations remains straightforward, i.e.,

$$A(\lambda) ⊜ [A^{-}(\lambda), A^{+}(\lambda)] ⊜ [M_{A(\lambda)} - R_{A(\lambda)}, M_{A(\lambda)} + R_{A(\lambda)}] \quad (3)$$

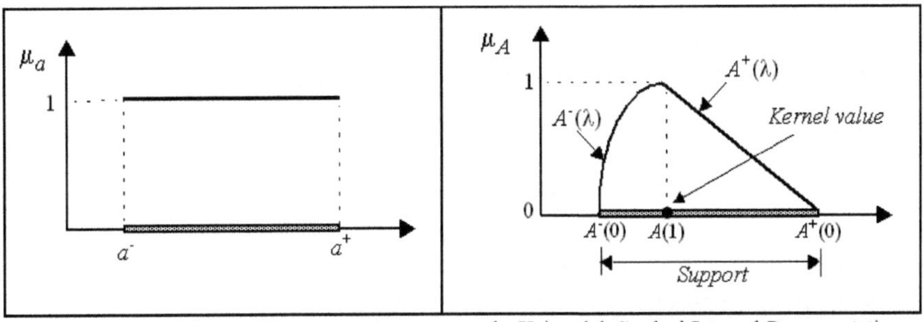

a. Conventional Interval Representation b. Unimodal Gradual Interval Representation

Fig. 1. Conventional and Unimodal Gradual Interval Representations

In this paper, the set of gradual intervals is denoted by $\mathbb{Q}(\mathbb{R})$ ($\mathbb{Q}^*(\mathbb{R})$ for gradual intervals which do not contain zero in their support). It is important to note here that $A(\lambda)$ is a gradual interval if and only if the condition $R_{A(\lambda)} \geq 0$ is respected. In other words, $A^{-}(\lambda)$ and $A^{+}(\lambda)$ have to form an ordered pair of gradual numbers ($A^{-}(\lambda) \leq A^{+}(\lambda)$).

2.2. Fuzzy and Gradual Arithmetic Operations

For two gradual intervals $A(\lambda)⊜(M_{A(\lambda)}, R_{A(\lambda)})$ and $B(\lambda)⊜(M_{B(\lambda)}, R_{B(\lambda)})$, all the equations developed previously for conventional intervals remain valid for gradual ones where a and b are respectively replaced by $A(\lambda)$ and $B(\lambda)$. For example, the division operator given by equation (14) in Part I becomes:

$$\text{for } A \in \mathbb{Q}(\mathbb{R}) \text{ and } B \in \mathbb{Q}^*(\mathbb{R})\text{: } A(\lambda) ⊘ B(\lambda) ⊜ \delta_{B(\lambda)} ⊛ (A(\lambda) \otimes B(\lambda)); \quad (4)$$

$$\text{where: } \delta_{B(\lambda)} = (M^2_{B(\lambda)} - R^2_{B(\lambda)})^{-1}$$

3 Optimistic and Exact Inverse Operators

An extension of the proposed operators defined for intervals to gradual ones can be directly deduced:

3.1 The Operators ⊟ and ⊡ for Gradual Intervals

*A. **Proposition*** **1:** for $A(\lambda)\in\mathbb{Q}(\mathbb{R})$ and $B(\lambda)\in\mathbb{Q}(\mathbb{R})$, the operator ⊟ is defined by:

$$A(\lambda)\boxminus B(\lambda)\ominus(M_{A(\lambda)}, R_{A(\lambda)})\boxminus(M_{B(\lambda)}, R_{B(\lambda)})\ominus(M_{A(\lambda)}\text{-}M_{B(\lambda)}, R_{A(\lambda)}\text{-}R_{B(\lambda)})\ominus\Phi(\lambda) \tag{5}$$

The exact difference operator ⊟ produces a gradual interval if and only if:

$$R_{A(\lambda)} \geq R_{B(\lambda)}, \forall\lambda \in [0,1] \tag{6}$$

- ***Proof***

Equation (5) follows immediately from the Proof of the proposition 1 (see Part I) by replacing a and b by $A(\lambda)$ and $B(\lambda)$. In the same time by adopting the same principle as proposition 2 of the Part I, $\Phi(\lambda)$ is a gradual interval if and only if:

$$R_{\Phi(\lambda)} \geq 0 \Rightarrow R_{A(\lambda)} - R_{B(\lambda)} \geq 0 \Rightarrow R_{A(\lambda)} \geq R_{B(\lambda)} \quad ;\text{ for } \lambda \in [0,1]$$

B. Proposition **2:** For $A(\lambda)\in\mathbb{Q}(\mathbb{R})$ and $B(\lambda)\in\mathbb{Q}^*(\mathbb{R})$, the operator ⊡ is defined by:

$$A(\lambda)\boxdot B(\lambda) \ominus(M_{A(\lambda)}, R_{A(\lambda)})\boxdot(M_{B(\lambda)}, R_{B(\lambda)}) \ominus(M_{\Phi(\lambda)}, R_{\Phi(\lambda)}) \ominus\Phi(\lambda) \tag{7}$$

The functions $M_{\Phi(\lambda)}$ and $R_{\Phi(\lambda)}$ are given by the equation (20) of Part I, where the intervals a and b are respectively replaced by $A(\lambda)$ and $B(\lambda)$.
The exact operator ⊡ produces a gradual interval if and only if:

$$\left|RX_{A(\lambda)}\right| \geq \left|RX_{B(\lambda)}\right| ;\text{ for } \lambda \in [0,1] \tag{8}$$

- ***Proof***

In this case, by adopting the same principle as the proof of propositions 3 and 4 (in Part I), it can be stated that the equations (7) and (8) are obtained by replacing a and b by $A(\lambda)$ and $B(\lambda)$.

3.2 Overestimation between Operators

The overestimation errors detailed in section 4 of the Part I, for conventional intervals are directly extended to the gradual ones.

A. Overestimation between ⊖ and ⊟

For gradual intervals, it can be deduced that the overestimation error between the subtraction operators are given by the following equation:

$$\Delta_R(\lambda) = 2.R_{B(\lambda)}$$

B. Overestimation between ⊘ and ⊡

For gradual intervals, the overestimation error between the division operators is given by (see section 4 of the Part I):

If $\left|RX_{A(\lambda)}\right|>1$ and $M_{A(\lambda)}=0$: $\Delta_R(\lambda)=2.R_{A(\lambda)}R_{B(\lambda)}.\delta_{B(\lambda)}$

$$\text{If } \left|RX_{A(\lambda)}\right|>1 \text{ and } M_{A(\lambda)}\neq 0 : \begin{cases}\left|\Delta_M(\lambda)\right|=2.\left|M_{A(\lambda)}\right|R_{B(\lambda)}.\delta_{B(\lambda)}\\ \Delta_R(\lambda)=2.R_{A(\lambda)}R_{B(\lambda)}.\delta_{B(\lambda)}\end{cases}$$

$$\text{If } \left|RX_{A(\lambda)}\right|\leq 1 \; : \; \Delta_{RX}(\lambda)=2\left|RX_{B(\lambda)}\right|.(1-RX^2_{A(\lambda)})/(1-RX^2_{A(\lambda)}.RX^2_{B(\lambda)})$$

It is obvious that when λ =1, the overestimation errors are equal to 0.

3.3 Remarks

- It is important to note here that the inverse of a gradual interval *A* according to the operator ⊞̈ is undefined. Indeed, the condition (8) is violated witch means that the obtained result is not a gradual interval. The single case where condition (8) holds, i.e. *A* is invertible, is for *A* being a crisp number.
- It is important to note here that when the conditions (6) and (8) are not respected the obtained results can not be represented by gradual intervals.
- It can be stated that the proposed difference operator ⊟is an adapted version of the Hukuhara difference definition for gradual intervals. Indeed, the Hukuhara difference of two sets $A \in C$ and $B \in C$, if it exists, is a set $Z \in C$ such that $A = B + Z$, where *C* is the family of all nonempty convex compact subsets [11]. The Hukuhara difference is unique but a necessary condition for its existence is that *A* contains a translate $\{z\} + B$. In this case, a translation of this result to intervals, means that $R_{A(\lambda)} \geq R_{B(\lambda)}$ [11], [19], [20].
- If the two gradual intervals *A* and *B* are symmetric, then the conditions (6) and (8) are reduced to:

$$R_{A(0)} \geq R_{B(0)} \; : \text{for the operator } \boxminus$$

$$\text{and : } \left|RX_{A(0)}\right| \geq \left|RX_{B(0)}\right| \; : \text{for the operator } ⊞̈$$

4 Illustrative Examples

In this section, the subtraction and division operators are used. Of course, other operations can be defined and illustrated in a similar manner. In the illustrative examples, all computations are performed in the *MR* space. Moreover, for the sake of interpretation facility and as usually used in the fuzzy literature, the obtained results are plotted according to a translation to the *EF* space. Three illustration examples are considered in order to emphasize specific configurations of the gradual intervals.

***A. Example* 1:** Let us consider two gradual (fuzzy) intervals *A* and *B* given by:

$A(\lambda)⊜(M_{A(\lambda)}, R_{A(\lambda)})$ ⊜(4-λ, 3-3λ) ; and : $B(\lambda)$ ⊜ $(M_{B(\lambda)}, R_{B(\lambda)})$ ⊜(2, 1-λ)

Using the subtraction and the division operations, the following results are obtained:

- *Standard subtraction* : $A(\lambda)\ominus B(\lambda)$ ⊜(2-λ, 4-4λ)
- *Standard division* : $A(\lambda)⊘B(\lambda)$ ⊜ $\delta_{B(\lambda)}$⊛$(A(\lambda)\otimes B(\lambda))$;

where : $\delta_{B(\lambda)} = 1/(-\lambda^2+2\lambda+3)$

and $A(\lambda)\otimes B(\lambda) ⊜ (3\lambda^2-8\lambda+11,\ \lambda^2-11\lambda+10)$

- *Proposed subtraction*: $A(\lambda)\boxminus B(\lambda)$ ⊜ $(2-\lambda,\ 2-2\lambda)$
- *Proposed division:* $A(\lambda)\boxed{\div}B(\lambda)$ ⊜ $(M_{\Phi(\lambda)},\ R_{\Phi(\lambda)})$; where:

$$M_{\Phi(\lambda)} = (4-\lambda)/(2+\psi_{\Phi(\lambda)}.(1-\lambda)) \text{ with: } \psi_{\Phi(\lambda)} = (\lambda^2+\lambda-2)/(3\lambda^2-4\lambda-5)$$

$$\text{Thus: } M_{\Phi(\lambda)} = (3\lambda^2-4\lambda-5)/(\lambda^2-2\lambda-3) \text{ and: } R_{\Phi(\lambda)} = \psi_{\Phi(\lambda)}.M_{\Phi(\lambda)}$$

The obtained results are illustrated according to the *EF* representation as shown in Fig. 2.

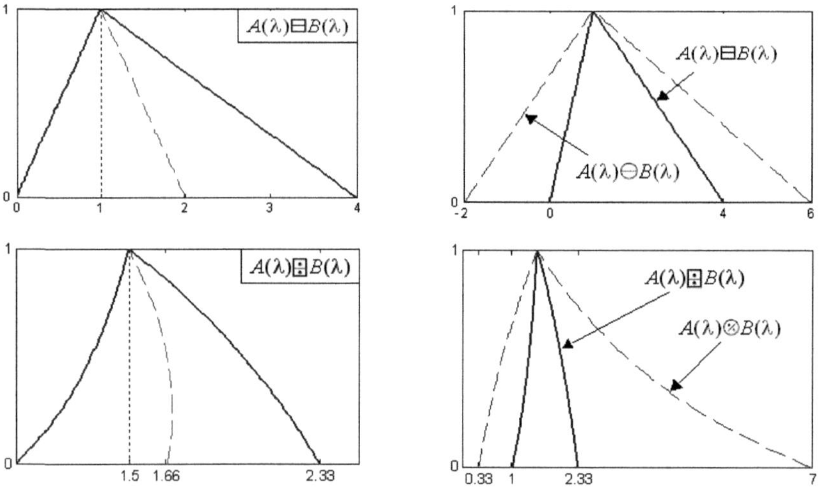

Fig. 2. The conventional and the proposed operators results in the *EF* space

It can be verified that $(A(\lambda)\boxed{\div}B(\lambda))\otimes B(\lambda)$ ⊜ $A(\lambda)$. The proposed operators $\boxminus$ and $\boxed{\div}$ are less imprecise than the standard ones $\ominus$ and $\oslash$. According to these results, it can be stated that maximum overestimation error is obtained for $\lambda = 0$ (see Fig. 3). This overestimation error can be quantified by Midpoint and Radius translations (see Fig. 3.a) or by the *RX* function as illustrated by the polar graph in Fig. 4.b.

In the opposite case when $\lambda = 1$, the operators give the same result, i.e. the division and subtraction for precise numbers.

More generally, the overestimation errors are given by:

$$\Delta_R(\lambda) = 2.R_{B(\lambda)} = 2-2\lambda \ : \text{between } \ominus \text{and } \boxminus$$

$$\Delta_{RX} = \frac{(1-\lambda)(8\lambda^2-10\lambda-7)}{2.25\lambda^4-9\lambda^3+12.5\lambda^2-\lambda-13.75} \ : \text{between } \oslash \text{and } \boxed{\div}$$

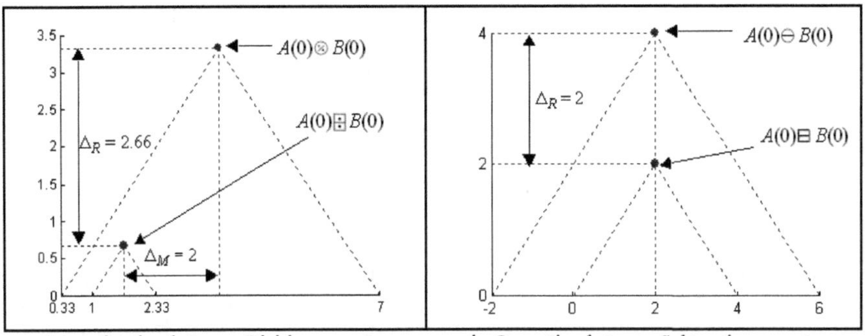

a. **Overestimation between Division operators** *b.* **Overestion between Subtraction operators**

Fig. 3. Overestimation between operators for $\lambda = 0$

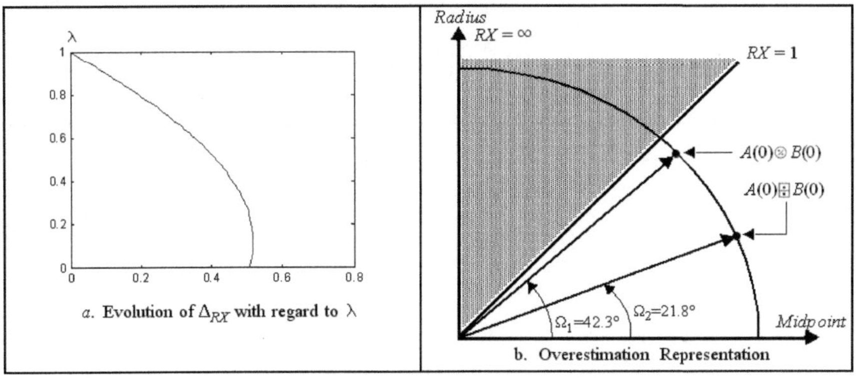

Fig. 4. Overestimation representation between ⊗and ⊡

In this case, it can be stated that if $\Delta_R(\lambda)$ is a linear function with regard to λ, the overestimation Δ_{RX} is a nonlinear one (see Fig. 4.a)

***B. Example* 2:** Let us consider two gradual (fuzzy) intervals *A* and *B* given by:

$$A(\lambda) ⊜ (M_{A(\lambda)}, R_{A(\lambda)}) ⊜ (-5\lambda^2 + 6\lambda + 4,\ 2\lambda^2 - 9\lambda + 7)$$

$$\text{and } B(\lambda) ⊜ (M_{B(\lambda)}, R_{B(\lambda)}) ⊜ (2,\ \cos(\alpha.\lambda))\ ;\ \text{with : } \phi = \pi/2$$

Using the subtraction and the division operations, the following results are obtained:

Indeed, the proposed operators are exact and less imprecise than the conventional ones. Indeed, the maximum overestimation between operators, obtained for $\lambda = 0$, is illustrated in Fig. 6. The evolution of Midpoint and Radius translations with regard to λ are given in Fig. 7.

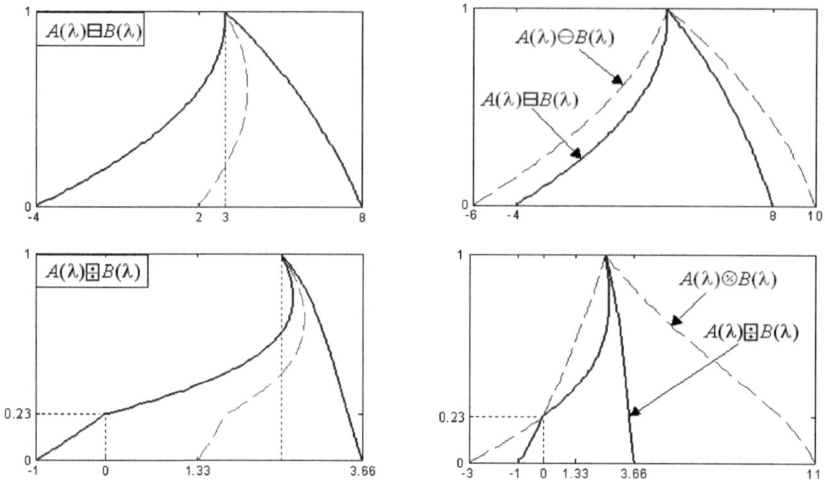

Fig. 5. The conventional and the proposed operators results in the *EF* space

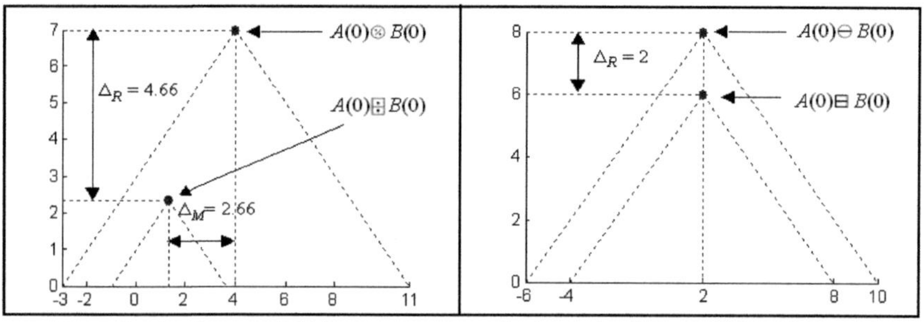

a. **Overestimation between Division operators** *b.* **Overestion between Subtraction operators**

Fig. 6. Overestimation between Operators for $\lambda = 0$

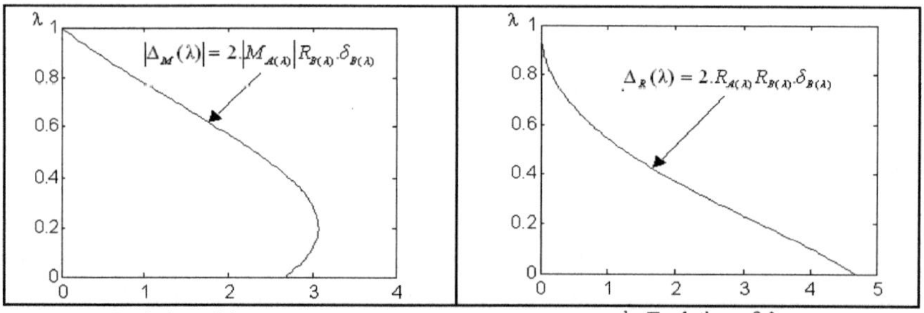

a. **Evolution of Δ_M** *b.* **Evolution of Δ_R**

Fig. 7. Evolutions of $|\Delta_M|$ and Δ_R

***C. Example* 3:** Let us consider two gradual intervals A and B given by:

$$A(\lambda) \circeq (M_{A(\lambda)}, R_{A(\lambda)}) \circeq (4\text{-}0.5.\sin(\lambda\pi), 2\text{-}2\lambda\text{+}0.5.\sin(\lambda\pi)) \text{ and}$$

$$B(\lambda) \circeq (M_{B(\lambda)}, R_{B(\lambda)}) \circeq (\text{-}2.75\text{-}0.25\lambda, 1.25\text{-}1.25\lambda)$$

Using the subtraction and the division operations, the following results are obtained:

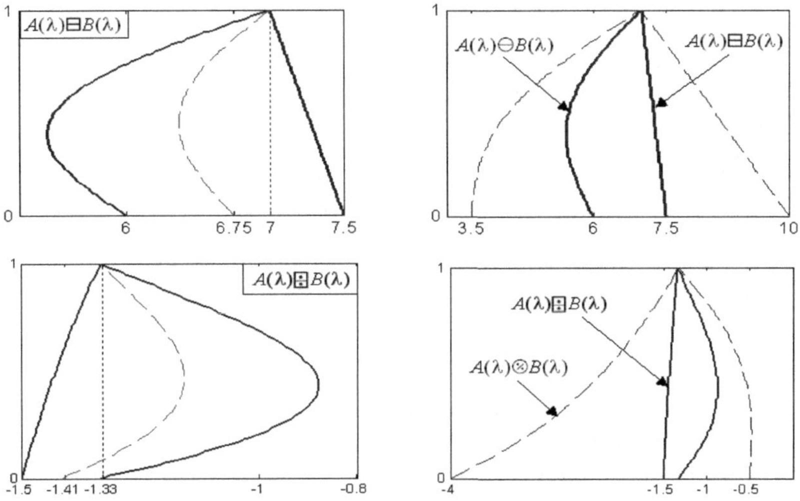

Fig. 8. The conventional and the proposed operators results in the *EF* space

In this case, the same remarks given in example 1 and 2 remain true. The only difference resides in the fact that the obtained intervals are purely gradual and cannot be represented by fuzzy ones.

5 Conclusion

This paper extends the interval exact operators developed in the Part I to gradual and fuzzy intervals with their existence conditions. Academic illustrative examples have been used for illustration. More complicated and realistic cases must be implemented. For example, the proposed optimistic operators may be used in the context of fuzzy inverse control and diagnosis methodologies, determining cluster centres for linguistic fuzzy C-means, fuzzy regression model inversion and reconstruction of inaccessible inputs, aggregation of Sugeno-like rule consequents, …

References

1. Bodjanova, S.: Alpha-bounds of fuzzy numbers. Information Sciences 152, 237–266 (2003)
2. Boukezzoula, R., Foulloy, L., Galichet, S.: Inverse Controller Design for Interval Fuzzy Systems. IEEE Transactions On Fuzzy Systems 14(1), 111–124 (2006)

3. Boukezzoula, R., Galichet, S., Foulloy, L.: MIN and MAX Operators for Fuzzy Intervals and their Potential Use in Aggregation Operators. IEEE Trans. on Fuzzy Systems 15(6), 1135–1144 (2007)
4. Dong, W.M., Wong, F.S.: Fuzzy weighted averages and implementation of the extension principle. Fuzzy Sets and Systems (21), 183–199 (1987)
5. Dubois, D., Prade, H.: Operations on fuzzy numbers. Journal of Systems Science (9), 613–626 (1978)
6. Dubois, D., Prade, H.: Gradual elements in a fuzzy set. Soft Comput. 12(2), 165–175 (2008)
7. Dubois, D., Kerre, E., Mesiar, R., Prade, H.: Fuzzy interval analysis. In: Dubois, D., Prade, H. (eds.) Fundamentals of Fuzzy Sets, The Handbooks of Fuzzy Sets Series, pp. 483–581. Kluwer, Boston (2000)
8. Fortin, J., Dubois, D., Fargier, H.: Gradual numbers and their application to fuzzy interval analysis. IEEE Trans. on Fuzzy Systems 16(2), 388–402 (2008)
9. Giachetti, R.E., Young, R.E.: A parametric representation of fuzzy numbers and their arithmetic operators. Fuzzy Sets and Systems 91(2), 185–202 (1997)
10. Giachetti, R.E., Young, R.E.: Analysis of the error in the standard approximation used for multiplication of triangular and trapezoidal fuzzy numbers and the development of a new approximation. Fuzzy Sets and Systems 91, 1–13 (1997)
11. Hukuhara, M.: Integration des applications measurables dont la valeur est compact convexe. Funkcialaj Ekvacioj 10, 205–223 (1967)
12. Kasperski, A., Zielinski, P.: Using Gradual Numbers for Solving Fuzzy-Valued Combinatorial Optimization Problems. In: Proc. of the 12th international Fuzzy Systems Association world congress (IFSA), Cancun, Mexico, pp. 656–665 (2007)
13. Kaufmann, A., Gupta, M.M.: Introduction to fuzzy arithmetic: Theory and Applications. Van Nostrand Reinhold Company Inc., New York (1991)
14. Lodwick, W.A., Jamison, K.D.: Special issue: interfaces between fuzzy set theory and interval analysis. Fuzzy Sets and Systems 135(1), 1–3 (2003)
15. Mizumoto, M., Tanaka, K.: The four operations of arithmetic on fuzzy numbers. Systems Comput. Controls 7(5), 73–81 (1976)
16. Mizumoto, M., Tanaka, K.: Some properties of fuzzy numbers. In: Gupta, M.M., Ragade, R.K., Yager, R.R. (eds.) Advances in Fuzzy Sets theory and Applications, pp. 156–164. North-Holland, Amsterdam (1979)
17. Moore, R., Lodwick, W.: Interval analysis and fuzzy set theory. Fuzzy Sets and Systems 135(1), 5–9 (2003)
18. Oussalah, M., De Schutter, J.: Approximated fuzzy LR computation. Information Sciences 153, 155–175 (2003)
19. Stefanini, L., Bede, B.: Generalization of Hukuhara differentiability of interval-valued functions and interval differential equations. Nonlinear Analysis 71, 1311–1328 (2009)
20. Stefanini, L.: A generalization of Hukuhara difference and division for interval and fuzzy arithmetic. Fuzzy sets and systems (2009), doi:10.1016/j.fss.2009.06.009
21. Yager, R.R.: On the lack of inverses in fuzzy arithmetic. Fuzzy Sets and Systems, 73–82 (1980)
22. Zadeh, L.A.: Fuzzy Sets. J. Information and Control (8), 338–353 (1965)
23. Zadeh, L.A.: The concept of a linguistic variable and its application to approximate reasoning. Information Sci., Part I: 8, 199–249; Part II: 8, 301–357; Part III: 9, 43–80 (1975)

Model Assessment Using Inverse Fuzzy Arithmetic

Thomas Haag and Michael Hanss

Institute of Applied and Experimental Mechanics, University of Stuttgart, Stuttgart, Germany
{haag,hanss}@iam.uni-stuttgart.de
http://www.iam.uni-stuttgart.de

Abstract. A general problem in numerical simulations is the selection of an appropriate model for a given real system. In this paper, the authors propose a new method to validate, select and optimize mathematical models. The presented method uses models with fuzzy-valued parameters, so-called comprehensive models, that are identified by the application of inverse fuzzy arithmetic. The identification is carried out in such a way that the uncertainty band of the output, which is governed by the uncertain input parameters, conservatively covers a reference output. Based on the so identified fuzzy-valued model parameters, a criterion for the selection and optimization is defined that minimizes the overall uncertainty inherent to the model. This criterion does not only consider the accuracy in reproducing the output, but also takes into account the size of the model uncertainty which is necessary to cover the reference output.

1 Introduction

A well-known problem in the numerical simulation of real-world systems is the question of how detailed the structure of the model has to be chosen in order to appropriately represent reality. If the structure of the model does not exhibit any simplifications of the reality, classical, crisp-valued model parameters are adequate to describe the real system, and a conventional model-updating procedure can be used for their identification. Due to computational limitations, however, it is often necessary to neglect certain phenomena in the modeling phase, such as high frequency dynamics or nonlinearities. This leads to simplified and, from a computational perspective, less expensive models, but obviously also to specific modeling errors when crisp parameters are used as the optimal ones. In order to cover these modeling errors in simplified models, uncertain parameters can be used, representing a so-called comprehensive model and providing a conservative prediction of the real system behavior. Based on the fact that the inherent uncertainty is a consequence of model simplifications, it can be classified as epistemic [1]. As probability theory may not be appropriate to effectively represent epistemic uncertainties [2], the alternative strategy of quantifying epistemic uncertainties by fuzzy numbers [3,4] is pursued in this paper. The propagation of

E. Hüllermeier, R. Kruse, and F. Hoffmann (Eds.): IPMU 2010, Part II, CCIS 81, pp. 461–470, 2010.

the uncertainty through the system, i.e. the evaluation of the model with fuzzy-valued parameters, is performed by the use of fuzzy arithmetic on the basis of the transformation method [4,5].

As an extension of the transformation method of fuzzy arithmetic, a special method to estimate the uncertain parameters of a simplified model on the basis of the output of an advanced model, or based on a single measurement signal is proposed in [6,7]. The proposed method uses inverse computations that are based on the data of the transformation method. This implies that an inversion of the model equations is not needed, which enables the method to be used with existing software, e.g. commercial finite element codes. By the computation of a so-called feasibility criterion, the identification procedure is limited to regions where reliable solutions are available. In this paper, the identification of multiple input parameters based on a single measurement output is presented and the resulting underdetermined problem is solved by using a constrained optimization procedure that minimizes the uncertainty in the input parameters. The major advancement in this paper is the definition and application of a model assessment criterion which is based on comprehensive models and inverse fuzzy arithmetic.

2 Inverse Fuzzy Arithmetical Identification

All fuzzy-valued computations in this paper are performed by the use of the transformation method, a detailed description of which can be found in [5,8]. The fundamental idea of the presented identification method is to identify the uncertain parameters of a simulation model by using measurement values of a real system.

This section is organized according to the steps that are needed for the identification: First, the notation and the procedure for the feedforward computation are explained. Second, the construction of the fuzzy-valued output quantities from the reference output is clarified. And third, the updated input parameters are computed, using an inverse fuzzy arithmetical approach.

2.1 Evaluation of Fuzzy-Parameterized Models

In general, a fuzzy-parameterized model consists of three key components:

1. A set of n independent fuzzy-valued model parameters $\widetilde{p}_i$ with the membership functions $\mu_{\widetilde{p}_i}(x_i)$, $i = 1, 2, \ldots, n$ (see Fig. 1(a)).
2. The model itself, which can be interpreted as a set of N generally nonlinear functions f_r, $r = 1, 2, \ldots, N$, that perform some operations on the fuzzy input variables $\widetilde{p}_i$, $i = 1, 2, \ldots, n$.
3. A set of N fuzzy-valued output parameters $\widetilde{q}_r$ with the membership functions $\mu_{\widetilde{q}_r}(z_r)$, $r = 1, 2, \ldots, N$, that are obtained as the result of the functions f_r.

Thus, a fuzzy-parameterized model can in general be expressed by a system of equations of the form

$$\widetilde{q}_r = f_r(\widetilde{p}_1, \widetilde{p}_2, \ldots, \widetilde{p}_n), \quad r = 1, 2, \ldots, N. \tag{1}$$

As a pre-condition for the application of inverse fuzzy arithmetic, the invertibility of the system, i.e. its unique solution for the uncertain model parameters $\widetilde{p}_i$, $i = 1, 2, \ldots, n$, has to be guaranteed. For this reason, only those models are considered in this paper where the output variables $\widetilde{q}_1, \widetilde{q}_2, \ldots, \widetilde{q}_N$ are strictly monotonic with respect to each of the model parameters $\widetilde{p}_1, \widetilde{p}_2, \ldots, \widetilde{p}_n$. This allows the uncertain model to be simulated and analyzed by simply applying the transformation method in its reduced form.

2.2 Definition of Fuzzy Outputs

To identify the uncertain input parameters of a model that is capable of representing a real system, a set of fuzzy-valued outputs $\widetilde{q}_r$ needs to be defined from the crisp outputs of the real system. The nominal values $\overline{z}_r$ of the fuzzy values $\widetilde{q}_r$ must correspond to the nominal output of the simplified model. The worst-case deviation from this nominal value is given by the output of the real system. Consequently, one value of the bounds of the support of $\widetilde{q}_r$ is set to the measurement value z_r^* while the other one is set to the nominal value. For the other μ-levels, a linear interpolation between $\mu = 0$ and $\mu = 1$ is chosen. This procedure leads to a single-sided, triangular fuzzy number, as visualized in Fig. 1(b).

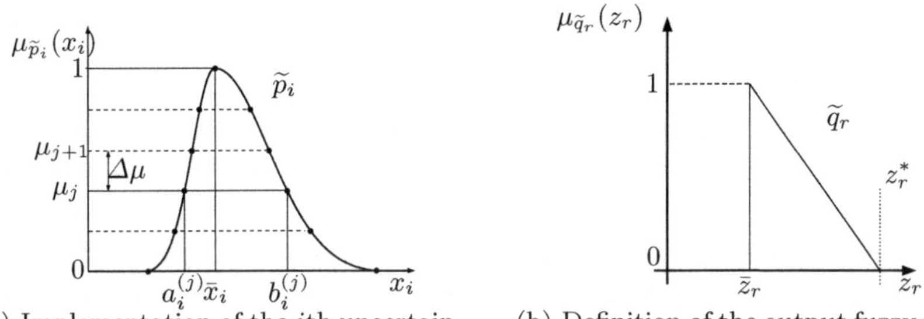

(a) Implementation of the ith uncertain parameter as a fuzzy number $\widetilde{p}_i$ decomposed into intervals (α-cuts)

(b) Definition of the output fuzzy number $\widetilde{q}_r$ based on a reference value

Fig. 1.

2.3 Inverse Fuzzy Arithmetic

With regard to the structure of fuzzy-parameterized models as defined in (1), the main problem of inverse fuzzy arithmetic consists in the identification of the uncertain model parameters $\widetilde{p}_1, \widetilde{p}_2, \ldots, \widetilde{p}_n$ on the basis of given values for the output variables $\widetilde{q}_1, \widetilde{q}_2, \ldots, \widetilde{q}_N$. In the case $N < n$, the identification problem is underdetermined, whereas it is overdetermined for $N > n$.

In the present paper, the case where an underdetermined inverse problem ($N^* = k, k < n$) has to be solved many times is considered. In the following, the variable S denotes the number of times for which the underdetermined inverse computation has to be performed.

To successfully solve the inverse fuzzy arithmetical problem, the following scheme can be applied, consisting of an appropriate combination of the simulation and the analysis parts of the transformation method:

1. Determination of the nominal values $\breve{\overline{x}}_1(s), \breve{\overline{x}}_2(s), \ldots, \breve{\overline{x}}_n(s)$: Owing to (1), the nominal values $\overline{x}_i(s)$ of the real model parameters $\widetilde{p}_i(s)$, $i = 1, 2, \ldots, n$, and the nominal values $\overline{z}_r(s)$ of the output variables $\widetilde{q}_r(s)$, $r = 1, 2, \ldots, k$, are related by the system of equations
$$\overline{z}_r(s) = f_r\left(\overline{x}_1(s), \overline{x}_2(s), \ldots, \overline{x}_n(s)\right), \quad r = 1, 2, \ldots, k. \tag{2}$$
Starting from the k given values $\overline{z}_r(s)$ in the inverse problem, estimations $\breve{\overline{x}}_i(s)$ of the n nominal values of the unknown fuzzy-valued model parameters $\breve{\widetilde{p}}_i(s)$, $i = 1, 2, \ldots, n$ can be determined either by analytically solving and optimizing (2) for $\overline{x}_i(s)$, $i = 1, 2, \ldots, n$, or by numerically solving and optimizing the system of equations using a certain iteration procedure.
2. Computation of the gain factors: For the determination of the single-sided gain factors $\eta_{ri+}^{(j)}(s)$ and $\eta_{ri-}^{(j)}(s)$, which play an important role in the inverse fuzzy arithmetical concept (see (4)), the model has to be simulated for some initially assumptive uncertain parameters $\widetilde{p}_i^*(s)$, $i = 1, 2, \ldots, n$, using the transformation method of fuzzy arithmetic. The nominal values of $\widetilde{p}_i^*(s)$ have to be set equal to the just computed values $\breve{\overline{x}}_i(s)$, $i = 1, 2, \ldots, n$, and the assumed uncertainty should be set to a large enough value, so that the expected real range of uncertainty in $\breve{\widetilde{p}}_i(s)$ is preferably covered.
3. Assembly of the uncertain parameters $\breve{\widetilde{p}}_1(s), \breve{\widetilde{p}}_2(s), \ldots, \breve{\widetilde{p}}_n(s)$: Recalling the representation of a fuzzy number in its decomposed form (Fig. 1(a)), the lower and upper bounds of the intervals of the fuzzy parameters $\breve{\widetilde{p}}_i(s)$ at the $(m+1)$ levels of membership μ_j shall be defined as $\breve{a}_i^{(j)}(s)$ and $\breve{b}_i^{(j)}(s)$, and the bounds of the given output values $\widetilde{q}_r(s)$ as $\breve{c}_r^{(j)}(s)$ and $\breve{d}_r^{(j)}(s)$. The interval bounds $\breve{a}_i^{(j)}(s)$ and $\breve{b}_i^{(j)}(s)$, which finally provide the membership functions of the unknown model parameters $\breve{\widetilde{p}}(s)_i$, $i = 1, 2, \ldots, n$, can then be determined through the following equation, where the dependency on s is left out for clarity:
$$\left[\begin{array}{c|c|c|c} H_{11}^{(j)} & H_{12}^{(j)} & \cdots & H_{1n}^{(j)} \\ \hline H_{21}^{(j)} & H_{22}^{(j)} & \cdots & H_{2n}^{(j)} \\ \hline \vdots & \vdots & \vdots & \vdots \\ \hline H_{k1}^{(j)} & H_{k2}^{(j)} & \cdots & H_{kn}^{(j)} \end{array}\right] \begin{bmatrix} \breve{a}_1^{(j)} - \breve{\overline{x}}_1 \\ \breve{b}_1^{(j)} - \breve{\overline{x}}_1 \\ \breve{a}_2^{(j)} - \breve{\overline{x}}_2 \\ \breve{b}_2^{(j)} - \breve{\overline{x}}_2 \\ \vdots \\ \breve{a}_n^{(j)} - \breve{\overline{x}}_n \\ \breve{b}_n^{(j)} - \breve{\overline{x}}_n \end{bmatrix} = \begin{bmatrix} c_1^{(j)} - \overline{z}_1 \\ d_1^{(j)} - \overline{z}_1 \\ c_2^{(j)} - \overline{z}_2 \\ d_2^{(j)} - \overline{z}_2 \\ \vdots \\ c_k^{(j)} - \overline{z}_k \\ d_k^{(j)} - \overline{z}_k \end{bmatrix} \tag{3}$$
$$\text{with } H_{ri}^{(j)} = \frac{1}{2}\begin{bmatrix} \eta_{ri-}^{(j)}\left(1 + \operatorname{sgn}(\eta_{ri-}^{(j)})\right) & \eta_{ri+}^{(j)}\left(1 - \operatorname{sgn}(\eta_{ri+}^{(j)})\right) \\ \eta_{ri-}^{(j)}\left(1 - \operatorname{sgn}(\eta_{ri-}^{(j)})\right) & \eta_{ri+}^{(j)}\left(1 + \operatorname{sgn}(\eta_{ri+}^{(j)})\right) \end{bmatrix}$$

$$i = 1, 2, \ldots, n\,, \quad r = 1, 2, \ldots, k\,, \quad j = 0, 1, \ldots, m-1\,.$$

The values $\breve{a}_i^{(m)} = \breve{b}_i^{(m)}$, $i = 0, 1, \ldots, n$, are already determined by the nominal values $\bar{\breve{x}}_i$.

Equation (3) is solved for the unknown quantities $\breve{a}_i^{(j)}(s), \breve{b}_i^{(j)}(s)$. By the assumption of k being smaller than n, the system of equations in (3) still possesses some degrees of freedom which are used to minimize the resulting uncertainty in the input parameters that are to be identified. Thus, (3) is solved by additionally minimizing

$$u(s) = U(s)^{\mathrm{T}} W(s) U(s) \quad \text{with} \tag{4}$$
$$U^{\mathrm{T}} = \left[\breve{a}_1^{(j)} - \bar{\breve{x}}_1,\, \breve{b}_1^{(j)} - \bar{\breve{x}}_1,\, \breve{a}_2^{(j)} - \bar{\breve{x}}_2,\, \breve{b}_2^{(j)} - \bar{\breve{x}}_2, \ldots, \breve{a}_n^{(j)} - \bar{\breve{x}}_n,\, \breve{b}_n^{(j)} - \bar{\breve{x}}_n\right].$$

The weighting matrix $W(s)$ realizes a standardization of the entries of $U(s)$ with respect to their dimensions and their modal values.

To extend the methodology to the globally overdetermined case where $N = S\,k$, two further steps are necessary. First, the inverse problem is solved for each s, for which a reliable solution can be expected as described above. Second, one final set of uncertain parameters needs to be derived from the solution for all $s = 1, 2, \ldots, S$. The extraction of those inverse problems where a reliable solution can be expected is based on the magnitudes of the elements of the matrix $H^{(j)}(s)$. In order to derive the one set of uncertain input parameters $\breve{\widetilde{p}}_i$ that is representative for all values of s, for each level of membership μ_j the maximum uncertainty, i.e. the union of all intermediate uncertain input parameters over all s is determinded. Details on these two steps are given in [7].

To verify the identified model parameters $\breve{\widetilde{p}}_1, \breve{\widetilde{p}}_2, \ldots, \breve{\widetilde{p}}_n$, the model equations (1) can be re-simulated by means of the transformation method, using $\breve{\widetilde{p}}_1, \breve{\widetilde{p}}_2, \ldots, \breve{\widetilde{p}}_n$ as the fuzzy input parameters.

In order to provide a quantitative measure for the quality of the identified model, the measure of total uncertainty Λ is defined in the following. The general idea of its definition is that the ideal model minimizes as well the uncertainty of the output that covers the reference output, as the uncertainty of the model parameters that cause the uncertain output.

In [5], the relative imprecision of a general fuzzy number $\widetilde{v}$ with modal value $\bar{v}$ is defined as

$$\mathrm{imp}_{\bar{v}}(\widetilde{v}) = \frac{1}{2m\bar{v}} \sum_{j=0}^{m-1} \left[b_i^{(j)} - a_i^{(j)} + b_i^{(j+1)} - a_i^{(j+1)}\right]. \tag{5}$$

The relative imprecision $\mathrm{imp}_{\bar{v}}(\widetilde{v})$ corresponds to the area between the membership function $\mu_{\widetilde{v}}(x)$ and the x-axis, which is normalized by the modal value $\bar{v}$. It quantifies the size of the uncertainty that is inherent to a fuzzy number $\widetilde{v}$. Based on the motivation given above, the total uncertainty Λ is defined as

$$\Lambda = \sum_{i=1}^{n} \left|\mathrm{imp}_{\bar{p}_i}(\widetilde{p}_i)\right| + \sum_{r=1}^{k} \left[\frac{1}{S} \sum_{s=1}^{S} \left(\left|\mathrm{imp}_{\bar{q}_r(s)}(\widetilde{q}_r(s))\right|\right)\right]. \tag{6}$$

The first summation captures the size of all input uncertainties whereas the second summation accounts for the mean uncertainty of the outputs.

3 Model Assessment

3.1 Preliminary Example

In this section, the presented method is applied to the rather simple mathematical problem of approximating the fourth-order polynomial

$$y(x) = 1 + x + x^2 + x^3 + x^4 \tag{7}$$

by a polynomial of lower degree $d < 4$

$$\widetilde{y}(x) = \sum_{l=0}^{d} \widetilde{a}_l x^l, \tag{8}$$

but with fuzzy-valued coefficients $\widetilde{a}_l$, $l = 0, 1, \ldots, d$. The modal values of the fuzzy coefficients $\widetilde{a}_l$ are computed through a best fit to the reference output $y(x)$, their uncertainties by the application of inverse fuzzy arithmetic. The output $\widetilde{y}(x)$ of the comprehensive model (8) with uncertain fuzzy-valued coefficients $\widetilde{a}_l$ is capable of entirely covering the reference output $y(x)$.

Figure 2 shows the approximations of the fourth-order polynomial (7) with three lower-order polynomials ($d = 1, 2, 3$) as well as with a polynomial of proper order ($d = 4$). On the left hand side, the fuzzy coefficients $\widetilde{a}_l$ are shown. On the right hand side, the uncertain outputs are plotted, where the degree of membership is visualized by color. A dark color corresponds to a high degree of membership, while a lighter color marks a decreasing degree of uncertainty. Each of the subfigures is labeled with the approximating model type and the measure of total uncertainty Λ. For the latter, the contribution of each summand is given in detail. The first $d + 1$ summands correspond to the uncertainty of the fuzzy model parameters $\widetilde{a}_0, ..., \widetilde{a}_d$, whereas the last summand corresponds to the output uncertainty.

The fourth-order polynomial is shown in Fig. 2(d) and, obviously, no uncertainty is needed to cover the reference output. In Figs. 2(a)-2(c), the degree of the approximation polynomial increases while the uncertainty in the coefficients decreases. Obviously, the quality of the model increases with increasing order of the polynomial as a smaller uncertainty in the model parameters is needed to cover the modeling error. The introduced measure Λ of the total uncertainty of the model, which is given below the plots, is obviously capable of quantifying this fact in a systematic way as the values of Λ decrease continously.

3.2 Application: Linearization of a Quadratic Function

As an application example, five different comprehensive linearizations of the fourth-order polynomial described in the previous section (see (7)) are compared.

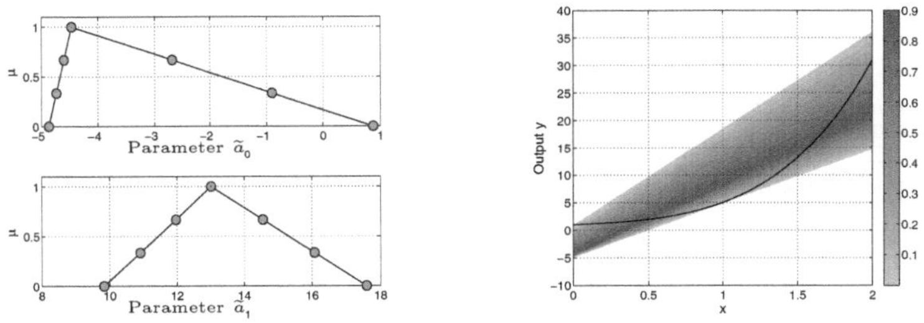

(a) Linear: $\Lambda = 64.3627 + 29.7902 + 49.6853 = 143.8382$

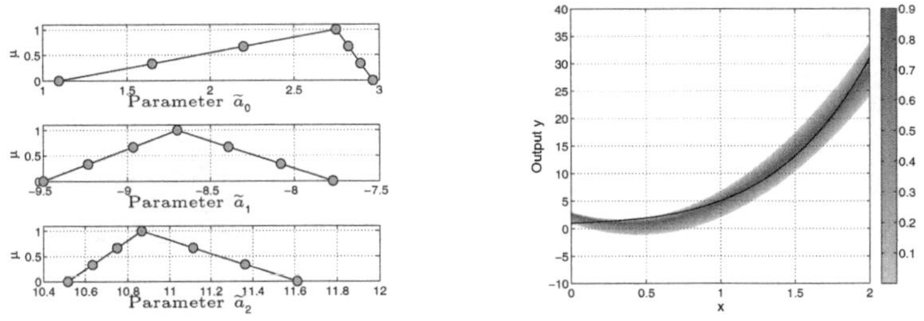

(b) Quadratic: $\Lambda = 33.9261 + 9.9192 + 5.0364 + 10.8632 = 59.745$

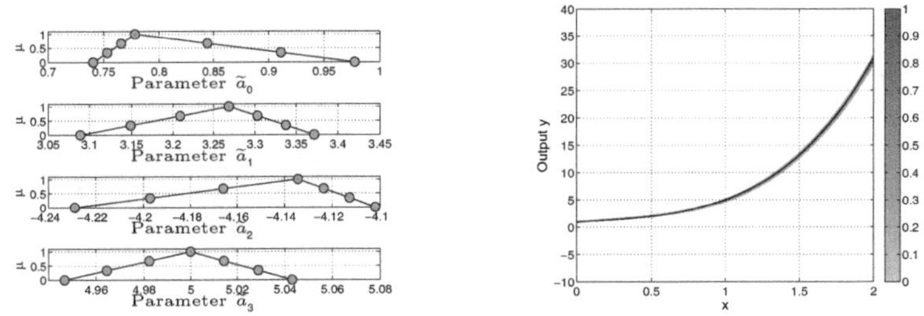

(c) Cubic: $\Lambda = 15.2032 + 4.3207 + 1.5335 + 0.96314 + 1.1726 = 23.1932$

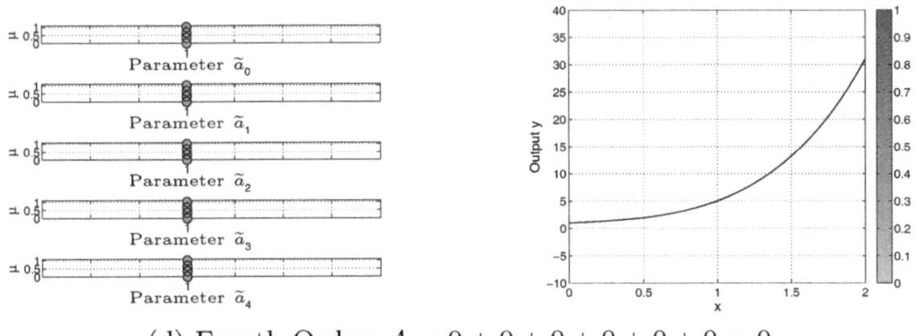

(d) Fourth Order: $\Lambda = 0 + 0 + 0 + 0 + 0 + 0 = 0$

Fig. 2. Approximations of the fourth-order polynomial with uncertain lower-order polynomials

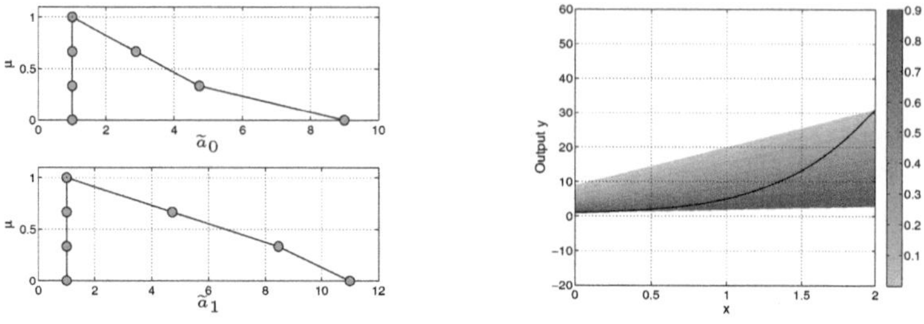

(a) L_1: $\Lambda = 320 + 540 + 127.9947 = 987.9947$

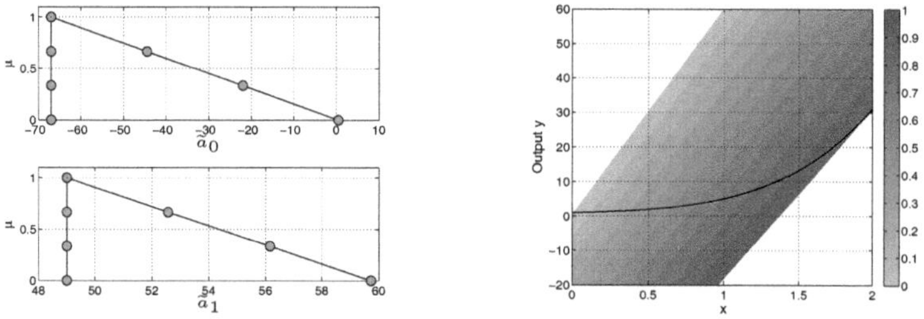

(b) L_2: $\Lambda = 50.3854 + 10.9578 + 92.39 = 153.7332$

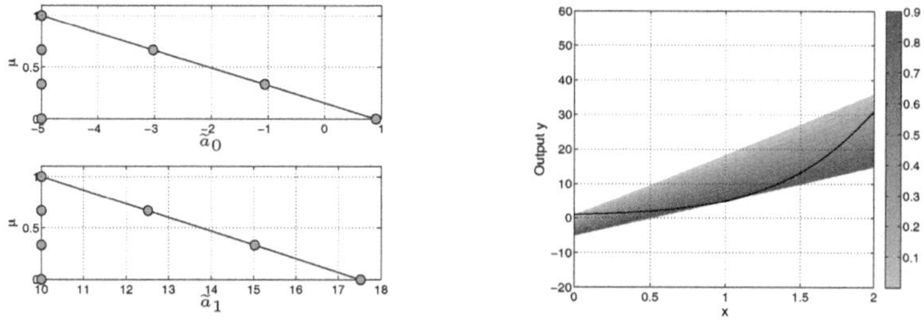

(c) L_3: $\Lambda = 59.0774 + 37.6471 + 56.0475 = 152.7719$

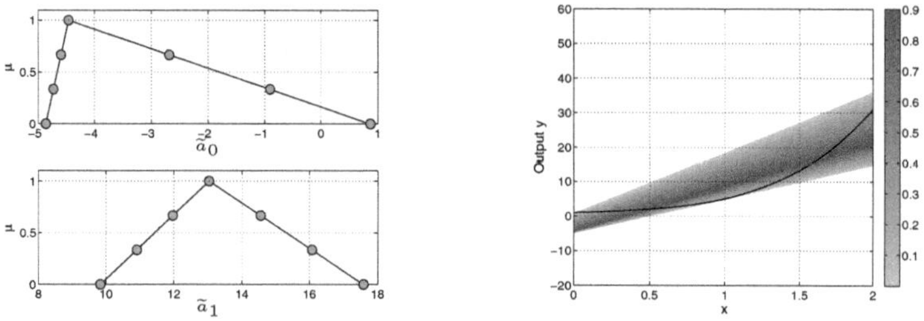

(d) L_4: $\Lambda = 64.3627 + 29.7902 + 49.6853 = 143.8382$

Fig. 3. Comprehensive linearizations of the fourth-order polynomial

All these linearizations are of the form

$$\widetilde{y} = \widetilde{a}_0 + \widetilde{a}_1 x, \tag{9}$$

where $\widetilde{a}_0$, $\widetilde{a}_1$ are fuzzy parameters. The five different linearizations are labeled L_1 to L_5 and differ in the way the modal values of the parameters $\widetilde{a}_0$, $\widetilde{a}_1$ are determined. The modal values of the linearizations L_1 to L_3 are generated by computing the Jacobian linearizations of the reference output (7) for the three points with $x = 0$, $x = 1$ and $x = 2$, respectively. For the linearization L_4, the best fit in terms of the mean-squared error of the output is computed and modal values of the fuzzy parameters $\widetilde{a}_0$, $\widetilde{a}_1$ are derived therefrom. The modal values of the linearization L_5 are computed by an optimization which minimizes the total uncertainty Λ of the model.

The uncertainty of the two fuzzy parameters $\widetilde{a}_0$, $\widetilde{a}_1$ are determined by inverse fuzzy arithmetic and the resulting input parameters lead to an uncertain output that in all cases covers the reference output in a conservative way. The identified uncertain input parameters are shown on the left hand side of the Figs. 3 and 4, whereas the uncertain outputs with the reference output are shown on the right hand side.

For each of the models L_1 to L_5, the corresponding measure Λ of the total uncertainty of the model is given below the plots. The measure Λ provides a mean to assess and compare the different models and is used to generate the optimal model L_5. A low value of Λ corresponds to a small total uncertainty which is aspired in the authors' opinion as it does not only focus on minimizing the deviation of the output but also the uncertainty of the model parameters. In engineering applications, the model parameters usually possess a physical meaning and their uncertainty can be influenced directly through adequate measures, which does not hold for the deviation of the output. For the current example, the relative impresicion of the model parameters is significantly smaller for the optimized model L_5 than for the best-fit model L_4. The averaged relative imprecision of the output, though, is only marginally larger.

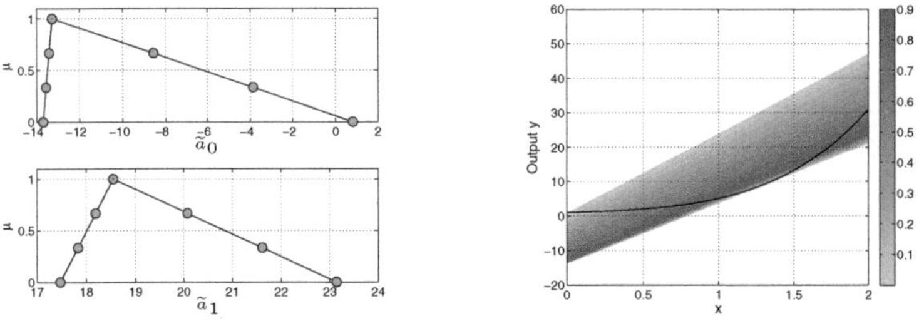

(a) L_5: $\Lambda = 54.6963 + 15.2567 + 57.1033 = 127.0564$

Fig. 4. Optimal linearization of the fourth-order polynomial

4 Conclusions

In this paper, a criterion to assess the quality of a model is defined and verified on two mathematical examples. Unlike the conventional way of proceeding, which focuses on the $\mathcal{L}_2$-norm of the output deviation only, the presented quality criterion also takes into account the uncertainty of the model parameters which are the source of the output deviation assuming a special model structure. Thereby, models can be assessed and optimized more comprehensively than this is possible with the rather narrow view of optimizing the output deviation only. For engineering applications, for example, the quantification of the model uncertainties, which cause the output deviation, enable the engineer to launch adequate measures in the actuator domain, rather than in the output domain, where this is impossible.

References

1. Oberkampf, W.L.: Model validation under both aleatory and epistemic uncertainty. In: Proc. of NATO AVT-147 Symposium on Computational Uncertainty in Military Vehicle Design, Athens, Greece (2007)
2. Hemez, F.M., Booker, J.M., J.R.L.: Answering the question of sufficiency: How much uncertainty is enough? In: Proc. of The 1st International Conference on Uncertainty in Structural Dynamics – USD 2007, Sheffield, UK (2007)
3. Zadeh, L.A.: Fuzzy sets. Information and Control 8, 338–353 (1965)
4. Kaufmann, A., Gupta, M.M.: Introduction to Fuzzy Arithmetic. Van Nostrand Reinhold, New York (1991)
5. Hanss, M.: Applied Fuzzy Arithmetic – An Introduction with Engineering Applications. Springer, Berlin (2005)
6. Hanss, M.: An approach to inverse fuzzy arithmetic. In: Proc. of the 22nd International Conference of the North American Fuzzy Information Processing Society – NAFIPS 2003, Chicago, IL, USA, pp. 474–479 (2003)
7. Haag, T., Reuß, P., Turrin, S., Hanss, M.: An inverse model updating procedure for systems with epistemic uncertainties. In: Proc. of the 2nd International Conference on Uncertainty in Structural Dynamics, Sheffield, UK, pp. 116–125 (2009)
8. Hanss, M.: The transformation method for the simulation and analysis of systems with uncertain parameters. Fuzzy Sets and Systems 130(3), 277–289 (2002)

New Tools in Fuzzy Arithmetic with Fuzzy Numbers

Luciano Stefanini

DEMQ - Department of Economics and Quantitative Methods
University of Urbino "Carlo Bo", Italy
lucste@uniurb.it

Abstract. We present new tools for fuzzy arithmetic with fuzzy numbers, based on the parametric representation of fuzzy numbers and new fuzzy operations, the generalized difference and the generalized division of fuzzy numbers. The new operations are described in terms of the parametric LR and LU representations of fuzzy numbers and the corresponding algorithms are described. An application to the solution of simple fuzzy equations is illustrated.

1 Parametric Fuzzy Numbers and Fundamental Fuzzy Calculus

In some recent papers (see [4], [5]), it is suggested the use of monotonic splines to model LU-fuzzy numbers and derived a procedure to control the error of the approximations. By this approach, it is possible to define a parametric representation of the fuzzy numbers that allows a large variety of possible types of membership functions and is very simple to implement. Following the LU-fuzzy parametrization, we illustrate the computational procedures to calculate the generalized difference and division of fuzzy numbers introduced in [6] and [7]; the representation is closed with respect to the operations, within an error tolerance that can be controlled by refining the parametrization.

A general *fuzzy set* over $\mathbb{R}$ is usually defined by its membership function $\mu : \mathbb{R} \longrightarrow [0,1]$ and a fuzzy set u of $\mathbb{R}$ is uniquely characterized by the pairs $(x, \mu_u(x))$ for each $x \in \mathbb{R}$; the value $\mu_u(x) \in [0,1]$ is the membership grade of x to the fuzzy set u. Denote by $\mathcal{F}(\mathbb{R})$ the collection of all the fuzzy sets over $\mathbb{R}$. Elements of $\mathcal{F}(\mathbb{R})$ will be denoted by letters u, v, w and the corresponding membership functions by μ_u, μ_v, μ_w.

The *support* of u is the (crisp) subset of points of $\mathbb{R}$ at which the membership grade $\mu_u(x)$ is positive: $supp(u) = \{x|x \in \mathbb{X}, \mu_u(x) > 0\}$. We always assume that $supp(u) \neq \emptyset$. For $\alpha \in]0,1]$, the $\alpha-$level cut of u (or simply the $\alpha-cut$) is defined by $[u]_\alpha = \{x|x \in \mathbb{R}, \mu_u(x) \geq \alpha\}$ and for $\alpha = 0$ by the closure of the support $[u]_0 = cl\{x|x \in \mathbb{R}, \mu_u(x) > 0\}$. The *core* of u is the set $core(u) = \{x|x \in \mathbb{R}, \mu_u(x) = 1\}$ and we say that u is normal if $core(u) \neq \emptyset$.

Well-known properties of the $level-cuts$ are:

$$[u]_\alpha \subseteq [u]_\beta \text{ for } \alpha > \beta \text{ and } [u]_\alpha = \bigcap_{\beta<\alpha} [u]_\beta \text{ for } \alpha \in]0,1]$$

E. Hüllermeier, R. Kruse, and F. Hoffmann (Eds.): IPMU 2010, Part II, CCIS 81, pp. 471–480, 2010.

and $\mu_u(x) = \sup\{\alpha | \ \alpha \in [0,1] \text{ for which } x \in [u]_\alpha\}$.

A particular class of fuzzy sets $u \in \mathcal{F}(\mathbb{R})$ is when the support is a convex set and the membership function is quasi-concave:

Definition 1. *Consider $u \in \mathcal{F}(\mathbb{R})$, and assume that supp(u) is a convex set; we say that the membership function μ_u is quasi-concave if $\mu_u((1-t)x' + tx'') \geq \min\{\mu_u(x'), \mu_u(x'')\}$ for every $x', x'' \in supp(u)$ and $t \in [0,1]$. Equivalently, μ_u is quasi-concave if the level sets $[u]_\alpha$ are convex sets for all $\alpha \in [0,1]$.*

Definition 2. *A fuzzy quantity is a fuzzy set $u \in \mathcal{F}(\mathbb{R})$ with the properties:*
1. *$core(u) \neq \emptyset$*
2. *$[u]_\alpha$ are non empty compact convex sets for all $\alpha \in [0,1]$.*

If the core is a singleton, we have a fuzzy number; if the core is an interval with positive length, we have a fuzzy interval. We will denote by $\mathcal{F}$ the space of fuzzy quantities (numbers or intervals). The $\alpha - cuts$ of a fuzzy quantity are non empty, compact intervals of the form $[u]_\alpha = [u_\alpha^-, u_\alpha^+] \subset \mathbb{R}$.

Definition 3. *An LR-fuzzy quantity (number or interval) u has membership function of the form*

$$\mu_u(x) = \begin{cases} L(\frac{a-x}{b-a}) & if \ \ a \leq x \leq b \\ 1 & if \ \ b \leq x \leq c \\ R(\frac{x-c}{d-c}) & if \ \ c \leq x \leq d \\ 0 & otherwise \end{cases}$$

where $L, R : [0,1] \to [0,1]$ are two non-increasing shape functions such that $R(0) = L(0) = 1$ and $R(1) = L(1) = 0$. If $b = c$ we obtain a fuzzy number.

The usual notation for an LR-fuzzy quantity is

$$u = \langle a, b, c, d \rangle_{L,R} \text{ for an interval, and } u = \langle a, b, c \rangle_{L,R} \text{ for a number.}$$

We refer to the two functions $L(.)$ and $R(.)$ as the left and right branches (shape functions) of u, respectively.

On the other hand, the *level-cuts* of a fuzzy number are "nested" closed intervals and this property is the basis for the LU representation (L for lower, U for upper).

Definition 4. *An LU-fuzzy quantity (number or interval) u is completely determined by any pair $u = (u^-, u^+)$ of functions $u^-, u^+ : [0,1] \longrightarrow \mathbb{R}$, defining the end-points of the $\alpha - cuts$, satisfying the three conditions:(i) $u^- : \alpha \longrightarrow u_\alpha^- \in \mathbb{R}$ is a bounded monotonic nondecreasing left-continuous function $\forall \alpha \in]0,1]$ and right-continuous for $\alpha = 0$;(ii) $u^+ : \alpha \longrightarrow u_\alpha^+ \in \mathbb{R}$ is a bounded monotonic non-increasing left-continuous function $\forall \alpha \in]0,1]$ and right continuous for $\alpha = 0$;(iii) $u_\alpha^- \leq u_\alpha^+ \ \forall \alpha \in [0,1]$.*

If $u_1^- < u_1^+$ we have a fuzzy interval and if $u_1^- = u_1^+$ we have a fuzzy number. We refer to the functions $u_{(.)}^-$ and $u_{(.)}^+$ as the *lower* and *upper* branches on u, respectively. In particular, if the two branches $u^-_{(.)}$ and $u^+_{(.)}$ are continuous invertible functions then $\mu_u(.)$ is formed by two continuous branches, the left being the increasing inverse of $u^-_{(.)}$ on $[u^-_0, u^-_1]$ and the right the decreasing inverse of $u^+_{(.)}$ on $[u^+_1, u^+_0]$. The notation $u_\alpha = [u^-_\alpha, u^+_\alpha]$, $\alpha \in [0,1]$ denotes explicitly the $\alpha - cuts$ of u. If the functions are monotonic and invertible, we have

$$u^-_\alpha = a - (b-a)L^{-1}(\alpha) \text{ and } u^+_\alpha = c + (d-c)R^{-1}(\alpha).$$

The simplest fuzzy quantities have linear branches (in the LR or LU representations); in the linear case, a trapezoidal fuzzy interval will be denoted by $u = \langle a, b, c, d\rangle$, where $a \leq b \leq c \leq d$. A triangular fuzzy number, is denoted by $u = \langle a, b, c\rangle$, where $a \leq b \leq c$.

Both the left-right shape functions $L, R : [0,1] \to [0,1]$ of the LR representation and the lower-upper branches $u^-, u^+ : [0,1] \to [a,b] \subset \mathbb{R}$ of the LU model are defined on the same standard domain $[0,1]$. We will obtain them as monotonic interpolators on a finite decomposition of the interval $[0,1]$ into N subintervals ($N+1$ points):

$$0 = \alpha_0 < \alpha_1 < ..... < \alpha_{i-1} < \alpha_i < .... < \alpha_N = 1.$$

Without loosing generality, we will use the same subdivision for all the shape functions L, R, u^- or u^+.

Denote the N subintervals by $I_i = [\alpha_{i-1}, \alpha_i]$, $i = 1, 2, ..., N$. For simplicity of notation, we will consider $N = 1$ (i.e. a single interval in the $\alpha-$decomposition) and we omit the subscript i. The monotonic functions u^-_α and u^-_α for $\alpha \in [0,1]$ are then characterized by the values and the slopes at $\alpha = 0$ and $\alpha = 1$:

$$\begin{gathered} u^-_\alpha = p^-(\alpha) \text{ and } u^+_\alpha = p^+(\alpha)\ ,\ \alpha \in [0,1] \\ p^-(0) = u^-_0,\ p^{-\prime}(0) = d^-_0,\ p^-(1) = u^-_1,\ p^{-\prime}(1) = d^-_1 \\ p^+(0) = u^+_0,\ p^{+\prime}(0) = d^+_0,\ p^+(1) = u^+_1,\ p^{+\prime}(1) = d^+_1. \end{gathered} \tag{1}$$

For a fuzzy number we have $u^-_0 \leq u^-_1$ (i.e. the data are increasing) then $d^-_0 \geq 0$ and $d^-_1 \geq 0$ are required; analogously, $u^+_0 \geq u^+_1$ (i.e. the data are decreasing) then $d^+_0 \leq 0$ and $d^+_1 \leq 0$ are required. In particular, $u_0 = u_1 \Leftrightarrow d_0 = d_1 = 0$.

The final step is to consider families of parametric increasing functions $s : [0,1] \to [0,1]$ as general standardized models for u^-_α and u^+_α; we require the conditions $s(0) = 0$, $s(1) = 1$ and the Hermite conditions $s'(0) = \beta_0$ and $s'(1) = \beta_1$. To refer different model functions, we explicit the slope parameters and write $s(.;\beta_0,\beta_1) : [0,1] \to [0,1]$ or $t \to s(t;\beta_0,\beta_1)$.

Examples of standardized parametric functions $s(t;\beta_0,\beta_1)$ are (see [4]):

o (2,2)-rational spline:

$$s_{rat22}(t;\beta_0,\beta_1) = \frac{t^2 + \beta_0 t(1-t)}{1 + (\beta_0 + \beta_1 - 2)t(1-t)}; \tag{2}$$

o mixed exponential spline:

$$s_{mixed}(t;\beta_0,\beta_1) = \frac{1}{k}[t^2(3-2t) + \beta_0 - \beta_0(1-t)^k + \beta_1 t^k] \tag{3}$$

where $k = 1 + \beta_0 + \beta_1$.

A fuzzy number u is obtained by representing the lower and the upper branches u_α^- and u_α^+ using (1) in combination with the model functions (2) or (3).

On the trivial decomposition of the interval $[0,1]$, with $N=1$ (without internal points) and $\alpha_0=0, \alpha_1=1$, u can be represented by a vector of 8 components

$$u=(u_0^-, d_0^-, u_0^+, d_0^+; u_1^-, d_1^-, u_1^+, d_1^+) \tag{4}$$

where $u_0^-, d_0^-, u_1^-, d_1^-$ are used for the lower branch u_α^-, and $u_0^+, d_0^+, u_1^+, d_1^+$ for the upper branch u_α^+; by the application of a monotonic interpolator on the whole interval $\alpha \in [0,1]$ we write

$$\begin{aligned} u_\alpha^- &= p^-(\alpha) = u_0^- + (u_1^- - u_0^-)s(\alpha; \beta_0^-, \beta_1^-) \\ u_\alpha^+ &= p^+(\alpha) = u_0^+ + (u_1^+ - u_0^+)s(\alpha; \beta_0^+, \beta_1^+) \end{aligned}$$

where the parameters β_0^-, β_1^- and β_0^+, β_1^+ are determined such that $\beta_i^\pm(u_1^\pm - u_0^\pm) = d_i^\pm$ for $i=0,1$ (if $u_1^- = u_0^-$ or $u_1^+ = u_0^+$ we fix $\beta_i^- = 0$ or $\beta_i^+ = 0$, respectively).

More generally, a parametric representation of a fuzzy number on a decomposition $0=\alpha_0<\alpha_1<...<\alpha_N=1$ can be written as:

$$\begin{array}{rcl} u & = & (u_{0,i}^-, d_{0,i}^-, u_{1,i}^-, d_{1,i}^-; u_{0,i}^+, d_{0,i}^+, u_{1,i}^+, d_{1,i}^+)_{i=1,...,N} \\ & \Updownarrow & \\ u_\alpha & = & [p_i(t_\alpha; u_{0,i}^-, \widetilde{d}_{0,i}^-, u_{1,i}^-, \widetilde{d}_{1,i}^-), p_i(t_\alpha; u_{0,i}^+, \widetilde{d}_{0,i}^+, u_{1,i}^+, \widetilde{d}_{1,i}^+)]_{i=1,2,...,N} \end{array} \tag{5}$$

where the functions $p_i(t_\alpha; u_{0,i}^-, \widetilde{d}_{0,i}^-, u_{1,i}^-, \widetilde{d}_{1,i}^-)$ and $p_i(t_\alpha; u_{0,i}^+, \widetilde{d}_{0,i}^+, u_{1,i}^+, \widetilde{d}_{1,i}^+)$ are obtained by the described monotonic models using the indicated data, with $\widetilde{d}_{k,i} = d_{k,i}(\alpha_i - \alpha_{i-1})$, $i=0,1$ and $t_\alpha = \frac{\alpha-\alpha_{i-1}}{\alpha_i-\alpha_{i-1}}$ for $\alpha \in [\alpha_{i-1}, \alpha_i]$. For $N \geq 1$ we have a total of $8N$ parameters $u_{0,1}^- \leq u_{1,1}^- \leq u_{0,2}^- \leq u_{1,2}^- \leq ... \leq u_{0,N}^- \leq u_{1,N}^-$, $d_{k,i}^- \geq 0$ defining the increasing lower branch u_α^- and $u_{0,1}^+ \geq u_{1,1}^+ \geq u_{0,2}^+ \geq u_{1,2}^+ \geq ... \geq u_{0,N}^+ \geq u_{1,N}^+$, $d_{k,i}^+ \leq 0$ defining the decreasing upper branch u_α^+ (obviously, also $u_{1,N}^- \leq u_{1,N}^+$ is required).

A simplification of (5) can be obtained by requiring continuous or differentiable branches; in the first case, $u_{1,i}^- = u_{0,i+1}^-$ and $u_{1,i}^+ = u_{0,i+1}^+$ for $i = 1,2,...,N-1$ while, to have differentiability, also the conditions $d_{k,i}^- = d_{k,i+1}^-$, $d_{k,i}^+ = d_{k,i+1}^+$ are required. For the two cases we then have $6N+2$ or $4N+4$ parameters, respectively.

In the following, we will consider only the differentiable case, for which we use the representation:

$$u = (\alpha_i; u_i^-, d_i^-, u_i^+, d_i^+)_{i=0,1,...,N} \tag{6}$$

with the data

$$u_0^- \leq u_1^- \leq ... \leq u_N^- \leq u_N^+ \leq u_{N-1}^+ \leq ... \leq u_0^+ \tag{7}$$

and the slopes

$$d_i^- \geq 0, d_i^+ \leq 0. \tag{8}$$

Denote by $\mathbb{F}_N = \left\{ u | \; u = (u_{0,i}^-, d_{0,i}^-, u_{1,i}^-, d_{1,i}^-; u_{0,i}^+, d_{0,i}^+, u_{1,i}^+, d_{1,i}^+)_{i=1,...,N} \right\}$ the set of LU-fuzzy numbers with a uniform decomposition $\alpha_i = i/N$. $\mathbb{F}_N$ is a $(8N)$-dimensional space and we can see that any fuzzy number (with a finite number of discontinuities) can be approximated using (5) by choosing a fine decomposition $\{\alpha_i; i = 0, ..., N\}$ with N sufficiently large and the parameters $u_{0,i}^-$, $d_{0,i}^-$, $u_{1,i}^-$, $d_{1,i}^-$, $u_{0,i}^+$, $d_{0,i}^+$, $u_{1,i}^+$, $d_{1,i}^+$ to satisfy interpolation and Hermite-type conditions on each subinterval. It follows that the family $\mathbb{F} = \bigcup_{N \geq 1} \mathbb{F}_N$ is dense with respect to all fuzzy numbers (with at most a countable number of discontinuities).

Also an LR fuzzy number can be obtained by using (2) or (3) as the shape functions with the parameters

$$u_{LR} = (u_{0,L}, d_{0,L}, u_{0,R}, d_{0,R}; \quad u_{1,L}, d_{1,L}, u_{1,R}, d_{1,R})$$

provided that $u_{0,L} \leq u_{1,L} \leq u_{1,R} \leq u_{0,R}$ and $d_{0,L}, d_{1,L} \geq 0$, $d_{0,R}, d_{1,R} \leq 0$. The last two pairs of parameters give the slopes (first derivatives) $d_{0,L}, d_{1,L}$ of the membership function μ at the points $x = u_{0,L}$ and $x = u_{1,L}$ and $d_{0,R}, d_{1,R}$ at the points $x = u_{1,R}$ and $x = u_{0,R}$, respectively. More generally, an LR membership function can be obtained on a discretization $0 = \alpha_0 \leq \alpha_1 \leq ... \leq \alpha_N = 1$ by $4(N+1)$ parameters

$$LR : \mu = \alpha_i \rightarrow (u_{i,L}, d_{i,L}, u_{i,R}, d_{i,R}) \; , \; \; i = 0, 1, ..., N \tag{9}$$

where $u_{i,L}, d_{i,L}$, $u_{i,R}, d_{i,R}$ are related to the N pieces $(i = 1, 2, ..., N)$ by

$$\begin{cases} L_i(t) = \alpha_i + (\alpha_{i-1} - \alpha_i) s(t; \beta_{i,0}^L, \beta_{i,1}^L) \\ R_i(t) = \alpha_i + (\alpha_{i-1} - \alpha_i) s(t; \beta_{i,0}^R, \beta_{i,1}^R) \end{cases} , \;\; t \in [0, 1],$$

and the parameters $\beta_{i,0}^L, \beta_{i,1}^L$ and $\beta_{i,0}^R, \beta_{i,1}^R$ are determined such that the slopes of the membership function at the nodes are the given values $d_{i,L}, d_{i,R}$

$$\begin{cases} (\alpha_{i-1} - \alpha_i) \beta_{i,0}^L = -(u_{i,L} - u_{i-1,L}) d_{i,L}, \\ (\alpha_{i-1} - \alpha_i) \beta_{i,1}^L = -(u_{i,L} - u_{i-1,L}) d_{i-1,L}, \\ (\alpha_{i-1} - \alpha_i) \beta_{i,0}^R = (u_{i-1,R} - u_{i,R}) d_{i,R}, \text{ and} \\ (\alpha_{i-1} - \alpha_i) \beta_{i,1}^R = (u_{i-1,R} - u_{i,R}) d_{i-1,R}. \end{cases}$$

The membership function is then

$$\mu(x) = \begin{cases} L_i(\frac{u_{i,L} - x}{u_{i,L} - u_{i-1,L}}) & \text{if } \; u_{i-1,L} \leq x \leq u_{i,L} \; i = 1, 2, ..., N \\ 1 & \text{if } \; u_{N,L} \leq x \leq u_{N,R} \\ R_i(\frac{x - u_{i,R}}{u_{i-1,R} - u_{i,R}}) & \text{if } \; u_{i,R} \leq x \leq u_{i-1,R} \; i = 1, 2, ..., N \\ 0 & \text{otherwise} \end{cases}$$

with the following conditions (to have a proper fuzzy number) $u_{i-1,L} \leq u_{i,L}$, $u_{i,R} \leq u_{i-1,R}$, $u_{N,L} \leq u_{N,R}$, $d_{i,L} \geq 0$ and $d_{i,R} \leq 0$.

2 Generalized Fuzzy Difference

In the fuzzy or in the interval arithmetic contexts, equation $u = v + w$ is not equivalent to $w = u - v = u + (-1)v$ or to $v = u - w = u + (-1)w$ and this has motivated the introduction of the following Hukuhara difference (see [6], [7]).

Definition 5. *Given $u, v \in \mathcal{F}$, the H-difference is defined by*

$$u \ominus_H v = w \Longleftrightarrow u = v + w.$$

Clearly, $u \ominus_H u = \{0\}$; if $u \ominus_H v$ exists, it is unique.

The gH-difference for fuzzy numbers is defined as follows:

Definition 6. *Given $u, v \in \mathcal{F}$, the gH-difference is the fuzzy quantity $w \in \mathcal{F}$, if it exists, such that*

$$u \ominus_{gH} v = w \Longleftrightarrow \begin{cases} (i) \;\; u = v + w \\ or\; (ii) \;\; v = u + (-1)w \end{cases}; \tag{10}$$

If $u \ominus_{gH} v$ and $u \ominus_H v$ exist, $u \ominus_H v = u \ominus_{gH} v$; if (i) and (ii) are satisfied simultaneously, then w is a crisp quantity. Also, $u \ominus_{gH} u = u \ominus_H u = \{0\}$.

The conditions of the existence of $w = u \ominus_{gH} v$ are

$$[w]_\alpha = [w_\alpha^-, w_\alpha^+] = [u]_\alpha \ominus_{gH} [v]_\alpha \text{ and}$$
$$\begin{cases} w_\alpha^- = \min\{u_\alpha^- - v_\alpha^-, u_\alpha^+ - v_\alpha^+\} \\ w_\alpha^+ = \max\{u_\alpha^- - v_\alpha^-, u_\alpha^+ - v_\alpha^+\} \end{cases}.$$

provided that w_α^- is nondecreasing, w_α^+ is nonincreasing and $w_1^- \leq w_1^+$; in particular, for $\alpha \in [0, 1]$,

$$\begin{array}{ll} (1) \begin{cases} w_\alpha^- = u_\alpha^- - v_\alpha^- \\ w_\alpha^+ = u_\alpha^+ - v_\alpha^+ \end{cases} & \text{if } len([u]_\alpha) \leq len([v]_\alpha) \\ (2) \begin{cases} w_\alpha^- = u_\alpha^+ - v_\alpha^+ \\ w_\alpha^+ = u_\alpha^- - v_\alpha^- \end{cases} & \text{if } len([u]_\alpha) \geq len([v]_\alpha) \end{array} \tag{11}$$

where $len([u]_\alpha) = u_\alpha^+ - u_\alpha^-$ is the length of the $\alpha - cuts$ of u (similarly $len([v]_\alpha)$ for v).

Remark 1. A similar construction has been proposed by S. Markov (see e.g. [1,2,3]) in the setting of interval analysis; his inner-difference is defined by first introducing the inner-sum of two intervals (or convex compact sets) A and B by

$$A +^- B = \begin{cases} X \text{ if } X \text{ solves } (-A) + X = B \\ Y \text{ if } Y \text{ solves } (-B) + Y = A \end{cases} \tag{12}$$

and then

$$A -^- B = A +^- (-B). \tag{13}$$

It is not difficult to see that $A \ominus_{gH} B = A -^- B$.

The monotonicity of $u_\alpha^- - v_\alpha^-$ and $u_\alpha^+ - v_\alpha^+$ according to (1) or (2) in (11) is an important condition for the existence of $u \ominus_{gH} v$ and is to be verified explicitly as in fact it may not be satisfied. Consider $[u]_\alpha = [5 + 4\alpha, 11 - 2\alpha]$ and $[v]_\alpha = [12 + 3\alpha, 19 - 4\alpha]$; then $u_\alpha^- - v_\alpha^- = -6 + \alpha$, $u_\alpha^+ - v_\alpha^+ = -8 + 2\alpha$ and $u_\alpha^+ - v_\alpha^+ < u_\alpha^- - v_\alpha^-$ but $u_\alpha^- - v_\alpha^-$ is not decreasing as required.

If the gH-differences $[u]_\alpha \ominus_{gH} [v]_\alpha$ do not define a proper fuzzy number, we can use the nested property of the α−cuts and obtain a proper fuzzy number by

$$[u \ominus_g v]_\alpha := cl \bigcup_{\beta \geq \alpha} ([u]_\beta \ominus_{gH} [v]_\beta) \text{ for } \alpha \in [0, 1]. \tag{14}$$

As each gH-difference $[u]_\beta \ominus_{gH} [v]_\beta$ exists for $\beta \in [0, 1]$ and (14) defines a proper fuzzy number, it follows that $z = u \ominus_g v$ can be considered as a generalization of Hukuhara difference for fuzzy numbers, existing for any u, v.

The LU-fuzzy parametrization of $z = u \ominus_g v$ can be obtained by choosing a partition $0 = \alpha_0 < \alpha_1 < ... < \alpha_N = 1$ of $[0, 1]$ and, from $[w_i^-, w_i^+] = [u]_{\alpha_i} \ominus_{gH} [v]_{\alpha_i}$ by the following backward iteration

$$z_N^- = w_N^-, \quad z_N^+ = w_N^+ \tag{15}$$

$$\text{For } k = N - 1, ..., 0: \begin{cases} z_k^- = \min\{z_{k+1}^-, w_k^-\} \\ z_k^+ = \max\{z_{k+1}^+, w_k^+\} \end{cases}. \tag{16}$$

3 Generalized Fuzzy Division

We define the gH-division as follows:

Definition 7. *Let $u, v \in \mathcal{F}$ have $\alpha -$ cuts $[u]_\alpha = [u_\alpha^-, u_\alpha^+]$, $[v]_\alpha = [v_\alpha^-, v_\alpha^+]$, with $0 \notin [v]_\alpha$ $\forall \alpha \in [0, 1]$. The gH-division $\div_{gH}$ is the operation that calculates the fuzzy number $w = u \div_{gH} v \in \mathcal{F}$ having level cuts $[w]_\alpha = [w_\alpha^-, w_\alpha^+]$ (here $[w]_\alpha^{-1} = [\frac{1}{w_\alpha^+}, \frac{1}{w_\alpha^-}]$) defined by*

$$[u]_\alpha \div_{gH} [v]_\alpha = [w]_\alpha \iff \begin{cases} (i) \ [u]_\alpha = [v]_\alpha [w]_\alpha \\ or\ (ii) \ [v]_\alpha = [u]_\alpha [w]_\alpha^{-1} \end{cases}, \tag{17}$$

provided that w is a proper fuzzy number, where the multiplications between intervals are performed in the standard interval arithmetic setting.

The existence of $w = u \div_{gH} v$ requires that the level cuts $[w]_\alpha = [w_\alpha^-, w_\alpha^+]$ be computed according to the following rules (see [6]):

If ($u_\alpha^+ < 0$ and $v_\alpha^+ < 0$) or ($u_\alpha^- > 0$ and $v_\alpha^- > 0$) then $w_\alpha^- = \min\{\frac{u_\alpha^-}{v_\alpha^-}, \frac{u_\alpha^+}{v_\alpha^+}\}$ and $w_\alpha^+ = \max\{\frac{u_\alpha^-}{v_\alpha^-}, \frac{u_\alpha^+}{v_\alpha^+}\}$;

If ($u_\alpha^+ < 0$ and $v_\alpha^- > 0$) or ($u_\alpha^- > 0$ and $v_\alpha^+ < 0$) then $w_\alpha^- = \min\{\frac{u_\alpha^-}{v_\alpha^+}, \frac{u_\alpha^+}{v_\alpha^-}\}$ and $w_\alpha^+ = \max\{\frac{u_\alpha^-}{v_\alpha^+}, \frac{u_\alpha^+}{v_\alpha^-}\}$;

If $v_\alpha^+ < 0$ and $u_\alpha^- \leq 0 \leq u_\alpha^+$ then $w_\alpha^- = \frac{u_\alpha^+}{v_\alpha^-}$ and $w_\alpha^+ = \frac{u_\alpha^-}{v_\alpha^-}$;

If $v_\alpha^- > 0$ and $u_\alpha^- \leq 0 \leq u_\alpha^+$ then $w_\alpha^- = \frac{u_\alpha^-}{v_\alpha^+}$ and $w_\alpha^+ = \frac{u_\alpha^+}{v_\alpha^+}$.

The fuzzy gH-division $\div_{gH}$ is well defined if the $\alpha - cuts$ $[w]_\alpha$ are such that $w \in \mathcal{F}$ (i.e. if w_α^- is nondecreasing, w_α^+ is nonincreasing, $w_1^- \leq w_1^+$).

If the gH-divisions $[u]_\alpha \div_{gH} [v]_\alpha$ do not define a proper fuzzy number, we can proceed similarly to what is done in section 3.3 and obtain an approximated fuzzy division with $\alpha - cuts$

$$[u \div_g v]_\alpha := cl \bigcup_{\beta \geq \alpha} ([u]_\beta \div_{gH} [v]_\beta). \tag{18}$$

The LU-fuzzy version of $z = u \div_g v$ on a partition $0 = \alpha_0 < \alpha_1 < ... < \alpha_N = 1$ of $[0,1]$ is obtained using $[w_i^-, w_i^+] = [u]_{\alpha_i} \div_{gH} [v]_{\alpha_i}$ and a backward procedure identical to (15)-(16).

4 Computation of Generalized Difference and Division

The computation of gH-difference and gH-division for fuzzy numbers are performed easily in the LU and LR parametric representations. We will detail the LU setting (the LR setting is analogous).

Given two fuzzy numbers $u, v \in \mathcal{F}$ in LU-parametric representation

$$u = (\alpha_i; u_i^-, \delta u_i^-, u_i^+, \delta u_i^+)_{i=0,1,...,N}$$
$$v = (\alpha_i; v_i^-, \delta v_i^-, v_i^+, \delta v_i^+)_{i=0,1,...,N}$$

we compute the fuzzy gH-difference $w = u \ominus_{gH} v$ in LU parametric form $w = (\alpha_i; w_i^-, \delta w_i^-, w_i^+, \delta w_i^+)_{i=0,1,...,N}$ as follows:

Algorithm 1: *LU-Fuzzy gH-difference*
for $i = 0, ..., N$
$m_i = u_i^- - v_i^-$, $p_i = u_i^+ - v_i^+$, $dm_i = \delta u_i^- - \delta v_i^-$, $dp_i = u_i^+ - v_i^+$
if $m_i \geq p_i$ **then**
$\begin{cases} w_i^- = p_i, \ \delta w_i^- = \max\{0, dp_i\} \\ w_i^+ = m_i, \ \delta w_i^+ = \min\{0, dm_i\} \end{cases}$
else
$\begin{cases} w_i^- = m_i, \ \delta w_i^- = \min\{0, dm_i\} \\ w_i^+ = p_i, \ \delta w_i^+ = \max\{0, dp_i\} \end{cases}$
endif
end

The algorithm for the fuzzy gH-division $w = u \div_{gH} v$ is the following:
Algorithm 2: *LU-Fuzzy gH-division*
for $i = 0, ..., N$
if $(u_i^+ < 0$ **and** $v_i^+ < 0)$ **or** $(u_i^- > 0$ **and** $v_i^- > 0)$ **then**
$\begin{cases} m_i = u_i^- / v_i^-, \ dm_i = (v_i^- \delta u_i^- - u_i^- \delta v_i^-)/(v_i^-)^2 \\ p_i = u_i^+ / v_i^+, \ dp_i = (v_i^+ \delta u_i^+ - u_i^+ \delta v_i^+)/(v_i^+)^2 \end{cases}$
elseif $(u_i^+ < 0$ **and** $v_i^- > 0)$ **or** $(u_i^- > 0$ **and** $v_i^+ < 0)$ **then**

$$\begin{cases} m_i = u_i^-/v_i^+, \; dm_i = (v_i^+ \delta u_i^- - u_i^- \delta v_i^+)/(v_i^+)^2 \\ p_i = u_i^+/v_i^-, \;\; dp_i = (v_i^- \delta u_i^+ - u_i^+ \delta v_i^-)/(v_i^-)^2 \end{cases}$$

elseif ($u_i^- \le 0$ **and** $v_i^+ < 0$ **and** $u_i^+ \ge 0$) **then**

$$\begin{cases} m_i = u_i^-/v_i^-, \; dm_i = (v_i^- \delta u_i^- - u_i^- \delta v_i^-)/(v_i^-)^2 \\ p_i = u_i^+/v_i^-, \;\; dp_i = (v_i^- \delta u_i^+ - u_i^+ \delta v_i^-)/(v_i^-)^2 \end{cases}$$

else

$$\begin{cases} m_i = u_i^-/v_i^+, \; dm_i = (v_i^+ \delta u_i^- - u_i^- \delta v_i^+)/(v_i^+)^2 \\ p_i = u_i^+/v_i^+, \;\; dp_i = (v_i^+ \delta u_i^+ - u_i^+ \delta v_i^+)/(v_i^+)^2 \end{cases}$$

endif

if $m_i \ge p_i$ **then**

$$\begin{cases} w_i^- = p_i, \;\; \delta w_i^- = \max\{0, dp_i\} \\ w_i^+ = m_i, \; \delta w_i^+ = \min\{0, dm_i\} \end{cases}$$

else

$$\begin{cases} w_i^- = m_i, \; \delta w_i^- = \min\{0, dm_i\} \\ w_i^+ = p_i, \;\; \delta w_i^+ = \max\{0, dp_i\} \end{cases}$$

endif

end

If the algorithms 1. or 2. will not produce a proper fuzzy number (i.e. the produced LU-representation has non-monotonic w_i^- or w_i^+), the algorithm to adjust the solution according to (15)-(16) is the following:

Algorithm 3: *Adjust LU-Fuzzy g-difference or g-division z from the result w of Algorithm 1 or Algorithm 2.*

$z_N^- = w_N^-$, $\delta z_N^- = \delta w_N^-$, $z_N^+ = w_N^+$, $\delta z_N^+ = \delta w_N^+$

for $i = N-1, ..., 0$

if $z_{i+1}^- \le w_i^-$

then $z_i^- = z_{i+1}^-$, $\delta z_i^- = 0$

else $z_i^- = w_i^-$, $\delta z_i^- = \delta w_i^-$

endif

if $z_{i+1}^+ \ge w_i^+$

then $z_i^+ = z_{i+1}^+$, $\delta z_i^+ = 0$

else $z_i^+ = w_i^+$, $\delta z_i^+ = \delta w_i^+$

endif

end

Applications of the generalized difference and division in the field of interval linear equations and interval differential equations are described in [6] and [7].

Here, for given fuzzy numbers u, v, w with $0 \notin [u]_0$ we consider the fuzzy equation

$$ux + v = w \tag{19}$$

and solve it by

$$x_{gH} = (w \ominus_{gH} v) \div_{gH} u \tag{20}$$

or, more generally, using the (approximated) g-difference and g-division

$$x_g = (w \ominus_g v) \div_g u. \tag{21}$$

Clearly, equation (19) is interpreted here in a formal sense as in fact the found (unique) solution x_{gH} from (20) will not necessarily satisfy (19) exactly. But, taking into account the two cases in (10) and (17), it is possible to see that one of the following four cases is always satisfied (see [6]):

1. substitution $x = x_{gH}$ satisfies exactly $ux + v = w$,
2. substitution $x = x_{gH}$ satisfies exactly $v = w - ux$,
3. x_{gH}^{-1} exists and substitution $y = x_{gH}^{-1}$ satisfies exactly $u \div_{gH} y + v = w$,
4. x_{gH}^{-1} exists and substitution $y = x_{gH}^{-1}$ satisfies exactly $v = w - u \div_{gH} y$.

The following two examples are obtained using the LU-parametrization with $N = 5$ and a uniform decomposition of $[0, 1]$. The data u, v and w are triangular fuzzy numbers with linear membership functions. The solution is obtained by computing $z = w \ominus_{gH} v$ (Algorithm 1) and $x = z \div_{gH} u$ (Algorithm 2).

Ex. 1: Consider $u = \langle 1, 2, 3\rangle$, $v = \langle -3, -2, -1\rangle$, $w = \langle 3, 4, 5\rangle$. The membership function of x_{gH} is illustrated in Figure 1. It satisfies cases 3. and 2.

Ex. 2: Consider $u = \langle 8, 9, 10\rangle$, $v = \langle -3, -2, -1\rangle$, $w = \langle 3, 5, 7\rangle$. The solution (20) is illustrated in Figure 2; it satisfies case 1.

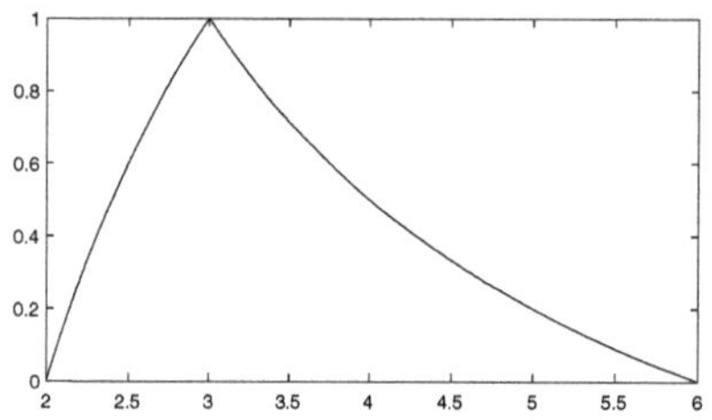

Fig. 1. Solution x_{gH} for Ex. 1

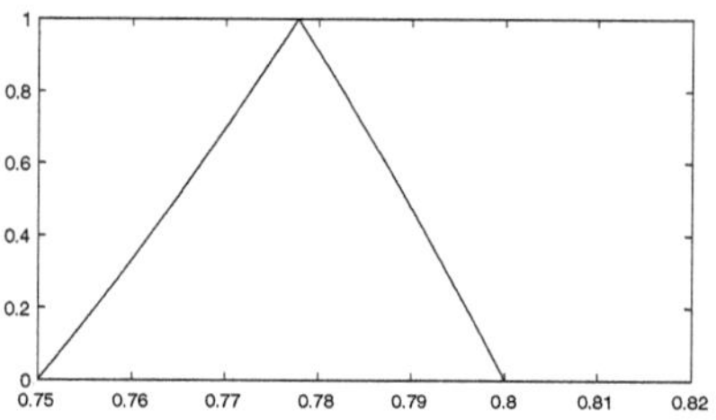

Fig. 2. Solution x_{gH} for Ex. 2

References

1. Markov, S.: A non-standard subtraction of intervals. Serdica 3, 359–370 (1977)
2. Markov, S.: Extended interval arithmetic. Compt. Rend. Acad. Bulg. Sci. 30(9), 1239–1242 (1977)
3. Markov, S.: On the Algebra of Intervals and Convex Bodies. Journal of Universal Computer Science 4(1), 34–47 (1998)
4. Stefanini, L., Sorini, L., Guerra, M.L.: Parametric representations of fuzzy numbers and application to fuzzy calculus. Fuzzy Sets and Systems 157, 2423–2455 (2006)
5. Stefanini, L., Sorini, L., Guerra, M.L.: Fuzzy Numbers and Fuzzy Arithmetic. In: Pedrycz, W., Skowron, A., Kreynovich, V. (eds.) Handbook of Granular Computing, ch. 12. J. Wiley & Sons, Chichester (2009)
6. Stefanini, L.: A generalization of Hukuhara difference and division for interval and fuzzy arithmetic. Fuzzy Sets and Systems 161, 1564–1584 (2009)
7. Stefanini, L., Bede, B.: Generalized Hukuhara differentiability of interval-valued functions and interval differential equations. Nonlinear Analysis 71, 1311–1328 (2009)

Application of Gaussian Quadratures in Solving Fuzzy Fredholm Integral Equations

M. Khezerloo*, T. Allahviranloo, S. Salahshour,
M. Khorasani Kiasari, and S. Haji Ghasemi

Department of Mathematics, Science and Research Branch,
Islamic Azad University, Tehran, Iran
khezerloo_m@yahoo.com

Abstract. In this paper first of all the integral term in the fuzzy Fredholm integral equation (FFIE) is approximated by one of the Gaussian methods. FFIE is transformed to a dual fuzzy linear system that it can be approximated by the method that proposed in [7]. In the special case, Chebyshev-Gauss quadrature is applied to approximate the mentioned integral.

Keywords: Gaussian quadrature; Chebyshev-Gauss quadrature; Fuzzy Fredholm integral equation; Dual fuzzy linear system; Nonnegative matrix.

1 Introduction

The fuzzy differential and integral equations are important part of the fuzzy analysis theory. Park et al. [9] have considered the existence of solution of fuzzy integral equation in Banach space and Subrahmaniam and Sudarsanam [13] proved the existence of solution of fuzzy functional equations. Allahviranloo et al. [1, 2, 3] iterative methods for finding the approximate solution of fuzzy system of linear equations (FSLE) with convergence theorems. Abbasbandy et al. have discussed LU decomposition method, for solving fuzzy system of linear equations in [5]. They considered the method in spatial case when the coefficient matrix is symmetric positive definite. Wang et al. presented an iterative algorithm for dual linear system of the form $X = AX + Y$, where A is real $n \times n$ matrix, the unknown vector X and the constant Y are all vectors consisting of n fuzzy numbers in [15].

The rest of the paper is organized as follows:

In Section 2, Ming Ma et al. [7] proposed method is brought. In Section 3, the main section of the paper, we introduce Gaussian quadratures and then Chebyshev-Gauss quadrature for solving Fredholm integral equation is proposed and discussed in the section 4. The proposed idea is illustrated by one example in details in Section 5. Finally conclusion is drawn in Section 6.

* Corresponding author.

E. Hüllermeier, R. Kruse, and F. Hoffmann (Eds.): IPMU 2010, Part II, CCIS 81, pp. 481–490, 2010.

2 Preliminaries

Definition 1. *[7], The fuzzy linear system*

$$BX = AX + Y \tag{2.1}$$

is called a dual fuzzy linear system where $A = (a_{ij})$, $B = (b_{ij})$, $1 \le i, j \le n$, are crisp coefficient matrix and X, Y are vectors of fuzzy numbers.

Theorem 1. *[7], Let $A = (a_{ij})$, $B = (b_{ij})$, $1 \le i, j \le n$, are nonnegative matrices. The dual fuzzy linear system (2.1) has a unique fuzzy solution if and only if the inverse matrix of $B - A$ exists and has only nonnegative entries.*

Consider the dual fuzzy linear system (2.1), and transform its $n \times n$ coefficient matrix A and B into $(2n) \times (2n)$ matrices as in the following:

$$\begin{aligned}
\underline{y}_1 &+t_{11}\underline{x}_1 + \cdots + t_{1n}\underline{x}_n + t_{1,n+1}(-\overline{x}_1) + \cdots + t_{1,2n}(-\overline{x}_n) \\
&= s_{11}\underline{x}_1 + \cdots + s_{1n}\underline{x}_n + s_{1,n+1}(-\overline{x}_1) + \cdots + s_{1,2n}(-\overline{x}_n) \\
&\vdots \\
\underline{y}_n &+t_{n1}\underline{x}_1 + \cdots + t_{nn}\underline{x}_n + t_{n,n+1}(-\overline{x}_1) + \cdots + t_{n,2n}(-\overline{x}_n) \\
&= s_{n1}\underline{x}_1 + \cdots + s_{nn}\underline{x}_n + s_{n,n+1}(-\overline{x}_1) + \cdots + s_{n,2n}(-\overline{x}_n) \\
-\overline{y}_1 &+t_{n+1,1}\underline{x}_1 + \cdots + t_{n+1,n}\underline{x}_n + t_{n+1,n+1}(-\overline{x}_1) + \cdots + t_{n+1,2n}(-\overline{x}_n) \\
&= s_{n+1,1}\underline{x}_1 + \cdots + s_{n+1,n}\underline{x}_n + s_{n+1,n+1}(-\overline{x}_1) + \cdots + s_{n+1,2n}(-\overline{x}_n) \\
&\vdots \\
-\overline{y}_n &+t_{2n,1}\underline{x}_1 + \cdots + t_{2n,n}\underline{x}_n + t_{2n,n+1}(-\overline{x}_1) + \cdots + t_{2n,2n}(-\overline{x}_n) \\
&= s_{2n,1}\underline{x}_1 + \cdots + s_{2n,n}\underline{x}_n + s_{2n,n+1}(-\overline{x}_1) + \cdots + s_{2n,2n}(-\overline{x}_n)
\end{aligned} \tag{2.2}$$

where s_{ij} and t_{ij} are determined as follows:

$$\begin{aligned}
&b_{ij} \ge 0, \; s_{ij} = b_{ij}, \qquad s_{i+n,j+n} = b_{ij} \\
&b_{ij} < 0, \; s_{i,j+n} = -b_{ij}, \; s_{i+n,j} = -b_{ij} \\
&a_{ij} \ge 0, \; t_{ij} = a_{ij}, \qquad t_{i+n,j+n} = a_{ij} \\
&a_{ij} < 0, \; t_{i,j+n} = -a_{ij}, \; t_{i+n,j} = -a_{ij}
\end{aligned} \tag{2.3}$$

while all the remaining s_{ij} and t_{ij} are taken zero.

The following theorem guarantees the existence of a fuzzy solution for a general case.

Theorem 2. *[7], The dual fuzzy linear system (2.1) has a unique fuzzy solution if and only if the inverse matrix of $S - T$ exists and nonnegative.*

In [4], Allahviranloo proved that $(S - T)_{ij} \geq 0$ is not necessary condition for an unique fuzzy solution of fuzzy linear system.

3 Gaussian Quadratures for Solving Fuzzy Fredholm Integral Equations

Definition 2. *The sequence $\{\Psi_n(x)\}$ of polynomials, with degree of n, is called orthogonal function over an interval $[a, b]$ when,*

$$\int_a^b w(x)\ \Psi_i(x)\ \Psi_j(x)dx = \begin{cases} \gamma_i \neq 0, & i = j \\ 0, & i \neq j \end{cases} \tag{3.4}$$

where $w(x)$ is weight function.

Definition 3. *The general form of Gaussian quadratures is as follows:*

$$\int_a^b w(x)\ \tilde{p}(x)dx = \sum_{j=1}^{n} H_j \tilde{p}(a_j) + E \tag{3.5}$$

where $w(x)$ does not appear on right-hand side of (3.5), a_j's are the zeros of orthogonal polynomial of degree n on $[a, b]$, H_j is weight of the Gaussian quadrature, $\tilde{p}(x)$ is a fuzzy-valued function ($\tilde{p} : (a, b) \to E$ where E is the set of all fuzzy numbers) and E is the error.

Let the coefficient of x^n in $\Psi_n(x)$ be A_n, then we can easily obtain

$$H_j = \frac{-A_{n+1}\gamma_n}{A_n \Psi_{n+1}(a_j)\Psi_n'(a_j)}, \qquad j = 1, \ldots, n \tag{3.6}$$

where it is dependent on orthogonal polynomials and is independent on $\tilde{p}(x)$.

If we ignore this error, then we have

$$\int_a^b w(x)\ \tilde{p}(x)dx \simeq \sum_{j=1}^{n} H_j \tilde{p}(a_j) \tag{3.7}$$

Now, consider the fuzzy Fredholm integral equation of second kind

$$\tilde{\phi}(x) = \tilde{f}(x) + \int_a^b K(x, y)\tilde{\phi}(y)dy, \qquad a \leq x \leq b \tag{3.8}$$

where $K(x, y)$ is a crisp kernel, $\tilde{f}(x)$ and $\tilde{\phi}(x)$ are fuzzy-valued functions.

We are going to obtain the fuzzy solution of the fuzzy Fredholm integral equation (3.8) by using any kind of Gaussian quadratures.

For approximation of integral term of Eq. (3.8), we must generate weight function $w(x)$. So, we have

$$\begin{aligned}\tilde{\phi}(x) &= \tilde{f}(x) + \int_a^b K(x,y)\tilde{\phi}(y)dy \\ &= \tilde{f}(x) + \int_a^b \frac{w(y)}{w(y)} K(x,y)\tilde{\phi}(y)dy \\ &= \tilde{f}(x) + \int_a^b w(y)\frac{1}{w(y)} K(x,y)\tilde{\phi}(y)dy \\ &= \tilde{f}(x) + \int_a^b w(y)F(x,y)\tilde{\phi}(y)dy \\ &\simeq \tilde{f}(x) + \sum_{j=1}^n F(x,a_j)\tilde{\phi}(a_j)\end{aligned} \tag{3.9}$$

where $F(x,y) = K(x,y)/w(y)$ and $a \leq x \leq b$.

There are various methods to apply such method like Laguerre-, Hermit-, Legendre-, Chebyshev-Gauss and etc. For instance, we use Chebyshev-Gauss quadrature. Moreover, we should convert $[a,b]$ to $[-1,1]$ for applying our method which will be done by changing variable, easily.

4 Chebyshev-Gauss Quadrature for Solving Fuzzy Fredholm Integral Equations

Given the fuzzy Fredholm integral equation of second kind

$$\tilde{\phi}(x) = \tilde{f}(x) + \int_{-1}^1 K(x,y)\tilde{\phi}(y)dy, \qquad -1 \leq x \leq 1 \tag{4.10}$$

in solving the integral equation with a kernel $K(x,y)$ and the fuzzy-valued function $\tilde{f}(x)$, the problem is typically to find the function $\tilde{\phi}(x)$.
In this section, we try to solve Eq. (4.10) by using Chebyshev-Gauss quadrature. So, we consider the crisp function $\frac{1}{(1-y^2)^{1/2}}$ and we have

$$\tilde{\phi}(x) = \tilde{f}(x) + \int_{-1}^1 K(x,y)\left(\frac{\frac{1}{(1-y^2)^{1/2}}}{\frac{1}{(1-y^2)^{1/2}}}\right)\tilde{\phi}(y)dy \tag{4.11}$$

We suppose $F(x,y) = K(x,y).(1-y^2)^{1/2}$. So by using Eq. (3.7), we get

$$\begin{aligned}\tilde{\phi}(x) &= \tilde{f}(x) + \int_{-1}^1 K(x,y)\left(\frac{\frac{1}{(1-y^2)^{1/2}}}{\frac{1}{(1-y^2)^{1/2}}}\right)\tilde{\phi}(y)dy \\ &= \tilde{f}(x) + \int_{-1}^1 \frac{1}{(1-y^2)^{1/2}} F(x,y)\tilde{\phi}(y)dy \\ &= \tilde{f}(x) + \frac{\pi}{n}\sum_{j=1}^n F(x,a_j)\tilde{\phi}(a_j)\end{aligned} \tag{4.12}$$

Now, Eq. (4.12) is transformed to the system by selecting several points in $[-1,1]$. In this study the selected points a_j, $j = 1,\ldots,n$ are zeros of Chebyshev polynomial of degree n. Therefore, Eq. (4.12) can be writen as follows:

$$\tilde{\phi}(a_i) = \tilde{f}(a_i) + \frac{\pi}{n}\sum_{j=1}^n F(a_i,a_j)\tilde{\phi}(a_j), \qquad i = 1,\ldots,n \tag{4.13}$$

Since the Eq. (4.13) is a dual system, so we are going to obtain the solution of the dual fuzzy linear system

$$X = Y + AX \tag{4.14}$$

where

$$A = \frac{\pi}{n}\begin{bmatrix} F(a_1,a_1) & F(a_1,a_2) & \cdots & F(a_1,a_n) \\ F(a_2,a_1) & F(a_2,a_2) & \cdots & F(a_2,a_n) \\ \vdots & \vdots & & \vdots \\ F(a_n,a_1) & F(a_n,a_2) & \cdots & F(a_n,a_n) \end{bmatrix}$$

and

$$X = \begin{bmatrix} \tilde{\phi}(a_1) \\ \tilde{\phi}(a_2) \\ \vdots \\ \tilde{\phi}(a_n) \end{bmatrix}, \quad Y = \begin{bmatrix} \tilde{f}(a_1) \\ \tilde{f}(a_2) \\ \vdots \\ \tilde{f}(a_n) \end{bmatrix}$$

It is obvious that A is a $n \times n$ crisp matrix and X, Y are vectors of fuzzy numbers.

In [7], Ming Ma et al. have proposed a method for solving dual fuzzy linear system and we use this method as well. By solving system (4.14), we obtain $\tilde{\phi}(a_i)$, $i = 1, \ldots, n$ and then approximate fuzzy-valued function $\tilde{\phi}(x)$ by using an interpolation method.

5 Numerical Examples

Example 1. Consider fuzzy Fredholm integral equation (4.10). Suppose

$$\tilde{f}(x) = (r, 2-r)x$$

$$K(x,y) = x + y$$

$$n = 2, \quad a_1 = -\sqrt{2}/2 = -a_2$$

Then,

$$F(x,y) = (x+y)\sqrt{1-y^2}$$

Using Eq. (4.12), we obtain

$$\tilde{\phi}(x) = (r, 2-r)x + \frac{\pi}{2}\sum_{j=1}^{n}\left((x+a_j)\sqrt{1-a_j^2}\right)\ \tilde{\phi}(a_j)$$

From Eq. (4.13),

$$\tilde{\phi}(a_i) = (r, 2-r)a_i + \frac{\pi}{2}\sum_{j=1}^{2}\left((a_i+a_j)\sqrt{1-a_j^2}\right)\ \tilde{\phi}(a_j), \quad i = 1, 2$$

So,

$$\begin{bmatrix} \tilde{\phi}(\sqrt{2}/2) \\ \tilde{\phi}(-\sqrt{2}/2) \end{bmatrix} = \begin{bmatrix} \tilde{f}(\sqrt{2}/2) \\ \tilde{f}(-\sqrt{2}/2) \end{bmatrix} + \frac{\pi}{2} \begin{bmatrix} 1 & 0 \\ 0 & -1 \end{bmatrix} \begin{bmatrix} \tilde{\phi}(\sqrt{2}/2) \\ \tilde{\phi}(-\sqrt{2}/2) \end{bmatrix}$$

Using Ming Ma et al. proposed method we get

$$\begin{bmatrix} \underline{\phi}(\sqrt{2}/2) \\ \underline{\phi}(-\sqrt{2}/2) \\ -\overline{\phi}(\sqrt{2}/2) \\ -\overline{\phi}(-\sqrt{2}/2) \end{bmatrix} = \begin{bmatrix} r\sqrt{2}/2) \\ (r-2)\sqrt{2}/2) \\ (r-2)\sqrt{2}/2) \\ r\sqrt{2}/2) \end{bmatrix} + \frac{\pi}{2} \begin{bmatrix} 1\,0\,0\,0 \\ 0\,0\,0\,1 \\ 0\,0\,1\,0 \\ 0\,1\,0\,0 \end{bmatrix} \begin{bmatrix} \underline{\phi}(\sqrt{2}/2) \\ \underline{\phi}(-\sqrt{2}/2) \\ -\overline{\phi}(\sqrt{2}/2) \\ -\overline{\phi}(-\sqrt{2}/2) \end{bmatrix}$$

Then,

$$\begin{bmatrix} \tilde{\phi}(\sqrt{2}/2) \\ \tilde{\phi}(-\sqrt{2}/2) \end{bmatrix} = \begin{bmatrix} \frac{\sqrt{2}}{2-\pi}(r, 2-r) \\ \frac{\sqrt{2}}{2-\pi}(r-2, -r) \end{bmatrix}$$

If we apply fuzzy Lagrange interpolation method for $n = 2$, we obtain

$$\tilde{\phi}(x) \simeq \tilde{p}_2(x) = \frac{2x - \sqrt{2}}{-2\sqrt{2}}\, \tilde{\phi}(-\sqrt{2}/2) + \frac{2x + \sqrt{2}}{2\sqrt{2}}\, \tilde{\phi}(\sqrt{2}/2)$$

Now, consider $n = 3$. Then, $a_1 = -\sqrt{3}/2 = -a_3$ and $a_2 = 0$. Using these points we have

$$\begin{bmatrix} \tilde{\phi}(-\sqrt{3}/2) \\ \tilde{\phi}(0) \\ \tilde{\phi}(\sqrt{3}/2) \end{bmatrix} = \begin{bmatrix} \tilde{f}(-\sqrt{3}/2) \\ \tilde{f}(0) \\ \tilde{f}(\sqrt{3}/2) \end{bmatrix} + \frac{\pi}{3} \begin{bmatrix} -\sqrt{3}/2 & -\sqrt{3}/2 & 0 \\ -\sqrt{3}/4 & 0 & \sqrt{3}/4 \\ 0 & \sqrt{3}/2 & \sqrt{3}/2 \end{bmatrix} \begin{bmatrix} \tilde{\phi}(-\sqrt{3}/2) \\ \tilde{\phi}(0) \\ \tilde{\phi}(\sqrt{3}/2) \end{bmatrix}$$

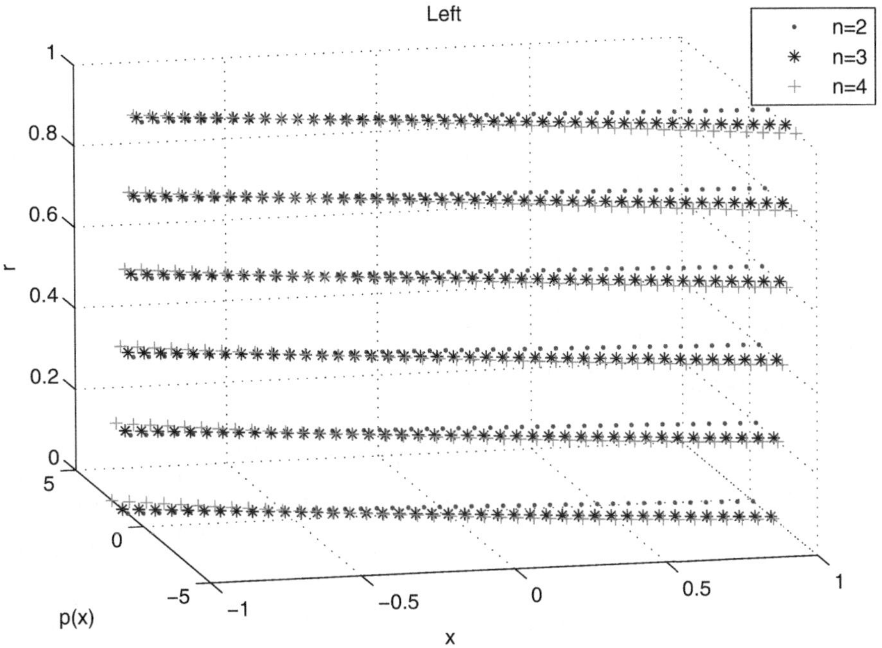

Fig. 1. Approximation of $\underline{p}_n(x)$

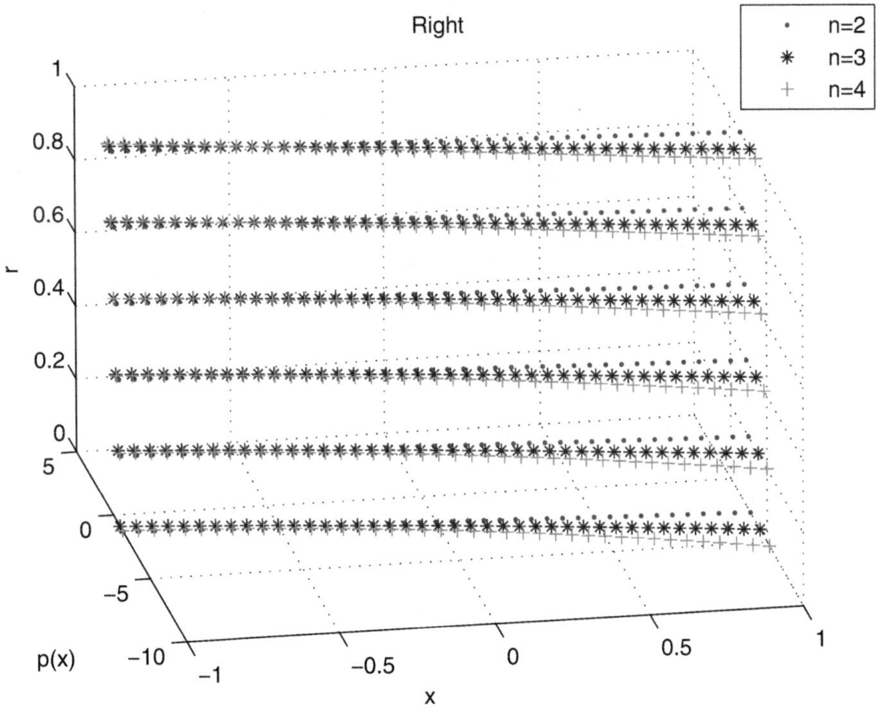

Fig. 2. Approximation of $\overline{p}_n(x)$

Using Ming Ma et al. proposed method we get

$$\begin{bmatrix} \underline{\phi}(-\sqrt{3}/2) \\ \underline{\phi}(0) \\ \underline{\phi}(\sqrt{3}/2) \\ -\overline{\phi}(-\sqrt{3}/2) \\ -\overline{\phi}(0) \\ -\overline{\phi}(\sqrt{3}/2) \end{bmatrix} = \begin{bmatrix} \sqrt{3}/2)(r-2) \\ 0 \\ \sqrt{3}/2r \\ \sqrt{3}/2r \\ 0 \\ \sqrt{3}/2(r-2) \end{bmatrix}$$

$$+\frac{\pi}{3}\begin{bmatrix} 0 & 0 & 0 & \sqrt{3}/2 & \sqrt{3}/2 & 0 \\ 0 & 0 & \sqrt{3}/4 & \sqrt{3}/4 & 0 & 0 \\ 0 & \sqrt{3}/2 & \sqrt{3}/2 & 0 & 0 & 0 \\ \sqrt{3}/2 & \sqrt{3}/2 & 0 & 0 & 0 & 0 \\ \sqrt{3}/4 & 0 & 0 & 0 & 0 & \sqrt{3}/4 \\ 0 & 0 & 0 & 0 & \sqrt{3}/2 & \sqrt{3}/2 \end{bmatrix} \begin{bmatrix} \underline{\phi}(-\sqrt{3}/2) \\ \underline{\phi}(0) \\ \underline{\phi}(\sqrt{3}/2) \\ -\overline{\phi}(-\sqrt{3}/2) \\ -\overline{\phi}(0) \\ -\overline{\phi}(\sqrt{3}/2) \end{bmatrix}$$

So,

$$\begin{bmatrix} \underline{\phi}(-\sqrt{3}/2) \\ \underline{\phi}(0) \\ \underline{\phi}(\sqrt{3}/2) \\ -\overline{\phi}(-\sqrt{3}/2) \\ -\overline{\phi}(0) \\ -\overline{\phi}(\sqrt{3}/2) \end{bmatrix} = \begin{bmatrix} -\sqrt{3}\times 0.3789(r-2) - \sqrt{3}\times 0.3067r \\ \sqrt{3}\times 0.0407(r-2) - \sqrt{3}\times 0.6624r \\ \sqrt{3}\times 0.3965(r-2) - \sqrt{3}\times 1.082r \\ -\sqrt{3}\times 0.3067(r-2) - \sqrt{3}\times 0.3789r \\ -\sqrt{3}\times 0.6624(r-2) + \sqrt{3}\times 0.0407r \\ -\sqrt{3}\times 1.082(r-2) + \sqrt{3}\times 0.3965r \end{bmatrix}$$

If we apply fuzzy Lagrange interpolation method for $n = 3$, we obtain

$$\begin{aligned} \tilde{p}_3(x) &= \left(-2x/\sqrt{3}\right)\left((2x-\sqrt{3})/(-2\sqrt{3})\right)\tilde{\phi}(-\sqrt{3}/2) \\ &+ \left((2x+\sqrt{3})/\sqrt{3}\right)\left((2x-\sqrt{3})/(-\sqrt{3})\right)\tilde{\phi}(0) \\ &+ \left((2x+\sqrt{3})/(2\sqrt{3})\right)\left(2x/\sqrt{3}\right)\tilde{\phi}(\sqrt{3}/2) \end{aligned}$$

and for $n = 4$,

$$\tilde{\phi}(x) = ((x - 0.9239)/ - 1.8478)\,((x + 0.3827)/ - 0.5412)\,((x - 0.3827)/ - 1.3066)\,\tilde{\phi}(-0.9239)$$
$$+ ((x + 0.9239)/ - 1.8478)\,((x + 0.3827)/1.3066)\,((x - 0.3827)/0.5412)\,\tilde{\phi}(0.9239)$$
$$+ ((x + 0.9239)/0.5412)\,((x - 0.9239)/ - 1.3066)\,((x - 0.3827)/ - 0.7654)\,\tilde{\phi}(-0.9239)$$
$$+ ((x + 0.9239)/1.3066)\,((x - 0.9239)/ - 0.5412)\,((x + 0.3827)/0.7654)\,\tilde{\phi}(-0.9239)$$

Fig. 1 and Fig. 2 show the $\underline{p}_n(x)$ and $\overline{p}_n(x)$, respectively, the at $n = 2$, $n = 3$ and $n = 4$.

6 Conclusion

In this work, the integral term in the fuzzy Fredholm integral equation (FFIE) was approximated by one of the Gaussian methods. FFIE was transformed to a dual fuzzy linear system that it can be approximated by the method that proposed in [7]. In the special case, Chebyshev-Gauss quadrature was applied to approximate the mentioned integral.

References

[1] Allahviranloo, T.: Successive over relaxation iterative method for fuzzy system of linear equations. Applied Mathematics and Computation 162, 189–196 (2005)

[2] Allahviranloo, T.: The Adomian decomposition method for fuzzy system of linear equations. Applied Mathematics and Computation 163, 553–563 (2005)

[3] Allahviranloo, T., Ahmady, E., Ahmady, N., Shams Alketaby, K.: Block Jacobi two-stage method with Gauss-Sidel inner iterations for fuzzy system of linear equations. Applied Mathematics and Computation 175, 1217–1228 (2006)

[4] Allahviranloo, T.: A comment on fuzzy linear system. Fuzzy Sets and Systems 140, 559 (2003)

[5] Abbasbandy, S., Ezzati, R., Jafarian, A.: LU decomposition method for solving fuzzy system of linear equations. Applied Mathematics and Computation 172, 633–643 (2006)

[6] Babolian, E., Sadeghi Goghary, H., Abbasbandy, S.: Numerical solution of linear Fredholm fuzzy integral equations of the second kind by Adomian method. Applied Mathematics and Computation 161, 733–744 (2005)

[7] Friedman, M., Ming, M., Kandel, A.: Duality in fuzzy linear systems. Fuzzy Sets and Systems 109, 55–58 (2000)

[8] Friedman, M., Ming, M., Kandel, A.: Fuzzy linear systems. Fuzzy Sets and Systems 96, 201–209 (1998)

[9] Park, J.Y., Kwun, Y.C., Jeong, J.U.: Existence of solutions of fuzzy integral equations in Banach spaces. Fuzzy Sets and Systems 72, 373–378 (1995)

[10] Park, J.Y., Jeong, J.U.: A note on fuzzy functional equations. Fuzzy Sets and Systems 108, 193–200 (1999)

[11] Park, J.Y., Lee, S.Y., Jeong, J.U.: On the existence and uniqueness of solutions of fuzzy Volterra-Fredholm integral equuations. Fuzzy Sets and Systems 115, 425–431 (2000)

[12] Park, J.Y., Jeong, J.U.: The approximate solutions of fuzzy functional integral equations. Fuzzy Sets and Systems 110, 79–90 (2000)
[13] Subrahmaniam, P.V., Sudarsanam, S.K.: On some fuzzy functional equations. Fuzzy Sets and Systems 64, 333–338 (1994)
[14] Wang, K., Zheng, B.: Symmetric successive overrelaxation methods for fuzzy linear systems. Applied Mathematics and Computation 175, 891–901 (2006)
[15] Wang, X., Zhong, Z., Ha, M.: Iteration algorithms for solving a system of fuzzy linear equations. Fuzzy Sets Syst. 119, 121–128 (2001)
[16] Zadeh, L.A.: The concept of a linguistic variable and its application to approximate reasoning. Information Sciences 8, 199–249 (1975)

Existence and Uniqueness of Solutions of Fuzzy Volterra Integro-differential Equations

S. Hajighasemi[1], T. Allahviranloo[2], M. Khezerloo[2], M. Khorasany[2], and S. Salahshour[3]

[1] Roudehen Branch, Islamic Azad University, Tehran, Iran
[2] Department of Mathematics, Science and Research Branch, Islamic Azad University, Tehran, Iran
[3] Department of Mathematics, Mobarakeh Branch, Islamic Azad University, Mobarakeh, Iran

Abstract. In this paper, we will investigate existence and uniqueness of solutions of fuzzy Volterra integrro-differential equations of the second kind with fuzzy kernel under strongly generalized differentiability. To this end, some new results are derived for Hausdorff metric.

Keywords: Fuzzy integro-differential equations, Fuzzy valued functions, Hausdorff metric.

1 Introduction

The fuzzy differential and integral equations are important part of the fuzzy analysis theory and they have the important value of theory and application in control theory.

Seikkala in [14] defined the fuzzy derivative and then, some generalizations of that, have been investigated in [4,11,12,13,15,17]. Consequently, the fuzzy integral which is the same as that of Dubois and Prade in [5], and by means of the extension principle of Zadeh, showed that the fuzzy initial value problem $x'(t) = f(t, x(t)), x(0) = x_0$ has a unique fuzzy solution when f satisfies the generalized Lipschitz condition which guarantees a unique solution of the deterministic initial value problem. Kaleva [7] studied the Cauchy problem of fuzzy differential equation, characterized those subsets of fuzzy sets in which the Peano theorem is valid. Park et al. in [9] have considered the existence of solution of fuzzy integral equation in Banach space and Subrahmaniam and Sudarsanam in [16] have proved the existence of solution of fuzzy functional equations.

Bede et.al in [2,3] have introduced a more general definition of the derivative for fuzzy mappings, enlarging the class of differentiable.

Park and Jeong [8,10] studied existence of solution of fuzzy integral equations of the form

$$x(t) = f(t) + \int_0^t f(t, s, x(s))ds, \quad t \geq 0,$$

where $f(t)$ and $x(t)$ are fuzzy valued functions and k is a crisp function on real numbers.

E. Hüllermeier, R. Kruse, and F. Hoffmann (Eds.): IPMU 2010, Part II, CCIS 81, pp. 491–500, 2010.

In this paper, we study the existence and uniqueness of the solution of fuzzy Volterra integro-differential of the form

$$\begin{cases} x'(t) = f(t) + \int_0^t k(t,s)g(x, x(s))ds, & t \geq 0 \\ x(0) = \tilde{c}, \quad \tilde{c} \in E \end{cases} \tag{1}$$

where $x(t)$ is an unknown fuzzy set-valued mapping, the kernel $k(t,s)$ is determined fuzzy set-valued mapping and Riemann integral is used [3].

This paper is organized as following:
In section 2, the basic concepts are given which are used throughout the paper. In section 3, the existence and uniqueness of solutions of fuzzy Volterra integro-differential equations of the second kind are investigated using Hausdorff metric properties and strongly generalized differentiability. Finally, conclusion and future research is drawn in section 4.

2 Preliminaries

Let $P(\Re)$ denote the family of all nonempty compact convex subsets of $\Re$ and define the addition and scalar multiplication in $P(\Re)$ as usual. Let A and B be two nonempty bounded subsets of $\Re$. The distance between A and B is defined by the Hausdorff metric,

$$d(A, B) = \max\{\sup_{a\in A} \inf_{b\in B} \|a - b\|, \sup_{b\in B} \inf_{a\in A} \|a - b\|\},$$

where $\|.\|$ denotes the usual Euclidean norm in $\Re$. Then it is clear thet $(P(\Re), d)$ becomes a metric space. Puri and Ralescu [11] have proved $(P(\Re), d)$ is complete and separable. Let $I = [0, a] \subset \Re$ be a closed and bounded interval and denote $E = \{u : \Re^n \to [0, 1] | u \quad \text{satisfies} \quad (i) - (iv) \quad \text{below}\}$,
where

(i) u is normal, i.e., there exists an $x_0 \in \Re$ such that $u(x_0) = 1$,
(ii) u is fuzzy convex,
(iii) u upper semicontinuous,
(iv) $[u]^0 = cl\{x \in \Re | u(x) > 0\}$ is compact.

For $0 < \alpha \leq 1$ denote $[u]^\alpha = \{x \in \Re | u(x) \geq \alpha\}$. Then from (i)-(iv), it follows that the α-level set $[u]^\alpha \in P(\Re)$ for all $0 < \alpha \leq 1$.
Also, set E is named as the set of all fuzzy real numbers. Obviously $\Re \subset E$.

Definition 1. *An arbitrary fuzzy number u in the parametric form is represented by an ordered pair of functions $(\underline{u}, \overline{u})$ which satisfy the following requirements:*

(i) $\overline{u} : \alpha \longrightarrow \overline{u}^\alpha \in \Re$ *is a bounded left-continuous non-decreasing function over* $[0, 1]$,
(ii) $\underline{u} : \alpha \longrightarrow \underline{u}^\alpha \in \Re$ *is a bounded left-continuous non-increasing function over* $[0, 1]$,

(iii) $\underline{u}^\alpha \leq \overline{u}^\alpha, \quad 0 \leq \alpha \leq 1.$

Let $D : E \times E \to \Re^+ \cup \{0\}$ be defined by

$$D(u, v) = \sup_{0 \leq \alpha \leq 1} d([u]^\alpha, [v]^\alpha),$$

where d is the Hausdorff metric defined in $(P(\Re), d)$. Then D is a metric on E. Further, (E, D) is a complete metric space [7,11].

Definition 2. *A mapping $x : I \to E$ is bounded, if there exists $r > 0$ such that*

$$D(x(t), \tilde{0}) < r \quad \forall t \in I.$$

Also, one can easily proved the following statements:

(i) $D(u + w, v + w) = D(u, v)$ for every $u, v, w \in E$,
(ii) $D(u + v, \tilde{0}) \leq D(u, \tilde{0}) + D(v, \tilde{0})$ for every $u, v \in E$,
(iii) $D(u \tilde{*} v, \tilde{0}) \leq D(u, \tilde{0}) D(v, \tilde{0})$ for every $u, v, w \in E$ where the fuzzy multiplication $\tilde{*}$ is based on the extension principle that can be proved by α-cuts of fuzzy numbers $u, v \in F$ and $\lambda \in \Re$,
(iv) $D(u + v, w + z) \leq D(u, w) + D(v, z)$ for u, v, w, and $z \in E$.

Definition 3. *(see [6]). Let $f : \Re \to E$ be a fuzzy valued function. If for arbitrary fixed $t_0 \in \Re$ and $\epsilon > 0$, a $\delta > 0$ such that*

$$|t - t_0| < \delta \Rightarrow D(f(t), f(t_0)) < \epsilon,$$

f is said to be continuous.

Definition 4. *Consider $u, v \in E$. If there exists $w \in E$ such that $u = v + w$, then w is called the H-difference of u and v, and is denoted by $u \ominus v$.*

Definition 5. *[3]. Let $f : (a, b) \to E$ and $t \in (a, b)$. We say that f is strongly generalized differentiable at t_0, if there exists an element $f'(t_0) \in E$, such that*

(i) for all $h > 0$ sufficiently small, $\exists f(t_0 + h) \ominus f(t_0), \exists f(t_0) \ominus f(t_0 - h)$ and the limited (in the metric D):

$$\lim_{h \to 0} \frac{f(t_0 + h) \ominus f(t_0)}{h} = \lim_{h \to 0} \frac{f(t_0) \ominus f(t_0 - h)}{h} = f'(t_0)$$

or

(ii) for all $h > 0$ sufficiently small, $\exists f(t_0) \ominus f(t_0 + h)$, $\exists f(t_0 - h) \ominus f(t_0)$ and the following limits hold (in the metric D):

$$\lim_{h \to 0} \frac{f(t_0) \ominus f(t_0 + h)}{-h} = \lim_{h \to 0} \frac{f(t_0 - h) \ominus f(t_0)}{-h} = f'(t_0)$$

or

(iii) *for all* $h > 0$ *sufficiently small,* $\exists f(t_0+h) \ominus f(t_0)$, $\exists f(t_0-h) \ominus f(t_0)$ *and the following limits hold (in the metric D):*

$$\lim_{h\to 0} \frac{f(t_0+h)\ominus f(t_0)}{h} = \lim_{h\to 0} \frac{f(t_0-h)\ominus f(t_0)}{-h} = f'(t_0)$$

or

(iv) *for all* $h > 0$ *sufficiently small,* $\exists f(t_0) \ominus f(t_0+h)$, $\exists f(t_0) \ominus f(t_0-h)$ *and the following limits hold (in the metric D):*

$$\lim_{h\to 0} \frac{f(t_0)\ominus f(t_0+h)}{-h} = \lim_{h\to 0} \frac{f(t_0)\ominus f(t_0-h)}{h} = f'(t_0)$$

It was proved by Puri and Relescu [11] that a strongly measurable and integrably bounded mapping $F: I \to E$ is integrable (i.e., $\int_I F(t)dt \in E$).

Theorem 1. *[16]. If* $F: I \to E$ *is continuous then it is integrable.*

Theorem 2. *[16]. Let* $F, G: I \to E$ *be integrable and* $\lambda \in \Re$*. Then*

(i) $\int_I (F(t)+G(t))dt = \int_I F(t)dt + \int_I G(t)dt$,
(ii) $\int_I \lambda F(t)dt = \lambda \int_I F(t)dt$,
(iii) $D(F,G)$ *is integrable,*
(iv) $D(\int_I F(t)dt, \int_I G(t)dt) \leq \int_I D(F(t),G(t))dt$.

Theorem 3. *[3]. For* $t_0 \in \Re$*, the fuzzy differential equation* $x' = f(t,x)$, $x(t_0) = x_0 \in E$*, where* $f: \Re \times E \to E$ *is supposed to be continuous, is equivalent to one of the integral equations*

$$x(t) = x_0 + \int_{t_0}^{t} f(s,x(s))ds, \quad \forall t \in [t_0,t_1]$$

or

$$x(0) = x(t) + (-1).\int_{t_0}^{t} f(s,x(s))ds, \quad \forall t \in [t_0,t_1]$$

on some interval $(x_0,x_1) \subset \Re$*, under the strong differentiability condition, (i) or (ii), respectively. Here the equivalence between two equation means that any solution of an equation is a solution for the other one.*

Lemma 1. *[1]. If the H-difference for arbitrary* $u,v \in E$ *and* $u,w \in E$ *exists, we have*

$$D(u\ominus v, u\ominus w) = D(v,w), \quad \forall u,v,w \in E$$

Lemma 2. *If the H-difference for arbitrary* $u,v \in E$ *exists, we have*

$$D(u\ominus v, \tilde{0}) = D(u,v), \quad \forall u,v \in E$$

Lemma 3. *If the H-difference for arbitrary* $u,v \in E$ *and* $w,z \in E$ *exists, we have*

$$D(u\ominus v, w\ominus z) \leq D(u,w) + D(v,z), \quad \forall u,v,w,z \in E.$$

3 Existence Theorem

We consider the following fuzzy Volterra integro-differential equation

$$\begin{cases} x'(t) = f(t) + \int_0^t k(t,s)g(x,x(s))ds, & t \geq 0 \\ x(0) = \tilde{c}, \quad \tilde{c} \in E, \end{cases} \tag{2}$$

where $f : [0,a] \to E$ and $k : \Delta \to E$, where $\Delta = (t,s) : 0 \leq s \leq t \leq a$, and $g : [0,a] \times E \to E$ are continuous.

Case I. Let us consider $x(t)$ is a (i)-differentiable function, of Theorem 3, we get the following:

$$x(t) = \tilde{c} + \int_0^u f(t)dt + \int_0^u \int_0^t k(t,s)g(s,x(s))dsdt, \;\; 0 \leq t \leq a, \;\; 0 \leq u \leq 1, \;\; \tilde{c} \in E, \tag{3}$$

then we have the corresponding theorem:

Theorem 4. *Let a and L are positive numbers. Assume that Eq.(3) satisfies the following conditions:*

(i) $f : [0,a] \to E$ *is continuous and bounded i.e., there exists* $R > 0$ *such that* $D(f(t),\tilde{0}) \leq R$.
(ii) $k : \Delta \to E$ *is continuous where* $\Delta = (t,s) : 0 \leq s \leq t \leq a$ *and there exists* $M > 0$ *such that* $\int_0^t D(k(t,s),\tilde{0})ds \leq M$.
(iii) $g : [0,a] \times E \to E$ *is continuous and satisfies the Lipschitz condition, i.e.,*

$$D(g(t,x(t)),g(t,y(t)) \leq LD(x(t),y(t)), \quad 0 \leq t \leq a, \tag{4}$$

where $L < M^{-1}$ *and* $x,y : [0,a] \to E$.
(iv) $g(t,\tilde{0})$ *is bounded on [0,a].*

Then, there exists an unique solution $x(t)$ *of Eq.(3) on* $[0,a]$ *and the successive iterations*

$$\begin{aligned} &x_0(t) = \tilde{c} + \textstyle\int_0^u f(t)dt, \\ &x_{n+1}(t) = \tilde{c} + \textstyle\int_0^u f(t)dt + \int_0^u \int_0^t k(t,s)g(s,x_n(s))dsdt, \\ &\quad 0 \leq t \leq a, \quad 0 \leq u \leq 1, \quad \tilde{c} \in E \end{aligned} \tag{5}$$

are uniformly convergent to x(t) on $[0,a]$.

Proof. It is easy to see that all $x_n(t)$ are bounded on $[0,a]$. Indeed $x_0 = f(t)$ is bounded by hypothesis. Assume that $x_{n-1}(t)$ is bounded, we have

$$\begin{aligned} D(x_n(t),\tilde{0}) &\leq D(\tilde{c},\tilde{0}) + \textstyle\int_0^u D(f(t),\tilde{0})dt + \int_0^u \int_0^t D(k(t,s)g(s,x_{n-1}(s)),\tilde{0})dsdt \\ &\leq D(\tilde{c},\tilde{0}) + Ru + \textstyle\int_0^u ((\sup_{0\leq t\leq a} D(g(t,x_{n-1}(t)),\tilde{0})) \int_0^t D(k(t,s),\tilde{0})ds)dt \end{aligned} \tag{6}$$

Taking every assumptions into account

$$\begin{aligned} D(g(t,x_{n-1}(t)),\tilde{0}) &\leq D(g(t,x_{n-1}(t)),g(t,\tilde{0}))+D(g(t,\tilde{0}),\tilde{0}) \\ &\leq LD(x_{n-1}(t),\tilde{0})+D(g(t,\tilde{0}),\tilde{0}). \end{aligned} \tag{7}$$

Since, $u \leq 1$ we obtain that $x_n(t)$ is bounded. Thus, $x_n(t)$ is a sequence of bounded functions on $[0,a]$. Next we prove that $x_n(t)$ are continuous on $[0,a]$. For $0 \leq t_1 \leq t_2 \leq a$, we have the following:

$$\begin{aligned} D(x_n(t_1),x_n(t_2)) &\leq D(\textstyle\int_0^u f(t_1)dt_1, \int_0^u f(t_2)dt_2)+D(\tilde{c},\tilde{c}) \\ &+D(\textstyle\int_0^u \int_0^{t_1} k(t_1,s)g(s,x_{n-1}(s))dsdt, \int_0^u \int_0^{t_2} k(t_2,s)g(s,x_{n-1}(s))dsdt) \\ &\leq \textstyle\int_0^u D(f(t_1),f(t_2))dt \\ &+D(\textstyle\int_0^u \int_0^{t_1} k(t_1,s)g(s,x_{n-1}(s))dsdt, \int_0^u \int_0^{t_1} k(t_2,s)g(s,x_{n-1}(s))dsdt) \\ &+D(\textstyle\int_0^u \int_{t_1}^{t_2} k(t_2,s)g(s,x_{n-1}(s))dsdt,\tilde{0}) \\ &\leq D(f(t_1),f(t_2)) \\ &+\textstyle\int_0^{t_1} D(k(t,s)g(s,x_{n-1}(s)),\tilde{0})D(k(t_2,s)g(s,x_{n-1}(s)),\tilde{0})ds \\ &+\textstyle\int_{t_1}^{t_2} D(k(t_2,s),\tilde{0})D(g(s,x_{n-1}(s)),\tilde{0})ds \\ &\leq D(f(t_1),f(t_2)) \\ &+\textstyle\int_0^u((\sup_{0\leq t\leq a} D(g(t,x_{n-1}(t)),\tilde{0}))\int_0^{t_1} D(k(t_1,s),k(t_2,s))ds)dt \\ &+\textstyle\int_0^u((\sup_{0\leq t\leq a} D(g(t,x_{n-1}(t)),\tilde{0}))\int_{t_1}^{t_2} D(k(t_2,s),\tilde{0})ds)dt. \end{aligned}$$

By hypotheses and (7), we have $D(x_n(t_1),x_n(t_2)) \to 0$ as $t_1 \to t_2$.

Thus the sequence $x_n(t)$ is continuous on $[0,a]$.

Relation (4) and its analogue corresponding to $n+1$ will give for $n \geq 1$:

$$\begin{aligned} D(x_{n+1}(t),x_n(t)) &\leq \textstyle\int_0^u \int_0^t D(k(t,s),\tilde{0})D(g(s,x_n(s)),g(s,x_{n-1}))dsdt \\ &\leq \textstyle\int_0^u(\sup_{0\leq t\leq a} D(g(s,x_n(s)),g(s,x_{n-1}))ds \int_0^t D(k(t,s),\tilde{0})ds)dt \\ &\leq \textstyle\int_0^u(ML\sup_{0\leq t\leq a} D(x_n(t),x_{n-1}(t)))dt \end{aligned}$$

Thus, we get the following:

$$\sup_{0\leq t\leq a} D(x_{n+1}(t),x_n(t)) \leq \int_0^u ML \sup_{0\leq t\leq a} D(x_n(t),x_{n-1}(t))dt \tag{8}$$

For $n=0$, we have

$$\begin{aligned} D(x_1(t),x_0(t)) &\leq \textstyle\int_0^u \int_0^t D(k(t,s),\tilde{0})D(g(s,f(s)),\tilde{0})dsdt \\ &\leq \textstyle\int_0^u(\sup_{0\leq t\leq a} D(g(t,f(t)),\tilde{0}) \int_0^t D(k(t,s),\tilde{0})ds)dt \end{aligned} \tag{9}$$

So, we obtain

$$\sup_{0\le t\le a} D(x_1(t), x_0(t)) \le uMN,$$

where $N = \sup_{0\le t\le a} D(g(t, f(t)), \tilde{0})$ and $u \in [0, a]$. Moreover, we derive

$$\sup_{0\le t\le a} D(x_{n+1}(t), x_n(t)) \le u^{n+1}L^n M^{n+1} N \tag{10}$$

which shows that the series $\sum_{n=1}^{\infty} D(x_n(t), x_{n-1}(t))$ is dominated, uniformly on $[0, a]$, by the series $uMN \sum_{n=0}^{\infty} (uLM)^n$.

But (4) and $u \le 1$ guarantees the convergence of the last series, implying the uniform convergence of the sequence $x_n(t)$. If we denote $x(t) = lim_{n\to\infty} x_n(t)$, then $x(t)$ satisfies (3). It is obviously continuous on $[0, a]$ and bounded.

To prove the uniqueness, let $y(t)$ be a continuous solution of (3) on $[0, a]$. Then,

$$y(t) = f(t) + \int_0^t k(t, s)g(s, y(s))ds \tag{11}$$

From (4), we obtain for $n \ge 1$,

$$\begin{aligned}
D(y(t), x_n(t)) &\le \textstyle\int_0^u \int_0^t D(k(t, s), \tilde{0}) D(g(s, y(s)), g(s, x_{n-1}(t)))dsdt \\
&\le \textstyle\int_0^u (\sup_{0\le t\le a} D(g(t, y(t)), g(t, x_{n-1}(s))) \int_0^t D(k(t, s), \tilde{0})ds)dt \\
&\le \textstyle\int_0^u (LM \sup_{0\le t\le a} D(y(t), x_{n-1}(t)))dt \\
&\ \ \vdots \\
&\le \textstyle\int_0^u ((uLM)^n \sup_{0\le t\le a} D(y(t), x_0(t)))dt
\end{aligned}$$

Since $LM < 1, \quad u \le 1$

$$lim_{n\to\infty} x_n(t) = y(t) = x(t), \quad 0 \le t \le a,$$

Case II. If $x(t)$ be (ii)-differentiable, of theorem (3) we have the following:

$$x(t) = \tilde{c} \ominus (-1). \left(\int_0^u f(t)dt + \int_0^u \int_0^t k(t, s)g(s, x(s))dsdt \right), 0 \le t \le a, 0 \le u \le 1,\ \tilde{c} \in E \tag{12}$$

Theorem 5. *Let a and L be positive numbers. Assume that Eq.(12) satisfies the following conditions:*

(i) $f : [0, a] \to E$ *is continuous and bounded i.e., there exists* $R > 0$ *such that* $D(f(t), \tilde{0}) \le R$.

(ii) $k : \Delta \to E$ *is continuous where* $\Delta = (t, s) : 0 \le s \le t \le a$ *and there exists* $M > 0$ *such that* $\int_0^t D(k(t, s), \tilde{0})ds \le M$.

(iii) $g : [0, a] \times E \to E$ *is continuous and satisfies the Lipschitz condition, i.e.,*

$$D(g(t, x(t)), g(t, y(t)) \le LD(x(t), y(t)), \quad 0 \le t \le a, \tag{13}$$

where $L < M^{-1}$ *and* $x, y : [0, a] \to E$.

(iv) $g(t, \tilde{0})$ *is bounded on [0,a].*

then there exists a unique solution $x(t)$ *of Eq.(12) on* $[0, a]$ *and the successive iterations*

$$x_0(t) = f(t)$$

$$x_{n+1}(t) = \tilde{c} \ominus (-1). \int_0^u f(t)dt \ominus (-1). \int_0^u \int_0^t k(t, s)g(s, x_n(s))dsdt, \tag{14}$$

$$0 \le t \le a, \quad 0 \le u \le 1, \quad \tilde{c} \in E$$

are uniformly convergent to x(t) on $[0, a]$.

Proof. It is easy to see that all $x_n(t)$ are bounded on $[0, a]$. Indeed, $x_0 = f(t)$ is bounded by hypothesis. Assume that $x_{n-1}(t)$ is bounded from lemma 2 we have

$$\begin{aligned} D(x_n(t), \tilde{0}) &\le D(\tilde{c}, \tilde{0}) + D(\textstyle\int_0^u f(t)dt, \tilde{0}) + \int_0^u \int_0^t D(k(t, s), \tilde{0}) D(g(s, x_{n-1}(s)), \tilde{0})dsdt \\ &\le D(\tilde{c}, \tilde{0}) + Ru + \textstyle\int_0^u ((\sup_{0 \le t \le a} D(g(t, x_{n-1}(t)), \tilde{0})) \int_0^t D(k(t, s), \tilde{0})ds)dt \end{aligned} \tag{15}$$

Taking every assumptions into account

$$\begin{aligned} D(g(t, x_{n-1}(t)), \tilde{0}) &\le D(g(t, x_{n-1}(t)), g(t, \tilde{0})) + D(g(t, \tilde{0}), \tilde{0}) \\ &\le LD(x_{n-1}(t), \tilde{0}) + D(g(t, \tilde{0}), \tilde{0}), \end{aligned} \tag{16}$$

Since, $u \le 1$ we obtain that $x_n(t)$ is bounded. Thus, $x_n(t)$ is a sequence of bounded functions on $[0, a]$. Next we prove that $x_n(t)$ are continuous on $[0, a]$. For $0 \le t_1 \le t_2 \le a$, from Lemma 1 and Lemma 2 we have

$$\begin{aligned} D(x_n(t_1), x_n(t_2)) &\le \textstyle\int_0^u D(f(t_1), f(t_2))dt \\ &+ D(\textstyle\int_0^u \int_0^{t_1} k(t_1, s)g(s, x_{n-1}(s))dsdt, \int_0^u \int_0^{t_1} k(t_2, s)g(s, x_{n-1}(s))dsdt) \\ &+ D(\textstyle\int_0^u \int_{t_1}^{t_2} k(t_2, s)g(s, x_{n-1}(s))dsdt, \tilde{0}) \\ &\le D(f(t_1), f(t_2)) \\ &+ \textstyle\int_0^u ((\sup_{0 \le t \le a} D(g(t, x_{n-1}(t)), \tilde{0})) \int_0^{t_1} D(k(t_1, s), k(t_2, s))ds)dt \\ &+ \textstyle\int_0^u ((\sup_{0 \le t \le a} D(g(t, x_{n-1}(t)), \tilde{0})) \int_{t_1}^{t_2} D(k(t_2, s), \tilde{0})ds)dt. \end{aligned}$$

By hypotheses and (16), we have $D(x_n(t_1), x_n(t_2)) \to 0$ as $t_1 \to t_2$. Thus the sequence $x_n(t)$ is continuous on $[0, a]$. Relation (13) and its analogue corresponding to $n + 1$ will give for $n \ge 1$ similar proving previous theorem, and Lemmas 1-3 we get the following:

$$\sup_{0 \le t \le a} D(x_{n+1}(t), x_n(t)) \le \int_0^u ML \sup_{0 \le t \le a} D(x_n(t), x_{n-1}(t))dt \tag{17}$$

For $n = 0$, we obtain

$$D(x_1(t), x_0(t)) = \leq \int_0^u (\sup_{0 \leq t \leq a} D(g(t, f(t)), \tilde{0}) \int_0^t D(k(t, s), \tilde{0}) ds) dt \qquad (18)$$

So, we have

$$\sup_{0 \leq t \leq a} D(x_1(t), x_0(t)) \leq uMN,$$

where $N = \sup_{0 \leq t \leq a} D(g(t, f(t)), \tilde{0})$ and $u \in [0, a]$. Moreover, from (17), we derive

$$\sup_{0 \leq t \leq a} D(x_{n+1}(t), x_n(t)) \leq u^{n+1} L^n M^{n+1} N \qquad (19)$$

which shows that the series $\sum_{n=1}^{\infty} D(x_n(t), x_{n-1}(t))$ is dominated, uniformly on $[0, a]$, by the series $uMN \sum_{n=0}^{\infty} (uLM)^n$. However, Eq. (13) and $u \leq 1$ guarantees the convergence of the last series, implying the uniform convergence of the sequence $x_n(t)$. If we denote $x(t) = lim_{n \to \infty} x_n(t)$, then $x(t)$ satisfies (12). It is obviously continuous on $[0, a]$ and bounded.

Uniqueness of solution be asserted, similar proving previous theorem and By using Lemmas 1-3, which ends the proof of theorem.

4 Conclusion

In this paper, we proved the existence and uniqueness of solution of fuzzy Volterra integro-differential equations under strongly generalized differentiability. Also, we used fuzzy kernels to obtain such solutions which is the first attempt in the fuzzy literature in our best knowledge.

For future research, we will prove fuzzy fractional Vloterra integro-differential equations using strongly generalized differentiability.

References

1. Allahviranloo, T., Kiani, N.A., Barkhordari, M.: Toward the existence and uniqueness of solutions of second-order fuzzy differential equations. Information Sciences 179, 1207–1215 (2009)
2. Bede, B., Rudas, I.J., Attila, L.: First order linear fuzzy differential equations under generalized differentiability. Information Sciences 177, 3627–3635 (2007)
3. Bede, B., Gal, S.G., Generalizations of the differentiability of fuzzy-number-valued functions with applications to fuzzy differential equations. Fuzzy Sets and Systems 151, 581–599 (2005)
4. Chalco-cano, Y., Roman-Flores, H.: On new solution of fuzzy differential equations. Chaos, Solitons and Fractals 38, 112–119 (2008)
5. Dubois, D., Prade, H.: Towards fuzzy differential calculus, Part I: integration of fuzzy mappings, class of second-order. Fuzzy sets and Systems 8, 1–17 (1982)
6. Friedman, M., Ma, M., Kandel, A.: Numerical solution of fuzzy differential and integral equations. Fuzzy Sets and System 106, 35–48 (1999)

7. Kaleva, O.: The Cauchy problem for fuzzy differential equations. Fuzzy sets and Systems 35, 389–396 (1990)
8. Park, J.Y., Jeong, J.U.: A note on fuzzy functional equations. Fuzzy Sets and Systems 108, 193–200 (1999)
9. Park, J.Y., Kwun, Y.C., Jeong, J.U.: Existence of solutions of fuzzy integral equations in Banach spaces. Fuzzy Sets and Systems 72, 373–378 (1995)
10. Park, J.Y., Lee, S.Y., Jeong, J.U.: On the existence and uniqueness of solutions of fuzzy Volterra-Fredholm integral equuations. Fuzzy Sets and Systems 115, 425–431 (2000)
11. Puri, M.L., Ralescu, D.A.: Differentials of fuzzy functions. J. Math. Anal. Appl. 91, 552–558 (1983)
12. Rodriguez-Munize, L.J., Lopez-Diaz, M.: Hukuhara derivative of the fuzzy expected value. Fuzzy Sets and Systems 138, 593–600 (2003)
13. Rodriguez-Lipez, R.: Comparison results for fuzzy differential equations. Information Sciences 178, 1756–1779 (2008)
14. Seikkala, S.: On the fuzzy initial value problem. Fuzzy Sets and Systems 24, 319–330 (1987)
15. Stefanini, L.: On the generalized LU-fuzzy derivative and fuzzy differential equations. In: IEEE International Conference on Fuzzy Systems, art. no. 4295453 (2007)
16. Subrahmaniam, P.V., Sudarsanam, S.K.: On some fuzzy functional equations. Fuzzy Sets and Systems 64, 333–338 (1994)
17. Zhang, D., Feng, W., Qiu, J.: Global existence of solutions to fuzzy Volterra integral equations. ICIC Express Letters 3, 707–711 (2009)

Expansion Method for Solving Fuzzy Fredholm-Volterra Integral Equations

S. Khezerloo[1,⋆], T. Allahviranloo[2], S. Haji Ghasemi[2], S. Salahshour[2], M. Khezerloo[2], and M. Khorasan Kiasary[2]

[1] Department of Mathematics, Karaj Branch, Islamic Azad University, Karaj, Iran
tofigh@allahviranloo.com
[2] Department of Mathematics, Science and Research Branch, Islamic Azad University, Tehran, Iran

Abstract. In this paper, the fuzzy Fredholm-Volterra integral equation is solved, where expansion method is applied to approximate the solution of an unknown function in the fuzzy Fredholm-Volterra integral equation and convert this equation to a system of fuzzy linear equations. Then we propose a method to solve the fuzzy linear system such that its solution is always fuzzy vector. The method is illustrated by solving several examples.

Keywords: Expansion method; Fuzzy Fredholm-Volterra Integral Equations; Linear fuzzy system, Fuzzy number.

1 Introduction

The fuzzy differential and integral equations are important part of the fuzzy analysis theory. Park et al. [10] have considered the existence of solution of fuzzy integral equation in Banach space and Subrahmaniam and Sudarsanam [14] proved the existence of solution of fuzzy functional equations. Park and Jeong [11, 12] studied existence of solution of fuzzy integral equations of the form

$$x(t) = f(t) + \int_0^t f(t, s, x(s))ds, \quad 0 \leq t$$

where f and x are fuzzy-valued functions ($f, x : (a, b) \to E$ where E is the set of all fuzzy numbers) and k is a crisp function on real numbers. In [13] they studied existence of solution of fuzzy integral equations of the form

$$x(t) = f(t) + \int_0^t f(t, s, x(s))ds + \int_0^a g(t, s, x(s))ds, \quad 0 \leq t \leq a$$

Babolian et al. [5], proposed a numerical method for solving fuzzy Fredholm integral equation. In present, we try to approximate the solution of fuzzy Fredholm-Volterra integral equation.

⋆ Corresponding author.

E. Hüllermeier, R. Kruse, and F. Hoffmann (Eds.): IPMU 2010, Part II, CCIS 81, pp. 501–511, 2010.

Allahviranloo et al. [1, 2, 3] studied iterative methods for finding the approximate solution of fuzzy system of linear equations (FSLE) with convergence theorems. Abbasbandy et al. have discussed LU decomposition method, for solving fuzzy system of linear equations in [4]. They considered the method in spatial case when the coefficient matrix is symmetric positive definite.

In Section 2, the basic concept of fuzzy number operation is brought. In Section 3, the main section of the paper, a fuzzy Fredholm-Volterra integral equation is solved by the expansion method. A new method for solving the fuzzy linear system is proposed and discussed in section 4. The proposed ideas are illustrated by some examples in Section 5. Finally conclusion is drawn in Section 6.

2 Basic Concepts

There are various definitions for the concept of fuzzy numbers ([6, 8])

Definition 1. *An arbitrary fuzzy number u in the parametric form is represented by an ordered pair of functions (u^-, u^+) which satisfy the following requirements:*

1. $u^- : \alpha \mapsto u^-_\alpha \in \Re$ *is a bounded left-continuous non-decreasing function over* $[0,1]$.
2. $u^+ : \alpha \mapsto u^+_\alpha \in \Re$ *is a bounded left-continuous non-increasing function over* $[0,1]$.
3. $u^-_\alpha \leq u^+_\alpha, \quad 0 \leq \alpha \leq 1$.

A crisp number r is simply represented by $u^-_\alpha = u^+_\alpha = r, \quad 0 \leq \alpha \leq 1$. If $u^-_1 < u^+_1$, we have a fuzzy interval and if $u^-_1 = u^+_1$, we have a fuzzy number. In this paper, we do not distinguish between numbers or intervals and for simplicity we refer to fuzzy numbers as interval. We also use the notation $u_\alpha = [u^-_\alpha, u^+_\alpha]$ to denote the α-cut of arbitrary fuzzy number u. If $u = (u^-_\alpha, u^+_\alpha)$ and $v = (v^-_\alpha, v^+_\alpha)$ are two arbitrary fuzzy numbers, the arithmetic operations are defined as follows:

Definition 2. *(Addition)*

$$u + v = (u^-_\alpha + v^-_\alpha, u^+_\alpha + v^+_\alpha) \tag{2.1}$$

and in the terms of α-cuts

$$(u + v)_\alpha = [u^-_\alpha + v^-_\alpha, u^+_\alpha + v^+_\alpha], \quad \alpha \in [0,1] \tag{2.2}$$

Definition 3. *(Subtraction)*

$$u - v = (u^-_\alpha + v^+_\alpha, u^+_\alpha - v^-_\alpha) \tag{2.3}$$

and in the terms of α-cuts

$$(u - v)_\alpha = [u^-_\alpha - v^+_\alpha, u^+_\alpha - v^-_\alpha], \quad \alpha \in [0,1] \tag{2.4}$$

Definition 4. *(Scalar multiplication)*
For given $k \in \Re$

$$ku = \begin{cases} (ku_\alpha^-, ku_\alpha^+), & k > 0 \\ (ku_\alpha^+, ku_\alpha^-), & k < 0 \end{cases} \tag{2.5}$$

and

$$(ku)_\alpha = [\min\{ku_\alpha^-, ku_\alpha^+\}, \max\{ku_\alpha^-, ku_\alpha^+\}] \tag{2.6}$$

In particular, if $k = 1$, we have

$$-u = (-u_\alpha^+, -u_\alpha^-)$$

and with α-cuts

$$(-u)_\alpha = [-u_\alpha^+, -u_\alpha^-], \quad \alpha \in [0, 1]$$

Definition 5. *(Multiplication)*

$$uv = ((uv)_\alpha^-, (uv)_\alpha^+) \tag{2.7}$$

and

$$\begin{aligned} (uv)_\alpha^- &= \min\{u_\alpha^- v_\alpha^-, u_\alpha^- v_\alpha^+, u_\alpha^+ v_\alpha^-, u_\alpha^+ v_\alpha^+\} \\ (uv)_\alpha^+ &= \max\{u_\alpha^- v_\alpha^-, u_\alpha^- v_\alpha^+, u_\alpha^+ v_\alpha^-, u_\alpha^+ v_\alpha^+\}, \quad \alpha \in [0, 1] \end{aligned} \tag{2.8}$$

Definition 6. *(Division)*
If $0 \notin [v_0^-, v_0^+]$

$$\frac{u}{v} = ((\frac{u}{v})_\alpha^-, (\frac{u}{v})_\alpha^+) \tag{2.9}$$

and

$$\begin{aligned} (\tfrac{u}{v})_\alpha^- &= \min\{\tfrac{u_\alpha^-}{v_\alpha^-}, \tfrac{u_\alpha^-}{v_\alpha^+}, \tfrac{u_\alpha^+}{v_\alpha^-}, \tfrac{u_\alpha^+}{v_\alpha^+}\} \\ (\tfrac{u}{v})_\alpha^+ &= \max\{\tfrac{u_\alpha^-}{v_\alpha^-}, \tfrac{u_\alpha^-}{v_\alpha^+}, \tfrac{u_\alpha^+}{v_\alpha^-}, \tfrac{u_\alpha^+}{v_\alpha^+}\}, \quad \alpha \in [0, 1] \end{aligned} \tag{2.10}$$

Definition 7. *[9]. An arbitrary fuzzy number u are represented by a vector of 8 component of the interval* $[0, 1]$ *without internal points, i.e.* $\alpha = 0$ *and* $\alpha = 1$*, as follows:*

$$u = (u_0^-, d_0^-, u_1^-, d_1^-, u_0^+, d_0^+, u_1^+, d_1^+) \tag{2.11}$$

where

$$u_0^- = u^-(0), \quad u_1^- = u^-(1), \quad u_0^+ = u^+(0), \quad u_1^+ = u^+(1)$$

and

$$d_0^- = (u^-)'(0), \quad d_1^- = (u^-)'(1), \quad d_0^+ = (u^+)'(0), \quad d_1^+ = (u^+)'(1)$$

By definition (1), it is clear that $d_0^- > 0$, $d_1^- > 0$, $d_0^+ < 0$ and $d_1^+ < 0$. For an arbitrary trapezoidal fuzzy number u, we have

$$d_0^- = d_1^- = (u^-)', \quad d_0^+ = d_1^+ = (u^+)'$$

and if $u_1^- = u_1^+$, then u is an triangular fuzzy number.

A crisp real number a and a crisp interval $[a,b]$ have the forms $(a,0,a,0,a,0,a,0)$ and $(a,0,a,0,b,0,b,0)$, respectively.

Let u and v be two fuzzy numbers in form

$$u = (u_0^-, d_0^-, u_1^-, d_1^-, u_0^+, d_0^+, u_1^+, d_1^+), \quad v = (v_0^-, e_0^-, v_1^-, e_1^-, v_0^+, e_0^+, v_1^+, e_1^+)$$

Definition 8. *(Addition) , [9].*

$$u+v = (u_0^- + v_0^-, d_0^- + e_0^-, u_1^- + v_1^-, d_1^- + e_1^-, u_0^+ + v_0^+, d_0^+ + e_0^+, u_1^+ + v_1^+, d_1^+ + e_1^+) \tag{2.12}$$

Definition 9. *(Scalar multiplication) , [9]. For given $k \in \Re$*

$$ku = \begin{cases} (ku_0^-, kd_0^-, ku_1^-, kd_1^-, ku_0^+, kd_0^+, ku_1^+, kd_1^+), & k > 0 \\ (ku_0^+, kd_0^+, ku_1^+, kd_1^+, ku_0^-, kd_0^-, ku_1^-, kd_1^-), & k < 0 \end{cases} \tag{2.13}$$

In particular, if $k = -1$ then

$$(-u_0^+, -d_0^+, -u_1^+, -d_1^+, -u_0^-, -d_0^-, -u_1^-, -d_1^-)$$

and subtraction has been defined by

$$u - v = u + (-v)$$

Note that proposed representation in definition (7) is exact for the fuzzy numbers having left and right branches described by polynomial up to the third degree; so, the approximated arithmetic operations produce exact results for the low-degree left and right functions and, in particular, the trapezoidal and triangular fuzzy numbers.(see[9])

Guerra et al. [9] used the Hermit approximation to obtain the (approximated) membership function and they suggested how to go from the α-cuts to the membership function and vice versa.

3 Expansion Method for Solving Fuzzy Fredholm-Volterra Integral Equation

Consider the fuzzy Fredholm-Volterra integral equation (FF-VIE) of second kind

$$\tilde{f}(x) = \mu\tilde{\phi}(x) + \lambda \int_0^1 K(x,y)\tilde{\phi}(y)dy + \lambda \int_0^x F(x,z)\tilde{\phi}(z)dz \qquad 0 \le x \le 1 \tag{3.14}$$

where $K, F : [0,1] \times [0,1] \rightarrow \Re$ are arbitrary given kernel functions, $\tilde{\phi}, \tilde{f} : [0,1] \rightarrow E^n$ are a fuzzy-valued functions while $\tilde{\phi}$ is the unknown fuzzy-valued function to be determined and λ, μ are numerical parameters.

Here, in the present work we consider $K(x,y)$ and $F(x,z)$ are continuous in $0 \leq x \leq 1$ and for fixed nonnegative y, z over $[0,1]$ and known term $f(x)$ is a continuous function.

We are going to approximate fuzzy-valued function $\tilde{\phi}(x)$ by using a sequence of functions as $\{h_i\}_{i=1}^{\infty}$ in the following form:

$$\tilde{\phi}(x) = \sum_{i=0}^{\infty} \tilde{a}_i h_i(x) \tag{3.15}$$

where $\tilde{a}_i$, $i = 0, 1, 2, \ldots$ are fuzzy coefficients. It is clear that

$$\tilde{\phi}(x) = \sum_{i=0}^{n} \tilde{a}_i h_i(x) + E\tilde{\phi} \tag{3.16}$$

where $E\tilde{f}$ is the error.

By cancelation $E\tilde{\phi}$ in Eq. (3.16) and replacing in Eq. (3.14), we have:

$$\begin{aligned} \tilde{f}(x) &= \mu \textstyle\sum_{i=0}^{n} \tilde{a}_i h_i(x) + \lambda \int_0^1 K(x,y) \sum_{i=0}^{n} \tilde{a}_i h_i(y) dy + \lambda \int_0^x F(x,z) \sum_{i=0}^{n} \tilde{a}_i h_i(z) dz \\ &= \mu \textstyle\sum_{i=0}^{n} \tilde{a}_i h_i(x) + \lambda \sum_{i=0}^{n} \tilde{a}_i \int_0^1 K(x,y) h_i(y) dy + \lambda \sum_{i=0}^{n} \tilde{a}_i \int_0^x F(x,z) h_i(z) dz \\ & \qquad 0 \leq x \leq 1 \end{aligned} \tag{3.17}$$

Let $h_i(x) = x^i$, $i = 0, \ldots, n$, so

$$\begin{aligned} & h_i(x) \geq 0 \\ & r_i(x) = \textstyle\int_0^1 K(x,y) h_i(y) dy = \int_0^1 K(x,y) y^i dy \geq 0 \\ & p_i(x) = \textstyle\int_0^x F(x,z) h_i(z) dz = \int_0^x F(x,z) z^i dz \geq 0 \end{aligned} \tag{3.18}$$

If we suppose $\mu \geq 0$, $\lambda \geq 0$ then, we obtain

$$\begin{aligned} \tilde{f}(x) &= \textstyle\sum_{i=0}^{n} \tilde{a}_i (\mu h_i(x) + \lambda r_i(x) + \lambda p_i(x)) \\ &= \textstyle\sum_{i=0}^{n} \tilde{a}_i m_i(x) \end{aligned} \tag{3.19}$$

Therefore,

$$\tilde{f}(x_j) = \sum_{i=0}^{n} \tilde{a}_i m_i(x_j), \qquad j = 1, \ldots, n+1 \tag{3.20}$$

To calculate the solution of Eq. (3.21), we try to solve linear fuzzy system

$$MA = Y \tag{3.21}$$

where

$$M = \begin{bmatrix} m_0(x_1) & m_1(x_1) & \cdots & m_n(x_1) \\ m_0(x_2) & m_1(x_2) & \cdots & m_n(x_2) \\ \vdots & \vdots & & \vdots \\ m_0(x_{n+1}) & m_1(x_{n+1}) & \cdots & m_n(x_{n+1}) \end{bmatrix}, \quad A = \begin{bmatrix} \tilde{a}_0 \\ \tilde{a}_1 \\ \vdots \\ \tilde{a}_n \end{bmatrix}, \quad Y = \begin{bmatrix} \tilde{f}(x_1) \\ \tilde{f}(x_2) \\ \vdots \\ \tilde{f}(x_{n+1}) \end{bmatrix}$$

It is obvious that M is a $(n+1) \times (n+1)$ crisp matrix and A, Y are $(n+1) \times 1$ fuzzy matrix. If M is nonsingular matrix, then

$$A = M^{-1} Y$$

4 Fuzzy Solution of Fuzzy Linear System

First of all we define the fuzzy linear system (FLS).

Definition 10. *The $n \times n$ linear system*

$$\begin{cases} a_{11}x_1 + a_{12}x_2 + \cdots + a_{1n}x_n = y_1 \\ a_{21}x_1 + a_{22}x_2 + \cdots + a_{2n}x_n = y_2 \\ \vdots \\ a_{n1}x_1 + a_{n2}x_2 + \cdots + a_{nn}x_n = y_n \end{cases} \tag{4.22}$$

is called a fuzzy linear system(FLS) where the coefficient matrix $A = [a_{ij}]_{i,j=1}^{n}$ is a crisp nonsingular matrix and

$$y_i = (y_{i0}^-, d_{i0}^-, y_{i1}^-, d_{i1}^-, y_{i0}^+, d_{i0}^+, y_{i1}^+, d_{i1}^+), \qquad 1 \le i \le n$$

is a fuzzy number.

The matrix form of the system (4.22) is

$$AX = Y$$

Let $B = A^{-1}$ is inverse of matrix A, so we have

$$X = A^{-1}Y = BY \tag{4.23}$$

therefore,

$$\begin{bmatrix} x_1 \\ x_2 \\ \vdots \\ x_n \end{bmatrix} = \begin{bmatrix} b_{11} & b_{12} & \cdots & b_{1n} \\ b_{21} & b_{22} & \cdots & b_{2n} \\ \vdots & \vdots & \vdots & \vdots \\ b_{n1} & b_{n2} & \cdots & b_{nn} \end{bmatrix} \begin{bmatrix} (y_{10}^-, d_{10}^-, y_{11}^-, d_{11}^-, y_{10}^+, d_{10}^+, y_{11}^+, d_{11}^+) \\ (y_{20}^-, d_{20}^-, y_{21}^-, d_{21}^-, y_{20}^+, d_{20}^+, y_{21}^+, d_{21}^+) \\ \vdots \\ (y_{n0}^-, d_{n0}^-, y_{n1}^-, d_{n1}^-, y_{n0}^+, d_{n0}^+, y_{n1}^+, d_{n1}^+) \end{bmatrix} \tag{4.24}$$

so,

$$\begin{cases} b_{11}(y_{10}^-, d_{10}^-, y_{11}^-, d_{11}^-, y_{10}^+, d_{10}^+, y_{11}^+, d_{11}^+) + b_{12}(y_{20}^-, d_{20}^-, y_{21}^-, d_{21}^-, y_{20}^+, d_{20}^+, y_{21}^+, d_{21}^+) \\ \quad + \cdots + b_{1n}(y_{n0}^-, d_{n0}^-, y_{n1}^-, d_{n1}^-, y_{n0}^+, d_{n0}^+, y_{n1}^+, d_{n1}^+) = x_1 \\ \vdots \\ b_{n1}(y_{10}^-, d_{10}^-, y_{11}^-, d_{11}^-, y_{10}^+, d_{10}^+, y_{11}^+, d_{11}^+) + b_{n2}(y_{20}^-, d_{20}^-, y_{21}^-, d_{21}^-, y_{20}^+, d_{20}^+, y_{21}^+, d_{21}^+) \\ \quad + \cdots + b_{nn}(y_{n0}^-, d_{n0}^-, y_{n1}^-, d_{n1}^-, y_{n0}^+, d_{n0}^+, y_{n1}^+, d_{n1}^+) = x_n \end{cases} \tag{4.25}$$

Using definition (9), it is clear that

$$x_i = (x_{i0}^-, e_{i0}^-, x_{i1}^-, e_{i1}^-, x_{i0}^+, e_{i0}^+, x_{i1}^+, e_{i1}^+), \qquad 1 \le i \le n \tag{4.26}$$

is a fuzzy number, therefore fuzzy linear system (4.22) always has a fuzzy solution.

5 Numerical Examples

Example 1. *Consider fuzzy Fredholm-Volterra integral equation (3.14). Suppose that*

$$\tilde{f}(x) = (0, 1, 1, 1, 2, -1, 1, -1)x$$

$$K(x, y) = xy$$

$$F(x, z) = x^2 z$$

$$\lambda = \mu = 1$$

Using Eqs. (3.17) and (3.18), we have

$$\begin{aligned} (0, 1, 1, 1, 2, -1, 1, -1)x &= \sum_{i=0}^{n} \tilde{a}_i x^i + \sum_{i=0}^{n} \tilde{a}_i \int_0^1 xy^{i+1} dy + \sum_{i=0}^{n} \tilde{a}_i \int_0^x x^2 z^{i+1} dz \\ &= \sum_{i=0}^{n} \tilde{a}_i x^i + \sum_{i=0}^{n} \tfrac{1}{i+2}\, \tilde{a}_i\, x + \sum_{i=0}^{n} \tfrac{1}{i+2}\, \tilde{a}_i\, x^{i+4} \\ &= \sum_{i=0}^{n} \tilde{a}_i \left(x^i + \tfrac{1}{i+2}\, x + \tfrac{1}{i+2}\, x^{i+4}\right) \end{aligned}$$

Now, let $n = 2$. By substitution $x_1 = 0.25, x_2 = 0.5, x_3 = 0.75$, we obtain

$$\begin{bmatrix} 1.1270\ 0.3337\ 0.1251 \\ 1.2813\ 0.6771\ 0.3789 \\ 1.5332\ 1.0791\ 0.7945 \end{bmatrix} \begin{bmatrix} \tilde{a}_0 \\ \tilde{a}_1 \\ \tilde{a}_2 \end{bmatrix} = \begin{bmatrix} (0, 0.25, 0.25, 0.25, 0.5, -0.25, 0.25, -0.25) \\ (0, 0.5, 0.5, 0.5, 1, -0.5, 0.5, -0.5) \\ (0, 0.75, 0.75, 0.75, 1.5, -0.75, 0.75, -0.75) \end{bmatrix}$$

Since $|M| \neq 0$, *then*

$$\begin{bmatrix} \tilde{a}_0 \\ \tilde{a}_1 \\ \tilde{a}_2 \end{bmatrix} = \begin{bmatrix} (-3.0465, 3.0116, -0.0349, 3.0116, 2.9766, -3.0116, -0.0349, -3.0116) \\ (-14.4829, 15.4772, 0.9943, 15.4772, 16.4715, -15.4772, 0.9943, -15.4772) \\ (-16.4929, 16.1538, -0.3391, 16.1538, 15.8147, -16.1538, -0.3391, -16.1538) \end{bmatrix}$$

So, the approximation of solution is

$$\begin{aligned} \tilde{\phi}(x) &= (-3.0465, 3.0116, -0.0349, 3.0116, 2.9766, -3.0116, -0.0349, -3.0116) \\ &+(-14.4829, 15.4772, 0.9943, 15.4772, 16.4715, -15.4772, 0.9943, -15.4772)x \\ &+(-16.4929, 16.1538, -0.3391, 16.1538, 15.8147, -16.1538, -0.3391, -16.1538)x^2 \end{aligned}$$

Example 2. *Consider fuzzy Fredholm-Volterra integral equation (3.14). Suppose that*

$$\begin{aligned} \tilde{f}(x) &= (0, 1, 1, 1, 2, -1, 1, -1) \sin \pi x \\ K(x, y) &= \sin \tfrac{\pi x}{2} \\ F(x, z) &= xz \\ \lambda &= \mu = 1 \end{aligned}$$

Using Eqs. (3.17) and (3.18), we have

$$\begin{aligned} (0, 1, 1, 1, 2, -1, 1, -1) \sin \pi x &= \textstyle\sum_{i=0}^{n} \tilde{a}_i x^i + \sum_{i=0}^{n} \tilde{a}_i \int_0^1 \sin \frac{\pi x}{2} y^i dy + \sum_{i=0}^{n} \tilde{a}_i \int_0^x x z^{i+1} dz \\ &= \textstyle\sum_{i=0}^{n} \tilde{a}_i x^i + \sum_{i=0}^{n} \frac{1}{i+1} \tilde{a}_i \sin \frac{\pi x}{2} + \sum_{i=0}^{n} \frac{1}{i+2} \tilde{a}_i x^{i+3} \\ &= \textstyle\sum_{i=0}^{n} \tilde{a}_i \left(x^i + \frac{1}{i+1} \sin \frac{\pi x}{2} + \frac{1}{i+2} x^{i+3} \right) \end{aligned}$$

Now, let $n = 2$. *By substitution* $x_1 = 0, x_2 = 0.5, x_3 = 1$, *we obtain*

$$\begin{bmatrix} 1 & 0 & 0 \\ 1.7696 & 0.8744 & 0.4935 \\ 2.5 & 1.8333 & 1.5833 \end{bmatrix} \begin{bmatrix} \tilde{a}_0 \\ \tilde{a}_1 \\ \tilde{a}_2 \end{bmatrix} = \begin{bmatrix} (0, 0, 0, 0, 0, 0, 0, 0) \\ (0, 1, 1, 1, 2, -1, 1, -1) \\ (0, 0, 0, 0, 0, 0, 0, 0) \end{bmatrix}$$

Since $|M| \neq 0$, *then*

$$\begin{bmatrix} \tilde{a}_0 \\ \tilde{a}_1 \\ \tilde{a}_2 \end{bmatrix} = \begin{bmatrix} (0, 0, 0, 0, 0, 0, 0, 0) \\ (0, 3.3009, 3.3009, 3.3009, 6.6018, -3.3009, 3.3009, -3.3009) \\ (-7.6442, 3.8221, -3.8221, 3.8221, 0, -3.8221, -3.8221, -3.8221) \end{bmatrix}$$

So, the approximation of solution is

$$\tilde{\phi}(x) = (0, 3.3009, 3.3009, 3.3009, 6.6018, -3.3009, 3.3009, -3.3009)x$$
$$+(-7.6442, 3.8221, -3.8221, 3.8221, 0, -3.8221, -3.8221, -3.8221)x^2$$

Example 3. *Consider fuzzy Fredholm-Volterra integral equation (3.14). Suppose that*

$$\tilde{f}(x) = (0, 1, 1, 1, 2, -1, 1, -1)x^2 + (1, 1, 2, 1, 4, -0.5, 2, -0.5)$$
$$K(x, y) = x + y$$
$$F(x, z) = x$$
$$\lambda = \mu = 1$$

Using Eqs. (3.17) and (3.18), we have

$$(0, 1, 1, 1, 2, -1, 1, -1)x^2 + (1, 1, 2, 1, 4, -0.5, 2, -0.5)$$
$$= \sum_{i=0}^{n} \tilde{a}_i x^i + \sum_{i=0}^{n} \tilde{a}_i \int_0^1 (x + y)y^i dy + \sum_{i=0}^{n} \tilde{a}_i \int_0^x xz^i dz$$
$$= \sum_{i=0}^{n} \tilde{a}_i x^i + \sum_{i=0}^{n} \tilde{a}_i \left(\frac{1}{i+1}\, x + \frac{1}{i+2}\right) + \sum_{i=0}^{n} \frac{1}{i+1}\, \tilde{a}_i\, x^{i+2}$$
$$= \sum_{i=0}^{n} \tilde{a}_i \left(x^i + \frac{1}{i+1}\, x + \frac{1}{i+2} + \frac{1}{i+1}\, x^{i+2}\right)$$

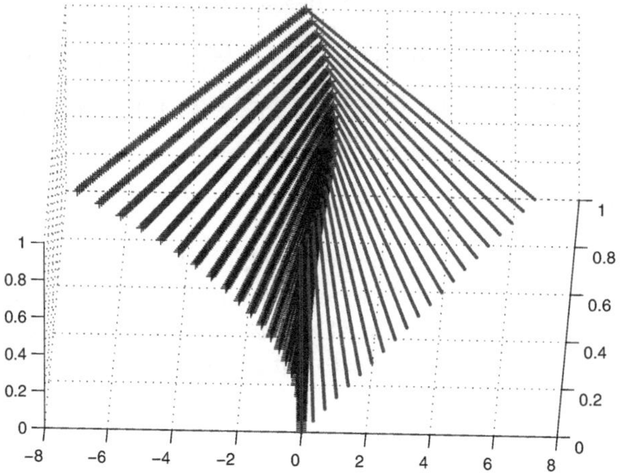

Fig. 1. Approximation of $\tilde{\phi}(x)$

Now, let $n = 3$. *By substitution* $x_1 = 0, x_2 = 0.25, x_3 = 0.5, x_4 = 1$, *we obtain*

$$\begin{bmatrix} 1.5 & 0.3333 & 0.25 & 0.2 \\ 1.8125 & 0.7161 & 0.3971 & 0.2784 \\ 2.25 & 1.1458 & 0.6875 & 0.4578 \\ 3.5 & 2.3333 & 1.9167 & 1.7 \end{bmatrix} \begin{bmatrix} \tilde{a}_0 \\ \tilde{a}_1 \\ \tilde{a}_2 \\ \tilde{a}_3 \end{bmatrix}$$

$$= \begin{bmatrix} (1, 1, 2, 1, 4, -0.5, 2, -0.5) \\ (1, 17/16, 33/16, 17/16, 33/8, -9/16, 33/16, -9/16) \\ (1, 5/4, 9/4, 5/4, 9/2, -3/4, 9/4, -3/4) \\ (1, 2, 3, 2, 6, -3/2, 3, -3/2) \end{bmatrix}$$

Since $|M| \neq 0$, *then*

$$\begin{bmatrix} \tilde{a}_0 \\ \tilde{a}_1 \\ \tilde{a}_2 \\ \tilde{a}_3 \end{bmatrix} = \begin{bmatrix} (-0.3689, 1.4795, 1.5814, 1.4795, 4.3875, -1.0516, 1.5814, -1.0516) \\ (-36.8450, 18.6653, -1.0258, 18.6653, 34.3608, -18.8816, -1.0258, -18.8816) \\ (-115.0260, 61.1495, -0.5528, 61.1495, 112.9612, -61.6291, -0.5528, -61.6291) \\ (-82.4059, 45.6518, 0.54, 45.6518, 83.9875, -45.401, 0.54, -45.401) \end{bmatrix}$$

So, the approximation of solution is

$$\begin{aligned} \tilde{\phi}(x) &= (-0.3689, 1.4795, 1.5814, 1.4795, 4.3875, -1.0516, 1.5814, -1.0516) \\ &+(-36.8450, 18.6653, -1.0258, 18.6653, 34.3608, -18.8816, -1.0258, -18.8816)x \\ &+(-115.0260, 61.1495, -0.5528, 61.1495, 112.9612, -61.6291, -0.5528, -61.6291)x^2 \\ &+(-82.4059, 45.6518, 0.54, 45.6518, 83.9875, -45.401, 0.54, -45.401)x^3 \end{aligned}$$

6 Conclusion

In this work, we considered a fuzzy Fredholm-Volterra integral equation and tried to approximate its solution by expansion method. Therefore, we proposed a new model for solving a system of n fuzzy linear equations with n variables. It was proved that the solution vector of fuzzy linear system always is a fuzzy solution.

References

[1] Allahviranloo, T.: Successive over relaxation iterative method for fuzzy system of linear equations. Applied Mathematics and Computation 162, 189–196 (2005)

[2] Allahviranloo, T.: The Adomian decomposition method for fuzzy system of linear equations. Applied Mathematics and Computation 163, 553–563 (2005)

[3] Allahviranloo, T., Ahmady, E., Ahmady, N., Shams Alketaby, K.: Block Jacobi two-stage method with Gauss-Sidel inner iterations for fuzzy system of linear equations. Applied Mathematics and Computation 175, 1217–1228 (2006)
[4] Abbasbandy, S., Ezzati, R., Jafarian, A.: LU decomposition method for solving fuzzy system of linear equations. Applied Mathematics and Computation 172, 633–643 (2006)
[5] Babolian, E., Sadeghi Goghary, H., Abbasbandy, S.: Numerical solution of linear Fredholm fuzzy integral equations of the second kind by Adomian method. Applied Mathematics and Computation 161, 733–744 (2005)
[6] Dubois, D., Prade, H.: Towards fuzzy differential calculus: Part 3, differentiation. Fuzzy Sets and Systems 8, 225–233 (1982)
[7] Friedman, M., Ming, M., Kandel, A.: Fuzzy linear systems. Fuzzy Sets and Systems 96, 201–209 (1998)
[8] Gal, S.G.: Approximation theory in fuzzy setting. In: Anastassiou, G.A. (ed.) Handbook of Analytic-Computational Methods in Applied Mathematics, pp. 617–666. Chapman Hall & CRC Press (2000)
[9] Guerra, M.L., Stefanini, L.: Approximate fuzzy arithmetic operation using monotonic interpolation. Fuzzy Sets and Systems 150, 5–33 (2005)
[10] Park, J.Y., Kwun, Y.C., Jeong, J.U.: Existence of solutions of fuzzy integral equations in Banach spaces. Fuzzy Sets and Systems 72, 373–378 (1995)
[11] Park, J.Y., Jeong, J.U.: A note on fuzzy functional equations. Fuzzy Sets and Systems 108, 193–200 (1999)
[12] Park, J.Y., Lee, S.Y., Jeong, J.U.: On the existence and uniqueness of solutions of fuzzy Volterra-Fredholm integral equuations. Fuzzy Sets and Systems 115, 425–431 (2000)
[13] Park, J.Y., Jeong, J.U.: The approximate solutions of fuzzy functional integral equations. Fuzzy Sets and Systems 110, 79–90 (2000)
[14] Subrahmaniam, P.V., Sudarsanam, S.K.: On some fuzzy functional equations. Fuzzy Sets and Systems 64, 333–338 (1994)
[15] Wang, K., Zheng, B.: Symmetric successive overrelaxation methods for fuzzy linear systems. Applied Mathematics and Computation 175, 891–901 (2006)
[16] Zadeh, L.A.: The concept of a linguistic variable and its application to approximate reasoning. Information Sciences 8, 199–249 (1975)

Solving Fuzzy Heat Equation by Fuzzy Laplace Transforms

Soheil Salahshour and Elnaz Haghi

Department of Mathematics, Islamic Azad University,
Mobarakeh Branch, Mobarakeh, Iran

Abstract. In this paper, we solve the fuzzy heat equations under strongly generalized H-differentiability by fuzzy Laplace transforms. To this end, the original fuzzy heat equation is converted to the corresponding fuzzy two point boundary value problem (FBVP) based on the fuzzy Laplace transform. Then, we will solve the obtained FBVP using characterization theorem. Finally, some numerical examples are given to illustrate the utility of the fuzzy Laplace transform method.

Keywords: Fuzzy heat equations, Fuzzy Laplace transforms, Strongly generalized H-differentiability, Characterization theroem.

1 Introduction

The subject of partial differential equations holds an exciting and special position in mathematics. Problems involving time t as one independent variable lead usually to parabolic or hyperbolic equations.

The simplest parabolic equation, $\frac{\partial u}{\partial t} = c\frac{\partial^2 u}{\partial x^2}$, derives from the theory of heat conduction and its solution gives, for example, the temperature u at a distance x units of length from one end of a thermal insulated bar after t seconds of heat conduction.

In real world, knowledge about dynamical systems modeled by partial differential equations, specially in heat equations, is often incomplete or vague. These vagueness may be appeared in each part of heat equations like initial condition, boundary condition or etc. So, if we want to solve heat equations in sense of real conditions, we have to use interval or fuzzy computations. However, it is obvious that the theory of fuzzy computations (fuzzy differential equations, fuzzy calculations and etc.) is richer and is more coincided with real applications than interval computations.

Buckley et. al [6] introduced the fuzzy partial differential equations. Then, Allahviranloo [2] proposed difference method for solving fuzzy partial differential equations(FPDEs).

The technique of direct and inverse F-transform and approximating properties of them are described in [11,12]. Recently, Allahviranloo and Barkhordari in [3] proposed fuzzy Laplace transforms for solving first order fuzzy differential equations under generalized H-differentiability. However, one can see more

E. Hüllermeier, R. Kruse, and F. Hoffmann (Eds.): IPMU 2010, Part II, CCIS 81, pp. 512–521, 2010.

references about theory of fuzzy differential equations and numerical solution of them in [1,5,8,10,14,15,16].

The motivation of this paper is constructed on solving fuzzy heat equation using fuzzy Laplace transform. It seems that, this is one of the first attempt to solve such well-known fuzzy partial differential equations under strongly generalized H-differentiability in fuzzy literature.

This paper is organized as follows:

In section 2 preliminaries are given, then fuzzy partial generalized H-derivatives are given in section 3. Moreover, the fuzzy heat equation is presented and solved with fuzzy Laplace transform method in section 4. Some illustrative examples are solved in order to show the ability of proposed method in section 5 and the conclusion is drawn in section 6.

2 Preliminaries

We now recall some definitions which are needed throughout the paper. By R, we denote the set of all real numbers.

Definition 1. *A fuzzy number u in parametric form is a pair $(\underline{u}, \overline{u})$ of functions $\underline{u}(r)$, $\overline{u}(r)$, $0 \leq r \leq 1$, which satisfy the following requirements:*

1. *$\underline{u}(r)$ is a bounded non-decreasing left continuous function in $(0, 1]$, and right continuous at 0,*
2. *$\overline{u}(r)$ is a bounded non-increasing left continuous function in $(0, 1]$, and right continuous at 0,*
3. *$\underline{u}(r) \leq \overline{u}(r)$, $0 \leq r \leq 1$.*

We recall that for $a < b < c$ which $a, b, c \in R$, the triangular fuzzy number $u = (a, b, c)$ determined by a, b, c is given such that $\underline{u}(r) = a + (b - a)r$ and $\overline{u}(r) = c - (c - b)r$ are the endpoints of the r-level sets, for all $r \in [0, 1]$.

The Hausdorff distance between fuzzy numbers given by $D : E \times E \longrightarrow R_+ \bigcup\{0\}$,

$$D(u, v) = \sup_{r\in[0,1]} \max\left\{|\underline{u}(r) - \underline{v}(r)|, |\overline{u}(r) - \overline{v}(r)|\right\},$$

It is easy to see that D is a metric in E and has the following properties (see [13])

(i) $D(u \oplus w, v \oplus w) = D(u, v), \quad \forall u, v, w \in E,$

(ii) $D(k \odot u, k \odot v) = |k|D(u, v), \quad \forall k \in R, u, v \in E,$

(iii) $D(u \oplus v, w \oplus e) \leq D(u, w) + D(v, e), \quad \forall u, v, w, e \in E,$

(iv) (D, E) is a complete metric space.

Definition 2. *Let $x, y \in E$. If there exists $z \in E$ such that $x = y + z$, then z is called the H-difference of x and y, and it is denoted by $x \ominus y$.*

Now we investigate the partial lateral type of H-derivatives for fuzzy-valued functions $f = f(x, y)$ with respect to x as following similar one dimensional case [7]:

Definition 3. *A function $f : (a,b) \times (a,b) \longrightarrow E$ is differentiable at x_0 with respect to x, if there exists a $\frac{\partial f(x,y)}{\partial x}|_{x=x_0} \in E$ such that*
(1) for all $h > 0$ sufficiently near to 0, $\exists f(x_0+h,y)\ominus f(x_0,y)$, $\exists f(x_0,y)\ominus f(x_0-h,y)$ and the limits(in the metric D)

$$\lim_{h\longrightarrow 0+}\frac{f(x_0+h,y)\ominus f(x_0,y)}{h}=\lim_{h\longrightarrow 0+}\frac{f(x_0,y)\ominus f(x_0-h,y)}{h}=\frac{\partial}{\partial x}f(x_0,y)$$

(2) for all $h < 0$ sufficiently near to 0, $\exists f(x_0+h,y)\ominus f(x_0,y)$, $\exists f(x_0,y)\ominus f(x_0-h,y)$ and the limits(in the metric D)

$$\lim_{h\longrightarrow 0-}\frac{f(x_0+h,y)\ominus f(x_0,y)}{h}=\lim_{h\longrightarrow 0-}\frac{f(x_0,y)\ominus f(x_0-h,y)}{h}=\frac{\partial}{\partial x}f(x_0,y)$$

Moreover, we can define partial lateral type of H-differentiability with respect to y as following:

Definition 4. *A function $f : (a,b) \times (a,b) \longrightarrow E$ is differentiable at y_0 with respect to y, if there exists a $\frac{\partial f(x,y)}{\partial y}|_{y=y_0} \in E$ such that*
(1) for all $k > 0$ sufficiently near to 0, $\exists f(x,y_0+k)\ominus f(x,y_0)$, $\exists f(x,y_0)\ominus f(x,y_0-k)$ and the limits(in the metric D)

$$\lim_{k\longrightarrow 0+}\frac{f(x,y_0+k)\ominus f(x,y_0)}{k}=\lim_{k\longrightarrow 0+}\frac{f(x,y_0)\ominus f(x,y_0-k)}{k}=\frac{\partial}{\partial y}f(x,y_0)$$

(2) for all $k < 0$ sufficiently near to 0, $\exists f(x,y_0+k)\ominus f(x,y_0)$, $\exists f(x,y_0)\ominus f(x,y_0-k)$ and the limits(in the metric D)

$$\lim_{k\longrightarrow 0-}\frac{f(x,y_0+k)\ominus f(x,y_0)}{k}=\lim_{k\longrightarrow 0-}\frac{f(x,y_0)\ominus f(x,y_0-k)}{k}=\frac{\partial}{\partial y}f(x,y_0)$$

Also, we can define second-order partial lateral type of H-derivatives for fuzzy-valued function $f = f(x,y)$ as following:

Definition 5. *function $f : (a,b) \times (a,b) \longrightarrow E$ is differentiable of the second-order with respect to x, if there exists a $\frac{\partial^2 f(x,y)}{\partial x^2}|_{x=x_0} \in E$ such that*
(1) for all $h > 0$ sufficiently near to 0, $\exists \frac{\partial}{\partial x}f(x_0+h,y)\ominus\frac{\partial}{\partial x}f(x_0,y)$, $\exists\frac{\partial}{\partial x}f(x_0,y)\ominus\frac{\partial}{\partial x}f(x_0-h,y)$ and the limits(in the metric D)

$$\lim_{h\longrightarrow 0+}\frac{\frac{\partial}{\partial x}f(x_0+h,y)\ominus\frac{\partial}{\partial x}f(x_0,y)}{h}=\lim_{h\longrightarrow 0+}\frac{\frac{\partial}{\partial x}f(x_0,y)\ominus\frac{\partial}{\partial x}f(x_0-h,y)}{h}=\frac{\partial^2}{\partial x^2}f(x_0,y)$$

(2) for all $h < 0$ sufficiently near to 0, $\exists \frac{\partial}{\partial x}f(x_0+h,y)\ominus\frac{\partial}{\partial x}f(x_0,y)$, $\exists\frac{\partial}{\partial x}f(x_0,y)\ominus\frac{\partial}{\partial x}f(x_0-h,y)$ and the limits(in the metric D)

$$\lim_{h\longrightarrow 0-}\frac{\frac{\partial}{\partial x}f(x_0+h,y)\ominus\frac{\partial}{\partial x}f(x_0,y)}{h}=\lim_{h\longrightarrow 0-}\frac{\frac{\partial}{\partial x}f(x_0,y)\ominus\frac{\partial}{\partial x}f(x_0-h,y)}{h}=\frac{\partial^2}{\partial x^2}f(x_0,y).$$

Please notice that in each case of differentiability, for sake of simplicity, we say f is (i)-differentiable if it is satisfied in the first form (1) of differentiability in each mentioned differentiability and we say f is (ii)-differentiable if it is satisfied in the second form (2) of differentiability in each mentioned differentiability.

Before starting the main goal of paper, we propose characterization theorem for second-order fuzzy two point boundary value problems which is connection between FBVPs and crisp BVPs.

Theorem 1. *(Characterization theorem). Let us consider the following fuzzy two point boundary value problem*

$$\begin{cases} u''(t) = f(t,u,u'), \quad t \in [t_0, P] \\ u(t_0) = u_0 \in E \\ u(P) = B \in E \end{cases} \tag{1}$$

where $f : (a,b) \times E \times E \longrightarrow E$ *is such that*
(i) $[f(t,x,y)]^r = \left[f(t, \underline{x}(r), \overline{x}(r), \underline{y}(r), \overline{y}(r))\right]$;

(ii) $\underline{f}$ *and* $\overline{f}$ *are absolutely continuous on* $[t_0, T]$ *with respect to* r ;

(iii) There exists $M : I \longrightarrow R^+$ *such that* $\forall (t,u,v) \in I \times E \times E$

$$\left|\frac{\partial}{\partial r} f_1(t,u,v,r)\right|, \left|\frac{\partial}{\partial r} f_2(t,u,v,r)\right| \leq M(r), \qquad a.e.\ r \in I = [0,1]$$

(iv) Let $B = (B_1, B_2) \in E$ *where* B_1 *and* B_2 *are absolutely continuous on* I, *and*

$$\left|\frac{\partial}{\partial r} B_1(r)\right|, \left|\frac{\partial}{\partial r} B_2(r)\right| \geq (b-a)^2 M(r), \qquad a.e., r \in I$$

Then fuzzy BVP (1) is equivalent to the following BVPs:

$$\begin{cases} \underline{u}''(t;r) = \underline{f}(t, \underline{x}(r), \overline{x}(r), \underline{y}(r), \overline{y}(r)), \ 0 \leq r \leq 1, \\ \overline{u''}(t;r) = \overline{f}(t, \underline{x}(r), \overline{x}(r), \underline{y}(r), \overline{y}(r)), \ 0 \leq r \leq 1, \\ \underline{u}(t_0;r) = \underline{u}_0(r), \ \overline{u}(t_0;r) = \overline{u}_0(r); \ 0 \leq r \leq 1, \\ \underline{u}(P;r) = \underline{u}_P(r), \overline{u}(P;r) = \overline{u}_P(r) \ 0 \leq r \leq 1, . \end{cases} \tag{2}$$

Proof. In [9], proved that using conditions $(i) - (iv)$ we can obtain unique solution of FBVP (1). The proof of equivalence of original FBVP and related deterministic BVP is completely similar to proof of Theorem 2 in [4].

3 The Fuzzy Heat Equation

In this section, we consider solutions of the fuzzy heat equations with real diffusion constant under strongly generalized H-differentiability (or lateral type of H-differentiability).

The temperature $u = u(x, y)$ of a thin rod , or bar, of constant cross-section and homogeneous material, lying along the axis and completely insulated laterally, may be modeled by the one-dimensional heat equation

$$\frac{\partial u}{\partial y} = c\frac{\partial^2 u}{\partial x^2} \tag{3}$$

with the fuzzy initial condition

$$u(x, 0) = u_0 \in E, \quad x \in [0, P] \tag{4}$$

and the fuzzy boundary conditions:

$$u(0, y) \in E \ \ at\ x = 0 \ \ and\ y > 0, \tag{5}$$

$$u(P, y) \in E \ \ at\ x = P\ and\ y > 0 \tag{6}$$

where c is constant diffusion such that in this paper we will consider without loss of generality $c = 1$. Although, for future research we will discuss on fuzzy heat equation with complex fuzzy number diffusion constant c.

4 The Fuzzy Laplace Transform Method

We consider the fuzzy-valued function $u = u(x, y)$, where $y \geq 0$ is a time variable. Denote by $\mathtt{U}(x, s)$ the fuzzy Laplace transform of u with respect to t, that is to say

$$\mathtt{U}(x, s) = \mathtt{L}\{u(x, y)\} = \int_0^\infty e^{-sy}u(x, y)dy \tag{7}$$

Indeed, we can present the above definition for fuzzy Laplace transform based on the r-cut representation of fuzzy-valued function u as following:

$$\mathtt{U}(x, s) = \mathtt{L}\{u(x, y)\} = [\underline{l}\{u(x, y; r)\}, \overline{l}\{u(x, y; r)\}], \quad 0 \leq r \leq 1$$

where

$$\underline{l}\{u(x, y; r)\} = l\{\underline{u}(x, y; r)\} = \int_0^\infty e^{-sy}\underline{u}(x, y; r)dy, \quad 0 \leq r \leq 1,$$

$$\overline{l}\{u(x, y; r)\} = l\{\overline{u}(x, y; r)\} = \int_0^\infty e^{-sy}\overline{u}(x, y; r)dy, \quad 0 \leq r \leq 1.$$

For applying the fuzzy Laplace transform method, we have to suppose some assumption as following:

Assumption 1: $$\mathtt{L}\left(\frac{\partial u}{\partial x}\right) = \int_0^\infty e^{-sy}\frac{\partial}{\partial x}u(x, y)dy = \frac{\partial}{\partial x}\int_0^\infty e^{-sy}u(x, y)dy = \frac{\partial}{\partial x}\mathtt{U}(x, s), \tag{8}$$

In other words, " The transform of the derivative is the derivative of the transform".

Assumption 2: $$\lim_{x\longrightarrow x_0}\int_0^\infty e^{-sy}u(x,y)dy = \int_0^\infty e^{-sy}u(x_0,y)dy, \tag{9}$$

That is,

$$\lim_{x\longrightarrow x_0} \mathtt{U}(x,s) = \mathtt{U}(x_0,s) \tag{10}$$

In (7) it is convenient to write

$$\frac{\partial}{\partial x}\mathtt{U}(x,s) = \frac{d}{dx}\mathtt{U}(x,s) = \frac{d\mathtt{U}}{dx}$$

Since our parameter s can be treated like a constant with respect to H-differentiation involved.

A second H-derivative version of (6) results in the expression

$$\mathtt{L}\left(\frac{\partial^2 u}{\partial x^2}\right) = \frac{\partial^2 \mathtt{U}}{\partial x^2} \tag{11}$$

Note that, based on the partial lateral type of H-differentiability of fuzzy-valued function $u = u(x,y)$, we get :

CaseA$_1$. Let us consider u and $\frac{\partial u}{\partial x}$ are (i)-differentiable or u and $\frac{\partial u}{\partial x}$ are (ii)-differentiable, then we get the following:

$$\mathtt{L}\left(\frac{\partial^2 u(x,y)}{\partial x^2}\right) = \frac{\partial^2 \mathtt{U}(x,s)}{\partial x^2} = \left[\frac{\partial^2 \underline{\mathtt{U}}(x,s;r)}{\partial x^2}, \frac{\partial^2 \overline{\mathtt{U}}(x,s;r)}{\partial x^2}\right], \quad 0 \le r \le 1 \tag{12}$$

CaseA$_2$. Let us consider u is (i)-differentiable and $\frac{\partial u}{\partial x}$ is (ii)-differentiable or u is (ii)-differentiable and $\frac{\partial u}{\partial x}$ is (i)-differentiable, then we get the following:

$$\mathtt{L}\left(\frac{\partial^2 u(x,y)}{\partial x^2}\right) = \frac{\partial^2 \mathtt{U}(x,s)}{\partial x^2} = \left[\frac{\partial^2 \overline{\mathtt{U}}(x,s;r)}{\partial x^2}, \frac{\partial^2 \underline{\mathtt{U}}(x,s;r)}{\partial x^2}\right], \quad 0 \le r \le 1 \tag{13}$$

CaseB$_1$. Let us consider u is differentiable in the first form (1) in Definition 4 with respect to y, then [3]:

$$\mathtt{L}\left(\frac{\partial u}{\partial y}\right) = s\mathtt{L}\{u(x,y)\} \ominus u(x,0) = s\mathtt{U}(x,s) \ominus u(x,0) \tag{14}$$

CaseB$_2$. Let us consider u is differentiable in the second form (2) in Definition 4 , then [3]:

$$\mathtt{L}\left(\frac{\partial u}{\partial y}\right) = -u(x,0) \ominus (-s\mathtt{L}\{u(x,y)\}) \ominus = -u(x,0) \ominus (-s\mathtt{U}(x,s)) \tag{15}$$

The fuzzy Laplace transform method applied to the solution of fuzzy heat equation consists of first applying the fuzzy Laplace transform to the both sides of equation. This will result in a FBVP involving $\mathtt{U}$ as a function of the single variable x. Moreover, since the boundary conditions also express u as a function of t, we take the fuzzy Laplace transform of boundary conditions as well. Indeed, we will see that the original fuzzy heat equation is converted to the FBVP. So, by solving such FBVP and applying inverse of fuzzy Laplace transform, we can get the solution of fuzzy heat equation.

By taking the fuzzy Laplace transform of Eq. (3) yields:

$$\mathtt{L}\left\{\frac{\partial u}{\partial y}\right\} = \mathtt{L}\left\{c\frac{\partial^2 u}{\partial x^2}\right\} \tag{16}$$

Then, by taking fuzzy Laplace transform of boundary conditions we get:

$$\mathtt{L}\{u(0,y)\} = \mathtt{U}(0,s),\ \mathtt{L}\{u(P,y)\} = \mathtt{U}(P,s) \tag{17}$$

Consequently, usage of lateral type of H-differentiability, leads to obtain the following crisp systems in order to obtain solution of original FPDE

Cases $(\mathbf{A_1}\&\mathbf{B_1})$ or $(\mathbf{A_2}\&\mathbf{B_2}) \equiv$

$$\begin{cases} \frac{d^2\underline{\mathtt{U}}(x,s;r)}{dx^2} = s\underline{\mathtt{U}}(x,s;r) - \underline{u}(x,0;r), \frac{d^2\overline{\mathtt{U}}(x,s;r)}{dx^2} = s\overline{\mathtt{U}}(x,s;r) - \overline{u}(x,0;r), \\ \underline{\mathtt{L}}\{u(0,y;r)\} = \underline{\mathtt{U}}(0,s;r), \overline{\mathtt{L}}\{u(0,y;r)\} = \overline{\mathtt{U}}(0,s;r), \\ \underline{\mathtt{L}}\{u(P,y;r)\} = \underline{\mathtt{U}}(P,s;r), \overline{\mathtt{L}}\{u(P,y;r)\} = \overline{\mathtt{U}}(P,s;r) \end{cases} \tag{18}$$

Cases $(\mathbf{A_1}\&\mathbf{B_2})$ or $(\mathbf{A_2}\&\mathbf{B_1}) \equiv$

$$\begin{cases} \frac{d^2\underline{\mathtt{U}}(x,s;r)}{dx^2} = -\overline{u}(x,0;r) + s\overline{\mathtt{U}}(x,s;r), \frac{d^2\overline{\mathtt{U}}(x,s;r)}{dx^2} = -\underline{u}(x,0;r) + s\underline{\mathtt{U}}(x,s;r), \\ \underline{\mathtt{L}}\{u(0,y;r)\} = \underline{\mathtt{U}}(0,s;r), \overline{\mathtt{L}}\{u(0,y;r)\} = \overline{\mathtt{U}}(0,s;r), \\ \underline{\mathtt{L}}\{u(P,y;r)\} = \underline{\mathtt{U}}(P,s;r), \overline{\mathtt{L}}\{u(P,y;r)\} = \overline{\mathtt{U}}(P,s;r) \end{cases} \tag{19}$$

The solutions of BVPs (18) and (19) are denoted respectively by $\underline{\mathtt{U}}$ and $\overline{\mathtt{U}}$. Then, by taking the inverse of fuzzy Laplace transform, we can get the solution of fuzzy heat equation as following:

$$u(x,y;r) = [\underline{u}(x,y;r), \overline{u}(x,y;r)] = \left[\mathtt{L}^{-1}\{\underline{\mathtt{U}(x,s;r)}\}, \mathtt{L}^{-1}\{\overline{\mathtt{U}(x,s;r)}\}\right],\ 0 \le r \le 1, \tag{20}$$

provided that $[\mathtt{L}^{-1}\{\underline{\mathtt{U}}\}, \mathtt{L}^{-1}\{\overline{\mathtt{U}}\}]$ define a fuzzy-valued function.

5 Examples

In this section, we will consider some illustrative examples with fuzzy initial and boundary conditions under lateral type of H-differentiability.

Example 1. Consider the following fuzzy heat equation

$$\frac{\partial u}{\partial y}=\frac{\partial^2 u}{\partial x^2},\quad 0<x<P,\ y>0 \tag{21}$$

with

$$I.C.\ \ u(x,0)=u_0\in E,\ \ B.Cs:=\begin{cases}(i)\ \frac{\partial u(0,y)}{\partial x}=\widetilde{0}\ \ (i.e.,\ left\ end\ insulated),\\ (ii)\ u(P,y)=u_1\in E,\ \ (\ len\{u_1\}\geq len\{u_0\})\end{cases}$$

So, by applying fuzzy Laplace transform method which is discussed in detail in previous section, we get the following:

Cases ($\mathbf{A_1}$&$\mathbf{B_1}$) or ($\mathbf{A_2}$&$\mathbf{B_2}$). Taking the fuzzy Laplace transform gives

$$\frac{d^2\mathsf{U}(x,s;r)}{dx^2}=\left[\frac{d^2\underline{\mathsf{U}}(x,s;r)}{dx^2},\frac{d^2\overline{\mathsf{U}}(x,s;r)}{dx^2}\right]=sU\ominus u_0.$$

Then,

$$\mathsf{U}(x,s)=c_1\cosh(\sqrt{s}x)+c_2\sinh(\sqrt{s}x)+\frac{u_0}{s},$$

and by (ii), $c_2=0$, so that

$$\mathsf{U}(x,s)=c_1\cosh(\sqrt{s}x)+\frac{u_0}{s}.$$

We find by (ii) that

$$\mathsf{U}(P,s)=\frac{u_1}{s}=c_1\cosh(\sqrt{s}P)+\frac{u_0}{s},$$

and so

$$c_1=\frac{u_1\ominus u_0}{s\cosh\sqrt{s}P}.$$

Therefore

$$\mathsf{U}(x,s)=\frac{(u_1\ominus u_0)\cosh\sqrt{s}x}{s\cosh\sqrt{s}P}+\frac{u_0}{s}.$$

Taking the inverse, gives

$$u(x,y;r)=u_0+4\left(\frac{u_1\ominus u_0}{\pi}\right)\sum_{n=1}^{\infty}\frac{(-1)^n}{2n-1}\exp(-\frac{(2n-1)^2\pi^2 x}{4P^2})\times\cos(\frac{2n-1}{2P}).\pi x.$$

Example 2. Consider the following fuzzy heat equation

$$\frac{\partial u}{\partial y}=\frac{\partial^2 u}{\partial x^2} \tag{22}$$

with

$$I.C.\ \ (i)\ u(x,0)=[1+r,3-r],\ \ B.Cs:=\begin{cases}(ii)\ u(0,y)=[r-1,1-r],\\ (iii)\ \lim_{x\longrightarrow\infty}u(x,y)=[1+r,3-r]\end{cases}$$

Then, we get the following:

Cases $(\mathbf{A_1}\&\mathbf{B_2})$ or $(\mathbf{A_2}\&\mathbf{B_1})$

$$\begin{cases} \frac{d^2\overline{\mathrm{U}}(x,s;r)}{dx^2} = s\underline{\mathrm{U}}(x,s;r) - (1+r), \\ \underline{\mathrm{L}}\{u(0,y;r)\} = \underline{\mathrm{U}}(0,s;r) = \frac{r-1}{s}, \\ \lim_{x\longrightarrow\infty} \underline{\mathrm{L}}\{u(x,y;r)\} = \lim_{x\longrightarrow\infty} \underline{\mathrm{U}}(x,s;r) = \frac{r+1}{s} \\ \frac{d^2\underline{\mathrm{U}}(x,s;r)}{dx^2} = s\overline{\mathrm{U}}(x,s;r) - (3-r), \\ \overline{\mathrm{L}}\{u(0,y;r)\} = \overline{\mathrm{U}}(0,s;r) = \frac{1-r}{s}, \\ \lim_{x\longrightarrow\infty} \overline{\mathrm{L}}\{u(x,y;r)\} = \lim_{x\longrightarrow\infty} \overline{\mathrm{U}}(x,s;r) = \frac{3-r}{s} \end{cases} \tag{23}$$

By solving system (23), we get the following:

$$\underline{u}(x,y;r) - 2 * erfc\left(\frac{x}{2\sqrt{y}}\right) + (1+r), \quad 0 \leq r \leq 1.$$

$$\overline{u}(x,y;r) = -2 * erfc\left(\frac{x}{2\sqrt{y}}\right) + (3-r), \quad 0 \leq r \leq 1.$$

6 Conclusion

In this paper, we investigated the fuzzy Laplace transform method to solve fuzzy heat equation under lateral type of H-differentiability or strongly generalized H-differentiability. To this end, we proposed the partial fuzzy derivatives with respect to independent variables x and y. Then, usage of fuzzy Laplace transforms, leads to translate the original FPDE to the corresponding fuzzy two point boundary value problem. To this end, some characterization theorem is given to connect the FBVP and the deterministic system of BVPs. Hence, obtaining the solution of original FPDE is equivalent to obtain solution of corresponding crisp BVPs.

Also, for future research we will consider solutions of some fuzzy partial differential equations, like as fuzzy hyperbolic differential equations and fuzzy elliptic differential equations under strongly generalized H-differentiability.

References

1. Abbasbandy, S., Allahviranloo, T., Lopez-Pouso, O., Nieto, J.J.: Numerical methods for fuzzy differential inclusions. Computer and Mathematics With Applications 48, 1633–1641 (2004)
2. Allahviranloo, T.: Difference methods for fuzzy partial differential equations. CMAM 2, 233–242 (2002)

3. Allahviranloo, T., Barkhordari Ahmadi, M.: Fuzzy Laplace transforms. Soft. Comput. 14, 235–243 (2010)
4. Bede, B.: Note on Numerical solutions of fuzzy differential equations by predictor-corrector method. Information Sciences 178, 1917–1922 (2008)
5. Bede, B., Gal, S.G.: Generalizations of the differentiability of fuzzy-number-valued functions with applications to fuzzy differential equations. Fuzzy Sets and Systems 151, 581–599 (2005)
6. Buckley, J.J., Feuring, T.: Introduction to fuzzy partial differential equations. Fuzzy Sets and Systems 105, 241–248 (1999)
7. Chalco-Cano, Y., Roman-Flores, H.: On new solutions of fuzzy differential equations. Chaos, Solitons and Fractals 38, 112–119 (2006)
8. Chang, S.S.L., Zadeh, L.: On fuzzy mapping and control. IEEE trans. System Cybernet 2, 30–34 (1972)
9. Chen, M., Wu, C., Xue, X., Liu, G.: On fuzzy boundary value problems. Information Scinece 178, 1877–1892 (2008)
10. Friedman, M., Ming, M., Kandel, A.: Numerical solution of fuzzy differential and integral equations. Fuzzy Sets and System 106, 35–48 (1999)
11. Perfilieva, I.: Fuzzy transforms: Theory and Applications. Fuzzy Sets and Systems 157, 993–1023 (2006)
12. Perfilieva, I., De Meyer, H., De Baets, B.: Cauchy problem with fuzzy initial condition and its approximate solution with the help of fuzzy transform. In: WCCI 2008, Proceedings 978-1-4244-1819-0, Hong Kong, pp. 2285–2290. IEEE Computational Intelligence Society (2008)
13. Puri, M.L., Ralescu, D.: Fuzzy random variables. J. Math. Anal. Appl. 114, 409–422 (1986)
14. Lakshmikantham, V., Nieto, J.J.: Differential equations in metric spaces, An introduction and an application to fuzzy differential equations. Dyn. Contin. Discrete Impuls. Syst. Ser. A: Math. Anal. 10, 991–1000 (2003)
15. Nieto, J.J., Rodriguez-Lopez, R.: Hybrid metric dynamical systems with impulses. Nonlinear Anal. 64, 368–380 (2006)
16. Nieto, J.J., Rodriguez-Lopez, R., Georgiou, D.N.: Fuzzy differential systems under generalized metric spaces approach. Dynam. Systems Appl. 17, 1–24 (2008)

A New Approach for Solving First Order Fuzzy Differential Equation

Tofigh Allahviranloo and Soheil Salahshour

Department of Mathematics, Islamic Azad University,
Mobarakeh Branch, Mobarakeh, Iran

Abstract. In this paper, a new approach for solving first order fuzzy differential equations (FDEs) with fuzzy initial value is considered under strongly generalized H-differentiability. In order to obtain solution of FDE, we extend the 1-cut solution of original problem. This extension is constructed based on the allocating some unknown spreads to 1-cut solution, then created value is replaced in the original FDE. However obtaining solutions of FDE is equivalent to determine the unknown spreads while 1-cut solution is derived via previous step (in general, 1-cut of FDE is interval differential equation). Moreover, we will introduce three new set solutions for FDEs based on the concepts of united solution set, tolerable solution set and controllable solution set. Indeed, our approach is designed to obtain such new solution sets while one of them has pessimistic/optimitic attitude. Finally, some numerical examples are solved to illustrate the approach.

Keywords: Fuzzy differential equation (FDE), Strongly generalized H-differentiability, Interval differential equation, United solution set (USS), Tolerable solution set (TSS), Controllable solution set (CSS).

1 Introduction

The topic of fuzzy differential equations (FDEs) has been rapidly growing in recent years. Kandel and Byatt [10,11] applied the concept of fuzzy differential equation (FDE) to the analysis of fuzzy dynamical problems. The FDE and the initial value problem (Cauchy problem) were rigorously treated by O. Kaleva [8], S. Seikkala [13] and by other researchers (see [3,5,7,13]). The numerical methods for solving fuzzy differential equations are investigated by [1,2].

The idea of the presented approach is constructed based on the extending 1-cut solution of original FDEs. Obviously, 1-cut of FDE is interval differential equation or ordinary differential equation. If 1-cut of FDE be an interval differential equation, we solve it with Stefanini et. al's method [14], otherwise, solving ordinary differential equation is done as usual. Consequently, we try to fuzzify obtained 1-cut solution in order to determine solutions of original FDE under strongly generalized H-differentiability. To this end, some unknown spreads are allocated to the 1-cut solution. Then, by replacing such fuzzified value into original FDE and also based on the type of differentiability, we can get the spreads of solution of FDE.

E. Hüllermeier, R. Kruse, and F. Hoffmann (Eds.): IPMU 2010, Part II, CCIS 81, pp. 522–531, 2010.

Moreover, we extend the concepts of united solution set, tolerable solution set and controllable solution set for fuzzy differential equations. So, we will find solutions of first order fuzzy differential equation which are placed in mentioned solution sets. Thus, we define three type of spreads while one of them is linear combination of the others. Such spread and related solution has pessimistic/optimistic attitude which is a new point of view to numerical solution of FDEs. Clearly, such property will allow to the decision maker to inference or analyze the system in the real senses based on the pessimistic or optimistic desires. It seems that proposed method has flexible structure in order to obtain numerical solutions of FDEs in different attitude.

The structure of paper is organized as follows:

In section 2, some basic definitions and results which will be used later are brought. In section 3, first order fuzzy differential equations is introduced and the proposed approach is given in detail. Moreover, concepts of united solution set, tolerable solution set and controllable solution set are introduced and discussed in detail in section 4, then the proposed technique is illustrated by solving several examples in section 5, and Concluding remarks are drawn in section 6.

2 Preliminaries

An arbitrary fuzzy number with an ordered pair of functions $(\underline{u}(r), \overline{u}(r))$, $0 \leq r \leq 1$, which satisfy the following requirements is represented [9].

Definition 1. *A fuzzy number u in parametric form is a pair $(\underline{u}, \overline{u})$ of functions $\underline{u}(r)$, $\overline{u}(r)$, $0 \leq r \leq 1$, which satisfy the following requirements:*

1. *$\underline{u}(r)$ is a bounded non-decreasing left continuous function in $(0, 1]$, and right continuous at 0,*
2. *$\overline{u}(r)$ is a bounded non-increasing left continuous function in $(0, 1]$, and right continuous at 0,*
3. *$\underline{u}(r) \leq \overline{u}(r)$, $0 \leq r \leq 1$.*

Let E be the set of all upper semicontinuous normal convex fuzzy numbers with bounded r-level intervals. It means that if $v \in E$ then the r-level set

$$[v]^r = \{s|\ v(s) \geq r\}, \quad 0 < r \leq 1,$$

is a closed bounded interval which is denoted by

$$[v]^r = [\underline{v}(r), \overline{v}(r)].$$

For arbitrary $u = (\underline{u}, \overline{u}), v = (\underline{v}, \overline{v})$ and $k \geq 0$, addition $(u+v)$ and multiplication by k as $(\underline{u+v}) = \underline{u}(r) + \underline{v}(r)$, $(\overline{u+v}) = \overline{u}(r) + \overline{v}(r)$, $(\underline{ku})(r) = k\underline{u}(r)$, $(\overline{ku})(r) = k\overline{u}(r)$, are defined.

The Hausdorff distance between fuzzy numbers given by $D : E \times E \longrightarrow R_+ \bigcup\{0\}$,

$$D(u, v) = \sup_{r\in[0,1]} \max\{|\underline{u}(r) - \underline{v}(r)|, |\overline{u}(r) - \overline{v}(r)|\},$$

where $u = (\underline{u}(r), \overline{u}(r))$, $v = (\underline{v}(r), \overline{v}(r)) \subset R$ is utilized (see [4]). Then, it is easy to see that D is a metric in E and has the following properties (see [12])
(i) $D(u \oplus w, v \oplus w) = D(u, v), \quad \forall u, v, w \in E,$
(ii) $D(k \odot u, k \odot v) = |k| D(u, v), \quad \forall k \in R, u, v \in E,$
(iii) $D(u \oplus v, w \oplus e) \leq D(u, w) + D(v, e), \quad \forall u, v, w, e \in E,$
(iv) (D, E) is a complete metric space.

Definition 2. *[9]. Let $f : R \to E$ be a fuzzy valued function. If for arbitrary fixed $t_0 \in R$ and $\epsilon > 0$, a $\delta > 0$ such that*

$$|t - t_0| < \delta \Rightarrow D(f(t), f(t_0)) < \epsilon,$$

f is said to be continuous.

Definition 3. *Let $x, y \in E$. If there exists $z \in E$ such that $x = y + z$, then z is called the H-difference of x and y, and it is denoted by $x \ominus y$.*

In this paper we consider the following definition of differentiability for fuzzy-valued functions which was introduced by Bede et. al [4] and investigate by Chalco-Cano et. al [6]).

Definition 4. *Let $f : (a, b) \to E$ and $x_0 \in (a, b)$. We say that f is strongly generalized H-differentiable at x_0, If there exists an element $f'(x_0) \in E$, such that*

(1) for all $h > 0$ sufficiently near to 0, $\exists f(x_0 + h) \ominus f(x_0)$, $\exists f(x_0) \ominus f(x_0 - h)$ and the limits(in the metric D)

$$\lim_{h \longrightarrow 0+} \frac{f(x_0 + h) \ominus f(x_0)}{h} = \lim_{h \longrightarrow 0+} \frac{f(x_0) \ominus f(x_0 - h)}{h} = f'(x_0)$$

or

(2) for all $h < 0$ sufficiently near to 0, $\exists f(x_0 + h) \ominus f(x_0)$, $\exists f(x_0) \ominus f(x_0 - h)$ and the limits(in the metric D)

$$\lim_{h \longrightarrow 0-} \frac{f(x_0) \ominus f(x_0 + h)}{h} = \lim_{h \longrightarrow 0-} \frac{f(x_0 - h) \ominus f(x_0)}{h} = f'(x_0)$$

In the special case when f is a fuzzy-valued function, we have the following result.

Theorem 1. *[6]. Let $f : R \to E$ be a function and denote $f(t) = (\underline{f}(t, r), \overline{f}(t, r))$, for each $r \in [0, 1]$. Then*
(1) if f is differentiable in the first form (1) in Definition 4, then $\underline{f}(t, r)$ and $\overline{f}(t, r)$ are differentiable functions and $f'(t) = (\underline{f}'(t, r), \overline{f}'(t, r))$

(2) if f is differentiable in the second form (2) in Definition 4, then $\underline{f}(t, r)$ and $\overline{f}(t, r)$ are differentiable functions and $f'(t) = (\overline{f}'(t, r), \underline{f}'(t, r))$.

The principal properties of the H-derivatives in the first form (1), some of which still holds for the second form (2), are well known and can be found in [8] and some properties for the second form (2) can be found in [6].

Notice that we say fuzzy-valued function f is I-differentiable if satisfy in the first form (1) in Definition 4, and we say f is II-differentiable if satisfy in the second form (2) in Definition 4.

3 First Order Fuzzy Differential Equations

In this section, we consider the following first order fuzzy differential equation:

$$\begin{cases} y'(t) = f(t, y(t)), \\ \widetilde{y}(t_0) = \widetilde{y}_0, \end{cases} \tag{1}$$

where $f : [a, b] \times E \longrightarrow E$ is fuzzy-valued function, $\widetilde{y}_0 \in E$ and strongly generalized H-differentiability is also considered which is defined in Definition 4.

Now, we describe our propose approach for solving FDE (1). In the beginning, we shall solve FDE (1) in sense of 1-cut as following:

$$\begin{cases} (y')^{[1]}(t) = f^{[1]}(t, y(t)), \\ y^{[1]}(t_0) = y_0^{[1]}, \ t_0 \in [0, T] \end{cases} \tag{2}$$

If Eq.(2) be a crisp differential equation we can solve it as usual, otherwise, if Eq.(2) be an interval differential equation we will solve it by Stefanini et. al's method which is proposed and discussed in [14]. Notice that the solution of differential equation (2) is presented with notation $y^{[1]}(t)$. Then, some unknown left spread $\alpha_1(t; r)$ and right spread $\alpha_2(t; r)$ are allocated to the 1-cut solution for all $0 \leq r \leq 1$. So, this approach leads to obtain

$$y(t) = [\underline{y}(t; r), \overline{y}(t; r)] = \left[\underline{y^{[1]}(t)} - \alpha_1(t; r), \overline{y^{[1]}}(t) + \alpha_2(t; r)\right] \tag{3}$$

as unknown solution of original FDE (1), then Eq.(3) is replaced into FDE (1). Hence, we have the following:

$$y'(t) = \left[\underline{y^{[1]}}(t) - \alpha_1(t; r), \overline{y^{[1]}}(t) + \alpha_2(t; r)\right]' =$$

$$\left[\underline{f}(t, \underline{y^{[1]}}(t) - \alpha_1(t; r), \overline{y^{[1]}}(t) + \alpha_2(t; r)), \overline{f}(t, \underline{y^{[1]}}(t) - \alpha_1(t; r), \overline{y^{[1]}}(t) + \alpha_2(t; r))\right].$$

Please notice that, we assumed the considered spreads and 1-cut solution are differentiable. Consequently, based on type of differentiability we have two following cases:

Case I. Suppose that $y(t)$ in Eq.(3) is I-differentiable, then we get:

$$y'(t) = \left[(\underline{y^{[1]}})'(t) - \alpha_1'(t; r), (\overline{y^{[1]}})'(t) + \alpha_2'(t; r)\right] \tag{4}$$

where, $\alpha_1'(t;r) = \frac{\partial \alpha_1(t;r)}{\partial t}$ and $\alpha_2'(t;r) = \frac{\partial \alpha_2(t;r)}{\partial t}$ for all $0 \le r \le 1$. Consider Eq.(4) and original FDE (1), then we have the following for all $r \in [0,1]$:

$$\begin{cases} (\underline{y^{[1]}})'(t) - \alpha_1'(t;r) = \underline{f}(t, \underline{y^{[1]}}(t) - \alpha_1(t;r), \overline{y^{[1]}}(t) + \alpha_2(t;r)), \ t_0 \le t \le T \\ (\overline{y^{[1]}})'(t) + \alpha_2'(t;r) = \overline{f}(t, \underline{y^{[1]}}(t) - \alpha_1(t;r), \overline{y^{[1]}}(t) + \alpha_2(t;r)), \ t_0 \le t \le T \end{cases} \tag{5}$$

Moreover, we modified fuzzy initial value $\widetilde{y}_0$ in terms of unknown left and right spreads and 1-cut solution. Consider fuzzy initial value $y(t_0) = [\underline{y}(t_0;r), \overline{y}(t_0;r)]$ for all $0 \le r \le 1$, then we can rewrite the lower and upper functions $\underline{y}(t_0;r)$ and $\overline{y}(t_0;r)$, respectively as following:

$$\begin{cases} \underline{y}(t_0;r) = \underline{y^{[1]}}(t_0) - \alpha_1(t_0;r), \\ \overline{y}(t_0;r) = \overline{y^{[1]}}(t_0) + \alpha_2(t_0;r), \end{cases} \tag{6}$$

Thus, Eq.(5) and Eq.(6) simultaneously lead to obtain the following ODEs:

$$\begin{cases} (\underline{y^{[1]}})'(t) - \alpha_1'(t;r) = \underline{f}(t, \underline{y^{[1]}}(t) - \alpha_1(t;r), \overline{y^{[1]}}(t) + \alpha_2(t;r)), \\ (\overline{y^{[1]}})'(t) + \alpha_2'(t;r) = \overline{f}(t, \underline{y^{[1]}}(t) - \alpha_1(t;r), \overline{y^{[1]}}(t) + \alpha_2(t;r)), \\ \underline{y}(t_0;r) = \underline{y^{[1]}}(t_0) - \alpha_1(t_0;r), \\ \overline{y}(t_0;r) = \overline{y^{[1]}}(t_0) + \alpha_2(t_0;r), \end{cases} \tag{7}$$

Clearly, in above ODEs (7), only left and right spreads $\alpha_1(t;r)$ and $\alpha_2(t;r)$ are unknown parameters. So, ODEs (7) can be rewritten as following:

$$\begin{cases} \alpha_1'(t;r) = H_1(t, \alpha_1(t;r), \alpha_2(t;r)), \ 0 \le r \le 1, \ t \in [0,T] \\ \alpha_2'(t;r) = H_2(t, \alpha_1(t;r), \alpha_2(t;r)), \ 0 \le r \le 1, \ t \in [0,T], \\ \alpha_1(t_0;r) = \underline{y^{[1]}}(t_0) - \underline{y}(t_0;r),, \ 0 \le r \le 1, \ t_0 \in [0,T], \\ \alpha_2(t_0;r) = \overline{y}(t_0;r) - \overline{y^{[1]}}(t_0),, \ 0 \le r \le 1, \ t_0 \in [0,T] \end{cases} \tag{8}$$

Indeed, we will find spreads $\alpha_1(t;r)$ and $\alpha_2(t;r)$ by solving ODEs (8). Hence, solution of original FDE (1) is derived based on the obtained spreads and 1-cut solution as follows:

$$y(t) = [\underline{y}(t;r), \overline{y}(t;r)]$$

where for all $0 \le r \le 1$ and $t \in [0,T]$ such that

$$\underline{y}(t;r) = \underline{y^{[1]}}(t) - \alpha_1(t;r), \quad \overline{y}(t;r) = \overline{y^{[1]}}(t) + \alpha_2(t;r)$$

Case II. Suppose that $y(t)$ in Eq.(3) is II-differentiable, then we get:

$$y'(t) = \left[(\overline{y^{[1]}})'(t) + \alpha_2'(t;r), (\underline{y^{[1]}})'(t) - \alpha_1'(t;r) \right] \tag{9}$$

Similarly, ODEs (7) can be rewritten in sense of II-differentiability as following:

$$\begin{cases} (\overline{y^{[1]}})'(t) + \alpha_2'(t;r) = \underline{f}(t, y^{[1]}(t) - \alpha_1(t;r), y^{[1]}(t) + \alpha_2(t;r)), \\ (\underline{y^{[1]}})'(t) - \alpha_1'(t;r) = \overline{f}(t, y^{[1]}(t) - \alpha_1(t;r), y^{[1]}(t) + \alpha_2(t;r)), \\ \underline{y}(t_0;r) = \underline{y^{[1]}}(t_0) - \alpha_1(t_0;r), \\ \overline{y}(t_0;r) = \overline{y^{[1]}}(t_0) + \alpha_2(t_0;r), \end{cases} \tag{10}$$

Since, only unknown parameters in ODEs (10) are $\alpha_1(t;r)$ and $\alpha_2(t;r)$, we can rewrite (10) in terms of $\alpha_1(t;r)$ and $\alpha_2(t;r)$ and their derivatives. So, we have the following:

$$\begin{cases} \alpha_2'(t;r) = H_1(t, \alpha_1(t;r), \alpha_2(t;r)), \ 0 \le r \le 1, \ t \in [0,T] \\ \alpha_1'(t;r) = H_2(t, \alpha_1(t;r), \alpha_2(t;r)), \ 0 \le r \le 1, \ t \in [0,T], \\ \alpha_1(t_0;r) = \underline{y^{[1]}}(t_0) - \underline{y}(t_0;r),, \ 0 \le r \le 1, \ t_0 \in [0,T], \\ \alpha_2(t_0;r) = \overline{y}(t_0;r) - \overline{y^{[1]}}(t_0),, \ 0 \le r \le 1, \ t_0 \in [0,T] \end{cases} \tag{11}$$

Finally, by solving above ODEs (11) unknown spreads are determined and follows we can derive solution of original FDE (1) in sense of II-differentiability by using

$$\underline{y}(t;r) = \underline{y^{[1]}}(t) - \alpha_1(t;r), \quad \overline{y}(t;r) = \overline{y^{[1]}}(t) + \alpha_2(t;r) \tag{12}$$

for all $0 \le r \le 1$.
Notice that solution of original FDE (1) is assumed fuzzy-valued function and under such assumption we determined the unknown left and right spreads $\alpha_1(t;r)$ and $\alpha_2(t;r)$. However, we will check that obtained spreads lead to derive fuzzy-valued function as solution of original FDE (1).

Theorem 2. *Suppose that left spread $\alpha_1(t;r)$ and right spread $\alpha_2(t;r)$ are obtained from ODEs (8) or (11). Then the following affirmations are equivalent:*

(1) *$\alpha_1(t;r)$ and $\alpha_2(t;r)$ are nonincreasing positive functions for all $0 \le r \le 1$, $t_0 \le t \le T$,*
(2) *$y(t)$ is fuzzy-valued function.*

4 Three New Solution Sets for FDEs

In this section, we try to extend concepts of united solution set (USS), tolerable solution set(TSS) and controllable solution set (CSS) to the theory of fuzzy differential equations.

Definition 5. *Let us consider FDE (1), then united solution set, tolerable solution set and controllable solution set are defined, respectively, as following:*

$$Y_{\exists\exists} = \{y(t) |\ y'(t) \cap f(t, y(t)) \neq \emptyset, \quad t \in [a,b]\}, \tag{13}$$

$$Y_{\forall\exists} = \{y(t) |\ y'(t) \subseteq f(t, y(t)), \quad t \in [a,b]\}, \tag{14}$$

$$Y_{\exists\forall} = \{y(t) |\ y'(t) \supseteq f(t, y(t)), \quad t \in [a,b]\}. \tag{15}$$

Subsequently, we try to obtain solution of FDE which are placed in TSS or CSS. To this end, some discussions are given to construct a solution of FDE such that it has pessimistic/optimistic attitude where pessimistic attitude is happened in TSS and optimistic attitude is placed in CSS. Clearly, in this sense, we will obtain a connected solution between TSS and CSS. However, this approach is coincide with real application while decision maker can obtain interested solution and can inference the systems in general cases.

Let us consider left and right spreads $\alpha_1(t;r), \alpha_2(t;r)$ are derived similar to previous section. Then, we define some spreads as following:

$$\alpha^-(t;r) = \min\{\alpha_1(t;r), \alpha_2(t;r)\}, \quad t_0 \le t \le T, \; 0 \le r \le 1, \tag{16}$$

$$\alpha^+(t;r) = \max\{\alpha_1(t;r), \alpha_2(t;r)\}, \quad t_0 \le t \le T, \; 0 \le r \le 1, \tag{17}$$

$$\alpha^\lambda(t;r) = \lambda\ \alpha^+(t;r) + (1-\lambda)\ \alpha^-(t;r), \quad t_0 \le t \le T, \; \lambda \in [0,1], \; 0 \le r \le 1. \tag{18}$$

Also, we define new solutions corresponding to each spreads (16)-(18), respectively, as following:

$$y^-(t) = \left[\underline{y^{[1]}}(t) - \alpha^-(t;r), \overline{y^{[1]}}(t) + \alpha^-(t;r)\right], \tag{19}$$

$$y^+(t) = \left[\underline{y^{[1]}}(t) - \alpha^+(t;r), \overline{y^{[1]}}(t) + \alpha^+(t;r)\right], \tag{20}$$

$$y^\lambda(t) = \left[\underline{y^{[1]}(t)} - \alpha^\lambda(t;r), \overline{y^{[1]}}(t) + \alpha^\lambda(t;r)\right], \tag{21}$$

Proposition 1. *Let us consider the spreads (16)-(17) and corresponding solutions (19)-(20), then we have the following:*
(1) $y^-(t) \in TSS$
(2) $y^+(t) \in CSS$

Proposition 2. *Let us consider $y^\lambda(t)$ which is defined by (21). Also, suppose that $\{\lambda_k\}_{k=0}^{\infty}$ is a nondecreasing consequence with initial value $\lambda_0 = 0$ such that $\lambda_k \longrightarrow 1$ when $k \longrightarrow \infty$. Then*

$$y^{\lambda_k}(t) = y^-(t) \in TSS \longrightarrow y^+(t) \in CSS, \quad \lambda_0 = 0, \; k \longrightarrow \infty$$

Proposition 3. *Let us consider $y^\lambda(t)$ which is defined by (21). Also, suppose that $\{\lambda_k\}_{k=0}^{\infty}$ is a nonincreasing consequence with initial value $\lambda_0 = 1$ such that $\lambda_k \longrightarrow 0$ when $k \longrightarrow \infty$. Then*

$$y^{\lambda_k}(t) = y^+(t) \in CSS \longrightarrow y^-(t) \in TSS, \quad \lambda_0 = 1, \; k \longrightarrow \infty$$

5 Examples

In this section, some examples are given to illustrate the technique. Notice that Example 1 is solved under I-differentiability and Example 2 is considered under II-differentiability.

Example 1. Let us consider the following FDE

$$\begin{cases} y'(t) = y(t), \\ y(0;r) = [1+r, 5-2r], \; 0 \le r \le 1 \end{cases} \tag{22}$$

Based on the proposed approach, 1-cut system is derived as follows:

$$\begin{cases} (y^{[1]})'(t) = y^{[1]}(t), \\ y^{[1]}(0) = [2,3], \end{cases} \tag{23}$$

Above interval differential equation is solved by Stefanini et. al's method [14] as follows:

$$\begin{cases} (\underline{y^{[1]}})'(t) = \underline{y^{[1]}}(t), \\ (\overline{y^{[1]}})'(t) = \overline{y^{[1]}}(t), \\ \underline{y^{[1]}(0)} = 2, \\ \overline{y^{[1]}(0)} = 3, \end{cases} \tag{24}$$

Then, solution of Eq.(24) is obtained $y(t) = [2e^t, 3e^t]$. Based on ODEs (8), we get:

$$\begin{cases} (\underline{y^{[1]}})'(t) - \alpha_1'(t;r) = \underline{y^{[1]}}(t) - \alpha_1(t;r), \\ (\overline{y^{[1]}})'(t) + \alpha_2'(t;r) = \overline{y^{[1]}}(t) - \alpha_2(t;r), \\ \alpha_1(0;r) = \underline{y^{[1]}}(0) - \underline{y}(0;r) = 1 - r, \\ \alpha_2(0;r) = \overline{y}(0;r) - \overline{y^{[1]}}(0) = 2(1-r), \end{cases} \tag{25}$$

Hence, ODEs (25) is rewritten based on the left and right unknown spreads as follows:

$$\begin{cases} \alpha_1'(t;r) = \alpha_1(t;r), \\ \alpha_2'(t;r) = \alpha_2(t;r), \\ \alpha_1(0;r) = 1 - r, \\ \alpha_2(0;r) = 2(1-r), \end{cases} \tag{26}$$

By solving ODEs (26), we get the spreads as following:

$$\alpha_1(t;r) = \alpha_1(0;r)\ e^t = (1-r)e^t$$

$$\alpha_2(t;r) = \alpha_2(0;r)\ e^t = (2-2r)e^t$$

Finally, solution of original FDE (22)

$$y(t) = [(1+r)\ e^t, (5-2r)e^t].$$

Clearly, our approach is coincide with the results of Bede et. al [3], Chalco-Cano et.al [6] and similar papers. It seems that proposed method has new point of view to solve FDE based on extending the 1-cut solution.

Now, we derive new solutions which are placed in CSS or TSS. Additionally, some pessimistic/optimistic solution is obtained, that is connected solution between TSS and CSS. So, by applying Eqs.(16)-(18) we have the following:

$$\alpha^-(t;r) = \min\{(1-r)\ e^t, (2-2r)e^t\} = (1-r)\ e^t, \quad \forall t \in [t_0, T],,\ 0 \le r \le 1,$$

$$\alpha^+(t;r) = \max\{(1-r)\ e^t, (2-2r)e^t\} = (2-2r)\ e^t, \quad \forall t \in [t_0, T],,\ 0 \le r \le 1,$$

$$\alpha^\lambda(t;r) = \lambda\alpha^+(t;r) + (1-\lambda)\alpha^-(t;r) = (1-\lambda)(1-r)e^t, \quad \forall t \in [t_0, T]$$

Therefore, corresponding solutions for above spreads are achieved as:

$$y^-(t;r) = [(1+r)e^t, (4-r)e^t],$$
$$y^+(t;r) = [2re^t, (5-2r)e^t],$$
$$y^\lambda(t;r) = [(1+\lambda+r(1-\lambda))e^t, (4-\lambda+r(\lambda-1))e^t].$$

It easy to see that $y^-(t) \in \texttt{TSS}$, $y^+(t) \in \texttt{CSS}$ and $y^\lambda(t)$ is pessimistic/optimistic solution for each $\lambda \in [0,1]$.

Example 2. Let us consider the following FDE

$$\begin{cases} y'(t) = -y(t), \\ y(0;r) = [1+r, 5-r], \ 0 \leq r \leq 1 \end{cases} \tag{27}$$

1-cut solution of above FDE is derived via Stefanini et al's method as $y^{[1]}(t) = [2e^{-t}, 4e^{-t}]$. Similar to Example 1, the original FDE is transformed to the following ODEs

$$\begin{cases} \alpha_1'(t;r) = -\alpha_1(t;r), \\ \alpha_2'(t;r) = -\alpha_2(t;r), \\ \alpha_1(0;r) = 1-r, \\ \alpha_2(0;r) = 1-r, \end{cases} \tag{28}$$

By solving ODEs (28), we get spreads

$$\alpha_1(t;r) = \alpha_1(0;r)\ e^{-t} = (1-r)e^{-t}$$

$$\alpha_2(t;r) = \alpha_2(0;r)\ e^{-t} = (1-r)e^{-t}$$

Finally, solution of original FDE (27) is derived

$$y(t) = [(1+r)\ e^{-t}, (5-r)e^{-t}].$$

Analogously, we determine new spreads based on Eqs.(16)-(18) as following:

$$\alpha^-(t;r) = \alpha^+(t;r) = (1-r)\ e^{-t}, \quad \forall t \in [t_0, T],\ 0 \leq r \leq 1,$$

$$\alpha^\lambda(t;r) = \lambda\alpha^+(t;r) + (1-\lambda)\alpha^-(t;r) = (1-r)e^{-t}, \quad \forall t \in [t_0, T], \quad \forall\lambda \in [0,1]$$

Then, we obtained relation solutions:

$$y^-(t;r) = y^+(t;r) = y^\lambda(t;r) = \left[(1+r)e^{-t}, (5-r)e^{-t}\right], \forall t \in [t_0, T], \forall\lambda \in [0,1],\ 0 \leq r \leq 1.$$

6 Concluding Remarks

In this paper, we proposed a new appraoch for solving first order fuzzy differential equations under strongly generalized H-differentiability. The main part of proposed technique is extending 1-cut solution of original FDEs by allocating some unknown spreads. Moreover, we extended concepts of united solution set, tolerable solution set and controllable solution set for theory of FDEs. Besides, proposed approach can adapted in order to obtain TSS or CSS. Clearly, TSS or CSS are approximated solutions generally while decision maker could inference and analyze real systems based on the connected solution between TSS and CSS which has pessimistic/optimistic attitude.

References

1. Abbasbandy, S., Allahviranloo, T., Lopez-Pouso, O., Nieto, J.J.: Numerical Method for Fuzzy Differential Inclusions. Computer and Mathematics with Applications 48, 1633–1641 (2004)
2. Allahviranloo, T., Kiani, N.A., Barkhordari, M.: Toward the existence and uniqueness of solutions of second-order fuzzy differential equations. Information Sciences 179, 1207–1215 (2009)
3. Bede, B., Rudas, I.J., Bencsik, A.L.: First order linear fuzzy differential equations under generalized differentiability. Information Sciences 177, 1648–1662 (2007)
4. Bede, B., Gal, S.G.: Generalizations of the differentiability of fuzzy-number-valued functions with applications to fuzzy differential equations. Fuzzy Sets and Systems 151, 581–599 (2005)
5. Buckley, J.J., Feuring, T.: Fuzzy differential equations. Fuzzy sets and Systems 110, 43–54 (2000)
6. Chalco-Cano, Y., Roman-Flores, H.: On new solutions of fuzzy differential equations. Chaos, Solitons and Fractals 38, 112–119 (2006)
7. Congxin, W., Shiji, S.: Existence theorem to the Cauchy problem of fuzzy differential equations under compactness-type conditions. Information Sciences 108, 123–134 (1998)
8. Kaleva, O.: Fuzzy differential equations. Fuzzy Sets and Systems 24, 301–317 (1987)
9. Friedman, M., Ming, M., Kandel, A.: Numerical solution of fuzzy differential and integral equations. Fuzzy Sets and System 106, 35–48 (1999)
10. Kandel, A.: Fuzzy dynamical systems and the nature of their solutions. In: Wang, P.P., Chang, S.K. (eds.) Fuzzy sets theory and Application to Policy Analysis and Information Systems, pp. 93–122. Plenum Press, New York (1980)
11. Kandel, A., Byatt, W.J.: Fuzzy differential equations. In: Proc. Internet. conf. Cybernetics and Society, Tokyo, November 1978, pp. 1213–1216 (1978)
12. Puri, M.L., Ralescu, D.: Fuzzy random variables. J. Math. Anal. Appl. 114, 409–422 (1986)
13. Seikkala, S.: On the fuzzy initial value problem. Fuzzy Sets and Systems 24, 319–330 (1987)
14. Stefanini, L., Bede, B.: Generalized Hukuhara differentiability of interval-valued functions and interval differential equations. Nonlinear Anal. 71, 1311–1328 (2009)

A Comparison Study of Different Color Spaces in Clustering Based Image Segmentation

Aranzazu Jurio*, Miguel Pagola, Mikel Galar,
Carlos Lopez-Molina, and Daniel Paternain

Dpt. Automática y Computación, Universidad Pública de Navarra,
Campus Arrosadía s/n, 31006 Pamplona, Spain
{aranzazu.jurio,miguel.pagola,mikel.galar,
carlos.lopez,daniel.paternain}@unavarra.es
http://giara.unavarra.es

Abstract. In this work we carry out a comparison study between different color spaces in clustering-based image segmentation. We use two similar clustering algorithms, one based on the entropy and the other on the ignorance. The study involves four color spaces and, in all cases, each pixel is represented by the values of the color channels in that space. Our purpose is to identify the best color representation, if there is any, when using this kind of clustering algorithms.

Keywords: Clustering; Image segmentation; color space; HSV; CMY; YUV; RGB.

1 Introduction

Segmentation is one of the most important tasks in image processing. The objective of image segmentation is the partition of an image into different areas or regions. These regions could be associated with a set of objects or labels. The regions must satisfy the following properties:

1. Similarity. Pixels belonging to the same region should have similar properties (intensity, texture, etc.).
2. Discontinuity. The objects stand out the environment and have clear contours or edges.
3. Connectivity. Pixels belonging the same object should be adjacent, i.e. should be grouped together.

Because of the importance of segmentation process, scientific community has proposed lots of methods and techniques to solve this problem [2,14].

Segmentation techniques can be divided in Histogram thresholding, Feature space clustering, Region-based approaches and Edge detection approaches. Color image segmentation attracts more and more attention mainly due to the following reasons: (1) color images can provide more information than gray level images; (2) the power of personal computers is increasing rapidly, and PCs can be

* Corresponding author.

E. Hüllermeier, R. Kruse, and F. Hoffmann (Eds.): IPMU 2010, Part II, CCIS 81, pp. 532–541, 2010.

used to process color images now [7]. Basically, color segmentation approaches are based on monochrome segmentation approaches operating in different color spaces. Color is perceived by humans as a combination of tristimuli R (red), G (green), and B (blue) which are usually called three primary colors. From R,G, B representation, we can derive other kinds of color representations (spaces) by using either linear or nonlinear transformations. There exist several works trying to identify which is the best color space to represent the color information, but there is not a common opinion about which is the best choice. However some papers identify the best color space for a specific task.

In [6] the authors present a complete study of the 10 most common and used colour spaces for skin colour detection. They obtain that HSV is the best one to find skin colour in an image. A similar study with 5 different colour spaces is made in [8] prooving that the polynomial SVM classifier combined with HSV colour space is the best approach for the classification of pizza toppings. For crop segmentation, in order to achieve real-time processing in real farm fields, Ruiz-Ruiz *et al.* [16] carry out a comparison study between RGB and HSV models, getting that the best accuracy is achieved with HSV representation. Although most authors use HSV in image segmentation, some works are showing that other color spaces are also useful [1,15].

When using any of the typical color spaces is not enough, some authors define a new kind of color spaces by selecting a set of color components which can belong to any of the different classical color spaces. Such spaces, which have neither psycho-visual nor physical color significance, are named hybrid color spaces [17].

Cluster analysis or clustering is the assignment of a set of observations into subsets (called clusters) so that observations in the same cluster are similar in some sense. Clustering is a method of unsupervised learning and a common technique for statistical data analysis used in many fields, including machine learning, data mining, pattern recognition, image analysis and bioinformatics. Image segmentation is also a topic where clustering techniques have been widely applied [9,5,13,11].

A cluster is a collection of objects which are similar between them and are dissimilar to the objects belonging to other clusters. Therefore, as within any clustering technique we must measure the distance or the similarity between objects, in color image segmentation it is very important to define which color space is going to be used because such measure will be defined within said space.

Clustering techniques can provide methods whose results satisfy the three properties demanded to segmented images. In this case the objects will be the pixels, and each pixel can be defined by its color, texture information, position, etc. In our experiments the features that identify each pixel are only the values of its three components in the selected color space.

This work is organized as follows: We begin recalling the different color spaces. In section 3 we present the two clustering algorithms that will be used in the segmentation process. Next in experimental results, we present the settings of the experiment and the results obtained. Finally we show some conclusions and future research.

2 Color Spaces

A color space is a tool to visualize, create and specify the color. For computers color is an excitation of three phosphors (blue, red, and green) and for a printing press color is a reflectance and absorbance of cyan, magenta, yellow and black inks on the paper. A color space is the representation of three attributes used to describe a color. A color space is also a mathematical representation of our perception [1].
We can distinguish between these two clases:

- Hardware oriented: They are defined according to the properties of the optical instruments to show the color, like TV, LCD screens or printers. Typical examples are RGB, CMY, YUV (it is the PAL/European standard for YIQ).
- User oriented: Based on human perception of colors by hue, saturation and brightness. Hue represents the wave length of the perceived color, the saturation or croma indicates the quantity of white light present in the color and the brightness or value the intensity of the color. Typical examples are: HLS, HCV, HSV, HSB and MTM, L*u*v*, L*a*b* y L*C*h*.

In this work we are going to compare four color spaces in image segmentation. These ones are RGB, CMY, HSV and YUV.

2.1 RGB

An RGB color space can be understood as all possible colors that can be made from three colourants for red, green and blue. The main purpose of the RGB color model is for the sensing, representation, and display of images in electronic systems, such as televisions and computers, though it has also been used in conventional photography. Although the RGB is the most used model to acquire digital images, it is said that it is not adequate for color image analysis. We are going to use this color space as the reference.

2.2 CMY

The CMY (Cyan, Magenta, Yellow) color model is a subtractive color model used in printing. It works by masking certain colors on typically white background, it means, absorbing particular wavelengths of light. Cyan is the opposite of red (it absorbs red color), magenta is the opposite of green and yellow is the opposite of blue. The conversion from RGB to CMY is:

$$\begin{array}{ll} C' = 1 - R & C = min(1, max(0, C' - K')) \\ M' = 1 - G & M = min(1, max(0, M' - K')) \\ Y' = 1 - B & Y = min(1, max(0, Y' - K')) \\ K' = min(C', M', Y') & \end{array} \qquad (1)$$

2.3 HSV

The HSV color model is more intuitive than the RGB color model. In this space, hue (H) represents the color tone (for example, red or blue), saturation (S) is the amount of color (for example, bright red or pale red) and the third component (called intensity, value or lightness) is the amount of light (it allows the distinction between a dark color and a light color).

If we take the HSV color space in a cone representation, the hue is depicted as a three-dimensional conical formation of the color wheel. The saturation is represented by the distance from the center of a circular cross-section of the cone, and the value is the distance from the pointed end of the cone.

Let R, G, B ∈ [0,1] be the red, green, and blue coordinates of a RGB image, *max* be the greatest of R, G, and B, and *min* be the lowest. In equation 2 it is shown how to transform this image into HSV space.

$$H = \begin{cases} 0, & \text{if } \max = \min \\ (60^\circ \times \frac{G-B}{\max - \min} + 360^\circ) \bmod 360^\circ, & \text{if } \max = R \\ 60^\circ \times \frac{B-R}{\max - \min} + 120^\circ, & \text{if } \max = G \\ 60^\circ \times \frac{R-G}{\max - \min} + 240^\circ, & \text{if } \max = B \end{cases} \tag{2}$$

$$S = \begin{cases} 0, & \text{if } \max = 0 \\ \frac{\max - \min}{\max} = 1 - \frac{\min}{\max}, & \text{otherwise} \end{cases} \tag{3}$$

$$V = \max \tag{4}$$

2.4 YUV

YUV color model imitates human vision. Term YUV designates a whole family of so called luminance (Y) and chrominance (UV) color spaces. In this work, we use YCbCr, which is an standard color space for digital television systems. To convert a RGB image into YUV space it is used the following expression:

$$\begin{bmatrix} Y \\ U \\ V \end{bmatrix} = \begin{bmatrix} 0,299 & 0,587 & 0,114 \\ -0,147 & -0,289 & 0,436 \\ 0,615 & -0,515 & -0,100 \end{bmatrix} \begin{bmatrix} R \\ G \\ B \end{bmatrix} \tag{5}$$

3 Clustering Algorithms

Among fuzzy clustering methods, the fuzzy c-means (FCM) method [2] is one of the most popular methods. One important issue in fuzzy clustering is identifying the number and initial locations of cluster centers. In classical FCM algorithm, these initial values are specified manually. But there exist another type of clustering algorithms that automatically determine the number of clusters and the location of cluster centers by the potential of each data point. Yao et al. in [18] proposed a clustering method based on the entropy measure instead of the potential measure. Also [10] we have proposed an improvement of such algorithm based on the Ignorance functions [4] that also segment images without selecting the initial number of clusters.

3.1 Entropy Based Fuzzy Clustering Algorithm

The basis of EFC is to find the elements which, if they are supposed to be the center of the cluster, then the entropy of the total set of elements is the lowest. This entropy is calculated for each element taking into account the similarity of that element with all the elements left ($S(x_i, x_j)$), with the following expression:

$$E(x_i) = -\sum_{j \in X}^{j \neq i} (S(x_i, x_j) log_2 S(x_i, x_j) + (1 - S(x_i, x_j)) log_2 (1 - S(x_i, x_j))) \quad (6)$$

Such a way the algorithm first selects the element with lowest entropy as the center of the first cluster. Once it is selected, it is deleted from the center candidates list. Also, the elements whose distance to the cluster center is lower than a a given threshold (β) are deleted. Once those elements are deleted from the candidates list, the element with lowest entropy is taken as the center of the second cluster. The process is repeated until the candidates list is empty.

Given a set T with N data, the algorithm is outlined as follows:

1. Calculate the entropy of each $x_i \in T$, for $i = 1, \dots, N$.
2. Choose $x_{i_{Min}}$ achieving the lowest entropy.
3. Delete from T, $x_{i_{Min}}$ and all the data whose distance to it is smaller than β.
4. If T is not empty, go to step 2.

We must notice that it is not possible to choose a priori the number of clusters in which the algorithm must split the data. The user must modify the value of threshold β to obtain the number of desired clusters.

3.2 Ignorance Based Clustering Algorithm

In [10] we propose a modification of the EFC algorithm. We replace the similarity between elements by restricted equivalence functions. In addition we use ignorance functions instead of entropy functions so, for us, the center of the cluster is the element which causes that the partition of the data has the lowest ignorance. With these two modifications we improve the results of the EFC, and solve some problems that it has with symmetrical data.

The ignorance functions estimate the uncertainty that exists when there are two membership functions. However, in this case we want to calculate the total ignorance of a set of elements by means of their membership degree to a cluster. If we are completely sure that an element is the center of the cluster, then we have no ignorance. If the membership of the element to the cluster is 0.5, then we say that we have total ignorance. Therefore we can deduce from a general ignorance function (please see theorem 2 of [4]) the following expression to calculate the ignorance associated to a single element:

$$Ig(x) = 4(1 - x)x \tag{7}$$

Given a set T with N data, the ignorance algorithm is as follows:

1. Calculate the ignorance of each $x_i \in T$, for $i = 1, \ldots, N$.
   1.1. Calculate the restricted equivalence between each pair of data.

$$Eq(x_i, x_j) = M(REF(x_{i1}, y_{x1}), REF(x_{i2}, y_{x2}), \ldots, REF(x_{in}, x_{jn})) \\ \text{for all } j = 1..N \text{ where } j \neq i \tag{8}$$

   1.2. Calculate the ignorance of each pair of data:

$$Ig(Eq(x_i, x_j)) = (1 - Eq(x_i, x_j)) * Eq(x_i, x_j) \tag{9}$$

   1.3. Calculate the ignorance of each datum.

$$I_T(x_i) = \frac{\sum_{j=1}^{N} Ig(Eq(x_i, x_j))}{N} \tag{10}$$

2. Choose $x_{i_{Min}}$ achieving the lower ignorance.
3. Delete from T, $x_{i_{Min}}$ and all the data whose distance to it is smaller than β.
4. If T is not empty, go to step 2.

4 Experimental Results

In this section we present the experimental study that we have done to discover which is the best color space to use in image segmentation based on clustering. We take four natural images with their ideal segmentation (see figure 1). These segmentations have been manually calculated into three different areas taking into account the image dataset [12]. These areas have been segmented following a color and object representation criteria. Each area, in the ideal segmented image (see figure 1), has been colored with the mean color of the pixels that belong to it.

For this set of images we execute the two algorithms, Ignorance based clustering (section 3.2) and Entropy based fuzzy clustering (section 3.1), four times, each one with a different color space RGB, CMY, HSV and YUV. In clustering algorithms each pixel is an element represented by three parameters, so each x_i is a vector with three values. These values vary in every execution, representing each color channel of the selected color space.

For the Ignorance based clustering we have selected the following expression of equivalence:

$$Eq(x_i, x_j) = (1 - |x_i^3 - x_j^3|)^3 \tag{11}$$

and for the Entropy based we have selected the expression of similarity proposed in the original work [18]:

$$S(x_i, x_j) = e^{-\alpha D(x_i, x_j)} \tag{12}$$

where $\alpha = -ln(0.5)/\bar{D}$ and D is the Minkowski distance with $p = 1$.

Fig. 1. Original images and ideal segmentation

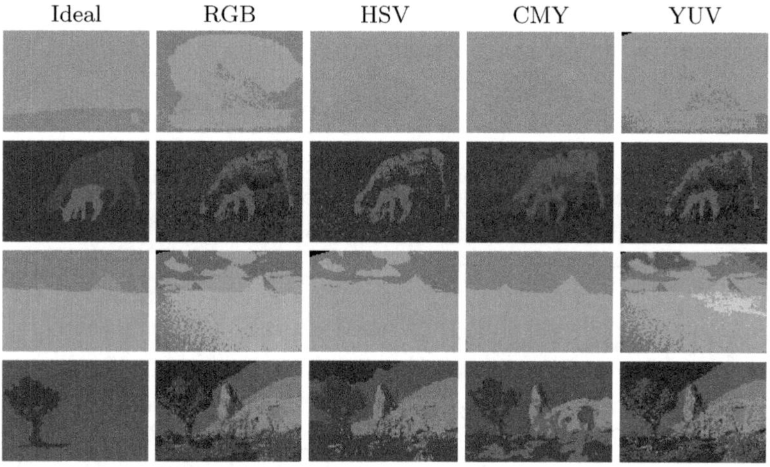

Fig. 2. Segmented images obtained with the ignorance based clustering

In our experiment we evaluate four different color spaces: CMY, RGB, YUV and HSV. In figures 2 and 3 we show the best segmented images obtained for every image and every color space for each algorithm. For a quantitative comparison we present table 1. In these tables we show the similarity between the ideal image and these segmented images using the following equation:

$$SIM(A,B) = \frac{1}{3 \times N \times M} \sum_{c \in \{R,G,B\}} \sum_{i} \sum_{j} 1 - |A_{ijc} - B_{ijc}| \quad (13)$$

being N and M the number of rows and columns of the image, where A is the segmented image obtained, B is the ideal one and A_{ijc} is the intensity of the pixel located in the i-th row in the j-th columns and the c-th channel of the image A. As every region of the image is coloured with the mean colour of that region, the more likeness are both mean colours, the greater is the similarity between those pixels. This similarity has been chosen because it fulfills the six properties demanded for a global comparison measure of two images[3].

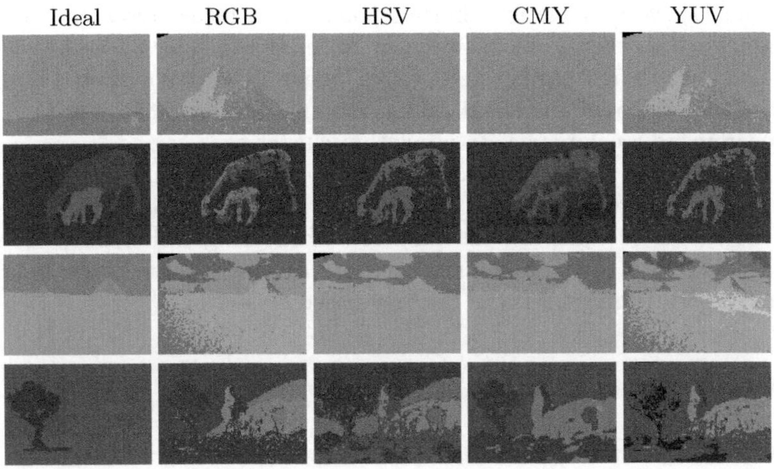

Fig. 3. Segmented images obtained with the entropy based clustering

Table 1. Similarities between the ideal images and best segmented images

Ignorance based clustering				
Image	CMY	RGB	YUV	HSV
1	0.9802	0.9396	0.9719	0.9818
2	0.9724	0.9567	0.9575	0.9717
3	0.9965	0.9246	0.9314	0.9687
4	0.9408	0.8886	0.8851	0.9012
Mean	**0.9724**	0.9273	0.9364	0.9558

Entropy based clustering				
Image	CMY	RGB	YUV	HSV
1	0.9801	0.9704	0.9715	0.9812
2	0.9712	0.968	0.9694	0.9723
3	0.9703	0.9278	0.993	0.9687
4	0.9355	0.9026	0.9222	0.9222
Mean	**0.9642**	0.9422	0.9640	0.9611

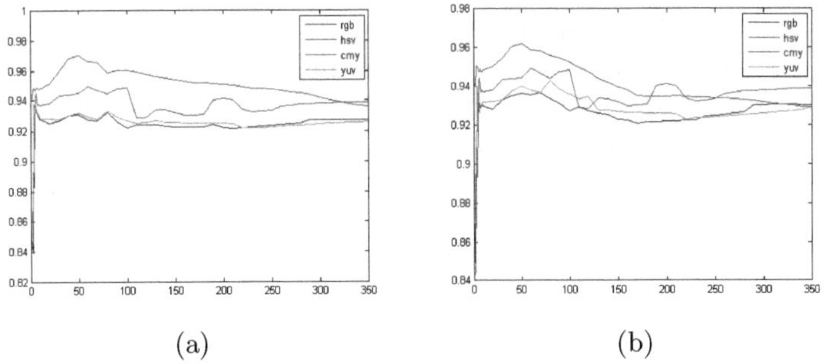

Fig. 4. Average similarity with respect the threshold value. (a) Ignorance based and (b) Entropy based clustering. Each line represent the average similarity for the set fo images.

We can see that the CMY space is the one which obtain better results in both algorithms.

As we have explained before, both algorithms have a threshold value, which will be a key in the number of final clusters. The selection of the the best threshold for every image is a difficult point and a future research line. In our first approach to this problem, we want to select the color space in which the influence of the threshold is the lowest. Therefore we have executed both algorithms for 45 different threshold values, ranging from 0 to 350. Such a way, we can recommend a color space to use within clustering.

In figure 4(a) we show the mean performance of the ignorance based clustering obtained for different threshold values using the four color spaces. It is clear that the threshold value has less influence when using CMY. But best results are obtained with CMY. Similar conclusions can be obtained form figure 4(b) where the algorithm used is the Entropy based clustering.

5 Conclusions and Future Research

In this work we have studied four color spaces for image segmentation based on clustering. These spaces are RGB, HSV, CMY and YUV. The clustering algorithms we have worked with depend on a threshold value. In this sense, we have also studied the importance of this value in the final segmented image.

Our experiments have revealed that the best results are obtained in most cases in the CMY color space. HSV also provides good results. Besides, CMY is the color space in which the quality of the segmented image is higher for any threshold. In the ignorance based algorithm this space is the best with a big difference while in the entropy based one, it is followed closer by YUV and HSV. So, we can conclude that the correct space to use is the CMY.

However, this is a preliminary study and it must be enlarged with more images. They must include different kind of images, like real images, synthetic images, etc. It must also be enlarged with different ideal segmentations for each image. As ground truth segmentations are not unique, the most siutable color space could change for different ideal solutions.

In the future, we will construct an automatic method to choose the best threshold in this kind of clustering algorithms. Besides, we want to extend this study by incorporating more color spaces, like L*a*b, YIB or LSLM and more clustering algorithms, like FCM.

Acknowledgments. This research was partially supported by the Grant TIN2007-65981.

References

1. Alata, O., Quintard, L.: Is there a best color space for color image characterization or representation based on Multivariate Gaussian Mixture Model? Computer Vision and Image Understanding 113, 867–877 (2009)
2. Bezdek, J.C., Keller, J., Krisnapuram, R., Pal, N.R.: Fuzzy Models and algorithms for pattern recognition and image processing. In: Dubois, D., Prade, H. (Series eds.). The Handbooks of Fuzzy Sets Series. Kluwer Academic Publishers, Dordrecht (1999)

3. Bustince, H., Pagola, M., Barrenechea, E.: Construction of fuzzy indices from fuzzy DI-subsethood measures: Application to the global comparison of images. Information Sciences 177, 906–929 (2007)
4. Bustince, H., Pagola, M., Barrenechea, E., Fernandez, J., Melo-Pinto, P., Couto, P., Tizhoosh, H.R., Montero, J.: Ignorance functions. An application to the calculation of the threshold in prostate ultrasound images. Fuzzy Sets and Systems 161(1), 20–36 (2010)
5. Celenk, M.: A Color Clustering Technique for Image Segmentation. Computer Vision Graphics and Image Processing 52(2), 145–170 (1990)
6. Chaves-González, J.M., Vega-Rodríguez, M.A., Gómez-Pulido, J.A., Sánchez-Pérez, J.M.: Detecting skin in face recognition systems: A colour spaces study. Digital Signal Process. (2009), doi:10.1016/j.dsp.2009.10.008
7. Cheng, H.D., Jiang, X.H., Sun, Y., Wang, J.: Color image segmentation: advances and prospects. Pattern Recognition 34(12), 2259–2281 (2001)
8. Du, C.-J., Sun, D.-W.: Comparison of three methods for classification of pizza topping using different colour space transformations. Journal of Food Engineering 68, 277–287 (2005)
9. Lo, H., Am, B., Lp, C., et al.: A Comparison of Neural Network and Fuzzy Clustering-Techniques in Segmenting Magnetic-Resonance Images of the Brain. IEEE Transactions on Neural Networks 3(5), 672–682 (1992)
10. Jurio, A., Pagola, M., Paternain, D., Barrenechea, E., Sanz, J., Bustince, H.: Ignorance-based fuzzy clustering algorithm. In: Ninth International Conference on Intelligent Systems Design and Applications, pp. 1353–1358 (2009)
11. Jurio, A., Pagola, M., Paternain, D., Lopez-Molina, C., Melo-Pinto, P.: Interval-valued restricted equivalence functions applied on Clustering Techniques. In: 13rd International Fuzzy Systems Association World Congress and 6th European Society for Fuzzy Logic and Technology Conference (IFSA-EUSFLAT 2009) (2009)
12. Martin, D., Fowlkes, C., Tal, D., Malik, J.: A database o f human segmented natural images and its application to evaluating segmentation algorithms and measuring ecological statistics. In: Proc. 8th Int'l. Conf. Computer Vision, July 2001, vol. 2, pp. 416–423 (2001)
13. Nam, I., Salamah, S., Ngah, U.: Adaptive Fuzzy Moving K-Means Clustering Algorithm For Image Segmentation. IEEE Transactions on Consumer Electronics 55(4), 2145–2153 (2009)
14. Pal, N.R., Pal, S.K.: A review of image segmentation techniques. Pattern recognition 26, 1277–1294 (1993)
15. Pagola, M., Ortiz, R., Irigoyen, I., Bustince, H., Barrenechea, E., Aparicio-Tejo, P., Lamsfus, C., Lasa, B.: New method to assess barley nitrogen nutrition status based on image colour analysis: Comparison with SPAD-502. Computers and Electronics in Agriculture 65(2), 213–218 (2009)
16. Ruiz-Ruiz, G., Gómez-Gil, J., Navas-Gracia, L.M.: Testing different color spaces based on hue for the environmentally adaptive segmentation algorithm (EASA). Computers and Electronics in Agriculture 68(1), 88–96 (2009)
17. Vandenbroucke, N., Macaire, L., Postaire, J.G.: Color image segmentation by pixel classification in an adapted hybrid color space. Application to soccer image analysis. Computer Vision and Image Understanding 90(2), 190–216 (2003)
18. Yao, J., Dash, M., Tan, S.T., Liu, H.: Entropy-based fuzzy clustering and fuzzy modeling. Fuzzy Sets Syst. 113(3), 381–388 (2000)

Retrieving Texture Images Using Coarseness Fuzzy Partitions

Jesús Chamorro-Martínez[1], Pedro Manuel Martínez-Jiménez[1,*], and Jose Manuel Soto-Hidalgo[2]

[1] Department of Computer Science and Artificial Intelligence, University of Granada
{jesus,pedromartinez}@decsai.ugr.es
[2] Department of Computer Architecture, Electronics and Electronic Technology, University of Córdoba
jmsoto@uco.es

Abstract. In this paper, a *Fuzzy Dominant Texture Descriptor* is proposed for semantically describing an image. This fuzzy descriptor is defined over a set of fuzzy sets modelling the "coarseness" texture property. Concretely, fuzzy partitions on the domain of coarseness measures are proposed, where the number of linguistic labels and the parameters of the membership functions are calculated relating representative coarseness measures (our reference set) with the human perception of this texture property. Given a "texture fuzzy set", its dominance in an image is analyzed and the dominance degree is used to obtain the image texture descriptor. Fuzzy operators over these descriptors are proposed to define conditions in image retrieval queries. The proposed framework makes database systems able to answer queries using texture-based linguistic labels in natural language.

1 Introduction

For analyzing an image several kind of features can be used. From all of them, texture is one of the most popular and, in addition, one of the most difficult to characterize due to its imprecision. For describing texture, humans use vague textural properties like *coarseness/fineness*, *orientation* or *regularity* [1,2]. From all of them, the *coarseness/fineness* is the most common one, being usual to associate the presence of fineness with the presence of texture. In this framework, a *fine* texture corresponds to small texture primitives (e.g. the image in figure 1(A)), whereas a *coarse* texture corresponds to bigger primitives (e.g. the image in figure 1(I)).

There are many measures in the literature that, given an image, capture the fineness (or coarseness) presence in the sense that the greater the value given by the measure, the greater the perception of texture [3]. However, given a certain measure value, there is not an immediate way to decide whether there is a fine texture, a coarse texture or something intermediate; in other words, there is not a textural interpretation.

* This work was supported by Spanish research programme Consolider Ingenio 2010: MIPRCV (CSD2007-00018).

E. Hüllermeier, R. Kruse, and F. Hoffmann (Eds.): IPMU 2010, Part II, CCIS 81, pp. 542–551, 2010.

Fig. 1. Some examples of images with different degrees of fineness

To face this problem, fuzzy logic has been recently employed for representing the imprecision related to texture. In many of these approaches, fuzzy logic is usually applied just during the process, being the output a crisp result [4,5]. Other approaches try to model the texture and its semantic by means of fuzzy sets defined on the domain of a given texture measure. In this last framework, some proposals model the texture property by means of an unique fuzzy set [6], and other approaches define fuzzy partitions providing a set of linguistic terms [7,8].

Focusing our study in the last type of approaches, two questions need to be faced for properly defining a fuzzy partition: (i) the number of linguistic labels to be used, and (ii) the parameters of the membership functions associated to each fuzzy set (and, consequently, the kernel localization). However, these questions are not treated properly in the literature. Firstly, the number of fuzzy sets are often chosen arbitrarily, without take into account the capability of each measure to discriminate between different categories. Secondly, in many of the approaches, just an uniform distribution of the fuzzy sets is performed on the domain of the measures, although it is well known that measure values corresponding to representative labels are not distributed uniformly. In addition, from our knowledge, none of the fuzzy approaches in the literature consider the relationship between the computational feature and the human perception of texture, so the labels and the membership degrees do not necessarily will match with the human assessments.

In this paper, we propose a fuzzy partition taking into account the previous questions. Firstly, in order to select the number of linguistic labels, we analyze the ability of each measure to discriminate different coarseness categories. For this purpose, data about the human perception of fineness is collected by means of a pool. This information is also used to localize the position and size of the kernel of each fuzzy set, obtaining a fuzzy partition adapted to the human perception of coarseness-fineness.

Moreover, we propose to apply the obtained fuzzy partition for texture image retrieval. The current image retrieval systems are based on features, such as color, texture or shape, which are automatically extracted from images. In this framework, a very important point to take into account is the imprecision in the feature descriptions, as well as the store and retrieval of that imprecise data. To deal with this vagueness, some interesting approaches introduce the use of fuzzy logic in the feature representation and in the retrieval process [9,10]. These fuzzy approaches also allow to perform queries on the basis of linguistic terms, avoiding one of the drawbacks of the classical image retrieval systems, where the queries have to be defined on the basis of images or sketches similar to the one we are searching for. This way, the proposed fuzzy partition will be used to describe images in terms of their texture coarseness and the queries will be performed by using linguistic labels.

The rest of the paper is organized as follows. In section 2 we present our methodology to obtain the fuzzy partition. In section 3 a Fuzzy Dominant Texture Descriptor is proposed in order to apply the obtained fuzzy partition to texture image retrieval. Results are shown in section 4, and the main conclusions and future work are sumarized in section 5.

2 Fuzzy Partitions for Coarseness

As it was pointed, there is not a clear perceptual interpretation of the value given by a fineness measure. To face this problem, we propose to define a fuzzy partition on the domain of a given fineness measure. For this purpose, several questions will be faced: (i) what reference set should be used for the fuzzy partition, (ii) how many fuzzy sets will compound the partition, and (iii) how to obtain the membership functions for each fuzzy set.

Concerning the reference set, we will define the partition on the domain of a given coarseness-fineness measure. From now on, we will note $\mathcal{P} = \{P_1, \ldots, P_K\}$ the set of K measures analyzed in this paper, Π_k the partition defined on the domain of $\mathcal{P}_k$, N_k the number of fuzzy sets which compounds the partition Π_k, and $\mathcal{T}_k^i$ the i-th fuzzy set in Π_k. In this paper, the set $\mathcal{P} = \{P_1, \ldots, P_K\}$ is formed by the $K = 17$ measures shown in the first column of table 1. It includes classical statistical measures, frequency domain approaches, fractal dimension analysis, etc. All of them are automatically computed from the texture image.

With regard to the number of fuzzy sets which compounds the partition, we will analyze the ability of each measure to distinguish between different degrees of fineness. This analysis will be based on how the human perceives the fineness-coarseness. To get information about human perception of fineness, a set of images covering different degrees of fineness will be gathered. These images will be used to collect, by means of a pool, human assessments about the perceived fineness. From now on, let $\mathcal{I} = \{I_1, \ldots, I_N\}$ be the set of N images representing fineness-coarseness examples, and let $\Gamma = \{v^1, \ldots, v^N\}$ be the set of perceived fineness values associated to $\mathcal{I}$, with v^i being the value representing the degree of fineness perceived by humans in the image $I_i \in \mathcal{I}$. We will use the texture image set and the way to obtain Γ described in [11].

Table 1. Result obtained by applying the algorithm proposed in [11]

Measure	N_k	Classes	$\bar{c}_5 \pm \Psi_5$	$\bar{c}_4 \pm \Psi_4$	$\bar{c}_3 \pm \Psi_3$	$\bar{c}_2 \pm \Psi_2$	$\bar{c}_1 \pm \Psi_1$
Correlation [3]	5	{1,2-4,5-6,7-8,9}	0.122±0.038	0.403±0.0272	0.495±0.0225	0.607±0.0133	0.769±0.0210
ED [12]	5	{1,2,3-5,6-8,9}	0.348±0.0086	0.282±0.0064	0.261±0.0063	0.238±0.0066	0.165±0.0061
Abbadeni [13]	4	{1,2-6,7-8,9}	-	5.672±0.2738	9.208±0.4247	11.12±0.2916	25.23±1.961
Amadasun [1]	4	{1,2-6,7-8,9}	-	4.864±0.271	7.645±0.413	9.815±0.230	19.62±1.446
Contrast [3]	4	{1,2-5,6-8,9}	-	3312±265.5	2529±295.5	1863±94.84	790.8±129.4
FD [14]	4	{1,2,3-8,9}	-	3.383±0.0355	3.174±0.0282	2.991±0.0529	2.559±0.0408
Tamura [2]	4	{1,2-6,7-8,9}	-	1.540±0.0634	1.864±0.0722	2.125±0.0420	3.045±0766
Weszka [15]	4	{1,2-6,7-8,9}	-	0.153±0.0064	0.113±0.0093	0.099±0.0036	0.051±0.0041
DGD [16]	3	{1,2-8,9}	-	-	0.020±0.0010	0.038±0.0017	0.091±0.0070
FMPS [17]	3	{1,2-8,9}	-	-	0.256±0.0477	0.138±0.0122	0.0734±0.0217
LH [3]	3	{1,2-8,9}	-	-	0.023±0.0010	0.052±0.0025	0.127±0.0096
Newsam [18]	3	{1,2-6,7-9}	-	-	0.1517±0.0425	0.2654±0.0466	0.4173±0.0497
SNE [19]	3	{1,2-8,9}	-	-	0.879±0.0182	0.775±0.0087	0.570±0.0232
SRE [20]	3	{1,2-8,9}	-	-	0.995±0.00026	0.987±0.00066	0.966±0.0030
Entropy [3]	2	{1,2-9}	-	-	-	9.360±0.124	8.656±0.301
Uniformity[3]	2	{1,2-9}	-	-	-	$1.3_{E^{-4}} \pm 2.6_{E^{-5}}$	$3.9_{E^{-4}} \pm 1.9_{E^{-4}}$
Variance[3]	1	-	-	-	-	-	-

Using the data about human perception, and the measure values obtained for each image $I_i \in \mathcal{I}$, we will apply a set of multiple comparison tests in order to obtain the number of fineness degrees that each measure can discriminate (section 2.1). In addition, with the information given by the tests, we will define the fuzzy sets which will compound the partition (2.2).

2.1 Distinguishability Analysis of the Fineness Measures

As it was expected, some measures have better ability to represent fineness-coarseness than the others. To study the ability of each measure to discriminate different degrees of fineness-coarseness (i.e. how many classes can P_k actually discriminate), we propose to analyze each $P_k \in \mathcal{P}$ by applying a set of multiple comparison tests following the algorithm shown in [11]. This algorithm starts with an initial partition[1] and iteratively joins clusters until a partition in which all classes are distinguishable is achieved. In our proposal, the initial partition will be formed by the 9 classes used in our poll (where each class will contain the images assigned to it by the majority of the subjects), as δ the Euclidean distance between the centroids of the involved classes will be used, as ϕ a set of 5 multiple comparison tests will be considered (concretely, the tests of Scheffé, Bonferroni, Duncan, Tukey's least significant difference, and Tukey's honestly significant difference [21]), and finally the number of positive tests to accept distinguishability will be fixed to $NT = 3$.

From now on, we shall note as $\Upsilon_k = C_1^k, C_2^k, \ldots, C_{N_k}^k$ the N_k classes that can be discriminated by P_k. For each C_i^k, we will note as $\bar{c}_i^k$ the class representative value. In this paper, we propose to compute $\bar{c}_i^k$ as the mean of the measure values in the class C_i^k.

[1] Let us remark that this partition is not the "fuzzy partition". In this case, the elements are measure values and the initial clusters the ones given by the pool.

Table 1 shows the parameters obtained by applying the proposed algorithm with the different measures considered in this paper. The second column of this table shows the N_k classes that can discriminate each measure and the third column shows how the initial classes have been grouped. The columns from fourth to eighth show the representative values $\bar{c}_r^k$ associated to each cluster.

2.2 The Fuzzy Partitions

In this section we will deal with the problem of defining the membership function $\mathcal{T}_k^i(x)$ for each fuzzy set $\mathcal{T}_k^i$ compounding the partition Π_k. As it was explained, the number of fuzzy sets will be given by the number of categories that each measure can discriminate (shown in Table 1).

In this paper, trapezoidal functions are used for defining the membership functions. In addition, a fuzzy partition in the sense of Ruspini is proposed. Figure 2 shows some examples of the type of fuzzy partition used. To establish the localization of each kernel, the representative value $\bar{c}_i^k$ will be used (in our case, the mean). Concretely, this value will be localized at the center position of the kernel.

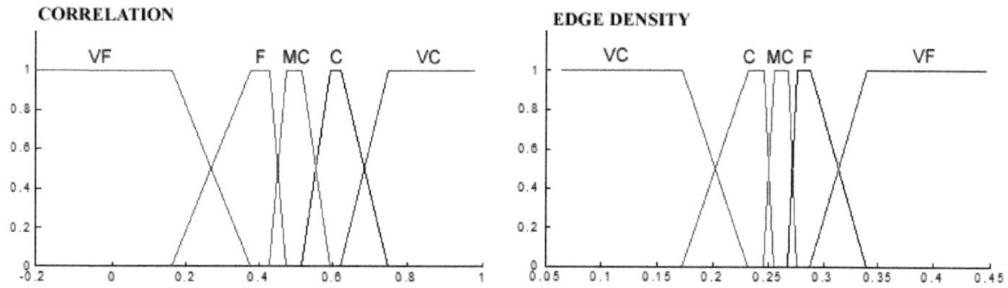

Fig. 2. Fuzzy partitions for the measures Correlation and Edge Density. The linguistic labels are VC = very coarse, C = coarse, MC = medium coarse, F = fine, VF = very fine.

To establish the size of the kernel, we propose a solution based on the multiple comparison tests used in section 2.1. As it is known, in these tests confidence intervals around the representative value of each class are calculated (being accomplished that these intervals do not overlap for distinguishable classes). All values in the interval are considered plausible values for the estimated mean. Based on this idea, we propose to set the kernel size as the size of the confidence interval.

The confidence interval CI_i^k for the class C_i^k is defined as

$$CI_i^k = \bar{c}_i^k \pm \Psi_i^k \tag{1}$$

where $\Psi_i^k = 1.96\bar{\sigma}_i^k / \sqrt{||C_i^k||}$, with $\bar{c}_i^k$ being the class representative value and $\bar{\sigma}_i^k$ being the estimated standard deviation for the class. Table 1 shows the values Ψ_i^k for each measure and each class.

Thus, the trapezoidal function that is used for defining the membership functions has the form

$$\mathcal{T}_k^i(x) = \begin{cases} 0 & x < a_k^i \quad or \quad x > d_k^i \\ \frac{x-a_k^i}{b_k^i-a_k^i} & a_k^i \le x \le b_k^i \\ 1 & b_k^i \le x \le c_k^i \\ \frac{d_k^i-x}{d_k^i-c_k^i} & c_k^i \le x \le d_k^i \end{cases} \tag{2}$$

with $a_k^i = c_k^{i-1}$, $b_k^i = \bar{c}_i^k - \Psi_i^k$, $c_k^i = \bar{c}_i^k + \Psi_i^k$ and $d_k^i = b_k^{i+1}$. It should be noticed that $a_k^1 = b_k^1 = -\infty$ and $c_k^{N_k} = d_k^{N_k} = \infty$.

Figure 2 shows the fuzzy partitions for the measures of correlation and ED (the ones with higher capacity to discriminate fineness classes).

3 Dominance-Based Fuzzy Texture Descriptor

As it was pointed, we propose to apply the obtained fuzzy partition to texture image retrieval. For describing semantically an image, the dominant textures will be used. In this section, a Fuzzy Dominant Texture Descriptor is proposed (section 3.2) on the basis of the dominance degree of a given texture (section 3.1).

3.1 Dominant Fuzzy Textures

Intuitively, a texture is dominant to the extend it appears frequently in a given image. As it is well known in the computer vision field, the histogram is a powerful tool for measuring the frequency in which a property appears in an image. Working with fuzzy properties suggests to extend the notion of histogram to "fuzzy histogram". In this sense, a fuzzy histogram will give us information about the frequency of each fuzzy property (texture in our case). In this paper, the counting will be performed by using the scalar sigma-count (i.e., the sum of membership degrees). Thus, for any fuzzy set T with membership function $T : X \rightarrow [0, 1]$, the fuzzy histogram is defined as[2]

$$h(T) = \frac{1}{NP} \sum_{x \in X} T(x) \tag{3}$$

with NP being the number of pixels. For texture properties, a window centered on each pixel will be used to calculate the measure value x.

Using the information given by the histogram, we will measure the "dominance" of a texture fuzzy set. Dominance is an imprecise concept, i.e., it is possible in general to find textures that are clearly dominant, textures that are clearly not dominant, and textures that are dominant to a certain degree, that depends on the percentage of pixels where the color/texture appears.

[2] In our case, this fuzzy set will correspond with the texture fuzzy set $\mathcal{T}_k$.

It seems natural to model the idea of dominance by means of a fuzzy set over the percentages given by $h(T)$, i.e., a fuzzy subset of the real interval $[0,1]$. Hence, we define the fuzzy subset "Dominant", denoted as Dom, as follows:

$$Dom(T) = \begin{cases} 0 & h(T) \leq u_1 \\ \frac{h(T)-u_1}{u_2-u_1} & u_1 \leq h(T) \leq u_2 \\ 1 & h(T) \geq u_2 \end{cases} \tag{4}$$

where u_1 and u_2 are two parameters such that $0 \leq u_1 < u_2 \leq 1$, and $h(T)$ is calculated by means of Eq. 3. We have intuitively fixed $u_1 = 0.2$ and $u_2 = 0.4$.

3.2 Fuzzy Dominant Texture Descriptor

On the basis of the dominance of textures, a new image descriptor is proposed for the "Texture Coarseness" property:

Definition 1. *Let $\mathcal{T}$ a finite reference universe of texture fuzzy sets. We define the* Fuzzy Dominant Texture Descriptor *as the fuzzy set*

$$FDTD = \sum_{T \in \mathcal{T}} Dom(T)/T \tag{5}$$

with $Dom(T)$ being the dominance degree of T given by Eq. 4.

3.3 Fuzzy Operators

Fuzzy operators over fuzzy descriptors are needed to define conditions in image retrieval queries. In this paper, the operators we proposed in [22] will be used.

The first one is the `FInclusion(A,B)` operator, which calculates the inclusion degree $A \subseteq B$, where, in our case, A and B are fuzzy texture descriptors. The calculus is done using a modification of the *Resemblance Driven Inclusion Degree* introduced in [23], which computes the inclusion degree of two fuzzy sets whose elements are imprecise.

The second one is the `FEQ(A,B)` operator, which calculates the resemblance degree between two fuzzy texture descriptors. The calculus is done by means of the *Generalized Resemblance between Fuzzy Sets* proposed in [23], which is based on the concept of double inclusion.

The described framework makes database systems able to answer queries based on the set of dominant textures within an image. Therefore, the user can define a fuzzy set of fuzzy textures (i.e, a descriptor) which must be included in, or resemble to, the descriptor of each image in the database. Each fuzzy texture in the fuzzy set can be defined by using the linguistic labels proposed in section 2.2, which makes possible to define queries using natural language.

4 Results

In this section, the dominance-based fuzzy texture descriptor proposed in section 3 will be applied to texture image retrieval in order to analyze its performance.

Fig. 3. Retrieval results in VisTex database using the linguistic label *very Coarse* as query

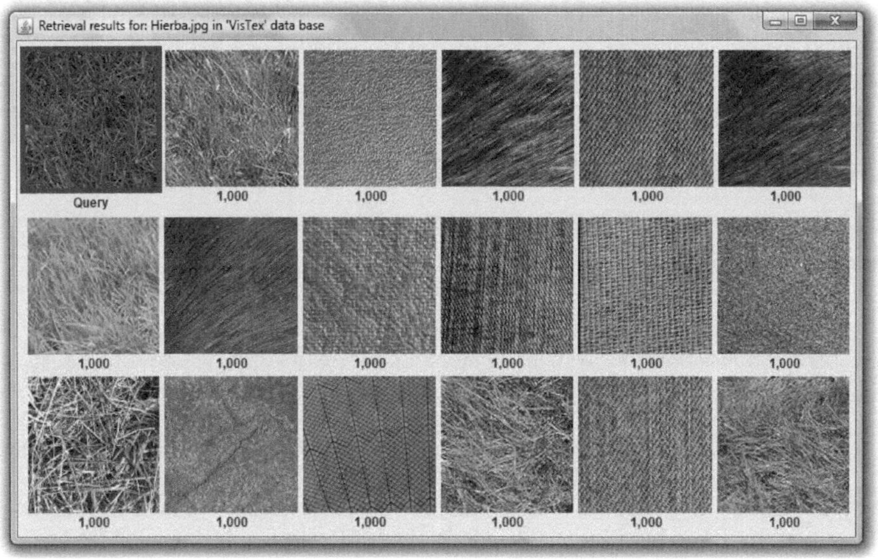

Fig. 4. Retrieval results in VisTex database using an image as query

The fuzzy partition defined for the measure *Correlation*, that has the highest capacity to discriminate fineness classes, will be used.

Figure 3 shows a retrieval example using the linguistic label *very coarse* as query. Fuzzy dominant texture descriptor $FDTD = 1/verycoarse$ has been used. The resemblance fuzzy operator described in Section 3.3 is used in this retrieval system. Figure 3 shows the retrieval results in VisTex database with resemblance degree 1. It can be noticed that the textures of all these images are perceived as very coarse.

Figure 4 shows an example where the query has been defined by an image, i.e. we are interested in getting images with a set of dominant textures similar to the one associated with the sample image (in this case, $FDTD = 1/veryfine$). The retrieval results with resemblance degree 1 are shown in Figure 4, and it can be noticed that the textures of all these images are perceived as very fine.

5 Conclusions

In this paper, a *Fuzzy Dominant Texture Descriptor* has been proposed for describing semantically an image. This fuzzy descriptor has been defined over a set of fuzzy sets modelling the "coarseness" texture property. Concretely, fuzzy partitions on the domain of coarseness measures have been proposed, where the number of linguistic labels and the parameters of the membership functions have been calculated relating representative coarseness measures with the human perception of this texture property. Given a "texture fuzzy set", we have proposed to analyze its dominance in an image and the dominance degree has been used to obtain the image texture descriptor. Fuzzy operators over these descriptors have been proposed to define conditions in image retrieval queries. The proposed framework has been applied to texture image retrieval in order to analyze its performance, obtaining satisfactory results.

References

1. Amadasun, M., King, R.: Textural features corresponding to textural properties. IEEE Transactions on Systems, Man and Cybernetics 19(5), 1264–1274 (1989)
2. Tamura, H., Mori, S., Yamawaki, T.: Textural features corresponding to visual perception. IEEE Transactions on Systems, Man and Cybernetics 8, 460–473 (1978)
3. Haralick, R.: Statistical and structural approaches to texture. Proceedings IEEE 67(5), 786–804 (1979)
4. Hanmandlu, M., Madasu, V.K., Vasikarla, S.: A fuzzy approach to texture segmentation. In: Proc. International Conference on Information Technology: Coding and Computing, vol. 1, pp. 636–642 (2004)
5. Barcelo, A., Montseny, E., Sobrevilla, P.: Fuzzy texture unit and fuzzy texture spectrum for texture characterization. Fuzzy Sets and Systems 158, 239–252 (2007)
6. Chamorro-Martinez, J., Galan-Perales, E., Soto-Hidalgo, J., Prados-Suarez, B.: Using fuzzy sets for coarseness representation in texture images. In: Proceedings IFSA 2007, pp. 783–792 (2007)

7. Kulkarni, S., Verma, B.: Fuzzy logic based texture queries for cbir. In: Proc. 5th Int. Conference on Computational Intelligence and Multimedia Applications, pp. 223–228 (2003)
8. Lin, H., Chiu, C., Yang, S.: Finding textures by textual descriptions, visual examples, and relevance feedbacks. Pattern Recognition Letters 24(14), 2255–2267 (2003)
9. Hsu, C.C., Chu, W., Taira, R.: A knowledge-based approach for retrieving images by content. IEEE Transactions on Knowledge and Data Engineering 8, 522–532 (1996)
10. Sanchez, D., Chamorro-Martinez, J., Vila, M.: Modelling subjectivity in visual perception of orientation for image retrieval. Information Processing and Management **39**(2) (2003) 251–266
11. Chamorro-Martinez, J., Galan-Perales, E., Sanchez, D., Soto-Hidalgo, J.: Modelling coarseness in texture images by means of fuzzy sets. In: International Conference on Knowledge-Based and Intelligent Information and Engineering Systems, vol. 2, pp. 355–362 (2006)
12. Canny, J.: A computational approach to edge detection. IEEE Transactions on Pattern Analysis and Machine Intelligence 8(6), 679–698 (1986)
13. Abbadeni, N., Ziou, N., Wang, D.: Autocovariance-based perceptual textural features corresponding to human visual perception. In: Proc. of 15th International Conference on Pattern Recognition, vol. 3, pp. 901–904 (2000)
14. Peleg, S., Naor, J., Hartley, R., Avnir, D.: Multiple resolution texture analysis and classification. IEEE Transactions on Pattern Analysis and Machine Intelligence (4), 518–523 (1984)
15. Weszka, J., Dyer, C., Rosenfeld, A.: A comparative study of texture measures for terrain classification. IEEE Transactions on Systems, Man and Cybernetics 6, 269–285 (1976)
16. Kim, S., Choi, K., Lee, D.: Texture classification using run difference matrix. In: Proc. of IEEE 1991 Ultrasonics Symposium, December 1991, vol. 2, pp. 1097–1100 (1991)
17. Yoshida, H., Casalino, D., Keserci, B., Coskun, A., Ozturk, O., Savranlar, A.: Wavelet-packet-based texture analysis for differentiation between benign and malignant liver tumours in ultrasound images. Physics in Medicine and Biology 48, 3735–3753 (2003)
18. Newsam, S., Kammath, C.: Retrieval using texture features in high resolution multi-spectral satellite imagery. In: Data Mining and Knowledge Discovery: Theory, Tools, and Technology VI, SPIE Defense and Security (April 2004)
19. Sun, C., Wee, W.: Neighboring gray level dependence matrix for texture classification. Computer Vision, Graphics and Image Processing 23, 341–352 (1983)
20. Galloway, M.: Texture analysis using gray level run lengths. Computer Graphics and Image Processing 4, 172–179 (1975)
21. Hochberg, Y., Tamhane, A.: Multiple Comparison Procedures. Wiley, Chichester (1987)
22. Chamorro-Martinez, J., Medina, J., Barranco, C., Galan-Perales, E., Soto-Hidalgo, J.: Retrieving images in fuzzy object-relational databases using dominant color descriptors. Fuzzy Sets and Systems 158(3), 312–324 (2007)
23. Marín, N., Medina, J., Pons, O., Sánchez, D., Vila, M.: Complex object comparison in a fuzzy context. Information and Software Technology 45(7), 431–444 (2003)

A Fuzzy Regional-Based Approach for Detecting Cerebrospinal Fluid Regions in Presence of Multiple Sclerosis Lesions

Francesc Xavier Aymerich[1,2], Eduard Montseny[2], Pilar Sobrevilla[3], and Alex Rovira[1]

[1] Unitat RM Vall Hebron (IDI). Hospital Vall Hebron, 08035 Barcelona, Spain
{xavier.aymerich,alex.rovira}@idi-cat.org
[2] ESAII Departament, Universitat Politècnica de Catalunya, 08034, Barcelona, Spain
{xavier.aymerich,eduard.montseny}@upc.edu
[3] MAII Departament, Universitat Politècnica de Catalunya, 08034, Barcelona, Spain
pilar.sobrevilla@upc.edu

Abstract. Magnetic Resonance Imaging (MRI) is an important paraclinical tool for diagnosing and following-up of Multiple Sclerosis (MS). The detection of MS lesions in MRI may require complementary information to filter false detections. Given that MS lesions cannot be located within cerebrospinal fluid (CSF), detection of this region is very helpful for our purpose. Although T1-weighted images are usually chosen to detect CSF regions, the gray level similarity between some MS lesions and CSF regions difficult this task. With the aim of discriminating CSF region within intracranial region, but considering aforementioned drawback, we propose a fuzzy-based algorithm that involves the regional analysis of the fuzzy information obtained from a previous local analysis. The proposed algorithm introduces location, shape and size constraints in CSF detection, and provides confidence degrees associated with the possibility of including MS lesion pixels.

Keywords: Magnetic resonance imaging, brain, multiple sclerosis, cerebrospinal fluid, fuzzy sets, regional analysis.

1 Introduction

Multiple sclerosis (MS) is a disease of the central nervous system (CNS) characterized by the destruction of the myelin that encases the axons. This demyelinating process results in an inhibition of neural transmission and causes the appearance of several clinical symptoms, such as motor and sensory disturbances, paralysis or visual alterations [1]. The diagnosis and follow-up of this disease involves the use magnetic resonance imaging (MRI), which is recognized as the imaging method of choice for examination of CNS disorders [2]. In the study of MS, MRI also plays a relevant role as diagnostic criteria [3][4].

Among all the different weighted images that can be obtained by MRI, T1-weighted images are the preferred choice for the analysis of the intracranial region. This is due to its high degree of anatomic information, and because they show good contrast between the parenchyma region and the cerebrospinal fluid (CSF) [5].

E. Hüllermeier, R. Kruse, and F. Hoffmann (Eds.): IPMU 2010, Part II, CCIS 81, pp. 552–561, 2010.

Although several algorithms have been proposed for the analysis of the intracranial region in T1-weighted images, most of them are focused on segmentation considering healthy volunteers [6][7] or other pathologies [7][8]. Consequently, these methods do not take into account the problems that the presence of MS lesions can introduce in the segmentation process.

In MS patients, the analysis of the intracranial region considering T1-weighted images is focused on tasks such as the measurement of brain volumes [9], the evaluation of black holes [10], and filtering false detections. Most of these tasks require the differentiation of the encephalic parenchyma in relation of its environment to achieve an accurate segmentation. In spin-echo or gradient-echo T1-weighted images, the similarity of gray-level values between CSF and MS hypointense lesions can introduce misclassifications of MS lesions as CSF instead of as parenchyma [11].

Recently we proposed an algorithm [12] for detecting CSF regions in presence of MS lesions considering a single T1-weighted scan. This algorithm carried out a local fuzzy analysis based on gray-level and texture features associated with CSF regions that allowed representing the intrinsic vagueness of the CSF features. However we observed the necessity of introducing a further analysis to take into consideration some location, shape and size constraints that the local analysis could not include. So, in this work we propose a regional fuzzy-based algorithm that allows taking into account aforementioned constraints. The use of fuzzy techniques will allow dealing with the vagueness of the CSF features, and will provide confidence degrees associated with the possibility that detections could correspond to MS lesions.

2 Definition of the CSF Regions

We considered axial slices acquired in a Siemens 1.5T Magnetom Vision MR System (Erlangen, Germany) using a T1-weighted spin-echo sequence (TR/TE/NEX/FA 667ms/14ms/1/70°) to cover a field of view of 250 mm in each 3 mm slice.

Based on the analysis of the anatomical structures of these images, in [12] we observed that CSF regions, such as the sulci or the ventricular regions, can be divided according to its width as follows:

1. Wide CSF regions (***WFR***): fluid regions whose width is equal to or greater than 5 pixels.
2. Narrow CSF regions (***NFR***): fluid regions whose width is lower than 5 pixels.

Due to MS lesions contiguous to CSF may be difficult to differentiate from CSF regions, we also differentiated inner and peripheral wide and narrow CSF regions. These regions were described based on gray level and texture features as follows:

1.1 Inner Wide Fluid Region (***IWFR***): region whose pixels show dark gray level –*dwgl*- and homogeneous texture –*ht*-.

1.2 Peripheral Wide Fluid Region (***PWFR***): region whose pixels show medium-dark gray level –*mdgl*- and micro-grainy texture –*mgt*-.

2.1 Inner Narrow Fluid Region (***INFR***): region whose pixels show dark gray level –*dngl*- and very micro-grainy texture –*vmgt*-.

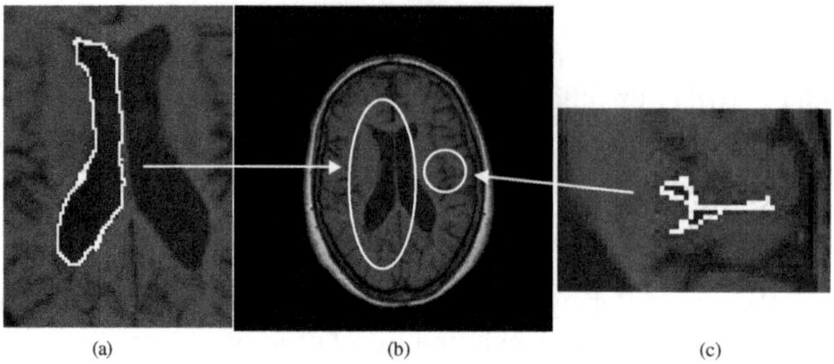

Fig. 1. Detail of the different CSF regions considered. (b) is the original image, (a) is a zoomed area that shows IWFR dark region surrounded by PWFR (solid white line), and (c) shows a zoomed area of the INFR dark regions surrounded by PNFR (solid white line).

2.2 Peripheral Narrow Fluid Region (***PNFR***): region whose pixels show medium-dark gray level *–mdgl-* and very micro-grainy texture *–vmgt-*.

Figure 1 depicts the locations of the regions above described. Image (a) shows that PWFR surround IWFR, whereas having a look at image (c) it can be appreciated that pixels in narrow regions labeled as PNFR do not require the proximity of INFR.

3 Local Analysis within Intracranial Region

Local analysis involved the study of perceptual features, gray level and texture, within intracranial region. This study required of two preprocessing tasks: Extraction of the intracranial region using a previously developed algorithm [13]; and normalization of the gray-level values to increase the gray-level uniformity among different slices.

Given the vagueness associated with the considered perceptual features, the local analysis carried out in [12] was based on the definition of the fuzzy sets F_{IWFR}, F_{PWFR}, F_{INFR} and F_{PNFR} associated with the four regions introduced at previous section, which were obtained aggregating the antecedent fuzzy sets –F_{dngl}, F_{dwgl}, F_{mdgl}, F_{vgt}, F_{gt} and F_{ht}- associated with the perceptual features *dngl*, *dwgl*, *mdgl*, *vgt*, *gt* and *ht* as follows:

$$\mu_{IWFR}(p_{ij}) = 0.7\,\mu_{dwgl}(p_{ij}) + 0.3\,\mu_{ht}(p_{ij}); \quad \mu_{PWFR}(p_{ij}) = 0.8\,\mu_{mdgl}(p_{ij}) + 0.2\,\mu_{gt}(p_{ij})\,. \tag{1}$$

$$\mu_{PWFR}(p_{ij}) = 0.8\,\mu_{mdgl}(p_{ij}) + 0.2\,\mu_{gt}(p_{ij}); \quad \mu_{PNFR}(p_{ij}) = \min(\mu_{mdgl}(p_{ij}), \mu_{vgt}(p_{ij}))\,. \tag{2}$$

Where μ_{dngl} and μ_{mdgl} were obtained through the evaluation of the normalized gray levels; μ_{dwgl} by assigning the mean gray level of pixels inside a 3x3 raster window to the central pixel; and texture membership functions, μ_{ht}, μ_{vgt}, and μ_{gt}, were obtained analyzing the differences among the gray levels of the central and surrounding pixels within square raster windows of size 3x3 for ht, 7x7 in the cases of vgt and gt.

4 Algorithm and Methodology

As we are interested in detecting CSF region within images containing MS lesions, the proposed algorithm takes into account CSF regional constraints to improve the results obtained by previously described local analysis. So, regional analysis consisted of obtaining new membership functions related to the fuzzy sets F_{IWFR}, F_{PWFR}, F_{INFR} and F_{PNFR} by considering the regional constraints of the regions given in section 2.

4.1 Obtaining Improved IWFR Membership Function

Analyzing the outcomes of local algorithm, it was appreciated that some IWFR pixels had low μ_{IWFR} values due to noise or image non-homogeneity. To avoid this problem it was considered that: "*A pixel p_{ij} is IWFR if it is surrounded by pixels of IWFR, although there can be some very small neighboring regions that are not IWFR*".

Given the characteristics of IWF regions, to implement previous property we considered four masks: A circular mask of 1.5 pixel radius, M_1; a circular ring, M_2, with radii 1.5 and 2.5 pixels; a circular ring, M_3, with inner and outer radii 1.5 and 3.2 pixels, divided into four regions ($\{M_3^k\}_{k=1}^4$), each having 7 pixels; and a circular mask, M_4, of 3.2 pixel radius divided into 4 inner regions ($\{M_{4I}^k\}_{k=1}^4$), with 2 pixels each, and 4 outer regions ($\{M_{4O}^k\}_{k=1}^4$) that match the ones of previous mask.

Then, once mask M_k is centered on pixel p_{ij}, $\mu_{M_k}(p_{ij})$ is obtained aggregating the membership values of the pixels covered by M_k using OWA operators [14] as follows:

- $W_{M_1} = (0,0,0,0,1,0,0,0)$ and $W_{M_2} = (0,0,0,0,0,0,0,0,0,1,0,0)$ are the weighting vectors applied for getting $\mu_{M_1}(p_{ij})$ and $\mu_{M_2}(p_{ij})$ respectively.
- Weighting vector $W_{M_3^k}$ =(0,0,0,0,0,1,0) are the used for obtaining the membership values $\mu_{M_3^k}(p_{ij})$ *($1 \le k \le 4$)*. Then, these values are aggregated by using W_{M_3} =(0,0,1,0) for obtaining $\mu_{M_3}(p_{ij})$.
- $W_{M_{4I}^k} = (0,1)$ and $W_{M_{4O}^k} = (0,0,0,0,0,1,0)$ are the weighting vectors applied for getting $\mu_{M_{4I}^k}(p_{ij})$ and $\mu_{M_{4O}^k}(p_{ij})$. Then, for each k *($1 \le k \le 4$)*, we get $\mu_{M_4^k}(p_{ij})$ using $W_{M_4^k} = (0,1)$, and finally these values are aggregated by using W_{M_4} = (0,0,1,0) for obtaining $\mu_{M_4}(p_{ij})$.

Then, the improved fuzzy set F_{IWFR} is given by the membership function:

$$\eta_{IWFR}(p_{ij}) = \begin{cases} \mu_{IWFR}(p_{ij}) & \text{if } \mu_{IWFR}(p_{ij}) > 0.75 \\ \max(\mu_{M_1}(p_{ij}), \mu_{M_2}(p_{ij}), \mu_{M_3}(p_{ij}), \mu_{M_4}(p_{ij})) & \text{otherwise} \end{cases} . \quad (3)$$

4.2 Obtaining Improved PWFR Membership Function

Regionally, a PWFR is a thin and closed region surrounding a IWF region. So, to improve the outcomes of local algorithm, we first considered the contiguity of PWFR and IWFR. To do it, for each pixel p_{ij} the membership values, $\mu_{IWFR}(p_{kl})$, of the pixels p_{kl} covered by a circular mask of 2 pixel radius (M_5), centered in p_{ij}, were aggregated using an OWA operator of weighting vector W_5 =(0,1,0,0,0,0,0,0,0,0,0,0,0). In this way we obtained a new membership value $\mu_{M_5}(p_{ij})$, that was used for getting the improved PWFR membership function given by equation (4).

The values of the parameters in this equation were obtained taking into account that a pixel of PWFR must show a *high enough* membership degree to F_{PWFR}, and F_{IWFR} in its neighborhood. Since a pixel of PWFR requires the presence of a minimum number of pixels belonging to IWFR in its near neighborhood showing high μ_{IWFR} value, we gave more relevance to the parameter associated with $\mu_{M_5}(p_{ij})$. Then, we obtained the values by heuristic analysis of the PWFR region at different locations. (An analogous process was applied for obtaining the parameters of next equations.)

$$\mu_{PFWR}(p_{ij}) = \begin{cases} 0.15\mu_{PWFR}(p_{ij}) + 0.85\mu_{M_5}(p_{ij}) & \text{if } \mu_{PWFR}(p_{ij}) > 0.4 \\ \mu_{PWFR}(p_{ij}) & \text{otherwise} \end{cases} . \quad (4)$$

Analyzing the results of the local algorithm, it was observed that the values of μ_{PWFR} and μ_{IWFR} were too low for some pixels adjacent to IWFR. To overcome this drawback we applied a morphological dilation -δ_{c1}- to previous values using a circular structuring element of radius 1. Then, the improved membership function defining F_{PWFR} is given by equation (5), where δ_{c1}^{IWFR} and δ_{c1}^{PWFR} are the values obtained by applying δ_{c1} on the values given by μ_{IWFR} and μ_{PWFR}, respectively.

$$\eta_{PFWR}(p_{ij}) = \begin{cases} \max(\mu_{PWFR}^{1}(p_{ij}), \delta_{c1}^{IWFR}(p_{ij})) & \text{if } \delta_{c1}^{IWFR}(p_{ij}) > 0.5 \\ \min(1-\eta_{IWFR}(p_{ij}), \delta_{c1}^{IWFR}(p_{ij})) & \text{if } \{\min(\delta_{c1}^{IWFR}(p_{ij}), \delta_{c1}^{PWFR}(p_{ij})) > 0.5 \text{ and } \mu_{PWFR}^{1}(p_{ij}) \le 0.5\} \\ \mu_{PWFR}^{1}(p_{ij}) & \text{otherwise} \end{cases} . \quad (5)$$

4.3 Obtaining Improved INFR Membership Function

Because of pixels within Inner Narrow Fluid regions can not be connected with central locations of wide fluid regions, to improve the outcomes of local algorithm we considered that: "*If pixel p_{ij} is INFR and is connected to any pixel that is PWFR or IWFR in a central location of the intracranial region, then p_{ij} is not INFR*".

To implement the central location property, given the set R_{cent} that contains the 60% of more inner pixels within the intracranial region, we defined the membership function $\mu_{WCR}(p_{ij})$ as equal to the maximum of $\eta_{IWFR}(p_{ij})$ and $\eta_{PWFR}(p_{ij})$ for all the pixels of R_{cent}, and equal to zero otherwise. For implementing the connectivity we considered the binary image I_B such that $I_B(i,j)=1$ if $\max(\mu_{WCR}(p_{ij}), \mu_{INFR}(p_{ij}))>0.5$, and $I_B(i,j)=0$ otherwise. Then, if $C(p_{ij})$ is the set of pixels 8-connected to p_{ij} in I_B, we define $mwr(p_{ij})$ as the mean value of μ_{WCR} evaluated on $C(p_{ij})$. Moreover we considered the connectivity function $CF(p_{ij},\mu_{WCR}, 0.5)$ that counts the number of pixels within a 8-neighborhood of p_{ij}, whose values in μ_{WCR} are greater than 0.5. Using these expressions the improved membership function is given by:

$$\eta_{INFR}(p_{ij}) = \begin{cases} \min(\mu_{INFR}(p_{ij}), 1 - mwr(p_{ij})) & \text{if } \{\mu_{INFR}(p_{ij}) > 0.5 \text{ and } CF(p_{ij}, \mu_{WCR}, 0.5) > 0\} \\ \mu_{INFR} & \text{otherwise} \end{cases} . \quad (6)$$

4.4 Obtaining Improved PNFR Membership Function

As a result of improvements on INFR, we need to review the peripheral narrow fluid regions in the proximity of pixels where $\eta_{INFR}(p_{ij})<\mu_{INFR}(p_{ij})$. Thus, being p_{ij} a pixel whose $\mu_{PNFR}(p_{ij})$ shows a high enough membership degree but there exist a pixel p_{kl} within a 3x3 raster window centered on p_{ij} such that $\eta_{INFR}(p_{kl})<\mu_{INFR}(p_{kl})$, then p_{ij} may not be PNFR. To do it, we applied the following expression:

$$\eta_{PNFR} = \begin{cases} \min(\eta_{INFR}(p_{ij}), \mu_{PNFR}(p_{ij})) & \textit{if} \;\; \mu_{PNFR}(p_{ij}) > 0.5 \textit{ and } CF(p_{ij}, \mu_{INFR} - \eta_{INFR}, 0) > 0 \\ \mu_{PNFR}(p_{ij}) & \textit{otherwise} \end{cases} . \tag{7}$$

4.5 Introduction of Confidence Degrees in the Detection

Since misclassification of MS lesion pixels as CSF pixels highly depends on pixel location, we considered that *exists confidence* in the detection of a pixel as CSF if it presents a low possibility of misclassification as MS lesion pixel based on its location.

The definition of *confidence* was based on the aggregation of the fuzzy sets F_{PWFR}, F_{IWFR}, F_{PNFR} and F_{INFR} considering location and size constraints of the regions involved, and the possibility that lesion is present at them. Taking into account that narrow and wide fluid regions have different regional characteristics, we defined four new fuzzy sets: F_{CNFR}, and F_{CWFR}, which provide the possibility of a pixel to be detected as CSF in areas of narrow and wide fluid regions where there exists confidence in the detection; and F_{NCNFR} and F_{NCWFR} that are associated with areas in which there exists a higher possibility of misclassification. The membership functions that define F_{CNFR} and F_{NCNFR} are given by:

$$\mu_{CNFR}(p_{ij}) = \begin{cases} \min(\max(\eta_{PNFR}(p_{ij}), \eta_{INFR}(p_{ij})), \mu_{MPNFR}(p_{ij})) \\ \qquad\qquad \textit{if} \begin{cases} CF(p_{ij}, \eta_{INFR}, 0.5) > 0 \quad \textit{and} \\ \max(\eta_{PNFR}(p_{ij}), \eta_{INFR}(p_{ij})) > 0.5 \end{cases} \\ \min(\eta_{PNFR}(p_{ij}), \eta_{INFR}(p_{ij}), \eta_{MPNFR}(p_{ij})) \qquad \textit{otherwise} \end{cases} . \tag{8}$$

$$\mu_{NCNFR} = \min(\max(\eta_{PNFR}, \eta_{INFR}), 1 - \mu_{CNFR}) . \tag{9}$$

where $\mu_{MPNFR}(p_{ij})=\max(\mu_{PNFR}(p_{ij}), \mu_{INFR}(p_{ij}))$ if p_{ij} belongs to a region, R, whose pixels satisfy that $\max(\mu_{PNFR}(\bullet), \mu_{INFR}(\bullet))>0.5$, and at least a 60% of the pixels within R are in the 20% most outer pixels of the intracranial region; and, $\mu_{MPNFR}(p_{ij})=0$ otherwise.

As previously said, unlike narrow fluid regions, in wide CSF regions the higher possibility of correct detection is based on size constraints. Thus, if I_{WFR} is the binary image such that $I_{WFR}(i,j)=1$ if p_{ij} belongs to a region with a minimum width of 5 pixels whose pixels satisfy that $\max(\eta_{PWFR}(p_{ij}), \eta_{PNFR}(p_{ij}))>0.5$, and $I_B(i,j)=0$ otherwise. Then, the membership functions defining F_{CWFR} and F_{NCWFR} are given by:

$$\mu_{CWFR} = \begin{cases} 0.4 & \textit{if} \;\; \eta_{IWFR}(i,j) > 0.5 \;\; \textit{and} \;\; I_{WFR}(i,j) = 0 \\ \eta_{IWFR} & \textit{otherwise} \end{cases} . \tag{10}$$

$$\mu_{NCWFR} = \begin{cases} 0.2 & \text{if } \eta_{IWFR}(i,j) > 0.5 \text{ and } I_{WFR}(i,j) = 1 \\ \eta_{PWFR} & \text{otherwise} \end{cases} . \tag{11}$$

Finally, the degrees of confidence and no confidence in the detection were given by aggregation of previous membership functions according following expressions:

$$\mu_{CFR}(p_{ij}) = \begin{cases} 0.5 & \text{if} \begin{cases} (\max(\mu_{CWFR}(p_{ij}), \mu_{CNFR}(p_{ij})) > 0.5) \text{ and} \\ (\max(\mu_{NCWFR}(p_{ij}), \mu_{NCNFR}(p_{ij})) > 0.5) \text{ and} \\ (\max(\mu_{CWFR}(p_{ij}), \mu_{CNFR}(p_{ij})) \leq \\ \quad \max(\mu_{NCWFR}(p_{ij}), \mu_{NCNFR}(p_{ij})) - 0.1) \end{cases} \\ \max(\mu_{CWFR}(p_{ij}), \mu_{CNFR}(p_{ij})) & \text{otherwise} \end{cases} . \tag{12}$$

$$\mu_{NCFR}(p_{ij}) = \begin{cases} 0.5 & \text{if} \begin{cases} (\max(\mu_{CWFR}(p_{ij}), \mu_{CNFR}(p_{ij})) > 0.5) \text{ and} \\ (\max(\mu_{NCWFR}(p_{ij}), \mu_{NCNFR}(p_{ij})) > 0{,}5) \text{ and} \\ (\max(\mu_{CWFR}(p_{ij}), \mu_{CNFR}(p_{ij})) > \\ \quad \max(\mu_{NCWFR}(p_{ij}), \mu_{NCNFR}(p_{ij})) - 0.1) \end{cases} \\ \max(\mu_{NCWFR}(p_{ij}), \mu_{NCNFR}(p_{ij})) & \text{otherwise} \end{cases} . \tag{13}$$

4.6 Defuzzification and Quality Measures

The defuzzification process focused on obtaining the CSF region applying α-cuts to CFR and NCFR fuzzy sets. Then, the crisp representation of the CSF was obtained adding the binary results obtained for CFR and NCFR.

To determine the appropriate α-cuts we defined a detection quality factor, Q_{DFR}, obtained evaluating the efficiency in the detection of CSF regions, E_{FRP}, and considering two reliability factors: R_{FRP}, related to the number of CSF pixels detected; related to the number of MS lesion pixels. So, if N_{DFRP} is the number of detected CSF pixels, N_{NDFRP} is the number of non-detected CSF pixels, N_{FDFR} is the number of false detections detecting CSF pixels, N_{DLP} is the number of detected MS lesion pixels, and N_{LP} is the total number of MS lesion pixels, the efficiency in the detection and the reliability factors were defined by equations (14) and (15). Then the quality factor, Q_{DFR}, is given by equation (16).

$$E_{FRP} = 2\frac{N_{DFRP}}{2N_{DFRP} + N_{NDFRP} + N_{FDFR}} . \tag{14}$$

$$R_{FRP} = \frac{N_{DFRP}}{N_{DFRP} + N_{NDFRP}} ; \quad R_{NDLP} = 1 - \frac{N_{DLP}}{N_{LP}} . \tag{15}$$

$$Q_{DFR} = 0.6\left(\frac{E_{FRP} + R_{FRP}}{2}\right) + 0.4 R_{NDLP} . \tag{16}$$

CSF and MS lesion masks were manually obtained using Dispimage software [15].

The α-cuts had to provide a commitment between correct detection of fluid regions and avoiding misclassification. So for selecting the more suitable we studied the results obtained for α-cuts in the interval [0.55, 0.95] for the training images.

5 Results

The proposed algorithm was evaluated considering a training set compounded by 6 images acquired from two patients, and a test set constituted by 138 images acquired from three patients. All images were acquired using the T1-weighted sequence described in section 2.

From the study of the results obtained applying nine α-cuts in the interval [0.55, 0.95] to the values of μ_{CFR} for the training images, we selected the value 0.55 because it provided the higher CFR quality detection outputs, without misclassifications as MS lesions. Then, we considered the union of the binary image obtained applying this á-cut to μ_{CFR} and the binary images obtained applying the nine α-cuts to μ_{NCFR}. Next, we obtained their quality indexes, Q_{DFR}, and analyzing the results we observed that the optimal value was gotten when the α-cut for μ_{NCFR} was 0.55.

Having a look at columns 3 and 4 of Table 1, it can be appreciated that the indexes related to detection of CSF pixels, E_{FRP} and R_{FRP}, obtained considering the union improved the results obtained for the CFR, whereas we detected a reduced number of MS pixels. Moreover, looking at column 6 of Table 1 can be appreciated that global quality Q_{DFR} achieved an improvement around of 10% when NCFR is considered.

It must be pointed out that the introduction of regional analysis represented a significant improvement of the results over local analysis in the CFR, because E_{FRP} and R_{FRP} quality indexes increased around of 10-16%. In the case of R_{NDLP} and Q_{DFR} quality indexes, their values kept in the same range with absence of misclassifications and a slight better Q_{DFR} in the regional analysis.

The results obtained for the test images are shown at Table 2. As can be appreciated, the values of all quality factors obtained for the regional analysis (rows 2 and 4) improved the obtained with local analysis (rows 3 and 5). Differences in relation to

Table 1. Quality results obtained considering the training set images after regional analysis

Region	α-cut	E_{FRP}	R_{FRP}	R_{NDLP}	Q_{DFR}
CFR	0.55	0.673	0.580	1.0	0.776
CFR∪NCFR	0.55	0.724	0.832	0.951	0.847

Table 2. Summary of quality results obtained for CFR∪NCFR considering the test set images after regional analysis

Region		E_{FRP}	R_{FRP}	R_{NDLP}	Q_{DFR}
CFR	Regional	0.557	0.430	0.999	0.696
	Local	0.469	0.338	0.999	0.642
CFR∪NCFR	Regional	0.726	0.708	0.925	0.800
	Local	0.696	0.651	0.886	0.758

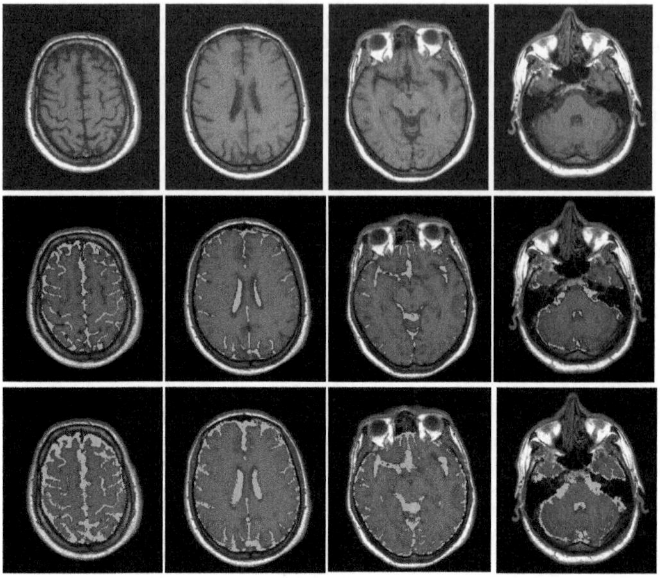

Fig. 2. Different levels of detection corresponding to different anatomical locations. (a) Original images. (b) Detection mask corresponding to CFR overlaid on the original images. (c) Detection mask corresponding to CFR∪NCFR overlaid on the original images.

the analysis of the images included in the training set showed, mainly, a reduction of CSF detection for CFR, whereas the capability of avoiding misclassifications was very close to the obtained for the training set. These results can be appreciated in Fig. 3 that shows some examples of results in the analysis of the test set corresponding to different anatomical locations.

To conclude, in this paper we have presented an algorithm which allows discriminating cerebrospinal fluid regions inside the intracranial region, providing confidence degrees that match with the possibility of including pixels associated to MS lesions. This work has focused on the introduction of a regional analysis in order to improve detections levels obtained after local analysis based on perceptual features. Thereby, the proposed algorithm has considered location, shape and size constraints, and has divided the results in function of the level of confidence in avoiding misclassification of MS pixels as CSF detections. The results show good CSF detection levels, and the values of the quality factors point out that CFR is free or practically free of misclassifications, whereas NCFR helped to improve CSF detection level without increasing significantly the number of misclassifications. The introduction of regional analysis has allowed improving both CSF detection levels and confidence in avoiding misclassifications in relation to previous local analysis. It must be also emphasized the improvements in values of R_{NDLP} quality factors and in the quality levels obtained considering the test set. Finally, the results obtained suggest that this algorithm, particularly images resulting of detecting CFR regions, can be applied to filter false detections of MS lesions due to misclassifications of these lesions as CSF. An improvement of the work here presented may be the introduction of an alternative approach to obtain the values of the parameters based on an optimization procedure; which would also help to carry out a robustness study.

Acknowledgments. This work has been partially supported by the Spanish CICYT project TIN2007-68063.

References

1. Edwards, M.K., Bonnin, J.M.: White Matter Diseases in Magnetic Resonance Imaging of the Brain and Spine. In: Atlas, S.W. (ed.) Magnetic Resonance Imaging of the Brain and Spine, pp. 467–477. Raven Press (1991)
2. Edelman, R.R., Hesselink, J.R.: Clinical Magnetic Resonance Imaging. W.B. Saunders Company, Philadelphia (1990)
3. McDonald, W.I., Compston, A., Edan, G., Goodkin, D., Hartung, H.P., Lublin, F.D., McFarland, H.F., Paty, D.W., Polman, C.H., Reingold, S.C., Sandberg-Wollheim, M., Sibley, W., Thompson, A., van den Noort, S., Weinshenker, B.Y., Wolinsky, J.S.: Recommended Diagnostic Criteria for Multiple Sclerosis: Guidelines from the International Panel on the Diagnosis of Multiple Sclerosis. Ann. Neurol. 50(1), 121–127 (2001)
4. Tintoré, M., Rovira, A., Río, J., Nos, C., Grivé, E., Sastre-Garriga, J., Pericot, I., Sánchez, E., Comabella, M., Montalban, X.: New Diagnostic Criteria for Multiple Sclerosis: Application in first Demyelinating Episode. Neurology 60(1), 27–30 (2003)
5. Miller, D.H., Barkhof, F., Frank, J.A., Parker, G.J.M., Thompson, A.J.: Measurement of Atrophy in Multiple Sclerosis: Pathological Basis, Methodological Aspects and Clinical Relevance. Brain 125(8), 1676–1695 (2002)
6. Boesen, K., Rehm, K., Schaper, K., Stoltzner, S., Woods, R., Lüders, E., Rottenberg, D.: Quantitative comparison of four brain extraction algorithms. Neuroimage 22(3), 1255–1261 (2004)
7. Lemieux, L., Hammers, A., Mackinnon, T., Liu, R.S.N.: Automatic Segmentation of the Brain and Intracranial Cerebrospinal Fluid in T1-weighted Volume MRI Scans of the Head, and its Application to Serial Cerebral and Intracranial Volumetry. Magn. Reson. Med. 49(5), 872–884 (2003)
8. Anbeek, P., Vincken, K.L., van Bochove, G.S., van Osch, M.J.P., van der Grond, J.: Probabilistic Segmentation of Brain tissue in MR imaging. Neuroimage 27(4), 795–804 (2005)
9. Sastre-Garriga, J., Ingle, G.T., Chard, D.T., Cercignani, M., Ramió-Torrentà, L., Miller, D.H., Thompson, A.J.: Grey and White Matter Volume Changes in Early Primary Progressive Multiple Sclerosis: a Longitudinal Study. Brain 128(6), 1454–1460 (2005)
10. Datta, S., Sajja, B.R., He, R., Wolinsky, J.S., Gupta, R.K., Narayana, P.A.: Segmentation and Quantification of Black Holes in Multiple Sclerosis. Neuroimage 29(2), 467–474 (2006)
11. Sharma, J., Sanfilipo, M.P., Benedict, R.H.B., Weinstock-Guttman, B., Munschauer, F.E., Bakshi, R.: Whole-Brain Atrophy in Multiple Sclerosis Measured by Automated versus Semiautomated MR Imaging Segmentation. AJNR 25(6), 985–996 (2004)
12. Aymerich, F.X., Montseny, E., Sobrevilla, P., Rovira, A.: FLCSFD- a Fuzzy Local-Based Approach for Detecting Cerebrospinal Fluid Regions in Presence of MS Lesions. In: ICME International Conference on Complex Medical Engineering, pp. 1–6 (2009)
13. Aymerich, F.X., Sobrevilla, P., Montseny, E., Rovira, A., Gili, J.: Automatic Segmentation of Intracraneal Region from Magnetic Resonance Images. Magnetic Resonance in Physics, Biology and Medicine 11(S1), 141 (2000)
14. Yager, R.R.: Families of OWA Operators. Fuzzy Sets and Systems 59, 125–148 (1993)
15. Plummer, D.L.: Dispimage: a Display and Analysis Tool for Medical Images. Riv. Neuroradiol. 19, 1715–1720 (1992)

Probabilistic Scene Models for Image Interpretation

Alexander Bauer

Fraunhofer Institute for Optronics, System Technologies
and Image Exploitation IOSB, Fraunhoferstr. 1, 76131 Karlsruhe, Germany
alexander.bauer@iosb.fraunhofer.de

Abstract. Image interpretation describes the process of deriving a semantic scene description from an image, based on object observations and extensive prior knowledge about possible scene descriptions and their structure. In this paper, a method for modeling this prior knowledge using probabilistic scene models is presented. In conjunction with Bayesian Inference, the model enables an image interpretation system to classify the scene, to infer possibly undetected objects as well as to classify single objects taking into account the full context of the scene.

Keywords: Image Interpretation, Image Understanding, High-level vision, Generative Models, Bayesian Inference, Relaxation Labeling, Importance Sampling.

1 Introduction

Many applications of computer vision aim at the automated interpretation of images as a basis for decision making and planning in order to perform a specific task. Image interpretation summarizes the process of creating a semantic scene description of a real-world scene from single or multiple images. A scene represents a spatio-temporal section of the real-word in terms of its physical objects, their properties and relations. The corresponding semantic scene description contains all task-relevant objects, properties and relations described at a task-relevant abstraction level.

In many applications of image interpretation, it is not sufficient to detect and classify objects based on its appearance alone (e. g. the existence of a building in the scene). Rather, higher-level semantic descriptions (e. g. the function of the building being a workshop) have to be inferred based on the spatial configuration of multiple objects and prior knowledge about possible scenes. Prior knowledge can also be useful to improve results of purely appearance-based object recognition methods by ruling out unlikely detections and focus the attention on likely occurring, but undetected objects.

The image interpretation problem has drawn scientific interest since the 80s in the fields of artificial intelligence and computer vision. The main challenges have been identified early [1], yet their solution has not been ultimately determined:

- **Knowledge representation** – How to model prior knowledge about possible scene descriptions?
- **Hypotheses Matching** – How to match possible scene descriptions against incomplete and erroneous detections of objects in the image?

E. Hüllermeier, R. Kruse, and F. Hoffmann (Eds.): IPMU 2010, Part II, CCIS 81, pp. 562–571, 2010.

- **Inference** – How to derive inferences from prior knowledge in order to improve and complete the scene description?

Probabilistic models, becoming more and more popular in computer vision, provide an intuitive way to model uncertainty, and the Bayes' Theorem provides a consistent framework for inference based on incomplete evidence. These properties of probabilistic methods have motivated the development of probabilistic scene models for image interpretation, meant to model prior knowledge about possible scenes using probability theory. The presented model design is targeted to the assisted interpretation of infrastructure facilities from aerial imagery [2], but it potentially generalizes to other image interpretation applications as well. This paper describes the scene model structure and how the main challenges of knowledge representation, hypotheses matching and inference can be tackled in a probabilistic framework for image interpretation. For better understanding, the contribution is illustrated on the application for the interpretation of airfield scenes.

2 Related Work

Early approaches to image interpretation aiming for the description of complex scenes were inspired by the advances in artificial intelligence in the 80s in the field of rule-based inference [1],[4],[5],[6]

From the 90s until today, probabilistic approaches and Bayesian inference have drawn attention from cognitive psychology as well as from the computer vision community. Since then, several probabilistic approaches for high level image interpretation have been proposed, of which only a few can be mentioned here. Rimey and Brown developed a system to control selective attention using Bayesian Networks and Decision Theory [7]. Lueders used Bayesian Network Fragments to model taxonomy and partonomy relations between scene objects to compute the most probable scene interpretation based on perceptive information [8]. A stochastic graph grammar in conjunction with a Markov Random Field has been used by Lin et al. to recognize objects which are composed of several parts with varying alignment and occurrence [9]. In [10] it was also applied to aerial imagery.

Following this current, the presented approach contributes to the efficient application of probability theory for the interpretation of images depicting complex scenes, such as they appear in remote sensing and aerial reconnaissance. In contrast to previous approaches, it is focused on the classification of objects at the functional level (see Section 1), rather than on detection and classification on the appearance level. It is able to improve and interpret results acquired from low-level methods such as automated object recognition algorithms and can be used to control their execution.

3 Bayesian Inference Applied to Image Interpretation

The Bayes' theorem provides a sound formalism to infer the distribution of a random variable given evidence in terms of uncertain observations and measurements. Everything that is required for Bayesian inference is to model the prior distribution of the random variable and to define a conditional probability of the observations given each realization of the variable. Applied to the image interpretation problem, the unknown random variable S represents the correct semantic description of the scene. Evidence collected from the image as a set of object observations is described by the random variable O. According to the Bayes' theorem, the updated posterior probability distribution can be calculated using

$$P(S = s_i \mid O = o_k) \propto P(O = o_k \mid S = s_i) \cdot P(S = s_i) \,. \tag{1}$$

For brevity, the term for the normalization of the distribution to 1 is omitted in (1). The prior distribution $P(S = s_i)$ is defined by a probabilistic scene model, modeling possible scene realizations and their typical object configurations in terms of their probability of occurrence. The conditional probability $P(O = o_k \mid S = s_i)$ models the uncertainty in the recognition of objects. Using the posterior distribution, several useful inferences can be calculated to direct the iterative development of the scene description, which will be explained in Section 6.

4 Representing Possible Scene Descriptions

A scene description describes a real-world scene in terms of its task-relevant physical objects. In real-world scenes, functionally related objects are often arranged in spatial relation to each other to form an object composition which is described as a new object.

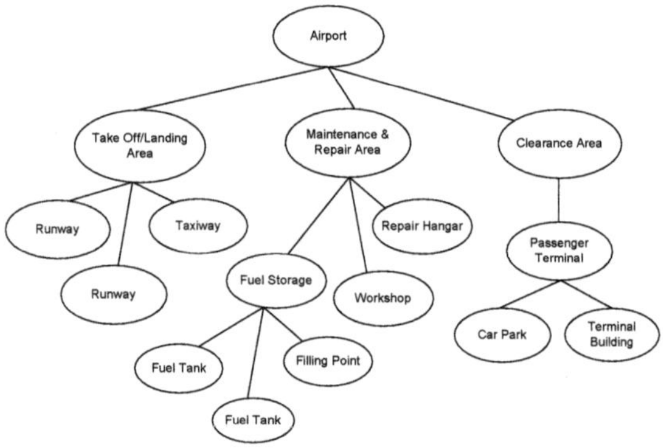

Fig. 1. Example of an interpretation tree for an airfield scene

For example on an airfield, buildings which are dedicated to the maintenance and repair of aircraft are composed to a "Maintenance & Repair Area". This leads to a natural description of a scene as a tree of objects, in which the edges define functional composition, as illustrated using an example in Fig. 1.

The resulting *interpretation tree* $s=(\Omega,F)$ describes all objects of a possible scene description in terms of *object nodes* $\omega_i \in \Omega$. The set of edges F represents their functional composition. Object nodes are associated with a particular *object class* Φ, written as $\omega_i \sim \Phi$. Object classes are concepts such as "Airport", "Runway", etc.

The number of interpretation trees possible to occur can be very large, due to the high variability of real-world scenes. A lot of variations results from different numbers of occurrence of objects of a single object class. Variations in the structure of real-world scenes are also likely, even if the objects in the scene serve a similar function. For example in the case of airfields, regional variations and the advances in design of airfields over decades resulted in very different object configurations. However, object class occurrence and scene structure are important characteristics of a scene and therefore have to be taken account for in the model. To tackle the complexity problem, an approximation method is proposed for inference in Section 6.

5 Modeling Prior Knowledge in a Probabilistic Scene Model

As mentioned in Section 3, the probabilistic scene model must provide a prior probability for each possible interpretation tree, i.e. a distribution of the random variable S, which represents the correct scene description of the currently investigated image. A second requirement on the scene model is that the acquisition of the model parameters must be tractable and comprehensible. As sufficient training data is hardly available for a complex domain, in most cases it will be necessary to consult a human expert to establish a comprehensive and useful scene model. Therefore a modeling syntax is chosen, which is inspired by the verbal description of object classes by a human expert. Nevertheless, learning can be implemented by estimating the conditional probabilities of the model from training data.

The scene model is defined as the set of interrelated *object class models* $M(\Phi)$, from which all possible interpretation trees can be generated. As an illustrative example, Fig. 2 shows some of the object class models and their relations necessary to model possible interpretation trees of an airfield. Three types of object class models are defined:

- *Composition models* (C-Models $M_C(\Phi)$) describe an object class in terms of a composition of other objects (e. g. the 'Airport' model in Fig. 2). Such object classes occur at the upper levels of the interpretation tree (such as "Runway Area", see Fig. 1). To represent the probability of all possible compositions, the distributions of the number of occurrences of each subordinate object model is defined in the compositional model. Assuming independence on the occurrence of different object models, it is sufficient to define the distribution for each single object class and to establish the joint probability distribution by multiplication. In the example shown in Fig. 2 the distributions are chosen to be uniform inside a reasonable interval, which simplifies the acquisition process in cooperation with an expert by using statements such as: "Airports have at least one runway, up to a maximum of 5 runways". However, more informative distributions can be used to incorporate more detailed prior knowledge, also taking into account dependencies between the occurrence probabilities of different object classes.
- *Taxonomy models* (T-Models $M_T(\Phi)$) define abstract object class models, which summarize different realizations of an abstract object class. For example the object class "Airfield" is further specified by the discrimination

of disjunctive subtypes of that object class, as depicted in Fig. 2. For each subtype, a probability is defined which represents the conditional probability $P(\omega_i \sim \Phi_j | \omega_i \sim \Phi_T)$ of an object node ω_i to be associated with the subtype discriminations Φ_j, given it is associated with the abstract object class Φ_T.

- *Atomic models* (A-Models $M_A(\Phi)$) define object classes which can be neither further discriminated by more specific object classes nor divided into sub-parts. In Fig. 2, all objects which are not described by a box are represented by an A-model.

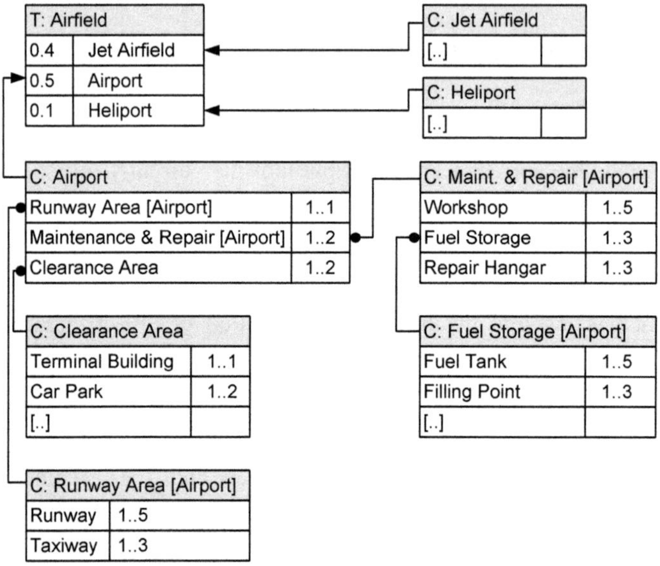

Fig. 2. Section of a scene model for airfield scenes. Abbreviations: T: Taxonomy Model, C: Composition Model. Figures in T-Models represent probabilities of different realizations, intervals in C-Models stand for uniform distributions of the number of occurrences for each subordinate object model.

The complete *scene model* $M(S)$ is defined by the tupel $M(S)=<M_C, M_T, M_A, \Phi_0>$ in terms of the sets of the three kinds of object class models and the root object class model Φ_0. From the scene model, all possible interpretation trees $s \in S$ and their corresponding prior probability $P(S = s)$ can be generated using the following algorithm:

1. Create object node ω_0 as the root node of the interpretation tree s and associate it with object class Φ_0. Initialize $P(S = s) := 1$.
2. For every object node ω_i associated with a T-Model, choose a subtype object class Φ_j and update $P(S = s)$ with $P(S = s) \cdot P(\omega_i \sim \Phi_j | \omega_i \sim \Phi_T)$ as defined in the T-model.
3. For every object node ω_i newly associated with a C-Model, choose a composition according to the C-Model description and create the object nodes of the composition. Update the scene prior probability $P(S = s)$ by multiplying with the composition probability according to the C-Model.

4. Repeat step 2 and 3 until no T-Model remains and all C-models have been treated in step 3.

Using the algorithm, for any given interpretation tree s, the corresponding prior probability $P(S = s)$ can be determined by choosing the respective subtype classes in step 2 and the respective compositions in step 3.

If the decisions in step 2 and 3 on the T-model subtype or the C-model composition are chosen randomly according to the corresponding discrete probability distribution defined in the models, the scene model draws samples from the prior probability $P(S = s)$. This fact is exploited for Monte-Carlo approximation in Section 6.

6 Matching Object Observations to Interpretation Trees

Object observations in the image can be either made by a human interpreter or by a computer vision system in a bottom-up process. In order to apply prior knowledge defined in the scene model of Section 4, it is necessary to determine a likelihood probability $P(O \mid S)$ for a set of observations given a candidate interpretation tree. To expressed the probability of mismatch in terms of the number of unmatched object observations n and the number of unmatched object nodes p, a heuristic likelihood function is chosen:

$$P(O = o_k \mid S = s_i) \propto \exp(-[\lambda \cdot n(o_k, s_i) + q(o_k, s_i)]) . \quad (2)$$

The parameter λ controls the balance of influence of both counts on the inference result. To determine the counts however, a matching between object observations and the object nodes of the interpretation tree has to be established. This has to be done based on the features of the observed objects and their spatial relations.

To approach this problem, the object nodes of the interpretation tree are rearranged as nodes of a graph with the connecting arcs representing their expected spatial relations. Expected features are represented as node attributes. Accordingly, object observations, their observed features and spatial relations are represented as a graph as well, based on a probabilistic object-oriented representation as described in [11]. This formulation relates the matching problem to general graph matching problems, which have been extensively studied in literature [12]. In many cases, good and efficient approximations to the NP-complete matching problem have been achieved using relaxation labeling [13]. It is an iterative graph matching method using heuristic compatibility coefficients. One of the most appealing reformulations of relaxation labeling has been presented by Christmas, Kittler and Petrou [14] by deriving compatibility coefficients and update function in accordance with probability theory. However, their formulation is only suitable for continuous spatial relations such as distance, but not on discrete locative expressions such as nearness and adjacency. If a formally derived probabilistic relaxation scheme can be found, it might be possible to derive a formal definition of the likelihood probability, for example based on graph-edit distance [15].

In the context of this paper, relaxation labeling using heuristic compatibility coefficients and the likelihood function (2) has been used.

7 Inference

If the posterior distribution $P(S = s_i \mid O = o_k)$ is established, manifold inferences can be calculated. A specific class of inferences can be expressed as the expectation of an indicator function of the scene description variable S

$$E_{S|o_k}\{I_\Psi(S)\}. \tag{3}$$

The indicator function takes the value 1 if a specific condition on the interpretation tree, the object observations or the corresponding matching holds true; otherwise it is defined to be zero.

Selecting the next object class to search for in the image is a task, which can be supported by defining the indicator function $I_\Phi(S)$ to represent the condition that an unmatched object instance of object class Φ exists in the interpretation tree S. The expectation of that indicator function is equal to the probability of occurrence of the object class. The occurrence probabilities can be used to guide the image interpretation process for efficient establishment of a complete scene description.

The same indicator function is useful to determine the distribution of the root node of the interpretation tree, representing the overall classification of the scene, for example in the case of airfields, if it is a military airfield or a civil airport.

To classify an observed object based on its features and taking into account the occurrence of other object observations and their spatial relations, the indicator function can be designed to resemble the condition that the object of interest has been matched to an interpretation node associated to a specific object class model Φ. This way, the probability for the object to be interpreted as being of object class Φ is determined.

If the number of possible interpretation trees is large, such as in a comprehensive model of airfield scenes, calculation of expectations becomes intractable. However, using Importance Sampling, a Monte-Carlo estimation method, approximations can be calculated [16]. As the generation algorithm described in Section 2 is able to generate samples from S according to the prior distribution, the prior distribution is used as proposal distribution for Importance Sampling. Respectively, the estimator for the expectation of a function $g(S)$ given the posterior distribution is defined as

$$\tilde{E}_{S|O}\{g(S)\} = \frac{\sum_{i=1}^{n} \omega(s_i) g(s_i)}{\sum_{i=1}^{n} \omega(s_i)} \tag{4}$$

using the weights

$$\omega(s_i) = \frac{P(S = s_i \mid O = o_k)}{P(S = s_i)} = P(O = o_k \mid S = s_i). \tag{5}$$

The estimator (4) is independent of the normalization of the weights, so the definition of the observation likelihood probability (2) does not need to be normalized.

By previous experiments, it was found that a sample size of 10.000 is sufficient to estimate the expectations at a reasonable accuracy [3]. The Java™ implementation of the estimator is able to generate and process 10.000 samples per second on an Intel™ 2.1 GHz Core 2 CPU. As the sampling distribution is independent on the observations, samples can be reused for different sets of observations and do not have to be redrawn for each recursion step. Therefore, after the initial generation of samples, the calculation time is well below one second, which is acceptable for the application in decision-support systems.

8 Experiments

To study the feasibility of the presented method, scene models for the interpretation of airfield and harbor image have been developed in cooperation with image interpretation experts, each involving about 50 different objects classes. As ground-truth for the evaluation, airfields and harbors scenes labeled from aerial images. As a first step, to compare the benefit of different modeling aspects (unary features, global scene context, local object context and spatial relations), the respective classification accuracy has been determined for objects in 10 different airfield scenes. The unary feature in this experiment was the appearance-level class (building, paved area or antenna). In order to incorporate spatial relations, the relaxation labeling of Rosenfeld et al. [13] was used and the compatibility coefficients were chosen to be 1 in the case that the object's distance was below a fixed threshold, zero in all other cases. Figure 3 displays the results, which show that the classification accuracy is significantly improved when taking into account prior knowledge about different scene realizations and using additional binary features such as spatial nearness for the association of objects nodes in the model and objects observations in the image.

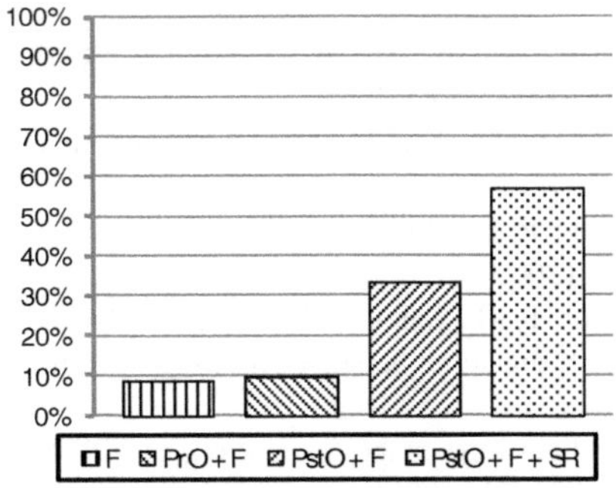

Fig. 3. Experimental result for the classification accuracy of single objects in airfield scenes using different levels of prior knowledge modeling. F: unary features and uniform prior occurrence probability, PrO+F: prior occurrence probability from scene model and unary features, PstO+F: posterior occurrence probability taking into account other objects in the scene and unary features, PstO+F+SR: Like PstO+F but using spatial relations (nearness) to objects in the local neighborhood in the relaxation scheme.

9 Conclusion and Outlook

For automated image interpretation, the problems of knowledge representation, hypotheses matching and inference have to be addressed, especially if higher-level semantic descriptions have to be extracted from the image. In this paper, probabilistic scene models are suggested to model prior knowledge about possible scene descriptions and their application in image interpretation using Bayesian inference is explained. Probabilistic scene models are defined in a human understandable way, allowing a human expert to determine the required parameters using compositional and taxonomic models even in the absence of training data. To match possible scene descriptions against object observations, a heuristic likelihood function is proposed and the use of relaxation labeling is suggested to establish a correspondence between object observations and a candidate scene description. Three exemplary inferences, which can be derived from the posterior distribution of scene descriptions given incomplete object observations, are proposed and their approximate calculation in the context of high-dimensional scene models using Importance Sampling is suggested. The evaluation of the object classification accuracy shows that using the proposed method, object classification significantly benefits from the consideration of prior knowledge about possible scene realizations and spatial relations between objects.

Future work will address the modeling of spatial relations in more detail for the application in aerial image interpretation of complex scenes such as airfields, harbors and industrial installations. Relaxation labeling methods will be evaluated and the integration of discrete locative expressions in the context of probabilistic relaxation will be investigated, based on a representative set of ground-truth labeled scenes. The benefit of an interactive decision-support system for image interpretation [2] based on the presented probabilistic scene models will then be evaluated on aerial images of airfields and harbors.

References

1. Matsuyama, T., Hwang, V.: SIGMA: A Knowledge-Based Aerial Image Understanding System. Plenum Press, New York (1990)
2. Bauer, A.: Assisted Interpretation of Infrastructure Facilities from Aerial Imagery. Proc. of SPIE 7481, 748105 (2009)
3. Bauer, A.: Probabilistic Reasoning on Object Occurrence in Complex Scenes. Proc. of SPIE, 7477A, 74770A (2009)
4. Russ, T.A., MacGregor, R.M., Salemi, B., Price, K., Nevatia, R.: VEIL: Combining Semantic Knowledge with Image Understanding. In: ARPA Image Understanding Workshop (1996)
5. Dillon, C., Caelli, T.: Learning image annotation: the CITE system. Journal of Computer Vision Research 1(2), 90–121 (1998)
6. Hanson, A., Marengoni, M., Schultz, H., Stolle, F., Riseman, E., Jaynes, C.: Ascender II: a framework for reconstruction of scenes from aerial images. In: Workshop Ascona 2001: Automatic Extraction of Man-Made Objects from Aerial and Space Images (III), pp. 25–34 (2001)
7. Rimey, R.D., Brown, C.M.: Control of Selective Perception Using Bayes Nets and Decision Theory. International Journal of Computer Vision 17, 109–173 (1994)

8. Lueders, P.: Scene Interpretation Using Bayesian Network Fragments. Lecture Notes in Economics and Mathematical Systems, vol. 581, pp. 119–130 (2006)
9. Lin, L., Wu, T., Porway, J., Xu, Z.: A Stochastic Graph Grammar for Compositional Object Representation and Recognition. Pattern Recognition 42(7), 1297–1307 (2009)
10. Porway, J., Wang, K., Yao, B., Zhu, S.C.: A Hierarchical and Contextual Model for Aerial Image Understanding. In: IEEE Conference on Computer Vision and Pattern Recognition, CVPR, pp. 1–8 (2008)
11. Bauer, A., Emter, T., Vagts, H., Beyerer, J.: Object-Oriented World Model for Surveillance Applications. In: Future security: 4th Security Research Conference Karlsruhe, Congress Center Karlsruhe, Germany, September 2009, pp. 339–345. Fraunhofer IRB Verl. Stuttgart (2009)
12. Conte, D., Foggia, P., Sansone, C., Vento, M.: Thirty Years of Graph Matching in Pattern Recognition. International Journal of Pattern Recognition and Artificial Intelligence 18(3), 265–298 (2004)
13. Rosenfeld, A., Hummel, R.A., Zucker, S.W.: Scene Labeling by Relaxation Operations. Systems, Man and Cybernetics 6(6), 420–433 (1976)
14. Christmas, W.J., Kittler, J., Petrou, M.: Structural Matching in Computer Vision using Probabilistic Relaxation. IEEE Transactions on Pattern Analysis and Machine Intelligence 17(8), 749–764 (1995)
15. Myers, R., Wilson, R.C., Hancock, E.R.: Bayesian Graph Edit Distance. IEEE Transactions on Pattern Analysis and Machine Intelligence 22(6), 628–635 (2000)
16. Koch, K.R.: Introduction to Bayesian Statistics, 2nd edn. Springer, Heidelberg (2007)

Motion Segmentation Algorithm for Dynamic Scenes over H.264 Video

Cayetano J. Solana-Cipres[1], Luis Rodriguez-Benitez[1], Juan Moreno-Garcia[2], and Luis Jimenez-Linares[1]

[1] ORETO Research Group, E.S. Informatica, University of Castilla-La Mancha, Paseo de la Universidad 4, 13071 Ciudad Real, Spain
[2] E.U.I.T. Industrial, University of Castilla-La Mancha, Avda. Carlos III s/n, 45071 Toledo, Spain

Abstract. In this paper, a novel algorithm to carry out the segmentation of moving objects in dynamic cameras is proposed. The developed system distinguishes what actions to execute in function of the environment conditions. In this way, the algorithm can segment objects in static and dynamic scenes and in ideal and noisy conditions. Therefore, the main target of this system is to cover the wider range of ambient situations. The segmentation algorithms have been developed for H.264 compressed domain because it is a modern encoder used in many modern multimedia applications and it can be decoded in real-time. Experimental results show promising performance in standard video sequences.

1 Introduction

Moving object segmentation is an essential component of an intelligent video surveillance system. Even more, segmentation could be considered as the foot of the pyramid in surveillance systems due to the fact that it has to support their next stages like object tracking, activity analysis or event understanding. In fact moving object segmentation has been considered one of the historical computer vision research issues for many years. Some techniques like background subtraction and temporal differences have thought out and there are efficient proposals to identify objects in real-time over scenes with a static background. However, segmentation of multiple objects in complex scenes is still an open field due to shadows, occlusions, illumination changes, dynamic backgrounds, poorly textured objects, and so on.

Furthermore, object segmentation applied to surveillance systems has three additional constraints. First, it is important to achieve very high accuracy in the detection, with the lowest possible false alarm rates and detection misses. Second, segmentation algorithms should work in real-time. And three, object segmentation architectures have to be adaptive, i.e., they should cover the wider range of possible ambient conditions: sunny and rainy days, at the day and at the night, to support contrast and illumination changes, ready to static and moving cameras. This research field is actually far from being completely solved, but the latest researches show suggestive results in dynamic scenes.

E. Hüllermeier, R. Kruse, and F. Hoffmann (Eds.): IPMU 2010, Part II, CCIS 81, pp. 572–581, 2010.

There has not been developed a whole surveillance system useful in all situations, but the appropriate union of different algorithms could solve partially the issue. For example, Chien et al. [2] propose an architecture with a baseline mode to static cameras and no light changing which can be complete with other three modules: shadow cancellation mode to delete shadows, global motion compensation mode to moving camera situations and adaptative threshold mode to decide automatically the required parameters. In [11] we propose a real-time moving object segmentation to static cameras which it is now extended to translation cameras and dynamic backgrounds.

The paper is organized as follows. Section 2 briefly reviews some related works of segmentation algorithms over dynamic scenes in H.264/AVC compressed domain. Section 3 describes the architecture of the segmentation approach. Later, in Section 4 experimental results are shown. Finally, conclusions and future works are described in Section 5.

2 Recent Works

Segmentation is a very difficult task in the presence of camera movement and dynamic background, therefore it is an open research field and several works have been proposed. There are several techniques at pixel level which have good performance: using region-based active contours [5], based on Markov Random Fields [3] or using a multi-cue descriptor with a variable bandwidth mean shift technique [1]. These pixel-level approaches have to fully decode the compressed video first, so they are quite accurate but cannot fulfill the requeriment of real-time applications. However, many multimedia communication applications have real-time requirement, therefore an efficient algorithm for automatic video segmentation is very desirable, and the video surveillance field is not an exception. Most real-time segmentation techniques that worked in MPEG-2 compressed domain were based on the motion vector field, not in luminance and chrominance information, hence it is no surprise that this is adopted in literature by H.264/AVC techniques. Note that the motion vectors (MVs) are two dimensional vectors which represent the pattern of movement between frames.

The first motion-level approach was developed by Zeng et al. [12], who present an algorithm that employs a block-based Markov Random Field model to segment moving objects from the MV field obtained directly from the H.264 bitstream. However, this approach is only applicable to video sequences with stationary background. Liu et al. [6] propose a later approximation where a real-time spatiotemporal segmentation is presented. In this case, spatial segmentation only exploits the MV field extracted from the H.264 compressed video. Regions are classified using the block residuals of global motion compensation and the projection is exploited for inter-frame tracking of video objects. Liu et al. [7] also propose an algorithm where the MVs are temporally and spatially normalized and then accumulated by an iterative backward projection to enhance salient motions and alleviate noisy MVs. Then the MV field is segmented into regions using a statistical region growing approach and finally the moving objects are extracted. In addition, Hong et al. [4] propose a H.264/AVC segmentation with works with 4x4 pixel blocks as the basic processing unit and a six-parameter global motion affine model. The architecture is divided in four stages: to approximate the global motion estimation, to assign weightings according to the macroblock (MB) partition modes, to apply a spatio-temporal refinement and to segment out the foreground blocks by an adaptative threshold. Finally, other two approaches have been presented. Mak and Cham [8] propose a real-time algorithm in H.264 to identify background motion model and foreground moving objects. They use an eight-parameter perspective motion model to represent the background model, a Markov Random Field (MRF) to model the foreground field and the coefficients of the residual frames to improve the segmentation results. In Poppe et al. [9], a moving object detection algorithm for video surveillance applications is explained. It is a background segmentation technique at MB level.

3 Architecture Overview

Figure 1 shows a general overview of the architecture. The H.264/AVC stream is the global input of the system. The frame information is processed by a H.264 decoder module, which extracts the useful information for the next stages of the system: the decision modes (DMs) and the motion vectors. Each one of these pieces of information will suffer a particular set of changes to obtain the required results. There are two main decision nodes which select different algorithms in function of the domain conditions. The first one is the nature of the camera, which can be static, translation or mobile. This paper is focused on translation cameras, whereas static cameras are analyzed in [11] and mobile cameras have not been dealed until now. The second decision node is the motion level presented in the scene and it distinguishes between a little, medium and a lot of motion (in Section 3.2 there is a discussion about this feature of the architecture). These decisions are automatically obtained from the video information by the system and will determine the quality of the segmentation performance in terms of both quality and processing time. The DMs and the MVs extracted from the H.264 stream are used into the segmentation system. A K-neighbor algorithm is applied to the DMs. This stage is described in [11] and obtains as output a matrix with the different motion areas of the scene. On the other hand, the MVs are processed in a different way depending on the type of camera. This decision node determines the *motion compensation*. If the camera presents any translation movement, then the system has to learn the information about the scenario through the *Velocity Labels Dynamic Update* module, which study the global motion of the background. This module is briefly explained in Section 3.1. If there is a translation camera and the named module has been executed (or the camera is static) then the fuzzification of the motion vectors is performed.

After fuzzification, linguistic motion vectors (LMVs) [10] are obtained. At this point, the system has to discriminate the motion degree of the scene by using a decision node. The *Motion Level* decision mode determines the way in which the *Pre-processing Filter* is executed. If there is a high level of motion, then the

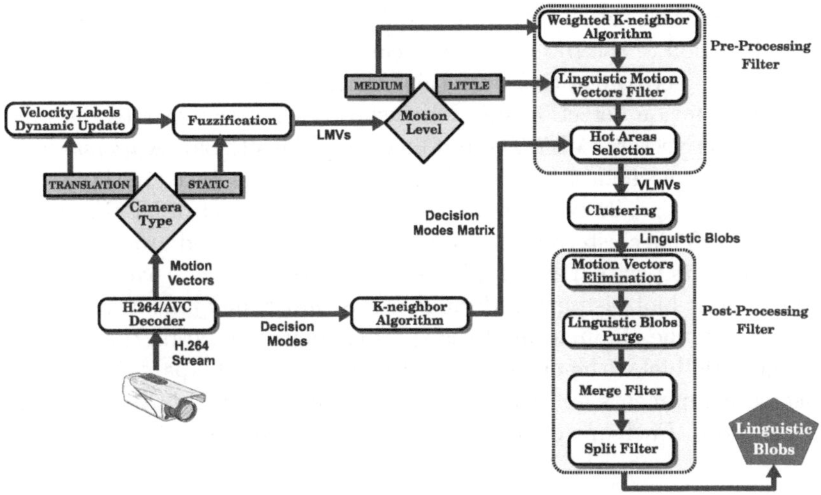

Fig. 1. Architecture overview

Fig. 2. Motion detection stage: (a) original frame, (b) decision modes matrix, (c) macroblocks encoded with decision modes higher than 4 and (d) macroblocks selected into the motion detection step

Weighted K-neighbor module has to be executed over the LMVs (this module is justified in Section 3.2). Then, a LMVs filter is executed to delete useless vectors, which are those with a minimum size and therefore without relevant movement. The next module consists in the selection of the LMVs belonging to hot areas in function of the decision modes matrix: only LMVs belonging to macroblocks codified with high DMs, as shown in Figure 2. In this moment, the motion detection of the frame has been carried out, so the next step consists in the moving object recognition from this information.

After the motion detection stage, a clustering algorithm groups the valid linguistic motion vectors (VLMVs) into linguistic blobs (LBs). A *linguistic blob* is the conceptual representation of each moving object presented in the scene characterized by its size, its position, its movement and the set of VLMVs belonging to the object. This clustering process is described in [11] and groups the VLMVs in function of its position and displacement, so occlusions are avoid. Finally, the set of linguistic blobs is purged through a *Post-processing filter* which is divided in four stages: the elimination of the spurious motion vectors, the purge of the valid linguistic blobs, the decreasing of the merge ratio by partitioning blobs in connected components and the decreasing of the split ratio by fusing together neighbor blobs. Each one of these steps are described next.

3.1 First Decision Node: Camera Type

The segmentation architecture uses four fuzzy variables to model the scenario: horizontal and vertical position and down and right velocity. The clustering algorithm groups the VMLVs in function of these parameters using euclidean distances, so the distribution of the labels of these linguistic variables is crucial to improve the segmentation performance. The position variables are fixed to incorporate expert knowledge into the scenario; for example, if there is a scenario with three lanes, the labels of the position variables can be adapted to this situation. In a similar way, the velocity variables should be updated to represent the movement of the camera in a process commonly called *motion compensation*.

Once motion vectors have been extracted, the system analyzes the type of the camera: static or with a translation motion. The architecture obtains the *global motion vector* $GMV = (gmv_x, gmv_y)$ and the variances σ_x and σ_y of the motion vectors of the frame to identify the type of the camera. The GMV is obtained as the average vector of the MVs field of a frame and the variances are calculated as the average difference between each MV and the GMV. This process is done in the first P-frame of each group of pictures (GOP). If both horizontal and vertical components of the GMV are near to zero, the camera should be static and the segmentation algorithm will use the initial velocity labels (Figure 3a). It was observed that most MVs are (0,1) and (1,0) in static cameras due to the static background, therefore we consider that there is no motion in areas with this slight motion and these are noisy MVs.

However, if at least one of the components (right or down motion) of the GMV is greater than 1, the camera is mobile, so the motion compensation process should be executed by redesigning the set of linguistic labels (Figure 3). Figure 3b shows the dynamic update of the fuzzy variable *right velocity* using GMV and σ_y. The value of the GMV marks the center value of the *No Motion* label and this set has a core equals to the variance σ_y (other dispersion measurements have been tried, but the results were worse). The neighbor labels (*Slow Right* and *Slow Left*) have a greater core than the *No Motion* label, and the *Normal* labels

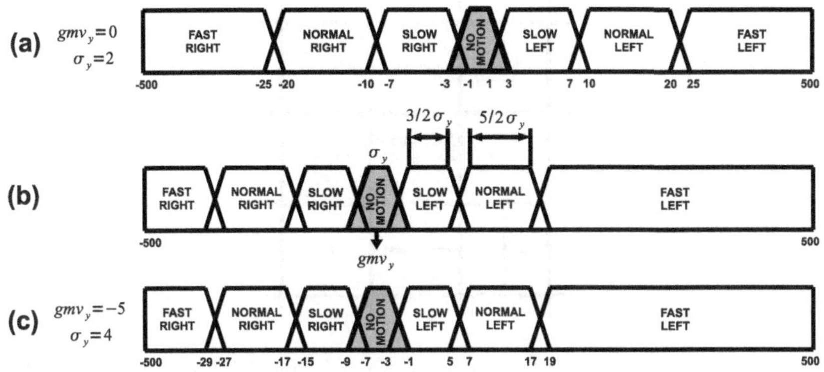

Fig. 3. Linguistic labels: (a) standard static camera, (b) general translation camera and (c) example of translation camera

still greater than them. The size of the labels have been empirically inferred to reach an optimal distribution. Figure 3c shows a real example of the *right velocity* variable into a mobile camera where $GMV = (gmv_x, -5)$ and $\sigma_y = 4$.

3.2 Second Decision Node: Motion Level

The motion degree of a camera can be different in function of external factors and, for that reason, a decision node is needed to distinguish different situations. This differentiation is done by analyzing the variances of the MVs (σ_x and σ_y). If there is no motion, the segmentation should be done with classical methods (background subtraction or temporal differences) because a change-detection segmentation will be unsatisfactory (these methods do not segment the regions without MVs). On the other side, if there is a lot of motion, for example due to moving clouds, shaking sheets or changing water, there will be unwanted noise in the frames and the segmentation could also be wrong. In the third case, a little motion degree, the results of the clustering algorithm will be satisfactory. Finally, if there is a medium motion level, it is a necessary step to delete noise.

In words of Poppe et al. [9], motion vectors are created in H.264/AVC from a coding perspective, so additional processing is needed to clean the noisy field. This fact is due to the MVs are created to optimally compress the video, not to optimally represent the real motion in the sequence. Besides, Hong et al. [4] inform that the MVs of MBs on frame boundaries become irregular for moving cameras. By starting from this assumption, a motion vector filter could be useful to reduce unwanted noise. Concretely, we propose a filter based on a *weighted K-neighbor algorithm*. Each new motion vector is calculated as the sum of three components: the first one is itself, the second component is the mean of the MVs belonging to the same MB and the last one is the mean of the MVs belonging to adjacent MBs. This algorithm is also weighted because the first component will have a value of the 50% of the final MV and the other two components will have

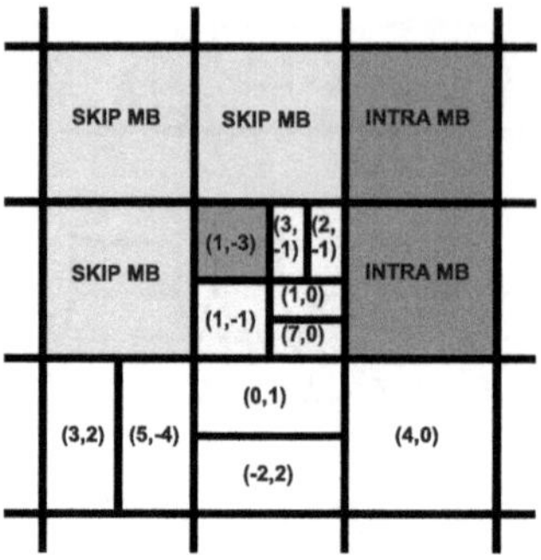

Fig. 4. Example of motion vectors of a partitioned frame

a value of 25% each one. This filter allows to reduce the encoding noise before selecting the useful vectors. Figure 4 shows a representative example in which the value of the blue motion vector after applying the filter is calculated:

$$\frac{1}{2}(1,-3)+\frac{1}{4}\frac{(14,-3)}{5}+\frac{1}{4}\frac{(10,1)}{5}=(1.7,-1.6)$$

3.3 Post-Processing Filter

Once clustering algorithm has grouped the motion vectors into the liguistic blobs, the segmentation system purges the results by applying a post-processing filter. This stage is divided in four steps. First, linguistic blobs with a size lower than a predefined threshold are deleted because they are considered noisy blobs generated as residual information into the clustering process. This threshold is experimentally obtained and its value is usually near 3 or 4 motion vectors. In the second step, the spurious MVs included into the LBs are deleted where a *spurious motion vector* is one that is not adjacent to any macroblock belonging to the LB. Next, merge ratio is decreased by partitioning blobs in connected components where a connected component is one that has all its vectors belonging to adjacent macroblocks. Finally, split ratio is decreased by fusing together neighbor blobs if two conditions are fullfilled: first, one of them is much bigger than the other blob, and second, they have a lot of adjacent macroblocks. These two conditions (difference of size and adjacency) are fixed by both thresholds in function of the scenary. Figure 5 shows specific examples of the different steps in Post-Processing stage; in left images, post-processing filter has not been used while the post-processing stage improves the segmentation performance in right images. In Figure 5a two blobs are deleted (blue and cyan) and two blobs are divided (red and magenta), so the five objects are well identified in right image. In Figure 5b both green and red blobs are merged to avoid split and cyan and magenta blobs are automatically divided to resolve merge problems. Finally, in Figure 5c it can be seen how cyan and green blobs are deleted because they are small and magenta and yellow blobs are divided to avoid merge.

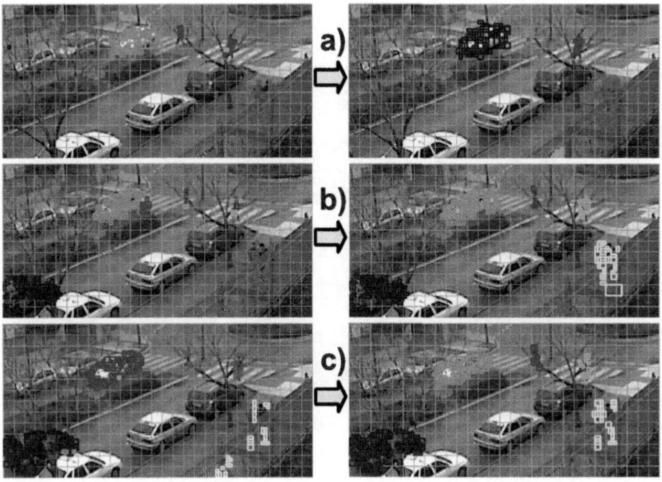

Fig. 5. Improvement of the segmentation method using the post-processing filter

4 Experimental Results

The segmentation approach has been evaluated on several video sequences which have been compressed using the H.264 encoder of JM 15.1. The encoder configuration has been set with *baseline profile* and *low complexity mode* on RD Optimization. The resolution of the tested videos goes from 320x240 to 640x480 pixels and this fact determines the processing time (Table 1). In this work a supervised, subjective method to evaluate the segmentation quality is used. Concretely, the segmentation performance is described through two measurements: the *detection possibility* (percentage of the real regions detected) and the *precision* (percentage of the detected objects corresponding to real objects). Therefore, high detection possibility means less miss-segmentation and high precision means less false segmentation. Besides, two measurements are defined to analyze the *computation time* and the *segmentation size*, i.e., the temporal and spatial requirements of the algorithm. Figure 6 shows three snapshots of one tested sequence where a camera is placed inside a mobile car; this camera allows to detect other vehicles

Table 1. Segmentation results

Measurement	Coast Guard	Highway	Mobile
Video resolution	352x288	320x240	640x480
Detection possibility	81.3 %	94.5 %	82.6 %
Precision	78.4 %	89.7 %	76.6 %
Computation time	35.7 ms/frame	23.3 ms/frame	76.2 ms/frame
Segmentation size	11.6 KB/frame	2.25 KB/frame	40.17 KB/frame

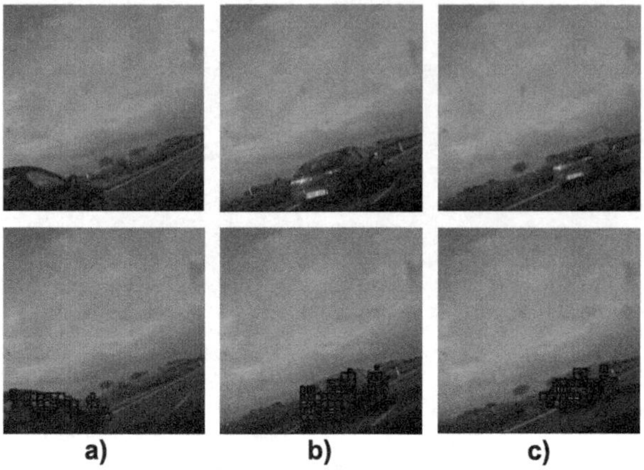

Fig. 6. *Highway* segmented frames: a) frame 112, b) frame 135 and c) frame 143

Fig. 7. *Coast Guard* segmented frames: a) frame 9, b) segmented frame 9, c) frame 199 and d) segmented frame 199

on the road. In Figure 7, there are two snapshots of original and segmented frames of the *Coast Guard* standard sequence.

5 Conclusion

In this paper, a real-time moving object segmentation algorithm for dynamic scenes has been proposed. This algorithm is fitted into a general segmentation architecture to cover the wider range of ambient and camera situations. The named architecture is focused on a video surveillance system, so it works in real-time and is adaptive in an automatic way in function of the environment conditions. Finally, the algorithm has been developed for H.264/AVC compressed domain due to its compression ratio and because it is a modern encoder used in many modern multimedia applications.

As future work is planned to extend the architecture to free-moving cameras by developing a new motion compensation algorithm. This algorithm should predict the next-frame camera motion to delete it, in this way the real motion of the objects could be distinguished before the motion detection and object identification stages. Another research line appears to identify objects when the motion degree is very low. In this case, there would be slightly moving objects or stopped objects. We propose to use a merge between classical pixel-level techniques based on motion or color and a buffer of image areas where motion has been detected.

Acknowledgments

This work was supported by the Council of Science and Technology of Castilla-La Mancha under FEDER Projects PIA12009-40, PII2I09-0052-3440 and PII1C09-0137-6488.

References

1. Bugeau, A., Perez, P.: Detection and segmentation of moving objects in complex scenes. Computer Vision and Image Understanding 113, 459–476 (2009)
2. Chien, S.Y., Huang, Y.W., Hsieh, B.Y., Ma, S.Y., Chen, L.G.: Fast video segmentation algorithm. IEEE Transactions on Multimedia 6, 732–748 (2004)
3. Cucchiara, R., Prati, A., Vezzani, R.: Real-time motion segmentation from moving cameras. Real-Time Imaging 10, 127–143 (2004)
4. Hong, W.D., Lee, T.H., Chang, P.C.: Real-time foreground segmentation for the moving camera based on H.264 video coding information. Proc. Future Generation Communication and Networking 01, 385–390 (2007)
5. Jehan-Besson, S., Barlaud, M., Aubert, G.: Region-based active contours for video object segmentation with camera compensation. In: Proc. IEEE International Conference on Image Processing, Thessaloniki, Greece, October 2001, pp. 61–64 (2001)
6. Liu, Z., Lu, Y., Zhang, Z.: Real-time spatiotemporal segmentation of video objects in the H.264 compressed domain. Journal of Visual Communication and Image Representation 18, 275–290 (2007)
7. Liu, Z., Shen, L., Zhang, Z.: An efficient compressed domain moving object segmentation algorithm based on motion vector field. Journal of Shanghai University 12, 221–227 (2008)
8. Mak, C.M., Cham, W.K.: Fast video object segmentation using Markov Random Field. In: Workshop on Multimedia Signal Processing, pp. 343–348 (2008)
9. Poppe, C., De Bruyne, S., Paridaens, T., Lambert, P., Van de Walle, R.: Moving object detection in the H.264/AVC compressed domain for video surveillance applications. Journal of Visual Communication and Image Rep. 20, 428–437 (2009)
10. Rodriguez-Benitez, L., Moreno-Garcia, J., Castro-Schez, J.J., Albusac, J., Jimnez-Linares, L.: Automatic objects behaviour recognition from compressed video domain. Image and Vision Computing 27, 648–657 (2009)
11. Solana-Cipres, C.J., Fdez-Escribano, G., Rodriguez-Benitez, L., Moreno-Garcia, J.: Real-time moving object segmentation in H.264 compressed domain based on approximate reasoning. Int. J. of Approximate Reasoning 51, 99–114 (2009)
12. Zeng, W., Du, J., Gao, W., Huang, Q.: Robust moving object segmentation on H.264/AVC compressed video using the block-based MRF model. Real-Time Imaging 11, 290–299 (2005)

Using Stereo Vision and Fuzzy Systems for Detecting and Tracking People

Rui Paúl*, Eugenio Aguirre, Miguel García-Silvente, and Rafael Muñoz-Salinas

Department of Computer Science and A.I., E.T.S. Ingeniería Informática,
University of Granada, 18071 Granada, Spain
Department of Computing and Numerical Analysis, E.P.S.,
University of Córdoba, Córdoba, Spain
{ruipaul,eaguirre,m.garcia-silvente}@decsai.ugr.es,
rmsalinas@uco.es

Abstract. This paper describes a system capable of detecting and tracking various people using a new approach based on stereo vision and fuzzy logic. First, in the people detection phase, two fuzzy systems are used to assure that faces detected by the OpenCV face detector actually correspond to people. Then, in the tracking phase, a set of hierarchical fuzzy systems fuse depth and color information captured by a stereo camera assigning different confidence levels to each of these information sources. To carry out the tracking, several particles are generated while fuzzy systems compute the possibility that some generated particle corresponds to the new position of people. The system was tested and achieved interesting results in several situations in the real world.

Keywords: People Tracking, Stereo Vision, Fuzzy systems, Particle Filtering, Color Information.

1 Introduction and Related Work

People detection and tracking can be done in various ways and with different kind of hardware. When computer vision is used, the system analyzes the image and searches for cues that provide important information in the detection of people. Those cues could be, for instance, morphological characteristics of the human body [1]. Due to illumination change problems some authors have opted to use dynamic skin color models [2].

In this work stereo vision has been used so 3D information could be extracted from the images. This information is relatively invariable with respect to illumination changes. In [3], the authors present a system capable of detecting and tracking several people. Their work is based on a skin detector, a face detector and the disparity map provided by a stereo camera. In the work of Grest and Koch [4] a particle filter [5] is also used to estimate the position of the person

* This work is supported by the FCT Scolarship SFRH/BD/22359/2005, Spanish MCI Project TIN2007-66367 and Andalusian Regional Government project P09-TIC-04813.

E. Hüllermeier, R. Kruse, and F. Hoffmann (Eds.): IPMU 2010, Part II, CCIS 81, pp. 582–591, 2010.

and create color histograms of the face and breast regions of that person and stereo vision to compute the real position of the person in the room. However, stereo and color were not integrated in the tracking process and they use cameras positioned in different parts of a room rather than one stereo camera. Moreno *et. al.* [6] present a system able to detect and track a single head using the Kalman filter. They combined color and stereo information but head color does not provide enough information to distinguish among different users. In [7] and [8], the authors present an approach to detect and track several people using *plan-view maps*. They use information provided by an *occupancy map* and a *height map* using the Kalman filter.

In our approach the problem is solved using a new approach based on a particle filter which generates particles that are evaluated by means of fuzzy logic. Although we also use depth and color information as sources of information, they are supplied to several hierarchically organized fuzzy systems. People tracking is done by generating different particles in the image and then computing their possibility to be part of a previous detected person using a fuzzy system approach. We opted for using fuzzy logic [9] in order to have the possibility of dealing with uncertainty and vagueness in a flexible manner so we can avoid possible restrictions when representing imprecision and uncertainty with probabilistic models. Furthermore, when using linguistic variables and rules to define the behavior of our system, it turns out to be more understandable and similar to the way humans represent and process knowledge.

2 People Detection and Tracking

Our system is based on a stereo camera (BumbleBee model) that allows us to extract not only visual information from the images but also depth information about most of the objects appearing in the images. By combining these two different types of information it is possible to achieve a more robust tracking than when using only one of them. If one of them fails, it is possible to keep track of a person by using the other and vice-versa.

2.1 People Detection

The detection of people begins with a face detector phase. This is done by using the face detector available in the OpenCV library [10], that is free to use and download. Although this detector is free, fast and able to detect people with different morphological faces, false positives can be found. The classifier outputs the rectangular region(s) of the faces detected in our RGB image. In order to reject possible false positives each of the detected face(s) have to pass two tests to assure that it belongs to a person.

The first test use the concept of the projection of the model of a person. Taking into account the usual size of a person we can estimate the projection of a person in our camera image, according to his or her distance to the camera and knowing the intrinsic parameters of the camera. From now on we will call the projection of the model of a person as R_p (standing for Region of Projection).

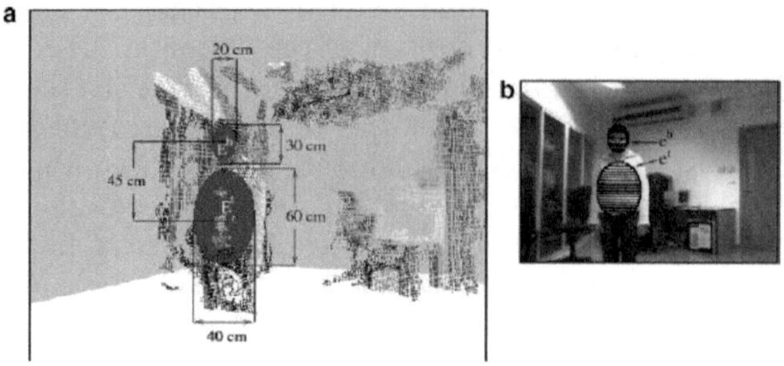

Fig. 1. (a) Model employed. (b) Projection of the model on the reference image.

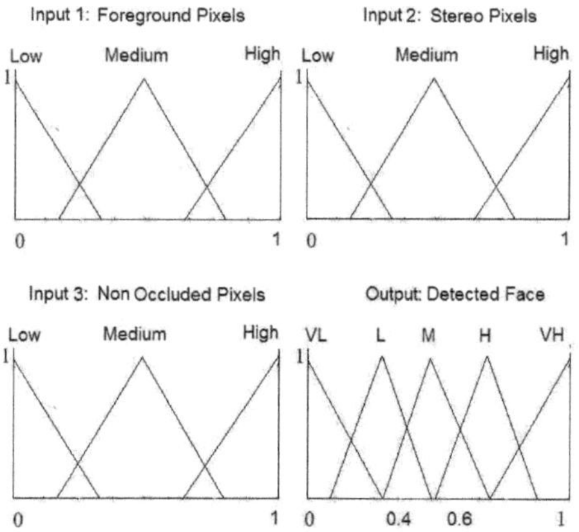

Fig. 2. Fuzzy sets for detecting faces with variables ForegroundPixels (ratio), StereoPixels (ratio), NonOccludedPixels (ratio) and DetectedFace

Fig.1 shows the region of projection in a stereo image and its corresponding reference image.

The goal of this test is to check whether inside R_p there are enough pixels respecting three conditions: they have to belong to the foreground (if they belong to the background they cannot be considered as being part of a person), they have disparity information (if there is a person in R_p then there should be a high number of pixels containing depth information) and they are not occluded (if most of the pixels inside R_p are occluded then R_p represents a region where visual and depth information, important for the tracking process, is not sufficient and consequently trustable).

These three measures are fuzzified by three linguistic variables labeled as $ForegroundPixels$, $StereoPixels$ and $NonOccludedPixels$, respectively (see Fig.2). Using these three variables as input variables to the Fuzzy System 1 shown by Table 1, the fuzzy output $DetectedFace$ is computed. Fuzzy System

Table 1. Rules for Fuzzy System 1

IF			THEN
ForegroundPixels	StereoPixels	NonOccludedPixels	DetectedFace
High	High	High	Very High
High	High	Medium	High
High	High	Low	Medium
High	Medium	High	High
...	...	...	...
Low	Low	Low	Very Low

1 and the rest of the fuzzy systems shown in this work use the Mamdani inference method. The defuzzified value of $DetectedFace$ indicates the possibility, from 0 to 1, whether region R_p is worth to contain a true positive face. If this value is higher than α_1, the detected face passes to the second and last test. The second test also checks whether R_p may contain a true positive face. However the idea is different now. If there is a person in that region, then pixels inside R_p should have approximately the same depth. Therefore the Fuzzy System 2 receives, as input, the difference between the average depth of R_p and the depth of the detected face as seen in Eq.1.

$$d = |Z - \frac{\sum_{j=1}^{n}(z_j)}{n}|. \tag{1}$$

where d is the difference we want to compute, Z the actual depth of the detected face, z_j the depth of the j pixel inside R_p and n the total number of pixels inside R_p. This value is fuzzified by the linguistic variable $AverageDifference$.

Fuzzy System 2 also receives the standard deviation of those pixels, fuzzified by the linguistic variable *StandardDeviation*, and the depth at which the face was detected, fuzzified by the linguistic variable *Depth*. Depth of the detected face is used to compute the confidence that we should assign to the values of the other variables. At farther distances, the uncertainty is higher. The output variable *SimilarDepth* is computed by Fuzzy System 2 and its defuzzified value is a value between 0 and 1 corresponding to the possibility that R_p contains pixels with depth similar to the depth of the detected face. In Fig.3 linguistic variables $AverageDifference$, *StandardDeviation*, *Depth* and *SimilarDepth* (output) are shown. In Table 2 it is possible to find examples of the rules defined for Fuzzy System 2.

Finally, if this value is higher than α_2, we assume that a person was detected and we assign a tracker for him or her. The values for parameters α_1 and α_2 have been experimentally tuned.

The rules and linguistic variables defined for other fuzzy systems in Section 2.2 are similar to the ones of Figures 2, 3 and Tables 1, 2 so that they are omitted in order to not exceed the allowed number of pages of this paper.

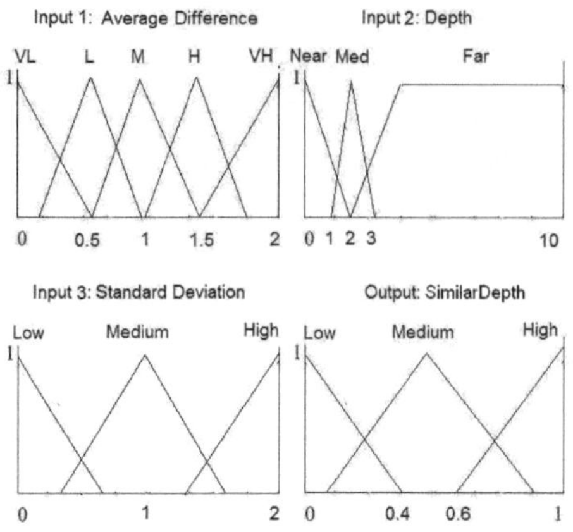

Fig. 3. Fuzzy sets for detecting faces with variables AverageDifference (meters), Depth (meters), StandardDeviation (meters) and SimilarDepth

Table 2. Rules for Fuzzy System 2

IF			THEN
AverageDifference	Depth	StandardDeviation	SimilarDepth
VL	Far	Low	High
VL	Far	Medium	High
VL	Far	High	Medium
L	Far	Low	High
...	...	...	...
VH	Near	High	Low

2.2 People Tracking

As we said before, a tracker is created for each person detected. The order of the trackers goes from the person that is closer to the camera to the person that is farther. The tracking process is divided into two parts. In the first one, a particle filter approach is used to generate and evaluate possible new positions for the person being tracked. This particle filter is based on the condensation algorithm, but particles are evaluated by means of fuzzy logic rather than using a probabilistic approach. After some experiments, 50 particles were considered to be sufficient to keep track of people without compromising performance. In the second one, the average position of all particles is computed. This average is an weighted average, based on the value of possibility $PossibilityP(i)$ of each previously generated particle. The average position $PersonPos_{(t)}$ is the new position of the person in 3D. We consider that the position of the person is his or her face position. There are as many trackers as people being tracked. So, this phase is repeated as many times as the number of people being tracked.

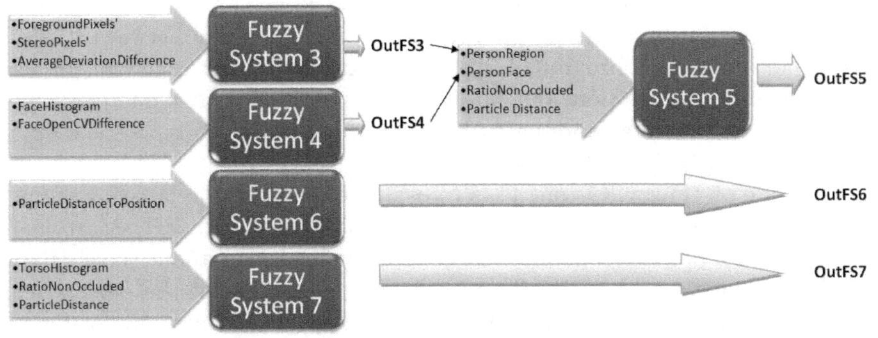

Fig. 4. Fuzzy Systems used to evaluate de overall quality of each generated particle. For each fuzzy system, the input linguistic variables are specified.

The generation of a new particle, is based on the previous position of the average particle in the previous frame $PersonPos_{(t-1)}$. The idea is to generate most particles in the surroundings of the previous and a few farther as people are not expected to move fast from frame to frame (the frame rate is 10 fps). The propagation model of the particles is based on the previous position of the person plus some δ value that follows a gaussian distribution with parameters $N(\mu = 0\,m, \sigma = 0.1\,m)$. After generating the set of particles we begin the process of evaluating the possibility ($PossibilityP(i)$) that each particle corresponds to the tracked person. The observation model for each particle is based on the output of different fuzzy systems as shown in Fig.4. We use a two layer fuzzy system approach to take into account the confidence level of the outputs of some of the fuzzy systems. This situation will be explained later when each of the fuzzy system is described. Finally, the overall result for each particle is given by $PossibilityP(i) = OutFS_5 * OutFS_6 * OutFS_7$ where $OutFS_i$ stands for the "ith" Fuzzy System defuzzified output and is a value between 0 and 1 (see Fig.4).

The goal of "Fuzzy System 3" is to evaluate the region of projection of some person $R_p(P_i)$ (see Fig.1) according to the depth of the current particle being evaluated (P_i). This evaluation will take into consideration only aspects related with the possibility that some object, similar to a person, is located in that region. The first step is to compute the area of $R_p(P_i)$. After obtaining this information we define three linguistic variables: $ForegroundPixels'$, $StereoPixels'$ and $AverageDeviationDifference$. $ForegroundPixels'$ and $StereoPixels'$ are defined in a similar way to $ForegroundPixels$, $StereoPixels$ at Section 2.1. $AverageDeviationDifference$ gives us information about the difference between the depth of P_i and the depth average of all pixels inside $R_p(P_i)$. This value is also fused with the standard deviation for those pixels. The reason for defining this variable is that, all pixels inside $R_p(P_i)$, should have approximately the same depth as P_i and should have approximately the same depth between them, as long as they belong to some person or object. These values will be the input to Fuzzy System 3 that will output a deffuzified value between 0 and 1. The higher amount of foreground, disparity pixels and lower difference in average and standard deviation, the closer the output is to 1. A value closer to 1 means that, in the area represented by $R_p(P_i)$, it is likely to have some object that could hypothetically be a person.

The scope of "Fuzzy System 4" is to evaluate face issues related to the person being tracked. We define two linguistic variables called $FaceHistogram$ and $FaceOpenCVDistance$. The first one contains information about the similarity between the face region of $R_p(P_i)$ and the face histogram of the person being tracked. As people from frame to frame (at a 15 fps frame rate) do not tend to move or rotate their face so abruptly, the histograms should be similar. We use the elliptical region of the face to create a color model [11]. We then measure the difference between the face histogram of region of $R_p(P_i)$ and the face histogram of the person being tracked. This difference is based on a popular measure between two color distributions: the Bhattacharyya coefficient [12]. This method gives us the similarity measure of two color models in the range $[0, 1]$. Values near 1 mean that both color models are identical. Values near 0 indicate that the distributions are different. An important feature of this method is that two color models can be compared even if they have been created using a different number of pixels. The second linguistic variable measures the distance between P_i and the position of the nearest face to P_i detected by the OpenCV face detector. Although OpenCV is not 100% accurate, most of time this information can be worth as it can tell if there is really a face near P_i. The deffuzified output of this fuzzy system is also a number between 0 and 1 where 1 is an optimal value.

The deffuzified outputs of "Fuzzy System 3" and "Fuzzy System 4" are then provided as input for another fuzzy system that we call "Fuzzy System 5". This fuzzy system allows us to measure the confidence of the outputs of Fuzzy Systems 3 and 4 based on occlusion and depth information. We define four linguistic variables called $PersonRegion$, $PersonFace$, $RatioNonOccluded$ and $ParticleDistance$ to compute the final output for Fuzzy System 5. $PersonRegion$ and $PersonFace$ have five linguistic labels Very Low, Low, Medium, High and Very High distributed in a uniform way into the interval $[0, 1]$ in a similar way to the membership functions of $AverageDifference$ shown by Fig. 3. Their inputs are the defuzzified outputs of "Fuzzy System 3" and "Fuzzy System 4" respectively. $RatioNonOccluded$ contains information about the ratio of non occluded pixels inside $R_p(P_i)$. The higher the number of non occluded pixels, the more confidence we have on the output values. In other words, the more pixels we can use from $R_p(P_i)$ to compute foreground, depth, average information and histogram the more trustable the outputs of "Fuzzy System 3" and "Fuzzy System 4". Finally $ParticleDistance$ has information about the distance of the particle evaluated (P_i). As errors in stereo information increase with distance, the farther the particle is located, the less trustable it is in means of depth information. The defuzzified output of "Fuzzy System 5" ($OutFS_5$) is also a number between 0 and 1. Higher values indicate a region with higher possibility to contain a person.

With respect to "Fuzzy System 6", this fuzzy system's goal is to evaluate whether P_i is likely to be the person being followed taking into consideration the distance to the previous location of the person (in the frame before). Due to the frame rate used, people from frame to frame are not expected to move significantly. Therefore, we define only one variable called $ParticleDistanceToPosition$ that contains information about the 3D distance between the 3D position of P_i and the 3D position of the currently tracked person ($PersonPos_{(t-1)}$). The deffuzified output will be, once again, a value between 0 and 1 represented by $OutFS_6$. An output equal to 1 means that P_i is located exactly in the same place where $PersonPos_{(t-1)}$ was located.

The last fuzzy system ("Fuzzy System 7") is related with torso information. Identically to "Fuzzy System 4" we also define a variable that translates the similarity between the torso histogram information of $R_p(P_i)$ and the histogram information of the torso of the person being tracked. This variable is called $TorsoHistogram$. We also use for this fuzzy system, the variables $RatioNonOccluded$ and $ParticleDistance$ analogously to "Fuzzy System 5". When doing this, we are adding a measure of confidence for the output which after its deffuzification is called $OutFS_7$ and has a value between 0 and 1.

As said before, all these outputs are multiplied and result on a final value between 0 and 1. Then an weighted average of the 3D position $PersonPos_{(t)}$ is computed by taking into consideration all the possibility values for the set of particles. A particle that has a possibility value closer to 1 will weight much more than one with a possibility value of 0. Its $R_p(P_i)$ is also added to an occlusion map, so the following trackers and the people detection's algorithm know, that there is already a person occupying that region. This occlusion map is reset every time a new frame is processed. The face and torso histograms are also updated.

3 Experimental Results

The system was tested in various scenarios and with different people. Videos were recorded with a resolution of 320x240 pixels and the system was tested with an Intel Core 2 Duo 3.17 Processor (only one processor is used for processing). The achieved operation frequency of our system was about 10 Hz in average depending on the number of people being tracked. As each tracked person implies a new tracker, processing time increases in average by 50 ms for each added tracker. We consider up to 4 people for the system to perform in real time with this kind of camera and processor.

We recorded 15 videos with 1, 2 and 3 people moving freely in a room. The set of videos provided over 15 minutes of recording with various people interacting freely. We could observe that, when either disparity or color information were not completely reliable, the system still kept track of the people. The average accuracy rate for tracking people in the test set was over 90%. This result was achieved after tuning different settings of the fuzzy systems. We think an higher rate could be achieved if these values keep being tuned.

In Fig.5 we show four frames taken from one of those videos, with both reference image and disparity image shown for each frame. In the disparity image,

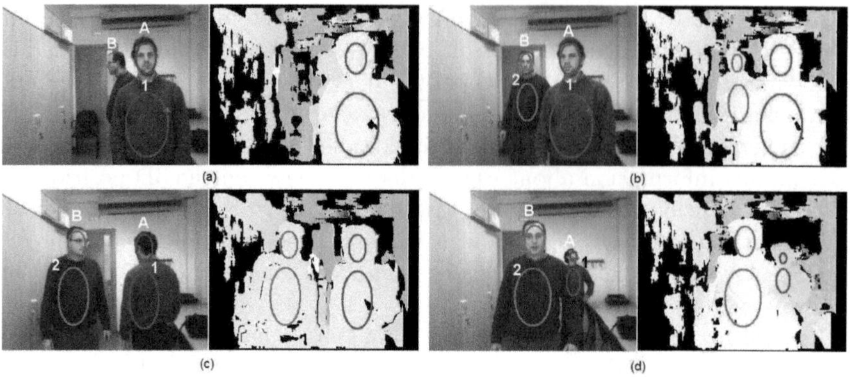

Fig. 5. Different frames taken from a video with 2 people being tracked

lighter areas represent shorter distances to the camera. In Fig.5(a) it is possible to see that the system detected person A (ellipse 1) while person B was not detected. In Fig.5(b) we can see that person B was detected (ellipse 2) since most of the pixels were visible. In Fig.5(c) it is possible to see that, although depth information for both people was very similar, the system could still keep an accurate track for each of the people. The reason for achieving this accuracy relies on color information that compensated the similarity of depth information. Finally in Fig.5(d) it is possible to see that, for person A, although part of his body was occluded, the system could still achieve an accurate tracking, based on disparity information rather than color information.

We would also like to mention that, when people cross their paths, the system manages to keep track of each person by making use of both depth and color information. However, situations in which two people dressed with the same colors and located at the same distance got very close, could originate that the system would confuse both targets. This issue is expected to be solved in a near future by providing more information sources to the system.

4 Conclusions and Future Work

The system proposed proved to work in real life situations, where people were interacting freely and occluded each other sometimes. The system was capable of detecting and tracking people based on fuzzy logic as it has proven in the past that it is an interesting tool for treating uncertainty and vagueness. A particle filter is used to generate particles that are evaluated using fuzzy logic instead of probabilistic methods. As we know, information supplied by sensors is commonly affected by errors, and therefore the use of fuzzy systems help us to deal with this problem. In our case, as stereo information is not 100% accurate, we may sometimes rely more on color information and solve that problem. On the other hand, we can easily manage unexpected situations as, for instance, sudden illumination changes, by giving more importance to stereo information. By setting up linguistic variables and rules that deal with this problem we achieved an efficient way of solving it. Also, when using fuzzy systems to represent knowledge, the complexity in understanding the system is substantially lower as this kind of knowledge representation is similar to the way the human being is used to represent its own knowledge. Furthermore, it allows an easy way of adding new features, just by adding more variables or fuzzy systems.

In this work, rules and linguistic variables are defined after testing different values in different experiments. As a future work, we would like to build a system capable of learning and therefore adjusting these parameters automatically.

References

1. Hirai, N., Mizoguchi, H.: Visual tracking of human back and shoulder for person following robot. In: IEEE/ASME International Conference on Advanced Intelligent Mechatronics, vol. 1, pp. 527–532 (2003)
2. Sigal, L., Sclaroff, S., Athitsos, V.: Skin color-based video segmentation under time-varying illumination. IEEE Transactions on Pattern Analysis and Machine Intelligence 26, 862–877 (2003)
3. Darrell, T., Gordon, G., Harville, M., Woodfill, J.: Integrated person tracking using stereo, color, and pattern detection. International Journal of Computer Vision 37, 175–185 (2000)
4. Grest, D., Koch, R.: Realtime multi-camera person tracking for immersive environments. In: IEEE Sixth Workshop on Multimedia Signal Processing, pp. 387–390 (2004)
5. Isard, M., Blake, A.: CONDENSATION-conditional density propagation for visual trackings. International Journal of Computer Vision 29, 5–28 (1998)
6. Moreno, F., Tarrida, A., Andrade-Cetto, J., Sanfeliu, A.: 3D real-time head tracking fusing color histograms and stereovision. In: International Conference on Pattern Recognition, pp. 368–371 (2002)
7. Harville, M.: Stereo person tracking with adaptive plan-view templates of height and occupancy statistics. Image and Vision Computing 2, 127–142 (2004)
8. Muñoz-Salinas, R., Aguirre, E., García-Silvente, M.: People Detection and Tracking using Stereo Vision and Color. Image and Vision Computing 25, 995–1007 (2007)
9. Yager, R.R., Filev, D.P.: Essentials of Fuzzy Modeling and Control. John Wiley & Sons, Inc., Chichester (1994)
10. Intel, OpenCV: Open source Computer Vision library, `http://www.intel.com/research/mrl/opencv/`
11. Birchfield, S.: Elliptical head tracking using intensity gradients and color histograms. In: IEEE Conf. Computer Vision and Pattern Recognition, pp. 232–237 (1998)
12. Kailath, T.: The divergence and Bhattacharyya distance measures in signal selection. IEEE Transactions on Communication Technology 15, 52–60 (1967)

Group Anonymity

Oleg Chertov and Dan Tavrov

Faculty of Applied Mathematics, National Technical University of Ukraine "Kyiv Polytechnic Institute", 37 Peremohy Prospekt, 03056 Kyiv, Ukraine
{chertov,kmudisco}@i.ua

Abstract. In recent years the amount of digital data in the world has risen immensely. But, the more information exists, the greater is the possibility of its unwanted disclosure. Thus, the data privacy protection has become a pressing problem of the present time.

The task of individual privacy-preserving is being thoroughly studied nowadays. At the same time, the problem of statistical disclosure control for collective (or group) data is still open.

In this paper we propose an effective and relatively simple (wavelet-based) way to provide group anonymity in collective data. We also provide a real-life example to illustrate the method.

Keywords: statistical disclosure control, privacy-preserving data mining, group anonymity, wavelet analysis.

1 Introduction

Year by year more and more data emerge in the world. According to the latest IDC research [1], the Digital Universe will double every 18 months, and the number of "security-intensive" information will grow from 30% to roughly 45% by the end of 2012. This means that the risk of privacy violation and confidential data disclosure rises dramatically. The threat doesn't only comprise the possibility of stealing a credit card and social security numbers, patient medical records and images, Internet commerce and other transaction data. It is popular nowadays to provide direct access to the depersonalized and non-aggregated primary data. E.g., one can easily gain access to the microfile data in census statistics and sociology. But, if these data aren't protected, individual anonymity can easily be violated. That has been clearly shown by Sweeney [2, p. 21]. Using Massachusetts voters data, she proved that knowing a person's birth date and full 9-digit ZIP code is enough to identify 97% of voters.

A problem of individual anonymity in primary data isn't a new one. It is being more or less successfully solved as one of the main tasks in privacy-preserving data mining. There are different statistical disclosure control methods [3] which guarantee:

- *k*-anonymity [4]. This means every attribute values combination corresponds to at least *k* respondents existing in the dataset sharing the combination;
- even more delicate *l*-diversity [5] and *t*-closeness [6].

At the same time a problem of providing group (collective) anonymity is still open [7]. By *group anonymity* we understand protecting such data features that can't be

E. Hüllermeier, R. Kruse, and F. Hoffmann (Eds.): IPMU 2010, Part II, CCIS 81, pp. 592–601, 2010.

distinguished by considering individual records only. E.g., we cannot protect the regional distribution of young unemployed females in terms of inidividual anonymity.

Either individual or group anonymity can be provided by introducing an acceptable level of uncertainty to the primary data. By making specific data records impossible to distinguish among the others, we guarantee required privacy-preserving.

When providing data anonymity, both group and individual, it is important to take into account the adversarial model which reflects what information is known to an adversary, and what is not.

In our work, we suppose that the potential adversary doesn't possess any other additional information except for the one contained in the primary data.

In general, there exist a variety of approaches to solving the group anonymity problem. In this paper, we will discuss a so-called extremum transition approach. Its main idea is to swap records with specific attribute values between the positions of their extreme concentrations and the other permitted ones.

Depending on the task definition, we can implement this approach by:

- swapping values of required attributes between respondents;
- transferring a respondent to be protected to the other place of living (to the other place of work). In most cases it is natural to transfer not only a single respondent, but the whole respondent's family as well;
- mere modifying the attribute values.

Of course, it's easy to provide group anonymity. All we need is a permission to move respondents between any possible places (as long as the population number on a particular territory remains stable). But such primary data deformation almost inevitably leads to considerable utility loss. Imagine that we want to transfer some respondents to a particular territory. But, there are not enough people of the same sex and age to fit our new "migrants". Obviously, such a transfer cannot be acceptable.

All this leads to a question. If we know what to modify to provide data group anonymity, what should we preserve to prevent data utility loss?

In general, it is possible to preserve *absolute* quantities only for a *particular* population category (all the population provided, respondents with required values of a certain attribute etc.). But, in many cases researches can be interested in the *relative* values rather than in the absolute ones. Let us consider some typical examples.

1. True quantity of military (or special service) officers can be absolutely confidential. This is also the case for their regional distribution. At the same time, information on their distribution by age, marital status or, say, wife's occupation can be very interesting and useful for sociologists.

2. In developing countries, there is usually no public statistics on a company's income. In this case, information on the company's income growth rates can serve as an important marker of its economic status and development prospective.

We come to conclusion that we need to preserve relations between strata, distribution levels, data ranges rather than the absolute values. But, it's not easy to alter data records with a particular attribute values combination and preserve proportional relations between all the other possible ones. Such a problem seems to be as complex as the k-anonymization problem. The latter, as stated in [8], is an NP-hard problem.

Certainly, there are different techniques that can aid in finding a balance between altering primary data and preventing utility loss. For instance, we can try to perform such data swapping that main statistical features such as data mean value and their standard deviation will persist. For example, in [11], a specific normalizing process to preserve these features is introduced.

But, in the current paper we propose to use wavelet transform (WT) instead. Surely, WT doesn't guarantee the persistance of all statistical data features (except for their mean value which will be discussed later in the paper), but it can preserve some information that can come in handy for specific studies.

Generally speaking, WT is an effective way to present a square-integrable signal by a basis obtained from certain wavelet and scaling functions providing its both time and frequency representation. We consider WT to be acceptable because:

- It splits primary data into approximation and multilevel details. To protect data, we can redistribute approximation values considering particular attribute values combinations. Besides, we can prevent utility loss by leaving details unchanged (or by altering them proportionally). In this case, proportional relations between different attribute values ranges will be preserved. To illustrate that, let's refer to [9]. In Russia, studying the responses to 44 public opinion polls (1994-2001) showed the following result. It turned out that details reflect hidden time series features which come in handy for near-term and medium-term social processes forecasting.
- We can use the fast Mallat's pyramid algorithm [10]. Its runtime complexity is $O(n)$, where n is the maximum wavelet decomposition level.
- WT is already being successfully and intensively used to provide individual anonymity [11].

Thereby, in this work we set and solve a following task. We want to provide group anonymity for depersonalized respondent data according to particular attribute combination. We propose to complete this task using WT. In this case, group anonymity is gained through redistributing wavelet approximation values. Fixing data mean value and leaving wavelet details unchanged (or proportionally altering them) preserves data features which might become useful for specific researches. Figuratively speaking, we change the relief (approximation) of a restricted area, but try to preserve local data distribution (details).

We would also like to admit that there is no feasible algorithm to restore primary data after modifying them using the proposed method.

2 Theoretic Background

2.1 General Group Anonymity Definitions

Let the microfile data be presented as Table 1. In this table, μ stands for the number of records (respondents), η stands for the number of attributes, r_i stands for the ith record, u_j stands for the jth attribute, z_{ij} stands for a microfile data element.

To provide group anonymity we need to decide first which attribute values and of what groups we would like to protect.

Let us denote by S_v a subset of a Cartesian product $u_{v_1} \times u_{v_2} \times ... \times u_{v_l}$ of Table 1 columns. Here, v_i, $i=\overline{1,l}$ are integers. We will call an element $s_k^{(v)} \in S_v$,

$k=\overline{1,l_v}$, $l_v \le \mu$ *a vital value combination* because such combinations are vital for solving our task. Respectively, each element of $s_k^{(v)}$ will be called *a vital value,* and u_{v_j} will be called *a vital attribute*.

Our task is to protect some of the vital value combinations. E.g., if we took "Age" and "Employment status" as vital attributes we could possibly be interested in providing anonymity for the vital value combination ("Middle-aged"; "Unemployed").

We will also denote by S_p a subset of microfile data elements corresponding to the pth attribute, $p \ne v_i$ $\forall i=\overline{1,l}$. Elements $s_k^{(p)} \in S_p$, $k=\overline{1,l_p}$, $l_p \le \mu$ will be called *parameter values*, whereas p^{th} attribute will be called *a parameter attribute* because it will be used for dividing microfile data into groups to be analyzed.

Table 1. Microfile data

	u_1	u_2	...	u_η
r_1	z_{11}	z_{12}	...	$z_{1\eta}$
r_2	z_{21}	z_{22}	...	$z_{2\eta}$
...	...	...	...	...
r_μ	$z_{\mu 1}$	$z_{\mu 2}$	...	$z_{\mu\eta}$

For example, if we took "Place of living" as a parameter attribute we could obtain groups of "Urban" and "Rural" residents.

After having defined both parameter and vital attributes and values, we need to calculate the quantities of respondents that correspond to a specific pair of a vital value combination and a parameter value

These quantities can be gathered into an array $q=(q_1,q_2,...,q_m)$ which we will call *a quantity signal.*

To provide group anonymity for the microfile we need to replace this quantity signal with another one: $\tilde{q}=(\tilde{q}_1,\tilde{q}_2,...,\tilde{q}_m)$. Also, we need to preserve specific data features.

First of all, we need to make sure that the overall number of records remains stable: $\sum_{i=1}^{m} q_i = \sum_{i=1}^{m} \tilde{q}_i$.

And, as it was mentioned in Section 1, we also need to preserve all the wavelet decomposition details of signal q up to some level k (or at least alter them proportionally).

Possible solution to the task is proposed in the following subsections.

2.2 General Wavelet Transform Definitions

In this subsection we will revise the WT basics which are necessary for the further explanation. For detailed information see [10].

Let us call an array $s=(s_1,s_2,...,s_m)$ of discrete values a signal.

Let a high-pass wavelet filter be denoted as $h=(h_1,h_2,...,h_n)$, and a low-pass wavelet filter be denoted as $l=(l_1,l_2,...,l_n)$.

To perform signal s one-level wavelet decomposition, we need to carry out following operations:

$$a_1 = s *_{\downarrow 2n} l;\ d_1 = s *_{\downarrow 2n} h. \tag{1}$$

In (**1**), a convolution (which is denoted by $*$) of s and l is taken, and then the result is being dyadically downsampled (denoted by $_{\downarrow 2n}$). Also, a_1 is an array of approximation coefficients, whereas d_1 is an array of detail coefficients.

To obtain approximation and detail coefficients at level k, we need to perform (**1**) on approximation coefficients at level $k-1$:

$$a_k = a_{k-1} *_{\downarrow 2n} l = (\underbrace{(s *_{\downarrow 2n} l)\ldots *_{\downarrow 2n} l}_{k\ times}); d_k = a_{k-1} *_{\downarrow 2n} h = ((\underbrace{(s *_{\downarrow 2n} l)\ldots *_{\downarrow 2n} l}_{k-1\ times}) *_{\downarrow 2n} h)\ . \tag{2}$$

We can always present an initial signal s as

$$s = A_k + \sum_{i=1}^{k} D_i\ . \tag{3}$$

Here, A_k is called an approximation at level k, and D_i is called a detail at level i.

Approximation and details from (**3**) can be presented as follows:

$$A_k = (\underbrace{(a_k *_{\uparrow 2n} l)\ldots *_{\uparrow 2n} l}_{k\ times});\ D_k = ((\underbrace{(d_k *_{\uparrow 2n} h) *_{\uparrow 2n} l)\ldots *_{\uparrow 2n} l}_{k\text{-}1\ times})\ . \tag{4}$$

In (**4**), a_k and d_k are being dyadically upsampled (which is denoted by $_{\uparrow 2n}$) first, and then convoluted with the appropriate wavelet filter.

As we can see, all A_k elements depend on the a_k coefficients.

According to Section 1, we need to somehow modify the approximation, and at the same time preserve the details. As it follows from (**4**), details do not depend on approximation coefficients. Thus, preserving detail coefficients preserves the details.

Respectively, to modify the approximation we have to modify corresponding approximation coefficients.

2.3 Obtaining New Approximation Using Wavelet Reconstruction Matrices

In [12], it is shown that WT can be performed using matrix multiplications. In particular, we can always construct a matrix such that

$$A_k = M_{rec} \cdot a_k \,. \tag{5}$$

For example, M_{rec} can be obtained by consequent multiplication of appropriate upsampling and convolution matrices.

We will call M_{rec} a wavelet reconstruction matrix (WRM).

Now, let us apply WRM to solve the problem stated in Subsection 2.1.

Let $q = (q_1, q_2, ..., q_m)$ be a quantity signal of length m. Let also $l = (l_1, l_2, ..., l_n)$ denote a low-pass wavelet filter.

Taking into consideration (**5**), all we need to do is to find new coefficients $\widehat{a}_k$. For example, they can be found by solving a linear programming problem with constraints obtained from matrix M_{rec}. Then, adding new approximation $\widehat{A}_k$ and all the details corresponding to q, we can get a new quantity signal $\widehat{q}$.

In many cases, adding $\widehat{A}_k$ can result in the negative values of a new signal $\widehat{q}$, which is totally unacceptable. In this case we can modify $\widehat{q}$ to make it non-negative (e.g., by adding to each element of $\widehat{q}$ a suitable value), and thus receive a new signal $\hat{q}$.

Another problem arises. The mean value of the resultant signal $\hat{q}$ will obviously differ from the initial one. To overcome this problem, we need to multiply it by such a coefficient that the result has the required mean value. Due to the algebraic properties of convolution, both resultant details' and approximation' absolute values will differ from the initial ones by that precise coefficient. This means that the details will be changed proportionally which totally suits our problem statement requirements. In result, we obtain our required signal $\tilde{q}$.

To illustrate this method we will consider a practical example.

3 Experimental Results

To show the method under review in action, we took the 5-Percent Public Use Microdata Sample Files from U.S. Census Bureau [13] corresponding to the 2000 U.S. Census microfile data on the state of California.

The microfile provides various information on more than 1,6 million respondents. We took a "Military service" attribute as a vital one. This attribute is a categorical one. Its values are integers from 0 to 4. For simplicity, we took one vital value combination consisting of only one vital value, i.e. "1". It stands for "Active duty".

We also took "Place of Work Super-PUMA" as a parameter attribute. This attribute is also a categorical one. Its values stand for different statistical area codes.

For our example, we decided to take the following attribute values as parameter ones: 06010, 06020, 06030, 06040, 06060, 06070, 06080, 06090, 06130, 06170, 06200, 06220, 06230, 06409, 06600 and 06700. These codes correspond to border, coastal and island statistical areas.

By choosing these exact attributes we actually set a task of protecting information on military officers' number distribution over particular Californian statistical areas.

According to Section 2, we need to construct an appropriate quantity signal. The simplest way to do that is to count respondents with appropriate pair of a vital value combination and a parameter value. The results are shown in Table 2 (the third row).

Let's use the second order Daubechies low-pass wavelet filter $l \equiv \left(\frac{1+\sqrt{3}}{4\sqrt{2}}, \frac{3+\sqrt{3}}{4\sqrt{2}}, \frac{3-\sqrt{3}}{4\sqrt{2}}, \frac{1-\sqrt{3}}{4\sqrt{2}} \right)$ to perform two-level wavelet decomposition (**2**) of a corresponding quantity signal (all the calculations were carried out with 12 decimal numbers, but we will present all the numeric data with 3 decimal numbers):

$a_2 = (a_2(1), a_2(2), a_2(3), a_2(4)) = (2272.128,\ 136.352,\ 158.422,\ 569.098)$.

Now, let us construct a suitable WRM:

$$M_{rec} = \begin{pmatrix} 0.637 & 0 & 0 & -0.137 \\ 0.296 & 0.233 & 0 & -0.029 \\ 0.079 & 0.404 & 0 & 0.017 \\ -0.012 & 0.512 & 0 & 0 \\ -0.137 & 0.637 & 0 & 0 \\ -0.029 & 0.296 & 0.233 & 0 \\ 0.017 & 0.079 & 0.404 & 0 \\ 0 & -0.012 & 0.512 & 0 \\ 0 & -0.137 & 0.637 & 0 \\ 0 & -0.029 & 0.296 & 0.233 \\ 0 & 0.017 & 0.079 & 0.404 \\ 0 & 0 & -0.012 & 0.512 \\ 0 & 0 & -0.137 & 0.637 \\ 0.233 & 0 & -0.029 & 0.296 \\ 0.404 & 0 & 0.017 & 0.079 \\ 0.512 & 0 & 0 & -0.012 \end{pmatrix}.$$

According to (**5**), we obtain a signal approximation:

$A_2 = (1369.821,\ 687.286,\ 244.677,\ 41.992,\ -224.98,\ 11.373,\ 112.86,\ 79.481,\ 82.24,\ 175.643,\ 244.757,\ 289.584,\ 340.918,\ 693.698,\ 965.706,\ 1156.942)$.

As we can see, according to the extremum transition approach, we have to lower the military men quantity in the 06700 area. At the same time, we have to raise appropriate quantities in some other areas. The particular choice either may depend on any additional goals to achieve or it can be absolutely arbitrary.

But, along with this, we have to avoid incidental raising of the other signal elements. We can achieve this by using appropriate constraints. Also, it is necessary to note down that there can possibly be some signal elements which do not play important role, i.e. we can change them without any restrictions.

To show how to formally express suitable constraints, we decided to raise the quantities in such central-part signal elements like 06070, 06080, 06090, 06130,

06170 and 06200; besides, we have chosen the first and the last three signal elements to lower their values.

Considering these requirements, we get the following constraints:

$$\begin{cases} 0.637\cdot\hat{a}_2(1)-0.137\cdot\hat{a}_2(4)\le 1369.821 \\ 0.296\cdot\hat{a}_2(1)+0.233\cdot\hat{a}_2(2)-0.029\cdot\hat{a}_2(4)\le 687.286 \\ 0.079\cdot\hat{a}_2(1)+0.404\cdot\hat{a}_2(2)+0.017\cdot\hat{a}_2(4)\le 244.677 \\ -0.137\cdot\hat{a}_2(1)+0.637\cdot\hat{a}_2(2)\ge -224.980 \\ -0.029\cdot\hat{a}_2(1)+0.296\cdot\hat{a}_2(2)+0.233\cdot\hat{a}_2(3)\ge 11.373 \\ 0.017\cdot\hat{a}_2(1)+0.079\cdot\hat{a}_2(2)+0.404\cdot\hat{a}_2(3)\ge 112.860 \\ -0.012\cdot\hat{a}_2(2)+0.512\cdot\hat{a}_2(3)\ge 79.481 \\ -0.137\cdot\hat{a}_2(2)+0.637\cdot\hat{a}_2(3)\ge 82.240 \\ -0.029\cdot\hat{a}_2(2)+0.296\cdot\hat{a}_2(3)+0.233\cdot\hat{a}_2(4)\ge 175.643 \\ 0.233\cdot\hat{a}_2(1)-0.029\cdot\hat{a}_2(3)+0.296\cdot\hat{a}_2(4)\le 693.698 \\ 0.404\cdot\hat{a}_2(1)+0.017\cdot\hat{a}_2(3)+0.079\cdot\hat{a}_2(4)\le 965.706 \\ 0.512\cdot\hat{a}_2(1)-0.012\cdot\hat{a}_2(4)\le 1156.942\,. \end{cases}$$

A possible solution is $\hat{a}_2 = (0,\ 379.097,\ 31805.084,\ 5464.854)$.

Using M_{rec} and **(5)**, we can get a new approximation:

$\hat{A}_2 = (-750.103,\ -70.090,\ 244.677,\ 194.196,\ 241.583,\ 7530.756,\ 12879.498,\ 16287.810,\ 20216.058,\ 10670.153,\ 4734.636,\ 2409.508,\ -883.021,\ 693.698,\ 965.706,\ -66.997)$.

Since our integral aim is to preserve signal details, we construct our masked quantity signal by adding a new approximation and primary details:

$\hat{q} = \hat{A}_2 + D_1 + D_2 = (-2100.924,\ -745.376,\ 153.000,\ 223.204,\ 479.563,\ 7598.383,\ 12773.639,\ 16241.328,\ 20149.818,\ 10764.510,\ 5301.879,\ 2254.924,\ -982.939,\ 14.000,\ 60.000,\ 3113.061)$.

As we can see, some signal elements are negative. Since quantities cannot be negative, we need to add to every signal's element an appropriate value, e.g. 2500:

$\hat{q} = (399.076,\ 1754.624,\ 2653.000,\ 2723.204,\ 2979.563,\ 10098.383,\ 15273.639,\ 18741.328,\ 22649.818,\ 13264.510,\ 7801.879,\ 4754.924,\ 1517.061,\ 2514.000,\ 2560.000,\ 5613.061)$.

Here, all the signal samples are non-negative. Therefore, the only requirement not fulfilled yet is the equality of corresponding mean values. To provide that, we need to multiply $\hat{q}$ by the coefficient $\sum_{i=1}^{16} q_i / \sum_{i=1}^{16} \hat{q}_i = 0.054$.

The resultant signal has the same mean value and wavelet decomposition details as the initial one. This can be checked-up through easy but rather cumbersome calculations.

Since quantities can be only integers, we need to round the signal. Finally, we get the required quantity signal $\tilde{q}$ (see Table 2, the fourth row).

As we can see, the masked data are completely different from the primary ones, though both mean value and wavelet decomposition details are preserved.

To finish the task, we need to compile a new microfile. It is always possible to do as long as there are enough records to modify vital values of. Anyway, we can always demand this when building-up linear programming problem constraints.

Table 2. Quantity signals for the U.S. Census Bureau microfile

Column number	1	2	3	4	5	6	7	8
Area code	06010	06020	06030	06040	06060	06070	06080	06090
Signal q	19	12	153	71	13	79	7	33
Signal $\tilde{q}$	22	95	144	148	162	549	831	1019

Column number	9	10	11	12	13	14	15	16
Area code	06130	06170	06200	06220	06230	06409	06600	06700
Signal q	16	270	812	135	241	14	60	4337
Signal $\tilde{q}$	1232	722	424	259	83	137	139	305

4 Conclusion and Future Research

In the paper, we have set the task of providing group anonymity as a task of protecting such collective data patterns that cannot be retrieved by analyzing individual information only.

We have proposed a wavelet-based method which aims at preserving the data wavelet details as a source of information on the data patterns and relations between their components with different frequencies, along with the data mean value. At the same time, the method actually provides group anonymity since an appropriate level of uncertainty is being introduced into the data (by modifying the wavelet approximation).

The method is relatively easy and can be implemented programatically. Also, the method is rather flexible and can yield various resultant data sets depending on the particular task definition. Moreover, it can be combined with any existing individual anonymity methods to gain the most efficiently protected datasets.

On the other hand, the method isn't acceptable in various cases because it doesn't guarantee that some statistical data features, such as standard deviation, persist.

In the paper, we only pointed out the problem of group anonymity. There remain many questions to answers and challenges to response. Among them we would especially like to stress on such ones:

- Using different wavelet bases can lead to obtaining different data sets.
- Modifying quantity signals isn't very useful for different real-life examples. In situations like protecting the regional distribution of middle-aged people the relative data such as ratios seem to be more important to protect.
- In general, it is not always easy to define parameter and vital sets to determine the records to redistribute. This procedure also needs to be studied thoroughly in the future.

References

1. Gantz, J.F., Reinsel, D.: As the Economy Contracts, the Digital Universe Expands. An IDC Multimedia White Paper (2009), http://www.emc.com/collateral/demos/microsites/idc-digital-universe/iview.htm
2. Sweeney, L.: Computational Disclosure Control: A Primer on Data Privacy. Ph.D. Thesis. Massachusetts Institute of Technology, Cambridge (2001)
3. Aggarwal, C.C., Yu, P.S. (eds.): Privacy-Preserving Data Mining: Models and Algorithms. Springer, New York (2008)
4. Sweeney, L.: k-anonymity: a Model for Protecting Privacy. International Journal on Uncertainty, Fuzziness and Knowledge-based Systems 10(5), 557–570 (2002)
5. Machanavajjhala, A., Kifer, D., Gehrke, J., Venkitasubramaniam, M.: *l*-diversity: Privacy Beyond *k*-anonymity. ACM Transactions on Knowledge Discovery from Data 1(1) (2007)
6. Li, N., Li, T., Venkatasubramanian, S.: *t*-closeness: Privacy Beyond *k*-anonymity and *l*-diversity. In: 23rd International Conference on Data Engineering, pp. 106–115. IEEE Computer Society, Washington (2007)
7. Chertov, O., Pilipyuk, A.: Statistical Disclosure Control Methods for Microdata. In: International Symposium on Computing, Communication and Control, pp. 338–342. IACSIT, Singapore (2009)
8. Meyerson, A., Williams, R.: General *k*-anonymization is Hard. Technical Report CMU-CS-03-113, Carnegie Mellon School of Computer Science (2003)
9. Davydov, A.: Wavelet-analysis of the Social Processes. Sotsiologicheskie issledovaniya, 11, 89–101 (2003) (in Russian), http://www.ecsocman.edu.ru/images/pubs/2007/10/30/0000315095/012.DAVYDOV.pdf
10. Mallat, S.: A Wavelet Tour of Signal Processing. Academic Press, New York (1999)
11. Liu, L., Wang, J., Zhang, J.: Wavelet-based Data Perturbation for Simultaneous Privacy-Preserving and Statistics-Preserving. In: 2008 IEEE International Conference on Data Mining Workshops, pp. 27–35. IEEE Computer Society, Washington (2008)
12. Strang, G., Nguyen, T.: Wavelet and Filter Banks. Wellesley-Cambridge Press, Wellesley (1997)
13. U.S. Census 2000. 5-Percent Public Use Microdata Sample Files, http://www.census.gov/Press-Release/www/2003/PUMS5.html

Anonymizing Categorical Data with a Recoding Method Based on Semantic Similarity

Sergio Martínez, Aida Valls, and David Sánchez

Departament d'Enginyeria Informàtica i Matemàtiques, Universitat Rovira i Virgili
Avda. Països Catalans, 26, 43007 Tarragona, Spain
{sergio.martinezl,aida.valls,david.sanchez}@urv.cat

Abstract. With the enormous growth of the Information Society and the necessity to enable access and exploitation of large amounts of data, the preservation of its confidentiality has become a crucial issue. Many methods have been developed to ensure the privacy of numerical data but very few of them deal with textual (categorical) information. In this paper a new method for protecting the individual's privacy for categorical attributes is proposed. It is a masking method based on the recoding of words that can be linked to less than *k* individuals. This assures the fulfillment of the *k*-anonymity property, in order to prevent the re-identification of individuals. On the contrary to related works, which lack a proper semantic interpretation of text, the recoding exploits an input ontology in order to estimate the semantic similarity between words and minimize the information loss.

Keywords: Ontologies, Data analysis, Privacy-preserving data-mining, Anonymity, Semantic similarity.

1 Introduction

Any survey's respondent (i.e. a person, business or other organization) must be guaranteed that the individual information provided will be kept confidential. Statistical Disclosure Control discipline aims at protecting statistical data in a way that it can be released and exploited without publishing any private information that could be linked with or identify a concrete individual. In particular, in this paper we focus on the protection of microdata, which consists on values obtained from a set of respondents of a survey without applying any summarization technique (e.g. publishing tabular data or aggregated information from multiple queries) [1].

Since data collected from statistical agencies is mainly numerical, several different anonymization methods have been developed for masking numerical values in order to prevent the re-identification of individuals [1]. Textual data has been traditionally less exploited, due to the difficulties of handling non-numerical values with inherent semantics. In order to simplify its processing and anonymization, categorical values are commonly restricted to a predefined vocabulary (i.e. a bounded set of modalities). This is a serious drawback because the list of values is fixed in advance and, consequently, it tends to homogenise the sample. Moreover, the masking methods for categorical data do not usually consider the semantics of the terms (see section 2). Very

E. Hüllermeier, R. Kruse, and F. Hoffmann (Eds.): IPMU 2010, Part II, CCIS 81, pp. 602–611, 2010.

few approaches have considered semantics in some degree. However, they require the definition of ad-hoc structures and/or total orderings of data before anonymizing them. As a result, those approaches cannot process unbounded categorical data. This compromises their scalability and applicability. Approximate reasoning techniques may provide interesting insights that could be applied to improve those solutions [2]. As far as we know, the use of methods specially designed to deal with uncertainty has not been studied in this discipline until now.

In this work, we extend previous methods by dealing with unbounded categorical variables which can take values from a free list of linguistic terms (i.e. potentially the complete language vocabulary). That is, the user is allowed to write the answer to a specific question of the survey using any noun phrase. Some examples of this type of attributes can be "Main hobby" or "Most preferred type of food".

Unbounded categorical variables provide a new way of obtaining information from individuals, which has not been exploited due to the lack of proper anonymization tools. Allowing a free answer, we are able to obtain more precise knowledge of the individual characteristics, which may be interesting for the study that is being conducted. However, at the same time, the privacy of the individuals is more critical, as the disclosure risk increases due to the uniqueness of the answers.

In this paper, an anonymization technique for this kind of variables is proposed. The method is based on the replacement or recoding of the values that may lead to the individual re-identification. This method is applied locally to a single attribute. Attributes are usually classified as *identifiers* (that unambiguously identify the individual), *quasi-identifiers* (that may identify some of the respondents, especially if they are combined with the information provided by other attributes), *confidential* outcome attributes (that contain sensitive information) and *non-confidential* outcome attributes (the rest). The method proposed is suitable for quasi-identifier attributes.

In unbounded categorical variables, textual values refer to concepts that can be semantically interpreted with the help of additional knowledge. Thus, terms can be interpreted and compared from a semantic point of view, establishing different degrees of similarity between them according to their meaning (e.g. for hobbies, *treking* is more similar to *jogging* than to *dancing*). The estimation of semantic similarity between words is the basis of our recoding anonymization method, aiming to produce higher-quality datasets and to minimize information loss.

The computation of the semantic similarity between terms is an active trend in computational linguistics. That similarity must be calculated using some kind of domain knowledge. Taxonomies and, more generally ontologies [3], which provide a graph model where semantic relations are explicitly modelled as links between concepts, are typically exploited for that purpose (see section 3). In this paper we focus on similarity measures based on the exploitation of the taxonomic relations of ontologies.

The rest of the paper is organized as follows. Section 2 reviews methods for privacy protection of categorical data. Section 3 introduces some similarity measures based on the exploitation of ontologies. In section 4, the proposed anonymization method is detailed. Section 5 is devoted to evaluate our method by applying it to real data obtained from a survey at the National Park "*Delta del Ebre*" in Catalonia, Spain. The final section contains the conclusions and future work.

2 Related Work

A register is a set of attribute values describing an individual. Categorical data is composed by a set of registers (i.e. records), each one corresponding to one individual, and a set of textual attributes, classified as indicated before (identifiers, quasi-identifiers, confidential and non-confidential). The anonymization or masking methods of categorical values are divided in two categories depending on their effect on the original data [4]:

- *Perturbative*: data is distorted before publication. They are mainly based on data swapping (exchanging the values of two different records) or the addition of some kind of noise, such as the replacement of values according to some probability distribution (PRAM) [5], [6] and [7].
- *Non-perturbative*: data values are not altered but generalized or eliminated [8], [4]. The goal is to reduce the detail given by the original data. This can be achieved with the local suppression of certain values or with the publication of a sample of the original data which preserves the anonymity. Recoding by generalization is also another approach, where several categories are combined to form a new and less specific value.

Anonymization methods must mask data in a way that disclosure risk is ensured at an enough level while minimising the loss of accuracy of the data, i.e. the information loss. A common way to achieve a certain level of privacy is to fulfil the *k-anonymity* property [9]. A dataset satisfies the *k*-anonymity if, for each combination of attribute values, there exist at least *k-1* indistinguishable records in the dataset. On the other hand, low information loss guarantees that useful analysis can be done on the masked data.

With respect to recoding methods, some of them rely on hierarchies of terms covering the categorical values observed in the sample, in order to replace a value by another more general one.

Samariti and Sweeney [10] and Sweeney [9] employed a generalization scheme named Value Generalization Hierarchy (VGH). In a VGH, the leaf nodes of the hierarchy are the values of the sample and the parent nodes correspond to terms that generalize them. In this scheme, the generalization is performed at a fixed level of the hierarchy. The number of possible generalizations is the number of levels of the tree. Iyengar [11] presented a more flexible scheme which also uses a VGH, but a value can be generalized to different levels of the hierarchy; this scheme allows a much larger space of possible generalizations. Bayardo and Agrawal [12] proposed a scheme which does not require a VGH. In this scheme a total order is defined over all values of an attribute and partitions of these values are created to make generalizations. The problem is that defining a total order for categorical attributes is not straightforward.

T.Li and N. Li [13] propose three generalization schemes: Set Partitioning Scheme (SPS), in which generalizations do not require a predefined total order or a VGH; each partition of the attribute domain can be a generalization. Guided Set Partitioning Scheme (GSPS) uses a VGH to restrict the partitions that are generated. Finally, the

Guided Oriented Partition Scheme (GOPS) includes also ordering restrictions among the values.

The main problem of the presented approaches is that either the hierarchies or the total orders are build ad-hoc for the corresponding data value set (i.e. categorical values directly correspond to leafs in the hierarchy), hampering the scalability of the method when dealing with unbounded categorical values. Moreover, as hierarchies only include the categorical data values observed in the sample, the resulting structure is very simple and a lot of semantics needed to properly understand the word's meaning is missing. As a result, the processing of categorical data from a semantic point of view is very limited. This is especially critical in non-hierarchy-based methods, which do not rely on any kind of domain knowledge and, in consequence, due to their completely lack of word understanding, they have to deal with categorical data from the point of view of Boolean word matching.

3 Ontology-Based Semantic Similarity

In general, the assessment of concept's similarity is based on the estimation of semantic evidence observed in a knowledge resource. So, background knowledge is needed in order to measure the degree of similarity between concepts.

In the literature, we can distinguish several different approaches to compute semantic similarity according to the techniques employed and the knowledge exploited to perform the assessment.

The most classical approaches exploit structured representations of knowledge as the base to compute similarities. Typically, subsumption hierarchies, which are a very common way to structure knowledge [3], have been used for that purpose. The evolution of those basic semantic models has given the origin to ontologies. Ontologies offer a formal, explicit specification of a shared conceptualization in a machine-readable language, using a common terminology and making explicit taxonomic and non-taxonomical relationships [14]. Nowadays, there exists massive and general purpose ontologies like WordNet [15], which offer a lexicon and semantic linkage between the major part of English terms (it contains more than 150,000 concepts organized into is-a hierarchies). In addition, with the development of the Semantic Web, many domain ontologies have been developed and are available through the Web [16].

From the similarity point of view, taxonomies and, more generally, ontologies, provide a graph model in which semantic interrelations are modeled as links between concepts. Many approaches have been developed to exploit this geometrical model, computing concept similarity as inter-link distance.

In an is-a hierarchy, the simplest way to estimate the distance between two concepts c_1 and c_2 is by calculating the shortest *Path Length* (i.e. the minimum number of links) connecting these concepts (1) [17].

$$dis_{pL}(c_1,c_2) = \min \# \; of \; is-a \; edges \; connecting \; c_1 \; and \; c_2 \tag{1}$$

Several variations of this measure have been developed such as the one presented by Wu and Palmer [18]. Considering that the similarity between a pair of concepts in an upper level of the taxonomy should be less than the similarity between a pair in a lower level, they propose a path-based measure that also takes into account the depth of the concepts in the hierarchy (2).

$$sim_{w\&p}(c_1,c_2)\frac{2\times N_3}{N_1+N_2+2\times N_3},\tag{2}$$

where N_1 and N_2 are the number of is-a links from c_1 and c_2 respectively to their Least Common Subsumer (LCS), and N_3 is the number of is-a links from the LCS to the root of the ontology. It ranges from 1 (for identical concepts) to 0.

Leacock and Chodorow [19] also proposed a measure that considers both the shortest path between two concepts (in fact, the number of nodes N_p from c_1 to c_2) and the depth D of the taxonomy in which they occur (3).

$$sim_{l\&c}(c_1,c_2)=-\log(N_p/2D)\tag{3}$$

There exist other approaches which also exploit domain corpora to complement the knowledge available in the ontology and estimate concept's *Information Content* (IC) from term's appearance frequencies. Even though they are able to provide accurate results when enough data is available [20], their applicability is hampered by the availability of this data and their pre-processing. On the contrary, the presented measures based uniquely on the exploitation of the taxonomical structure are characterized by their simplicity, which result is a computationally efficient solution, and their lack of constraints as only an ontology is required, which ensures their applicability. The main problem is their dependency on the degree of completeness, homogeneity and coverage of the semantic links represented in the ontology [21]. In order to overcome this problem, classical approaches rely on WordNet's *is-a* taxonomy to estimate the similarity. Such a general and massive ontology, with a relatively homogeneous distribution of semantic links and good inter-domain coverage is the ideal environment to apply those measures [20].

4 Categorical Data Recoding Based on Semantic Similarity

Considering the poor semantics incorporated by existing methods for privacy preserving of categorical values, we have designed a new local method for anonymization based on the semantic processing of, potentially unbounded, categorical values.

Aiming to fulfill the *k*-anonymity property but minimizing the information loss of textual data, it is proposed a recoding method based on the replacement of some values of one attribute by the most semantically similar ones. The basic idea is that, if a value does not fulfilling the *k*-anonymity, it will be replaced by the most semantically similar value on the same dataset. This decreases the number of different values. The process is repeated until the whole dataset fulfils the desired *k*-anonymity. The rationale for this replacement criterion is that if categorical values are interpreted at a conceptual level, the way to lead to the least information loss is to change those values in a way that the semantics of the record – at a conceptual level – is preserved. In order to ensure this, it is crucial to properly assess the semantic similarity/distance between categorical values. Path-length similarities introduced in the previous section have been chosen because they provide a good estimation of concept alikeness at a very low computational cost [19], which is important when dealing with very large datasets, as it is the case of inference control in statistical databases [1].

As categorical data are, in fact, text labels it is also necessary to morphologically process them in order to detect different lexicalizations of the same concept (e.g. singular/plural forms). We apply a stemming algorithm to both text labels of categorical attributes and ontological labels in order to compare words from their morphological root.

The inputs of the algorithm are: a dataset consisting on a single attribute with categorical values (an unbounded list of textual noun phrases) and *n* registers (*r*), the desired level of *k*-anonymity and the reference ontology.

$$
\begin{aligned}
&\textit{Algorithm Ontology-based recoding (dataset, k, ontology)}\\
&r_i' := stem\,(r_i)\,\forall\; i\; in\; [1 \ldots n]\\
&while\,(there\ are\ changes\ in\ the\ dataset)\,do\\
&\quad for\,(i\ in\ [1 \ldots n])\ do\\
&\qquad m := count\,(r_j' = r_i')\,\forall\; j\; in\; [1 \ldots n]\\
&\qquad if\,(m < k)\ then\\
&\qquad\quad r'_{Max} := argMax\,(similarity\,(r_i', r_j', ontology))\,\forall\; j\; in\; [1 \ldots n],\, r_i' \neq r_j'\\
&\qquad\quad r_p' := r'_{Max}\ \forall\; p\; in\; [1 \ldots n],\, r_p' = r_i'\\
&\qquad end\ if\\
&\quad end\ for\\
&end\ while
\end{aligned}
$$

The recoding algorithm works as follows. First, all words of dataset are stemmed, so that, two words are considered equal if their morphological roots are identical. The process iterates for each register r_i of the dataset. First, it checks if the corresponding value fulfils the *k*-anonymity by counting its occurrences. Those values which occur less than *k* times do not accomplish *k*-anonymity and should be replaced. As stated above, the ideal word to replace another one (from a semantic point-of-view) is the one that has the greatest similarity (i.e. the least distant meaning). Therefore, from the set of words that already fulfill the minimum *k*-anonymity, the most similar to the given one according to the employed similarity measure and the reference ontology is found and the original value is substituted. The process finishes when no more replacements are needed, meaning that the dataset fulfills the *k*-anonymity property.

It is important to note that, in our method, categorical values may be found at any taxonomical level of the input ontology. So, in comparison to hierarchical generalization methods introduced in section 2, in which labels are always leafs of the ad-hoc hierarchy and terms are always substituted by hierarchical subsumers, our method replaces terms for the nearest one in the ontology, regardless being a taxonomical sibling (i.e. the same taxonomical depth), a subsumer (i.e. a higher depth) or an specialization (i.e. lower depth), provided that those appear more frequently in the sample (i.e. they fulfill the *k*-anonymity).

5 Evaluation

In order to evaluate our method, we used a dataset consisting on textual answers retrieved from polls made by "*Observatori de la Fundació d'Estudis Turístics Costa Daurada*" at the Catalan National Park "*Delta del Ebre*". The dataset consists on a sample of the answers of the visitors to the question: *What has been the main reason to visit Delta del Ebre?*. As answers are open, the disclosure risk is high, due to the heterogeneity of the sample and the presence of uncommon answers, which are easily identifiable. The test collection has 975 individual registers and 221 different responses, 84 of them are unique (so they can be used to re-identify the individual), while the rest have different amount of repetitions (as shown in Table 1).

Table 1. Distribution of answers in the evaluation dataset (975 registers in total)

Number of repetitions	1	2	3	4	5	6	7	8	9	11	12	13	15	16	18	19	Total
Number of different responses	84	9	6	24	23	37	12	1	2	7	5	1	5	2	2	1	**221**
Total amount of responses	84	18	18	96	115	222	84	8	18	77	60	13	75	32	36	19	**975**

The three similarity measures introduced in section 3 have been implemented and WordNet 2.1 has been exploited as the input ontology. As introduced in section 2, WordNet has been chosen due to its general purpose scope (which formalizes in an unbiased manner concept's meaning) and its high coverage of semantic pointers. To extract the morphological root of words we used the Porter Stemming Algorithm [22].

Our method has been evaluated for the three different similarity measures presented in section 2, in comparison to a random substitution (i.e. a substitution method that consists on replacing each sensible value by a random one from the same dataset so that the level of *k*-anonymity is increased) The results obtained for the random substitution are the average of 5 executions. Different levels of *k*-anonymity have been tested.

The quality of the anonymization method has been evaluated from two points of view. On one hand, we computed the information loss locally to the sample set. In order to evaluate this aspect we computed the Information Content (IC) of each individual of each categorical value after the anonymization process in relation to the IC of the original sample. IC of a categorical value has been computed as the inverse to its probability of occurrence in the sample (4). So, frequently appearing answers had less IC than rare (i.e. more easily identifiable) ones.

$$IC(c) = -\log p(c) \tag{4}$$

The average of the IC value for each answer is subtracted to the average IC of the original sample in order to obtain a quantitative value of information loss with regards to the distribution of the dataset. In order to minimize the variability of the random substitution, we averaged the results obtained for five repetitions of the same test. The results are presented in Figure 1.

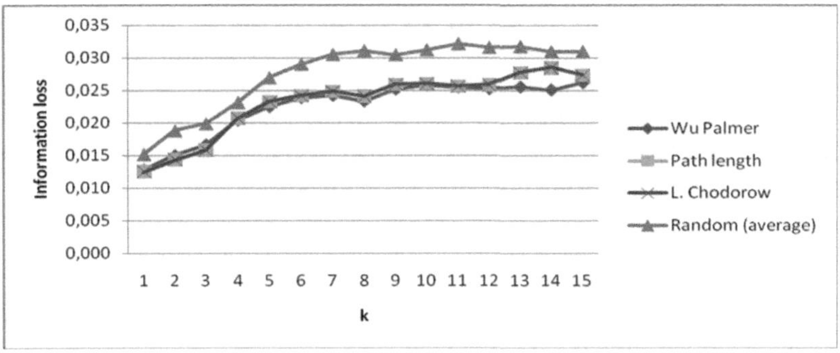

Fig. 1. Information loss based on local IC computation

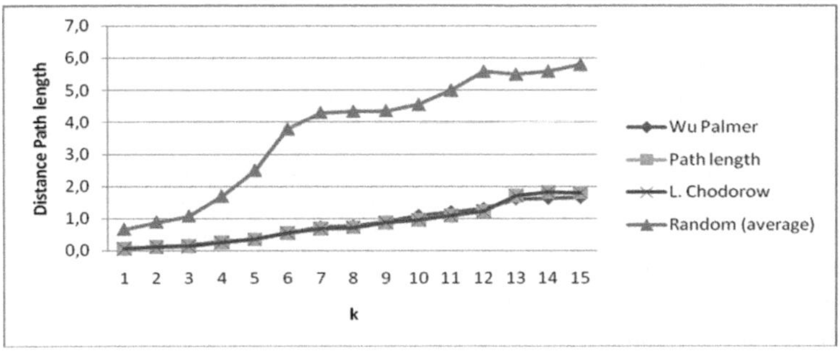

Fig. 2. Semantic distance of the anonymized dataset

To evaluate the quality of the masked dataset from a semantic point of view, we measured how different is the replaced value to the original one with respect to their meaning. This is an important aspect from the point of view of data exploitation as it represents a measure of up to which level the semantics of the original record are preserved. So, we computed the averaged semantic distance from the original dataset and the anonymized one using the Path Length similarity measure in WordNet. Results are presented in Figure 2.

Analyzing the figures we can observe that our approach is able to improve the random substitution by a considerable margin. This is even more evident for a high *k*-anonymity level. Regarding the different semantic similarity measures, they provide very similar and highly correlated results. This is coherent, as all of them are based on the same ontological features (i.e. absolute path length and/or the taxonomical depth) and, even though similarity values are different, the relative ranking of words is very similar. In fact, Path length and Leacock and Chorodow measures gave identical results as the later is equivalent to the former but normalized to a constant factor (i.e. the absolute depth on the ontology). Evaluating the semantic distance in function of the level of *k*-anonymity one can observe a linear tendency with a very smooth growth. This is very convenient and shows that our approach performs well regardless the desired level of anonymization.

The local information loss based on the computation of the averaged IC with respect to the original dataset follows a similar tendency. In this case, however, the information loss tends to stabilize for k values above 9, showing that the best compromise between the maintenance of the sample heterogeneity and the semantic anonymization have been achieved with k=9. The random substitution performs a little worse, even though in this case the difference is much less noticeable (as it tends to substitute variables in a uniform manner and, in consequence, the original distribution of the number of different responses tends to be maintained).

6 Conclusions

On the process of anonymization it is necessary to achieve two main objectives: on one hand, to satisfy the desired k-anonymity to avoid the disclosure, preserving the confidentiality and, on the other hand, to minimize the information loss to maintain the quality of the dataset. This paper proposes a method of local recoding for categorical data, based on the estimation of semantic similarity between values. As the meaning of concepts is taken into account, the information loss can be minimized.

The method uses the explicit knowledge formalized in wide ontologies (like Wordnet) to calculate the semantic similarity of the concepts, in order to generate a masked dataset that preserves the meaning of the answers given by the respondents.

In comparison with the existing approaches for masking categorical data based on generalization of terms, our approach avoids the necessity of constructing ad-hoc hierarchies according to data labels. In addition, our method is able to deal with unbounded attributes, which can take values in a textual form.

The results presented show that with a level of anonymity up to 6, the semantics of the masked data is maintained 3 times more than with a naive approach. Classical information loss measure based on information content also shows an improvement of the ontology-based recoding method.

After this first study, we plan compare our method with the existing generalization masking methods mentioned in section 2, in order to compare the results of the different anonymization strategies. For this purpose, different information loss measures will be considered. Finally, we plan extend the method for global recoding, where different attributes are masked simultaneously.

Acknowledgements

Thanks are given to “Observatori de la Funcació d’Estudis Turístics Costa Daurada” and “Parc Nacional del Delta de l’Ebre (Departament de Medi Ambient i Habitatge, Generalitat de Catalunya)” for providing us the data collected from the visitors of the park. This work is supported the Spanish MEC (projects ARES – CONSOLIDER INGENIO 2010 CSD2007-00004 – and eAEGIS – TSI2007-65406-C03-02). Sergio Martínez Lluís is supported by the Universitat Rovira i Virgili predoctoral research grant.

References

1. Domingo-Ferrer, J.: A survey of inference control methods for privacy-preserving data mining. In: Aggarwal, C.C., Yu, P.S. (eds.) Privacy-Preserving Data Mining: Models and Algorithms. Advances in Database Systems, vol. 34, pp. 53–80. Springer, Heidelberg (2008)

2. Bouchon-Meunier, B., Marsala, C., Rifqi, M., Yager, R.R.: Uncertainty and Intelligent Information Systems. World Scientific, Singapore (2008)
3. Gómez-Pérez, A., Fernández-López, M., Corcho, O.: Ontological Engineering, 2nd printing, pp. 79–84. Springer, Heidelberg (2004)
4. Willenborg, L., De Eaal, T.: Elements of Statistical Disclosure Control. Springer, New York (2001)
5. Guo, L., Wu, X.: Privacy preserving categorical data analysis with unknown distortion parameters. Transactions on Data Privacy 2, 185–205 (2009)
6. Gouweleeuw, J.M., Kooiman, P., Willenborg, L.C.R.J., DeWolf, P.P.: Post randomization for statistal disclousure control: Theory and implementation. Research paper no. 9731 (Voorburg: Statistics Netherlands) (1997)
7. Reiss, S.P.: Practical data-swapping: the first steps. ACM Transactions on Database Systems 9, 20–37 (1984)
8. Xu, J., Wang, W., Pei, J., Wang, X., Shi, B., Wai-Chee Fu, A.: Utility-based anonymization using local recoding. In: Proceedings of the 12th ACM SIGKDD international conference on Knowledge discovery and data mining, Philadelphia, PA, USA, pp. 785–790 (2006)
9. Sweeney, L.: k-anonymity: a model for protecting privacy. International Journal on Uncertainty, Fuzziness and Knowledge-based Systems 10(5), 557–570 (2002)
10. Samarati, P., Sweeney, L.: Protecting privacy when disclosing information: k-anonymity and its enforcement through generalization and suppression, Technical Report SRI-CSL-98-04, SRI Computer Science Laboratory (1998)
11. Iyengar, V.S.: Transforming data to satisfy privacy constraints. In: Proceedings of the 8th ACM SIGKDD Int. Conf. on Knowledge Discovery and Data Minig (KDD), pp. 279–288 (2002)
12. Bayardo, R.J., Agrawal, R.: Data privacy through optimal k-anonymization. In: Proceedings of the 21st International Conference on Data Engineering (ICDE), pp. 217–228 (2005)
13. Li, T., Li, N.: Towards optimal k-anonymization. Data & Knowledge Engineering 65, 22–39 (2008)
14. Guarino, N.: Formal Ontology in Information Systems. In: Guarino, N. (ed.) 1st Int. Conf. on Formal Ontology in Information Systems, pp. 3–15. IOS Press, Trento (1998)
15. Fellbaum, C.: WordNet: An Electronic Lexical Database. MIT Press, Cambridge (1998)
16. Ding, L., Finin, T., Joshi, A., Pan, R., Cost, R.S., Peng, Y., Reddivari, P., Doshi, V., Sachs, Swoogle, J.: A Search and Metadata Engine for the Semantic Web. In: Proc. 13th ACM Conference on Information and Knowledge Management, pp. 652–659. ACM Press, New York (2004)
17. Rada, R., Mili, H., Bichnell, E., Blettner, M.: Development and application of a metric on semantic nets. IEEE Transactions on Systems, Man and Cybernetics 9(1), 17–30 (1989)
18. Wu, Z., Palmer, M.: Verb semantics and lexical selection. In: Proc. 32nd annual Meeting of the Association for Computational Linguistics, New Mexico, USA, pp. 133–138 (1994)
19. Leacock, C., Chodorow, M.: Combining local context and WordNet similarity for word sense identification. In: Fellbaum (ed.) WordNet: An electronic lexical database, pp. 265–283. MIT Press, Cambridge (1998)
20. Jiang, J., Conrath, D.: Semantic similarity based on corpus statistics and lexical taxonomy. In: Proc. Int. Conf. on Research in Computational Linguistics, Japan, pp. 19–33 (1997)
21. Cimiano, P.: Ontology Learning and Population from Text. In: Algorithms, Evaluation and Applications, Springer, Heidelberg (2006)
22. Porter: An algorithm for suffix stripping. Program 14(3), 130–137 (1980)

Addressing Complexity in a Privacy Expert System

Siani Pearson

HP Labs, Long Down Avenue, Stoke Gifford, Bristol, BS34 8QZ, UK
Siani.Pearson@hp.com

Abstract. This paper shows how a combination of usability and heuristics can be used to reduce complexity for privacy experts who create and maintain the knowledge base of a decision support system. This system helps people take privacy into account during decision making without being overwhelmed by the complexity of different national and sector-specific legislation.

Keywords: privacy, decision support, usability, knowledge engineering.

1 Introduction

Privacy management for multinational companies is challenging due to the complex web of legal requirements and movement of data and business operations to cost-effective locations. Privacy requirements need to be addressed by dispersed teams, within the context of a variety of business processes, in a global context, and increasingly by people with little knowledge of privacy who have to handle personal data as part of their job. Organisational privacy rulebooks often run into hundreds of pages, and so it is not practical to expect employees' to know all of this information. A decision support system (DSS) can help with this problem by addressing the complexity of compliance requirements for end users, and particularly by assisting individuals who are not experts in privacy and security to find out what to do and highlight where they might not be compliant or where their behaviour poses risks.

Such a tool will have a knowledgebase (KB) that needs to be created and updated by experts on an ongoing basis. These experts can potentially be trained in this process, but they will usually be non-IT staff and may find it difficult to handle complex representations. Therefore, there is a need to address the complexity of the KB updating process. In this paper we explain our approach to this issue, which centres on provision of a novel user interface that facilitates arduous knowledge creation and maintenance tasks and reduces the need for training. Our approach is influenced by Alan Kay's maxim that 'Simple things should be simple, complex things should be possible' [1]. In a 'simple mode' for knowledge maintenance, heuristics are used to hide much of the complexity of the underlying representations from end users, and to fill in appropriate settings within the rules to allow creation of a basic functioning rulebase in a non-complex way. In our case study we have focused on privacy but this approach could be used for a number of other domains; however, privacy is a particularly suitable domain because of the contextual nature of privacy advice.

E. Hüllermeier, R. Kruse, and F. Hoffmann (Eds.): IPMU 2010, Part II, CCIS 81, pp. 612–621, 2010.

2 A Privacy Decision Support System

Our DSS is an expert system that captures data about business processes to determine their compliance. The tool supplies individuals who handle data with sufficient information and guidance to ensure that they design their project in compliance.

There are two types of user: end users (who fill in a questionnaire from which a report is generated), and domain experts (who create and maintain the KB). When an end user uses the DSS, they are initially taken through a series of customised questions and, based on their answers, a compliance report is automatically generated. They can use the tool in an educational 'guidance' mode, where their input is not logged, or alternatively in an 'assessment' mode where a report is submitted that scores the project for a list of risk indicators and a record is retained in the database. Where an issue has been identified, guidance is offered online that links into the external information sources and checklists and reminders are provided. In addition to this user perspective, the system provides a domain expert perspective which is a knowledge management interface for KB creation and update.

2.1 The Underlying Rule Representation

The DSS uses a rules engine, for which two types of rules are defined:

1. **question rules:** these automatically generate questions, in order to allow more subtlety in customisation of the questionnaire to the end user's situation
2. **domain rules:** these generate an output report for end users and potentially also for auditing purposes (with associated checklist, indication of risk, etc.)

All these rules have the general form: **when** condition **then** action.

The DSS uses a set of intermediate variables (IMs) to encode meaningful information about the project and drive the questionnaire, e.g. the IM 'project has transborder data flow' indicates that the current context allows transborder data flow.

The questionnaire maps to a tripartite graph structure as illustrated in Figure 1. The left nodes are monotonic expressions involving (question, answer) pairs. The middle partition consists of intermediate nodes that are semantically meaningful IMs. The right set of nodes represents "new" question(s) that will be asked. The question rules map to lines in Figure 1: they have as their conditions a monotonic expression (i.e. Boolean expression built up using *&* and *v* as logical operators) in IMs and/or (question, answer) pairs and as actions, directives to ask the user some questions or to set some IMs. The domain rules' condition is a Boolean expression in a set of IMs and answers to questions (cf. the conditions column of Figure 1) and they generate as their actions the content of the output report. See [2] for further details.

Further complexity in the rule expressions arises from the following system features, intended to enhance the end user experience:

- Customised help can be provided, by means of using rules where trigger conditions involve (question, answer) pairs and/or IMs and the inference engine is run to determine the appropriate help
- Subsections allow display of questions related to more complex knowledge

- The parameter "breadth first" (BF) or "depth first" (DF) attached to a question controls whether it is added in a 'drill down' fashion, i.e. immediately after the question which led to triggering it, or appended at the end of the list of questions
- An IM expression can trigger a set of questions instead of just one within a rule. In that case the order of questions specified by the expert user in the rule is respected when this block of questions is shown

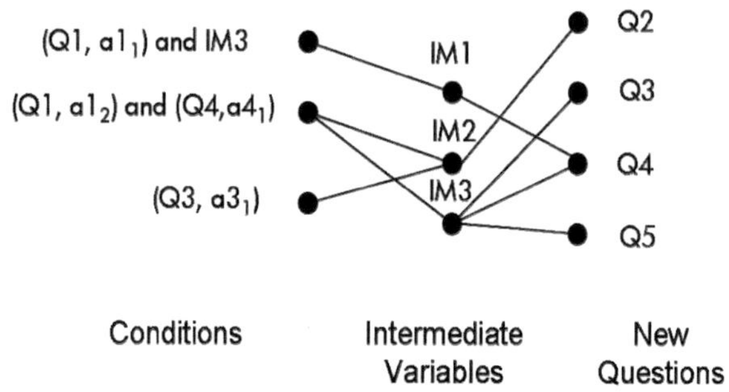

Fig. 1. A representation of the questionnaire using tripartite graphs

Let us consider a simple example of the underlying representation, in DRL format (although the rules can automatically be converted to XML format). Assume that an end user is answering a questionnaire, and that the question "*Is data confined to one country?*" is answered "No". This (question, answer) pair is added to working memory and as consequence the following question rule is triggered, asserting a new IM "Inv_Transborder_Data_Flow":

```
rule "IMR21" when QA (id == 48, value == "No") then
insert(new IM("Inv_Transborder_Data_Flow","Yes")); end
```

When the previous IM is asserted to working memory it triggers the following question rule which adds three new questions to the questionnaire:

```
rule "QR17" salience 1000
when IM (name == "Inv_Transborder_Data_Flow", value == "Yes")
then AddToDisplayList_DF(current, currentQuestion, new long[]
{49, 50, 51}); end
```

The initial (question, answer) pair will also generate a new parameter instance: "Data confined to one country" with value "No". When this parameter instance is added to the working memory of the privacy engine it triggers the following domain rule:

```
rule "Data confined to one country"
when ParameterInstance ( name == "Data confined to one
country" , value == "No"  )
then report.addRule(new RuleFacade().findById(50)); end
```

This rule adds a *Rule* object to the list of rules of the report. The rule will show a yellow flag (to indicate the seriousness of the issue) in the risk indicator "Transborder Data Flow" with the reason: "*Transborder data flow is involved in the project.*" More broadly, domain rules can generate as actions other items to be included within the report: a checklist entry which describes what the user should do about the issue raised in this rule; a link to more information.

2.2 Our Implementation: HP Privacy Advisor (HPPA)

HP Privacy Advisor (HPPA) is a DSS of the form described above that supports enterprise accountability: it helps an organisation to ensure privacy concerns are properly and proactively taken into account in decision making in the businesses as well as provide some assurance that this is case [2]. HPPA analyses projects' degree of compliance with HP privacy policy, ethics and global legislation, and integrates privacy risk assessment, education and oversight. Our implementation uses the production rule system Drools [3] for the rules engine and this is run after each question is answered by the user. Since the domain is focused on privacy, we refer to the domain rules as 'privacy rules'.

Several different methods were used for end user testing, and reactions to the tool have been overwhelmingly positive. We have also had validation from privacy experts when learning to use the KB management UIs that the simple mode described in the following section was very helpful, and have undergone a number of iterative improvements to the prototype based upon their suggestions in order to build up a privacy KB. In particular, these experts have entered privacy knowledge using these UIs into the tool that encodes the information from the 300-page HP privacy rulebook.

3 Simple Mode: Simplifying KB Maintenance

With regard to the DSS described in the previous section, the following issue relating to KB maintenance needs to be addressed: how can a non-IT person deal with the complex rule representation and create questionnaires and rules in an easy way?

We found the 'expert mode' screens initially implemented within HPPA for creating and editing question rules and privacy rules too complex for a non-trained person. These screens exposed the representation of the rules in a DRL-type format, as illustrated in Section 2, and also more complex editing that included customised help, tooltips and warnings, question sections, tagging, DF or BF generation of questions, etc. The complexity of this was particularly an issue as the domain experts usually do not have a technical background.

Hence we needed to find a reasonably simple means to update the rules in the KB that would work in the majority of cases and that can be used without the need for training or manipulating the underlying Drools representations. We foresaw two categories of domain experts: those who can carry out simple KB changes and build new questionnaires and those able to fine-tune the rules in the system. We designed a 'simple mode' for the former that could also be used by the latter. Our approach was to combine intuitive UIs with heuristics that hide the underlying complexity, as follows.

3.1 Usability Aspects

For the question rules, we designed a closer link between authoring and the finished questionnaire. The authoring environment resembles the questionnaire in layout, and the authoring vocabulary is closer to the vocabulary of use (i.e. not rules and variables but questions and answers): if you answer A then you are asked the follow up question B, and so on. The previewing of question sequences allows users to quickly switch from previewing a question in a sequence to editing that same question. The input screens for the privacy rules were also simplified.

We decided to restrict the interface for 'simple mode' to a small set of possible constructs. We actively fought against 'feature-creep', taking our goal for this mode as an interface that is restrictive. We had to balance restrictions against increased ease of learning and use: users can always enlist help or undertake training to achieve more complex goals, using the expert screens.

3.2 Heuristics

Analysis of our KB helped focus attention on the 'simple' tasks which make up the majority of the rules which are actually likely to be written by privacy experts, e.g.:

- Most questions had answers 'yes', 'no' and 'do not know'
- Most question-setting IMs had a trigger condition of the form: "When QA(id==ID, value==Value)"
- Most privacy rules had a trigger condition of the form: "When Parameter is Value"

The simplified UIs focus on making it easy to do these tasks; heuristics are used to hide the complexity of the underlying representation. In general they enable translation of the user requirements coming from the UIs into the machine readable formats of the rules discussed in Section 2. Thereby, Drools representations and IMs are not exposed to the simple user, and the corresponding 'simple' rules are built up by the system. There is no differentiation in the KB about rules derived from expert or simple mode, and this is instead derived from analysing those formats that can be manipulated by simple mode: if a privacy rule is created from the expert screens and has a complex trigger condition then the user is directed to switch to the expert mode to view and edit the rule; otherwise it may be edited within simple mode.

Examples of heuristics used include the following:

- governing whether the rules generated are BF or DF. For instance, when building the questionnaire, users can add follow-on questions; the BF and DF rules are separated by using the section information stored within the follow-up question itself. Questions in the same section as the parent question are made to be DF and questions from different sections are paired and saved as BF mode question rules.
- analysing whether rules need to be combined in order to express more than one follow-up question being generated
- generating IMs when questions are created in simple mode in order to automatically create the corresponding question rules

In addition, the following mechanisms are used:

- inheriting tags (used in order to identify subject domains) from higher levels in the questionnaire hierarchy (although the user can override this)
- maintaining a list of 'incomplete' nodes within the questionnaire 'tree' that the user should return to in order to complete the questionnaire. For example, if all answers to a question have follow-up questions defined or are marked as complete then the question is removed from the 'incomplete' list

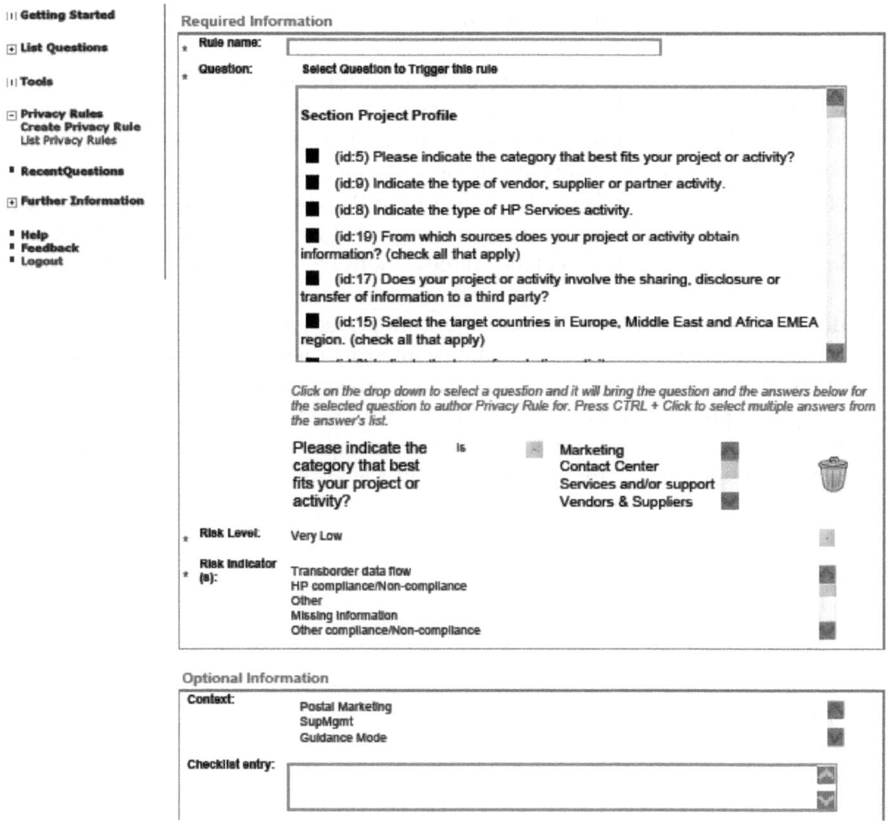

Fig. 2. Create Privacy Rule in Simple Mode

- preventing the user defining recursive chains when building up the questionnaire by checking there is no duplication of questions in each path

Despite the use of such mechanisms, we found that there are some aspects of the underlying system whose complexity is difficult to avoid and where the resulting solution could still be confusing to the user, notably:

- There is a need to distinguish between 'guidance' and 'assessment' mode mappings. Our solution was to categorise the rules into three modes that can be selected by users: 'guidance', 'assessment' or 'both': this obviates the need to find out the intersection or union of the mappings that exist in both modes. The active mode can be selected in the 'list questions' screen with the default selection as 'assessment'. Hence, if the context is set as 'assessment' then all the filters are done for that mode, so all question rules are checked for the mode selected before modifying them and the rules for 'guidance' or 'both' are not changed.
- Certain edits could cause major ramifications for other rules: for example, if a user edits a question (for example, amending answer text) that has follow-up questions defined then it is difficult to predict whether or not to keep or break the

corresponding links, and to what extent to highlight the effect on the privacy rules that might be triggered – directly or indirectly – by the original question but not the amended version. It is difficult to come up with a heuristic to decide accurately whether the associated rules should be amended or deleted, and so the user should be involved in this decision. Our solution to this was to show a notification to the user in simple mode that this affects the associated rules if they want to make this change, before they make it. They then have the choice whether this edit is automatically propagated throughout the rules, or whether to check the consequences via the expert mode, where the detailed ramifications on the other rules are displayed.

As discussed in Section 4, the translation from privacy laws to human-readable policies to machine-readable policies cannot be an exact one. We assume that the privacy expert is able to express in a semi-formalised manner corporate privacy policies or similar prescriptive rules that can be input directly via the UIs and then we automatically encode these into the system rules. Corporate privacy policies would already be close to a suitable form: for example, as illustrated in Figure 2, the 'simple mode'

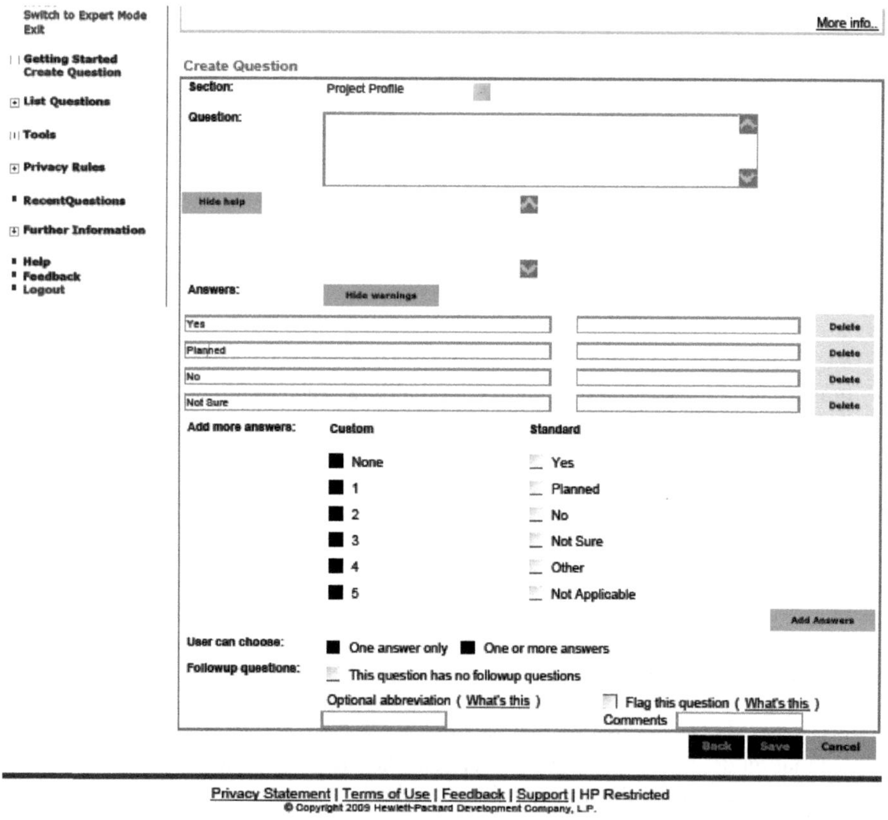

Fig. 3. Create Questionnaire Rule in Simple Mode

Fig. 4. List Privacy Rules in Simple Mode

input required to create a privacy rule is: a rule description; the question and answer(s) that triggers the output; what the output is (i.e. the risk level, risk indicator and optional information). A similar approach is taken for screens that allow editing.

Figure 3 illustrates how a simplified approach can be provided to enable generation of question rules. Additional screens allow creation and linkage of follow-on questions, editing question rules, listing questions (and subsets of the KB e.g. tagged questions), simple mode help and previewing the questionnaire (in the sense of stepping through paths of the questionnaire to try it out); for space reasons we are unable to display these UIs in this paper. The system can also highlight parts of the questionnaire that are unfinished, so that the user can complete these. Figure 4 shows how the privacy rules KB may be viewed in an intuitive form.

A number of open issues remain and we are working to refine our solutions. For example, all kinds of questions in natural language are allowed. Therefore, the system cannot automatically identify duplication of questions that are semantically equivalent but syntactically different. We do solve a restricted from of this problem by requesting the user to check a box when editing questions to indicate whether or not the new content is semantically equivalent to the old content, and hence enabling us to maintain the relationships between the corresponding rules in the former case.

4 Related Work

Policy specification, modelling and verification tools include EPAL [4], OASIS XACML [5], W3C P3P [6] and Ponder [7]. These policies, however, are at a different level to the ones we are dealing with in this paper, as for example they deal with operational policies, access control constraints, etc. and not a representation of country or context-specific privacy requirements. In addition they are targeted towards machine execution and the question of intermediate, human-actionable representation of policies has so far not been paid attention to in the policy research community. Related technologies in the Sparcle [8] and REALM projects [9] do not produce output useful for humans. OASIS LegalXML [10] has worked on creation and management of contract documents and terms, but this converts legal documents into an XML format that is too long to be human readable and not at the right level for the representation we need in our system. Breaux and Antón [11] have also carried out some work on how to extract privacy rules and regulations from natural language text. This type of work has a different focus then ours but could potentially be complementary in helping to populate the KB more easily. Translation of legislation/regulation to machine readable policies has proven very difficult, although there are some examples of how translations of principles into machine readable policies can be done, e.g. PISA project [12], P3P [6] and PRIME project [13].

The tool we have built is a type of expert system, as problem expertise is encoded in the data structures rather than the programs and the inference rules are authored by a domain expert. Techniques for building expert systems are well known [14]. A key advantage of this approach is that it is easier for the expert to understand or modify statements relating to their expertise. Our system can also be viewed as a DSS. Many different DSS generator products are available, including [15,16]. All use decision trees or decision tables which is not suitable for our use as global privacy knowledge is too complex to be easily captured (and elicited) via decision trees. Rule based systems and expert systems allow more flexibility for knowledge representation but their use demands great care: our rule representation is designed to have some important key properties such as completeness (for further details about the formal properties of our system, see [2]). There has also been some work on dynamic question generation in the expert system community [17,18] but their concerns and methods are very different.

Our research differs from preceding research in that we define an intermediate layer of policy representation that reflects privacy principles linked into an interpretation of legislation and corporate policies and that is human-actionable and allows triggering of customised privacy advice. The focus of this paper is novel use of a combination of heuristics and usability techniques to hide underlying system complexity from domain experts who create and maintain the KB.

5 Status and Conclusions

HPPA has transferred from HP Labs into a production environment and is being rolled out to HP employees in 2010. HPPA tackles complexity of international regulations, helping both expert and non-expert end users with identifying and addressing privacy requirements for a given context. Although our focus has been on privacy, this approach is applicable in a broader sense as it can also apply to other compliance areas, such as data retention, security, and export regulation.

In order to help privacy experts address the complexity of updating KBs in an expert system, a simple mode UI was implemented in HPPA in addition to expert mode screens. Both have been subject to recursive testing and improvement. We are currently working on allowing quarantine of rules built up in the simple mode, so that these can be run in test mode before being incorporated into the KB.

Acknowledgments. Simple mode benefitted from suggestions by L. Barfield, V. Dandamundi and P. Sharma. HPPA is a collaboration between an extended team.

References

1. Leuf, B., Cunningham, W.: The Wiki Way: Quick Collaboration on the Web. Addison-Wesley, Reading (2001)
2. Pearson, S., Rao, P., Sander, T., Parry, A., Paull, A., Patruni, S., Dandamudi-Ratnakar, V., Sharma, P.: Scalable, Accountable Privacy Management for Large Organizations. In: INSPEC 2009. IEEE, Los Alamitos (2009)
3. Drools, http://jboss.org/drools/
4. IBM: The Enterprise Privacy Authorization Language (EPAL), EPAL specification, v1.2 (2004), http://www.zurich.ibm.com/security/enterprise-privacy/epal/
5. OASIS: eXtensible Access Control Markup Language (XACML), http://www.oasis-open.org/committees/tc_home.php?wg_abbrev=xacml
6. Cranor, L.: Web Privacy with P3P. O'Reilly & Associates, Sebastopol (2002)
7. Damianou, N., Dulay, N., Lupu, E., Sloman, M.: The Ponder Policy Specification Language (2001), http://www-dse.doc.ic.ac.uk/research/policies/index.shtml
8. IBM: Sparcle project, http://domino.research.ibm.com/comm/research_projects.nsf/pages/sparcle.index.html
9. IBM: REALM project, http://www.zurich.ibm.com/security/publications/2006/REALM-at-IRIS2006-20060217.pdf
10. OASIS: eContracts Specification v1.0 (2007), http://www.oasis-open.org/apps/org/workgroup/legalxml-econtracts
11. Travis, D., Breaux, T.D., Antón, A.I.: Analyzing Regulatory Rules for Privacy and Security Requirements. IEEE Transactions on Software Engineering 34(1), 5–20 (2008)
12. Kenny, S., Borking, J.: The Value of Privacy Engineering, JILT (2002)
13. Privacy and Identity Management for Europe (2008), http://www.prime-project.org.eu
14. Russel, S., Norvig, P.: Artificial Intelligence – A Modern Approach. Prentice Hall, Englewood Cliffs (2003)
15. Dicodess: Open Source Model-Driven DSS Generator, http://dicodess.sourceforge.net
16. XpertRule: Knowledge Builder, http://www.xpertrule.com/pages/info_kb.htm
17. McGough, J., Mortensen, J., Johnson, J., Fadali, S.: A web-based testing system with dynamic question generation. In: Proc. Frontiers in Education Conference, Reno. IEEE, Los Alamitos (2001)
18. Bowen, J., Likitvivatanavong, C.: Question-Generation in Constraint-Based Expert Systems, http://ww.4c.ucc.ie

Privacy-Protected Camera for the Sensing Web

Ikuhisa Mitsugami[1], Masayuki Mukunoki[2], Yasutomo Kawanishi[2], Hironori Hattori[2], and Michihiko Minoh[2]

[1] Osaka University, 8-1, Mihogaoka, Ibaraki, Osaka 567-0047, Japan
mitsugami@am.sanken.osaka-u.ac.jp
[2] Kyoto University, Yoshida-Nihonmatsu, Sakyo, Kyoto 606-8501, Japan
{mukunoki,ykawani,hattori,minoh}@mm.media.kyoto-u.ac.jp
http://mm.media.kyoto-u.ac.jp/sweb/

Abstract. We propose a novel concept of a camera which outputs only privacy-protected information; this camera does not output captured images themselves but outputs images where all people are replaced by symbols. Since the people from this output images cannot be identified, the images can be opened to the Internet so that we could observe and utilize the images freely. In this paper, we discuss why the new concept of the camera is needed, and technical issues that are necessary for implementing it.

1 Introduction

In these days, many surveillance cameras are installed in our daily living space for several purposes; traffic surveillance, security, weather forecast, etc. Each of these cameras and its captured video are used only for its own purpose; traffic surveillance cameras are just for observing congestion degree of cars, and security cameras are just for finding suspicious people. The video may include various information other than that for the original purpose. If the video is shared among many persons through the Internet, the camera will become more convenient and effective. For example, we could get weather information from a traffic surveillance camera, congestion degree of shoppers in a shopping mall from a security camera, and so on. Considering these usages, we notice the usefulness of opening and sharing real-time sensory information on the Internet. The Sensing Web project[1,2], which was launched in the fall of 2007, proposes to connect all available sensors including the cameras to the Internet, and to open the sensory data to many persons in order that anyone can use the real-time sensory data for various purposes from anywhere.

On opening and sharing the sensory data, the most serious problem is privacy invasion of observed people. As long as the sensory data is closed in a certain system operated by an institution in the same way as most existing systems, the privacy information can be managed and controlled by the corresponding institution. We thus do not need to take care of the problem. On the other hand, in the case of the Sensing Web, the sensory data is opened to the public so that anyone can access any sensory data without any access managements.

E. Hüllermeier, R. Kruse, and F. Hoffmann (Eds.): IPMU 2010, Part II, CCIS 81, pp. 622–631, 2010.

Fig. 1. The output of a traditional security camera and privacy-protected camera

Especially, the video, which is the sensory data obtained by the cameras, contains rich information of the people and may cause the privacy invasion. In fact, a person in a video can be easily identified by his/her appearance features (face, motion, colors of cloths, etc.). The privacy invasion problem, therefore, has been a main obstacle against opening the sensory data.

In the Sensing Web project, we tackle the problem to realize an infrastructure where any sensory data are opened and shared. To overcome the problem, the privacy information has to be erased from the image before it is opened to the Internet. One of the ways to realize this privacy elimination is to mask the appearances of the people on the image. In fact, the Google Street View (GSV)[3] adopts this approach. Though this service offers not real-time sensory data but just the snapshots at a past moment, it faces the same problem as mentioned above. To overcome this problem, each person in the captured image is detected and masked automatically in the case of the GSV. This operation can be executed using a human detection technique. However, as the technique does not works perfectly, some people cannot be masked correctly and their privacy accordingly cannot be protected; when a person is detected in a wrong position or not detected, the mask is overlaid on a wrong position or is not overlaid at all, and as a result the person is left unmasked and clearly appeared in the output image. We thus propose another approach to overcome the privacy invasion problem based on a novel idea; the image of the camera is reconstructed by generating a background image without any people, and overlaying symbols at the positions of the corresponding people on the generated image. This idea is implemented as a new concept of the camera that we call a "Henshin" camera , which means a privacy-protected camera (Fig.1). In the case of this camera, even if the human detection does not work well, it just causes the rendering of the symbol on the different position or the lack of the character, but never causes the privacy invasion.

For realizing this privacy-protected camera, we need two techniques; a human detection and a background image generation. The former one has been studied for a long time, and there are a lot of existing studies. They are mainly categorized into two types of approaches; background subtraction methods[4,5] and human detection methods[6,7]. In this paper, we use a HOG-based human detection[7], which is known as a method that works robustly even when the luminosity of the scene changes frequently. On the other hand, the latter one has to be well considered. Although it looks just a conventional issue at a glance, it is indeed much different from the many existing methods for the background generations. Considering the concept of the privacy-protected camera, we have to design the background generation method ensuring that people would never appear in the output image even if a person stops for quite a long time in the scene, which is treated as not the foreground person but the background by the most methods. Besides, our method has to generate the background image as verisimilar to the truth as possible, because we would like to know various information of the observed area from this background image. Especially, lighting condition by the sun is helpful to know the weather. Therefore, such kind of information has to be well reconstructed. Considering the above discussions, this paper proposes a novel background generation technique which preserves the shadow accurately in outdoor scene while ensuring that a person never appears in the image. This technique is realized by collecting the images for super long term, categorizing them by time, and analyzing them using the eigenspace method.

2 Background Generation Using Long-Term Observation

2.1 Traditional Background Generation Methods

If a human detection performs perfectly and all the people in the image thus can be erased, the whole image except the people regions can be used as the accurate background image. However, when the people exist, the corresponding regions would be left as blanks so that the background image cannot be always fully generated. In addition, there has been no ideal method for the human detection, which is apparent to see that there are still many challenges for this topic. Therefore, in terms of the privacy protection, we must not directly use each image which is observed by the camera. We have to take an analytic approach by collecting many images for a certain period of time.

Calculating median or average of each pixel of the image sequence is a simple approach to generate the background image. In order to follow the background changing, the term of the image sequence is usually not very long. However, people who stop for the term appear in the generated image, which causes the privacy invasion. For generating the background avoiding privacy invasion, we have to analyze images collected during much longer term than people might stop. On the other hand, in terms of reconstructing the background image as similar to the truth as possible especially from the viewpoint of the lighting condition by the sun, such analytic approaches with long term image sequence do not perform well; they cannot follow immediately the sudden and frequent changes of the strength of the sunlight, because the generated image is influenced by the many images in the past. Such approaches cannot fulfill the demand for applying to the privacy-protected camera.

The eigenspace method is often used to analyze huge amount of data. We apply this method to the images collected by long term surveillance. Using the eigenspace method[8], we can analytically reconstruct the background image from the current image captured by the camera that may contain some people. This is achieved by the following process.

First, the eigenvectors $e_1, e_2, \cdots, e_k$ (sorted in descending order by their contributions) are calculated from a number of images by the principal component analysis (PCA). As the eigenspace defined by these eigenvectors indicates the variation of the image sequence, the background image x_t^b can be estimated from observed image x_t using this eigenspace; x_t^b is calculated by the following equation:

$$x_t^b \approx Ep = EE^T x_t \tag{1}$$

where p describes the corresponding point in the eigenspace and $E = |e_1, e_2, \cdots, e_k|$ is an orthonormal matrix. We have to use the images each of which may contain people. Note that the appearance of the people should have less influence than the variation of the background, as the people are usually much smaller than the size of the image and each of them moves randomly and is observed for just a short term. Thus, even if we use such images, we can get the eigenspace which includes no influence of people by using only s ($s << k$) eigenvectors to reconstruct the background. We use the orthonormal matrix $E' = |e_1, e_2, \cdots, e_s|$ instead of E to estimate the background image x_t^b (Fig.2). x_t^b is calculated by the following equation:

$$x_t^b \approx E'p' = E'E'^T x_t \tag{2}$$

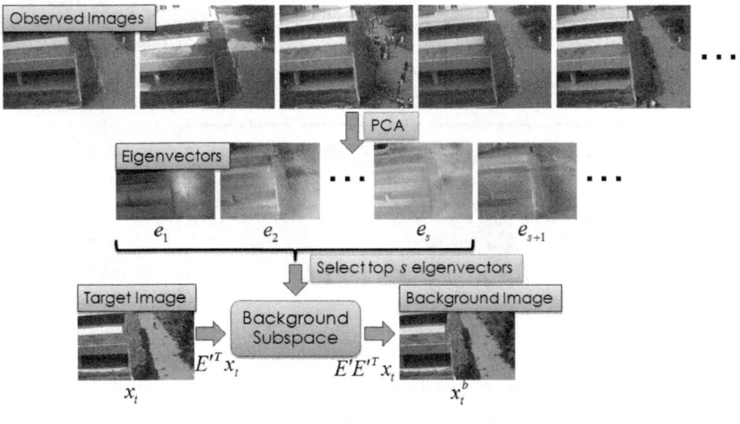

Fig. 2. The output of a traditional security camera and privacy-protected camera

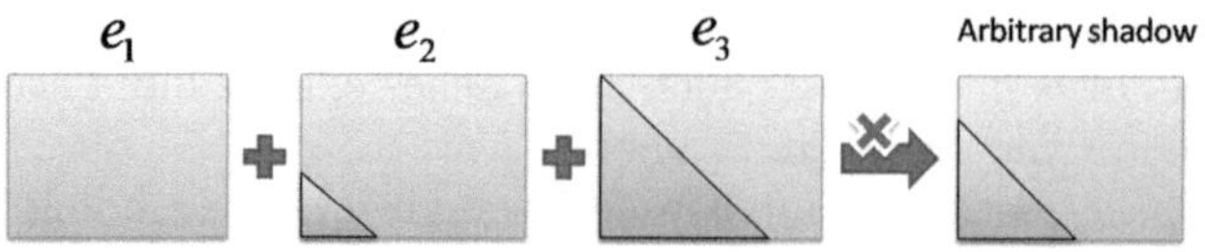

Fig. 3. Various shadow edges cannot be described as linear sum of the small number of the eigenspace

where p' describes the corresponding point in the eigenspace. It means that even when we use the image which may contain people, we expect to get the similar result to the case using the images without any people in them when we choose only such the small number of the eigenvectors.

Nevertheless, another problem still exist; the lighting condition may not be able to reconstructed by such the small (s) dimensional eigenspace.

2.2 Eigenspace Method with Classification by Observed Time

In outdoor scene, there are sharp shadow edges in the observed images and the position of the shadow edges are shifted gradually caused by the solar position. When we generate the background which includes various shadows in the scene, we need eigenvectors in the very high dimensional eigenspace which correspond to each position of shadow edges. When the dimension s is small as discussed in the previous section, such the eigenvectors corresponding to the moving shadow

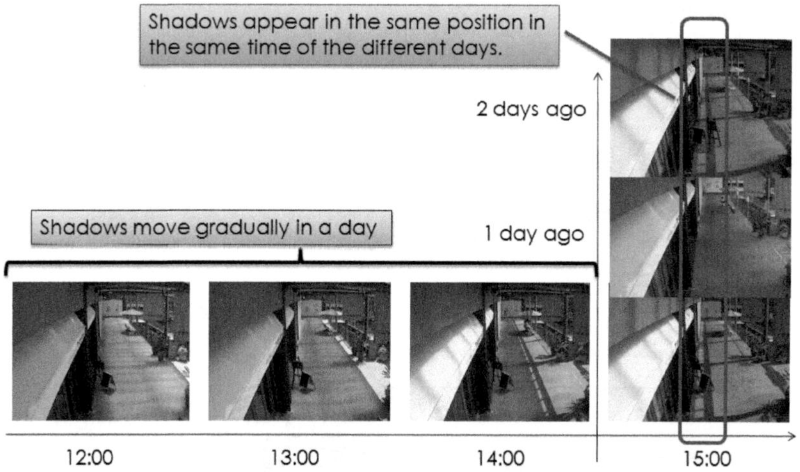

Fig. 4. Examples of the shadow positions. Shadow moves gradually in a day, and appears in the similar position in the images which were observed at the same time of the different days.

edges may be neglected. We, therefore, could not generate the background image keeping such various shadow edges by linear sum of top eigenvectors (Fig.3).

Our method relies on the fact that the shadows appear in the similar position in the same time even in another day. We collect huge amount of images by super long term surveillance and classify them into image sets according to their capture time. The images of each set are expected to have similar spatial appearance of shadows. Fig.4 shows the shadow appearances. Looking at the images observed in a day, there are shadows in the different position. On the other hand, we can see that the shadows appear in the similar position when we look at the images observed in 15 o'clock of different days.

Thus, the procedure of the proposed method is as follow. First, at a time t of a day d, we get a target image $x_{d,t}$. We then classify the observed images $x_{d,t}$

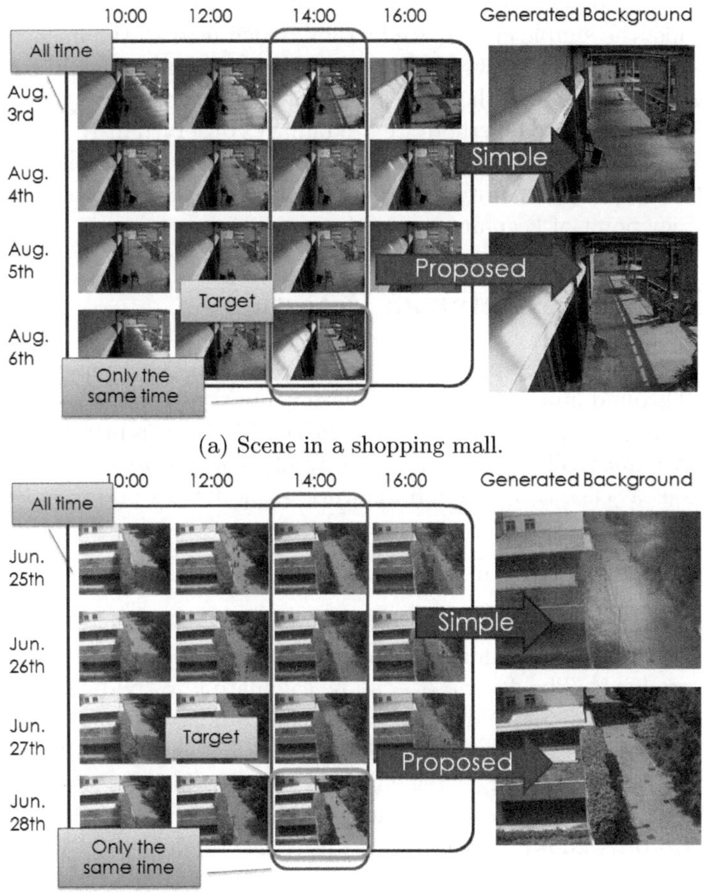

(a) Scene in a shopping mall.

(b) Scene in our university.

Fig. 5. Examples of the background image generation

according to their observed time t, and we can get an image set $x_{d-1,t}, x_{d-2,t}, \cdots$ which were observed at the time t of the different days $d-1, d-2, \cdots$. We refer each of the image sets as I_t. Finally, we apply the eigenspace method to the image set I_t, and then we can get the background keeping lighting conditions for each target image $x_{d,t}$ which is a raw image that may contain some people in it.

To show the effectiveness of our method, we experimented in some outdoor scenes. We generated the background images by the simple eigenspace method and the proposed method. We used the following two scenes:

- Scene 1: The input images and the results are shown in Fig.5(a). The images are collected from 1st to 31st in August at a shopping mall.
- Scene 2: The input images and the results are shown in Fig.5(b). The images are collected from 1st to 31st in August at our university.

For each scene, to generate background images of a certain day we applied the two methods: the simple eigenspace method which uses all the images for calculating the eigenspace, and the proposed method which first classifies the images into the sets according to the time and then applies the eigenspace method to the set. For the proposed method, we used images observed in 15 seconds around 14:00 of everyday. Comparing those two results, it is visually confirmed that the proposed method can generate better background image than the traditional one from the viewpoint of keeping the lighting condition.

3 Scene-Adaptive Human Detection

For realizing this privacy-protected camera, a human detection method as well as the background image generation mentioned in the previous section is important. For this purpose, the HOG-based method[7] is currently known to show good performance. This paper therefore fundamentally relies on this method, but proposes a scene adaptation framework to modify its performance.

3.1 HOG-Based Human Detection Method

First, we introduce the HOG descriptor defined in [7]. For a human image, local appearance and shape within the image are described by the distribution of intensity gradients. The descriptors can be calculated by dividing the image into small connected regions, which are called cells, and compiling a histogram of gradient for the pixels within each cell. The local histograms are then combined and normalized across a larger region of the image, which is called a block. The HOG descriptor denotes the combination of the histograms in the whole image. This descriptor is known to show good invariance to changes in illumination or shadowing.

The method then applies supervised learning technique to judge whether an image is of a human or not. Binary classifiers such as the Support Vector Machine(SVM) are often used. Once trained on images containing people, the classifier become able to make decisions regarding the appearance of a human.

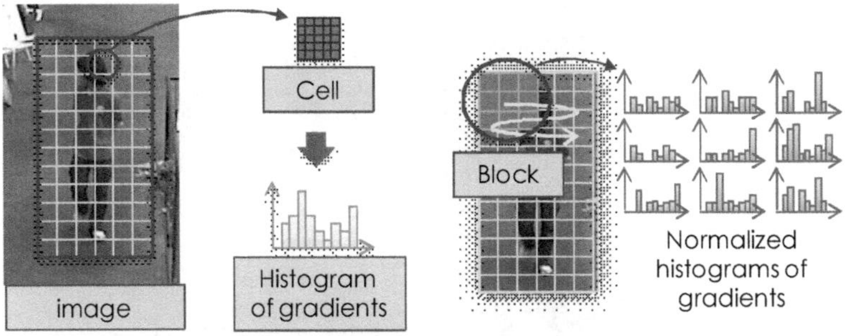

Fig. 6. The HOG descriptor

3.2 Scene Adaptation of HOG-Based Human Detector

Though in fact the HOG-based human detector mentioned in Section 3.1 runs quite well, it often causes the following errors; (i) it occasionally detects a false area since it might mistakenly detect a background area whose pattern is incidentally similar to human, and (ii) it sometimes spoils a human region when it fails to acquire enough features in that region. These failures are caused by the fact that the detector does not use information of the observed scene; the detector uses knowledge about human/non-human patterns contained in the images captured by other cameras and of other scenes. That is why it may mistakenly detect a pattern which is of the background but accidentally looks similar to a human, or it may sometimes spoil a human in the image.

The proposed method improves the performance by using additional information specific to each camera, which can be obtained by judging true or false detections from these results. This judgment cannot be achieved by using information only about each frame. However, it can be achieved by analyzing time series of detected results, because it is different between true and false detection. By this analysis we judge true or false detections and acquire additional information specific to each camera, and they are then used additionally in the supervised learning to update the detector, as shown in Fig.7.

The proposed method also improves how to search people in the image. The normal detector usually searches people through the whole image with all possible sizes and orientations, since it does not use any scene-specific information. On the other hand, the proposed method is designed to obtain relation between the sizes and orientations and positions of people specific in the scene by automatic camera calibration using the true detections. This modification is effective not only for the processing cost but also for the detection accuracy, because the patterns which are of the background but accidentally look similar to people can be eliminated from the candidate of the judgement.

Fig.8 shows the results of the existing and proposed methods. It is confirmed that the proposed method works more effectively; people are accurately detected throughout the image sequences while failure samples are much reduced.

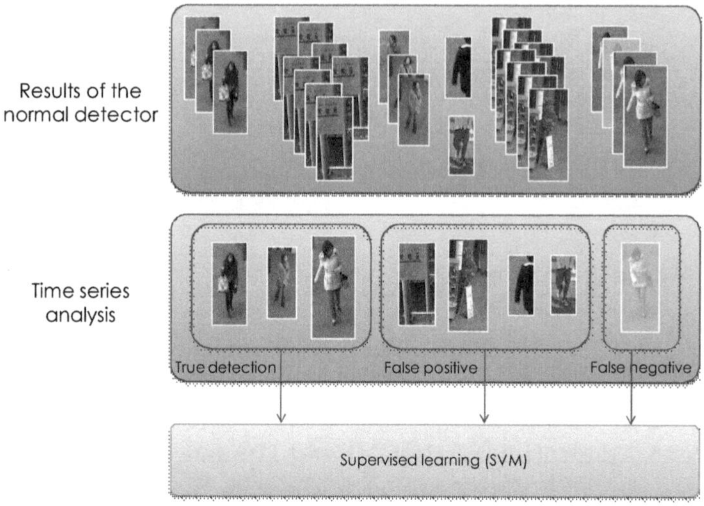

Fig. 7. Time series analysis for additional learning

Results of the existing method.

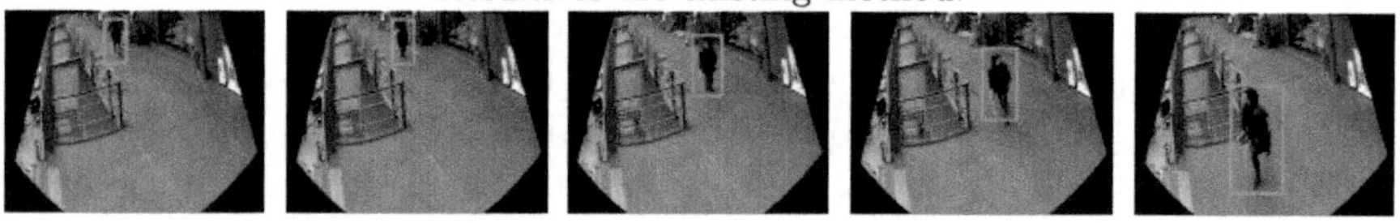

Results of the proposed method.

Fig. 8. Experimental results

4 Conclusion

In this paper, we proposed a novel concept of a camera, named the "Henshin" camera, which outputs privacy-protected information. This camera outputs only images where all people are replaced by symbols. The images thus can be opened and shared on the Internet so that we could observe and utilize the images freely.

For realizing this privacy-protected camera, we proposed a new background generation method and a human detection method. The former one was designed so as to ensure that people would never appear in the image even if the people stop for quite a long time in the scene. It was realized by collecting the images for super long term, categorizing them by time, and analyzing them using the eigenspace method. The latter one was based on the HOG human detection method but extended to adapt the detector to each scene so as to improve the accuracy of the detection. This is also realized by collecting the detected results from the past images.

Our future work contains pervasiveness of the concept of the privacy-protected and social investigation of its acceptability from the viewpoint of the privacy. The improvement of the proposed method is also one of the important topics.

References

1. Minoh, M., Kakusho, K., Babaguchi, N., Ajisaka, T.: Sensing Web Project - How to handle privacy information in sensor data. In: Proceedings of International Conference on Information Processing and Management of Uncertainty in Knowledge-Based Systems (IPMU), pp. 863–869 (2008)
2. Minoh, M.: Sensing Web: Concept and Problems (Special Issue Sensing Web). Journal of Japanese Society for Artificial Intelligence (2009) (in Japanese)
3. Google: Google Street View (2008), http://maps.google.com
4. Haritaoglu, I., Harwood, D., Davis, L.S.: Hydra: Multiple people detection and tracking using silhouettes. In: IEEE Workshop on Visual Surveillance, vol. 0, p. 6 (1999)
5. Jabri, S., Duric, Z., Wechsler, H., Rosenfeld, A.: Detection and location of people in video images using adaptive fusion of color and edge information. In: International Conference on Pattern Recognition, vol. 4, p. 4627 (2000)
6. Chen, Y.T., Chen, C.S.: Fast human detection using a novel boosted cascading structure with meta stages. IEEE Transactions on Image Processing 17(8), 1452–1464 (2008)
7. Dalal, N., Triggs, B.: Histograms of oriented gradients for human detection. In: IEEE Computer Society Conference on Computer Vision and Pattern Recognition, CVPR 2005, vol. 1, pp. 886–893 (2005)
8. Oliver, N., Rosario, B., Pentland, A.: A Bayesian computer vision system for modelling human interactions. In: Christensen, H.I. (ed.) ICVS 1999. LNCS, vol. 1542, pp. 255–272. Springer, Heidelberg (1998)

Bayesian Network-Based Approaches for Severe Attack Prediction and Handling IDSs' Reliability

Karim Tabia and Philippe Leray

LINA/COD CNRS UMR 6241 - Ecole Polytechnique de Nantes
{Karim.Tabia,philippe.Leray}@univ-nantes.fr

Abstract. Probabilistic graphical models are very powerful modeling and reasoning tools. In this paper, we propose efficient Bayesian network-based approaches for two major problems in alert correlation which plays an important role in nowadays computer security infrastructures. While the use of multiple intrusion detection systems (IDSs) and complementary approaches is highly recommended to improve the overall detection rates, this inevitably rises huge amounts of alerts most of which are redundant and false alarms. The aim of this work is twofold: Firstly, we propose an approach based on Bayesian multi-nets which allow to take advantage of local influence relationships in order to improve the prediction of severe attacks. Secondly, we propose to handle the reliability of IDSs by considering the uncertainty relative to the triggered alerts. Experimental studies carried out on real and recent IDMEF alerts produced by the *de facto* network-based IDS Snort shows significant improvements with respect to standard Bayesian approaches. More particularly, the handling of IDSs' reliability significantly reduces the false alarm rate which represents a crucial issue for intrusion detection development.

Keywords: Bayesian multi-nets, alert correlation, IDSs' reliability.

1 Introduction

Intrusion detection systems (IDSs) [11] act as burglar alarms and aim at detecting any malicious action compromising integrity, confidentiality or availability of computer and network resources or services . In practice, it is highly recommended to deploy several security products and solutions in order to increase the overall detection rates. However, the use of multiple IDSs and security components further multiplies the huge amounts of triggered alerts which cannot be manually analyzed by the security administrators. Alert correlation is needed to cope with these quantities of alerts. There are mainly three objectives targeted by alert correlation approaches: i) Reducing the number of triggered alerts by eliminating redundant ones [8], ii) multi-step attack detection where the different alerts may correspond to the execution of an attack plan consisting in several steps [10] and iii) prioritizing the triggered alerts according to the criteria and preferences of the security administrators [2].

E. Hüllermeier, R. Kruse, and F. Hoffmann (Eds.): IPMU 2010, Part II, CCIS 81, pp. 632–642, 2010.

In this paper, we are concerned with *predicting* severe attacks which often are the final step in multi-step attacks. More precisely, we are interested in anticipating severe attacks in order to prevent them by taking the appropriate countermeasures (such as preventing the suspected user from following his attack). Alerts triggered by IDSs are often associated with severity levels indicating the dangerousness of the detected attack. There are several attacks which are carried out by programs and attack scripts which perform predefined/frequent sequences of malicious actions leading to the dangerous one. For instance, an attacker may try to perform an IP scan, then a port scan and launch a dictionary attack to get a local access and finally, perform a buffer-overflow attack in order to gain root privileges. In this paper, our objective is to efficiently predict severe attacks with minimum intervention of the administrator. Our choice is motivated by the fact that most of low/medium severity attacks such as scans are inoffensive while several high severity level attacks are often preceded by preparatory inoffensive attacks. For instance, several worms spread by first scanning vulnerable machines then execute an exploit.

In this paper, we propose a probabilistic graphical model for severe attack prediction based on Bayesian multi-nets which allow modeling local influence relationships [5] and offer several benefits for our alert correlation problem. Moreover, we propose an approach for handling IDSs' reliability since these systems are known to trigger large amounts of false alerts. This approach is based on Pearl's virtual evidence [12] for reasoning with uncertain evidence in the framework of probabilistic models. As we will see in our experimental studies, the main benefits of our approaches are better prediction rates with minimum false alert rates while requiring minimum expert knowledge.

2 Basic Backgrounds

This section provides the basic backgrounds on alert correlation and Bayesian multi-net classifiers.

2.1 Alert Correlation

An alert is a message generated by un IDS when an attack is detected. It often contains an identification/name of the detected activity, its class, a severity level, the IP address of the attacker, the IP address of the victim, etc. Most of IDSs can report alerts in IDMEF format[1] which is the intrusion detection message exchange format enabling inter-operability among IDSs and other security tools.Alert correlation [8][7] consists in analyzing the alerts triggered by one or multiple IDS sensors in order to provide a *synthetic* and *high-level* view of the *interesting* malicious events targeting the information system.

Alert correlation approaches can be grouped into similarity-based approaches [8], predefined attack scenarios [10], pre and post-conditions of individual attacks

[1] `http://www.ietf.org/rfc/rfc4765.txt`

[7] and statistical approaches [15]. Note that most works on multi-step attack detection heavily rely on expert knowledge. In this paper, we are interested in severe attack detection which is a variant of multi-step attack recognition.

2.2 Bayesian Network Classifiers

Bayesian networks are powerful graphical models for representing and reasoning with uncertain and complex information [9]. They consist of a graphical component DAG (Directed Acyclic Graph) allowing an easy representation of the domain knowledge in the form of an influence network (vertices represent events while edges represent *dependence* relations between these events), and a quantitative probabilistic one allowing to specify the uncertainty relative to relationships between domain variables using conditional probability tables (CPTs). Bayesian network-based classification is a particular kind of probabilistic inference ensured by computing the greatest a posteriori probability of the class variable given the instance to classify. Namely, having an instance of the attribute vector $a_1a_2..a_n$ (observed variables $A_0=a_0$, $A_1=a_1$,.., $A_n=a_n$), it is required to find the most plausible class value c_k ($c_k \in$ C=$\{c_1, c_2,..,c_m\}$) for this observation. The maximum a posteriori classification rule can be written as follows:

$$Class = argmax_{c_k \in C}(p(c_i/a_1a_2..a_n)), \tag{1}$$

where term $p(c_i/a_1a_2..a_n)$ denotes the posterior probability of having the class instance c_i given the evidence $a_1a_2..a_n$. This probability is computed using Bayes rule as follows:

$$p(c_i/a_1a_2..a_n) = \frac{p(a_1a_2..a_n/c_i) * p(c_i)}{p(a_1a_2..a_n)} \tag{2}$$

In practice, the denominator of Equation 2 is ignored because it does not depend on the different classes. Equation 2 means that posterior probabilities are proportional to likelihood of the evidence and class prior probabilities while the evidence probability is just a normalizing constant. Note that most works use naive or semi-naive Bayesian network classifiers such as TAN (Tree Augmented Naive Bayes) and BAN (Bayesian Network Augmented Naive Bayes) [5] which make strong assumptions in order to simplify the classifier's structure learning from data. The other Bayesian network classifiers require more general structure learning and parameter estimation.

2.3 Bayesian Multi-net Classifiers

A standard Bayesian network classifier is defined as a unique network which encodes the dependence relationships existing in the training data. However, the relationships are not the same over the different classes. More particularly, in our severe attack prediction application, each severe attack is statistically correlated only with a small and specific set of other alerts most of the time because several attacks are undertaken by worms and scripts executing the same malicious events. For instance, the attack *WEB-IIS CodeRed v2 root.exe access* whose Snort signature identifier (*sid*) is 1256 is correlated in our data set only with alerts having *sid*=2 (*DOUBLE DECODING ATTACK*), 18(*WEBROOT DIRECTORY TRAVERSAL*) and 1002 (*WEB-IIS cmd.exe access*) because this worm uses a directory traversal attack (causing alert with *sid* 18 and 2) in order to access the *cmd.exe* executable on MS Windows systems. As we will see in our experimental studies, local correlation modeling leads to better likelihood estimation for the classification task. Note that most structure learning algorithms identify correlations using statistical tests, consequently when the classes are imbalanced the obtained network encodes mainly the correlations relative to the *majority* class. In a multi-net classifier, each class instance c_i is associated with a network N_{c_i} encoding only local influence relationships learnt from training instances belonging exclusively to the class c_i. The prior frequencies relative to the classes can be encoded by a root node C representing the class variable [4] or just by a local probability distribution as in [5]. Note that the computational complexity of learning a multi-net is the same as learning a single network. The generic procedure for learning a multi-net from empirical data is provided in Algorithm 1 while multi-net based classification is provided by Algorithm 2.

Input: Data set D of labeled training examples, Structure learning algorithm and its parameters
Output: Multi-net
Begin

1. Partition the training set D into subsets S_i where each S_i contains the data belonging only to class c_i.
2. For each training subset S_i,
   (a) Learn a Bayesian network N_{c_i} on S_i
      i. Learn the structure of N_{c_i} on S_i,
      ii. Learn the parameters of N_{c_i} on S_i,
   (b) Compute the frequency of class instance c_i.

ALGORITHM 1. Bayesian multi-net classifier learning

Input: Bayesian multi-net classifier, the instance to classify $a_1..a_n$
Output: Most probable class instance $c_i \in C$
Begin

1. For each possible class instance c_i,
   (a) Compute the likelihood of the evidence, namely $p_{N_{c_i}}(a_1..a_n)$.
   (b) Combine the likelihood with a priori frequency of class c_i, namely compute $p_{N_{c_i}}(a_1..a_n)*p(c_i)$.
2. Return the class instance having the utmost a posteriori probability degree, namely $argmax_{c_i \in D_c}(p_{N_{c_i}}(a_1..a_n)*p(c_i))$.

ALGORITHM 2. Classification based on a multi-net classifier

Authors in [5] argue that multi-net classifiers are as effective as the best state of the art classifiers while they are more expressive and less complex than BANs. Moreover, multi-nets allow to update the model with minimum modifications. For instance, in order to add a new class, one has just to learn an additional network for the added class and update $P(C)$. One can also update existent models by repeating the learning of local networks of the updated classes. Finally, multi-net classifiers offer the opportunity to choose a specific learning algorithm and select the most appropriate features for each class.

3 Multi-net Based Classifier for Predicting Severe Attacks

Bayesian network-based approaches are widely used in many areas of computer security. More particularly, Bayesian network classifiers are used in intrusion detection in several works such as [14]. In alert correlation, a Bayesian approach is used in [15] for alert fusion. In [1], the authors use naive and TAN classifiers for detecting attack plans and severe attacks. Note that all the works on detecting multi-step and severe attacks either use naive models or use the same model over all the target classes. Note also that to the best of our knowledge, there is no work addressing the IDSs reliability handling. In the following, we propose a multi-net based approach allowing to take advantage of local dependencies in order to better model the correlations of each target class.

In this work, severe attack detection is viewed as a classification problem: given a sequence of alerts $Alert_1$, $Alert_2$,..,$Alert_k$, we want to determine if this alert sequence *plausibly* will lead or be followed by a severe attack $Attack_i$. Here the attribute variables are the alerts with low/medium severity levels while the class variable C is associated with the different severe attacks we want to predict.

1. Predictors (attribute variables): The set of predictors (observed variables) is composed of the set of relevant alerts for predicting the severe attacks. Namely, with each relevant alert $Alert_i$, we associate an attribute variable A_i whose domain is $\{0, 1\}$ where the value 0 means that alert $Alert_i$ was not observed in the analyzed sequence while value 1 denotes the fact that the alert $Alert_i$ has been reported. The duration of alert sequences can be fixed experimentally or manually set by the expert. Note that in this paper, the predictors refer to alerts with *low* or *medium* severity level corresponding to inoffensive and benign events. The relevant predictors can be selected according to the experts knowledge or statistically by feature selection methods.

2. Class variable: The class variable C represents the severe attacks variable whose domain involves all the severe attacks $Attack_1$,.., $Attack_n$ to predict and another class instance $NoSevereAttack$ representing alert sequences that are not followed by severe attacks.

In addition to better modeling of low/medium severity level alert relationships, the main advantages of this multi-net based approach are: i) *Minimum expert intervention*: the expert has just to identify the severe attacks he wants to predict and use feature selection methods to select the relevant features for predicting these attacks on historical alert logs. ii) *Easy deployment:* one has just to preprocess the alerts logs in real-time to extract the predictors and perform real-time severe attack prediction. iii) *Easy update:* we can add (resp. discard) a new (resp. an existing) severe attack with minimum alterations on the existing multi-net.

4 Handling the IDSs' Reliability

The most important problem users of IDSs face is the one of false alerts which correspond to legitimate activities that have been mistakenly reported as malicious by the IDS. Indeed, nowadays IDSs are well-known to trigger high false alarm rates. For instance, the well-known Snort[2] IDS indicates for each attack, whether false alerts could be triggered. In an experimental evaluation of Snort

[2] `www.snort.org`

IDS [13], the authors concluded that 96% of the triggered alerts are false positives. Hence, taking into account the reliability of IDSs is an interesting issue for the prediction of severe attacks. For instance, if it is known that the 90% of alerts reporting a malicious event triggered by a given IDS are false, then this information should be taken into account if such alerts should be exploited as inputs by the alert correlation tool.

4.1 Handling IDSs' Reliability for Predicting Severe Attacks

Pearl's virtual evidence method [12] offers a natural way for handling and reasoning with uncertain evidence[3] in the framework of probabilistic networks. In this method, the uncertainty indicates the confidence on the evidence: to what extent the evidence is believed to be true. In our context, if an IDS triggers an alert and we know (from past experience for example) that this event is false alert in 95% of the cases then we are in presence of uncertain evidence. The main idea of Pearl's virtual evidence method is to recast the uncertainty relative to the uncertain evidence E on some *virtual* sure event η: the uncertainty regarding E is specified as the likelihood of η in the context of E.

In order to apply this method for efficiently predicting severe attacks, we must first assess the IDSs' reliability by means of empirical evaluations (an expert can examine for each alert type triggered by an IDS, the proportion of true/false alerts). An expert can also subjectively (by experience) fix the reliability of the IDSs composing his intrusion detection infrastructure.

Now, after assessing the reliability of the IDSs in triggering the alerts A_1,..,A_n, the handling of the uncertainty regarding an alert sequence, proceeds as follows:

1. For each alert variable A_i, add a child variable R_i as a virtual evidence recasting the uncertainty bearing on A_i. The domain of R_i is D_{R_i}=$\{0, 1\}$ where the value 0 is used to recast the uncertainty regarding the case A_i=0 (alert A_i was not triggered) while the value 1 is used to take into account the uncertainty in the case A_i=1 (alert A_i was triggered).
2. Each conditional probability distribution $p(R_i/A_i)$ encodes the reliability that the observed values (triggered alerts) are actually true attacks. For example, the probability $p(R_i{=}1/A_i{=}1)$ denotes the probability that the observation R_i=1 is actually due to a real attack.

The example of Figure 1 gives a tree-augmented naive Bayes network augmented with five nodes R_1,R_2,R_3,R_4,R_5 for handling the uncertainty relative to variables A_1,A_2,A_3,A_4,A_5 respectively. Henceforth, the observed variables are R_1,R_2,R_3,R_4,R_5 while variables A_1,A_2,A_3,A_4,A_5 are associated with the actual malicious/normal activities and they cannot be directly observed.

When analyzing an alert sequence $r_1 r_2 .. r_n$ (an instance of observation variables R_1,..,R_n), we compute $argmax_{c_i}(p(c_k/r_1..r_n))$ in order to predict severe

[3] In this paper, we use the term *uncertain evidence* to denote observations provided by unreliable sources (here IDSs).

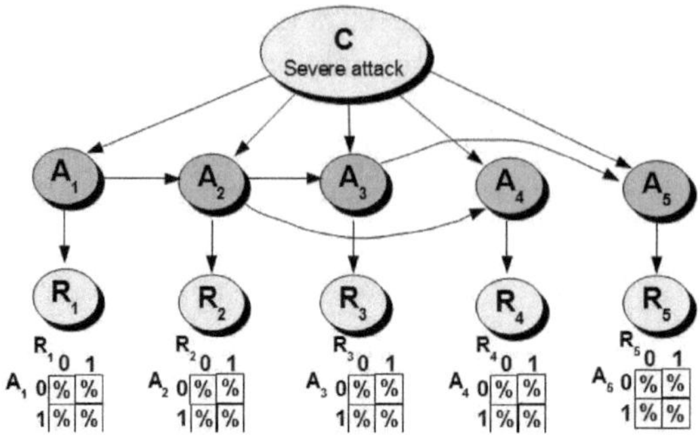

Fig. 1. Example of a Bayesian network classifier handling the reliability of inputs

attacks. Note that in practice it is less complicated to assess the false/true positive rates than assessing the false/true negative rates which requires analyzing the whole activities (for example, all the network traffic) in order to evaluate the proportion of attacks that were not detected by the IDSs.

5 Experimental Studies

Our experimental studies are carried on real and recent IDMEF Snort alert log files collected on a university campus. These alert logs represent three months activity collected during 2007 within the framework of PLACID project[4].

IDMEF alert preprocessing. In order to preprocess IDMEF alerts into CSV data that can be used for training our models, we developed an alert preprocessor which takes as input IDMEF alert log files and preprocessing options and outputs alert sequences in CSV format. Note that our preprocessing tool is used in off-line mode to provide the labeled data for training the Bayesian network models. In prediction mode, the tool preprocesses in real-time sequences of alerts generated by the IDSs and submits the preprocessed alerts for analysis.

Training and testing sets. In order to evaluate the effectiveness of our multi-net classifier and Pearl's virtual evidence method for handling IDSs reliability, we first preprocessed the first month of collected alerts in order to build the training data set and the second month to build the testing one. Table 1 provides details on the severe attacks we used in our experimentations. Among the severe attacks detected by Snort, we selected 9 Web-based severe attacks to predict on the basis of the alerts that often precede/prepare these severe attacks. All these severe attacks target either Web servers or related web-based applications. Such attacks may result in arbitrary code execution and full control of the targeted system. As for selecting the set of relevant predictors for our severe attacks, we first extracted all the existing alerts involving the same victims as the severe

[4] `http://placid.insa-rouen.fr`

Table 1. Training and testing set distributions

Sid	Alert name	Training set Number	Training set %	Testing set Number	Testing set %
1091	WEB-MISC ICQ Webfront HTTP DOS	87	0,18%	6	0,01%
2002	WEB-PHP remote include path	50	0,10%	231	0,47%
2229	WEB-PHP viewtopic.php access	5169	10,42%	1580	3,20%
1012	WEB-IIS fpcount attempt	3	0,01%	10	0,02%
1256	WEB-IIS CodeRed v2 root.exe access	2	0,004%	3	0,01%
1497	WEB-MISC cross site scripting attempt	5602	11,30%	7347	14,90%
2436	WEB-CLIENT Microsoft wmf metafile access	145	0,29%	53	0,11%
1831	WEB-MISC jigsaw dos attempt	659	1,33%	153	0,31%
1054	WEB-MISC weblogic/tomcat .jsp view source...	3412	6,88 %	3885	7,88%

attacks then using the information gain selection feature procedure, we selected a subset of relevant features. Note that the feature extraction process is similar to the works of [1][3].

5.1 Results

In order to evaluate the effectiveness of our multi-net classifier, we compare it with a naive Bayes and a network classifier built using MWST algorithm [6] which is a score based structure learning algorithm that rapidly builds simple and efficient tree structures. As for the multi-net, we also used the MWST algorithm to build the networks representing each class of alert sequences. We trained the three classifiers on the same training set and evaluated them on the same testing set of Table 1. The results of this experimentation are given in Table 2. Table 2 compares the results of a naive Bayes classifier, a MWST classifier and our multi-net classifier with respect to their prediction rate and the false alarm rate. The results of Table 2 show that the multi-net outperforms both the naive Bayes and MWST classifiers regarding the overall prediction and the false alert rates. In particular, the multi-net classifier predicted 76,92% of the severe attacks at a false alarm rate of 1,58% (29 false alarms/day) while the naive Bayes (resp. MWST) classifier triggered 3.21% (58 false alarms/day) (resp. 3,10% (56 false alarms/day)). This performance is due to the better modeling of each class leading to better estimation of the likelihoods of the alert sequences to analyze.

Table 2. Experimental results of Naive Bayes, MWST and multi-net classifiers

Sid	Alert name	Naive Bayes	MWST	Multi-net
1091	WEB-MISC ICQ Webfront HTTP DOS	0%	0%	0%
2002	WEB-PHP remote include path	26,41%	26,84%	26,84%
2229	WEB-PHP viewtopic.php access	74,81%	74,94%	72,15%
1012	WEB-IIS fpcount attempt	10,00%	0%	0%
1256	WEB-IIS CodeRed v2 root.exe access	0%	0%	0%
1497	WEB-MISC cross site scripting attempt	95,62%	95,62%	95,92%
2436	WEB-CLIENT Microsoft wmf metafile access	60,38%	60,38%	56,60%
1831	WEB-MISC jigsaw dos attempt	54,90%	54,25%	46,41%
1054	WEB-MISC weblogic/tomcat .jsp view source...	41,36%	38,69%	47,77%
	Prediction rate	75,32%	74,53%	**76,92%**
	False alarm rate	3,21%	3,10%	**1,58%**

It is important to note that the three classifiers failed to predict some severe attacks mainly because of the imbalance of class frequencies in the training set. This is for instance the reason why the severe attack with *Sid*=1256 (*WEB-IIS CodeRed v2 root.exe access*) which is represented in the training set only by 2 instances was not predicted. Finally, note that the naive and MWST classifiers achieved comparable prediction rates but suffer from high false alarm rates which is a crucial issue when using IDSs.

In the following, we report our experimental results on handling the IDS' reliability especially for reducing the false alarm rate. In this experimentation, we implemented the virtual evidence method as follows:

Reliability assessment: for each alert A_i used as a predictor, we first checked in Snort's database whether the rule associated with this attack is known to produce false positives. In the positive case, we computed on a representative corpus of the training data set the proportion of alerts A_i which actually correspond to true attacks. Namely, we computed two parameters $p(A_i{=}1/Attack{=}True)$ and $p(A_i{=}1/Attack{=}False)$. Note that taking account false negatives is in our case impossible because we have not the original network traffic in order to check whether there are attacks which were not detected by Snort.

Severe attack prediction taking into account the IDS reliability: When an alert sequence is submitted for analysis, the prediction is performed on the Bayesian network where the alert variables A_i are augmented by virtual evidence nodes (observed variables) R_i to handle the reliability of inputs.

Table 3 gives the results of handling the reliability of Snort IDS producing the alert sequences we analyze. The results of Table 3 show that the VE-MWST classifier implementing the virtual evidence method for handling the reliability of Snort IDS achieves comparable prediction rates with respect to MWST multi-net classifier but significantly reduces the false alarm rate down to **0,74%** (the false alarm rate was decreased from 29 down to only 13 false alarms/day). Note that this result is performed by handling the true/false positive reliability relative to only three alerts (those having sid=882, sid=1288 and sid=1852) constituting the majority of false alerts triggered by Snort in our data sets (see [13] for an analysis of these false alarms triggered by Snort). Such results are very promising

Table 3. Experimental results of MWST and VE-MWST multi-net classifiers

Sid	Alert name	MWST multi-net	VE-MWST multi-net
1091	WEB-MISC ICQ Webfront HTTP DOS	0%	0%
2002	WEB-PHP remote include path	26,84%	25,97%
2229	WEB-PHP viewtopic.php access	72,15%	74,30%
1012	WEB-IIS fpcount attempt	0%	0%
1256	WEB-IIS CodeRed v2 root.exe access	0%	0%
1497	WEB-MISC cross site scripting attempt	95,62%	93,32%
2436	WEB-CLIENT Microsoft wmf metafile access	56,60%	56,60%
1831	WEB-MISC jigsaw dos attempt	56,41%	37,25%
1054	WEB-MISC weblogic/tomcat .jsp view source...	47,77%	41,83%
	Prediction rate	**76,92%**	73,88%
	False alarm rate	1,58%	**0,74%**

but require a rigorous reliability assessment and handling false negatives which are very time consuming tasks.

6 Conclusion

This paper dealt with Bayesian network modeling/reasoning in the field of alert correlation. More precisely, we proposed a Bayesian multi-net model for predicting severe attacks allowing to prevent them by undertaking the appropriate countermeasures. The paper provided two main contributions: i) It proposed a method allowing to partially cope with the class imbalance problem characterizing the severe attack prediction problem (and more generally, several classification problems). This model allows to encode the local correlations leading to better likelihood estimation. ii) It proposed a model for handling the false alarms by handling the IDSs reliability based on Pearl's virtual evidence method. Experimental studies shows that our contributions allow high predictions rates with best false alarm rates which is actually the main issue in current IDSs development. As future works, we will explore predictive model learning with latent variables in order to build models from training data sets obtained from unreliable sources. Another future direction consists in exploring some heuristics and methods for feature selection in order to build more effective multi-net classifiers.

Acknowledgements

This work is supported by the ANR SETIN 2006 PLACID project (`http://placid.insa-rouen.fr/`).

References

1. Benferhat, S., Kenaza, T., Mokhtari, A.: Tree-augmented naive bayes for alert correlation. In: 3rd conference on Advances in Computer Security and Forensics (ACSF 2008), July 2008, pp. 45–52 (2008)
2. Benferhat, S., Sedki, K.: Alert correlation based on a logical handling of administrator preferences and knowledg. In: International Conference on Security and Cryptography (SECRYPT 2008), Porto, Portugal, July 2008, pp. 50–56 (2008)
3. Bin, Z., Ghorbani, A.: Alert correlation for extracting attack strategies. I. J. Network Security 3(3), 244–258 (2006)
4. Cano, A., Castellano, J.G., Masegosa, A.R., Moral, S.: Methods to determine the branching attribute in bayesian multinets classifiers. In: Godo, L. (ed.) ECSQARU 2005. LNCS (LNAI), vol. 3571, pp. 932–943. Springer, Heidelberg (2005)
5. Cheng, J., Greiner, R.: Learning bayesian belief network classifiers: Algorithms and system. In: 14th Conference of the Canadian Society on Computational Studies of Intelligence, London, UK, pp. 141–151. Springer, Heidelberg (2001)
6. Chow, C., Liu, C.: Approximating discrete probability distributions with dependence trees. IEEE Transactions on Information Theory 14(3), 462–467 (1968)
7. Cuppens, F., Miège, A.: Alert correlation in a cooperative intrusion detection framework. In: IEEE Symposium on Security and Privacy, pp. 187–200, USA (2002)

8. Debar, H., Wespi, A.: Aggregation and correlation of intrusion-detection alerts. In: Lee, W., Mé, L., Wespi, A. (eds.) RAID 2001. LNCS, vol. 2212, pp. 85–103. Springer, Heidelberg (2001)
9. Jensen, F.V., Nielsen, T.D.: Bayesian networks and decision graphs (information science and statistics). Springer, Heidelberg (June 2007)
10. Ning, P., Cui, Y., Reeves, D.S.: Constructing attack scenarios through correlation of intrusion alerts. In: 9th ACM conference on Computer and communications security, pp. 245–254. ACM, New York (2002)
11. Patcha, A., Park, J.: An overview of anomaly detection techniques: Existing solutions and latest technological trends. Computer Networks 51(12), 3448–3470 (2007)
12. Pearl, J.: Probabilistic reasoning in intelligent systems: networks of plausible inference. Morgan Kaufmann Publishers Inc, San Francisco (1988)
13. Tjhai, G.C., Papadaki, M., Furnell, S., Clarke, N.L.: Investigating the problem of ids false alarms: An experimental study using snort. In: 23rd International Information Security Conference (SEC 2008), pp. 253–267 (2008)
14. Valdes, A., Skinner, K.: Adaptive, model-based monitoring for cyber attack detection. In: Debar, H., Mé, L., Wu, S.F. (eds.) RAID 2000. LNCS, vol. 1907, pp. 80–92. Springer, Heidelberg (2000)
15. Valdes, A., Skinner, K.: Probabilistic alert correlation. In: Lee, W., Mé, L., Wespi, A. (eds.) RAID 2001. LNCS, vol. 2212, pp. 54–68. Springer, Heidelberg (2001)

Structuring and Presenting the Distributed Sensory Information in the Sensing Web

Rin-ichiro Taniguchi, Atsushi Shimada,
Yuji Kawaguchi, Yousuke Miyata, and Satoshi Yoshinaga

Graduate School of Information Science and Electrical Engineering,
Kyushu University
744, Motooka, Nishi-ku, Fukuoka 819-0395 Japan
http://limu.ait.kyushu-u.ac.jp/

Abstract. In the Sensing Web[1], a variety of sensors are installed dispersively, and, from those sensors, we acquire various information of the real world events. Although we can acquire a certain kind of information from each of the sensors separately, such information is fragmentary and integration or structurization of sensory data captured by multiple sensors is quite important for us to acquire truly meaningful information of the real world. From this point of view, we have researched into organization and presentation of distributed sensory data in the Sensing Web project. In this paper, we will present our research activity, especially wide-area object tracking, and some of demonstrative experiments.

1 Introduction

In the Sensing Web[1], a variety of sensors are installed dispersively, and, from those sensors, we acquire various information of the real world events. Although we can acquire a certain kind of information from each of the sensors separately, such information is fragmentary, and integration or structurization of sensory data captured by multiple sensors is quite important for us to acquire truly meaningful information of the real world. From this point of view, we have researched into structurization and presentation of distributed sensory data in the Sensing Web project.

Various sensors can be handled in the Sensing Web, but video cameras are one of the most popular sensors. Many cameras have been arranged around roads, buildings, etc for monitoring car traffic and people's activities, and the number of such cameras are definitely increasing to assure the safety of people's activity in the current society. In many cases, the images captured by those cameras have been monitored by human observers. However, as the number of such cameras increases, the burden of the human observers increases. In addition, if human observers directly see the captured images, the privacy invasion problem can occur.

To solve these problems, there have been many approaches in which the real world events are automatically detected by computer analysis of the captured images. Based on the approaches, the load of the human observers does not increase in principle when the number of the cameras increases. Since the analysis results of different cameras can be easily shared via computer networks, monitoring of very wide areas can be constructed as well. It should be noted that the privacy invasion problem can be also relaxed since human observers do not watch the captured images directly.

E. Hüllermeier, R. Kruse, and F. Hoffmann (Eds.): IPMU 2010, Part II, CCIS 81, pp. 643–652, 2010.

2 Object Tracking in Wide Aria Environment

As a practical Sensing Web application, which is realized by integrating sensing data acquired by multiple sensors, we have developed an object tracking system in a wide area environment. Object tracking in wide area environments can be used in many social applications, such as traffic analysis, purchasing behavior analysis in large-scaled shopping malls, etc.

2.1 Basic Concept

Although there have been many researches into object tracking in multi-view environments [2,3,4,5,6], we have made the following assumptions in our research, which are specific to the Sensing Web characteristics.

Geometrical relationship among the sensors are not known. In the Sensing Web, several people can install sensors individually, and, therefore, when we use those sensors we can not expect to have geometrical relationship among all of those sensors.

The sensors are installed sparsely. Therefore, their views do not overlap with each other. This is mainly due to the cost problem of system construction.

Output of the sensors should be represented in a privacy-protected form. This is the most important issue in the Sensing Web. Especially, when cameras are used, the detailed appearance of target objects should not be visible in the sensor output, and, in our prototype system, we expect only positions and, at most, very simplified appearance information such as color histograms of detected object regions are accessible.

Not only cameras but sensors which can acquire object positions can be used. In the prototype system, we have used not only ordinary video cameras but laser range sensors. The laser range sensors do not acquire the texture information of the objects, which is preferable in case of protecting personal privacy.

As mentioned above, we assume that the geometrical configuration among sensors are not known and that their views are not overlapped. To realize object tracking in such circumstances, we have developed a system having the following features.

- Each object is individually tracked across multiple sensor views. Object identification is realized referring to the topology of the sensor views, or the abstract structure of object flows among sensor views.
- The topology of sensor views are automatically acquired on-line. We do not have to explicitly prepare an off-line training phase. Of course, at beginning, the tracking accuracy is relatively low, but when time goes the accuracy becomes higher. The important point is that when the pattern of object movement changes, we expect the system can adapt the changes according to the on-line estimation algorithm.

2.2 Sensor-View Topology

Our tracking method relies on "sensor-view topology," or how sensor-views are related to one another, and, first, we present how the sensor-view topology is represented and acquired in our system.

Definition of Sensor-view Topology. We suppose that an object disappearing at the position, or "exit-point," p_k in a sensor S_i re-appears at the position, or "entry-point," p_l in S_j. If there is a road, or a walk, between p_k and p_l, a certain number of objects disappearing at p_k re-appear at p_l, and we can expect that an object disappearing at p_k will re-appear at p_l with a certain probability. However, it is not easy to make the correspondence between the disappeared object and the re-appeared object, since the appearances of an object change in different sensors. To solve this issue, we suppose that the moving speed of objects does not change largely in the same pair of the exit-point p_k and the entry-point p_l. This means that we can predict the time duration t_{p_k,p_l} required for objects to move from p_k to p_l, and, based on t_{p_k,p_l}, we can make the correspondence. To handle not only the position but also the time duration, we define "exit-information" and "entry-information" of objects. The exit-information of an object consists of the position p_k in S_i and the time t_{p_k}, while the entry-information consists of the position p_l in S_j and the time t_{p_l}.

According to the above consideration, here, we define sensor-view connection as tuple of $\{\boldsymbol{p}_k, \boldsymbol{p}_l, t\}$, that an object disappearing at $\boldsymbol{p}_k = (x_k, y_k)$ in a sensor S_i re-appears at $\boldsymbol{p}_l = (x_l, y_l)$ in S_j after a time duration t. The sensor-view topology is defined as a set of sensor-view connections, observed in a given set of sensor views. When views of sensors S_i and S_j overlap, we can handle such situation by supposing the transit time can be a negative value.

For example, in Figure 1 (a), the rectangle represents an observing view of a video sensor S_1, the half circle represents that of a laser range sensor S_2, and arrows represent trajectories of objects which should be tracked. In this case, $\mathrm{OUT}_1 \rightarrow \mathrm{IN}_1$ (5 sec) is a sensor-view connection between S_1 and S_2.

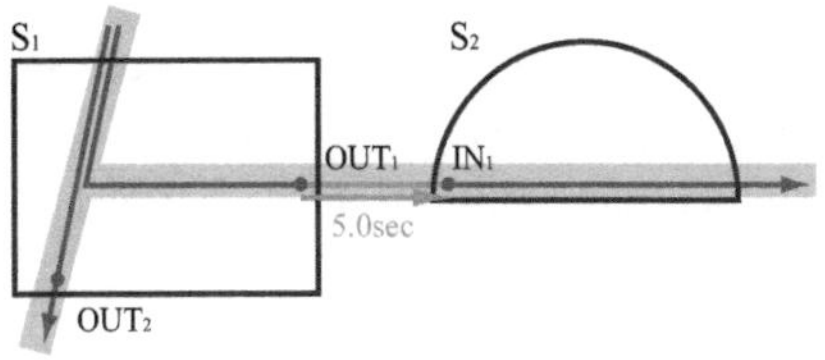

(a) An example of sensor-view connection.

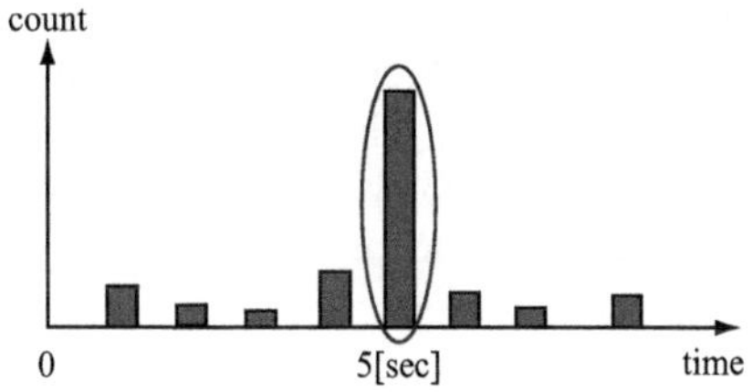

(b) A distribution of temporal pairs in estimating $S_1 \rightarrow S_2$.

Fig. 1. An example of estimation of a sensor-view connection

Estimation of Sensor-view Topology. We define a correct pair of exit/entry-information as true correspondence. Meanwhile, the other pairs are defined as false correspondence. When a number of objects are observed simultaneously, there are a lot of pairs of exit/entry-information. We have to extract the true correspondence from them. However, the object identification has not been achieved yet when the topology is being estimated. Therefore, we temporarily make exit-entry point pairs in all combinations, regardless of their correctness, and, we estimate the accurate topology based on the following uniformity relating to ture correspondence.

Spatial uniformity. Exit and entry-points are observed around some specific points.

Temporal uniformity. The moving speed of objects does not change largely from an exit-point to its corresponding entry-point.

For example, we suppose a simple example shown in Figure 1 (a), where the connection $S_1 \rightarrow S_2$ should be estimated. When all of the "exit/entry-information" are paired temporarily, a correct pair $\mathrm{OUT}_1 \rightarrow \mathrm{IN}_1$ and incorrect pairs, $\mathrm{OUT}_2 \rightarrow \mathrm{IN}_1$ etc, are obtained. Figure 1 (b) shows a typical histogram of the temporal pairs under the assumption of constant moving speed. The horizontal axis is the transit time of temporal pairs, and the vertical axis is the number of observation. We can find a peak of the distribution at 5 [sec], which corresponds to correct sensor-view connection, i.e., $\mathrm{OUT}_1 \rightarrow \mathrm{IN}_1$ with the transit time of 5 [sec].

The issue here is how to correctly estimate transition time when there are ambiguous correspondences having similar transit times. Therefore, to improve the accuracy, we can use additional features acquired from the observing scene if available. For example, in the case of video cameras, appearance features of objects, such as color histogram, becomes helpful information to estimate the topology.

When a connection of an exit-point in S_i and an entry-point in S_j is estimated, temporal connections are voted in a voting space $\boldsymbol{V}^{ij}$. A voted temporal connection is 5-dimensional vector: $\boldsymbol{B}^{ij} = (x^{\mathrm{OUT}}, y^{\mathrm{OUT}}, x^{\mathrm{IN}}, y^{\mathrm{IN}}, t)$, where $(x^{\mathrm{OUT}}, y^{\mathrm{OUT}})$, $(x^{\mathrm{IN}}, y^{\mathrm{IN}})$ are the coordinates of exit-point and entry-point, and t is transit time. We can give a weight w to each vote. When the weight is large value, it is highly possible that the exit/entry-information comes from the same object. In our experiment described later, we have used color histogram of objects if available, and we have implemented a hand-over mechanism of the histogram (see Figure 2). In this case, the w is calculated by histogram intersection.

Sensor-view connections are estimated whenever an object appears. Actually, temporal connections are classified by the Nearest Neighbor method, and each class corresponds to a sensor view connection. Mean μ_{tl} and variance σ_{tl} of transit times of each class, which corresponds to each sensor view connection, is used to identify objects in different sensor views.

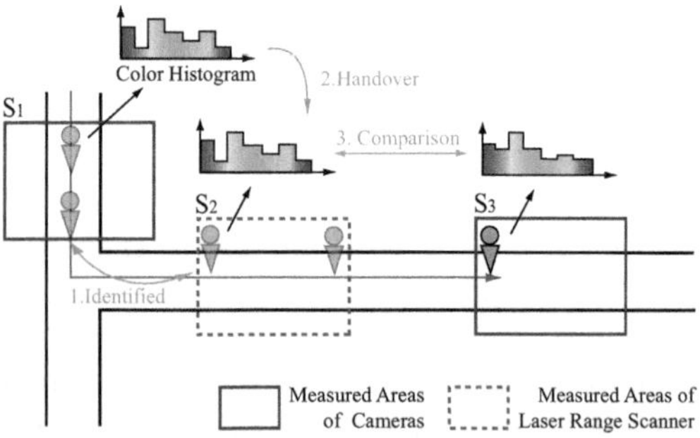

Fig. 2. Hand over of color information

2.3 Object Tracking across Multiple Sensors

Tracked objects in each sensor[1] are identified based on estimated sensor-view connections. To identify an object which passes in multiple observing views of sensors, we introduce a DoC (Degree of Confidence) of exit/entry-information pairs. A DoC is calculated in the following formula:

$$L(\mathrm{IN}, \mathrm{OUT}) = \frac{1}{\sqrt{2\pi}\sigma_t} \exp\left(\frac{(t-\mu_t)^2}{2\sigma_t^2}\right) \quad (1)$$

where μ_t and σ_t are mean and variance of transit time of a sensor-view connection which corresponds to the exit/entry-information pair. The processing flow of object identification is as follows:

Step1. When an object O_k appears and when its appearing point $(x^{\mathrm{IN}}, y^{\mathrm{IN}})$ is not close to any of entry-points of sensor-view connections, temporal sensor-view connections, which are acquired by combining the entry-information and all the recent exit-information, are calculated and put in the voting space. Then, the sensor view connection estimation process is executed.

Step2. When the appearing point is close to an entry-point of a sensor-view connection, we get a DoC $L()$ corresponding to the sensor-view connection. Then we search for an disappeared object which maximizes $L(\mathrm{IN}_{O_k}, \mathrm{OUT}_m)$ $(m = 1, \ldots, M)$, where OUT_m represents the exit-information of recent disappeared objects and M is the number of objects.

Step3. When the appearing object is identified, sensor-view connection which consists of the pair of corresponding exit/entry-information is voted to the voting space $\boldsymbol{V}^{ij}$. The amount of vote is $w = L(\mathrm{IN}_{O_k}, \mathrm{OUT}_m) \times W$ (W is constant when we only use exit/entry information.). Finally, the sensor-view connections are updated.

In our approach, sensor-view connections are estimated every time an object is observed. Therefore, estimation of connections and object tracking are achieved automatically without calibration of sensor's positions.

3 Implementation of Wide Area Tracking on Distributed System

When we construct a real-time object tracking system based on our proposed method, its computation cost, or it throughput, should be considered. The number of sensors which one computer can process at real-time is limited, and, in our implementation, one computer can process sensory data of at most two video cameras. Therefore, when a certain number of cameras are installed to observe a wide area environment, a distributed processing system should be constructed to achieve real-time processing.

In our algorithm, when an object moves between sensors S_i and S_j, we, at first, estimate sensor view connections from S_i to S_j and from S_j and S_i, and objects are tracked based on the estimated view connections. When we have n sensors we expect $n(n-1)$

[1] Object tracking in each sensor is realized by combining basic image analysis methods[7,8,9].

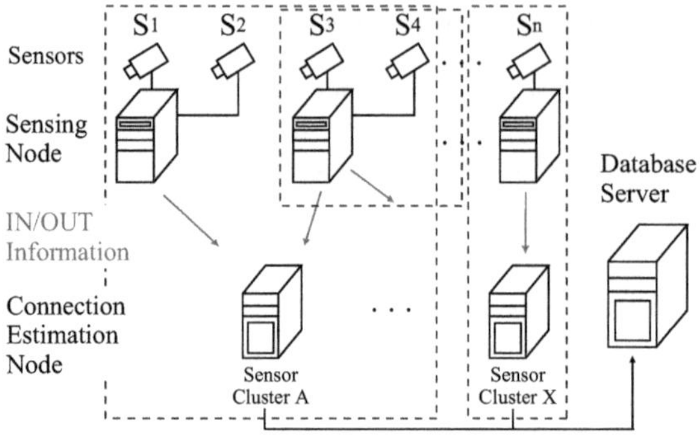

Fig. 3. Configuration of Distributed Object-tracking System

possible sensor view connections. Therefore, the number of connections becomes very large when the number of sensors is large, which drastically increases the computation cost. In such cases, we should introduce a hierarchical processing framework, where the sensors are grouped into several clusters, which are called "sensor clusters," and, in each of the sensor clusters, objects are tracked at first. Then, object tracking information acquired in different sensor clusters is integrated. An important point is that some of the sensors should be shared in different clusters so that acquired object tracking information can be tagged with one another among the clusters.

In our prototype system, we have three kinds of processing nodes: sensing nodes, connection estimation nodes, database server (see Figure 3). The functions of those nodes are summarized as follows.

Sensing node. In each sensing node, objects are tracked at real-time in each view of sensors connected the node. Processed result, or tracking result, of the node is represented in XML (see Figure 4), which consists of several tags: "time," "sensor," "object," "coordinates," and "color." The "sensor" tag has an attribute "id" which denotes the sensor in which the tracking result was acquired. The "object" tag has two attributes: one is "local_id" used in each sensor and the other is "global_id" which is a globally unique identifier. Note that the "global_id" is not available (actually N/A value is given) until object identification is finished among sensors. The values in the "coordinates" and "color" tags are calculated from the position of the object and its color histogram. Sensing nodes correspond to primitive elements of the Sensing Web, and their output are privacy preserved so that anyone can have access to the sensing nodes.

Connection estimation node. One connection estimation node is established in each sensor cluster, and it estimates sensor view topology among the sensors in its corresponding sensor cluster. Then, correspondence among tracking results acquired in the sensing nodes is established, and global object tracking in the sensor cluster is achieved. The acquired global object tracking is also represented in XML similar to the previous one.

```
<?xml version="1.0" encoding="UTF-8" ?>
<tracking_data>
    <time year="2010" month="1" day="15" hour="10" min="30" sec="20" msec="333">
        <sensor id="1">
            <object local_id="3" global_id="2">
                <coordinates>
                    <x_2d>44</x_2d>
                    <y_2d>55</y_2d>
                </coordinates>
                <color>14:421:421:324:0:0:···(512 bins)</color>
            </object>
            <object local_id="4" global_id="3">
                <coordinates>
                    <x_2d>48</x_2d>
                    <y_2d>50</y_2d>
                </coordinates>
                <color>14:21:41:32:51:0:···(512 bins)</color>
            </object>
        </sensor>
        <sensor id="2">
            .
            .
        </sensor>
    </time>
    <time •••>
      .
      .
    </time>
</tracking_data>
```

Fig. 4. An example of tracking result described in XML document

Database server. All the global object tracking results in all the sensor clusters are accumulated in this node, and matching of the tracking results are established. Any application program can retrieve the final tracking results from this database server.

4 Experimental Results

4.1 Experiments of Wide Area Object Tracking

We performed experiments of people tracking in a campus building using both of video sensors and laser range sensors. We have used 10 cameras and 2 laser range sensors, and Figure 5 shows a part of the sensor arrangement, i.e., sensors installed in the ground floor. In Figure 5, the green rectangles represent observing views of the video cameras, and the red rectangles represent observing views of laser range sensors. In this experiment, we observed the circumstance for about 5 hours and more than 400 people were observed.

Figure 6(a) shows result of sensor-view topology estimation, which was acquired after observing the circumstance for 10 to 15 minutes. Sensor-view topology was correctly estimated, and red lines in the figure indicate estimated sensor-view connections.

Then, we have evaluated the object tracking performance. Here, we used data collected from the sensors in the ground floor, i.e., from 5 video cameras and 2 laser range sensors, where about 350 people are observed. The performance was evaluated in terms of *recall* and *precision*.

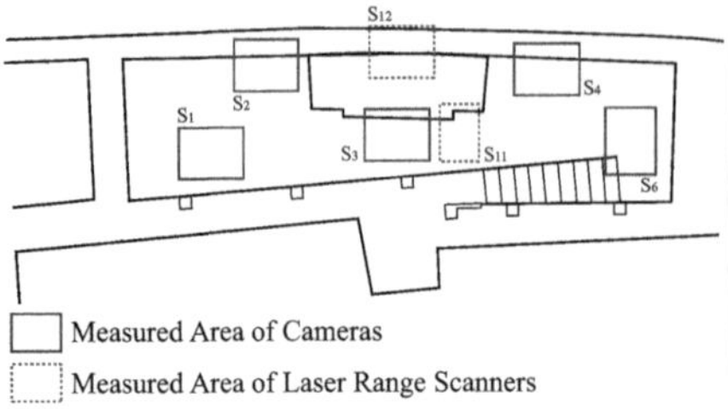

(a) Observed area on the ground floor

(b) Scene example

Fig. 5. Experimental circumstance

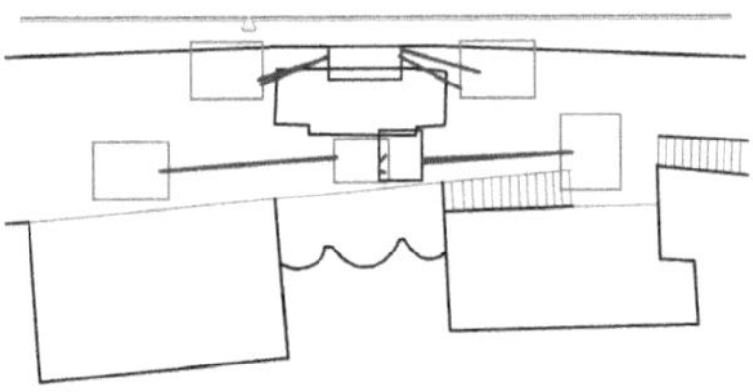

(a) Estimated topology (Figure 5(a))

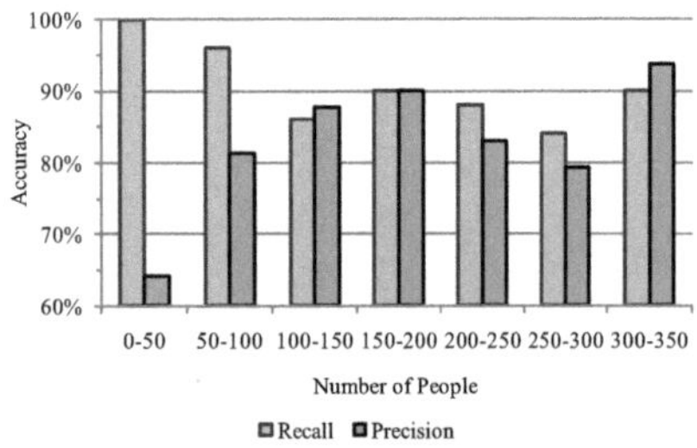

(b) Tracking accuracy

Fig. 6. Result of Object Tracking

Figure 6(b) shows how the performance changed as the number of observed people increased, where the recall and the precision in each observing period is illustrated. Here, the observing periods are represented in terms of the number of observed people, and, hence, their actual physical times are not the same. At beginning, the performance is not good, but it becomes better as the number of observations increases. Please note that color histogram information was used to evaluate object correspondence when object tracking results were available from video cameras[2]. When several people walking together disappear at the same time and when they re-appear at another entry point, those re-appeared people are possibly mis-identified. It sometimes happens especially when only exit and entry information is used to identify them. However, if we use information reflecting peoples appearance such as color histogram, this problem can largely relaxed.

4.2 Demonstrative Experiments in "Shin-Puh-Kan"

In the Sensing Web project, we have made demonstrative experiments in "Shin-Puh-Kan," a shopping mall in Kyoto. Our object tracking in a wide area environment has been incorporated in "Digital Diorama of Shin-Puh-Kan," which visualizes human activities in a 3D virtualized space[10]. Using the Digital Diorama, we can virtually see

[2] The similarity of the histogram was calculated based on histogram intersection.

(a) Shin-Puh-Kan

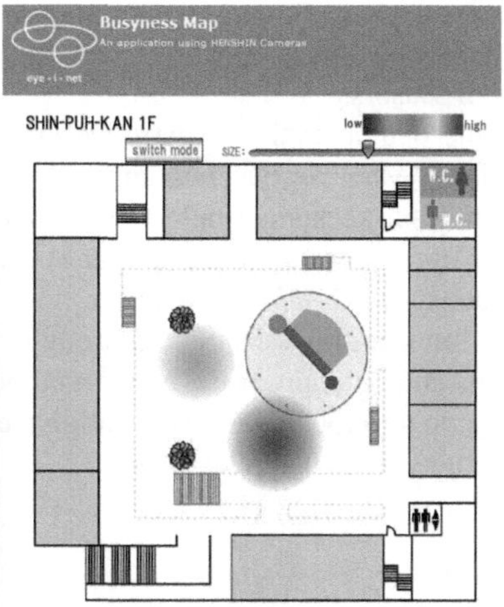

(b) NIGIWAI Map

Fig. 7. NIGIWAI (busyness) Map of Shin-Puh-Kan

the current scenes of the mall from any viewpoints at real-time. We have also developed "Shin-Puh-Kan NIGIWAI Map (Shin-Puh-Kan busyness map)," which visualizes the busyness of the mall by integrating object detection results acquired by video cameras installed in the mall (see Figure 7). We can see, at a glance, which part of the mall gathers many customers. From several questionnaires, we can see that the busyness map is convenient for customers of the mall, and that they want to have such busyness maps at complex shopping malls, large-scaled amusement parks, parking lots, etc.

5 Conclusion

We have presented our research into organizing distributed sensory information acquired in the Sensing Web. Especially, we have presented a system of object tracking in a wide area environment, which are observed by multiple sensors having non-overlapping views. The important feature of the Sensing Web is that accessible data from the sensors in the Sensing Web is privacy protected. In our prototype system, we have used video cameras and laser range sensors, and only positions of detected objects are used except that color histograms of the objects are additionally used in video cameras.

Our tracking system consists of estimation of the sensor-view topology of a given set of sensors and identification of objects appearing in different kinds of sensory data. By referring to the sensor-view topology, which represents the spatial and the temporal co-occurrence information of appearance/disappearance of objects, we can identify objects appeared in different sensor views. The sensor-view topology is, automatically and on-line, acquired by observing "exit/entry information" of all the objects in each sensor view and by finding co-occurrence in their "exit/entry information." We performed some experiments to evaluate our approach, and we found the topology was correctly estimated, and objects were correctly tracked across non-overlapping views of different kinds of sensors.

In the demonstrative experiments in a shopping mall, we have also constructed "NIGIWAI Map", or busyness map, of the mall, which is generated by integrating data from multiple sensors installed in the mall. This is a typical application to exhibit usefulness of the Sensing Web, and many customers of the mall feel it is convenient and interesting.

For future work, we should thoroughly evaluate the performance, in terms of accuracy and computation speed, of our method with more complicated sensor arrangement. We should develop better visualization mechanism to provide more convenient user interface for practical applications.

References

1. Minoh, M., Kakusho, K., Babaguchi, N., Ajisaka, T.: Sensing web project–how to handle privacy information in sensor data. In: Proceedings of International Conference on Information Processing and Management of Uncertainty in Knowledge-Based Systems (2008)
2. Zhao, H., Shibasaki, R.: A real-time system for monitoring pedestrians. In: Proceedings of IEEE Workshop on Applications of Computer Vision (2005)
3. Javed, O., Rasheed, Z., Shafique, K., Shah, M.: Tracking across multiple cameras with disjoint views. In: International Conference on Computer Vision 2003, pp. 952–957 (2003)
4. Cai, Y., Chen, W., Huang, K., Tan, T.: Continuously tracking objects across multiple widely separated cameras. In: Asian Conference on Computer Vision 2007, pp. 843–852 (2007)
5. Ukita, N.: Probabilistic-topological calibration of widely distributed camera networks. Machine Vision and Applications Journal 18(3-4), 249–260 (2007)
6. Makris, D., Ellis, T., Black, J.: Bridging the gaps between cameras. In: Conference on Computer Vision and Pattern Recognition 2004, vol. 2, pp. 205–210 (2004)
7. Tanaka, T., Shimada, A., Arita, D., Taniguchi, R.: Non-parametric background and shadow modeling for object detection. In: Proceedings of the 8th Asian Conference on Computer Vision, pp. 159–168 (2007)
8. Isard, M., Blake, A.: Condensation-conditional density propagation for visual tracking. International Journal on Computer Vision 29(1), 5–28 (1998)
9. Zhao, H., Chen, Y., Shao, X., Katabira, K., Shibasaki, R.: Monitoring a populated environment using single-row laser range scanners from a mobile platform. In: 2007 IEEE International Conference on Robotics and Automation, pp. 4739–4745 (2007)
10. Yamaguchi, R., Yamamoto, Y., Nitta, N., Ito, Y., Babaguchi, N.: Digital diorama: Adaptive 3d visualization system for indoor environments. In: Proceedings of International Workshop on Sensing Web (2007)

Evaluation of Privacy Protection Techniques for Speech Signals

Kazumasa Yamamoto and Seiichi Nakagawa

Toyohashi University of Technology,
Department of Computer Science and Engineering,
1-1 Hibarigaoka, Tenpaku-cho, Toyohashi, Aichi 441-8580, Japan
{kyama,nakagawa}@slp.cs.tut.ac.jp
http://www.slp.cs.tut.ac.jp/

Abstract. A ubiquitous networked society, in which all electronic equipment including "sensors" are connected to a network and are able to communicate with one another to share information, will shortly become a reality. Although sensor information is most important in such a network, it does include a large amount of privacy information and therefore it is preferable not to send raw information across the network. In this paper, we focus on privacy protection for speech, where privacy information in speech is defined as the "speaker's characteristics" and "linguistic privacy information." We set out to protect privacy information by using "voice conversion" and "deletion of privacy linguistic information from the results of speech recognition." However, since speech recognition technology is not robust enough in real environments, "speech elimination" technique is also considered. In this paper, we focus mainly on the evaluation of speech elimination and voice conversion.

Keywords: Privacy protection, Speech signal, Personal information, Speech elimination, Voice conversion.

1 Introduction

A ubiquitous networked society, in which all electronic equipment including "sensors" are connected to a network and are able to communicate with one another to share information, will shortly become a reality. Sensor information is very important in such a network, especially in virtual reality systems which give the illusion of actually being there.

We have been working on the project, "Contents Engineering for Social Use of Sensing Information" [1], which aims to collate real world observation content into a "Sensing Web." Sensor information is collected from a variety of sensors, such as obstacle sensors, video cameras, thermometers, microphones, and so on, which are installed at various locations for use not only by those who installed the sensor, but also by anyone in much the same way as information is disseminated on the World Wide Web.

On the Sensing Web, sensor information must either be filtered or privacy information encrypted, to ensure that the information can be used freely. Included

E. Hüllermeier, R. Kruse, and F. Hoffmann (Eds.): IPMU 2010, Part II, CCIS 81, pp. 653–662, 2010.

in this project is the development of techniques to provide privacy protection for information from a microphone (sound information) according to the user access level.

In this paper, we evaluate privacy protection techniques for speech signals. Note that we focus mainly on techniques for aspects of speech signal processing and do not describe issues related to speech recognition and linguistic processing.

2 Sound Information and Its Privacy Protection on the Sensing Web

Sound signals recorded by a sound sensor, such as a microphone, can be classified roughly into two categories:

- **Background (BG) sound / Environmental sound**
  Sound of the wind or vehicles, crowd noise, computer fan noise, background music, etc.
- **Speech**
  Human voice close to the microphone, which can be categorized further as "audible speech" or "non-audible speech," which is a human speech-like noise that comprises a mixture of human speech [2].

Generally, "speech" includes a great deal of privacy information, whereas "BG sound" does not. Although BG sound includes location information linked to where the speaker is, we do not treat this as privacy information because it does not relate to the speaker's individuality. "Non-audible speech" can also be treated as non-privacy information because we cannot associate it with any aspect of the speaker's individuality.

Speech conveys much information that can be used for biometric authentication [3]. Such information can be classified into three categories: linguistic information, para-linguistic information, and non-linguistic information. Para-linguistic information includes that which the speaker can purposefully control, other than linguistic information. It is mainly prosodic information and also includes the speaker's intention and behavior. On the other hand, non-linguistic information includes aspects of the speaker's individuality and emotion which the speaker cannot purposefully control, such as voice characteristics, gender, etc. Thus, all the non-linguistic information counts as privacy information.

Additionally, the data on the Sensing Web should contain:

- **Sound signals**
  Sound waves, which can be compressed.
- **Symbolized information**
  Results of speech recognition, speaker recognition, environmental sound recognition, and so on.

Both of these are required to protect privacy information.

We aim to develop a system, in which all speech privacy information can be protected (in other words, "encrypted") for the Sensing Web, and can then be decrypted according to the user access level. Additionally, this information must be available while encrypted (in a similar way, to an image with mosaic processing hiding only the faces in an image processing study). For this purpose, we use the following techniques [4]:

- **Voice Conversion**
  This technique is useful for protecting non-linguistic privacy information, i.e., individuality in speech signals. Individuality included in speech, such as voice characteristics or speaking habits, must be eliminated to ensure that the speaker cannot be identified.
- **Distant Speech Recognition**
  Linguistic privacy information must also be eliminated to protect privacy. To remove linguistic information from a speech signal, the time alignment of the information (words) in the speech signal must be known and the region must be replaced by another sound, such as a bleep. To do this, a highly accurate distant speech recognition technique for a real environment is required.
- **Speech Elimination**
  Since speech recognition based techniques are very expensive in terms of both computation and equipment investment, and are required to be highly accurate, we protect privacy information by eliminating only the speech from the recorded sound signals, resulting in only environmental sounds remaining. This protects all the privacy information included in the speech, making it useful as sound sensor information without privacy information.

In this paper, we focus on "Speech Elimination" and "Voice Conversion" techniques.

3 Privacy Protection by Speech Elimination

Speech signals include individuality information created simultaneously by voice characteristics, prosodic information, and linguistic individuality, such as the names of people. To eliminate linguistic individuality, a speech recognition technique is required to identify those words containing individual information. An elimination or substitution operation for the linguistic information is then needed. The required speech recognition based techniques, however, have very high computational cost and equipment investment, while the accuracy of current techniques for distant speech recognition in a real environment is not adequate for our purposes. On the other hand, BG sound is very important for understanding the surrounding environment in which the sensors are located. Therefore, we propose a "Speech Elimination System" to eliminate only speech from the sound sensor signals. Although many "noise elimination" techniques have been proposed for speech enhancement or speech recognition, to date, no "speech elimination" technique has been studied.

3.1 Speech Elimination Method

For speech elimination, we can simply use a noise suppression method, but exchanging the speech and noise components. However, it is difficult to apply noise

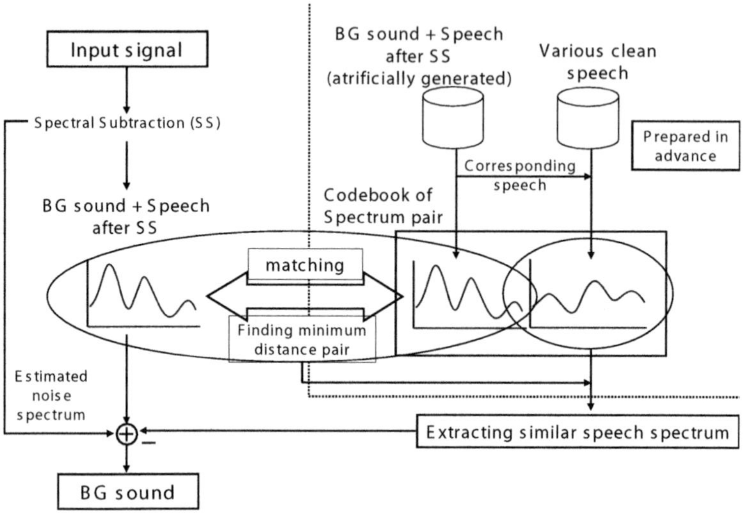

Fig. 1. Overview of speech elimination method

suppression methods directly, particularly those methods that assume stability of noise, since the speech signal is not a stable signal. We have thus proposed a vector quantization (VQ) based speech elimination technique [4], the procedure for which is described below:

1. Clean speech data and BG sound data are prepared as training data. The BG sound data are added to the clean speech data to make noisy speech data with a variety of SNRs.
2. The noisy speech data and clean speech data are analyzed in a spectral domain. Feature vectors are then generated by combining the noisy speech and clean speech amplitude spectra.
3. A VQ codebook is generated from the feature vectors using the LBG algorithm. For this process, the noisy speech amplitude spectrum is only used for VQ clustering.
4. Using the input sound signal (noisy speech) as the key, the codebook index is searched for the closest match to the input noisy speech by comparing with the noisy speech spectrum in the codebook.
5. A BG sound signal is synthesized using an overlap-add technique. The clean speech spectrum, taken from the codebook index obtained, is subtracted from the noisy speech spectrum, to produce the synthesized BG sound signal.

Fig. 1 illustrates the procedure for the speech elimination method.

3.2 Subjective Evaluation

A subjective evaluation was carried out through an experiment using the JNAS database [5]. The codebook was trained from speech data uttered by 103 male and 103 female speakers, with five sentences per speaker. As test data, we used ten sentences uttered by five male and five female speakers from newspaper articles selected from the JNAS database. Data from the test speakers were not included in the training data. The data were down sampled to 8 kHz from the original 16 kHz sampling. Conditions for speech analysis included a 32ms Hanning window (256 pts) and a 16 ms frame shift (128 pts). Restaurant noise from the AURORA-2 database [6] was used as BG sound, which was added to the clean speech at 20, 10, and 0dB SNRs for training the codebook. Dimensions of the code vector were 128 (for noisy speech) + 128 (for clean speech) (frequency bins) with a codebook size of 4096. In this experiment, we divided a spectral vector into four sub-vectors (32 dimensions each) to enlarge the size of the actual codebook. SNRs of the test data were set as 5dB, 0dB, and -5dB, while maintaining the background sound level, i.e., only the speech level was changed when adjusting the SNR. In total, 60 test sentences were used in the experiment (10 sentences × 3 SNR conditions × w/ and w/o elimination processing).

We used ten university students as subjects. They were requested to evaluate the "audibility/intelligibility" of the speech included in the background sound and the "naturality" of the background sound using a five-point evaluation scale. Subjects listened twice to all 60 test sentences presented in a random order.

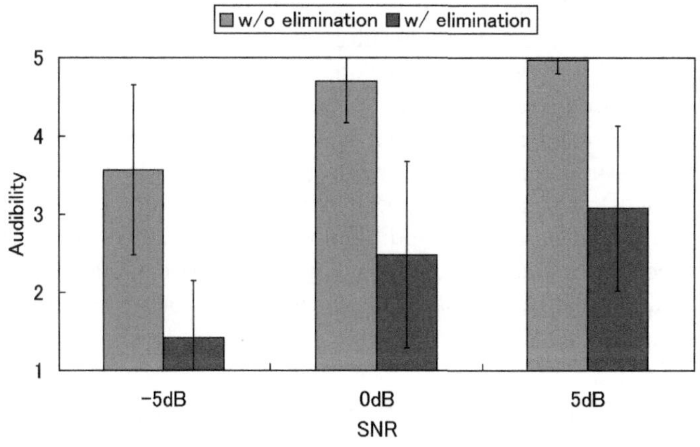

Fig. 2. Results of the audibility evaluation in speech elimination

Figs. 2 and 3 show the experimental results. Fig. 2 shows the results of the "audibility" evaluation of the speech included in the background sound. In the figure, "5" means "very easy to distinguish the words in the speech," whereas "1" means "very hard to distinguish the words in the speech." From the results, we can see that humans can distinguish the words almost perfectly with a 5dB SNR. Even with a -5dB SNR, humans can distinguish more than half of the words. However, using the speech elimination technique, even with a 5dB SNR, humans are unable to distinguish even half of the words, and it becomes more difficult with a -5dB SNR.

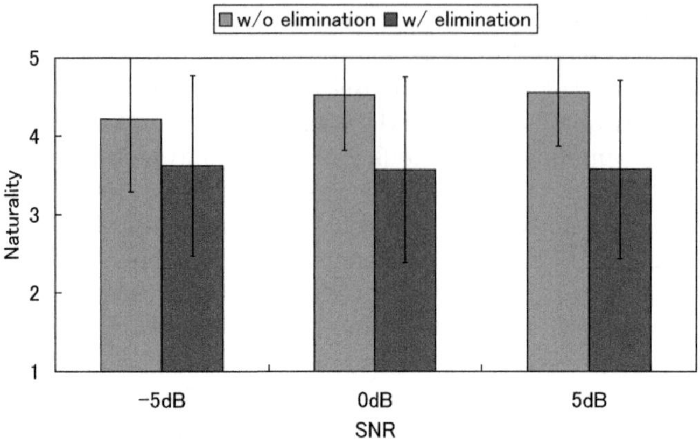

Fig. 3. Results of the naturality evaluation in speech elimination

On the other hand, Fig. 3 shows the results of evaluating the "naturality" of the sound. In the figure, "5" means a "very natural sound," whereas "1" means a "very unnatural sound." From these results we can see that the proposed technique maintains high naturality of background sound, although with slight degradation thereof. From these results, we can conclude that our proposed technique is useful for privacy protection.

Fig. 4 shows some results of the speech elimination. (a) and (c) show the original human speech waves and their spectra with background sound recorded in a university student restaurant, while (b) and (d) show the corresponding processed images, that is, with speech eliminated. We can see that the high amplitude components, namely audible speech, are removed from the original wave and spectrum including the frequency band between 1000 and 4000Hz.

Henceforth, we plan to evaluate the method using other background sounds, such as background music, etc.

4 Privacy Protection by Voice Conversion

Since speech includes various "speaker characteristics," which qualify as privacy information, it is not appropriate to publish such speech on the Sensing Web without some manipulation to protect the privacy information. In this regard, we use a voice conversion system to remove individuality included in speech signals and alter the speaker characteristics.

4.1 Voice Conversion Methods

Voice conversion systems have been studied for a long time, with many voice conversion methods being proposed.

Generally, speaker characteristics depend on the spectral peak positions, sharpness of the peaks, and formant frequencies, caused by the shape of the vocal tract,

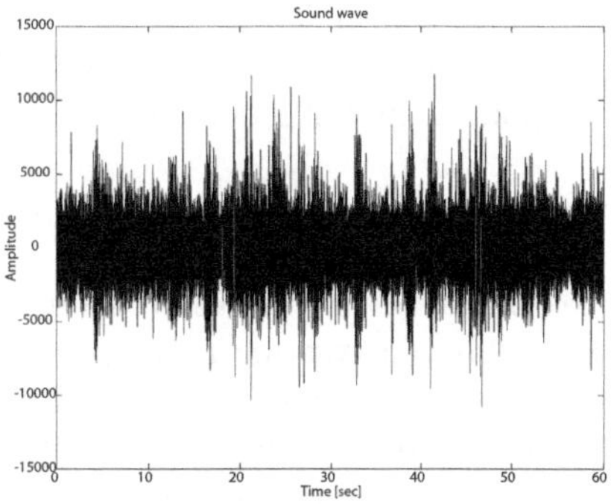

(a) Original sound wave with background noise

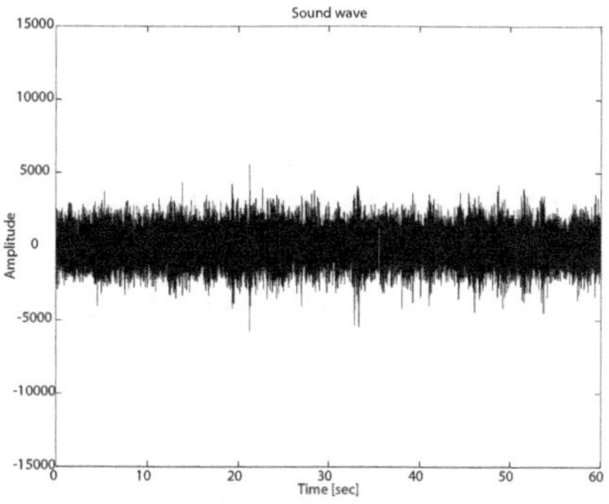

(b) Speech eliminated sound wave with background noise

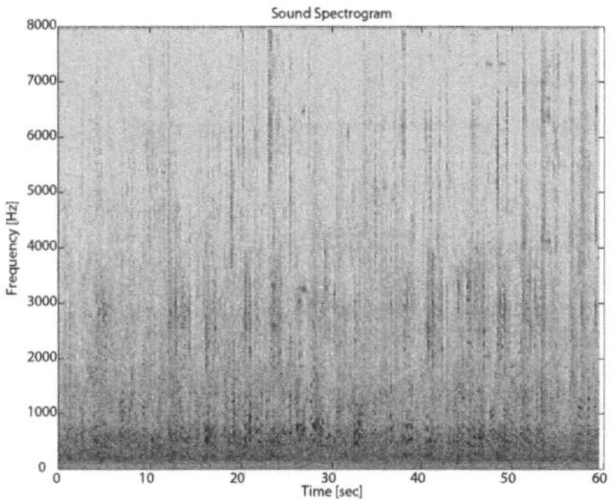

(c) Original sound spectrogram with background noise

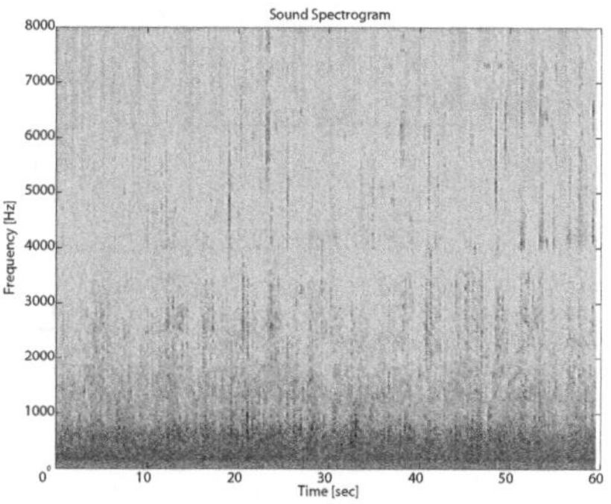

(d) Speech eliminated sound spectrogram with background noise

Fig. 4. Some results of the speech elimination

and the spectral slope, pitch, and accent caused by the sound source [7]. Traditional voice conversion techniques are based on changes in these parameters. One of the most popular voice conversion methods is extreme pitch conversion, commonly used in TV programs. In this method, to avoid reconstructing the original speech signal, the speech is converted and mixed using multiple scale pitch factors.

Recently, the main focus of voice conversion has relied on spectral transformation. A method using a mapping codebook was initially proposed [8]. This was then expanded to a statistical method using Gaussian Mixture Models (GMM) [9], which was in turn expanded using Eigen Voice (EV) [10].

4.2 Our Proposed System

First, in order to change the speaker characteristics in real-time, we attempted to change the spectral peak frequency, spectral peak sharpness, and spectral slope [7]. However, all these spectral modifications provided insufficient conversion quality. We then constructed a GMM-based voice conversion system [9] with a Mel-LPC analysis frontend [11] and MLSA filtering synthesis [12], as GMM-based voice conversion systems are able to map an unspecified person's voice to a specified person's voice.

In this study, we need a voice conversion system that works robustly in real (noisy) environments. The system must be able to separate the speech and noise signals, convert only the speech, and keep the original noise. Fig. 5 shows a block diagram of our proposed voice conversion system. In our system, input speech is sent to both the "speech eliminator" described in the previous section and the Voice Activity Detection (VAD) block. Next, noise suppression is done, simply by means of the Spectral Subtraction (SS) method. Thereafter, GMM-based voice conversion is performed. Finally, the converted speech and the speech eliminated background sound are mixed and output.

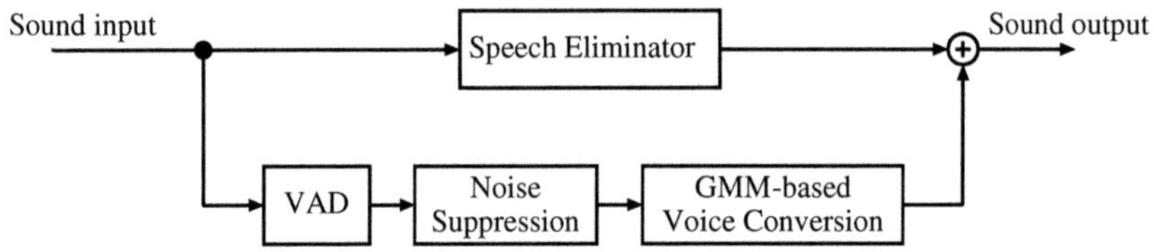

Fig. 5. Voice conversion system with speech elimination system

4.3 Subjective Evaluation

This experiment was also conducted using the JNAS database. The GMM was trained from the speech data uttered by one target male speaker and 29 other male speakers with 50 sentences per speaker. As test data we used ten sentences uttered by five male speakers (two sentences per speaker) from newspaper articles selected from the JNAS database. Data from test speakers were not included in the training data. Speech analysis conditions included a 16kHz sampling frequency, 25ms Hamming window (400 pts) and 10 ms frame shift (160 pts). The Mel-LPC analysis order was 16 and the dimensionality of the Mel-LPC cepstra was 20. Restaurant noise from the AURORA-2 database was once again used as BG sound. SNRs of the test data were set as 10dB and 0dB, while keeping the background sound constant. In total, 20 test sentence pairs were used in the experiment (5 speakers $\times$ 2 sentences $\times$ 2 SNR conditions).

The same subjects were used as in the previous experiment. They were requested to evaluate the "synthesized speech quality," "difference in voice characteristics from original voice," "difference in speech characteristics from original speech," and "audibility/intelligibility of speech" using a five-point evaluation scale. Subjects first listened to the original speaker's utterance, and then to the voice-converted utterance. They then compared the two utterances in evaluating each item.

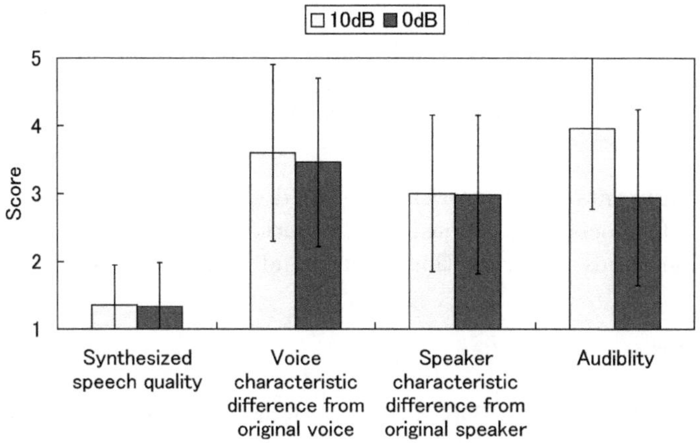

Fig. 6. Results of voice conversion evaluation

Fig. 6 gives the results of the evaluation, showing the "Score" which gives the measure of "synthesized speech quality" (with "5" being the best), different degrees of "voice characteristic difference from the original voice" and "speech characteristic difference from the original speech" (where "5" means "quite different from the original speaker" and "1" means "the same as the original speaker"), and degree of "audibility" (where "5" means "very audible").

Based on the results, the performance is not acceptable. The main reason for this is the poor performance of pitch estimation and voiced/unvoiced sound source selection in the noisy environment. In particular, as can be seen from the results, the quality of the synthesized speech was unsatisfactory, although this method does work in real-time. We need to improve the technique whilst retaining both real-time processing and good voice quality.

As one of our goals, we not only need a simple many-to-one voice conversion system, but also to keep information of the number of speakers. For example, when two people are speaking, we need a two-to-two voice conversion system. Consequently, we use a speaker diarization technique and speaker clustering technique. In future, we aim to improve the GMM-based voice conversion system in [10], and to evaluate the method combining spectral operation and the STRAIGHT vocoder system [13].

5 Summary

In this paper, we discussed the evaluation of speech elimination and voice conversion techniques for speech privacy protection on the Sensing Web. According to the experimental results, the proposed speech elimination technique worked well. However, the voice conversion system did not provide adequate performance in a noisy background environment, due to the lack of robustness of pitch estimation in the noisy environment.

The developed techniques, including the speech elimination technique, are not yet fully established, with the result that we need to continue developing new techniques for privacy protection, including speech processing methods, as well as speech recognition techniques and linguistic processing techniques.

Acknowledgments. This work was carried out as a part of the "Contents Engineering for Social Use of Sensing Information" project sponsored by Special Coordination Funds for Promoting Science and Technology, MEXT.

References

1. Minoh, M., Kakusho, K., Babaguchi, N., Ajisaka, T.: Sensing Web Project - How to handle privacy information in sensor data. In: Proc. 12th International Conference on Information Processing and Management Uncertainty in Knowledge-Based Systems (IPMU 2008), pp. 863–869 (2008)
2. Kobayashi, D., Kajita, S., Takeda, K., Itakura, F.: Extracting speech features from human speech-like noise. In: Proc. ICSLP 1996, vol. 1, pp. 418–421 (1996)
3. Impedovo, D., Refice, M.: Multiple speaker models and their combination in access control tasks. Journal of Information Assurance and Security 4(4), 346–353 (2009)
4. Yamamoto, K., Nakagawa, S.: Privacy protection for speech information. Journal of Information Assurance and Security 5(1), 284–292 (2010)
5. Ito, K., et al.: JNAS: Japanese speech corpus for large vocabulary continuous speech recognition research. Journal of the Acoustical Society of Japan (E) 20(3), 199–206 (1999)
6. Hirsch, H., Pearce, D.: The AURORA experimental framework for the performance evaluation of speech recognition systems under noisy conditions. In: Proc. ISCA ITRW ASR 2000 on Automatic Speech Recognition: Challenges for the next Millennium (2000)
7. Childers, D.G., Yegnanarayana, B., Wu, K.: Voice conversion: factors responsible for quality. In: Proc. ICASSP 1985, pp. 748–751 (1985)
8. Arslan, L.M., Talkin, D.: Voice conversion by codebook mapping of line spectral frequencies and excitation spectrum. In: Proc. EUROSPEECH 1997, pp. 1347–1350 (1995)
9. Stylianou, Y., Cappe, O., Moulines, E.: Continuous probabilistic transform for voice conversion. IEEE Trans. on Speech and Audio Processing 6(2), 131–142 (1998)
10. Toda, T., Black, A.W., Tokuda, K.: Voice conversion based on maximum-likelihood estimation of spectral parameter trajectory. IEEE Trans. on Audio, Speech, and Language Processing 15(8), 2222–2235 (2007)
11. Matsumoto, H., Moroto, M.: Evaluation of Mel-LPC cepstrum in a large vocabulary continuous speech recognition. In: Proc. ICASSP 2001, vol. 1, pp. 117–120 (2001)
12. Imai, S., Sumita, K., Furuichi, C.: Mel log spectrum approximation (MLSA) filter for speech synthesis. Electronics and Communications in Japan (Part I: Communications) 66(2), 10–18 (1983)
13. Kawahara, H.: STRAIGHT, Exploration of the other aspect of VOCODER: Perceptually isomorphic decomposition of speech sounds. Acoustic Science and Technology 27(6), 349–353 (2006)

Digital Diorama: Sensing-Based Real-World Visualization

Takumi Takehara[1], Yuta Nakashima[2], Naoko Nitta[2], and Noboru Babaguchi[2]

[1] School of Engineering, Osaka University
[2] Graduate School of Engineering, Osaka University
{takehara,nakashima,naoko,babaguchi}@nanase.comm.eng.osaka-u.ac.jp

Abstract. Many sensors around the world are consistently collecting the real-time real-world data. The data streams captured by these sensors can give us an idea of what is going on in a specific area; however, it is not easy for humans to understand their spatial and temporal relationships by just looking at them independently. This paper proposes to construct Digital Diorama, a three-dimensional view where viewers can see at a glance how people are moving around the monitored space without violating their privacy, by integrating multiple data streams captured by stationary cameras and RFID readers in real time. Digital Diorama realizes such real-world visualization with the following features: 1) view control, 2) real-time camera image superimposition, and 3) privacy control. We have demonstrated that Digital Diorama for a shopping center was able to present the current positions of persons and real-time camera images in approximately 1 frame per second.

1 Introduction

In recent years, many sensors such as stationary and mobile cameras, microphones, GPS receivers, and RFID readers are distributed around the world for monitoring purposes. For example, many cameras are often installed in stations, airports, shopping centers, etc., for the purpose of crime deterrence and investigations. So far, the information obtained from these sensors have been used only by authorized persons. If such information can be collected together and become open to public through a sensor network, more beneficial services can be provided. However, as the number of sensors increases, it becomes harder for humans to relate the separate information streams and to grasp the whole picture of the monitored space. For example, when we look at multiple camera images simultaneously, it is hard to understand the spatial and temporal relations among the objects in the camera images.

Many methods have been proposed to present a comprehensible view by integrating multiple camera images. Ikeda et al. [1] used two vertically-aligned omni-directional stereo and a laser range finder to construct a three-dimensional geometry model of the monitored space. The images of stationary objects captured by the omni-directional stereo are mapped to the three-dimensional geometry model so that the spatial continuity of the images can be understood.

E. Hüllermeier, R. Kruse, and F. Hoffmann (Eds.): IPMU 2010, Part II, CCIS 81, pp. 663–672, 2010.

Sawhney et al. also proposed Video Flashlights [2], which renders multiple live video images over a three-dimensional geometry model in real time. Further, Sebe et al. proposed to detect moving objects from camera images and to apply their textures to the billboards located at the positions of the moving objects [3]. Girgensohn et al. proposed DOTS [4] which offers a three-dimensional view of a building aiming at following persons of interest. Moving objects are considered as persons and the positions of tracked persons, which are obtained by video analysis, are marked in the three-dimensional geometry model by displaying their textures. The viewpoint is switched automatically to follow the person and his texture obtained from the camera which is closest to the viewpoint is shown as a billboard facing the viewpoint, so that viewers can intuitively understand how the person moves through the monitored space. Similarly, Haan et al. [5] targeted on following persons of interest in multiple camera images and proposed a three-dimensional interface, where multiple video streams are selected, transformed, and blended to provide a smooth transition between camera images rendered in the three-dimensional geometry model.

As Wang et al. have suggested in [6], embedding camera images in three-dimensional geometry models as described above should improve the viewers' performance in monitoring or tracking tasks. Similarly, targeting on a public space monitored by stationary cameras and RFID readers, this paper proposes to construct a three-dimensional view called Digital Diorama [7], which visualizes how persons move around the space in a comprehensible way by integrating the captured information. Especially, we focus on the following three issues: 1) in real life situation, only a limited number of sensors can be installed in the space, 2) the privacy information of persons should not be presented without their consent, and 3) Digital Diorama can be viewed over the networks simultaneously by different persons with different requests. In addressing these issues, Digital Diorama selectively presents captured real-time information such as real-time camera images and the privacy information of persons on a three-dimensional view according to the viewers' requests: the viewer's ID and a set of view and gaze points.

2 Digital Diorama

Digital Diorama is designed for a public space equipped with two types of sensors: stationary cameras and RFID readers. There are generally two types of objects in the space: stationary objects such as floors or walls and moving objects such as persons. Here, all moving objects are considered as persons. The cameras provide the visual information and the positions of these objects, while RFID readers provide the identity of the persons carrying RFID tags and their rough positions. By collecting and integrating these information from the sensors in real time, Digital Diorama visualizes how persons move around the space in a three-dimensional view.

Since only a limited number of cameras can be installed in a public space in real life, the space can be visualized only with the limited amount of real-time visual information. In addition, in order to protect the privacy of persons in a public space, the privacy information of persons such as their appearances should not be disclosed without their consent. Furthermore, since Digital Diorama can be accessed simultaneously by the general public over the networks, a specific view which meets an individual need should flexibly be constructed. Focusing on these issues, Digital Diorama firstly presents the three-dimensional view of only the stationary objects with the visual information prepared in advance as a basic view, and then selectively presents the information obtained from the sensors on the basic view depending on the viewer. As shown in Fig. 1, the viewer i can give a request R_i consisting of his/her ID id_i and from and around where he/she would like to view the space, as a set of view and gaze points S_i. Then, the view V_i is constructed by selectively presenting the information obtained from sensors on the basic view with the following three features.

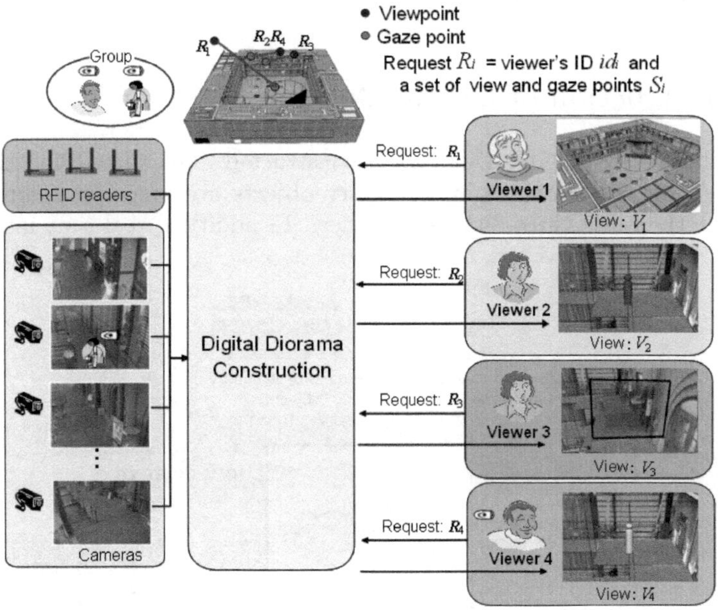

Fig. 1. Digital Diorama

View control: Arbitrary areas of interest can be viewed by moving the view and gaze points. Fig. 1 shows how Digital Diorama presents the distant view of a public space to Viewer 1, while it presents a close-up view of one person in the same space to Viewer 2 according to the view and gaze points specified by each viewer.

Real-time camera image superimposition: Real-time camera images are superimposed on the basic view to visualize the real-time visual information of stationary objects. These images are seamlessly presented only when the viewpoints are at the camera positions. Fig. 1 shows how a superimposed camera image is presented to Viewer 3 when the corresponding camera position is specified as the viewpoint.

Privacy control: Each person is presented on the basic view at his/her position obtained from camera images in real time. Representing every person by an anonymous human-shaped bar presents the positions of persons without violating their privacy. Moreover, a specific person can be represented differently with his/her consent, so that part of his/her privacy information is disclosed. For example, his/her position can be recognized by changing the color of the corresponding human-shaped bar. Therefore, assuming that group members are registered in advance, Digital Diorama represents the persons registered in the same group with the viewer differently according to the viewer ID. Fig. 1 shows how Digital Diorama presents the same person differently to Viewers 2 and 4, so that only Viewer 4 can identify the person as his group member.

3 Construction of Digital Diorama

Fig. 2 shows how Digital Diorama is constructed. The three-dimensional geometry model and textures of stationary objects are prepared beforehand to construct the basic three-dimensional view. In addition, real-time information

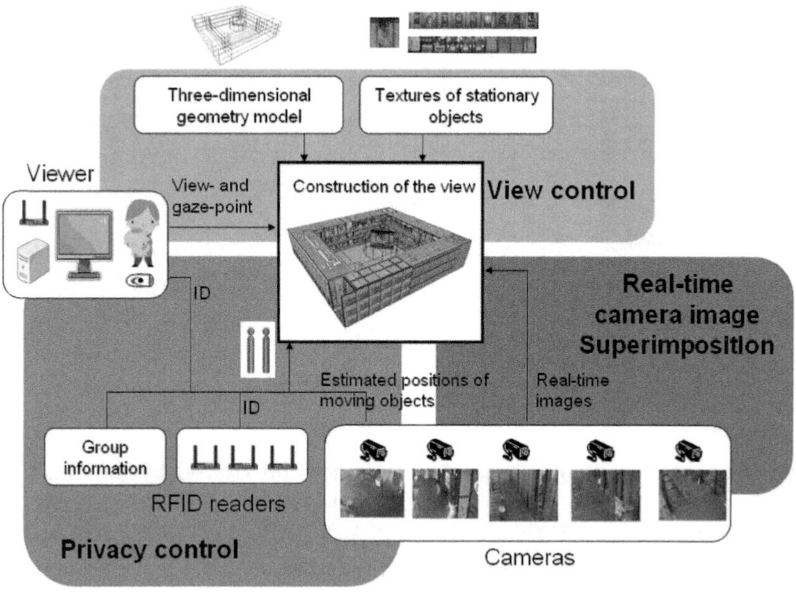

Fig. 2. Overview of Digital Diorama construction

obtained from the sensors is selectively presented with the following three features: view control, real-time camera image superimposition, and privacy control. The details of each feature are described in the following subsections.

3.1 View Control

The viewer is able to view the monitored space by specifying the arbitrary view and gaze points. The viewpoint corresponds to the position of viewer's eyes in the three-dimensional model and the gaze point corresponds to the position over the line of sight. This enables the viewer to view areas of interest freely.

As shown in Fig. 3, the viewpoint can be moved by the viewer on the surface of the sphere whose center is the current gaze point and radius equals the distance between the view and gaze points. Similarly, the gaze point can be moved on the surface of the sphere whose center is the current viewpoint and radius equals the distance between the view and gaze points. Moreover, the viewpoint can also be moved forward or backward along the line determined by the view and gaze points.

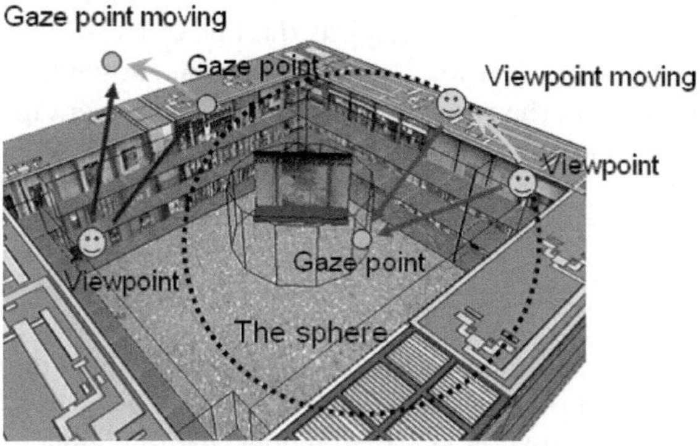

Fig. 3. How to move the view and gaze points

3.2 Real-Time Camera Image Superimposition

The textures of stationary objects such as walls or floors are prepared beforehand because there is not enough number of cameras installed in the monitored space to capture all the stationary objects in real life. Therefore, stationary objects placed after the texture preparation do not appear in Digital Diorama. Furthermore, illumination changes due to time and weather conditions cannot be reflected. By superimposing real-time camera images, real-time visual information can be provided to viewers within the confined space of cameras' fields of view.

As illustrated in Fig. 4, a camera image can be superimposed by arranging the camera image in the three-dimensional model so that each corner point of

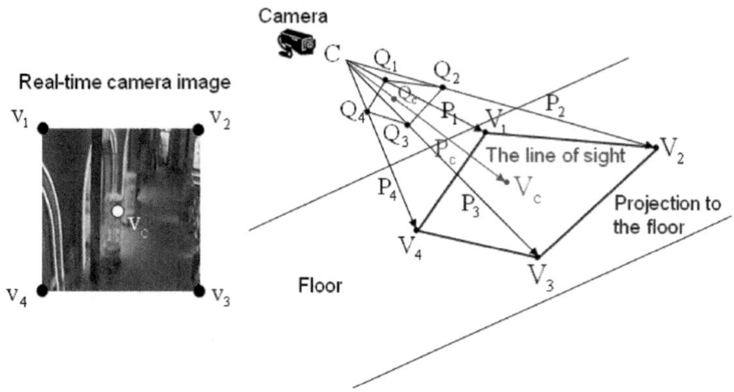

Fig. 4. The positions in the three-dimensional model to display the real-time camera image

the superimposed image $\mathbf{v}_i$ ($i = 1, 2, 3, 4$) is on the line determined by their corresponding points in the three-dimensional model, $\mathbf{V}_i$, and the given camera position $\mathbf{C}$, and is on the plane vertical to the line of the sight of the camera.

$\mathbf{V}_i$ can be calculated as follows. Let H denote the two-dimensional projective matrix, which represents the relationship between the camera image coordinates and the floor coordinates in the three-dimensional model. $\mathbf{V}_i = (X_i, Y_i, Z_i)$ is determined as $X_i = X'_i/Z'_i$, $Y_i = Y'_i/Z'_i$, where $\mathbf{V}'_i = (X'_i, Y'_i, Z'_i)$ is obtained from the following equation using H,

$$\mathbf{V}'_i = H\mathbf{v}_i. \tag{1}$$

Z_i is the height of the floor and is determined from the three-dimensional geometry model of the monitored space.

$\mathbf{V}_\mathrm{c}$, the point in the three-dimensional model corresponding to the image center point $\mathbf{v}_\mathrm{c}$, is calculated likewise. The line of sight of the camera is represented by the line determined by $\mathbf{C}$ and $\mathbf{V}_\mathrm{c}$. An arbitrary vector $\mathbf{x}$, which is vertical to the line, satisfies the equation,

$$\mathbf{x} \cdot \mathbf{P}_\mathrm{c} = 0, \tag{2}$$

where $\mathbf{P}_\mathrm{c} = \mathbf{V}_\mathrm{c} - \mathbf{C}$. Since each corner point of the superimposed image is on the line determined by $\mathbf{C}$ and $\mathbf{V}_i$ and is on the plane vertical to the line of sight of the camera, the following equation is obtained.

$$(t_i\mathbf{P}_i - \alpha\mathbf{P}_\mathrm{c}) \cdot \mathbf{P}_\mathrm{c} = 0, \tag{3}$$

where $\mathbf{P}_i = \mathbf{V}_i - \mathbf{C}$ and α is a constant which determines the distance between the camera position and the superimposed image. By solving this equation with respect to t_i, the corner points of the superimposed image in the three-dimensional model, $\mathbf{Q}_i$, are obtained as

$$\mathbf{Q}_i = \frac{\alpha|\mathbf{P}_\mathrm{c}|^2}{\mathbf{P}_i \cdot \mathbf{P}_\mathrm{c}}\mathbf{P}_i + \mathbf{C}. \tag{4}$$

The real-time camera image is displayed in the quadrilateral consisting of $\mathbf{Q}_i$ $(i = 1, 2, 3, 4)$.

Let us note that selecting a camera in Digital Diorama can move the viewpoint automatically to the camera position, since the superimposed image can be displayed seamlessly only when the viewpoint is exactly at the same position as the camera position. To show the viewers each camera position and the area captured by the camera, Digital Diorama displays the view frustum of each camera, which indicates the field of view of the camera. The view frustum is determined by $\mathbf{V}_i$ $(i = 1, 2, 3, 4)$ and the camera position $\mathbf{C}$.

3.3 Privacy Control

The persons can be detected from camera images and image processing techniques can be applied to the detected regions in camera images to hide various types of privacy information. For example, Chinomi et al. proposed PriSurv [8], a video surveillance system, where the appearances of persons are changed to various types of representation such as dot, bar, and edge, each of which reveals the existence, the height, and the shape of the persons respectively. Chen et al. also proposed to represent persons by edge motion history [9], so that their activity can be recognized without violating their privacy. Applying these ideas, Digital Diorama represents every person in the monitored space by an anonymous three-dimensional human-shaped bar on the basic view to present only his/her current position in the monitored space without disclosing his/her other privacy information.

Moreover, privacy information of specific persons can also be disclosed with their consent. To this end, we assume that some persons can carry RFID tags and be registered as a group in advance. The group members are stored as group information. RFID readers are used to detect the IDs and the rough positions of the persons carrying RFID tags in the monitored space. The positions of the persons with RFID tags are determined by matching the rough positions of RFID tags obtained from RFID readers to the positions of persons estimated from camera images. After viewer identification, by referring to the group information, group members of the viewer are presented differently from persons outside the group, selectively disclosing the privacy information of the persons in the monitored space. For example, changing the color of the corresponding human-shaped bar discloses the positions of group members of the viewer. Such privacy control enables us to provide various applications such as searching for a lost child.

As for the privacy issue of persons in camera images used in Section 3.2, image processing techniques for privacy protection of persons in the images discussed above [8] [9] can be applied so that the images are displayed without disclosing the appearances of persons.

4 Implementation

We installed 10 cameras and 11 RFID readers on one floor in a shopping center. Table 1 shows the specification of the PC on which Digital Diorama was implemented. We used an RFID reader for viewer identification and SONY DUALSHOCK 2 game pad with two analog sticks, each of which is used to move the viewpoint and the gaze point respectively, for easy and intuitive view control.

We firstly stored real-time information from the sensors, i.e., positions of persons and real-time camera images, in databases connected to the PC through networks. By obtaining the real-time information repeatedly from the databases, positions of persons and real-time camera images in Digital Diorama are

Table 1. PC specification

OS	Windows XP Professional Service Pack3
CPU	Intel Xeon 3.73 GHz
RAM	3.00 GBytes
GPU	ATI FireGL v3400
Display Size	1280 × 1024
Graphics API	OpenGL 2.0

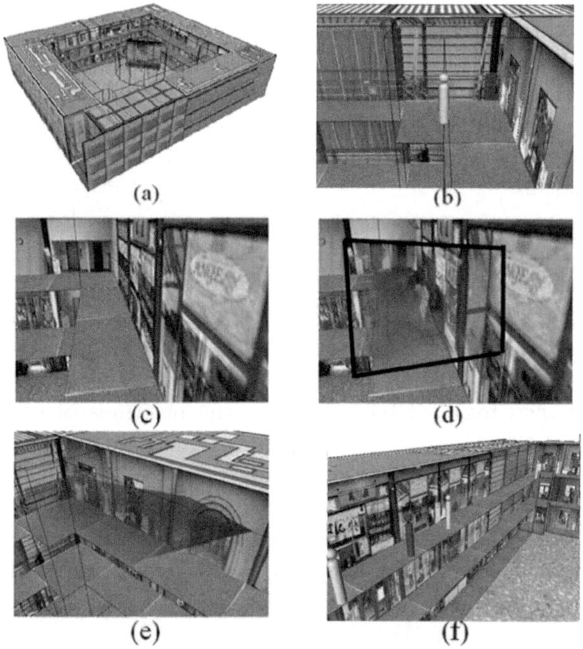

Fig. 5. Example views of each feature. (a), (b) View control. (c) A basic view from a camera position. (d) The view with the superimposed camera image. (e) The view frustums of a camera. (f) Privacy control.

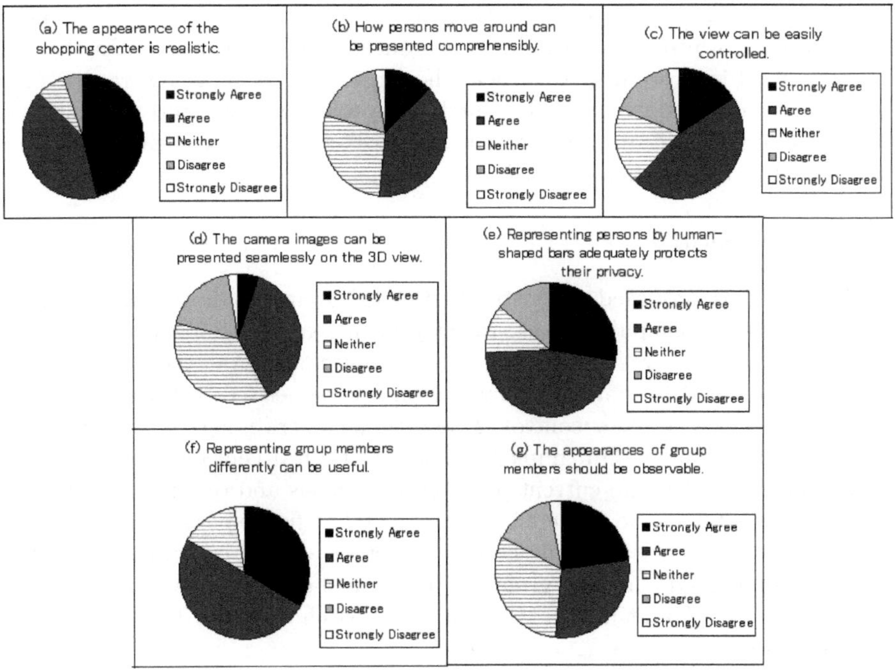

Fig. 6. Result of the questions

updated. This requires time for accessing or searching the databases. As a result of such implementation, the three-dimensional view was reconstructed according to the view and gaze points specified by the viewer in 380.9 frames per second, while only positions of persons were obtained from the database in 21.7 times per second and both the positions of persons and real-time camera images were obtained from the databases in 0.987 times per second. Thus, in Digital Diorama, the positions of persons and the real-time camera images are updated in about 1 frame per second.

Fig. 5 shows example views of each feature. The view can be controlled using the game pad from an overhead view to a close-up view of one person as shown in Fig. 5 (a) and (b). On a basic view from a camera position as shown in Fig. 5 (c), the real-time camera image is superimposed and seamlessly presented as shown in Fig. 5 (d). The positions and the fields of view of the cameras can be shown by displaying the view frustums of the cameras as shown in Fig. 5 (e). By identifying the viewer using an RFID reader, the group member of the viewer is represented by a human-shaped bar of a different color as shown in Fig. 5 (f), disclosing his/her position.

To evaluate the usability of Digital Diorama, we carried out a questionnaire composed of 7 questions in the shopping center, and obtained responses from 40 men and women in their teens to over sixties. Fig. 6 shows each question and the result. A majority of subjects responded positively to the appearance of the shopping center, the ease of view control, and our privacy control, showing expectations toward disclosing the appearance of group members. On the contrary, less subjects were satisfied with how the camera images were presented on the three-dimensional view. For further improvement, accurate estimation of positions and identification of persons are necessary to show their appearances and real-time camera images need to be presented more naturally.

5 Conclusion

In this paper, we proposed to construct Digital Diorama, a comprehensible three-dimensional view of a public space, by integrating real-time information obtained from stationary cameras and RFID readers. Information from the sensors are selectively presented to viewers depending on the requests of the viewers by the following features: 1) view control, 2) real-time camera image superimposition, and 3) privacy control. Our implementation of Digital Diorama for a shopping center indicated that the current positions of persons and real-time camera images are presented to the viewers in approximately 1 frame per second. Our main future work is to revise the method to present real-time camera images. This work was supported partly by grant funding from Japan Science and Technology Agency and by a Grant-in-Aid for scientific research from the Japan Society for the Promotion of Science.

References

1. Ikeda, S., Miura, J.: 3D Indoor Environment Modeling by a Mobile Robot with Omnidirectional Stereo and Laser Range Finder. In: IEEE/RSJ IROS, pp. 3435–3440 (2006)
2. Sawhney, H.S., Arpa, A., Kumar, R., Samarasekera, S., Aggarwal, M., Hsu, S., Nister, D., Hanna, K.: Video Flashlights - Real Time Rendering of Multiple Videos for Immersive Model Visualization. In: ACM EGWR, pp. 157–168 (2002)
3. Sebe, I.O., Hu, J., You, S., Neumann, U.: 3D Video Surveillance with Augmented Virtual Environments. In: ACM IWVS, pp. 107–112 (2003)
4. Girgensohn, A., Kimber, D., Vaughan, J., Yang, T., Shipman, F., Turner, T., Rieffel, E., Wilcox, l., Chen, F., Dunnigan, T.: DOTS: Support for Effective Video Surveillance. ACM Multimedia, 423–432 (2007)
5. de Haan, G., Scheuer, J., de Vries, R., Post, F.H.: Egocentric Navigation for Video Surveillance in 3D Virtual Environments. In: 3DUI, pp. 103–110 (2009)
6. Wang, Y., Krum, D.M., Coelho, E.M., Bowman, D.A.: Contextualized Videos: Combining Videos with Environment Models to Support Situational Understanding. IEEE TVCG 13(6), 1568–1575 (2007)
7. Takehara, T., Nakashima, Y., Nitta, N., Babaguchi, N.: Digital Diorama: Real-Time Adaptive Visualization of Public Spaces. SPC, 2 pages (2009)
8. Chinomi, K., Nitta, N., Ito, Y., Babaguchi, N.: Prisurv: Privacy Protected Video Surveillance System Using Adaptive Visual Abstraction. In: Satoh, S., Nack, F., Etoh, M. (eds.) MMM 2008. LNCS, vol. 4903, pp. 144–154. Springer, Heidelberg (2008)
9. Chen, D., Chang, Y., Yan, R., Yang, J.: Tools for Protecting the Privacy of Specific Individuals in Video. EURASIP Journal on Advances in Signal Processing 2007, 9 pages (2007)

Personalizing Public and Privacy-Free Sensing Information with a Personal Digital Assistant

Takuya Kitade[1], Yasushi Hirano[2], Shoji Kajita[3], and Kenji Mase[1]

[1] Nagoya University, Graduate School of Information Science, Nagoya 464-8603, Japan
[2] Yamaguchi University, Graduate School of Medicine, Ube 755-8611, Japan
[3] Nagoya University, Information and Communication Technology Services, Nagoya 464-8601, Japan
{kitade,kajita,mase}@nagoya-u.jp,
yhirano@yamaguchi-u.ac.jp

Abstract. Mobile devices have been used, and various sensors have been installed in public places. Ordinary people might benefit if they could utilize these sensors for private use. However, privacy issues must be addressed. We developed an application using public sensing data and conducted open experiments with it at a shopping mall. The results suggest that using public sensors for personal use is beneficial.

Keywords: Location-based services, Privacy protection, Sensing data.

1 Introduction

Sensors are found in many public places, including security cameras, magnetometers, and thermometers. However, each sensor currently exists independently of all other sensors. In the future, such sensors are expected to be utilized in and to constitute sensor networks, but this use evokes privacy concerns, such as unexpected exposure of personal activities. Privacy information in sensor data acquisition must be managed[1].

The popularity of mobile phones and Personal Digital Assistants (PDA) continues to increase. If people exploit various sensors for personal use with these terminals, they can augment the real world with existing sensors. Some previous research has investigated augmenting the real world [2]. It is crucial to determine locations in real space in Augmented Reality. Today, there are many varieties of location-based applications [3-6]. Location information has a user's context, so it provides convenience to users, and location information will become more and more useful in the near future.

In the following sections, we propose a location-based application with a PDA and report the evaluation results under a real environment.

2 Application

We developed a location-based comprehensive experimental project named Eye-i-net Personal using a PDA for mobile use at a Kyoto shopping mall in a three-story

E. Hüllermeier, R. Kruse, and F. Hoffmann (Eds.): IPMU 2010, Part II, CCIS 81, pp. 673–679, 2010.

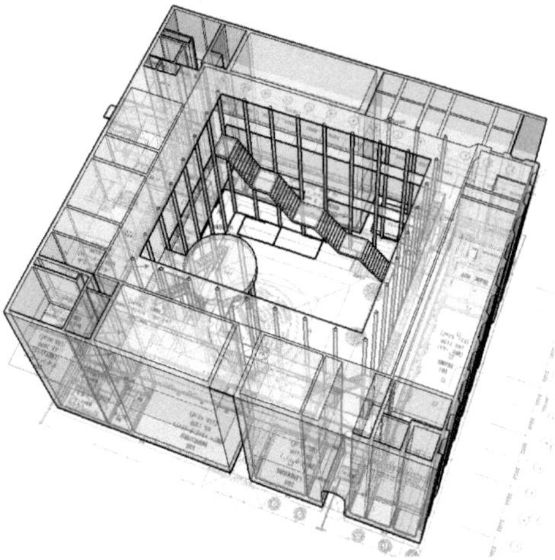

Fig. 1. 3D exterior appearance 1

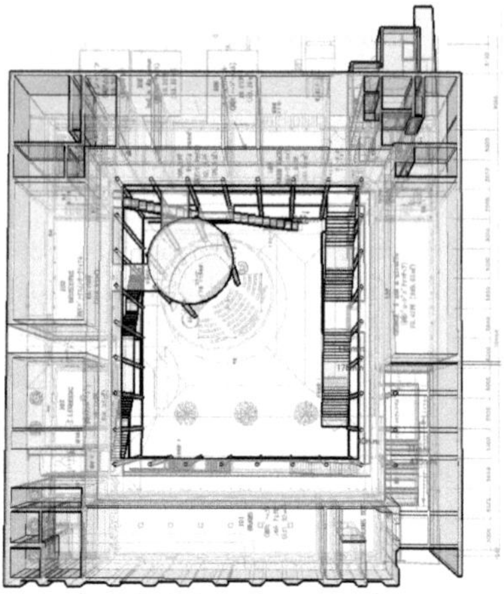

Fig. 2. 3D exterior appearance 2

Fig. 3. Interior photo 1

Fig. 4. Interior photo 2

building with a stairwell (Fig. 1-4). Our aim is to utilize various public sensors for personal use and to propose a location-based application using sensing data. Privacy protection is an important requirement of the project [7].

In this study, we regarded a Wi-Fi communication device as a sensor.

2.1 Components

We describe the design of the application components and their aims below.

- Interactive floor maps

Users can browse different floor maps by flicking on the PDA along with individual store web pages by tapping store icons. User's current location and store rating information are also displayed on the maps.

- Wi-Fi-based location estimation

Fig. 5. 1st floor map

Fig. 6. 2nd floor map

Fig. 7. 3rd floor map

User location is estimated with Wi-Fi signal strengths [8] and is shown on the display (Fig. 6). The PDA receives the signal strengths of the open Wi-Fi access points and transmits the signal data to the privacy protected user activity server to estimate user location. Only allowed users can access their own location information to cooperatively utilize the user location information with controlled privacy protection.

- Location-based rating

As an application of Wi-Fi-based location estimation, a location rating function is also provided. User recommendations represented by stars can be posted at the estimated user location. Ratings are quickly reflected on the floor map, and users can see the feedback of other users in real time. This provides users with opportunities to communicate with each other.

- Sound mixing

As another application of location estimation, we present a sound mixing game that provides users with a collection of preset sound clips hidden in virtual spaces connected to the user location. Users can play the music and enjoy hunting for the sounds and the resulting route with the mixed music. Users can identify their own routes by music and share the varieties with others.

- Photo shooting with public cameras

This function takes pictures with public cameras with a monitor arranged as a picture kiosk in the mall (See next section for details). Users can take their own picture by entering a four-digit number given randomly to their monitors; the number is used as a one-time ID to personalize the camera. Faces are automatically hidden from the public, but displayed normally on the PDA. We believe that public users will take personal pictures with such public video cameras as security cameras, web cameras, and so on.

3 Open Experiments

We conducted five-hour open events of Eye-i-net Personal with iPod touches that we loaned to visitors on Sep. 19, 2009 and Nov. 15, 2009. Users participated in the event and were asked to use the application as long as they want. We collected questionnaire responses during the experiments from all users, who also received a gift certificate worth about 4 US dollars.

3.1 Implementation

We utilized Wi-Fi signal strength for user location estimation and set public cameras in the mall. In this section, we explain each purpose.

3.1.1 Application Implementation

We developed Eye-i-net Personal for iPod touch, one of the most popular PDAs for which it is easy to develop applications. In the first event, we used 20 iPod touches. In the second event, we used 30. The iPod touch devices were loaned to mall visitors.

3.1.2 Wi-Fi Location Estimation

User location is estimated with Wi-Fi signal strengths, because GPS was not appropriate for location estimation at the mall, which is a closed space. We installed Wi-Fi access points only for the project in advance for precise location estimation and set a total of 16 Wi-Fi access points on each corner and in the middle of the mall's ceiling on the first and the third floors. We eventually utilize Wi-Fi signal strengths for location estimation. A location estimation algorithm was determined by referring to [8].

3.1.3 Placement of Picture Kiosks

We propose a location-based application of public anonymized sensing data using personal devices. Erecting picture kiosks is an attempt to use public anonymized sensing data for personal use. We arranged three picture kiosks on each floor (Fig. 8 and 9).

Fig. 8. Picture kiosk

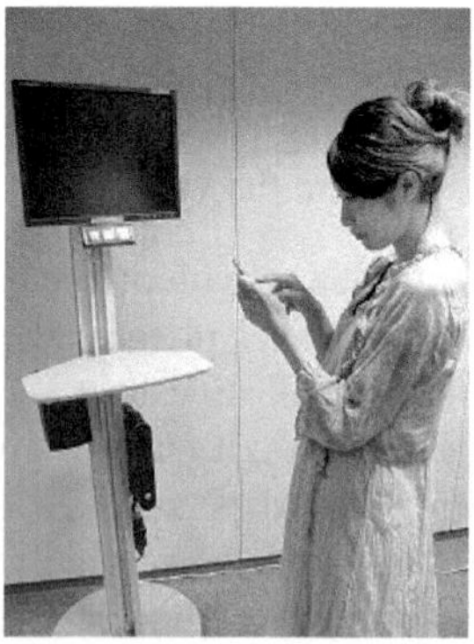

Fig. 9. Picture kiosk usage

3.2 Experiments Results

The number of subject users was 26 and 36 in each event, respectively (Fig. 10 and 11). In the first event, 56 photos were taken by 24 users with the picture kiosks. In the second event, 72 photos were taken by 33 users. Here are some photo samples (Fig. 12).

Fig. 10. Invitation during the event 1

Fig. 11. Invitation during the event 2

Fig. 12. User photo samples at picture kiosks

3.2.1 Questionnaire Results

We collected questionnaire responses from all users. Some questionnaire results are given as follows.

These results, especially Fig. 13, show that ordinary people may utilize public sensors for personal use; improving services about privacy protection will make public sensors usable for people (Fig. 14 and 15). Fig. 16 suggests that security cameras can be applied for private cameras.

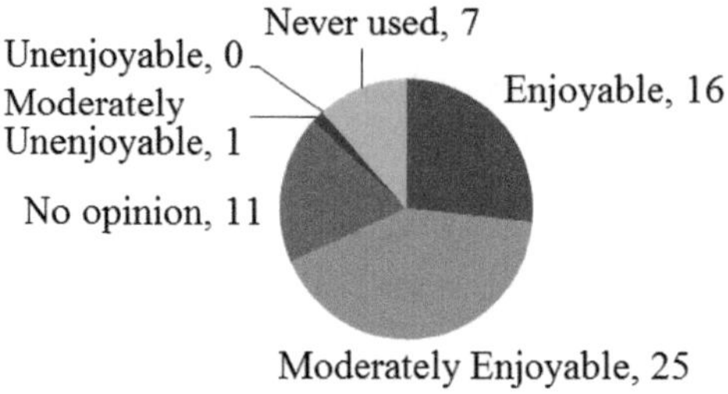

Fig. 13. Taking photos with public cameras

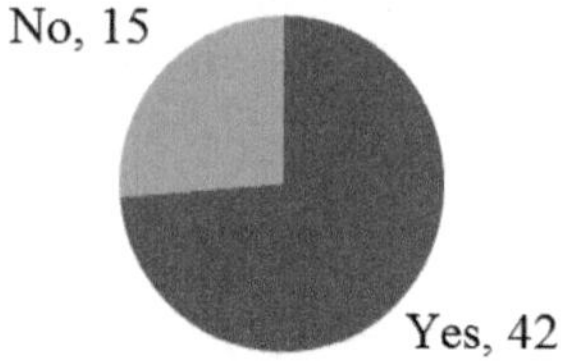

Fig. 14. Did photos of others encourage you to take photos?

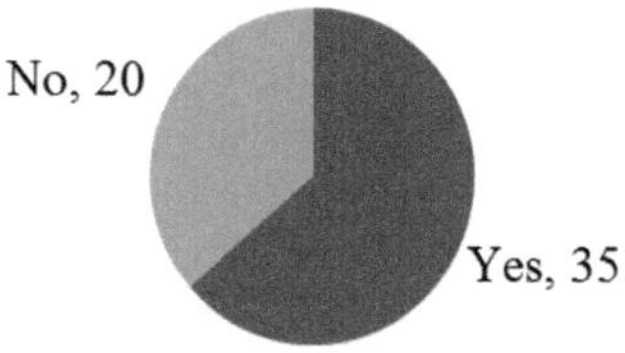

Fig. 15. Does hiding other faces soften your resistance to taking photos?

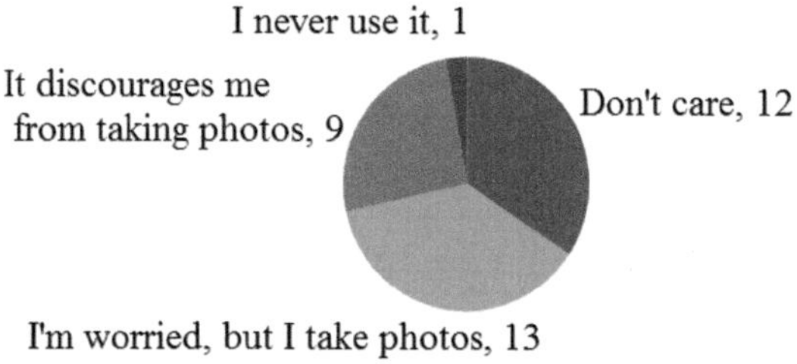

Fig. 16. How do you feel if picture kiosks always record?

4 Conclusions

The popularity of mobile devices continues to increase, and many kinds of sensors have been installed in public places. If ordinary people can utilize these sensors for private use, it is meaningful. However, privacy information management must be considered. We developed an application called Eye-i-net Personal and conducted two open experiments with it with iPod touches loaned to visitors in a shopping mall. Our results suggest that ordinary people think that public sensor utilization for personal use is useful.

One of the solutions to privacy problems is utilizing anonymized sensing data for personal use. That is, only allowed users get their own information; other users can only see anonymized sensing data. This corresponds to location-based rating in our experiments. It is better that users can find their own data in a large quantity of anonymous data; a searching method is required, and this is future work.

Acknowledgments. This research was supported in part by grant funding from Japan Scienceand Technology Agency (Sensing Web) and NICT projects.

References

1. Minoh, M., Kakusho, K., Babaguchi, N., Ajisaka, T.: Sensing Web Project - How to handle privacy information in sensor data-. In: Proceedings of IPMU 2008, 12th International Conference on Information Processing and Management Uncertainty in Knowledge-Based Systems, Málaga, pp. 863–869 (2008)
2. Mistry, P., Maes, P.: SixthSense: a wearable gestural interface. In: SIGGRAPH ASIA 2009: ACM SIGGRAPH ASIA 2009 Art Gallery & Emerging Technologies: Adaptation, p. 85. ACM, Yokohama (2009)
3. Google Latitude, `http://google.com/latitude`
4. brightkite, `http://brightkite.com/`
5. Sekai Camera, `http://sekaicamera.com/`
6. foursquare, `http://foursquare.com/`
7. Kitade, T., Niwa, K., Koyama, Y., Naito, K., Iwasaki, Y., Kawaguchi, N., Hirano, Y., Kajita, S., Mase, K.: A location-based application with public anonymized sensor data for personal use. In: International Conference on Security Camera Network, Privacy Protection and Community Safety 2009, SPC (2009)
8. Mase, T., Hirano, Y., Kajita, S., Mase, K.: Improving Accuracy of WLAN-Based Location Estimation by Using Recursive Estimation. In: 11th International Symposium on Wearable Computers, ISWC 2007, pp. 117–118 (2007)

The Open Data Format and Query System of the Sensing Web

Naruki Mitsuda and Tsuneo Ajisaka

Faculty of Systems Engineering, Wakayama University,
Wakayama City 640-8510, Japan
{manda,ajisaka}@sys.wakayama-u.ac.jp

Abstract. For the observation of the real world, many sensors are set to obtain information automatically and precisely. It is necessary to establish openness of access to these sensors for utilizing them effectively more. We design a software architecture of the Sensing Web which is one of such open sensor networks and arrange the open data format of sensor information. We also design a query system which performs matching of demand and supply in the Sensing Web.

1 Introduction

The Sensing Web collects information from various sensors and provides it for open use. The Sensing Web is different from most of web applications based on human-made documents. It is also different from control/embedded systems in its openness. Control/embedded systems are event-driven and those events are originated from sensors in various dynamic environments.

Though sensor data is a key factor shared by both the Sensing Web and control/embedded systems, those software architectures are totally different. A control/embedded system is watching events and invokes functions specified by a combination of events and states of its target machine or environment. The Sensing Web is not watching specific events but continuously collects and selects sensor data captured by multiple sensors. The Sensing Web user requires to view and know how an environment is situated, not to operate a specific machine. Another key factor of the Sensing Web is therefore to support flexible combination of various sensor data to make applications user friendly.

Openness always requires matching of demand and supply. It is generally not straightforward because of difficulty of matching the intention of a requirement and the meaning of an information source. Comparing to ordinary web applications of which source is human-made documents, the distance between the intention and the meaning is much closer in Sensing Web of which source is sensor data. So we try to provide good matching of demand and supply by presenting a format and semantics of data using in Sensing Web. Unlike ordinary web applications, information of time and space (location) is always sticky to all Sensing Web data. Sensing real view and sound of environments, privacy control is also indispensable during tracking data of this kind of systems.

E. Hüllermeier, R. Kruse, and F. Hoffmann (Eds.): IPMU 2010, Part II, CCIS 81, pp. 680–689, 2010.

This paper presents a software architecture that enables to implement Sensing Web with sound modularity, and a data format and its transformation support for matching of demand and supply.

2 Software Architecture of the Sensing Web

In order to implement the Sensing Web systematically, its architecture should be designed with sound modularity.

2.1 Components of the Sensing Web

The Sensing Web is an open ubiquitous sensor network which is used by many and unspecified users. The network node which constitutes it must be designed after taking its openness and flexibility into consideration. The sensor network is designed as Fig. 1 and consists of the following network nodes.

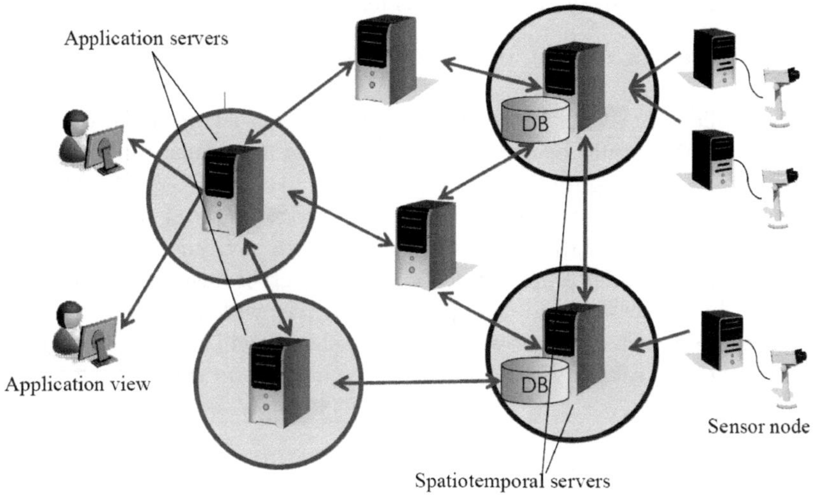

Fig. 1. Network architecture of the Sensing Web

Application view. A user operates application views directly on a browser or specialized client programs.

Application server. Application servers provide user services using the data obtained from sensors.

Sensor node. A sensor node is directly linked with each sensor, and performs an acquisition and publish of sensor data.

Middle server. A middle server processes the sensor data obtained from sensor nodes or other middle servers so that the data will be easy to utilize for other servers.

The discreteness of common processing logics is obtained by locating middle servers. Various applications can share these logics. The following can be considered as typical instances of middle servers.

Spatiotemporal server. This server unifies the data obtained from each sensor using spatiotemporal properties. For example, the data obtained from the sensors which subsist in a certain specific space is managed collectively.

Format conversion server. This server manages formatting rules of sensor information and carries out the automatic translation of the description of a sensor data according to application requests.

In Fig. 1, the Sensing Web network consists of several server nodes and each has a different role. It is because a loose coupling status is promoted by separating the stage of the information processing which matches the data obtained from sensors to an application users request. In addition, there is intention to reduce the load of each server because it is necessary to process the request of a huge quantity when the Sensing Web becomes a huge network in the future.

The flow of the Sensing Web use is as follows.

1. First, a user accesses an application screen and demands a required information.
2. The application server which received the request decides from what kind of sensor a data should be obtained, and requires an acquisition of a data to a suitable spatiotemporal server.
3. After analyzing the received request, the spatiotemporal server will collect and unify suitable data, if the managed sensors have enough data.
4. If those sensors do not have enough data, the server answers that the request cannot be processed, and the application server will try a different request.
5. If suitable information is obtainable, translation will be performed if necessary, and the user can be shown the information from the application server.

2.2 Control of Privacy Information

Since privacy information may be included in the data obtained from sensors, it is important to consider how to treat privacy in the Sensing Web.

This problem is solvable, if all privacy informations are removed when a sensor node discloses its data. However, originally it becomes settled by the relationship between the owner and user of the information whether privacy information could be disclosed[7]. An application of which convenience is increased by the exchange of privacy information may subsist.

In the Sensing Web, when exchanging privacy information, it was presupposed that the communication must be enciphered. Only the user who clears an authentication and gains a key shall decode the information. It is planned to control privacy information by practical use of the existing security technology.

This idea enables us to manage the organization of a data network, and control of privacy independently in a different layer, as shown in Fig. 2.

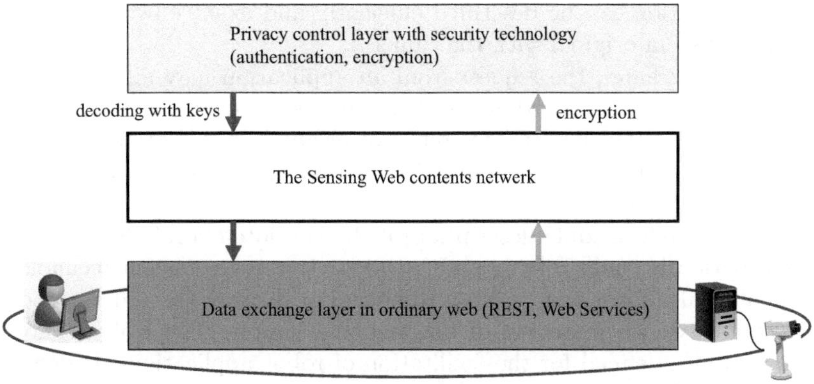

Fig. 2. Privacy control using security technology

3 The Open Data Format of the Sensing Web

In order to achieve open sharing of sensor information, the technology which matches sensor data and application requests needs to be realized. For that purpose, the metadata which shows the property of sensor data must be defined.

This research defines a metadata set as a tag library of XML[3]. The metadata for sensors, such as spatiotemporal elements, physical quantity properties, measurement accuracy, is restrictive. Therefore, the combination of a few tags can describe various sensor data. The compositional framework of information supply and demand is decided using the translation language for XML description. Furthermore, based on the syntax and semantics of the tag set, programming with tag libraries implements information supply and demand in detail.

Here, the extensibility which is one of the advantages of XML makes it possible to correspond to the variety of the sensors and the extensibility of services,which are the special features of the Sensing Web. Although the type of sensors or application services will increase in the future, it is expectable that an extension of tag sets stays in a fixed number of variations by repeating prudent arrangement of tags.

When the Sensing Web will spread widely in the future, the functional range of information definition and translation description will be expanded regarding reusability and maintainability. In addition, it will be necessary to also perform suitably the technical development for the improvement in reliability or efficiency.

3.1 Definition, Use, and Translation of Spatiotemporal Information

In the Sensing Web which obtains data directly from the sensors which observe the real world and utilizes them, it is important to preserve flexibility to the description schema of spatiotemporal information[6,5]. For example, when a camera expresses the location of objects which subsist in its field of view, a spatial information can be described efficiently and exactly by using a relative coordinate system original with the camera.

On the other hand, the request from an application may also be wanted to refer to another original coordinate system. In order to correspond to such status, while accepting definition and reference of original coordinate systems to any sensors or applications, it is required to be automatically convertible between each coordinate systems.

In this research, flexible description of the spatiotemporal elements for sensor information is enabled by applying the existing management technology for spatiotemporal information, named Robotic Localization[2], and incorporating this technology into matching of demand and supply. The Robotic Localization is a service defined for the facilitation of robot applications. It defines the framework of the treatment and expression of location information with many object-oriented models. Since its specification is not dependent on the type of robot applications, or the details of application logics, if the rules included in the specification are implemented with XML, it will become a technology which can be utilized for a spatiotemporal information management in the Sensing Web.

3.2 The Data Format of Sensor Information

This section shows the design strategy and example about representation of the sensor information used when a sensor node offers its data in the Sensing Web.

Since the Sensing Web is a sensor network with emphasis on openness, the flexibility and extensibility of the representation technique of symbolic data are important. We propose a XML tag set aiming at these properties, in consideration of the performance of sensors and the efficiency of data transmission. Since a usability changes depending on the organization of XML tags, it is necessary to decide the specification in consideration of the load of the data disclosure by persons who install sensors or servers.

Fig.3 is an example of representation of the data which sensor nodes supply. In this example, we assume that a camera can analyze its shot image taken for every definite period of time, and can detect the region of objects or persons in the images.

The type of sensor and the identifier given for every sensor are described in the line 2. The `<crs>` tag in the line 3 means "Coordinate Reference System", and here shows that location information is described with JGD2000 coordinate system. The `<frame>` tag in the line 4 expresses the temporal information given for every video frames which the camera cuts out. The value of `timestamp` attribute contained in this tag is recorded every moment. In the line 5, the identifier and category are given to an object (analysis target) recognized in the image. The `<coordinate>` tag in the line 6 expresses detailed location of the object with the latitude, longitude, and altitude on the coordinates reference system specified in the line 3. The size and color of the object are described in the line 10 - 11. The `<object>` tag is repeated the number of detected objects.

Fig. 4 is an example of a generalized description of sensor data. It is the data description which treats symbolic data, such as the temperature and humidity,

```
<sensing_data>
  <sensor id=“xxxxx“ category=“camera”/>
  <crs type=“JGD2000”/>
  <frame timestamp=“XX:YY:ZZ”/>
    <object id="0“ category=“human”>
      <coordinate>
        <lat>35.885623</lat><lng>139.801354</lng><alt>0.5</alt>
      </coordinate>
      <size>100</size>
      <color>******</color>
    </object>
    <object id="1“ cetegory=“car”>
    ....
  </frame>
</sensing_data>
```

Fig. 3. An example of representation for sensor information

```
<sensing_data>
  <sensor id=“xxxxx“ category=“camera”/>
  <crs type=“JGD2000”>
    <lat>35.885623</lat><lng>139.801354</lng><alt>0.5</alt>
  </crs>
  <frame timestamp=“XX:YY:ZZ”/>
    <object id="0“ category=“temperature”>
      <value>30.2</value>
    </object>
    <object id=“1“ category=“humidity”>
      <value>70.1</value>
    </object>
  </frame>
</sensing_data>
```

Fig. 4. Generalized representation of sensor information

as uniformed objects which are obtained in an analysis. In this example, location information is described at one place because it does not change for every symbolic data.

4 The Query System of the Sensing Web

This section shows the design strategy and example about the query system used when an application server requires sensor information. The query in the Sensing Web must be able to express the various requests to various data exactly.

In addition, it must be taking into consideration that the data schema for sensors differs from that for requests.

Furthermore, since the query itself should have openness, it is necessary to decide the specification of the query system with intelligible structure for users.

4.1 Description of Queries

Fig.5 is an example of a query description in case an application requires a sensor information of the Sensing Web. This query demands the latest temperature and humidity which are observed within 1 km centering on a certain point.

```
<query>
  <object_ref  name= “temp ” category=  “temperature  ”/>
  <object_ref  name= “hum” category=  “humidity ”/>
  <crs type= “JGD2000” range= “1000”>
    <lat>35.885623</lat>
    <lng>139.801354</lng>
    <alt>0.5</alt>
  </ crs>
  <schema>
    <frame timestamp=  “current ”/>
      <object ref=  “temp “/>
      <object ref=  “hum“/>
    </frame>
  </schema>
</query>
```

Fig. 5. An example of query description

The <object> tags in the line 2 and 3 specify the type of target information, and the type of target sensor using `refname` and `category` attributes. The <crs> tag in the line 4 specifies the coordinates reference system used in this query with the type property, and specifies the effective range. In the lines 5-7, the coordinates which take the center of the range are specified with its latitude, longitude, and altitude. The <schema> tag in the line 9 specifies the data format of response. The object elements included in the tag are replaced with the target information. Furthermore, according to the timestamp property of the <frame> tag in the line 10, it is required that the latest information should always be obtained.

In this query, since the description is divided into the part of the request and that of the format, it turns out that understandability is taken into consideration.

4.2 Format Conversion of Sensor Information

An application demands an ideal sensor information from a query system. However, the suitable information for an application cannot always be offered because of transition of the setting status of sensors, or an occurrence of errors. In addition, a system may be urged to rearrange the obtained information in any sequence using a fixed rule, or to process the sensor information received according to the query.

In order to correspond to such a status flexibly, the mechanism for performing the automatic format conversion of sensor information should be prepared. The relationship which subsists between the tag library used for queries and the library for representation of sensor data is extracted as a set of rules. The status where a part of extracted rules is not satisfied means the status which needs format conversion, and it will repeat the conversion until all the rules are satisfied.

5 Implementation of the Query System

In order to share a sensor information efficiently, the mechanism for matching demand and supply is indispensable. This mechanism combines the symbolic data obtained from sensors according to the query from an application, and returns suitable response.

5.1 Query Matching Process

Fig.6 expresses the matching process in the Sensing Web.The query used for a request is divided into the target part indicating the target data, and the schema part indicating the specification of structural transformation for formatting presentation of the received data. In response to the description of a target part, the query system selects and combines required data from sensor nodes, and unifies them in one XML document.

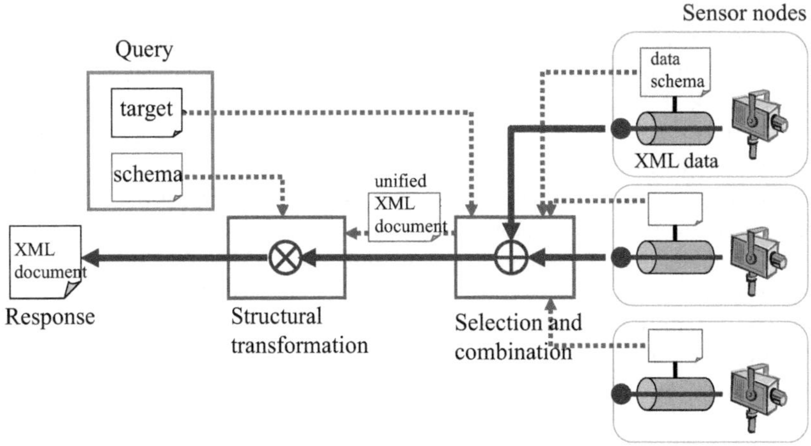

Fig. 6. Structure for matching demand and supply

Subsequently, the structure of the document and the schema part of the query are compared. Then, structural transformation is performed if necessary. The matching process is finished with providing a final response to the application.

XSLT[4] is used in order to describe rules extracted from the relationship between the XML tags. As mentioned above, the key technology of the matching process is the selection, combination, and transformation of XML documents. XSLT is suitable technology for these functions.

5.2 Management of Coordinate Systems

Applications using sensor information often use their own coordinate system, in order to implement their features efficiently.For example, the application that provides services in a limited space needs a relative coordination from a reference point in the space. Such application can describe coordinate values in details and manage them easily.

Also in query description, applications will desire to use their own coordinate systems. Therefore, the query system must provide the functionality to register application original coordinate systems. The application using its own coordinate system registers the coordinate system into the query system, before it publishes queries. And the application specifies the name of the registered coordinate system in queries.

When an application registers own relative coordinate system, it must define the translation logic between its coordinate system and other generic one, with the information of the reference point and axis.

6 Concluding Remarks

In order to make brokering demand and supply of information, metadata is required to describe and validate passing information. For information resources in general, what should be described as metadata is quite arbitrary and therefore the framework of description tends to be generic. It is mostly impossible to get a precise match required with such generic descriptions. Metadata of sensor data is limited to location, direction, time, physical units, precision, frequency, and several others, so that selection and combination of information sources is less difficult if requirements descriptions are properly stated.

This paper is centered for getting current information or that of a little bit earlier in time. Another utility of the Sensing Web is to search for patterns and trends of past data and estimation based on them. As sensor data is spawned continuously, information explosion would be more serious than the ordinary web unless filtering and selected storing is carefully controled and managed. Under the condition of good quantity control, data mining itself is rather optimistic because of the simplicity of data syntax and semantics.

References

1. Sensing Web: Content Engineering for Social Use of Sensing Information (2007), `http://www.imel1.kuis.kyoto-u.ac.jp/sweb/en/`
2. OMG: Robotic Localization Service (2008), `http://www.omg.org/docs/robotics/08-05-02.pdf`
3. W3C: Extensible Markup Language (XML) 1.1, 2nd edn. (2006), `http://www.w3.org/TR/2006/REC-xml11-20060816/`
4. W3C: XSL Transformations (XSLT) Version 2.0. (2007), `http://www.w3.org/TR/xslt20/`
5. Gupta, H., Zhou, Z., Das, S.R., Gu, Q.: Connected sensor cover: self-organization of sensor networks for efficient query execution. IEEE/ACM Trans. Netw. 14(1), 55–67 (2006)
6. Chang, S.K., Costagliola, G., Jungert, E.: Multi-sensor Information Fusion by Query Refinement. In: Proc. 5th International Conference on Recent Advances in Visual Information Systems, pp. 1–11 (2002)
7. Suko, Y.: Privacy Management in Information Delivery Using Social Network Structure. Keio Sfc Journal 4(1), 128–153 (2005) (in Japanese)

See-Through Vision: A Visual Augmentation Method for Sensing-Web

Yuichi Ohta, Yoshinari Kameda, Itaru Kitahara,
Masayuki Hayashi, and Shinya Yamazaki

Department of Intelligent Interaction Technologies, University of Tsukuba
Tennoudai 1-1-1, Tsukuba Science City, Ibaraki, Japan
{ohta,kameda,kitahara}@iit.tsukuba.ac.jp,
mhayashi@image.iit.tsukuba.ac.jp

Abstract. Many surveillance cameras are being installed throughout the environments of our daily lives because they effectively maintain safety and offer security to ordinary people. On the other hand, they can also cause serious discomfort and anxiety to the same people. This paper proposes a novel framework called See-Through Vision that utilizes surveillance cameras as a public sensing device by exploiting a state-of-the-art mixed-reality technique. Some advanced systems developed by us for See-Through Vision are introduced, and we discuss their advantages and how they actually maximize the benefits of surveillance cameras. Since See-Through Vision is a powerful tool that may violate the privacy of other people, we propose a sophisticated solution that can strike a good balance between the benefits and the drawbacks of such an approach. We propose privacy-safe See-Through Vision and demonstrate the system at a shopping mall in Kyoto.

Keywords: Sensing-Web, Outdoor Mixed-Reality, Visual Augmentation, Privacy Control.

1 Introduction

With increasing concerns about security, surveillance cameras continue to be ubiquitously installed in public places. When installing a surveillance camera, it is important to build a social consensus that reaches a good compromise between the security of the public space and the comfort of the monitored people. Although one crucial role of surveillance cameras is maintaining security, this benefit is not easily recognized since safety is invisible and access to such systems is not provided to the public. We are enhancing video surveillance systems with the additional role of letting people visually and intuitively understand the benefits of surveillance cameras.

This paper introduces our visual augmentation methods for surveillance cameras to carry out the above approach by utilizing a mixed-reality (MR) technique. Users can see and use extra visual information that is captured by surveillance cameras, appropriately transformed in MR fashion, and superimposed by an MR technique onto their view through their mobile display devices such as personal digital assistants (PDAs) or cell phones.

E. Hüllermeier, R. Kruse, and F. Hoffmann (Eds.): IPMU 2010, Part II, CCIS 81, pp. 690–699, 2010.

Fig. 1. Concept of See-Through Vision

We have been developing a See-Through Vision system [1] that allows users to observe such hidden areas as the blind corners of buildings (Fig. 1). It also helps people who might be the subject of surveillance cameras to easily understand the contents of the visual information and the properties of the capturing camera (e.g., resolution, field of view, and capturing angle) and thus decrease their level of discomfort.

2 See-Through Vision

A see-through display is implemented using surveillance cameras installed in public outdoor spaces. By calibrating surveillance cameras beforehand and calibrating a handheld device's camera on-line, objects in a hidden area can be displayed on the handheld device in MR style.

Although it is easy to estimate the pose of a handheld device's camera if artificial markers are available, there are few outdoor scenes where specially designed calibration markers such as checkerboards can be installed. Therefore, we propose a method to estimate the pose of the handheld device's camera without using artificial markers. Instead, we use the substructures of buildings as calibration markers. Usually, the shapes of substructures (called landmarks) can be obtained from CAD data of the building or the scene.

One of the critical problems of this approach is that the appearance of a landmark changes widely while the handheld device is moving. The major reasons for such appearance change are changes in light conditions, color fading, aging, etc. These phenomena cannot be prevented, even if relatively large and substantial parts of the buildings in a scene are set as landmarks. Therefore, we extract the live textures of landmarks from the videos of surveillance cameras to follow their texture changes.

2.1 Camera Registration

Our handheld device is equipped with a GPS, a digital compass, an inertia sensor, and a color video camera. The GPS and digital compass obtain the initial estimation of location and the orientation of the handheld device. The inertia sensor can track the pose change of the handheld device at high frequency. However, it is not adequate for precisely superimposing 3D CG models and live video onto the handheld device's camera images due to their problems of long-term instability such as drift.

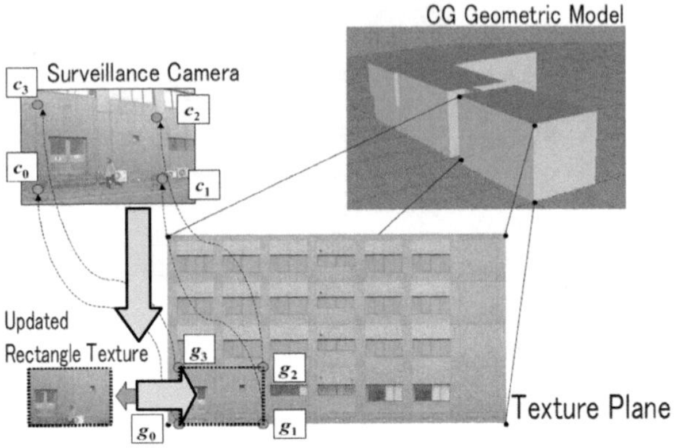

Fig. 2. Visualizing hidden areas; a rectangle on a wall becomes visible

We define landmarks in the outdoor scenes and use them as embedded natural markers. To accurately define the landmarks for image processing, they must be distinct in the scene. We adopt the substructures of buildings with good features for tracking [2] when they are captured on the handheld device's camera image. Choosing these substructures reduces the chance of false extraction in image processing. Each landmark has geometric information given by the CAD and/or CG data and pictorial information obtained with a surveillance camera in real time. The best surveillance camera, which actually takes pictorial information of a landmark, is selected based on the location relationship of the handheld device's camera, the landmark, and the surveillance camera.

2.2 Visualizing Hidden Areas

To visualize objects in hidden areas, we use a simple yet effective rectangle-based video warping method. Rectangles are set perpendicular to the ground in the hidden areas. Corresponding video segments are transmitted from an appropriate surveillance camera capturing the area to the handheld device, and the segments are warped based on the projection matrices given by the calibration results. Figure 2 shows an example of the visualization. In this case, a rectangle is set on the wall of a building adjacent to the hidden area. The system holds the geometric 3D points of the rectangle (g0…g3) and the corresponding 2D points (c0…c3) in the live video frame taken by the most appropriate surveillance camera. With these corresponding points, a homography matrix can be calculated to project the captured live video segment onto an image shown on the handheld device's display.

3 Advanced See-Through Vision

With basic See-Through Vision as described in the previous section, users can only observe a hidden area occluded by a building in front of them. On the other hand, if more surveillance cameras were installed here and there, richer visual information of a target scene would be available. In this section, we introduce our two types of advanced See-Through Vision systems designed with new functions that allow users to utilize visual information captured over a wider area.

3.1 See-Through Vision in Wider Areas

The virtual viewpoint motion shown in Fig. 3 is a novel interface for improving the visibility of hidden areas. Users can virtually get close to their desired area and see objects there in high resolution [3]. Since the virtual viewpoint can go through buildings, users can clearly see even objects that are in distant blind areas hidden by blindfolds such as buildings.

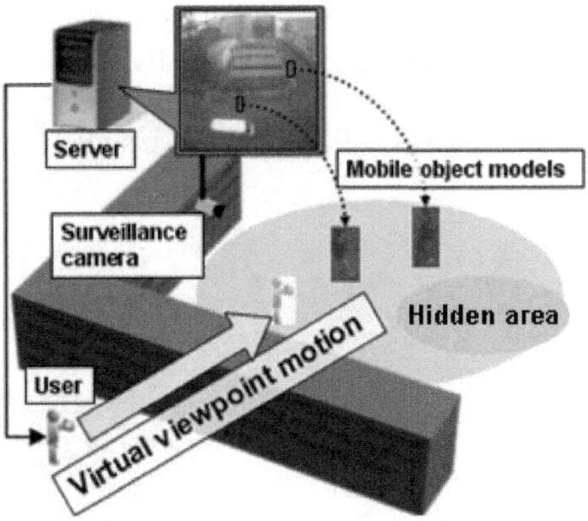

Fig. 3. See-Through Vision in Wide Area

3D Model of Background Region

3D information is necessary to generate a view from an arbitrary viewpoint. We acquire the 3D shapes of such static background regions as roads and buildings in advance. The texture is updated online by mapping segments of images captured by surveillance cameras while referring to the projective matrix of the surveillance cameras.

3D Model of Foreground Objects

The foreground objects are modeled by the billboard technique [4] with live texture captured by surveillance cameras. The object's billboard is put at a position estimated from the image of surveillance cameras.

Generally speaking, at least two cameras are necessary to estimate the 3D position of an object in a scene. However, surveillance cameras usually do not have overlapping areas among the cameras; consequently, we could not apply this method to our problem. Our system estimates the 3D positions of foreground objects using only a single camera while assuming that all of the objects stand perpendicular to the ground plane. Here, a foreground object is modeled by a rectangular polygon called a billboard (Fig. 4). By defining the billboard with a planner bounding box that encompasses the object, we can know the diagonal corners X_{lb} and X_{ru} of the billboard in 3D space. We assume that the height difference of X_{lb} and X_{ru} is the height of human beings, or h. Since the camera has already been calibrated, we can find the corresponding 2D points U_{lb} and U_{ru} in the camera image. By cutting out the image segment between U_{lb} and U_{ru} and then by projecting the cut segment onto the billboard in 3D virtual space, the user can see the handheld camera's synthesized image on his/her display with the objects superimposed at geometrically correct positions.

Figure 5 shows a sequence of snapshots depicting virtual viewpoint motion. Initially, a user stands far from the white building that hides a court that the user wants to see. Then the user starts virtually moving toward the white building (2nd picture), walks through the building, and then reaches the other side of the building (3rd picture), where the user virtually sees the court. The user can continue ahead to walk through the court (4th and 5th pictures) and then finally the user happens to encounter a person wearing a white T-shirt who is represented by a billboard; actually, the texture comes from a surveillance camera on-line, which is installed to monitor the court.

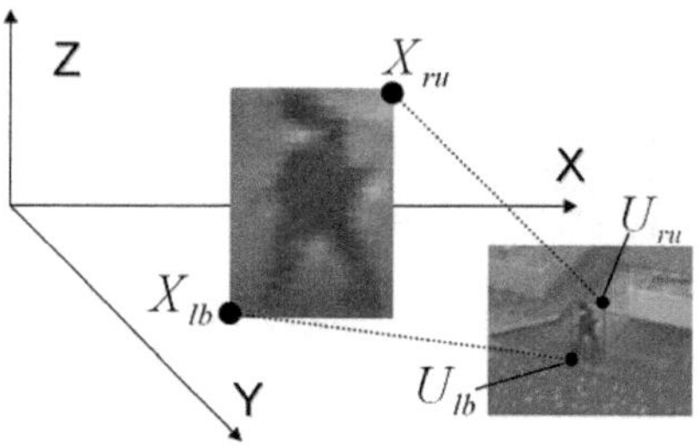

Fig. 4. 3D position estimation of a billboard

Fig. 5. Example of virtual viewpoint motion. A user's viewpoint comes through the building in front of him/her (left to right).

3.2 Smooth Video Hopping

As more and more surveillance cameras are installed in public spaces, they are going to be more accessible to the public, and people may want to see these videos. In such scenarios, people often find it difficult to understand which place they are seeing because the viewer sometimes does not understand the spatial relationship among cameras in a real space. Therefore, a sophisticated camera-switching method is needed to enable viewers to understand how the camera viewpoints change.

Smooth video hopping offers viewers the opportunity to intuitively understand the surrounding area by smoothly switching among multiple surveillance videos. It allows viewers to "hop" from one camera to another [5]. When a viewer is watching one camera, the viewer only sees the video taken directly by the camera. Then when the viewer wants to change the viewpoint to another camera, our system shows a pseudo-3D transition video sequence until the transition is complete. After the transition, the viewer can directly watch video taken from the new position. Since these transitions resemble flying over a scene from one viewpoint to another, viewers can easily understand the spatial relationship between the two cameras.

Transitions between Cameras
We created a pseudo-3D transition video of the hop from one camera to another using the projection mapping of live video textures onto such major static objects as buildings as well as the ground in a scene. An example of hopping video is shown in Fig. 6.

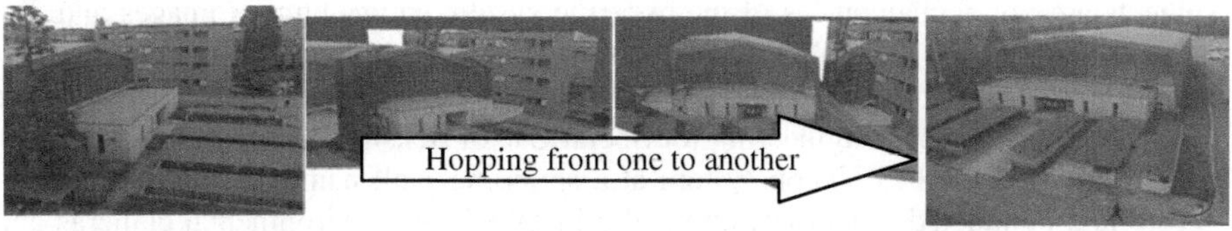

Fig. 6. Smooth Video Hopping for Surveillance Cameras: snapshots of a pseudo-3D transition between two surveillance cameras

We model only the major objects and the ground because our purpose is to assist viewers in understanding what they see when they change the viewpoint. We believe that detailed scene observation can only be done when they view real video from a camera's viewpoint. Since people are not sensitive to shapes when they fly over a scene, it is not important to reconstruct a precise 3D world during the transition in this application. The textures of the models are updated on-line.

The transition path between the two cameras is set by linearly interpolating the rigid translation matrices of the cameras; the intrinsic parameter matrix is also calculated using linear interpolation.

4 Privacy-Safe See-Through Vision

Since the See-Through Vision is powerful, it may violate the privacy of other people. Therefore, we need a solution that strikes a good balance between system benefits and expectations of privacy. We propose a Privacy-Safe See-Through Vision system (P-S Vision for short) that utilizes images of the surveillance cameras. This new and unique enhanced vision lets people directly recognize the good aspects of having the cameras installed. P-S Vision is privacy-safe because we designed the system to work only when privacy is not violated.

P-S vision is an extended system of the original See-Through Vision. The system consists of a mobile handheld device with a video camera and environmental cameras, and both types are connected via a sensor network given by [6]. Figure 7 shows an example situation of our scenario; there are some parasols on the ground level, and a user is at a higher level. If the user wants to see the spaces beneath the parasols to find free space available or to know where his/her acquaintances are resting, he/she directs the mobile device to a parasol. An environmental camera shooting beneath the parasol captures an image and extracts an appropriate image segment. Then, as shown in Fig. 7, they see objects behind the occluders on the screen. We demonstrated the system in a shopping mall in Kyoto. To operate, the system needs to know extrinsic parameters of both the environmental cameras and the mobile camera to synthesize the see-through vision images by an MR technique. We used ARToolkit [7] and devised indirect parameter estimation for placing the subjects at the right position.

Our system provides a privacy-safe vision. If the subjects of an environmental camera have some relationship with the user, in other words, if they share the same privacy level, the system shows clear and detailed images of the subjects obtained by surveillance cameras on-line (Fig. 8(a) and (b)). If no such relationship exists, i.e., the subjects are not acquaintances of the user, the system shows blurred images and displays human-shaped icons (Fig. 8(c)). This privacy-safe presentation requires an identification mechanism of subjects, which is given by the technological results of the work entitled "Content Engineering for Social Use of Sensing Information" [8].

We installed the P-S Vision system at a shopping mall named "Shin-puh-kan" in Kyoto and conducted a demonstration. We installed two environmental cameras and selected two parasols. Figure 9 is a snapshot of the demonstration. In this experiment,

Fig. 7. Snapshot image of See-Through Vision. User can observe invisible space using the images from environmental cameras.

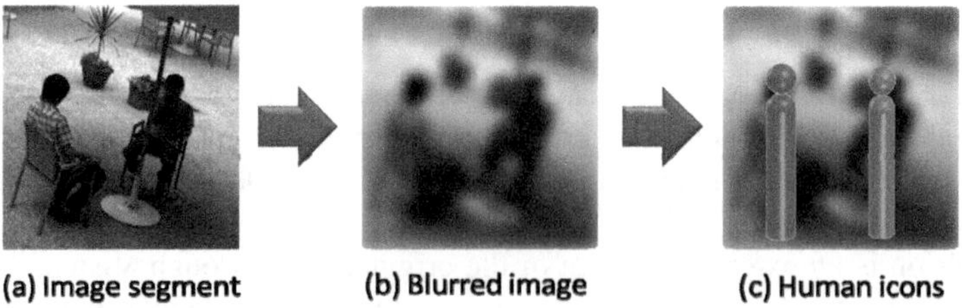

Fig. 8. Privacy-safe visualization. (a) An image segment captured by a surveillance camera. (b) Blurred image making recognition impossible. (c) Human-shaped icons.

privacy information and the number of subjects under the parasols were given manually. We asked users to evaluate the system in two aspects. One is an evaluation of the original See-Through Vision itself, and the other evaluation concerns the privacy-safe aspect. For the See-Through Vision evaluation, some users reported an uncomfortable feeling about the difference in view angle between the user and the surveillance cameras. We can ease the problem by using multiple cameras that shoot the same space from different positions and by selecting the camera that has the nearest view angle to the user's viewpoint.

In the situation of sharing the same privacy level with the people under the parasol, the system provided clear and detailed image segments (left of Fig. 9). Users could understand how many people were in the space. Most users also felt satisfied with the functionality of watching the people beneath the parasols. Some users complained that they could not clearly see the people beneath the parasol. One reason for this problem may be the smallness of the mobile device's monitor, preventing users from seeing the image in detail. The other reason would be the lighting conditions. Sometimes an image from an environmental camera would become too dark to distinguish its contents.While viewing the parasol with the privacy-safe service implemented, users expressed satisfaction with the method of visualization using blurring and human icons.

Fig. 9. Snapshot from the experiment. A user can have a clear view under the left parasol because they share the same privacy level with the subjects. On the other hand, the privacy of the people beneath the right parasol is protected.

We conducted an open demonstration like that shown in Fig. 9 at the shopping mall for three days last year and posed two questions to investigate how ordinary people felt about our technology for providing Privacy-Safe See-Through Vision.

Question-1: This technology enables users to look through walls and parasols so that they can see people, but these people are represented by stick shapes. Do you think this technology is convenient? Please mark 1 if you strongly agree, 3 if you feel neutral, and 5 if you strongly disagree.

Question-2: This technology enables users to look through walls and parasols so that they can see people, but these people are represented by stick shapes. Do you think this technology is threatening from the viewpoint of privacy protection? Please mark 1 if you think this is not at all threatening, 3 for neutral, and 5 if you strongly feel that it's threatening.

We obtained 215 responses for question-1 and 214 for question-2. Note that 1 is the most positive and 5 is the most negative, 3 is neutral, and * means no response (left blank). As shown in Fig.10, negative users marking 4 or 5 account for only 3.2% in question-1 and 19.1% in question-2. That implies that users are ready to accept the proposed privacy-safe see-through vision.

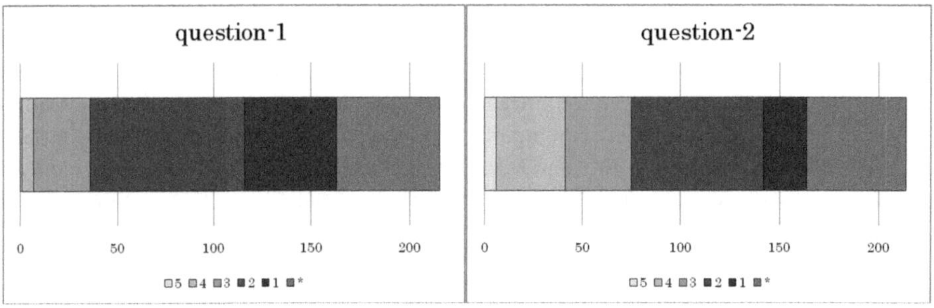

Fig. 10. Results of subjective evaluations for P-S Vision

5 Conclusion

We introduced our visual augmentation method for sensing-web, named See-Through Vision, by utilizing mixed-reality (MR) technology. Users can observe the visual information captured by surveillance cameras on the displays of mobile electronic devices such as PDAs or cell phones. The appearance captured by a surveillance camera is appropriately transformed and superimposed onto the observer's view by an MR technique. Advanced See-Through Vision systems were also introduced to emphasize the intuitive interface that helps users to gather information on the target environment. In order to provide a solution for privacy issues, we proposed our Privacy-Safe See-Through Vision. We confirmed the effectiveness of this approach through a demonstration experiment at a shopping mall in Kyoto.

References

1. Kameda, Y., Takemasa, T., Ohta, Y.: Outdoor See-Through Vision Utilizing Surveillance Cameras. In: ISMAR 2004, pp. 151–160 (2004)
2. Shi, J., Tomashi, C.: Good features to track. In: IEEE International Conference on Computer Vision and Pattern Recognition (CVPR 1994), pp. 593–600 (1994)
3. Yamazaki, S., Kitahara, I., Kameda, Y., Ohta, Y.: See-Through Vision in Wide Area with Virtual Viewpoint Motion. In: The 1st Korea-Japan Workshop on Mixed Reality, KJMR 2008 (2008)

4. Koyama, T., Kitahara, I., Ohta, Y.: Live mixed reality 3D video in soccer stadium. In: International Symposium on Mixed and Augmented Reality (ISMAR 2003), pp. 178–187 (2003)
5. Tsuda, T., Kitahara, I., Kameda, Y., Ohta, Y.: Smooth Video Hopping for Surveillance Cameras. In: SIGGRAPH 2006, Sketches (2006)
6. Minoh, M., Kakusho, K., Babaguchi, N., Ajisaka, T.: Sensing Web Project - How to Handle Privacy Information in Sensor Data. In: 12th International Conference on Information Processing and Management Uncertainty in Knowledge-Based Systems, pp. 863–869 (2008)
7. Kato, H., Billinghurst, M.: Marker Tracking and HMD Calibration for a Video-Based Augmented Reality Conferencing System. In: Proceedings of 2nd IEEE and ACM International Workshop on Augmented Reality, pp. 85–94 (1999)
8. Pentenrieder, K., Meier, P., Klinker, G., Gmbh, M.: Analysis of Tracking Accuracy for Single-Camera Square-Marker-Based Tracking. In: Third Workshop on Virtual and Augmented Reality of the GI-Fachgrouppe VR/AR, p. 4 (2006)

Manufacturing Virtual Sensors at Caterpillar, Inc.

Timothy J. Felty, James R. Mason, and Anthony J. Grichnik

Caterpillar Inc., Product Developement Center of Excellence,
100 N E Adams Street, Peoria, IL, 61629 USA
{Felty_Tim,Mason_James_X,Grichnik_Anthony_J}@cat.com

Abstract. Embedded software allows manufacturers to meet technological demands on their products more efficiently. The complexity of onboard electronic systems has grown to the point where embedded software is being used to replace or back up physical sensors in order to meet the engineering and system reliability requirements of Caterpillar and other manufacturers of large industrial equipment. This paper describes a process that allows Caterpillar to develop virtual sensors in a standard, repeatable manner that improves on conventional processes which are often inconsistent, inefficient, or both.

Keywords: Virtual Sensors, Intelligent Failover, Network Intelligence.

1 Virtual Sensors at Caterpillar: A History

The concept of virtual, or soft, sensors as models used in place of real sensors has existed for a relatively long time [1] [2]. At Caterpillar the first generation of these models were derived from physical relationships. An example would be using mass flow to determine the change in pressure across an orifice plate (Equation 1). These models were created to estimate values such as temperatures and pressures that were needed by the control system but were difficult or impossible to measure.

$$\Delta p = .5*(m^2/(\rho A^2 C^2)\ .^1 \tag{1}$$

Since the systems had grown more complex and Caterpillar engineers could no longer rely on a physics or chemistry based models, they turned to statistical and soft computing techniques. This switch was due to the lack of closed form, non-iterative models for certain physical phenomenon, such as combustion and emission-related processes. The first statistical method used was linear regression. This was due to the relative simplicity of the technique and the readily available software capable of performing the calculations.

Standard regression analysis was not able to provide the accuracy needed during the development of some virtual sensors. For these applications Caterpillar engineers turned to soft computing techniques, such as feed forward multilayer perceptron neural

[1] Where m is the mass flow rate, ρ is the density of the fluid, A is the cross-sectional area of the orifice, and C is the orifice flow coefficient.

E. Hüllermeier, R. Kruse, and F. Hoffmann (Eds.): IPMU 2010, Part II, CCIS 81, pp. 700–708, 2010.

networks. Off-the-shelf software was used to produce the neural networks, such as the MathWorks Neural Network Toolbox™ and StatSoft PROCEED™ [3] [4].

1.1 Virtual Sensor Implementation

The development and implementation of the control system is divided among several different groups at Caterpillar. Many divisions within Caterpillar develop software, including virtual sensors that will be included in the embedded software of the final product. All of these disparate groups have their own standards and procedures for developing, testing, and documenting software. The only commonality is that all the groups use Mathworks Simulink™ for development, and all target the same corporate hardware platform.

Distributed development raises issues ranging from different numerical representations, to storing training and testing data in incompatible formats, to using differing validation controls. Different numerical representations are slowly becoming less of an issue due to technology changes occurring with Caterpillar's electronic hardware. (A more in-depth discussion of fixed-pointed arithmetic is provided in an appendix.)

Each of these issues is not insurmountable but arises from a lack of a set methodology that applies to the entire enterprise. The differing validation controls reflect different philosophies on controls development. For instance, some groups try to validate models against a wide variety of machine hardware while others target a specific application. This becomes problematic when a group starts using a model developed for a specific application and another starts applying the same model to a different application. In these instances the model performance is more than likely inadequate but the root cause is unknown to the second group. In these cases virtual sensor technology is often portrayed as being unreliable or inadequate for meeting the needs of the project.

2 One-Off to Mass Production

Speaking mechanically, in the pre-industrialization era production was done in a one-off process. Craftsmen made parts as they were needed. This resulted in parts that were not strictly interchangeable and of varying quality between items made by the same craftsman (let alone by different craftsmen) in the same shop. The state of virtual sensor technology and utilization is much like pre-industrialization manufacturing. However, as the need for virtual sensors continues to increase, one-off production must give way to a faster and more efficient assembly line-like process.

There are many important requirements that must be met to transform virtual sensing from a custom product to a manufactured, industrialized process. The first requirement is that the engineer should be able to focus on the virtual sensor's application, as opposed to the data modeling required. The engineer should not have to concern themselves with the underlying techniques for building a virtual sensor. The second requirement is the process must be repeatable[2]. Third, the process should

[2] Models that are built using stochastic techniques or that have a termination based on error criterion may not be repeatable during a run. Neural networks built in successive runs may not be the same due to the randomness in starting weights. However, they should have similar error statistics.

provide verifiable feedback throughout the development. These requirements lead to a process that is both user centric and meets the needs of the enterprise.

Once the requirements of the process were laid out, the research team began to study what each of those requirements meant to a Caterpillar engineer in a production role. To accomplish this, the team questioned engineers involved with the existing effort to build virtual sensors. These discussions included data formats, workflow, and details related to the embedded implementation. Once the feedback from the engineers was received, the research team started to plan the implementation details.

To allow the engineers to use the process effectively, a guided GUI interface was chosen as the way to implement the software. The GUI would allow the engineer to receive information about progress within the process and to allow the engineer to visually inspect and confirm information as it is being generated.

To implement the software the research team decided to use off-the-shelf software as the backbone for the data modeling. StatSoft's Statistica™ and PROCEED software packages were chosen due the research group's familiarity with StatSoft's software. Statistica is scriptable, which allowed the research team to develop software that can interact with the user but leveraged the statistical functions needed for data modeling. Being able to leverage this functionality also gave the research group the advantage of reducing development time needed to implement the process.

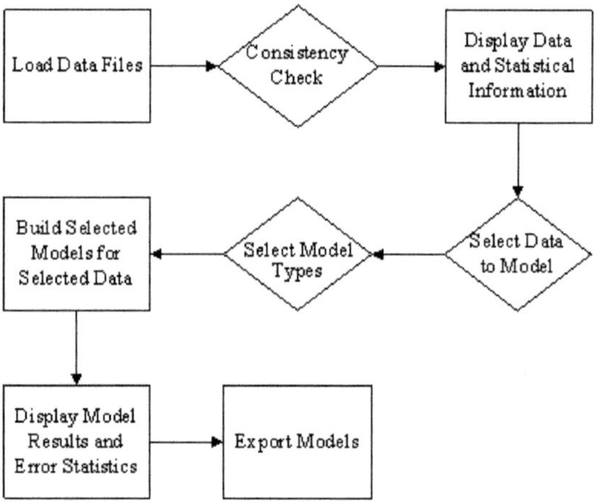

Fig. 1. Process flow diagram for manufacturing virtual sensors at Caterpillar

Figure 1 describes the process for building virtual sensors at Caterpillar. It begins with the software asking for two data sets to be loaded. The data sets should consist of columns that are sensor data channels to be used in the modeling process. The data sets are required to be in the Statistica proprietary *.sta format. Statistica can convert most spreadsheet and text data files into *.sta files; this is a necessary pre-processing step, but is not too difficult for the user. One of the data files will be used to provide training cases while the other will be used as a blind holdout sample [5]. Once loaded into memory, the data is checked for any gaps or non-numeric values. Gaps may be due to a sensor not being read, or the deletion of a data record by a user outside the process. Once the data files are confirmed to be intact the data is presented to the user for visual inspection. The data is presented as line plots, box plots, and a plot of the Mahalanobis Distance (MD) for each vector in the data set, as well as plots of other statistical comparisons. These plots allow the engineer to see if the training and testing data have similar characteristics. The line plot also allows the engineer to check the data for unreasonable values that might arise due to failures in either the mechanical hardware or data collection systems. The box plot allows the engineer to determine if the data gathered has an abundance of outliers that may also represent issues with the data quality. The MD plot allows the engineer to determine if a vector of data is similar to the training or testing data in MD space. This feedback requires the engineer to verify the data quality for both the training and testing data. While this verification is human dependent, it does force close examination of the data before modeling, as opposed to blindly building models on data that might be of low quality and wasting significant time.

After the data files have been checked and presented to the engineer, the engineer must input which data channels are to be modeled and which modeling techniques are to be considered for this application. The software then prepares the data and calls the appropriate modeling routines within Statistica and PROCEED. The process currently supports the creation of decision trees, linear regression, and feed-forward neural networks [6] in addition to other model types. Models are created with the training data are tested with the holdout data set. Since the engineer can choose to model the same data channel with different modeling techniques, each model is presented to the engineer along with its error statistics. The engineer is then able to choose the model they feel best meets their application needs based on model error, correlation, and model complexity [7]. Model complexity is an important consideration due to the limited storage space and computational ability on the embedded computing platforms used on our products[3].

Once the engineer has chosen which models to use for the virtual sensor applications, the software generates the models and exports them into a text file format. Self-protection models [8] are automatically generated along with the virtual sensor models and are output into separate text files. These self-protection models implement checks on the virtual sensors' input data within the overall control system architecture. The simplest check implemented is a bounds check on the upper and lower limits seen during the training phase of model building. The other automatically generated self-protection model is an MD check measuring the distance from the mean of the training set to the current input vector. These real-time checks when taken as a whole make up the virtual sensor network's self-protection systems, and prevent the virtual sensor network from making unqualified predictions. The traps raise a flag to the rest of the control software and signal the engineer to an error in the system, or that these conditions were not part of training and should be captured and used for refining the model in the future.

Another part of the software process then converts the models that are stored in a text format to their equivalent Simulink model files. Figures 2, 3, and 4 show examples of how the linear and neural network models appear in Simulink. These

[3] See Appendix: Hardware and Software on Cat Machines, for a discussion about storage and computational hardware used to run embedded software.

files are automatically converted to fixed-point math [9] for the user because fixed-point arithmetic is the default for embedded software at Caterpillar. The conversion to a fixed-point representation is accomplished by evaluating the extrema of the values that occur in the model and scaling the calculations accordingly.

After the virtual sensor models are converted into Simulink models, the engineer can use Matlab scripts to verify the accuracy of the models in Simulink. This verification in Simulink also creates a calibration certificate that contains information about the virtual sensor such as correlation, error statistics, and creation date and time. These certificates allow for traceability during embedded software integration.

By implementing the process as described, engineers at Caterpillar are able to develop virtual sensors in a consistent, repeatable process that is verifiable during all of its stages. This leaves only the requirement that the models should be of high quality. All of the model types except for neural networks can be shown to have minimal error for their individual types through analysis [10] [11]. For the development of neural networks, a large number of neural networks are created with varying activation functions and numbers of hidden neurons. The best performing neural network is returned and

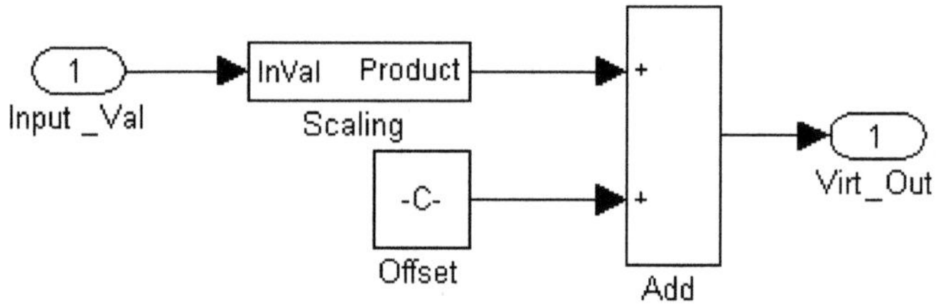

Fig. 2. Automatically generated linear regression model of a temperature sensor as it appears in Simulink. Variable names have been changed.

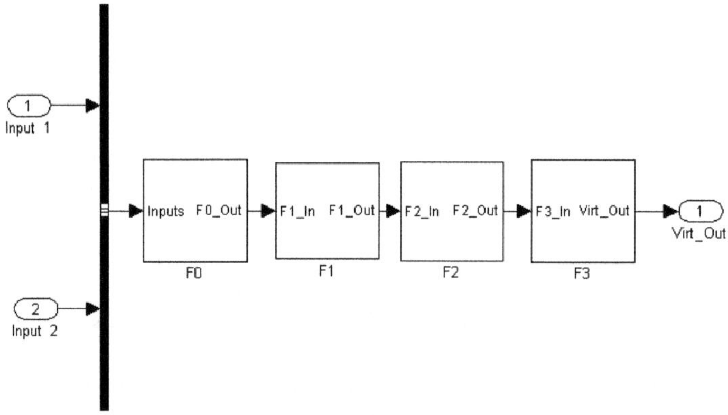

Fig. 3. Feed-forward neural network model with single hidden layer and input and output scaling layers as it appears in Simulink. This was a model for an emission product. Variable names have been changed.

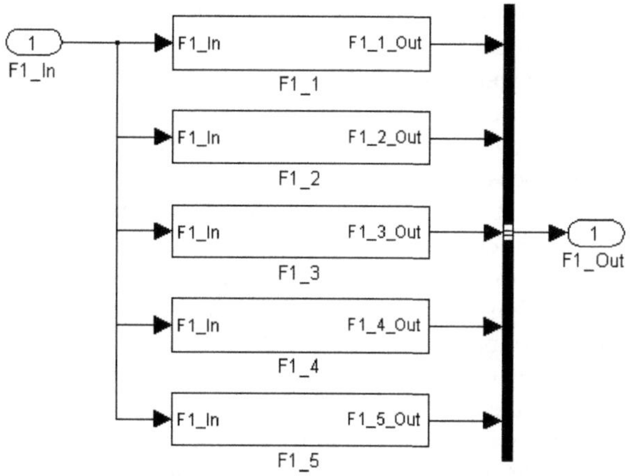

Fig. 4. Hidden layer of neural network. Each sub-block is a standard perceptron.

presented to the user for selection. While this does not ensure optimality, the returned neural networks generally have good error and correlation statistics and are very useful when other mathematical forms cannot adequately describe the desired response.

3 Applications and Benefits

Some groups within Caterpillar have adopted the process and the embedded software integration team has provided support in making the process accepted enterprise wide. Since the process and supporting software has been implemented, it has been used to virtualize temperature, pressure, and emission related values. These virtual sensors have been used to replace physical sensors as well as to provide failover capabilities. Both of these uses are beneficial; one reduces system cost while the other increases system reliability. The new process and software tools have proven to be more efficient in terms of development time. During development of the process the workflow was compared with the existing workflow to determine what kind of time savings were possible. The tasks and times required to build a virtual sensor from data collection to validation were estimated based on the experience of Caterpillar engineers. Using the same list of tasks it was estimated that the new process described in this paper reduces development time by a factor of 8. Much of the gains are due to less time being spent in model building and Simulink conversion.

As real-world comparison, a performance analysis group within Caterpillar wanted to build an emissions virtual sensor. Without the process it took six months of development time to train, test, and convert the model to fixed-point for inclusion within Caterpillar's embedded software. The new process was given the same task; build a virtual sensor to replace the same emission sensor but for a different machine application. The research team who developed the process and software were able to build the virtual sensor in a matter of days and were able to demonstrate the virtual sensor running on production hardware. During the demonstration the real sensor was disconnected and the control system was able to function properly based on the virtual sensor. Development time for this virtual sensor was reduced by a factor of 10 compared to conventional processes.

This turnaround time and the demonstration cemented the process as being the way forward for virtual sensor development within Caterpillar. As such the demand for the process and its associated technologies has grown since they were unveiled. Engineers have used the process and toolset to model the performance of engine systems for ratings development. Due to the defined output formats the models for different configurations can be interchanged leading to decrease in development time. The models have also found uses in optimization problems. The engineers responsible for system tuning have used high fidelity system response to optimize the control strategy with respect to fuel consumption, emissions and durability simultaneously.

The process and software have shown themselves to be useful outside the area of virtual sensors, such as design evaluation and optimization. To meet these additional needs the process was expanded so the models could be exported to other languages including Excel, C++, Matlab M-Script, and Java. Support for more formats can be added at a later date. By allowing one model to take on many forms, many different enterprise needs can be met from the same process.

4 Conclusion

The need for virtual sensors has grown beyond the one-off mentality that has plagued many soft computing technologies in the past. By developing and implementing a process Caterpillar hopes to use these soft computing techniques to improve their products, both in the manufacturing and research stages.

Future work will include adding support for more model types and more target languages. Model types under consideration include MARS and other splines, other neural network architectures, and non-linear regression. Target languages currently under consideration include Python, C#, IDL, and R.

References

1. Kramer, M.A.: Autoassociative Neural Networks. Computer & Chemical Engineering 16(4), 313–328 (1992)
2. Jet Propulsion Laboratory, `http://www.jpl.nasa.gov/news/releases/98/ford.html`
3. Grichnik, A., Seskin, M.: Dimensionality Reduction and Variable Selection. In: Proceedings of 2007 International Conference on Data Mining (DMIN 2007) (2007)
4. Grichnik, A., Seskin, M.: An Improved Metric for Robust Engineering. In: Proceedings of the Internation 2007 Conference on Scientific Computing (CSC 2007) (2007)
5. Caterpillar Inc., Grichnik, A., Mason, J., Felty, T.: Virtual Sensor Network (VSN) System and Method. US Patent Publication #2009/0119065
6. Ibid
7. Ibid
8. Ibid
9. Caterpillar, Inc., Grichnik, A., Mason, J., Felty, T.: Fixed-Point Virtual Sensor Control System and Method. US Patent #7593804

10. Ripley, B.: Pattern Recognition and Neural Networks. Cambridge University Press, Cambridge (1996)
11. Plackett, R.L.: Some Theorems in Least Squares. Biometrika 37, 149–157 (1950)
12. Freescale Semiconductor: MPC561/MPC563 Reference Manual (2005)
13. Freescale Semiconductor: RCPU RISC Central Processing Unit Reference Manual (1999)

Appendix: Hardware and Software on Cat Machines

Most Caterpillar products have multiple electronic control modules (ECMs). One of the ECMs is used to implement all of the engine controls. The engine ECM contains a Freescale MPC563 microcontroller running at 66MHz. The chip features 32KiB of internal RAM and 512KiB of internal Flash ROM [12]. The system provides an extra 512KiB of RAM and 4MiB of ROM. To conserve space and reduce runtime, Caterpillar control code uses fixed-point arithmetic. The MPC563 supports floating-point data types, but this increases the storage space and processing time requirements. The lower storage requirements are useful due to the fact that the onboard operating system requires space for itself, leaving less than the stated amounts to the control system.

By using fixed-point data types Caterpillar engineers can determine an appropriate tradeoff between representation accuracy and storage constraints. Caterpillar has a predefined list of fixed-point data types for engineers to use. As an example, Caterpillar has a data type for pressure with units of KPa. This data type is a signed 16-bit value with a radix of –5. This allows the engineer to represent pressures from –1024KPa to 1023.96875KPa with a precision of $1/32^{nd}$ of a KPa. A pressure value stored in this representation requires only 16 bits of storage, as opposed to 32 bits for a single precision number. This allows the engineers to save a considerable amount of storage space. Fixed-point data types make use of 8, 16, and 32 bit data types. This is a space savings of 75%, 50%, and 0%, respectively.

The control software runs close to real time, so computational speed becomes a major concern. On the MPC563 integer and floating-point units require a different amount of clock cycles to perform the same task. Table 1 shows the clock cycles needed to perform operations on integers and single precision numbers.

Table 1. Clock Cycles Required for Arithemetic Operations on MPC563 [13]

Function	Integer Clock Cycles	Single Prec. Clock Cycles
Add/Sub	1	4
Multiplication	2	4
Division	2 – 11	10
Shift	1	N/A
Logic/Compare	1	1
Convert Integer/Float	3	

For control applications this savings in execution time was deemed necessary. Most of the software within the control system involves lookups and accumulators, so the fixed-point data types do not have to be changed as data flows through the system.

However, once mathematical models started being included in the software data types could not be held static as the calculations were performed. For example, if temperature were to be predicted from a pressure and volume, the resulting value would have to be scaled to a temperature data type. This unfortunately eliminates most of the time savings provided by fixed-point data types. An addition between two values with different data types and the result having a third unique data type would require two shifts and one addition. For a multiplication with the same data types, the fixed-point implementation requires more clock cycles than the single precision implementation. The following C++ code shows how the addition and subtraction would be implemented.

Example implementation of fixed-point and floating point arithmetic in C++

```
void main()
{
   float x,y,z;
   int16 a,b,c;
   //a is signed 16, with 2^-8 scaling
   //b is signed 16, with 2^-2 scaling
   //c is signed 16, with 2^-3 scaling

   //set values for a,b,c,x,y,z
       .........

   //floating point add
   z = x + y;

   //floating point multiply
   z = x * y;

   //fixed-point add
   c = (a + (b<<6))>>5;

   //fixed-point multiply
   c = a * b;
   c = ((c < 0) && (c & 127)) + (c >> 7)) >> 3;
}
```

The floating-point addition and multiplication uses four clock cycles. Fixed-point addition with unique representations for each variable requires three clock cycles, and the multiplication requires a minimum of six and a maximum of eight clock cycles.

Modelling Low-Carbon UK Energy System Design through 2050 in a Collaboration of Industry and the Public Sector

Christopher Heaton and Rod Davies

Energy Technologies Institute, Holywell Building, Holywell Park, Loughborough, LE11 3UZ, United Kingdom

Abstract. Over the coming decades the renewal of our energy systems poses a significant technology challenge to all countries via a combination of drivers: resource availability, affordability, energy security and greenhouse gas (GHG) emission reduction to tackle global climate change. The Energy Technologies Institute (ETI) is a unique partnership between industry and Government with the objective of addressing the large-scale engineering challenges involved, through technology development and demonstration. The ETI has developed a distinctive energy system model to inform the identification and prioritisation of focus areas for technology development. The model places particular emphasis on the engineering system design, geographic representation, probabilistic treatment of uncertainty and backcasting from 2050 to the present day. A core objective is also to have a transparent modelling approach to the energy system. This paper describes the development and approach of the model and its use, both in focus area prioritisation and as a platform for dialogue between industry and policy makers.

Keywords: Energy system modelling, industrial applications, probabilistic and stochastic techniques.

1 Introduction

A large body of work has been directed towards energy system modelling in recent years. Market liberalisation has been one important historical factor, but increasingly the driving motivation is the need to address GHG reduction targets for global warming which necessarily imply significant change to our energy systems (see Figure 1). Other drivers include future uncertainty over primary resource availability, energy affordability and energy security. In this paper the term energy system refers to all activities which convert, transmit or consume energy in its various forms. We divide the system into four broad sectors: Power, Transport, Heat and Infrastructure. The first 3 of these are significant sources of direct emissions as shown in Figure 1, whereas Infrastructure refers to the storage, transmission and distribution of energy vectors.

The ETI [1] is a unique UK-based partnership between industry and Government with a budget of up to £1bn and the objective of addressing significant

E. Hüllermeier, R. Kruse, and F. Hoffmann (Eds.): IPMU 2010, Part II, CCIS 81, pp. 709–718, 2010.

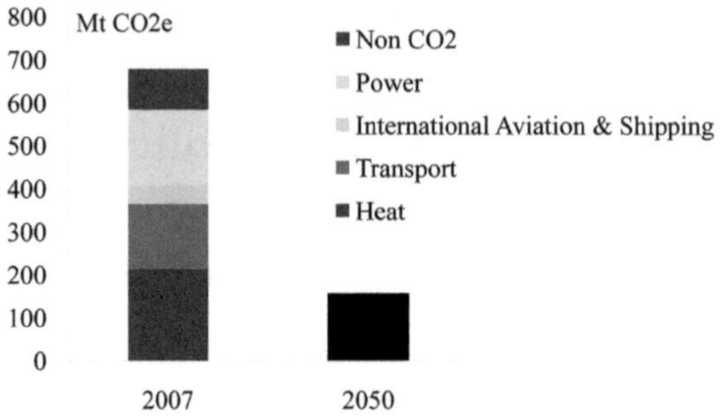

Fig. 1. Annual UK GHG emissions [2] (Mt CO2 equivalent). The UK is committed to an 80% reduction in 2050 compared to 1990 levels. It is widely believed that the limited reduction opportunities in other areas necessitate a reduction greater than 80% in the energy system.

technology challenges involved in delivering affordable, secure and sustainable energy to meet UK needs. It does this by technology and engineering development and demonstration, with the aim of building skills and capacity to implement new value-adding capability in the UK and internationally. The ETI membership currently comprises six global industrial companies: BP, Caterpillar, E.ON, EDF Energy, Rolls-Royce and Shell (with further recruitment ongoing), and a public sector membership which spans the relevant Government departments, the UK research councils and other interested bodies. The ETI focus is on complex large-scale engineering challenges that can only be addressed by bringing together the capabilities and market reach of a wide range of industrial and academic organisations. ETI projects are typically delivered by consortia involving both industrial and academic participants, and these consortia are supported by the resources of the ETI industrial members and UK Government.

The ETI performs an annual strategic analysis aimed at understanding how individual technologies can contribute to the performance of the UK energy system as a whole out to 2050 and beyond. This directly informs the ETI's project activities by identifying specific programme focus areas where ETI capabilities can make a difference. The rigorous and integrated approach to understanding how to engineer cost-effective and robust energy systems, and the pathways to them, is embodied in a new model of the UK energy system. The Energy Systems Modelling Environment (ESME) has been developed to address system design under uncertainty, filling an identified gap in the modelling landscape.

There are a number of existing platforms for energy system modelling, with varying emphasis on different aspects of the problem. Preeminent is the economic Market Allocation optimisation model (Markal) [3], originally begun in the early 1980s under the auspices of the International Energy Agency (IEA) Energy Technology Systems Analysis Programme, and subsequently developed into a family of models applied in different settings and used by various governments [4] and other bodies [5]. Other models have been developed with differing emphasis, such as the Stochastic Energy Deployment Model (SEDS) [6] which concentrates on a lightweight rapid formulation and stochastic simulations, but a complete review of all energy models is not possible here.

The needs of the ETI described above determined a set of distinctive requirements, and following a review of existing models it was concluded that none sufficiently addressed the key priorities of the ETI. In particular, given the ETI focus on engineering and technology on a distant time horizon of 2050, uncertainty over model input assumptions is a major issue. An early decision was therefore taken that the ETI needed a model taking account of uncertainty via statistical or stochastic methods in order to add robustness to the results. As well as being a reflection of the reality, an approach to uncertainty is also intended to provide insight on the range of outcomes for the energy system, rather than a single deterministic result.

Consequently in late 2008 the ETI began development of a new model. The project since has been led by ETI staff, with input from ETI stakeholders including: the ETI members, the Committee on Climate Change, the Carbon Trust and University College London. Additionally ETI has been assisted by two consulting companies: CRA International, working on scoping the functional specification and input assumptions, and Redpoint Energy, who supported the ETI on model formulation and development. The approach adopted is a Linear Program (LP) optimisation model for least-cost energy system designs with four key features:

- Engineering system focus: a bottom-up model of the energy system (macroeconomics and policy are not modelled).
- Geographical representation of supply and demand: capturing meaningful effects in energy system design, e.g. significant infrastructure costs
- Probabilistic approach: stochastic simulations to address inherent uncertainty in key 2050 assumptions, e.g. technology performance.
- Backcasting approach: a primary focus on the 2050 system design, followed by the pathway from the present day to the desired configuration in 2050.

In the rest of this paper we describe the modelling assumptions in more detail and the data inputs (Section 2), the platform and approach (Section 3), give some preliminary results to illustrate the functionality (Section 4) and finally some conclusions and discussion of further development (Section 5).

2 Input Assumptions and Data

The engineering system focus (first in the list above) has some important corollaries for the nature of the input assumptions and data required by ESME. Since the aim is to determine optimal deployment and configuration of technologies in the energy system, the cost and performance of each technology are important input data. The primary focus on engineering also leads to the position of macro-economics and government policy outside the model scope. In essence, ESME addresses the question of "what can technologies deliver?" and "what is the lowest cost system on a level playing field?" Government taxes, incentives and other policies which affect the price of energy technologies must therefore be absent from the input data. This may appear esoteric at first, but recall that the emphasis is on 2050; one would expect that all such policies will be significantly revised over the coming forty years, particularly in the context of a (potentially) radically different energy system and energy economy to today. A philosophical distinction is drawn between the aims of ESME and, say, a modelling tool for predicting with high confidence incremental evolution under the existing policies and economic conditions. The ESME approach allows the possibility of a two-stage process whereby an 'optimal' 2050 energy system design is identified in a policy vacuum, and then as a second stage appropriate policies to encourage evolution towards that design are determined, e.g. by using a predictive incremental model. It is hoped that this transparent approach will appeal to both policy makers and industry.

Macro-economic factors cannot, however, be removed completely. The approach taken here is to adopt five 'demand scenarios', each codifying information on: UK population size and distribution, macro-economic conditions and behavioural choices of the population. This information is translated into quantitative demand data, which, once the scenario is selected, is a static input to the optimisation procedure. In this context a demand is a requirement for a fundamental service that involves energy consumption. This includes, e.g. passenger transport, freight transport, heavy industrial activity, commercial floor space and domestic dwellings. These demands for fundamental services imply demand for fuels and energy vectors only when combined with technologies (e.g. electric or petrol cars). These technology choices are determined within the model by an optimisation procedure. The demand scenarios include one reference case and four further scenarios which test the extremes of future possibility.

The data specifying technology cost and performance from 2010 to 2050 is by volume the biggest input required, and it also naturally carries significant issues of commercial sensitivity. The industrial ETI members all recognise added value in an ETI dataset for ESME which pools their expertise across different sectors. However, their commercially sensitive views on the future cost and performance of technologies, which may be core to their business, must remain confidential. In response to this, the dataset was constructed by collecting data from the ETI member companies individually (and additionally from published and other external sources where available), which was kept confidential within ETI staff and then combined into an averaged and anonymous dataset. The ETI is the owner of the resultant averaged dataset, which can be shared and reviewed amongst the ETI members and stakeholders, and is used to drive ESME. An additional benefit of this approach is that the collection of views from multiple sources on each number can begin to identify the input parameters which have high uncertainty and inform the probability distributions to be assigned in those cases. Further, the model functionality allows for correlations to be specified between any two quantities, an optional and more subtle layer of data to specify.

3 Model Platform and Approach

The mathematical approach taken is an LP optimisation to find least-cost feasible 2050 energy system designs. The variables and constraints involved are described later in this section. In combination with the core LP, there is also pathway functionality for optimising both future energy system design (i.e. in 2050) and the pathway of technology deployments and retirements in intermediate years to achieve transition from the present day to the 2050 design. This is essentially an extension of the core LP, with modifications, into a larger LP optimisation problem. Finally, the probabilistic approach is achieved by Monte Carlo simulations which repeatedly sample probability distributions assigned to the input data and solve the modified optimisation problems.

The choice to restrict our formulation to a pure LP was taken for a number of reasons. The scope and resolution required by the core use within the ETI implies a large and complex optimisation problem (tens of thousands of variables and constraints), and yet the importance of stochastic simulations requires that a fast optimisation can be repeated many thousands of times in order to explore state space sufficiently. Further, we anticipated that within the stochastic simulations some very different optimal systems might arise far apart in the feasible space, and by the open-ended nature of the question being addressed we are interested to identify the qualitatively different or extreme optimals which may occur. These criteria fit well with the guaranteed convergence of an LP, and which also brings a degree of transparency and reproducibility which is valuable. Nevertheless the LP formulation imposes restrictions, and a nonlinear or mixed integer problem would allow greater flexibility to model nonlinear effects. This is deemed acceptable because the scope of ESME is very broad and it is naturally a high-level model of what in reality are very intricate complex systems. In a familiar way we approximate processes as linear at this high level, for which detailed lower-level modelling is not linear, and we must check *a posteriori* that the solution does not violate our approximations. For example, the cost of installing power stations will fail to scale linearly if the deployment involved is small, comparable with the size of a single installation.

For the ETI's purposes a complete high-level representation of the energy system involves the order of 100 technologies (see Table 1). A meaningful high level representation of UK geography was sought by splitting the country into 12 onshore regions (at which all aspects of supply and demand are specified), and 10 offshore regions (primarily sources of offshore renewable energy resources, and hence not greatly adding to computational complexity). The final resolution to specify is the temporal resolution. ESME does not attempt to model chronologically all activity within the energy system (such models are known as dispatch models). However, in order to approximately capture seasonal and diurnal effects on the energy system at a high level, ESME has a concept of six notional time periods: two seasons (currently configured as winter and summer), and three daytime periods (configured as 7 hours low-, 13 hours mid- and 4 hours peak-demand). With these it is possible to approximate seasonal and

Table 1. An indicative list of technologies in the ESME model (taken from the test dataset in January 2010)

Power	**Heat**	**Transport**	**Infrastructure**
Conventional Fossil Fuel (6)	Boilers (4)	Cars (5)	Transmission (4)
Renewables (10)	Heat Pumps (2)	Buses	Distribution (2)
Fossil Fuel + CCS (5)	Building insulations (3)	Trucks (3)	Storage (5)
Nuclear	Cookers (2)	Rail (3)	**Other**
Hydrogen	District Heating (4)	Aviation	Lighting (3)
Micro Generation (3)		Maritime	Other Conversion (10)
			Industry (8)
			Buildings (8)

diurnal patterns in demand for quantities such as heating and electricity, patterns which certainly impact on the design and operation of the energy system.

The structure of the optimisation problem is essentially a collection of commodity or mass balances for a set of 'products' which includes:

- The energy services specified in the demand scenario,
- Primary resources such as fuels and renewable energy sources,
- Intermediate products produced and consumed within the energy system such as electricity, hydrogen, etc.
- Emissions such as carbon dioxide

A mass balance must hold for each product, at each node and for each time period (though note a saving on complexity is made by specifying that some products need only be considered on an annual or seasonal basis as appropriate). Within the mass balance equations a large number of variables appear representing the utilisation of the various energy technologies, which typically act to convert one or more input products into one or more output products. The only exceptions to this are the special cases of transmission technologies (which transfer a product from one node to an adjacent node) and storage technologies (which transfer a product from one time period to another).

The key low-level constraint is that for any technology (conversion, transmission or storage) the utilisation is appropriately limited by the deployed capacity of the technology at the corresponding node. The objective function is system cost and thus comprises: all capital expenditure (from the technology deployment), operational costs (from the utilisation) and fuel costs. A number of higher level constraints are also applied, and which are likely of more interest to the model user or customer. These include:

- Emissions constraints to cap emissions at the target level
- Peak reserve margin constraint: requiring that sufficient electricity generating capacity is deployed to exceed peak demand by a specified margin. For this purpose capacity reliant on intermittent resources (e.g. wind turbines) is scaled down by a factor reflecting the probability of availability at peak demand.

- Flexibility constraint: requiring that sufficient flexible electricity generating capacity is deployed to respond to fluctuations in demand and/or intermittent supply. For this purpose capacities are scaled by a factor reflecting the reliable response time of each technology.
- Build rate constraints: constraints on the new deployment per annum for each given technology. This is primarily to reflect supply chain constraints of various sorts, and is particularly key for technologies which have not yet been developed and deployed at large scale.

This final constraint is particularly important in the broader context of the ETI's work, and early results indeed show sensitivity to the assumed rates at which emergent technologies can be deployed.

The platform for the model splits into three main parts: the formulation and solution of the core LP is performed in AIMMS [7], input and output data are held in a separate database, and a GUI is written in Microsoft Excel. Additionally, outputs can be exported to a geographical information system. The model operates in a data-driven manner, allowing flexibility to add or remove technologies, products or indeed almost any component. The LP complexity and solution time is therefore a function of the database size and the constraints imposed, but the typical benchmark for a single LP solution is of the order of 10-20s, sufficient for interactive use in deterministic mode and overnight runs in stochastic mode.

As mentioned above, the ETI decided to build ESME in late 2008, and the first version of the model was built and run during 2009. At the time of writing (early 2010) the core functionality described above has all been implemented and tested. Some items of functionality have been prototyped but remain to be completed, such as the pathway calculation which is currently being finalised. Incremental development will undoubtedly continue, but it is planned to compile and fix a definitive first version of ESME in mid 2010.

4 Illustrative Results

The ETI plans to formally publish an analysis of outputs from ESME later this year, when the model development and input data have been fully reviewed and approved. In this paper it is therefore not possible to present results which are definitive, or which represent the views of the ETI or its members. Instead, we display here some limited excerpts of results based on test data, purely for the purpose of demonstrating model operation and capability.

Figure 2 shows a subset of the results from one test run plotted on a UK map. The 12 onshore supply/demand regions can be seen clearly by the different shading on the background map. In this instance we have plotted the capacity of CCS generation (of which there are various types) indicated by the size of the blue circles and the deployment of associated pipelines shown by the red lines. In this example there are deemed to be two offshore locations for CO2 sequestration. Similar plots can be generated for almost all variables as supply, demand, technology deployment and utilisation are all specified by region.

Fig. 2. Illustrative example of geographical output. Each of 12 onshore regions is shaded green. The results shown are capacities of CCS generation of various types (blue circles) and associated pipeline network for the captured CO2 (red lines).

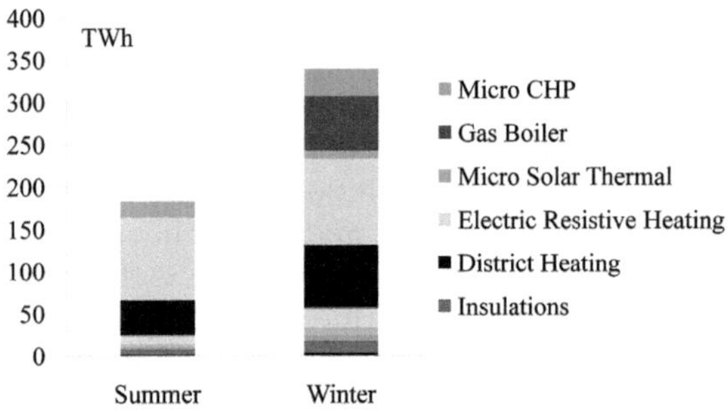

Fig. 3. Illustrative seasonal variation of heat production (space heat and hot water) in the test dataset

Figure 3 shows a simple example of the temporal variation in one aspect of the results from a test run, and illustrates a number of points in the modelling approach. In this run hot water demand is assumed constant throughout the year, but the demands for space heat are considerably higher in winter than summer. Note that gas boilers are here an example of a technology which is deployed and hence available for use all year round, however is only actually utilised in winter. Conversely the supply of heat from micro solar thermal diminishes in winter, because the performance of this technology is specified seasonally to reflect the reduced incident solar radiation in winter. Finally we note that in this instance the gas boiler was asserted in the input data to be overall the cheapest technology to produce heat. However, several more expensive technologies are seen deployed in Figure 3. This is due a combination of system-wide effects including the CO2 emission constraint, fuel costs, resource availability, transmission and distribution, seasonal and diurnal patterns etc., all of which affect the marginal costs of the alternative technologies.

Finally, Figure 4 shows an example of a deployment histogram generated from a test run in stochastic simulation mode. In this case probability distributions

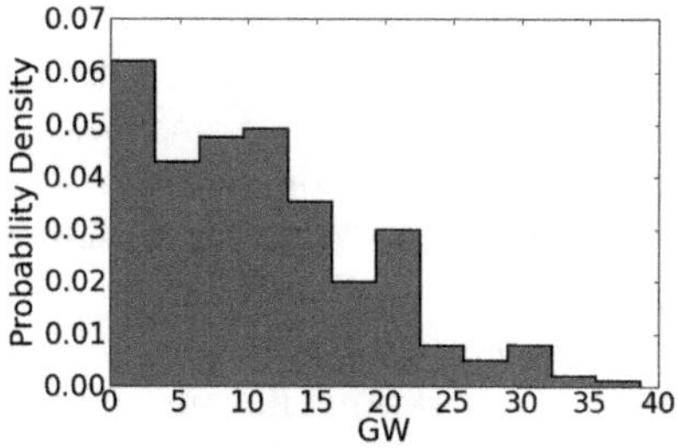

Fig. 4. Illustrative histogram of deployed capacity of H2 turbines, having assigned probability distributions to some of the technology costs in the test dataset

were assigned to a number of the technology costs in the test dataset, and a run of two thousand simulations was performed. The technology shown is a hydrogen turbine, which in most cases was used for supplying electricity demand peaks. The combusted hydrogen was generated elsewhere in the energy system from a number of sources (e.g. from coal or gas in combination with CCS, or from electrolysis), the quantities and relative proportions varying across the Monte Carlo simulations. Histograms similar to Figure 4 can be produced for any technology, and typically it may be interesting to inspect the histograms for a sector of related technologies, such as Heat or Transport. More detailed information is also encoded in the results, such as relationships and correlations between technologies across the system.

The distribution of the total energy system costs is another quantity which is of interest to study. One interesting option is to remove a given technology and then recompute the distribution of system costs, and hence to observe the implied benefit of having the technology in the available options. This is further evidence, in addition to the deployment histogram, which can inform the ETI's role in developing technologies with impact in 2050.

5 Conclusions and Discussion

Modelling is a useful tool in the energy sector, and is used to inform everything from immediate individual investment decisions to long term global policies. ESME is a new model developed by the ETI particularly aimed at studying system-level engineering on a distant time horizon of 2050. The result is a model with an engineering system focus, geographic representation of the energy system, probabilistic treatment of uncertainty in input assumptions and a back-casting approach. Early results from the model are promising, and it is planned that the ETI will finalise a first version of the model and dataset later in 2010, in order to publish findings based on ESME outputs.

Avenues for further work include a number of opportunities to add sophistication in the representation of the energy system, e.g. further security of supply considerations, or subgrid-scale modelling of local distribution networks, or of course increased resolution in space and/or time. All such possibilities however have to be carefully weighed against the wish to retain solution speed (key to enabling Monte Carlo simulations) and the transparency afforded by an intuitive and relatively lightweight formulation. On the stochastic front there are certainly opportunities to extract more relationship information from the large datasets produced via improved inference techniques. We have not explored any properties of the near-optimal solutions to the LP, and this too may reveal interesting results about robustness of various flavours of energy system. Following from the discussion in section 3, it is also interesting to speculate on the effect of departing from a pure LP optimisation. This could involve using an alternative optimiser with 'robustness' characteristics, or introducing some limited nonlinearity to improve the modelling of certain effects. In both cases a careful comparison of model performance and results, particularly on a large stochastic simulation, should be made to the existing baseline.

Finally, we remark that ESME is distinctive by virtue of the unique environment that the ETI offers in terms of collaboration between major industrial companies and the UK Government and broader public sector. In this respect the accompanying dataset which the ETI is developing in parallel has similar potential value as a common knowledge platform pooled from all the ETI stakeholders. The exact role that the modelling approach described can play within the ETI members and more broadly remains to be seen, but early reaction to the developing model has been positive, including from key Government stakeholders [8]. Ultimately it is hoped that ESME may provide a useful and dispassionate platform for dialogues, as well as being a tool for internal analysis. A longer term opportunity could also be to apply the model to other countries or regions, given the generic approach taken.

References

1. The Energy Technologies Institute (UK), `http://www.energytechnologies.co.uk`
2. Department for Energy and Climate Change, UK emissions statistics, `http://www.decc.gov.uk`
3. IEA Energy Technology Systems Analysis Programme's Markal model, `http://www.etsap.org/markal/main.html`
4. Department of Trade and Industry (UK): Meeting the Energy Challenge: A White Paper on Energy (2007), `http://www.berr.gov.uk/files/file39387.pdf`
5. Committee on Climate Change (UK): Building a low-carbon economy - the UK's contribution to tackling climate change (2008), `http://www.theccc.org.uk/pdf/TSO-ClimateChange.pdf`
6. Strategic Energy Analysis Centre of the National Renewable Energy Laboratory (USA), `http://www.nrel.gov/analysis`
7. Paragon Decision Technology Ltd. (NL), `http://www.aimms.com/aimms`
8. Beddington, P.J.: UK Government Chief Scientific Adviser: The ETI energy system model is an excellent tool. It will help identify key areas for strategic investment by ETI and, equally importantly it will help underpin and inform UK Government energy policy

A Remark on Adaptive Scheduling of Optimization Algorithms

Krisztián Balázs[1] and László T. Kóczy[1,2]

[1] Department of Telecommunications and Media Informatics, Budapest University of Technology and Economics, Hungary
balazs@tmit.bme.hu, koczy@tmit.bme.hu
[2] Institute of Informatics, Electrical and Mechanical Engineering, Faculty of Engineering Sciences, Széchenyi István University, Győr, Hungary
koczy@sze.hu

Abstract. In this paper the scheduling problem of optimization algorithms is defined. This problem is about scheduling numerical optimization methods from a set of iterative 'oracle-based' techniques in order to obtain an as efficient as possible optimization process based on the given set of algorithms.

Statements are formulated and proven about the scheduling problem and methods are proposed to solve this problem.

The applicability of one of the proposed methods is demonstrated through a simple fuzzy rule based machine learning example.

1 Introduction

There are a huge number of numerical optimization algorithms known from the literature. A large part of them is formed by the iterative 'oracle-based' (also known as 'black box') techniques evaluating the given objective function in each iteration to compute new states. They can be used in the field of complex optimization problems, because these techniques claim only a few assumptions about a given problem, thus they are rather general algorithms. A part of them are invented intuitively (e.g. steepest descent [1] and Levenberg-Marquardt [2] [3] methods), another part of them are inspired by natural processes (e.g. Genetic Algorithm [4] and Bacterial Evolutionary Algorithm [5]). However, there is a huge cost of this generality: there are no exact results which technique is how efficient in general, or even in a particular problem field at all.

Therefore there are only heuristics for deciding which ones to use and how to parameterize them. These heuristics are based on intuition and simulation results (e.g. [5] [6] [7], and many more).

There are often big differences in the difficulty of the algorithms, which result in different type of characteristics. A simpler technique can be faster, but less efficient and a more difficult one can be much slower, but much more efficient from iteration to iteration [7].

Often, in the early part of the optimization it is easy to reach better and better states in the problem space, but after a long iteration period it is quite difficult to find a better state. Thus, a simpler algorithm can perform better at the beginning of a numerical optimization process (as a more global search) due to its higher iteration speed, whereas

E. Hüllermeier, R. Kruse, and F. Hoffmann (Eds.): IPMU 2010, Part II, CCIS 81, pp. 719–728, 2010.

a more difficult algorithm can be a more appropriate choice at the end (as a more local search) due to its higher efficiency.

Therefore, it would be often useful to apply simpler algorithms at the beginning and more difficult ones at the end of a numerical optimization process. More generally, it would be desired to find out which optimization algorithm to use on a particular part of a problem. (This problem is more general than 'Which algorithm should be applied on an optimization problem?'). Rephrasing the desire: an effective schedule of the optimization algorithms should be found adaptively to the particular numerical optimization problem. (Obviously, not only optimization algorithms can be scheduled, but whole optimization architectures, or simply parameterizations, too.)

This problem defines a decision tree, where the levels, the edges and the vertices represent iteration levels, executions of optimization algorithms and states, respectively. It can be explored totally or partially. The former way of exploration has a huge computational demand (exponential), whereas the latter way has lower (either linear, depending on the size of the explored part).

In our work we define the scheduling problem in an exact way, we formulate and formally prove statements, furthermore we propose methods for giving an efficient schedule in linear time complexity under particular circumstances. Apparently, scheduling even in linear complexity also means a significant overhead in an optimization process, but to choose and to apply improper optimization algorithms or parameters on a problem or on a part of a problem may cost a lot more.

This paper will consider only maximization problems, minimization tasks can obviously be formulated as maximization problems, though.

The next section defines the scheduling problem and reflects on the necessity of assumptions. Section 3 makes some assumptions about the scheduling problem, proposes *Greedy scheduler* and discusses its efficiency. Section 4 proposes *Fast greedy scheduler* and discovers its relationship to the previous scheduling method. Section 5 presents a simple example simulation, a fuzzy rule based learning process, to demonstrate the usability of the theory introduced in this paper as well as the efficiency of Fast greedy scheduler. Finally, Section 6 draws some conclusions and finishes the paper.

2 Optimal Schedule of Optimization Algorithms

First of all some basic notations are defined that will be applied throughout this paper.

Definition 1. ***Basic notations***

1. S — *state space, containing the possible collective states of the optimization algorithms (including the positions in the domain of the given objective function)*
2. $\mathbf{x}_0 \in S$ — *starting state of the optimization*
3. $f_i : S \mapsto S$ — *change of the state caused by applying the ith algorithm one iteration long*
4. $t_i \in \mathbb{R}^+$ — *time cost of applying the ith algorithm one iteration long*
5. $\mathcal{F} = \{(f_i, t_i)\}$ — *set of optimization algorithms*
6. $m = |\mathcal{F}|$ — *cardinality of $\mathcal{F}$ (number of the members of the set)*
7. $<i>_0^n = i_0, i_1, \ldots, i_n$ — *sequence of indices ($\forall j$: $i_j \in \{1 \ldots m\}$)*

8. $\bigcirc_{j=j_0}^{n} f_{i_j} = f_{i_{j_0}} \circ f_{i_{j_0+1}} \circ \cdots \circ f_{i_n}$ — *composition of a sequence of functions (if* $j_0 > n$ *then* $\bigcirc_{j=j_0}^{n} f_{i_j}$ *is the identity function, i.e.* $\bigcirc_{j=j_0}^{n} f_{i_j}(\mathbf{x}) = \mathbf{x}$*)*
9. $R_j\{i_0, i_1, \ldots i_n\} = i$ *if* i *is the* j^{th} *minimal element in* $\{i_0, i_1, \ldots i_n\}$
10. $F : S \mapsto \mathbb{R}$ — *objective function (a greater objective function value denotes a closer position to the optima) – any monotonic increasing transformation of the real objective function may also be used, e.g. fitness function in evolutionary algorithms*
11. $T \in \mathbb{R}^+$ — *time limit (a given constant)*
12. $\phi_{F(\mathbf{x})^{(k-1)}}(F(f_1(\mathbf{x}))^{(k)} \ldots F(f_m(\mathbf{x}))^{(k)})$ — *the density function of the distribution of the objective function values after applying all of the optimization algorithms one iteration long at* $\mathbf{x}$ *(*$^{(k-1)}$ *and* $^{(k)}$ *only indicate that it is the objective function value after k-1 and k steps)*

A state $\mathbf{x}_1 \in S$ is *better* than another state $\mathbf{x}_2 \in S$, if $F(\mathbf{x}_1) > F(\mathbf{x}_2)$. (Similarly, *worse* can also be defined.)

During the optimization process the state $\mathbf{x} \in S$ stores (among other informations) the domain point having the best ever explored objective value and $F(\mathbf{x})$ gives back this value, hence the following condition always stands.

Definition 2. ***Objective Function Value Monotonity Condition***
C_{om}: $F(\bigcirc_{j=j_0}^{n_1} f_{i_j}(\mathbf{x})) \leq F(\bigcirc_{j=j_0}^{n_2} f_{i_j}(\mathbf{x}))$ *if* $n_1 \leq n_2$.

(On both sides there are subsequences of the same sequence of indices, but the subsequence on the left is at most as long as the subsequence on the right side.)

A sequence of indices means a schedule of the optimization algorithms. It can be evaluated by calling the 'oracle' (F objective function) in order to compare different schedules.

Definition 3. ***Evaluation of a sequence of indices***
In case of a given $\mathbf{x}_0 \in S$ *and* $<i>_0^n$ *the evaluation of* $<i>_0^n$ *is the computation of* $F(\bigcirc_{j=0}^{n} f_{i_j}(\mathbf{x}_0))$.

A sequence is *better* than another one, if its evaluation produces a greater value. (Similarly, *worse* can also be defined.)

The ith optimization algorithm is *locally best* at a given $\mathbf{x} \in S$ state, if $\forall j \in \{1 \ldots m\}$: $F(f_i(\mathbf{x})) \geq F(f_j(\mathbf{x}))$. (Similarly, *locally worst* can also be defined.)

After these basic definitions the main task can be formulated as follows.

Definition 4. ***Optimal schedule of optimization algorithms (time limited version)***
In case of a given $\mathcal{F}$, F *and* T *the optimal scheduling of optimization algorithms task is to determine a sequence* $<\nu>_0^l$ *(optimal sequence) for which* $\sum_{j=0}^{l} t_{\nu_j} \leq T$ *and for all* $<i>_0^n$ *for which* $\sum_{j=0}^{n} t_{i_j} \leq T$, $F(\bigcirc_{j=0}^{l} f_{\nu_j}(\mathbf{x}_0)) \geq F(\bigcirc_{j=0}^{n} f_{i_j}(\mathbf{x}_0))$ *holds.*

According to the definition, the optimal schedule is the one that reaches the greatest objective function value within the given time limit.

A sequence $<i>_0^n$ for which $\sum_{j=0}^{n} t_{i_j} \le T$ *fulfills the time condition.*

It is quite obvious to prove that in case of arbitrary $\mathbf{x}_0 \in S$ and $t_j \in \mathbb{R}^+$ values ($j \in \{1 \ldots m\}$) all of the sequences fulfilling the time condition must be evaluated to obtain an optimal sequence, if for all $\mathbf{x}_1, \mathbf{x}_2 \in S$ ($\mathbf{x}_1 \ne \mathbf{x}_2$) it cannot be determined whether $F(\mathbf{x}_1) \le F(\mathbf{x}_2)$ or $F(\mathbf{x}_1) \ge F(\mathbf{x}_2)$ holds without computing both $F(\mathbf{x}_1)$ and $F(\mathbf{x}_2)$.

Therefore, the number of evaluations is in the interval $[m^{\lfloor \frac{T}{max\{t_j\}} \rfloor}, m^{\lfloor \frac{T}{min\{t_j\}} \rfloor}]$ and the time of evaluations is in the interval $[m^{\lfloor \frac{T}{max\{t_j\}} \rfloor} min\{t_j\}, m^{\lfloor \frac{T}{min\{t_j\}} \rfloor} max\{t_j\}]$.

In terms of T this means exponential time complexity ($\Theta(m^T)$).

3 Greedy Scheduler

The previous statement indicates that there is need for some assumptions. In the following part of this paper two assumptions will be made about the given scheduling problem. The first one is the *Time Uniformity Condition* that expects algorithms having the same time costs.

Definition 5. ***Time Uniformity Condition***
$C_{tu}\colon \forall j \in \{1 \ldots m\}\colon t_j = t$

If C_{tu} is fulfilled, the intervals mentioned above contract into a single point, because $\forall j \in \{1, m\}\colon min\{t_j\} = t = max\{t_j\}$. Thus, the number of evaluations is $m^{\lfloor \frac{T}{t} \rfloor}$ and the time of evaluations is $m^{\lfloor \frac{T}{t} \rfloor} t$.

Since exponential time complexity holds if it cannot be determined whether $F(\mathbf{x}_1) \le F(\mathbf{x}_2)$ or $F(\mathbf{x}_1) \ge F(\mathbf{x}_2)$ holds without computing both $F(\mathbf{x}_1)$ and $F(\mathbf{x}_2)$, some condition about the relation of different $F(\mathbf{x})$-s would be desirable.

The second assumption is the *Strong Monotonicity Condition*. It requires that if a state is not worse than another one, the same relation will be held after applying the same algorithm on the states. This is definitely an irrational assumption about general scheduling problems, but it will be eased later.

Definition 6. ***Strong Monotonicity Condition***
C_{sm} : $\forall \mathbf{x}_1, \mathbf{x}_2 \in S$ *and* $\forall i \in \{1 \ldots m\}$: $F(\mathbf{x}_1) \le F(\mathbf{x}_2)$ *implies* $F(f_i(\mathbf{x}_1)) \le F(f_i(\mathbf{x}_2))$

The following lemma is based on the Strong Monotonicity Condition and will help in proving Theorem 1. (Although Theorem 1 could be proven without this lemma, some future considerations will be based on it.)

Lemma 1. *Suppose that C_{sm} holds and $<k>_0^l$ is a sequence fulfilling the following. For all $<i>_0^p$ sequence for which $\sum_{j=0}^{l} t_{k_j} \le \sum_{j=0}^{p} t_{i_j}$ there exists such a $h \le p$ for which $\sum_{j=0}^{l} t_{k_j} \le \sum_{j=0}^{h} t_{i_j}$ and $F(\bigcirc_{j=0}^{l} f_{k_j}(\mathbf{x}_0)) \ge F(\bigcirc_{j=0}^{h} f_{i_j}(\mathbf{x}_0))$. Furthermore there exists an optimal sequence $<\nu>_0^n$ so that $\sum_{j=0}^{l} t_{k_j} \le \sum_{j=0}^{n} t_{\nu_j}$. Then there exists an optimal sequence whose first l members are $<k>_0^l$.*

Proof. Since $\sum_{j=0}^{l} t_{k_j} \leq \sum_{j=0}^{n} t_{\nu_j}$, there exists such a $h \leq n$ for which $\sum_{j=0}^{l} t_{k_j} \leq \sum_{j=0}^{h} t_{\nu_j}$ and $F(\bigcirc_{j=0}^{l} f_{k_j}(\mathbf{x}_0)) \geq F(\bigcirc_{j=0}^{h} f_{\nu_j}(\mathbf{x}_0))$. Thus $T \geq \sum_{j=0}^{n} t_{\nu_j} = \sum_{j=0}^{h} t_{\nu_j} + \sum_{j=h+1}^{n} t_{\nu_j} \geq \sum_{j=0}^{l} t_{k_j} + \sum_{j=h+1}^{n} t_{\nu_j}$, furthermore $F(\bigcirc_{j=0}^{n} f_{\nu_j}(\mathbf{x}_0)) = F(f_{\nu_0} \circ f_{\nu_1} \circ \cdots \circ f_{\nu_n}(\mathbf{x}_0)) = F(f_{\nu_{h+1}} \circ f_{\nu_{h+2}} \circ \cdots \circ f_{\nu_n}(f_{\nu_0} \circ f_{\nu_1} \circ \cdots \circ f_{\nu_h}(\mathbf{x}_0))) \leq F(f_{\nu_{h+1}} \circ f_{\nu_{h+2}} \circ \cdots \circ f_{\nu_n}(f_{k_0} \circ f_{k_1} \circ \cdots \circ f_{k_l}(\mathbf{x}_0))) = F(\bigcirc_{j=0}^{l} f_{k_j} \circ \bigcirc_{j=h+1}^{n} f_{\nu_j}(\mathbf{x}_0))$, due to condition C_{sm}.

Therefore, $k_0, k_1, \ldots k_l, \nu_{h+1}, \nu_{h+2}, \ldots \nu_n$ is also an optimal sequence. □

At this point the first scheduling method, *Greedy scheduler* will be proposed. The main idea of this technique is to choose always the locally best optimization algorithm by evaluating all of the optimization algorithms at the actual state.

Definition 7. ***Greedy scheduler (GS)***
Greedy scheduler is a method producing a $<\nu>_0^n$ *sequence, for which:*
$\forall i \in \{0 \ldots n\}: \nu_i = j$, *where* $F(\bigcirc_{k=0}^{i-1} f_{\nu_k} \circ f_j(\mathbf{x}_0)) \geq F(\bigcirc_{k=0}^{i-1} f_{\nu_k} \circ f_h(\mathbf{x}_0)), \quad h \in \{1 \ldots m\}$

Theorem 1. *If both* C_{tu} *and* C_{sm} *hold and* $\mathbf{x}_0 \in S$ *is arbitrary, Greedy scheduler finds an optimal schedule within* $m\lfloor \frac{T}{t} \rfloor t$ *time.*

Proof. The technique of induction is applied:

1. Base ($i = 0$): $F(\bigcirc_{k=0}^{i-1} f_{\nu_k} \circ f_{\nu_0}(\mathbf{x}_0)) \geq F(\bigcirc_{k=0}^{i-1} f_{\nu_k} \circ f_h(\mathbf{x}_0))$, i.e. $F(f_{\nu_0}(\mathbf{x}_0)) \geq F(f_h(\mathbf{x}_0))$, $h \in \{1 \ldots m\}$ by definition and since for all j: $t_{\nu_0} = t = t_j$ and C_{sm} holds, the condition part of Lemma 1 is fulfilled (because $h = i$ is suitable). Thus there exists an optimal sequence whose first member is $\nu_0 = <\nu>_0^i$.
2. Induction step ($i > 0$): $F(\bigcirc_{k=0}^{i-1} f_{\nu_k} \circ f_{\nu_i}(\mathbf{x}_0)) \geq F(\bigcirc_{k=0}^{i-1} f_{\nu_k} \circ f_h(\mathbf{x}_0))$, $h \in \{1 \ldots m\}$ by definition, where $F(\bigcirc_{k=0}^{i-1} f_{\nu_k}(\mathbf{x}_0)) \geq F(\bigcirc_{k=0}^{i-1} f_{j_k}(\mathbf{x}_0))$ for all $<j>_0^{i-1}$ sequence.
   Since for all $<j>_0^i$ sequence $\sum_{l=0}^{i} t_{\nu_l} = i \cdot t = \sum_{l=0}^{i} t_{j_l}$ and C_{sm} holds, the condition part of Lemma 1 is fulfilled (because $h = i$ is suitable). Thus there exists an optimal sequence whose first i members are $<\nu>_0^i$.

The induction works until i becomes greater than the length of an optimal sequence. Then an optimal sequence is produced by GS.

Due to Definition 4, the length of the sequence is $\lfloor \frac{T}{t} \rfloor$. In each step all the m optimization algorithms are tested, therefore the number of evaluations is $m\lfloor \frac{T}{t} \rfloor$ and the time of evaluations is $m\lfloor \frac{T}{t} \rfloor t$.

□

In terms of T this means linear time complexity ($\Theta(mT)$).

The Strong Monotinicity Condition is far too bold for general optimal scheduling problems. Actually, in case of oracle-based optimization tasks, there is a very poor knowledge about the problem. Thus, any deterministic assumption is doubtful. Stochastic ones are rather suitable.

In the rest of this section the deterministic Strong Monotonicity Condition will be eased to a more general and intuitively more reasonable stochastic assumption. The following definition is necessary in order to formulate the eased assumption. The *expected value of chosen path* is an expected value calculated from $F(\mathbf{x})$ values as random variables. The result is the expected value of a schedule, where the upper indices shows, that iteration-by-iteration the $l_k^{th} \dots l_{n-1}^{th}$ locally worst optimization algorithms are applied.

Definition 8. ***Expected value of chosen path***

$$\underset{F(\mathbf{x})^{(k)}}{\overset{(l_k \dots l_{n-1})}{E}} F(\mathbf{x})^{(n)} = \int_{-\infty}^{\infty} \dots \int_{-\infty}^{\infty} \phi_{F(\mathbf{x})^{(k)}}(F(f_1(\mathbf{x}))^{(k+1)} \dots F(f_m(\mathbf{x}))^{(k+1)}) \cdot$$

$$\cdot \underset{R_{l_k}\{F_1((x)^{(k+1)}) \dots F_m((x)^{(k+1)})\}}{\overset{(l_{k+1} \dots l_{n-1})}{E}} F(\mathbf{x})^{(n)} \mathrm{d}F_1((x)^{(k+1)}) \dots \mathrm{d}F_m((x)^{(k+1)})$$

$$\underset{F(\mathbf{x})^{(n)}}{\overset{()}{E}} F(\mathbf{x})^{(n)} = F(\mathbf{x})^{(n)}$$

This definition implicitly assumes that the random variables of the optimization algorithms at a state are independent of the random variables of the optimization algorithms at another state. However, it does not assume that the random variable of an optimization algorithm at a state is independent of the random variable of another optimization algorithm at the same state.

The eased assumption is as follows. Advancing from a state along a path, where at the beginning the l_k^{th}, at the second step the l_{k+1}^{th} and so forth ... finally, at the last step the l_{n-1}^{th} locally worst algorithm is chosen, does not result a state being expectedly better than advancing from a not worse state.

Definition 9. *Lower Monotonicity Condition:*

$$C_{lm}: \underset{F(\mathbf{x}_1)^{(k)}}{\overset{(l_k \dots l_{n-1})}{E}} F(\mathbf{x}_1)^{(n)} \le \underset{F(\mathbf{x}_2)^{(k)}}{\overset{(l_k \dots l_{n-1})}{E}} F(\mathbf{x}_2)^{(n)}, \textit{ if } F(\mathbf{x}_1)^{(k)} \le F(\mathbf{x}_2)^{(k)}$$

Definition 10. *Upper Monotonicity Condition:*

$$C_{lm}: \underset{F(\mathbf{x})^{(k)}}{\overset{(l_k \dots l_i^{(1)} \dots l_{n-1})}{E}} F(\mathbf{x})^{(n)} \le \underset{F(\mathbf{x})^{(k)}}{\overset{(l_k \dots l_i^{(2)} \dots l_{n-1})}{E}} F(\mathbf{x})^{(n)}, \textit{ if } l_i^{(1)} \le l_i^{(2)}$$

Proposition 1. *If Lower Monotonicity Condition holds, Upper Monotonicity Condition also holds.*

Proof. The statement can be proven by substituting Definition 8 into Definition 10 and considering that $\underset{R_{l_i^{(1)}}\{F_1((x)^{(i+1)}) \dots F_m((x)^{(i+1)})\}}{\overset{(l_{i+1} \dots l_{n-1})}{E}} \le \underset{R_{l_i^{(2)}}\{F_1((x)^{(i+1)}) \dots F_m((x)^{(i+1)})\}}{\overset{(l_{i+1} \dots l_{n-1})}{E}}$,

because
$R_{l_i^{(1)}}\{F_1((x)^{(i+1)})\ldots F_m((x)^{(i+1)})\} \leq R_{l_i^{(2)}}\{F_1((x)^{(i+1)})\ldots F_m((x)^{(i+1)})\}$ and C_{lm} holds.

□

Corollary 1. *Based on Proposition 1, it is obvious to prove that if C_{lm} holds there is no method (among the ones choosing at most one optimization algorithm at each state) giving an expectedly better schedule for the optimal scheduling problem than Greedy scheduler, if C_{tu} holds.*

4 Fast Greedy Scheduler

Sometimes the optimization problem shows favorable properties and particular optimization algorithms can perform best during long periods. In this case the number of switches between the algorithms (n_{sw}) in an optimal schedule may be low.

To exploit this advantage of such problems, another version of the above discussed Greedy scheduler is proposed. If the scheduling problem has the mentioned favorable property, this scheduling method is faster, than the previous one.

The *Fast greedy scheduler* does not compare the optimization algorithms in each step, but only after a '*blind running*' time t_{br} ($\frac{t_{br}}{t} \in \mathbb{Z}^+$), while it applies the last locally best algorithm.

Definition 11. ***Fast greedy scheduler (FGS)***
Fast greedy scheduler is a method producing a $<\nu>_0^n$ sequence, for which $\forall i \in \{0\ldots n\}$:

$$\nu_i = \begin{cases} j & \textit{if } (i \cdot t) \textit{ mod } (t + t_{br}) = 0 \\ \nu_{i-1} & \textit{otherwise} \end{cases}$$

where j: $F(\bigcirc_{k=0}^{i-1} f_{\nu_k} \circ f_j(\mathbf{x}_0)) \geq F(\bigcirc_{k=0}^{i-1} f_{\nu_k} \circ f_h(\mathbf{x}_0))$, $h \in \{1\ldots m\}$

The following lemma gives an estimation for the computational time of FGS.

Lemma 2. *If C_{tu} and C_{sm} hold and $\mathbf{x}_0 \in S$ is arbitrary, then Fast greedy scheduler produces a $<k>_0^p$ sequence for which $F(\bigcirc_{j=0}^{p} f_{k_j}(\mathbf{x}_0)) \geq F(\bigcirc_{j=0}^{n} f_{\nu_j}(\mathbf{x}_0))$, where $<\nu>_0^n$ is an optimal sequence, in a time cost t_{FGS} for which:*

1. *If $n_{sw} \geq n = \lfloor\frac{T}{t}\rfloor$: $t_{FGS} \leq n(tm + t_{br})$*
2. *If $n_{sw} < n = \lfloor\frac{T}{t}\rfloor$: $t_{FGS} \leq n_{sw}(tm + t_{br}) + (n - n_{sw})t + \lceil\frac{(n-n_{sw})t}{t+t_{br}}\rceil(m-1)t$*

Proof. In worst case when FGS does not run on an optimal sequence the value of F does not change (because of C_{om}). Then FGS must run altogether n iterations long on optimal sequences during the whole scheduling process. In comparing phase FGS always chooses the locally best optimization algorithm. In blind running phase two cases can be distinguished:

1. *Inefficient blind running*: There is at least one switch between the algorithms in the optimal schedule. In this case FGS might not run any iterations on optimal sequences during the blind running phase.
2. *Efficient blind running*: During blind running FGS runs on an optimal sequence (because there are no switches between the algorithms in the optimal schedule). In this case $\frac{t_{br}}{t}$ is added to the number of iterations run on optimal sequences.

If $n_{sw} \geq n = \lfloor \frac{T}{t} \rfloor$, each comparing phase might be followed by an inefficient blind running phase. Then $t_{FGS} \leq n(tm + t_{br})$.

If $n_{sw} < n = \lfloor \frac{T}{t} \rfloor$, there is at most n_{sw} inefficient blind running phase. In this case the rest of the time is spent in comparing and efficient blind running phases, thus the rest of the wasted time is made by the $\lceil \frac{(n-n_{sw})t}{t+t_{br}} \rceil$ comparing phases: $t_{FGS} \leq n_{sw}(tm + t_{br}) + (n - n_{sw})t + \lceil \frac{(n-n_{sw})t}{t+t_{br}} \rceil (m-1)t$. □

Theorem 2 shows that if the number of switches between the optimization algorithms in the optimal schedule is suitably low, Fast greedy scheduler is more efficient than Greedy scheduler.

Theorem 2. *If C_{tu} and C_{sm} hold and $\mathbf{x}_0 \in S$ is arbitrary, then Fast greedy scheduler outperforms Greedy scheduler in the sense that while GS finds an optimal schedule for a scheduling problem, FGS finds a schedule that, although, does not solve the scheduling problem, is better (in the sense defined in Definition 3) than the one found by GS, if $n_{sw} < n$, furthermore:*

$$n_{sw} \leq \frac{\frac{t_{br}}{t}(mn - m - n + 1) + (1 - m)}{\frac{t_{br}}{t}(m + \frac{t_{br}}{t})}$$

Proof. The aim is to show that $t_{FGS} \leq t_{GS} = m\lfloor \frac{T}{t} \rfloor t$. The statement of the theorem can be achieved by rearranging the following inequality (obtained from the previous lemma):

$t_{FGS} \leq n_{sw}(tm + t_{br}) + (n - n_{sw})t + \lceil \frac{(n-n_{sw})t}{t+t_{br}} \rceil (m-1)t \leq n_{sw}(tm + t_{br}) + (n - n_{sw})t + (\frac{(n-n_{sw})t}{t+t_{br}} + 1)(m-1)t \leq mnt = t_{GS}$ □

5 Simulation Results

In this section a simulation run will be presented, which was carried out in order to demonstrate the usability of the recently discussed theory. The following simple example shows the efficiency of Fast greedy scheduler on a fuzzy rule based machine learning problem.

The problem was to learn (i.e. to approximate) the following six dimensional function (that was also applied by Nawa and Furuhashi to evaluate Bacterial Evolutionary Algorithm [5]): $f_{6dim} = \boldsymbol{x}_1 + \sqrt{\boldsymbol{x}_2} + \boldsymbol{x}_3\boldsymbol{x}_4 + 2e^{2(\boldsymbol{x}_5 - \boldsymbol{x}_6)}$ (where $\boldsymbol{x}_1, \boldsymbol{x}_2 \in [1,5]$, $\boldsymbol{x}_3 \in [0,4]$, $\boldsymbol{x}_4 \in [0,0.6]$, $\boldsymbol{x}_5 \in [0,1]$, $\boldsymbol{x}_6 \in [0,1.2]$).

The learning architecture applied Mamdani-inference [8], furthermore Bacterial Evolutionary Algorithm (BEA) [5] and Bacterial Memetic Algorithm (BMA) [6] as optimization algorithms to find proper parameters for the fuzzy rules.

In the simulation the parameters had the following values. The number of rules in the rule base was 4, the number of individuals in a generation was 5, the number of clones was 5 and 4 gene transfers were carried out in each generation. In case of BMA 4 iterations long gradient steps were applied. The number of training samples were 200 in the learning process. The whole optimization took 120 seconds long.

During the run, the fitness values of the best individuals were monitored in terms of time. These fitness values were calculated based on the Mean Squared Error (MSE) values (measured on the training samples) as follows: $F = \frac{10}{\text{MSE}+1}$.

The results are presented in Fig. 1 to get a better overview. The horizontal axis shows the elapsed computation time in seconds and the vertical axis shows the fitness values of the best individuals at the current time. In the figure the dashed line shows the result of the pure evolutionary algorithm (BEA), the dotted line presents the graph of the method using Levenberg-Marquardt technique (BMA) and the solid line denotes the values produced by FGS.

For FGS $\mathcal{F} = \{(f_1, t_1), (f_2, t_2)\}$ was defined as follows. In order to meet the Time Uniformity Condition, $t_1 = t_2 = t$ and since the fitness values may not increase during a long period, t was set to 20 seconds to have an effective comparing phase. Thus, due to the very differing time demands of the algorithms, f_1 denotes approximately 1070 iterations of BEA, whereas f_2 denotes 9 iterations of BMA. The blind running phase was also 20 seconds long ($t_{br} = t$).

As it can be observed in Fig. 1, after 120 second BMA gives only a slightly better result than BEA, due to the phenomenon mentioned in the Introduction, i.e. BEA (the simpler algorithm) was more effective in the early part of the optimization due to its higher iteration speed, whereas BMA (the more difficult technique) gave better performance after reaching a particular fitness level due to its higher efficiency.

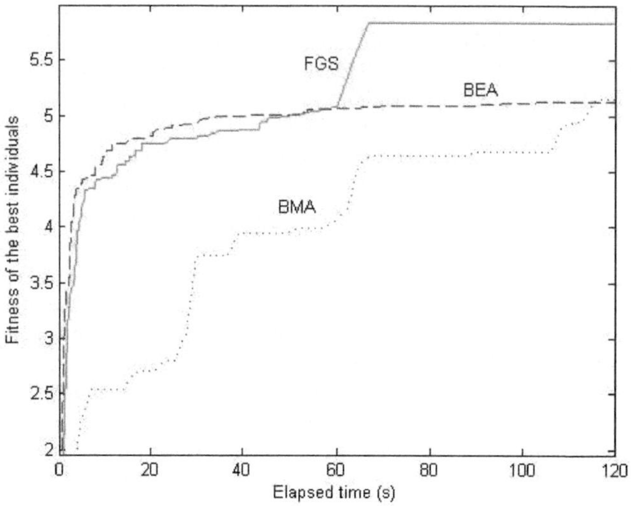

Fig. 1. Fitness of the best individuals (BEA - dashed line, BMA - dotted line, FGS - solid line)

However, using FGS and adaptively switching between the algorithms (i.e. scheduling them) the advantages of the different methods could be exploited by applying the proper technique in the different parts of the optimization process.

6 Conclusion

In this paper the theory of *adaptive scheduling of optimization algorithms* was introduced. In the frame of this theory two scheduling methods (Greedy and Fast greedy schedulers) have been proposed and the connection between them have been discussed.

The usability of this theory and the efficiency of Fast greedy scheduler have been demonstrated through an example simulation run.

There are a huge number of exciting open questions left about this field. What are reasonable assumptions for a set of problems? How could efficient schedulers by these assumptions be established? How should the parameters of a scheduler be set? And so forth...

Our future work aims to answer such questions and explore deeper the theory of this promising field as well as its applicability.

Acknowledgments

This paper was supported by the National Scientific Research Fund Grant OTKA K75711, a Széchenyi István University Main Research Direction Grant and the Social Renewal Operation Programme TÁMOP-4.2.2 08/1-2008-0021.

References

1. Snyman, J.A.: Practical Mathematical Optimization: An Introduction to Basic Optimization Theory and Classical and New Gradient-Based Algorithms. Springer, New York (2005)
2. Levenberg, K.: A method for the solution of certain non-linear problems in least squares. Quart. Appl. Math. 2(2), 164–168 (1944)
3. Marquardt, D.: An algorithm for least-squares estimation of nonlinear parameters. J. Soc. Indust. Appl. Math. 11(2), 431–441 (1963)
4. Holland, J.H.: Adaption in Natural and Artificial Systems. The MIT Press, Cambridge (1992)
5. Nawa, N.E., Furuhashi, T.: Fuzzy system parameters discovery by bacterial evolutionary algorithm. IEEE Transactions on Fuzzy Systems 7(5), 608–616 (1999)
6. Botzheim, J., Cabrita, C., Kóczy, L.T., Ruano, A.E.: Fuzzy rule extraction by bacterial memetic algorithms. In: Proceedings of the 11th World Congress of International Fuzzy Systems Association, IFSA 2005, Beijing, China, pp. 1563–1568 (2005)
7. Balázs, K., Botzheim, J., Kóczy, L.T.: Comparison of Various Evolutionary and Memetic Algorithms. In: Proceedings of the International Symposium on Integrated Uncertainty Management and Applications, IUM 2010, Ishikawa, Japan (2010) (accepted for publication)
8. Mamdani, E.H.: Application of fuzzy algorithms for control of simple dynamic plant. IEEE Proc. 121(12), 1585–1588 (1974)

An Adaptive Fuzzy Model Predictive Control System for the Textile Fiber Industry

Stefan Berlik and Maryam Nasiri

University of Siegen, Software Engineering Institute,
D-57068 Siegen, Germany
{berlik,nasiri}@informatik.uni-siegen.de

Abstract. Many process steps in the production of modern fibers and yarns are hallmarked by their high complexity and require thus a great know-how of the operating personnel. To support their work an adaptive fuzzy model predictive control system has been designed whose characteristics are sketched here. The system is build upon an expert specified rule base and comprises a data driven optimization component. Two disparate types of measures are collected and exploited for this: continuous available online measurements stemming from machine sensors and sporadic analyses from laboratory spot tests. Further key feature is an inferential control mechanism that allows for continuous control in absence of the primary values from the lab.

1 Introduction

Numerous machines and processes for production of fibers, yarns and textile structures require a great know-how of the operating personnel. This applies to the specific choice of the appropriate machine parameters at product changes or the introduction of new products or materials. Often, expensive trial-and-error experiments are necessary to establish meaningful machine settings. Suitable machine settings are not only important for economic reasons, but also for environmental protection.

Against the background of the increasing relocation of production facilities abroad, this expertise is often locally not available any longer. Therefore, systems that help the operator in finding the optimal setting of the machinery and equipment are becoming increasingly important. For many years now systems based on artificial neural networks are developed and used industrially for this. The big advantage of such systems lies in the ease of use. Their big disadvantage is that contexts that are analytically studied and thus quantifiable or able to be formulated in rules cannot be regarded. Thus often a significant part of the available operator's knowledge is not included in the recommended setting. In addition, the acquired knowledge is present only implicitly as a black box.

The use of systems based on fuzzy logic promise to remedy this [1]. It allows formulating fuzzy rules based on known coherencies. Through a combination of such rules the fuzzy system can make a statement on the optimal setting of the machine or system. In contrast to a prediction of an artificial neural network

E. Hüllermeier, R. Kruse, and F. Hoffmann (Eds.): IPMU 2010, Part II, CCIS 81, pp. 729–736, 2010.

this prediction is immediately replicable by the operator since the knowledge is present here explicitly. Its acceptance is therefore considerably higher.

A problem often to be encountered in practice are disparate types of measurements and different rates at which they occur. Frequently parameters defining the objective function are available only sporadic due to necessary analyses from laboratory spot tests. On the other hand online measurements stemming from machine sensors are continuously available; however their expressiveness might be unknown at first. Special care has to be taken in these situations.

For a machine in the mentioned environment an advanced control system shall be developed. To be able to use both, operator's expertise and knowledge stemming from measured data a fuzzy model shall shape the core of the system. Hence it seems reasonable to use a model predictive control scheme. To enable continuous optimization of the fuzzy model an adaptive system is developed [2]. The contribution of this paper is the development of a consistent draft of such a system and the presentation of first results of its partly implementation.

This paper is organized as follows. First, some related work is sketched in the next section. The architecture and some details of the system are discussed in Section 3. Section 4 outlines the current state and Section 5 finally presents our conclusions.

2 Related Work

A modern method for the control of complex processes is *model predictive control* (MPC), also referred to as *receding horizon control* (RHC)[3,4]. Model predictive control uses a time-discrete dynamic model of the process to be controlled to calculate its future states and output values as a function of the input signals. Using this prediction, in particular suitable input signals for desired output values can be found. While the model behavior will be predicated several steps ahead till a certain time frame, the input signal is usually only searched for the next time step and then the optimization is repeated. For the calculations of the next time step then the actual measured state is used, resulting in a feedback and thus a closed loop. Model predictive control technology offers a number of advantages that have made it one of the most widely used advanced control methods in the process industry: It can calculate control variables where classical nonlinear control techniques fail, is easily extended to multivariable problems, can take restrictions on the actuators into account, permits the operation near the constraints, allows a flexible specification of the objective function, delivers optimum control devices and is last not least model based.

Fuzzy set theory provides structured ways of handling uncertainties, ambiguities, and contradictions which made systems based on fuzzy set theory the approach of choice in many situations. Since its introduction in 1965, fuzzy set theory has found applications in a wide variety of disciplines. Modeling and control of dynamic systems belong to the fields in which fuzzy set techniques have received considerable attention, not only from the scientific community but also from industry. Their effectiveness together with their ease of use compared to systems based on classical two-valued logic paved the way to countless practical applications of fuzzy systems. Also in the textile industry fuzzy set techniques have successfully been used, e.g. in the area of textile technology, textile chemistry, textile software, and textile economy [5,6,7,8].

2.1 Fuzzy Model Predictive Control

Generally, a distinction is made between linear and nonlinear model predictive control. Linear model predictive control uses linear models for the prediction of system dynamics, considers linear constraints of the states and inputs, and a quadratic objective function. Nonlinear model predictive control is based on nonlinear models and can include non-linear constraints and / or general objective functions. When historically first linear model predictive control strategies have been examined and used, nowadays nonlinear control is of increasingly interest. A major reason for this is that many processes are inherently nonlinear and are often run over a wide operation range, linear regulators therefore ultimately achieve only a low control quality or stability problems may also arise. Particularly suitable for the creation of the nonlinear models and thus the core of the model predictive control is the fuzzy set theory.

Compared with already investigated controls based on neural networks, the use of fuzzy models for control of the machine offers significant advantages: Already existing expertise can be directly fed into the system, in contrast to neural networks the knowledge is explicitly, thus self-explanatory and the acceptance of the procedure higher.

The modeling and control of nonlinear systems using fuzzy concepts is described in [9]. Current methods for identification are data-extraction techniques on the one hand and the expertise on the other, but also mixed forms of both approaches. Basis of the data-driven approach is a clustering algorithm whereby fuzzy models are derived, preferably of the Takagi-Sugeno type. The obtained models form the basis of the desired nonlinear control, whether using the inversion of the model or model predictive control.

Characteristic of the project are the different rates at which the output variables are measured and the input variables can be varied. Generally, processes as such are termed *dual rate systems*. In the present case, the primary variables, i.e. the yarn parameters are available solely after laboratory tests and this only sporadically. For continuous adjustment of the process they are only suitable to a limited extent. What can be used in addition then is an approach known under the name *inferential control*. Online measured secondary variables of the process are in the absence of primary values used to estimate the actual disturbances affecting the process [10].

3 System Development

Figure 1 shows the interaction of the model predictive control system and the production process together with the appertaining data flow. It can be seen

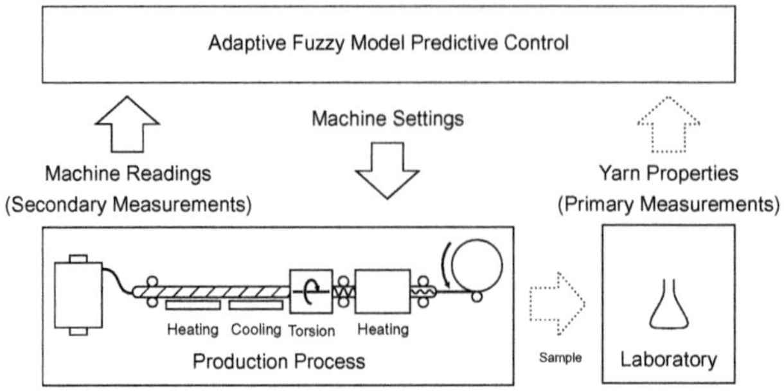

Fig. 1. Interaction of the control system and the production process

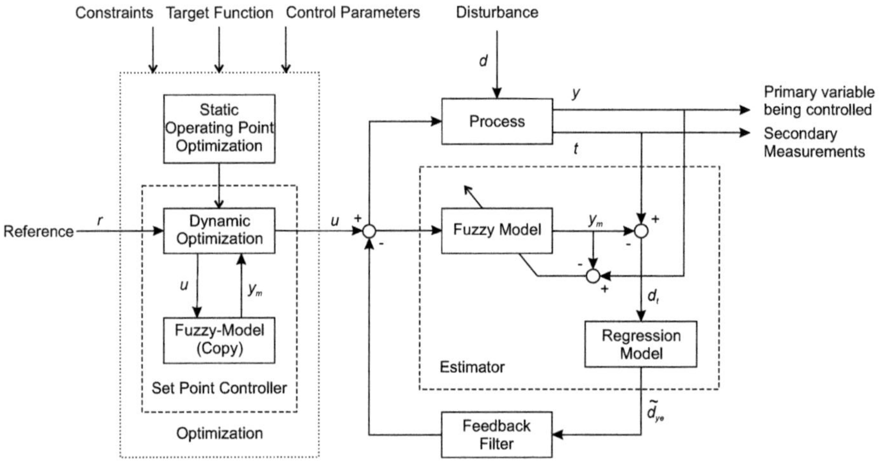

Fig. 2. Adaptive fuzzy model predictive control system with inferential control element

that the system operates the production process via the machine settings and receives feedback in form of machine readings and yarn properties. Both have to be considered adequately to improve overall system performance.

A more detailed view of the mentioned controller is given in Figure 2. On the left-hand side one can see the set point controller using the fuzzy model to find optimal machine settings together with the higher level static operating point optimization. On the right-hand side the estimator of the inferential control component is shown. Build upon the fuzzy model and the secondary measurements it estimates the disturbances on the primary variable. By feeding back the primary variable into the fuzzy model automatic adaptation of the model becomes possible. The following subsections describe the treated aspects in greater detail.

3.1 Development of the Initial Fuzzy Model

Basis of the initial fuzzy model is the acquired knowledge of a domain expert. By means of interviews, his knowledge is formulated as a linguistic fuzzy model in form of fuzzy rules and appropriate fuzzy sets. In combination these model the known relationships of the application domain. According to the expected complexity of the different control ranges of the fuzzy controller and the available data gaps in the data set are identified. Missing data sets are generated by practical experiments.

The collected data provide the alternative to automatically deduce a fuzzy model using clustering techniques - preferably in the form of Takagi-Sugeno type. After transformation of also the linguistic model into a Takagi-Sugeno model both can be checked by experts for consistency. Should significant differences appear the causes and effects are to question. Often this approach reveals previously unknown relationships and the process expert gains a deeper understanding of the process. Newly acquired knowledge is incorporated into the fuzzy model.

For fine tuning the initial system the authors developed an evolutionary optimization module based on the Covariance Matrix Adaptation Evolution Strategy [11,12]. The module was successfully applied in earlier projects, e.g. in the area of dyeing polyester fibers [13]. To verify that the resultant rule-based model maps the relationship between process parameters and yarn properties sufficiently precise, its forecast quality is reviewed for an independent data set [14].

3.2 Implementation of the Nonlinear Fuzzy Model Predictive Controller

Expertise in the field of yarn processing lies predominantly in the direction 'When machine setting, then yarn property', so according to the causal relationship. Consequently, the initial fuzzy system uses as inference direction a forward chaining of facts from the domain of machine settings to draw conclusion in the domain of yarn properties. To answer the more relevant question of appropriate machine settings for desired yarn properties the other inference direction is necessary. For this purpose with the model predictive control an appropriate solution is available that uses the controller structure known as *internal model control*, see Figure 3.

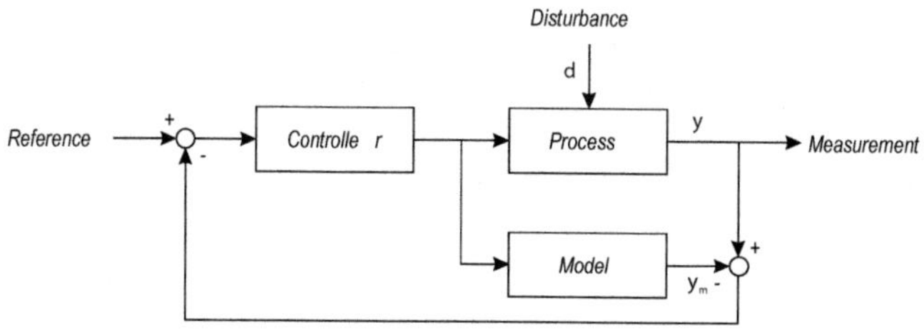

Fig. 3. Internal model control scheme

3.3 Integration of an Inferential Control

As already mentioned, the primary variables, i.e. the yarn parameters are available solely after laboratory tests and this only sporadically. For continuous adjustment of the process they are only suitable to a limited extent. Promising seems to be the additional use of an inferential system whereby an estimate of the influence of non-measurable disturbances on the controlled variables is returned to the control loop, see Figure 2. The aim of the inferential control scheme is therefore, to estimate the disturbance d even if there are no primary values y but only secondary values t available. This is a major difference to procedures such as Kalman filters that use the secondary values in order to predict the primary values and to regulate it - and thus in particular violate the feedforward nature of the inferential control. To be solved is essentially the identification problem of the form: Find an estimator for a given set of past primary values y and secondary values t, that estimates the influence of noise d_{ye} on the primary values from the influence on the secondary values d_t. Used for this purpose is first a linear regression model.

3.4 Extension to an Adaptive Fuzzy Model Predictive Control

The fuzzy model predictive control is based at first solely on the expert knowledge and a static set of observed historical data with which it supplies yarn parameter forecasts with respect to given machine settings. The sporadic incoming laboratory results on the other hand describe the effective, current correlation of the input and output variables, so this knowledge is usefully used to the continuous adaptation of the fuzzy system. Integration of an adaptation mechanism allows for two things: First, in the steady state of the process model discrepancies can be identified and eliminated. Ideally, it is possible to reconstruct an accurate, error-free process model over time. Furthermore, the model can adjust itself to (temporarily) changing process conditions and also determine optimal control variables in these situations. For this purpose the controller structure is extended so that the difference between model and process is returned as a new control variable into the fuzzy model.

3.5 Integration of a Nonlinear Estimator and a Static Working Point Optimization

The previously integrated inferential scheme can be extended to a nonlinear estimator - similar to the automated procedure that has been used to derive the initial TS fuzzy model. Necessary data for this will have been collected at this point already during former work packages of the project. Expected is an increase in the quality of the estimates on the error signals acting on the primary values. The new estimator is evaluated and compared with the linear estimator.

In addition to the previously discussed dynamic optimization of the operating point, i.e. the search of appropriate control variables for given yarn quality, the chosen architecture allows with only slight modifications also for a static operating point optimization. While the former is more likely to be assigned to the process control, the latter belongs to the plant management level. Although it is in our case not desired to improve the output parameters by multi-objective optimization, but rather to keep them constant, the process can be optimized with respect to higher goals, such as to increase throughput, minimize energy use, et cetera. For this purpose, simply the objective function has to be modified adequately and possibly extended by new constraints.

4 Current State

The presented system is still under construction and right now only partly implemented using the technical mathematical software Matlab. However, first results have been achieved. For the problem given, a proof-of-concept has been done with a prototypic rule base containing seven rules. The results the rule base gave for a set of test data are quite promising. Also the tuning module for the rule base works stable. Next steps are to incorporate the measured data by means of a clustering algorithm as described in Section 3.1. The model predictive control algorithm can then be applied to the machine. At this preliminary stage, it corresponds to the status quo of the control of the machine. As the feedback of the measured yarn parameters in the machine settings is sporadic, the control with lengthy periods rather equals to an open-loop control.

Consequently, the next milestone is the integration of an inferential control scheme in order to regulate the process in times without primary measures. Expected is a significant improvement in the management of the production process. The next aim is the structural extension of the control scheme to an adaptive fuzzy model predictive control to ensure continuous improvement of the underlying fuzzy model as well as an automatic adjustment to changing process conditions. Finally, the nonlinear estimator has to be introduced into the inferential scheme and the static operating point optimization has to be added.

5 Conclusion

Presented has been an architecture that addresses the design of an adaptive fuzzy model predictive control system, comprising a data driven optimization component supporting disparate types of measures. Since many production processes are characterized by producing dual rate data, including an inferential control mechanism to subsume laboratory spot tests - at least partly - by online available measurements is of great help. The cycle time of control can be reduced and costs saved.

References

1. Hopgood, A.A.: Intelligent Systems for Engineers and Scientists. CRC Press, Boca Raton (2001)
2. Cordón, O., Herrera, F., Hoffmann, F., Magdalena, L.: Genetic Fuzzy Systems. World Scientific Publishing Company, Singapore (2001)
3. Maciejowski, J.: Predictive Control with Constraints. Prentice Hall, Englewood Cliffs (2001)
4. Brosilow, C., Joseph, B.: Techniques of Model-Based Control. Prentice Hall, Englewood Cliffs (2002)
5. Thomassey, S., Happiette, M., Dewaele, N., Castelain, J.M.: A short and mean term forecasting system adapted to textile items' sales. The Journal of the Textile Institute 93, 95–104 (2002)
6. Lennox-Ker, P.: Using fuzzy logic to read image signatures. Textile Month (1997)
7. Kuo, C.F.J.: Using fuzzy theory to predict the properties of a melt spinning system. Textile Research Journal 74(3), 231–235 (2004)
8. Kim, S., Kumar, A., Dorrity, J., Vachtsevanos, G.: Fuzzy modeling, control and optimization of textile processes. In: Proceedings of the 1994 1st International Joint Conference of NAFIPS/IFIS/NASA, San Antonio, TX, USA, pp. 32–38 (1994)
9. Babuška, R.: Fuzzy Modeling for Control. Kluwer Academic Publishers, Norwell (1998)
10. Joseph, B.: A tutorial on inferential control and its applications. In: Proceedings of the American Control Conference, San Diego (June 1999)
11. Hansen, N., Ostermeier, A.: Adapting arbitrary normal mutation distributions in evolution strategies: The covariance matrix adaptation. In: IEEE International Conference on Evolutionary Computation, pp. 312–317 (1996)
12. Hansen, N., Ostermeier, A.: Completely derandomized self-adaptation in evolution strategies. Evolutionary Computation 9(2), 159–195 (2001)
13. Nasiri, M., Berlik, S.: Modeling of polyester dyeing using an evolutionary fuzzy system. In: Carvalho, J.P., Dubois, D., Kaymak, U., da Costa Sousa, J.M. (eds.) Proc. of the 2009 Conf. of the International Fuzzy Systems Association (IFSA) and the European Society for Fuzzy Logic and Technology (EUSFLAT), Lisbon, Portugal, July 20-24, pp. 1246–1251 (2009)
14. Siler, W., Buckley, J.J.: Fuzzy Expert Systems and Fuzzy Reasoning. Wiley-Interscience, Hoboken (2005)

Methodology for Evaluation of Linked Multidimensional Measurement System with Balanced Scorecard

Yutaka Kigawa[1], Kiyoshi Nagata[2], Fuyume Sai[2], and Michio Amagasa[2]

[1] Musashino Gakuin University, 860 Kamihirose, Sayama-city, Saitama, 350-1321, Japan
[2] Daito Bunka University, 1-9-1 Takashimadaira, Itabashi-ku, Tokyo, 175-8571, Japan

Abstract. Balanced Scorecard was articulated as a comprehensive framework for evaluating a company's business performance. Some of authors have proposed a multidimensional measurement system with BSC in which a structural modeling of four perspectives is constructed in the 1st stage, and the evaluation process in each perspective is performed in the 2nd stage, then the integrated value is calculated in the following stage. In that system, FSM based structural modeling method is applied for calculating the weight of measures in each perspective, then fuzzy inference mechanism and Choquet integrals are applied to have integrated values. In this paper, we propose a new methodology for evaluating the strategy oriented business performance of a company by means of linked structure of measurements in BSC framework. Fuzzy inference mechanism and FSM based structural modeling method also play an important role. We give a scheme of our system and show how it works by an illustrative example.

1 Introduction

In order to evaluate current companies' performance, Kaplan and Norton proposed the balanced scorecard (BSC) in view of multidimensional measurement of business performance in 1992 ([KN1]). BSC has four perspectives, "financial", "customer", "internal process", and "learning and growth", each of which has corresponding goal and several measures. Some of authors have proposed a system to aggregate them using FSM(Fuzzy Structural Modeling method) based structural modeling method, named the Modified Structural Modeling Method(MSMM), fuzzy inference mechanism, and Choque integral [MAS, CSAM]. In [KN1], although the four perspectives of BSC are linked with each other, only the bidirectional relationships between two of them are indicated as links. When we investigate the company's performance level at a fixed point of time, the evaluation values of each perspective can be aggregated subject to their independency. The proposed method in [MAS, CSAM] is this kind of evaluation and we call it a "static" evaluation.

The main concept of BSC is to evaluate company's performance level not only in financial figures but also in dimensions of three other categories in order to

E. Hüllermeier, R. Kruse, and F. Hoffmann (Eds.): IPMU 2010, Part II, CCIS 81, pp. 737–746, 2010.

accomplish its strategic objectives. This might be an epoch-making concept in those days when the finance related measures were recognized as almost unique measures to access the business performance.

Some companies adopted BSC as an assessment tool of their business performance, others consider it as a strategic measurement system [KN2]. These days, some of researchers make critical argument on its efficiency as strategic measurement system. We consider that an evaluation system of business performance headed to accomplish the determined business strategy should properly access the performance level of an activity without ignoring the effects operated by other activities. We call this kind of evaluation a "dynamic evaluation".

In this paper, we propose a methodology to perform the dynamic evaluation of business performance using fuzzy inference mechanism and MSMM in BSC framework. The rest of the paper is organized as follows: in the next section, we review BSC. A brief explanation of the "static" evaluation method is in the section 3. Then, the proposed methodological system is developed in section 4, and an illustrative example is in section 5. Finally, discussion and conclusion are coming up.

2 Balanced Scorecard and Its Strategic Implementation

The BSC is the concept that Robert Kaplan and David Norton devised to make up for a fault of conventional financial indices-centered management technique of business performance, typically consisting of following four processes.

1. Make a vision and a strategy clear and replace it with plain words,
2. link a strategic target and job performance evaluation index, and enforce common knowledge,
3. keep a consistency of plan, goal setting, strategic program,
4. promote strategic feedback and learning.

As for the word "balance", the balance of long-term and short-term business targets, the balance of job performance evaluation indices of financial and non-financial affairs, the balance of job performance evaluation indices in the past and in the future, the balance of a viewpoint of the outside and the inside. The business performance evaluation index of the BSC represents a company's job performance derived from their vision and strategy in the following four perspectives.

- The financial perspective:
  set indices that measure how the company earns financially to live up to the stakeholders' expectation.
- The customer perspective:
  set indices that measure how the company act for a customer to achieve their vision
- The business processes perspective:
  set indices that measure how the company builds superior business processes to improve the customer satisfaction and to achieve the financial target

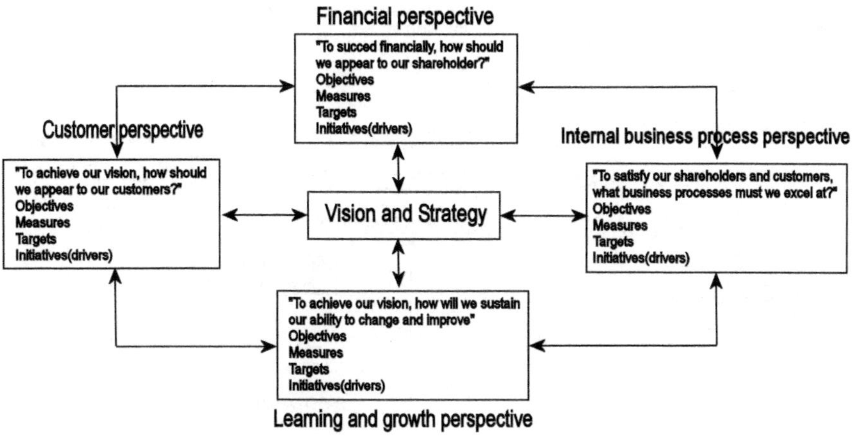

Fig. 1. The framework of BSC (source: [KN3])

- The learning and the growth perspective:
  set indices that measure how plans and educations make ability improvement both as an organization and as an individual to achieve the vision of the company

These perspectives above are totally linked with each other as in the Fig.1.

Although BSC was developed to achieve a company's business strategic objectives and might well work in some companies, such as Rockwater, Apple, Advanced Micro Devices, etc. (see [KN2]), it becomes that several shortcomings are discerned. For example in [IL], Christopher Itter and David Larcher pointed out the importance of link among four perspectives to see whether the company's activities are well performed heading for completion of the strategy. They also noted four feasible mistakes to be committed such as "not linking measures to strategy", "not validating the links", "not setting the right performance targets", and "measuring incorrectly". In [KN4], Kaplan and Norton proposed a method to implement a business strategy by mapping them into a linked actions chart, where BSC's four perspectives are hierarchically laid out and effective actions are allocated with links.

3 Multidimensional Measurement System with BSC

In [CSAM], Cui et. al proposed a measurement system of company's business performance by integrating multidimensional measures related to four perspectives of BSC using FSM base structural modeling method, fuzzy inference mechanism and Choque integrals. The Fig.2 describes the flow of the total system consists of four stages.

In the initial stage, stage A, a group of executives/managers indicates a consented structural model of possible measures in each of four perspectives. If they initially do not have one, Modified Structural Modeling Method (MSMM) can

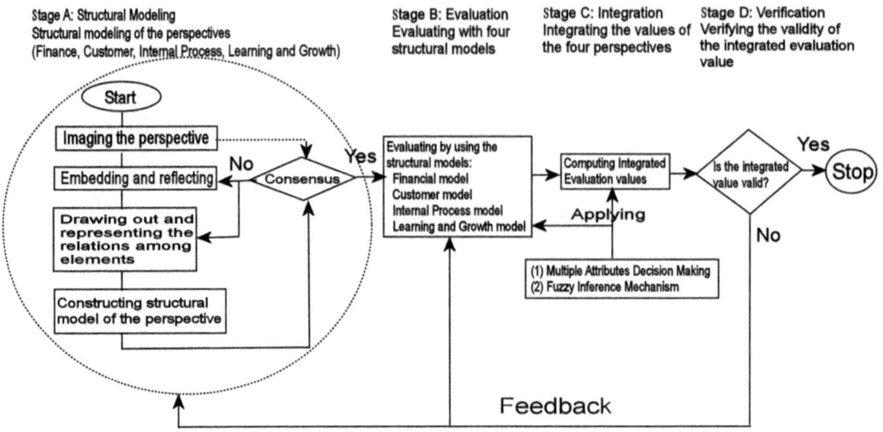

Fig. 2. The multidimensional measurement system with BSC without link

be applied to obtain a consented structural model of graphical image of items by multi-participants decision makers, see [NUCA]. In this stage, the structural models should be constructed to reflect not only the company's business strategy to accomplish the consented target, but also the company's organizational, financial, internal process and human resource managing status and concept.

In the stage B, the performance level in each of perspectives is evaluated from lower level to higher level. Except for the evaluation in financial perspective, the fuzzy inference mechanism is applied to obtain the higher levels' measurement evaluation value from lower ones, and the Choque integral with weights obtained as the components in the eigenvectors of fuzzy reachability matrices in MSMM is used to calculate the evaluation value in financial perspective.

The stage C is that where the four top level values of each perspective are integrated by means of fuzzy inference mechanism. If the resulted value seems to be valid in the stage D, then the system of process completes. If not, it needs to go back to one of previous stages.

This system might be good for evaluating the current business performance level in view of BSC's four perspectives. However, there are no explicit links between any two of measures in distinct perspectives. In order to evaluate the performance level aiming toward a definite strategy, we should take these links into account. We call these kind of evaluation a "dynamic evaluation" comparing to "static evaluation", and propose a "dynamic evaluation" system in the next section.

4 Evaluation Methodology of Linked Multidimensional Measurement System

Now we give our "dynamic" evaluation system with links between measures of four perspectives in order to achieve a definite strategic objective. Under the assumption that there is a hierarchy of four perspectives “learning and growth”, “internal process”, “customer”, “financial” in this order from down to top, we propose the following system consist of four phases and one preliminary one.

Phase0. Compose a group of some of executives, top managers, representatives of each department, strategy experts, etc., which we call a “strategy team”. The strategy team initially creates a vision, then make a strategy clear and replace it with plain words, as we see in the section2.

Phase1. This phase is corresponding to the stage A in Fig.2, where the current state of consented structural model is analyzed in the following steps.

Step1. Classify and embed possible measures (or activities) into each perspectives.

Step2. Construct a consented structural models of measures in each perspective.

Step3. Give the current scores to each measure.

Phase2. In this phase, the company's strategy is mapped into the structural model in the following steps.

Step1. For each perspective, determine the levels of measures to be considered as linked measures among perspectives. We call them the “critical level measures”.

Step2. Carefully noticing the influence caused by each critical measures, choose effective measures in the perspectives to achieve the strategic objective.

Step3. Construct a consented structural model of critical measures, where the values at (i, j)-component in the fuzzy structure matrix are given so as to represent the degree of influence of the i-th measure to the j-th one. In this step, we notice that links exist only between measures in distinct perspectives, because the relationships among measures within the same perspective are already considered in the step2 of phase1.

Step4. Find out all the measures related to the strategic objective and the path from measures in lower perspective to those in the higher one which we call the “critical path”.

After completing these steps, we might have the strategy oriented measures with links which we call the “critical linked measures”. If there are some measures seemed not to be related to the strategy objective, omit them from the “critical linked measures”, and we have targeted object to be evaluated.

Phase3. Evaluation of the performance level of each measure in the “critical linked measures” is executed from lower perspective to higher one in the following steps.

Step1. Set the perspective P=“internal process”, the second lowest perspective.

Step2. For any fixed measure in P, say m, find out all the measures in lower perspectives from which a link exists. Let m_i $(i = 1, ..., k)$ be such all measures in lower perspectives, define possible fuzzy inference rules such as
"If m is Low, m_1 is High,..., and m_k is Middle then m is Very High".
To define these rules, the eigenvectors calculated from the fuzzy matrix in the step3 of the phase2 might be helpful, since these values represent the importance degree of measures.

Step3. Perform the fuzzy inference mechanism using the scores given in the step3 of phase1.

Step4. If P is "financial" perspective, then quit the series of steps. If not, set P be one level higher perspective, and go to the step 2.

Phase4. Apply the multidimensional measurement system quoted in the previous section to see the total evaluation of the critical linking measures.

After completing all the phases of our system, we obtained several kinds of outputs as follows.

- Four structure models of measures with current scores in each perspective
- A strategy oriented structure model of critical measures across the perspectives with values assigned to the possible directed links ("critical linked measures"="critical leveled measures"+"critical path")
- Fuzzy inference rules among some of critical linked measures
- Dynamic evaluated scores for each of critical measures
- Total evaluation value of the organization's business performance for accomplishing the strategic objective

If the total evaluation value is satisfactory, then it is judged that the measures or activities are well performed to achieve the strategic objective. But if the value is not good, then need to find out some measures to be supplied more manpower or financial support, or to be improved in the process management. The values estimated in the phase3 are helpful to detect them.

5 Illustrative Example

Now we give an illustrative example to show how our system works for evaluating the business performance of a certain company X.

Phase0. The strategy team in X decided to improve their financial growth performance level in view of the company's strategic vision.

Phase1. In [CSAM], Cui et. al gave the list of measures to be evaluated, and constructed the consented structural models of each perspective. Using MSMM, the consented structural model of measures in financial perspective described in Fig.3 was obtained with four levels of measures. Top level measure is "finance" with estimated value 0.887 using Choque integral, and the second level measures are "productivity", "growth", "profitability", and "stability" of weight 0.23, 0.19, 0.30, and 0.27 respectively. The Fig.4, Fig.5, and Fig.6 also describe the consented structural model of measures in each perspective with the initial scores.

Phase2. Define the second level as the critical level except for the internal process perspective, and the third level for the internal process perspective. The target of dynamic evaluation is the measure "growth" in the financial perspective, encircled in Fig.3, and choose {Customer satisfaction(Cs),New product ratio(Nr),Market share growth rate(Ms)} from measures in the customer perspective, {Lead time shortening(Ls),Production cost reduction(Pc), Participating rate of QC(Qc), Fulfillment on an incentive system(Is), Organizational efficiency(Of)} from the internal process perspective, and {Employee

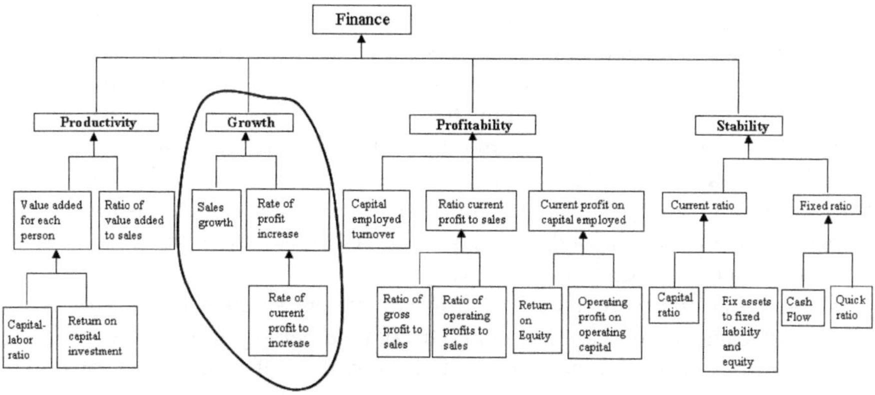

Fig. 3. The structural model of measures in financial perspective

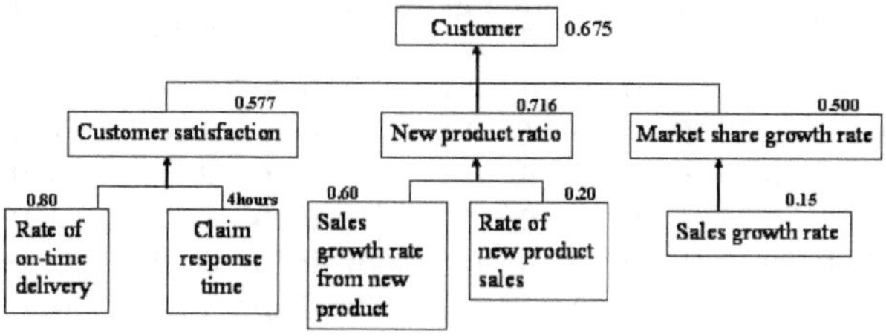

Fig. 4. The structural model of measures in customer perspective

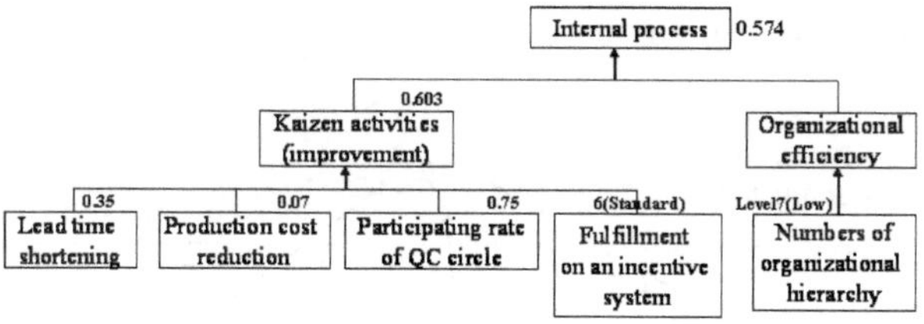

Fig. 5. The structural model of measures in internal process perspective

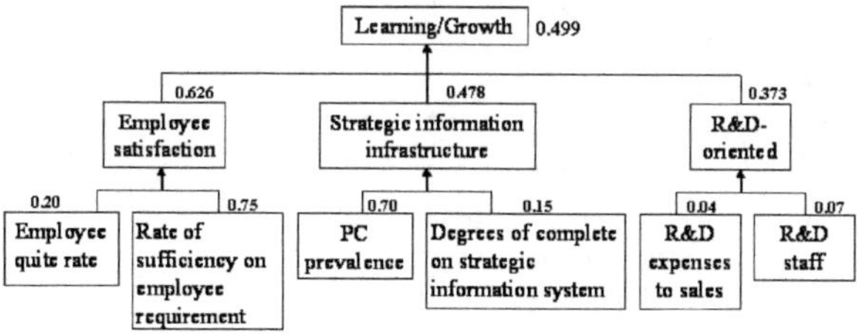

Fig. 6. The structural model of measures in learning and growth perspective

satisfaction(Es),Strategic information infrastructure(Si),R&D oriented(Ro)} from the learning and growth perspective. The fuzzy structure matrix representing the degree of influence caused by one measure to another one and the reachability matrix are shown in the table1.

When we set the threshold value $\alpha = 0.5$, the graphical image of critical linked measures with the influence degree is obtained as in the Fig.7.

Phase3. From the Fig.7, we define fuzzy inference rules from measures in lower perspective to one in higher perspective. As an example, we show how new score of the measure Is is calculated.

We see that the score is derived from its current score, the scores of Es and Si, and their influences. First define fuzzy membership functions of scores for these three measures, next give all the possible inference rules considering

Table 1. Fuzzy structure matrices:left is original matrix, right is reachability matrix

	Es	Si	Ro	Ls	Pc	Qc	Is	Of	Cs	Nr	Ms	Fg
Es	0	0	0	.6	.7	0	.6	0	.8	0	0	0
Si	0	0	0	.7	.4	0	.7	0	0	0	0	0
Ro	0	0	0	0	0	0	0	0	0	.7	.8	0
Ls	0	0	0	0	0	0	0	0	.8	0	0	0
Pc	.2	0	0	0	0	0	0	0	0	0	0	.4
Qc	0	0	0	0	0	0	0	0	0	.4	0	0
Is	0	0	0	0	0	0	0	0	0	0	.8	0
Of	0	0	0	0	0	0	0	0	0	0	.6	0
Cs	.3	0	0	0	0	0	0	0	0	0	0	.5
Nr	0	0	0	0	0	0	0	0	0	0	0	.7
Ms	0	0	0	0	0	0	0	0	0	0	0	.8
Fg	0	0	0	0	0	0	0	0	0	0	0	0

$\Longrightarrow$

Es	Si	Ro	Ls	Pc	Qc	Is	Of	Cs	Nr	Ms	Fg
.3	0	0	.6	.7	0	.6	0	.8	0	.6	.6
.3	0	0	.7	.4	0	.7	0	.7	0	.7	.7
0	0	0	0	0	0	0	0	0	.7	.8	.8
.3	0	0	.3	.3	0	.3	0	.8	0	.3	.5
.2	0	0	.2	.2	0	.2	0	.2	0	.2	.4
0	0	0	0	0	0	0	0	0	.4	0	.4
0	0	0	0	0	0	0	0	0	0	.8	.8
0	0	0	0	0	0	0	0	0	0	.6	.6
.3	0	0	.3	.3	0	.3	0	.3	0	.3	.5
0	0	0	0	0	0	0	0	0	0	0	.7
0	0	0	0	0	0	0	0	0	0	0	.8
0	0	0	0	0	0	0	0	0	0	0	0

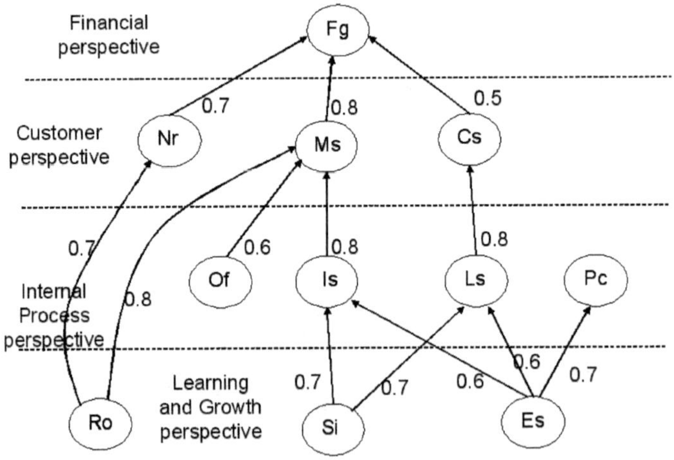

Fig. 7. The graphical structure model of measures across the perspectives

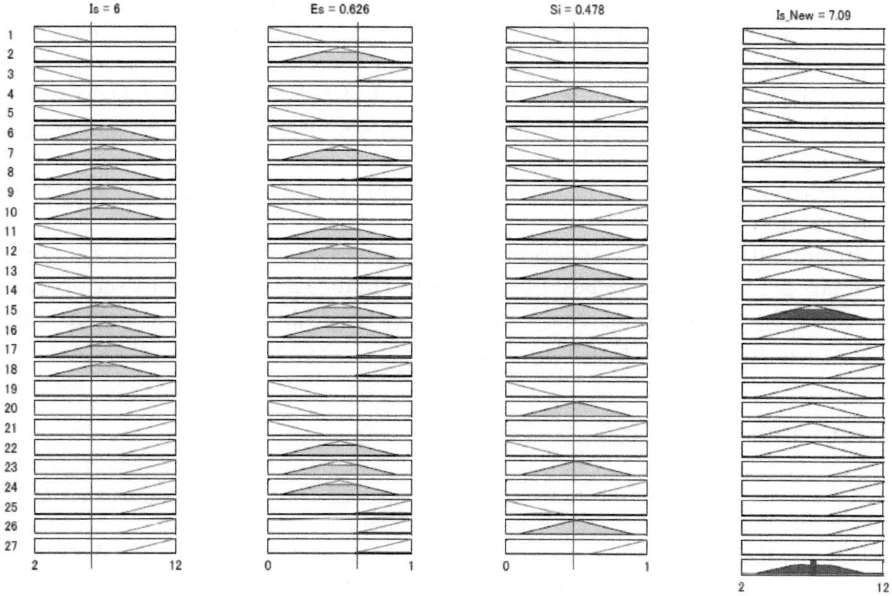

Fig. 8. The output of fuzzy inference mechanism

the meaning of measures and state of influences. Then the fuzzy inference mechanism is performed with Mamdani method to have the aggregation value for the new score of Is. The Fig.8 illustrates the aggregation process, and we have the new score 7.09 from the current score 6.
Performing the same process for {Ls, Si, Es } and for {Pc, Es}, we obtain the new scores for Ls and Pc. After that, go on the evaluation of measures in the customer perspective, and of those in the financial one. Then we have new scores for all the critical linked measures, and the total evaluation degree is to be calculated.

6 Discussion and Conclusion

In this paper, we propose a new methodology for evaluating strategy oriented business performance in BSC framework. We apply several fuzzy theory based methods, such as Modified Structural Modeling Method, fuzzy inference mechanism, in order to embed the uncertainty into the decision making process, which also gives our system the capability to adjust flexibilities caused by company's characteristics. The effectiveness of our methodological system highly depends on ability of members in the strategy team, however estimations and decisions are done by multi-participants in numerical and visual view. Thus correction or modification or adjustment is easy when new opinions or views are proposed.

We assumed the hierarchy of four perspectives in BSC and considered only the case that links of measures are directed from lower perspective to higher one. The inverse direction links might be possible, such as a financial related measure affects the employees quits rates, and we might need to reflect that type of linkages to our evaluation values. One idea to perform it is to continue the similar steps in the phase 3 cyclically even if P is financial perspective, and consider links from measures in P to other sort of measures. However, the target measures are usually in higher perspectives and cyclic iteration would cause the redundant evaluation problem. We should try to ameliorate our system by solving these problem.

In our future work, we consider that it is necessary to apply our system to evaluate some real companies' business performance level or at least set up more precise model of fictitious company. We also try to build the cost performance concept in it such as CPM from PERT.

References

[KN1] Kaplan, R.S., Norton, D.P.: The Balanced Scorecard-Measures that Drive Performance. Harvard Business Review 70(1), 71–79 (1992)

[KN2] Kaplan, R.S., Norton, D.P.: Putting the Balanced Scorecard to Work. Harvard Business Review 71(5), 134–149 (1993)

[KN3] Kaplan, R.S., Norton, D.P.: Balanced Scorecard. Harvard Business School Press (1996)

[KN4] Kaplan, R.S., Norton, D.P.: How to Implement a New Strategy Without Disrupting Your Organization. Harvard Business Review, 1–10 (March 2006)

[IL] Itter, C.D., Larcher, D.F.: Coming Up Short on Nonfinancial Performance Measurement. Harvard Business Review 81(11), 88–95 (2003)

[MAS] Matsuo, T., Amagasa, M., Suzuki, K.: Balanced Scorecard with Fuzzy Inference as a Performance Measurement. In: Proceedings of the 7th Asia Pacific Industrial Engineering and Management Systems Conference, Bangkok, Thailand, pp. 364–375 (2006)

[CSAM] Cui, D., Suzuki, K., Amagasa, M., Matsuo, T.: A Multidimensional Measurement System with Balanced Scorecard. In: Proceedings of 38th Annual Meeting Decision Sciences Institute, Phoenix, Arizona, pp. 5061–5066 (2007)

[NUCA] Nagata, K., Umezawa, M., Cui, D., Amagasa, M.: Modified Structural Modeling Method and Its Application -Behavior Analysis of Passengers for East Japan Railway Company. Journal of Industrial Engineering and Management Systems 7(3), 245–256 (2008)

Predictive Probabilistic and Possibilistic Models Used for Risk Assessment of SLAs in Grid Computing

Christer Carlsson and Robert Fullér

IAMSR, Åbo Akademi University,
Joukahainengatan 3-5A, Åbo, FI-20520, Finland
{christer.carlsson,robert.fuller}@abo.fi

Abstract. We developed a hybrid probabilistic and possibilistic technique for assessing the risk of an SLA for a computing task in a cluster/grid environment. The probability of success with the hybrid model is estimated higher than in the probabilistic model since the hybrid model takes into consideration the possibility distribution for the maximal number of failures derived from a resource provider's observations. The hybrid model showed that we can increase or decrease the granularity of the model as needed; we can reduce the estimate of the $P(S^* = 1)$ by making a rougher, more conservative, estimate o f the more unlikely events of $(M + 1, N)$ node failures. We noted that M is an estimate which is dependent on the history of the nodes being used and can be calibrated to 'a few' or to 'many' nodes.

Keywords: Grid computing, Service Level Agreement (SLA), Predictive probabilities, Predictive possibilities.

1 Introduction

There is an increasing demand for computing power in scientific and engineering applications which has motivated the deployment of high performance computing (HPC) systems that deliver tera-scale performance. Current and future HPC systems that are capable of running large-scale parallel applications may span hundreds of thousands of nodes.

The current highest processor count is 131K nodes according to top500.org [16]. For parallel programs, the failure probability of nodes and computing tasks assigned to the nodes has been shown to increase significantly with the increase in number of nodes. Large-scale computing environments, such as the current grids CERN LCG, NorduGrid, TeraGrid and Grid'5000 gather (tens of) thousands of resources for the use of an ever-growing scientific community. Many of these Grids offer computing resources grouped in clusters, whose owners may share them only for limited periods of time and Grids often have the problems of any large-scale computing environment to which is added that their middleware is still relatively immature, which contributes to making Grids relatively

E. Hüllermeier, R. Kruse, and F. Hoffmann (Eds.): IPMU 2010, Part II, CCIS 81, pp. 747–757, 2010.

unreliable computing platforms. Iosup et al [9] collected and present material from Grid'5000 which illustrates this contention. On average, resource availability (for a period of 548 days) in Grid'5000 on the grid level is 69% (±11.42), with a maximum of 92% and a minimum of 35%. Long et al [12] collected a dataset on node failures over 11 months from1139 workstations on the Internet to determine their uptime intervals; Planck and Elwasif [14] collected failure information for 16 DEC Alpha work-stations at Princeton University. Even if this is a typical local cluster of homogeneous processors the failure data shows similar characteristics as for the larger clusters. Schroeder and Gibson [17] analyse failure data (23 000 failure events) ars at Los Alamos National Laboratory (LANL). Their study includes root causes of failures, the mean time between failures, and the mean time to repair. They found that average failure rates differ wildly across systems, ranging from 20-1000 failures per year, mean repair time varies from less than an hour to more than a day. When node failures occur in the LANL, hardware was found to be the single largest cause (30-60%); software is the second largest contributor (with 5-24%), but in most systems the root cause remained undetermined for 20-30% of the failures (cf. [17]). They also found that the yearly failure rate varies widely across systems, ranging from 17 to an average of 1159 failures per year for several systems. The main reason for the differences was that the systems vary widely in size and that the nodes run different workloads.

Iosup et al [9] fit statistical distributions to the Grid'5000 data using maximum likelihood estimation (MLE) to find a best fit for each of the model parameters. They also wanted to find out if they can decide where (on which nodes or in which cluster) a new failure could/should occur. Since the sites are located and administered separately, and the network between them has numerous redundant paths, they found no evidence for any other assumption than that there is no correlation between the occurrence of failures at different sites. For the LANL dataset Schroeder and Gibson [17] studied the sequence of failure events and the time between failures as stochastic processes. This includes two different views of the failure process: (i) the view as seen by an individual node; (ii) the view as seen by the whole system. They found that the distribution between failures for individual nodes is well modelled by a Weibull or a Gamma distribution; both distributions create an equally good visual fit and the same negative log-likelihood. For the system wide view of the failures the basic trend is similar to the per node view during the same time.

A significant amount of the literature on grid computing addresses the problem of resource allocation on the grid (see, e.g., [Brandt [1], Czajkowski [7], Liu et al [11], Magana et al [13], and Tuecke [18]). The presence of disparate resources that are required to work in concert within a grid computing framework increases the complexity of the resource allocation problem. Jobs are assigned either through scavenging, where idle machines are identified and put to work, or through reservation in which jobs are matched and pre-assigned with the most appropriate systems in the grid for efficient workflow. In grid computing a resource provider [RP] offers resources and services to other Grid users based on agreed service level agreements [SLAs]. The research problem we have addressed is formulated as follows:

- o the RP is running a risk to be in violation of his SLA if one or more of the resources [nodes in a cluster or a Grid] he is offering to prospective customers will fail when carrying out the tasks
- o the RP needs to work out methods for a systematic risk assessment [RA] in order to judge if he should offer the SLA or not if he wants to work with some acceptable risk profile

In the context we are going to consider (a generic grid computing environment) resource providers are of various types which mean that the resources they manage and the risks they have to deal with are also different; we have dealt with the following RP scenarios (but we will report only on extracts due to space):

- RP1 manages a cluster of n_1 nodes (where $n_1 < 10$) and handles a few (< 5) computing tasks for a T;
- RP2 manages a cluster of n_2 nodes (where $n_2 < 150$) and handles numerous (≈ 100) computing tasks for a T; RP2 typically uses risk models building on stochastic processes (Poisson-Gamma) and Bayes modelling to be able to assess the risks involved in offering SLAs;
- RP3 manages a cluster of n_3 nodes (where $n_3 < 10$) and handles numerous (≈ 100) computing tasks for a T; if the computing tasks are of short duration and/or the requests are handled online RP3 could use possibility models that will offer robust approximations for the risk assessments;
- RP4 manages a cluster of n_4 nodes (where $n_4 < 150$) and handles numerous (≈ 100) computing tasks for a T; typically RP4 could use risk models building on stochastic processes (Poisson-Gamma) a nd Bayes modelling to assess the risks involved in offering SLAs; if the computing tasks are of short duration and/or the requests are handled online hybrid models which combine stochastic processes and Bayes modelling with possibility models could provide tools for handling this type of cases.

During the execution of a computing task the fulfilment of the SLA has the highest priority, which is why an RP often is using resource allocation models to safeguard against expected node failures. When spare resources at the RP's own site are not available outsourcing will be an adequate solution for avoiding SLA violations.

The risk assessment modelling for an SLA violation builds on the development of predictive probabilities and possibilities for possible node failures and combined with the availability of spare resources. The rest of the paper will be structured as follows: in Section 2 we will work out the basic conceptual framework for risk assessment, in Section 3 we will introduce the Bayesian predictive probabilities as they apply to the SLAs for RPs in grid computing, in Section 4 we will work out the corresponding predictive possibilities and show the results of the validation work we carried out for some RP scenarios; in Section 5 there is a summary and conclusions of the study.

2 Risk Assessment

There is no universally accepted definition of business risk but in the RP context we will understand risk to be a potential problem which can be avoided or mitigated (cf. [5] for references). The potential problem for an RP is that he has accepted an SLA and may not be able to deliver the necessary computing resources in order to fulfil a computing task within an accepted time frame T. *Risk assessment* is the process through which a resource provider tries to estimate the probability for the problem to occur within T and risk management the process through which a resource provider tries to avoid or mitigate the problem. In classical decision theory risk is connected with the probability of an undesired event; usually the probability of the event and its expected harm is outlined with a scenario which covers the set of risk, regret and reward probabilities in an expected value for the outcome. The typical statistical model has the following structure,

$$R(\theta, \delta(x)) = \int L(\theta, \delta(x)) f(x|\theta) \mathrm{d}x \qquad (1)$$

where L is a loss function of some kind, x is an observable event (which may not have been observed) an $\delta(x)$ is an estimator for a parameter θ which has some influence on the occurrence of x. The risk is the expected value of the loss function. The statistical models are used frequently because of the very useful tools that have been developed to work with large datasets. The statistical model is influenced by the modern capital markets theory where risk is seen as a probability measure related to the variance of returns (cf. Markowitz [14]. The analogy would be that an RP – handling a large number of nodes and a large number of computing tasks – will reach a steady state in his operations so that there will be a stable systematic risk (market risk') for defaulting on an SLA which he can build on as his basic assumption and then a (small') idiosyncratic risk which is situation specific and which he should estimate with some statistical models. We developed a hybrid probabilistic and possibilistic model to assess the success of computing tasks in a Grid. The model first gives simple predictive estimates of node failures in the next planning period when the under-lying logic is the Bayes probabilistic model for observations on node failures. When we apply the possibilistic model to a dataset we start by selecting a sample of k observations on node failures. Then we find out how many of these observations are different and denote this number by l; we want to use the two datasets to predict what the $k + 1$: st observation on node failures is going to be. The possibility model is used to find out if that number is going to be $0, 1, 2, 3, 4, 5, \ldots$ etc.; for this estimate the possibility model uses the most usual' numbers in the larger dataset and makes an estimate which is as close as possible' to this number. The estimate we use is a triangular fuzzy number, i.e. an interval with a possibility distribution. The possibility model turned out to be a faster and more robust estimate of the $k+1$: st observation and to be useful for online and real-time risk assessments with relatively small samples of data.

3 Predictive Probabilities

In the following we will use node failures in a cluster (or a Grid) as the focus, i.e. we will work out a model to predict the probabilities that n nodes will fail in a period covered by an SLA ($n = 0, 1, 2, 3, \ldots$). In the interest of space we have to do this by sketches as we deal with standard Bayes theory and modelling (cf. [5] for references).

The first step is to determine a probability distribution for t he number of node failures for a time interval $(t_1, t_2]$ by starting from some basic property of the process we need to describe. Typically we assume that the node failures represent a Poisson process which is non-homogenous in time and has a rate function $\lambda(t), t > 0$.

The second step is to determine a distribution for $\lambda(t)$ given a number of observations on node failures from r comparable segments in the interval $(t_1, t_2]$. This count normally follows a Gamma(α, β) distribution and the posterior distribution $p(\lambda_{t1,t2})$, given the count of node failures, is also a Gamma distribution according to the Bayes theory. Then, as we have been able to determine $\lambda_{t1,t2}$ we can calculate the predictive distribution for the number of node failures in the next time segment; Bayes theory shows that this will be a Poisson-Gamma distribution.

The third step is to realize that a computing task can be carried out successfully on a cluster (or a Grid) if all the needed nodes are available for the scheduled duration of the task. This has three components: (i) a predictive distribution on the number of nodes needed for a computing task covered by an SLA; (ii) a distribution showing the number of nodes available when an assigned set of nodes is reduced with the predicted number of node failures and an available number of reserve nodes is added (the number of reserve nodes is determined by the resource allocation policy of the RP); (iii) a probability distribution for the duration of the task.

The fourth step is to determine the probability of an SLA failure: p_1 (n nodes will fail for the scheduled duration of a task) $\times$ $(1 - p_2$ (m reserve nodes are available for the scheduled duration of a task)) if we consider only the systematic risk. We need to use a multinomial distribution to work out the combinations. Consider a Grid of k clusters, each of which contains n_c nodes, leading to the total number of nodes $n = \sum n_c$, where $c = 1, \ldots, k$, in the Grid. Let in the following $\lambda(t), t > 0$, denote generally a time non-homogeneous rate function for a Poisson process $N(t)$. We will assume that we have the RP4 scenario as our context, i.e. we will have to deal with hundreds of nodes and hundreds of computing tasks with widely varying computational requirements over the planning period for which we are carrying out the risk assessment.

The predictive distribution of the number of events is a Poisson-Gamma distribution, obtained by integrating the likelihood with respect to the posterior. Under the reference prior the predictive probability of having x events in the future on a comparable time interval equals

$$p\left(x \middle| r, \frac{1}{2} + \sum_{i=1}^{r} x_i, 1\right) = \frac{\Gamma(\sum_{i=1}^{r} x_i + x + 1/2) \times r^{\left(\frac{1}{2} + \sum_{i=1}^{r} x_i\right)}}{\Gamma(1/2 + \sum_{i=1}^{r} x_i) x! (r+1)^{\left(\frac{1}{2} + \sum_{i=1}^{r} x_i\right) + x}} \tag{2}$$

When a computing job begins execution in a cluster (or Grid) its successful completion will require a certain number of nodes to be available over a given period of time. To assess the uncertainty about the resource availability, we need to model both the distribution of the number of nodes and the length of the time required by the jobs. Given observed data on the number of nodes required by computing tasks, the posterior distribution of the probabilities $\mathbf{p}$ is available in an analytical form under a Dirichlet prior, and its density function can be written as

$$p(\mathbf{p} \mid \mathbf{w}) = \frac{\Gamma\left(\sum_{m=1}^{u} \alpha_m + w_m\right)}{\prod_{m=1}^{u} \Gamma(\alpha_m + w_m)} \times \prod_{m=1}^{u} p_m^{\alpha_m + w_m - 1}, \tag{3}$$

where w_m corresponds to the number of observed tasks utilising m nodes, α_m is the a priori relative weight of the m-th component in the vector $\mathbf{p}$, and $\mathbf{w}$ is the vector (w_m), where $m = 1, \dots, u$. The corresponding predictive distribution of the number of nodes required by a generic computing task in the future equals the Multinomial-Dirichlet distribution, which is obtained by integrating out the uncertainty about the multinomial parameters with respect to the posterior distribution. The Multinomial-Dirichlet distribution is in our notation defined as,

$$\begin{aligned} p(M = m^* | \mathbf{w}) &= \frac{\Gamma(\sum_{m=1}^{u} \alpha_m + w_m)}{\prod_{m=1}^{u} \Gamma(\alpha_m + w_m)} \frac{\prod_{m=1}^{u} \Gamma(\alpha_m + w_m + I(m = m^*))}{\Gamma(1 + \sum_{m=1}^{u} \alpha_m + w_m)} \\ &= \frac{\Gamma(\sum_{m=1}^{u} \alpha_m + w_m)}{\Gamma(\alpha_{m^*} + w_{m^*})} \frac{\Gamma(\alpha_{m^*} + w_{m^*} + 1)}{\Gamma(1 + \sum_{m=1}^{u} \alpha_m + w_m)}. \end{aligned} \tag{4}$$

By combining the above distributions, we will find the probability distribution for the number of nodes in use for computing tasks in a future time interval $(t_1, t_2]$, as the corresponding random variable equals the product XM.

To simplify the inference about the length of a task affecting a number of nodes we assume that the length follows a Gaussian distribution with expected value μ and variance σ^2. Obviously, it is motivated to have separate parameter sets for different types of tasks. Assuming the standard reference prior for the parameters, we obtain the predictive distribution for the expected time, T, used for a future computing task, in terms of the probability density for the expected time as follows,

$$\begin{aligned} &p(t|\bar{t}, ((b-1)/(b+1))s^2, b-1) = \\ &\frac{\Gamma(b/2)}{\Gamma(\frac{b-1}{2})\Gamma(1/2)} \left(\frac{1}{(b+1)s^2}\right)^{\frac{1}{2}} \left[1 + \frac{1}{b+1)s^2} (t - \bar{t})^2\right]^{-\frac{b}{2}}. \end{aligned} \tag{5}$$

The probability that a task lasts longer than any given time t equals $P(T > t) = 1 - P(T \leq t)$, where $P(T \leq t)$ is the cumulative distribution function

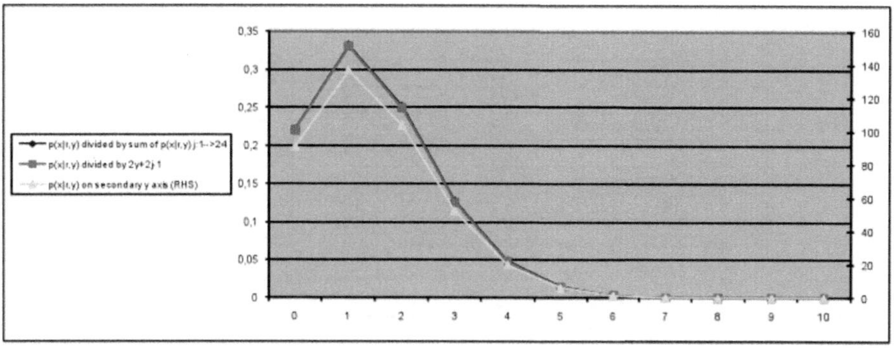

Fig. 1. Prediction of node failures [Prob of Failure: 0.1509012] with 136 time slots of LANL cluster data for computing tasks running for 9 days, 0 hours

(CDF) of the T-distribution. The value of the CDF can be calculated numerically using functions existing in most computing environments. However, it should also be noted that for a moderate to large b, the predictive distribution is well approximated by the Gaussian distribution with the mean $\bar{t}$ and the variance $s^2 \frac{(b+1)}{(b-3)}$. Consequently, if the Gaussian approximation is used, the probability $P(T \leq t)$ can be calculated using the CDF of the Gaussian distribution. We now consider the probability that a computing task will be successful. This happens as the sum P('none of the nodes allocated to the task fail') + $\sum_{m=1}^{m_{\max}} P$('m of the nodes allocated to the task fail & at least m idle nodes are available as reserves'). Here $m_{\max}$ is an upper limit for the number of failures considered. Note that we simplify the events below by considering the m failures to take place simultaneously. We then get

$$\begin{aligned} P(S=1) &= 1 - P(S=0) \\ &= 1 - \sum_{m=1}^{m_{\max}} P(m \text{ failures occur \& less than } m \text{ free nodes available}) \\ &\geq 1 - \sum_{m=1}^{m_{\max}} P(m \text{ failures occur}) P(\text{less than } m \text{ free nodes at any time point}) \end{aligned} \tag{6}$$

The probability P(m failures occur) is directly determined by the failure rate model discussed above. The other term, the probability P(less than m free nodes at any time point), is dependent on the resource allocation policy and the need of reserve nodes by the other tasks running simultaneously. The predictive probabilities model has been extensively tested and verified with data from the LANL cluster (cf. Shroeder-Gibson [17]). Here we collected results for the RP1 scenario where the RP is using a cluster with only a few nodes; the test runs have been carried out also for the scenarios RP2-4 with some variations to the results.

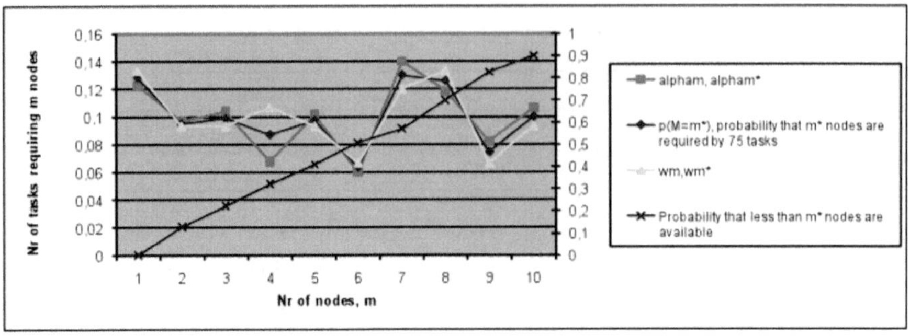

Fig. 2. Probability that less than m nodes will be available for computing tasks requiring < 10 nodes; the number of nodes randomly selected; 75 tasks simulated over 236 time slots; LANL

4 Predictive Possibilities

In this section we will introduce a hybrid method for simple predictive estimates of node failures in the next planning period when the underlying logic is the Bayes models for observations on node failures that we used for the RA models in section 3. This is essentially a standard regression model with parameters represented by triangular fuzzy numbers - typically this means that the parameters are intervals and represent the fact that the information we have is imprecise or uncertain. We can only sketch the model here, for details cf. [5]; the model builds on some previous results in [2], [3], [4] and [6]. We will take a sample of a dataset (in our case, a sample from the LANL dataset) which covers inputs and fuzzy outputs according to the regression model; let x_i, Y_i be a sample of the observed crisp inputs and fuzzy outputs of model (1). The main goal of fuzzy nonparametric regression is to estimate $F(x)$ at any $x \in \mathbb{D}$ from $(x_i, Y_i), i = 1, 2, ..., n$. And the membership function of an estimated fuzzy output should be as close as possible to that of the corresponding observations forming the fuzzy number, i.e. we should estimate $a(x), \alpha(x), \beta(x)$ for each $x \in \mathbb{D}$ so that we get a fit between the estimated Y and the observed Y' which is 'a closest fit'; here we will use Diamond's distance measure (cf. [8]). Let $A = (a, \alpha_1, \beta_1), B = (b, \alpha_2, \beta_2)$ be two triangular fuzzy numbers. Then the squared distance between A and B is defined by, $d^2(A, B) = (\alpha_1 - \alpha_2)^2 + (a - b)^2 + (\beta_1 - \beta_2)^2$. Let us now assume that the observed (fuzzy) output is $Y_i = (a, \alpha_i, \beta_i)$, then with the Diamond distance measure and a local linear smoothing technique we need to solve a locally weighted least-squares problem in order to estimate $F(x_0)$, for a given kernel K and a smoothing parameter h, where $K_h(|x_i - x_0|) = K\left(\frac{|x_i - x_0|}{h}\right)$. The kernel is a sequence of weights at x_0 to make sure that data that is close to x_0 will contribute more when we estimate the parameters at x_0 than those that are farther away, i.e. are relatively more distant in terms of the parameter h. Let $\hat{F}_{(i)}(x_i, h) = (\hat{a}_{(i)}(x_0, h), \hat{\alpha}_{(i)}(x_0, h), \hat{\beta}_{(i)}(x_0, h))$ be the predicted fuzzy regression function at input x_i. Compute $\hat{F}_{(i)}(x_i, h)$ for each x_i and let $CV(h) = \frac{1}{l}\sum_{i=1}^{l} d^2(Y_i, \hat{F}_{(i)}(x_i, h))$. We should select h_0 so that it is optimal in the following expression $CV(h_0) = \min_{h>0} CV(h)$. By solving this minimalisation problem, we get an estimate of $F(x)$ at x_0 by, $\hat{F}(x_0) = (\hat{a}(x_0), \hat{\alpha}(x_0), \hat{\beta}(x_0))$, and the following estimate of $F(x)$ at x_0, $\hat{F}(x_0) = (\hat{a}(x_0), \hat{\alpha}(x_0), \hat{\beta}(x_0)) = (\mathbf{e}_1^T \mathbf{H}(x_0, h)\mathbf{a}_Y, \mathbf{e}_1^T \mathbf{H}(x_0, h)\alpha_Y, \mathbf{e}_1^T \mathbf{H}(x_0, h)\beta_Y)$. We can use this model – which we have decided to call a predictive possibility model – to estimate a prediction of how many node failures we may have in the next planning period given that we have used statistics of the observed number of node failures to build a base of Poisson-Gamma distributions (for details cf. [6]). We will use the Fréchet-Hoeffding bounds for copulas to show a lower limit for the probability of success of a computing task in a cluster (or a Grid). Then we can rewrite (6) as,

$$
\begin{aligned}
P(\text{success}) \geq 1 - \sum_{m=1}^{m_{\max}} &P(m \text{ failures occur}) \\
\times \Bigg\{ &\sum_{j=1}^{t^*} P(\text{the length of a task} < j, \text{less-than-m-anytime}) \\
&+ P(\text{the length of a task} \geq t^*, \text{less-than-m-anytime}) \Bigg\}
\end{aligned}
$$

Let us introduce the notations $G(m) = P(\text{less-than-m-anytime})$ and $F(t) = P(\text{the duration of a task is less than t})$. Let t^* be chosen such that $1 - F(t^*) > 0.995$. Furthermore, denote the copula of F and G by H, where $H(t, m) = P(\text{the duration of a task is less than t, less-than-m-anytime})$. Then using the Fréchet-Hoeffding upper bound for copulas we find that,

$$
\begin{aligned}
&P(\text{success}) \geq 1 - \sum_{m=1}^{m_{\max}} P(m \text{ failures occur}) \\
&\times \Bigg\{ \sum_{j=1}^{t^*} \min \Bigg\{ \frac{\Gamma(b/2)}{\Gamma(\frac{b-1}{2})\Gamma(1/2)} \left(\frac{1}{(b+1)s^2}\right)^{\frac{1}{2}} \left[1 + \frac{1}{(b+1)s^2}(j - \bar{j})^2\right]^{-\frac{b}{2}}, \\
&P(\text{less than } m \text{ free nodes at any time point}) \Bigg\} \\
&+ \underbrace{P(\text{the length of a task is } \geq t^*, \text{less-than-m-anytime})}_{\leq 0.005} \Bigg\}
\end{aligned}
$$

Now we can use the new model as an alternative for predicting the number of node failures and use it as part of the Bayes model for predictive probabilities. In this way we will have hybrid estimates of the expected number of node failures - both probabilistic and possibilistic estimates. An RP may use either one estimate for his risk assessment or use a combination of both. We carried out a number of validation tests in order to find out (i) how well the predictive possibilistic

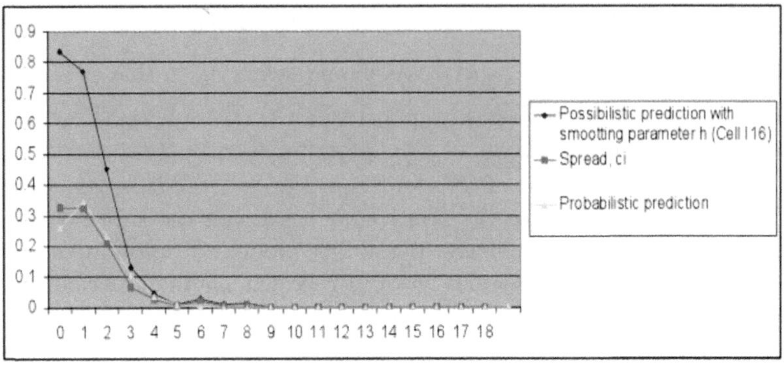

Fig. 3. 4 Possibilistic and probabilistic prediction of n node failures for a computing task with a duration of 8 days on 10 nodes; 153 time slots simulated

models can be fitted to the LANL dataset, (ii) what differences can be found between the probabilistic and possibilistic predictions and (iii) if these differences can be given reasonable explanations. The testing was structured as follows:

- 5 time frames for the possibilistic predictive model with a smoothing parameter from the smoothing function: $h = 382.54 \times \text{Nr of timeslots}^{-0.5325}$
- 5 feasibility adjustments from the hybrid possibilistic adjustment model to the probabilistic predictive model

The smoothing parameter h for the possibilistic models should be determined for the actual cluster of nodes, and in such a way that we get a good fit between the probabilistic and the possibilistic models. The approach we used for the testing was to experiment with combinations of h and then to fit a distribution to the results; the distribution could then be used for interpolation.

5 Summary and Conclusion

We developed a hybrid probabilistic and possibilistic technique for assessing the risk of an SLA for a computing task in a cluster/grid environment. The probability of success with a hybrid model is estimated higher than in the probabilistic model since the hybrid model takes into consideration the possibility distribution for the maximal number of failures derived from the RP's observations. The hybrid model showed that we can increase or decrease the granularity of the model as needed; we can reduce the estimate of the $P(S^* = 1)$ by making a rougher, more conservative, estimate of the more unlikely events of $(M + 1, N)$ node failures. We noted that M is an estimate which is dependent on the history of the nodes being used and can, of course, be calibrated to 'a few' or to many' nodes. The probabilistic models scale from 10 nodes to 1 00 nodes and then on to any (reasonable) number of nodes; in the same fashion also the possibilistic models scale to 100 nodes and then on to any (reasonable) number of nodes.

The RP can use both the probabilistic and the possibilistic models to get two alternative risk assessments and then (i) choose the probabilistic RA, (ii) the possibilistic RA or (iii) use the hybrid model for a combination of both RAs - this is a cautious/conservative approach.

References

1. Brandt, J.M., Gentile, A.C., Marzouk, Y.M., Pbay, P.P.: Meaningful Statistical Analysis of Large Computational Clusters (2005)
2. Carlsson, C., Fullér, R.: On Possibilistic Mean and Variance of Fuzzy Numbers. Fuzzy Sets and Systems 122, 315–326 (2001)
3. Carlsson, C., Fullér, R.: Fuzzy Reasoning in Decision Making and Optimization. Studies in Fuzziness and Soft Computing Series. Springer, Heidelberg (2002)
4. Carlsson, C., Fullér, R.: A Fuzzy Approach to Real Option Valuation. Fuzzy Sets and Systems 139, 297–312 (2003)
5. Carlsson, C., Weissman, O.: Advanced Risk Assessment, D4.1, The AssessGrid Project, IST-2005-031772, Berlin (2009)
6. Carlsson, C., Fullér, R., Mezei, J.: A lower limit for the probability of success of computing tasks in a grid. In: Proceedings of the Tenth International Symposium of Hungarian Researchers on Computational Intelligence and Informatics (CINTI 2009) Budapest, Hungary, pp. 717–722 (2009)
7. Czajkowski, K., Foster, I., Kesselman, C.: Agreement-Based Resource Management. Proceedings of the IEEE 93(3), 631–643 (2005)
8. Diamond, P.: Fuzzy least squares. Information Sciences 46, 141–157 (1988)
9. Iosup, A., Jan, M., Sonmez, O.O., Epema, D.H.J.: On the Dynamic Resource Availability in Grids. In: Proceedings of the 8th IEEE/ACM International Conference on Grid Computing (GRID 2007), pp. 26–33 (2007)
10. Lintner, J.: The Valuation of Risk Assets and the Selection of Risky Investments in Stock Portfolios and Capital Budgets. Review of Economics and Statistics 47, 13–37 (1965)
11. Liu, Y., Ngu, A.H.H., Zeng, L.: QoS Computation and Policing in Dynamic Web Service Selection. In: Proceedings of the 13th international World Wide Web (WWW 2004), pp. 66–73 (2004)
12. Long, D., Muir, R., Golding, R.: A Longitudinal Survey of Internet Host Reliability. In: 14th Symposium on Rel. Dist. Sys., pp. 2–9 (1995)
13. Magana, E., Serrat, J.: Distributed and Heuristic Policy-Based Resource Management System for Large-Scale Grids. In: Bandara, A.K., Burgess, M. (eds.) AIMS 2007. LNCS, vol. 4543, pp. 184–187. Springer, Heidelberg (2007)
14. Markowitz, H.M.: Portfolio Selection. Journal of Finance 7, 77–91 (1952)
15. Plank, J.S., Elwasif, W.R.: Experimental Assessment of Workstation Failures and Their Impact on Checkpointing Systems. In: 28th International Symposium on Fault-Tolerant Computing (1998)
16. Raju, N., Gottumukkala, N., Liu, Y., Leangsuksun, C.B.: Reliability Analysis in HPC Clusters. In: High Availability and Performance Computing Workshop (2006)
17. Schroeder, B., Gibson, G.A.: A Large-Scale Study of Failures in High-Performance Computing Systems. In: DSN 2006 Conference Proceedings, Philadelphia (2006)
18. Tuecke, S.: SNAP: A Protocol for Negotiating Service Level Agreements and Coordinating Resource Management in Distributed Systems. In: 8th Workshop on Job Scheduling Strategies for Parallel Processing, Edinburgh, Scotland (2002)

Author Index

GPSR Compliance

The European Union's (EU) General Product Safety Regulation (GPSR) is a set of rules that requires consumer products to be safe and our obligations to ensure this.

If you have any concerns about our products, you can contact us on ProductSafety@springernature.com

In case Publisher is established outside the EU, the EU authorized representative is:

Springer Nature Customer Service Center GmbH
Europaplatz 3
69115 Heidelberg, Germany

Batch number: 10704691

Printed by Printforce, the Netherlands